PAEDIATRIC CLINICAL PHARMACOLOGY

PAEDIATRIC CLINICAL PHARMACOLOGY

Edited by

Evelyne Jacqz-Aigrain
Imti Choonara

CRC Press
Taylor & Francis Group
Boca Raton London New York

CRC Press is an imprint of the
Taylor & Francis Group, an **informa** business

CRC Press
Taylor & Francis Group
6000 Broken Sound Parkway NW, Suite 300
Boca Raton, FL 33487-2742

First issued in paperback 2019

© 2010 by Taylor & Francis Group, LLC
CRC Press is an imprint of Taylor & Francis Group, an Informa business

No claim to original U.S. Government works

ISBN-13: 978-0-8247-2189-3 (hbk)
ISBN-13: 978-0-367-39133-1 (pbk)

Visit the Taylor & Francis Web site at
http://www.taylorandfrancis.com

and the CRC Press Web site at
http://www.crcpress.com

Preface

The treatment of children with medicinal products is an important scientific area. It is recognised that many medicines that are used extensively in paediatric patients are either unlicensed or off-label. This textbook will hopefully help paediatric health professionals to treat children with the most appropriate medicine, i.e. effectively with minimal side effects.

The first three chapters deal with the generic principles behind the use of medicines in children. This involves the science of paediatric clinical pharmacology, clinical trials and aspects relating to formulation development. The rest of the textbook deals with specific clinical areas encompassing the fetus, neonates and paediatrics.

The approach presented throughout the textbook is very much a European outlook. The textbook would not have been possible without the support of all the authors and in particular we would like to thank colleagues within the European Society for Developmental, Perinatal and Paediatric Pharmacology. Several of the chapters have been reprinted from the journal *Paediatric and Perinatal Drug Therapy* and we would like to thank the journal for permission to reprint these papers.

We trust that the textbook will help you to improve the medical management of children. If you feel that there are important areas of omission, please do not hesitate to contact either of us.

Evelyne Jacqz-Aigrain   *Imti Choonara*

Table of Contents

PREFACE   iii

DRUG EVALUATION IN CHILDREN   1

1.1 UNLICENCED AND OFF-LABEL DRUG USAGE IN
    CHILDREN IN EUROPE   3
    Sharon Conroy
    1.1.1   INTRODUCTION   3
    1.1.2   PAEDIATRIC INTENSIVE CARE   6
    1.1.3   NEONATAL INTENSIVE CARE   7
    1.1.4   GENERAL MEDICAL AND SURGICAL
            PAEDIATRIC IN-PATIENTS   7
    1.1.5   OUTPATIENTS AND PRIMARY CARE   10
    1.1.6   MISCELLANEOUS   13
    1.1.7   RISKS ASSOCIATED WITH UNLICENCED AND OFF-LABEL
            DRUG USE IN CHILDREN   15
            REFERENCES   16

1.2 EUROPEAN REGULATION AND THE EMEA   19
    Agnès Saint Raymond and Kalle Hoppu
    1.2.1   INTRODUCTION   19
    1.2.2   REGULATORY FRAMEWORK FOR MEDICINAL
            PRODUCTS IN THE EUROPEAN UNION   19
    1.2.3   THE EUROPEAN AGENCY FOR THE EVALUATION OF
            MEDICINAL PRODUCTS (EMEA)   20
    1.2.4   PAEDIATRIC MEDICINES AND THE EMEA   21
    1.2.5   THE PAEDIATRIC EXPERT GROUP (PEG)   22
    1.2.6   THE US DRUG REGULATION AND THE FOOD AND DRUG
            ADMINISTRATION   22
    1.2.7   CONCLUSION   23
            REFERENCES   24

1.3 A COMPARISON BETWEEN EUROPEAN
    AND AMERICAN LEGISLATION   25
    Joséphine Géraci Buiche
    1.3.1   THE USA AND ITS LEGISLATION RELATED TO
            PAEDIATRIC MEDICATION   25

1.3.2   EUROPE AND LEGISLATION CONCERNING
        PAEDIATRIC MEDICINES   29
1.3.3   THE EXAMPLE OF RARE DISEASES   30
1.3.4   A COMPARISON BETWEEN THE UNITED STATES
        AND EUROPE   31
        REFERENCES   33

1.4   CLINICAL TRIALS IN CHILDREN:
      GUIDANCE ICH E11   35
      Natalie Hoog-Labouret
      1.4.1   ICH E11, INTRODUCTORY NOTES   35
      1.4.2   THE CLASSES OF AGE   36
      1.4.3   EVALUATION   36
      1.4.4   PHARMACEUTICAL FORMS ADAPTED FOR THE CHILD   37
      1.4.5   CLINICAL TRIALS WITH CHILDREN   38
      1.4.6   THE DIRECTIVE 2001/20/CEE OF 4 APRIL 2001   38

1.5   ETHICAL ISSUES   41
      John McIntyre
      1.5.1   INTRODUCTION   41
      1.5.2   REGULATORY FRAMEWORK FOR NEW MEDICINES   41
      1.5.3   ETHICAL ISSUES IN RESEARCH: GENERAL GUIDELINES   42
      1.5.4   ETHICAL ISSUES IN RESEARCH: SPECIFIC GUIDELINES   42
      1.5.5   PROBLEM AREAS   43
      1.5.6   THE WAY FORWARD   46
      1.5.7   CONCLUSION   47
              REFERENCES   47

1.6   PRACTICAL ISSUES IN RELATION TO CLINICAL TRIALS   49
      Imti Choonara
      1.6.1   LOCATION   49
      1.6.2   WHICH MEDICINES?   49
      1.6.3   TIMING OF STUDIES   50
      1.6.4   WHICH PAEDIATRIC PATIENTS?   51
      1.6.5   STUDY DESIGN   51
      1.6.6   PHARMACOKINETICS   51
      1.6.7   NON-INVASIVE METHODS   52
      1.6.8   PHARMACODYNAMICS   52
      1.6.9   RECRUITMENT   52
      1.6.10 PHARMACOVIGILANCE   53
      1.6.11 CONCLUSIONS   53
              REFERENCES   53

1.7   SPECIFIC ISSUES IN RELATION TO CLINICAL TRIALS   55
      Corinne Alberti and Bruno Giraudeau
      1.7.1   INTRODUCTION   55
      1.7.2   "N OF 1" TRIAL   56
      1.7.3   SEQUENTIAL DESIGN   57

1.7.4   ADAPTIVE DESIGN: PLAY THE WINNER   60
1.7.5   THE BAYESIAN ANALYSIS   60
       REFERENCES   62

1.8   A REGISTER OF CLINICAL TRIALS IN CHILDREN   63
Chiara Pandolfini and Maurizio Bonati
       REFERENCES   71

1.9   EPIDEMIOLOGICAL ASPECTS OF DRUG USE   73
Antonio Clavenna and Maurizio Bonati
1.9.1   INTRODUCTION   73
1.9.2   DRUG UTILISATION PROFILE   73
1.9.3   ANTI-ASTHMATIC DRUGS   76
1.9.4   PSYCHOTROPIC MEDICATIONS   77
1.9.5   CONCLUSIONS   80
       REFERENCES   81

PAEDIATRIC CLINICAL PHARMACOLOGY   85

2.1   THE EFFECT OF AGE ON DRUG METABOLISM   87

Saskia de Wildt, Trevor N. Johnson and Imti Choonara
2.1.1   INTRODUCTION   87
2.1.2   GENERAL TRENDS IN THE DEVELOPMENT OF
       PHASE 1 DRUG METABOLISM IN CHILDREN   88
2.1.3   CYP3A4   88
2.1.4   CYP1A2   90
2.1.5   GLUCURONIDATION AND SULPHATION   92
2.1.6   CONCLUSIONS   94
       REFERENCES   95

2.2   APPROACHES TO STUDYING THE DEVELOPMENT
      OF DRUG METABOLISM IN CHILDREN   97
Trevor N. Johnson
2.2.1   INTRODUCTION   97
2.2.2   ENZYMES INVOLVED IN DRUG METABOLISM   98
2.2.3   APPROACHES TO STUDYING THE DEVELOPMENT OF
       DRUG METABOLISM IN HUMANS   100
2.2.4   CONCLUSIONS   118
       REFERENCES   119

2.3   DRUG DISPOSITION   123
Gérard Pons, Jean-Marc Treluyer and Elisabeth Rey
2.3.1   INTRODUCTION   123
2.3.2   ABSORPTION OF DRUGS   123
2.3.3   DRUG DISTRIBUTION   124
2.3.4   DRUG METABOLISM   125

	2.3.5	DRUG ELIMINATION	126
	2.3.6	PRACTICAL CONSEQUENCES	127
		REFERENCES	127

2.4 PHARMACOKINETICS AND PHARMACODYNAMICS   129
Noel E. Cranswick

	2.4.1	INTRODUCTION	129
	2.4.2	PHARMACOKINETICS	129
	2.4.3	DRUG TRANSFER	130
	2.4.4	SYSTEMIC AVAILABILITY	135
	2.4.5	VOLUME OF DISTRIBUTION	136
	2.4.6	CLEARANCE	137
	2.4.6	WHICH PRIMARY PARAMETERS	139
	2.4.7	SECONDARY PARAMETERS	139
	2.4.8	THERAPEUTIC RANGE	140
	2.4.9	APPLICATION OF PHARMACOKINETIC PRINCIPLES TO MULTIPLE DOSE REGIMENS	141
	2.4.10	PHARMACODYNAMICS	142
	2.4.11	COMPUTER PROGRAMMES	143
		ABBREVIATIONS	144
		REFERENCES	145
		SUGGESTED AND FURTHER READING	145

2.5 POPULATION PHARMACOKINETICS AND PHARMACODYNAMICS   147
Alison H. Thomson

	2.5.1	INTRODUCTION	147
	2.5.2	PHARMACOKINETICS	148
	2.5.3	PHARMACODYNAMICS	149
	2.5.4	THE POPULATION APPROACH	149
	2.5.5	PAEDIATRIC APPLICATIONS OF POPULATION PHARMACOKINETIC ANALYSIS	154
	2.5.6	CONCLUSIONS	158
		ACKNOWLEDGEMENTS	158
		REFERENCES	158

2.6 PHARMACOGENETICS   161
Evelyne Jacqz-Aigrain and Anders Rane

	2.6.1	DEFINITION	161
	2.6.2	POLYMORPHISMS OF DRUG METABOLISING ENZYMES	162
	2.6.3	POLYMORPHISM OF ABC TRANSPORTERS	169
	2.6.4	POLYMORPHISM OF TARGET PROTEINS	169
	2.6.5	METHODS IN PHARMACOGENETICS	169
	2.6.6	IMPLICATIONS OF PHARMACOGENETICS IN THERAPEUTICS	170
	2.6.7	PROSPECTS IN PHARMACOGENETICS IN DRUG TREATMENT OF CHILDREN	173
	2.6.8	CONCLUSIONS	174
		REFERENCES	175

2.7 THERAPEUTIC DRUG MONITORING 177
Rafael Gorodischer
2.7.1 INTRODUCTION 177
2.7.2 INDIVIDUALISATION OF DRUG THERAPY 177
2.7.3 ADVANTAGES AND LIMITATIONS OF TDM 180
2.7.4 PAEDIATRIC AND NEONATAL TDM 181
2.7.5 SPECIFIC DRUGS: SPECIAL CONSIDERATIONS 182
2.7.6 TDM IN DEVELOPING COUNTRIES 187
2.7.7 FUTURE DIRECTIONS 187
ABBREVIATIONS 187
REFERENCES 188

2.8 DRUG TOXICITY AND PHARMACOVIGILANCE 191
Imti Choonara
2.8.1 INTRODUCTION 191
2.8.2 DRUG TOXICITY IN THE NEONATE 191
2.8.3 DRUG TOXICITY IN CHILDREN 193
2.8.4 FORMULATION ERRORS 195
2.8.5 PHARMACOVIGILANCE 196
2.8.6 CONCLUSION 198
REFERENCES 198

2.9 POISONING IN CHILDHOOD 201
Michael Riordan and George Rylance
2.9.1 ASSESSMENT 201
2.9.2 MECHANISMS OF TOXICITY 202
2.9.3 PHARMACOKINETICS 204
2.9.4 INVESTIGATION 206
2.9.5 TREATMENT 206
2.9.6 SUPPORTIVE TREATMENT 209
2.9.7 ANTIDOTES 211
2.9.8 PREVENTION 215
REFERENCES 215

THE ADMINISTRATION OF MEDICINES TO CHILDREN 217

3.1 ROUTES OF ADMINISTRATION AND FORMULATIONS 219
Anthony J. Nunn
3.1.1 INTRODUCTION 219
3.1.2 EXCIPIENTS 220
3.1.3 GENERAL CONSIDERATIONS 222
3.1.4 ORAL ROUTE 223
3.1.5 INTRAVENOUS ROUTE 225
3.1.6 OTHER INJECTIONS 226
3.1.7 PULMONARY ROUTE 226
3.1.8 OTHER ROUTES OF ADMINISTRATION 227

3.1.9   EXTEMPORANEOUS PREPARATION   228
3.1.10 CONCLUSION   230
        REFERENCES   231

3.2 NOVEL APPROACHES TO THE ROUTE OF ADMINISTRATION   235
     Jörg Breitkreutz
     3.2.1   INTRODUCTION   235
     3.2.2   PARENTERAL ADMINISTRATION   236
     3.2.3   ORAL ADMINISTRATION   236
     3.2.4   RECTAL ADMINISTRATION   240
     3.2.5   TRANSDERMAL ADMINISTRATION   241
     3.2.6   ALTERNATIVE ROUTES OF ADMINISTRATION   242
             REFERENCES   243

3.3 MEDICATION ERRORS   245
     David H. Cousins
     3.3.1   PATIENT SAFETY, MEDICAL ERRORS AND
             ADVERSE EVENTS   245
     3.3.2   MEDICATION ERRORS, ADVERSE DRUG EVENTS
             AND ADVERSE DRUG REACTIONS   246
     3.3.3   THE INCIDENCE OF PAEDIATRIC MEDICATION ERRORS   249
     3.3.4   TYPES OF PAEDIATRIC MEDICATION ERROR   250
     3.3.5   CASE REPORTS   253
     3.3.6   METHODS TO PREVENT MEDICATION ERRORS
             ON CHILDREN   257
     3.3.7   CONCLUSION   262
             REFERENCES   263

3.4 COMPLIANCE OR CONCORDANCE IN CHILDREN?   265
     Sharon Conroy, Tony Nunn and Steve Tomlin
     3.4.1   COMPLIANCE OR CONCORDANCE IN CHILDREN?   265
     3.4.2   WHAT DO PARENT/CARERS THINK?   266
     3.4.3   CONSEQUENCES OF NON-COMPLIANCE   266
     3.4.4   FACTORS INCREASING NON-COMPLIANCE   267
     3.4.5   WHAT CAN WE DO TO HELP?   270
             REFERENCES   273

DRUGS IN PREGNANCY AND LACTATION   275

4.1 EPIDEMIOLOGY OF DRUG USE IN PREGNANCY   277
     Gideon Koren
     4.1.1   INTRODUCTION   277
     4.1.2   PREGNANCY-INDUCED CONDITIONS   278
     4.1.3   CHRONIC CONDITIONS   278
     4.1.4   STUDYING THE SAFETY OF DRUGS IN PREGNANCY   279
     4.1.5   CONCLUSION   280
             ACKNOWLEDGMENT   280
             REFERENCES   280

4.2   PLACENTAL TRANSFER OF DRUGS   281
      Raphaël Serreau and Evelyne Jacqz-Aigrain
      4.2.1   PLACENTAL TRANSFER MECHANISMS AND THEIR
              QUANTIFICATION   281
      4.2.2   PLACENTAL TRANSFER OF DRUGS - A CLASSIFICATION   284
      4.2.3   CONCLUSION   289
              REFERENCES   289

4.3   DRUG ADMINISTRATION DURING PREGNANCY-
      EVALUATION OF FETAL TOXICITY: ANIMAL STUDIES   291
      Pierre Guittin
      4.3.1   INTRODUCTION: THE NEED FOR ANIMAL
              TERATOLOGY STUDIES   291
      4.3.2   STUDY PLAN ENDPOINTS   292
      4.3.3   ANALYSIS OF RESULTS   294
      4.3.4   CONCLUSION   298
              REFERENCES   298

4.4   EVALUATION OF FETAL TOXICITY: HUMAN DATA   299
      Anders Rane
      4.4.1   INTRODUCTION   299
      4.4.2   ONTOGENIC DEVELOPMENT OF DRUG METABOLISING
              ENZYMES IN LABORATORY ANIMALS   300
      4.4.3   ONTOGENESIS OF DRUG METABOLISM IN HUMANS   301
      4.4.5   DRUG DISPOSITION IN PREGNANCY   302
      4.4.6   PASSAGE OF DRUGS ACROSS THE PLACENTA   302
      4.4.7   EFFECTS ON THE FETUS   304
      4.4.8   SUMMARY   306
              REFERENCES   306

4.5   FETAL PHARMACOLOGY AND THERAPY   309
      Evelyne Jacqz-Aigrain
      4.5.1   INTRODUCTION   309
      4.5.2   PHARMACOKINETICS DURING PREGNANCY   310
      4.5.3   EXAMPLES OF FETAL THERAPY   311
      4.5.4   MATERNAL-FETAL SURGERY   317
      4.5.5   CONCLUSION   318
              REFERENCES   318

4.6   DRUG TOXICITY DURING BRAIN DEVELOPMENT   321
      Pierre Gressens, Ignacio Sfaello and Philippe Evrard
      4.6.1   INTRODUCTION   321
      4.6.2   ETHANOL   323
      4.6.3   OPIATES   324
      4.6.4   COCAINE   326
      4.6.5   CAFFEINE   327
      4.6.6   SMOKING AND NICOTINE   328
      4.6.7   EPILEPSY AND ANTI-EPILEPTIC DRUGS   329

4.6.8   SELECTIVE SEROTONIN REUPTAKE INHIBITORS (SSRI)   331
        REFERENCES   331

4.7 NEONATAL EFFECTS OF DRUGS ADMINISTRATED
     DURING PREGNANCY   333
     John N. van den Anker
     4.7.1   INTRODUCTION   333
     4.7.2   ANALGESICS   333
     4.7.3   CORTICOSTEROIDS   335
     4.7.4   ANGIOTENSIN CONVERTING ENZYME INHIBITORS   336
     4.7.5   NONSTEROIDAL ANTI-INFLAMMATORY DRUGS   338
             REFERENCES   340

4.8 MANAGEMENT OF NEONATES EXPOSED TO
     ILLICIT DRUGS DURING PREGNANCY   343
     Claude Lejeune
     4.8.1   INTRODUCTION   343
     4.8.2   NEONATAL ABSTINENCE SYNDROME (NAS)   343
     4.8.3   SYMPTOMATIC TREATMENT   344
     4.8.4   MEDICAL TREATMENT   345
     4.8.5   USE OF OPIATE-DERIVED MEDICATIONS   345
     4.8.6   USE OF OTHER DRUGS   346
     4.8.7   CONCLUSIONS   347
             REFERENCES   347

4.9 MEDICATION AND BREAST FEEDING   349
     E. Rey and R. Serreau
     4.9.1   INTRODUCTION   349
     4.9.2   PHYSIOLOGY OF LACTATION   350
     4.9.3   PASSAGE OF DRUGS INTO BREAST MILK   350
     4.9.4   AMOUNT OF DRUG EXCRETED INTO BREAST MILK   352
     4.9.5   CLASSIFICATION OF MEDICATIONS BY CATEGORY
             OF RISK IN THE CASE OF BREASTFEEDING   353
     4.9.6   CONCLUSION: PRACTICAL RECOMMENDATIONS   355
             REFERENCES   357

4.10 CLINICAL DRUG TRIALS IN PREGNANCY:
     CHALLENGES AND SOLUTIONS   359
     Gideon Koren
     4.10.1 INTRODUCTION   359
     4.10.2 CHANGES IN DRUG DISPOSITION IN LATE PREGNANCY   360
     4.10.3 LEARNING FROM THE GLYBURIDE MILESTONE   360
     4.10.4 A PROPOSED FRAMEWORK FOR DRUG TRIALS
             IN PREGNANCY   361
             ACKNOWLEDGEMENT   362
             REFERENCES   362

MEDICINES IN NEONATES 363

5.1 PROBLEMS WITH MEDICINES IN THE
NEWBORN INFANT 365
John N. van den Anker
  5.1.1 INTRODUCTION 365
  5.1.2 DRUG ABSORPTION 366
  5.1.3 DRUG DISTRIBUTION 369
  5.1.4 DRUG METABOLISM 371
  5.1.5 RENAL EXCRETION 371
  REFERENCES 373

5.2 NEONATAL RDS-PRENATAL CORTICOSTEROIDS 377
Henry L. Halliday
  5.2.1 INTRODUCTION 377
  5.2.2 GLUCOCORTICOIDS USED IN CLINICAL TRIALS 378
  5.2.3 RANDOMISED CLINICAL TRIALS 379
  5.2.4 GUIDELINES FOR PRENATAL GLUCOCORTICOIDS 382
  5.2.5 COMBINATION THERAPIES 383
  5.2.6 CONCLUSIONS 384
  REFERENCES 384

5.3 NEONATAL RDS-SURFACTANT 387
Henry L. Halliday
  5.3.1 INTRODUCTION 387
  5.3.2 CHEMISTRY AND METABOLISM OF SURFACTANT 387
  5.3.3 BIOPHYSICAL AND PHYSIOLOGICAL EFFECTS
OF SURFACTANT 389
  5.3.4 SURFACTANT PREPARATIONS 389
  5.3.5 CLINICAL TRIALS 389
  5.3.6 TREATMENT REGIMENS 392
  5.3.7 ACUTE RESPONSES AND VENTILATOR SETTINGS 393
  5.3.8 ADVERSE EFFECTS 394
  5.3.9 NEW GENERATION SYNTHETIC SURFACTANTS 394
  5.3.10 CONCLUSIONS 395
  REFERENCES 395

5.4 NON-STEROIDAL ANTIINFLAMMATORY DRUGS
(NSAIDS), PROSTANOIDS AND THE DEVELOPMENTAL
KIDNEY: A SUBTLE COMPROMISE 397
J. P. Langhendries
  5.4.1 INTRODUCTION 397
  5.4.2 PROSTANOIDS ACTION ON THE KIDNEY THROUGH THEIR
SPECIFIC RECEPTORS 397
  5.4.3 NON-STEROIDAL ANTIINFLAMMATORY DRUGS (NSAIDS)
AND THE IMMATURE KIDNEY 399
  REFERENCES 401

5.5 PHARMACOLOGY OF ANALGESICS IN NEONATES   403
     John N. van den Anker and Dick Tibboel
     5.5.1 INTRODUCTION   403
     5.5.2 OPIOIDS   403
     5.5.3 PARACETAMOL   406
            REFERENCES   406

5.6 PREVENTION AND MANAGEMENT OF
     PAIN IN NEONATES   409
     E. Jacqz-Aigrain and K. J. S. Anand
     5.6.1 INTRODUCTION   409
     5.6.2 CONSENSUS STATEMENT   410
            REFERENCES   414

5.7 PHARMACOLOGY AND USE OF METHYLXANTHINES AND
     DOXAPRAM FOR THE TREATMENT OF NEONATAL APNOEA 417
     Marie-Jeanne Boutroy
     5.7.1 METHYLXANTHINES   417
     5.7.2 DOXAPRAM   420
     5.7.3 MISCELLANEOUS   421
     5.7.4 CONCLUSIONS   422
            REFERENCES   422

5.8 THERAPY FOR PERSISTENT DUCTUS ARTERIOSUS   425
     Bart Van Overmeire
     5.8.1 INTRODUCTION   425
     5.8.2 INDOMETHACIN   426
     5.8.3 IBUPROFEN   428
     5.8.4 ALTERNATIVE DRUGS   431
     5.8.5 SURGICAL TREATMENT   431
     5.8.6 CONCLUSION   432
            REFERENCES   432

5.9 ANTIBIOTICS IN NEONATES: THE NEED FOR
     A MORE RATIONAL APPROACH   435
     Jean-Paul Langhendries and John N. van den Anker
     5.9.1 INTRODUCTION   435
     5.9.2 BACTERIAL EPIDEMIOLOGY   436
     5.9.3 WHICH ANTIBIOTICS SHOULD BE PRESCRIBED?   436
     5.9.4 PHARMACOKINETICS (PK)   440
     5.9.5 PHARMACOKINETIC/PHARMACODYNAMIC (PK/PD)
            RELATIONSHIP   441
     5.9.6 ADMINISTRATION OF ANTIBIOTICS   444
            REFERENCES   446

5.10 THE USE OF ANTIEPILEPTIC DRUGS IN NEONATES   449
      Alexis Arzimanoglou   449
      5.10.1 INTRODUCTION   449
      5.10.2 CLINICAL AND EEG FEATURES OF NEONATAL SEIZURES   450

5.10.3 THE ROLE AND IMPACT OF ELECTRICAL DISCHARGES   451
5.10.4 AETIOLOGY   452
5.10.5 RECOGNISED EPILEPSY SYNDROMES IN NEONATES   454
5.10.6 MANAGEMENT OF NEONATAL SEIZURES   456
5.10.7 TREATMENT FOR ELECTROLYTE ABNORMALITIES   456
5.10.8 ANTIEPILEPTIC DRUG THERAPY   457
5.10.9 CONCLUSIONS   459
        REFERENCES   459

INFECTIONS   463

6.1 ANTIBACTERIAL AGENTS   465
Patricia Mariani-Kurkdjian and Edouard Bingen
     6.1.1 INTRODUCTION   465
     6.1.2 INHIBITORS OF CELL WALL SYNTHESIS   465
     6.1.3 INHIBITORS OF PROTEIN SYNTHESIS   471
     6.1.4 INHIBITORS OF NUCLEIC ACID SYNTHESIS   479
            TEXTBOOKS   483
            REFERENCES   483

6.2 URINARY TRACT INFECTIONS   485
Ulf Jodal
     6.2.1 CLINICAL PRESENTATION   485
     6.2.2 BACTERIA AND THE HOST REACTION   485
     6.2.3 RENAL DAMAGE   486
     6.2.4 ANTIBACTERIAL TREATMENT   486
            REFERENCES   488

6.3 BACTERIAL MENINGITIS   491
Antoine Bourillon and Edouard Bingen
     6.3.1 PATHOPHYSIOLOGY OF MENINGEAL INFECTION   491
     6.3.2 TREATMENT   492
     6.3.3 PREDICTIVE EFFICACY CRITERIA OF ANTIBIOTICS
            DURING MENINGITIS   494
     6.3.4 EPIDEMIOLOGY OF THE ORGANISMS INVOLVED
            IN BACTERIAL MENINGITIS   495
     6.3.5 THERAPEUTIC STRATEGY   498
            REFERENCES   500

6.4 ANTIVIRALS   503
Pierre Lebon
     6.4.1 INTRODUCTION   503
     6.4.2 MECHANISM OF ACTION   503
     6.4.3 INHIBITORS OF CELLULAR ENTRY   505
     6.4.4 INHIBITORS OF VIRAL EXCRETION: VIRAL INHIBITORS
            OF INFLUENZA A AND B   506
     6.4.5 INHIBITORS OF VIRAL MULTIPLICATION: INHIBITORS
            OF THE REPLICATION OF VIRAL NUCLEIC ACIDS   508

        6.4.6   RIBAVIRIN   511
                REFERENCES   512

6.5   ANTIRETROVIRAL DRUGS IN PAEDIATRIC PATIENTS   515
        Jean Marc Tréluyer
        6.5.1   INTRODUCTION   515
        6.5.2   THE HUMAN IMMUNODEFICIENCY VIRUS-HIV   516
        6.5.3   CLASSIFICATION OF ANTIRETROVIRAL DRUGS   516
        6.5.4   ANTIRETROVIRAL DRUGS IN NEONATES   518
        6.5.5   ANTIRETROVIRAL DRUGS IN INFANTS AND CHILDREN   522
        6.5.6   CONCLUSIONS   523
                REFERENCES   523

6.6   PROBLEMS OF DRUG THERAPY IN DEVELOPING COUNTRIES   527
        J. Brian S. Coulter
        6.6.1   INTRODUCTION   527
        6.6.2   ADMINISTRATION   527
        6.6.3   ESSENTIAL DRUG SCHEME   528
        6.6.4   DRUG RESISTANCE   528
        6.6.5   DRUG METABOLISM IN MALNUTRITION   529
                REFERENCES   530

6.7   IMMUNISATIONS AND DRUG THERAPY   531
        Robert Cohen
        6.7.1   INTRODUCTION   531
        6.7.2   ANTIBIOTIC TREATMENT   531
        6.7.3   SPACING OF DIFFERENT VACCINES   531
        6.7.4   CORTICOSTEROIDS   532
        6.7.5   IMMUNOSUPPRESSIVE THERAPY   534
        6.7.6   IMMUNOGLOBULIN AND BLOOD PRODUCTS   534
        6.7.7   ANTICOAGULANT THERAPY   535
                REFERENCES   535

CRITICAL CARE, NEUROLOGY AND ANALGESIA   537

7.1   NITRIC OXIDE   539
        Jean-Christophe Mercier and Anh Tuan Dinh-Xuan
        7.1.1   CLINICAL SETTING   539
        7.1.2   INHALED NITRIC OXIDE THERAPY   540
        7.1.3   PHARMACOLOGY OF NO   541
        7.1.4   INHALED NO THERAPY SIDE EFFECTS   542
        7.1.5   CONCLUSION   543
                REFERENCES   543

7.2   DRUG DISPOSITION DURING EXTRA-CORPOREAL
        MEMBRANE OXYGENATION (ECMO)   545
        Hussain Mulla
        7.2.1   INTRODUCTION   545

7.2.2 DRUG DISPOSITION DURING ECMO 545
REFERENCES 551

7.3 MUSCLE RELAXANTS 553
Stephen D. Playfor
7.3.1 INDICATIONS FOR NEUROMUSCULAR BLOCKADE 553
7.3.2 PHYSIOLOGY OF NEUROMUSCULAR BLOCKADE 554
7.3.3 COMMONLY USED NEUROMUSCULAR BLOCKING AGENTS 554
7.3.4 TOXICITY AND COMPLICATIONS OF
NEUROMUSCULAR BLOCKADE 557
REFERENCES 558

7.4 SEDATION 559
Stephen D. Playfor
7.4.1 THE AIMS OF SEDATION 559
7.4.2 COMMONLY USED SEDATIVE AND ANALGESIC AGENTS 560
7.4.3 TOXICITY AND COMPLICATIONS OF SEDATIVE AGENTS 562
7.4.4 ASSESSMENT OF SEDATION 563
REFERENCES 564

7.5 EPILEPSY AND ANTIEPILEPTIC DRUGS 565
Richard Appleton
7.5.1 INTRODUCTION 565
7.5.2 RECOGNITION OF EPILEPTIC SEIZURES 566
7.5.3 EPILEPTIC SEIZURES 566
7.5.4 EPILEPSIES AND EPILEPSY SYNDROMES 567
7.5.5 AETIOLOGY OF EPILEPSY 567
7.5.6 ASSOCIATED IMPAIRMENTS 568
7.5.7 PROGNOSIS OF EPILEPSY 568
7.5.8 DRUG TREATMENT OF EPILEPSY 568
7.5.9 WHEN TO START A DRUG? 569
7.5.10 WHICH DRUG AND IN WHAT DOSE? 569
7.5.11 WHEN TO CHANGE A DRUG OR ADD A SECOND DRUG? 571
7.5.12 MECHANISM OF ACTION OF ANTIEPILEPTIC DRUGS 571
7.5.13 WHEN TO STOP THE DRUG 575
7.5.14 TOXICITY 575
7.5.15 HOLISTIC TREATMENT 578
7.5.16 STATUS EPILEPTICUS (CONVULSIVE AND
NON-CONVULSIVE) 579
7.5.17 CONCLUSION 579
REFERENCES 580

7.6 SEDATIVES 581
William P. Whitehouse
7.6.1 INTRODUCTION 581
7.6.2 WHAT IS SEDATION? 582
7.6.3 WHAT IS SEDATION FOR? 584
7.6.4 CHOICE OF DRUG(S) 585

7.6.5	ADVERSE EFFECTS AND SAFETY	586
7.6.6	CONCLUSIONS	586
	REFERENCES	586

7.7 OPIOID ANALGESIC DRUGS   589
Greta M. Palmer and Brian J. Anderson

7.7.1	INTRODUCTION	589
7.7.2	NATURALLY OCCURRING OPIOIDS	590
7.7.3	SEMI-SYNTHETIC OPIOIDS	597
7.7.4	SYNTHETIC OPIOIDS	598
7.7.5	MIXED OPIOID AGONIST/ANTAGONISTS	604
	REFERENCES	605

7.8 NON-OPIOID ANALGESICS   613
Greta M. Palmer and Brian J. Anderson

7.8.1	TRAMADOL	613
7.8.2	CLONIDINE	615
7.8.3	KETAMINE	616
	REFERENCES	617

7.9 PARACETAMOL   621
Brian J. Anderson

7.9.1	MECHANISM OF ACTION	621
7.9.2	PHARMACODYNAMICS	621
7.9.3	PHARMACOKINETICS	622
7.9.4	TOXICITY	624
7.9.5	PARACETAMOL DOSING	625
	REFERENCES	626

IMMUNOSUPPRESSANTS, RHEUMATIC AND GASTROINTESTINAL TOPICS
CLINICAL SECTION (A)

629

8.1 PHARMACOLOGY AND USE OF IMMUNOSUPPRESSANTS   631
Evelyne Jacqz-Aigrain and Pierre Wallemacq

8.1.1	INTRODUCTION	631
8.1.2	PHARMACOLOGY AND USE OF IMMUNOSUPPRESSANTS	632
8.1.3	CONCLUSION	641
	REFERENCES	641

8.2 RHEUMATIC DISORDERS   643
A. G. Cleary and H. Venning

8.2.1	INTRODUCTION	643
8.2.2	TREATMENT OPTIONS	644
8.2.3	IMMUNISATION	651
	REFERENCES	651

8.3 TREATMENT OF INFLAMMATORY BOWEL DISEASES
IN CHILDREN 653
Jean Pierre Cezard and Jean Pierre Hugot
8.3.1 INTRODUCTION 653
8.3.2 TREATMENT 654
8.3.3 CONCLUSION 659
REFERENCES 659

8.4 GASTRO-OESOPHAGEAL REFLUX 661
Marc Bellaiche, Prévost Jantchou and Jean Pierre Cezard
8.4.1 INTRODUCTION 661
8.4.2 NON-PHARMACOLOGICAL THERAPIES 661
8.4.3 ANTACIDS 662
8.4.4 PROKINETICS 662
8.4.5 ACID SUPPRESSANTS 663
8.4.6 SURGICAL THERAPY 663
8.4.7 THERAPEUTIC ENDOSCOPIC PROCEDURES 664
8.4.8 CONCLUSION 664
REFERENCES 664

8.5 DIARRHOEA 665
Ulrich Meinzer and Marc Bellaiche
8.5.1 INTRODUCTION 665
8.5.2 DRUGS 665
8.5.3 CLASSIFICATION OF DRUGS 666
8.5.4 CONCLUSION 668
REFERENCES 669

8.6 INSULIN 671
Rachel M. Williams and David B. Dunger
8.6.1 INTRODUCTION 671
8.6.2 HUMAN INSULIN 672
8.6.3 INSULIN ANALOGUES 673
8.6.4 SIDE EFFECTS OF INSULIN THERAPY 678
8.6.5 STRATEGIES FOR INSULIN THERAPY 679
8.6.6 SUMMARY 681
REFERENCES 681

8.7 HUMAN GROWTH HORMONE 683
Paul Czernichow
8.7.1 GH PHARMACOLOGY 683
8.7.2 GH AS A THERAPEUTIC AGENT 684
8.7.3 SIDE EFFECTS 685
REFERENCES 686

8.8 ANTIPSYCHOTROPIC DRUGS 687
Marie France LeHeuzey
8.8.1 INTRODUCTION 687
8.8.2 STIMULANT MEDICATIONS 687

8.8.3   ANTIDEPRESSANTS   690
8.8.4   NEUROLEPTICS AND ATYPICAL ANTIPSYCHOTICS   694
8.8.5   CONCLUSIONS   697
         REFERENCES   699

RESPIRATORY, ENDOCRINE, CARDIAC, AND RENAL TOPICS
CLINICAL SECTION (B)

                                                                  701

9.1   ASTHMA   703
Kristine Desager
      9.1.1   CLINICAL SETTING   703
      9.1.2   THERAPY   703
      9.1.3   PHARMACOLOGY AND SIDE EFFECTS OF
               INHALED CORTICOSTEROIDS   706
      9.1.4   PRACTICAL PROPOSAL: TREATMENT OF ASTHMA   709
               REFERENCES   710

9.2   CYSTIC FIBROSIS   711
Alan Smyth
      9.2.1   CLINICAL SETTING   711
      9.2.2   PHARMACOLOGY AND DRUG THERAPY   712
      9.2.3   TREATMENT OF PULMONARY INFECTION   713
      9.2.4   CONCLUSION   717
               REFERENCES   717

9.3   MANAGEMENT OF HYPERTENSIVE EMERGENCIES
      IN CHILDREN   719
Peter Houtman
      9.3.1   INTRODUCTION   719
      9.3.2   HYPERTENSION   720
      9.3.3   MANAGEMENT   721
      9.3.4   DRUG THERAPY   722
      9.3.5   CONCLUSION   724
               REFERENCES   725

9.4   HEART FAILURE   727
Beat Friedli
      9.4.1   INTRODUCTION   727
      9.4.2   DIURETICS   728
      9.4.3   DIGOXIN   730
      9.4.4   ANGIOTENSIN CONVERTING ENZYME (ACE) INHIBITORS   731
      9.4.5   OTHER VASODILATORS   733
      9.4.6   BETA-BLOCKERS   734
      9.4.7   SPIRONOLACTONE   735
               REFERENCES   736

9.5　DIURETICS　737
Jean-Pierre Guignard
　9.5.1　INTRODUCTION　737
　9.5.2　RENAL FUNCTION　737
　9.5.3　BODY FLUID HOMEOSTASIS　739
　9.5.4　CLINICAL USE OF DIURETICS　739
　9.5.5　CLASSIFICATION OF DIURETICS　742
　9.5.6　FILTRATION DIURETICS　742
　9.5.7　OSMOTIC DIURETICS　746
　9.5.8　INHIBITORS OF CARBONIC ANHYDRASE　746
　9.5.9　LOOP DIURETICS　747
　9.5.10　DISTAL DIURETICS　749
　9.5.11　K$^+$-SPARING DRUGS　750
　9.5.12　NEW DEVELOPMENTS　752
　　　REFERENCES　753

9.6　ENURESIS　755
Pierre Cochat and Behrouz Kassaï
　9.6.1　INTRODUCTION　755
　9.6.2　CLINICAL PRESENTATION　756
　9.6.3　PATHOPHYSIOLOGY　758
　9.6.4　MANAGEMENT　759
　9.6.5　CONCLUSION　763
　　　REFERENCES　763

9.7　STEROID SENSITIVE NEPHROTIC SYNDROME　765
Nicholas J. A. Webb
　9.7.1　INTRODUCTION　765
　9.7.2　DRUG THERAPY OF THE PRESENTING EPISODE OF
　　　　NEPHROTIC SYNDROME　766
　9.7.3　MANAGEMENT OF INITIAL RELAPSES OF SSNS　767
　9.7.4　MANAGEMENT OF FREQUENTLY RELAPSING AND
　　　　STEROID DEPENDENT SSNS　767
　9.7.5　LEVAMISOLE　770
　9.7.6　ALKYLATING AGENTS　770
　9.7.7　CICLOSPORIN　772
　9.7.8　NEWER AGENTS　773
　9.7.9　PROGNOSIS OF SSNS　773
　　　REFERENCES　774

9.8　CANCER CHEMOTHERAPY FOR PAEDIATRIC
　　　MALIGNANCIES　775
Gilles Vassal
　9.8.1　DACTINOMYCIN (ACTINOMYCIN D)　776
　9.8.2　ASPARAGINASE　776
　9.8.3　ALKYLATING AGENTS　777
　9.8.4　CYCLOPHOSPHAMIDE　778
　9.8.5　IFOSFAMIDE　779

9.8.6	PLATINUM SALTS	780
9.8.7	CARBOPLATIN	781
9.8.8	CYTARABINE	782
9.8.9	ANTHRACYCLINES	783
9.8.10	ETOPOSIDE	784
9.8.11	IMATINIB MESYLATE	785
9.8.12	6-MERCAPTOPURINE (6-MP)	786
9.8.13	METHOTREXATE	787
9.8.14	TEMOZOLOMIDE	789
9.8.15	VINCRISTINE AND OTHER SPINDLE POISONS	790
	REFERENCES	791
INDEX		793

Drug evaluation in children

1.1 Unlicenced and off-label drug usage in children in Europe

Sharon Conroy

Academic Division of Child Health, University of Nottingham, Derbyshire Children's Hospital, Uttoxeter Road, Derby, DE22 3NE, UK

1.1.1 INTRODUCTION

Most drugs used in adults are high quality products with a licence for their use. In children however, we often need to use medicines that are unlicenced (UL), or licenced medicines in ways not covered by the licence (off label) (OL). We cannot guarantee the safety, efficacy or quality of these drugs, as they have not been tested in clinical trials in the paediatric population for the indications and methods by which they are used. Many reasons are given by the pharmaceutical industry for not testing drugs in children, e.g. difficulties in studying drugs in children, ethical problems and consent issues. However, the real reason is that it is time consuming and expensive to perform trials in children, and without a legal obligation and/or financial incentives to do so then they are unlikely to happen.

Between January 1995 and April 1998, 45 drugs were licenced through the European Medicines Evaluation Agency (EMEA), 29 were of potential use in children, but only 10 of these (34%) were licenced for paediatric use [1]. A further study showed that the situation to Sept 2001 had not changed. Only 47 (33%) of 141 potentially useful medicines had been licenced by the EMEA for use in children [2]. Many of the medicines licenced were vaccines (with a large paediatric market) and anti-HIV drugs (for which high financial incentives for development had been made available). Entries in a Dutch drug information compendium showed similar disappointing figures [3]. Of 1157 drugs potentially useful in children, 29% were licenced for use in all children, 22% in some paediatric age/weight groups and the rest gave no or little paediatric information or had disclaimers against use in children.

The use of UL and OL medicines does not imply poor medical or pharmacy practice, it is done when there is no alternative. European legislation makes provision for doctors to prescribe, pharmacists to dispense and nurses to administer UL and OL medicines [4]. Licensing principally regulates the pharmaceutical industry in their marketing of drugs.

Unlicenced medicines

UL medicines must be used when there is no commercially available, suitable formulation of the drug to allow administration of the neonatal or paediatric dose, in a form that the child can take. Pharmacists therefore have to provide the drug by one of the following means:

Extemporaneously dispensed medicines

This may involve crushing tablets, opening capsules, or weighing the 'raw' drug and suspending the resulting powder in an agent to produce a liquid formulation; or diluting with a bulking agent to a specific strength and supplying as a powder in a sachet, or as a capsule for the patient to swallow or to open and give the contents in food or milk.

Such liquids, capsules and powders are regularly produced in large volumes in hospital pharmacy departments throughout Europe [5]. This may have to be done with little information regarding bioavailability of the drug from the final product, the physical, chemical and microbial stability of the preparation, and a lack of methods to prepare a palatable formulation. Such products are unlikely to be subjected to quality assurance procedures. Preparations often have short shelf lives, requiring fresh supplies to be made on a frequent basis, inconvenient to a busy family and potentially affecting concordance. Adequate communication between hospital and community to share information on such formulations is essential to avoid interruptions in therapy.

Purchase of UL formulations

UL formulations may be purchased from a 'specials' pharmaceutical manufacturer. These products may have undergone quality assurance processes and be supplied with a certificate guaranteeing the desired standard. Examples are caffeine injection and oral solution.

Importing drugs licenced in other countries

Licenced products do not have the same availability between countries, even within Europe. For 75% of medicines requiring extemporaneous dispensing in one country, a suitable licenced alternative is available in another European country, North America or Australia [5,6]. Drugs licenced in other countries may be imported, but issues surrounding importation and free movement of medicines between countries make this difficult and sometimes impossible.

'Named patient supplies'

Pharmaceutical companies may supply UL drugs on a 'named-patient' basis, e.g. captopril 2mg tablets in the UK. Low strength tablets of mercaptopurine were a preferable alternative to extemporaneous liquids for children who could swallow small tablets, until their supply was discontinued by the manufacturer [7,8]. This was disappointing and has led to many problems. UK guidelines have now been issued for the conduct of notification of product discontinuations, which may reduce the risk of this happening again [9].

Use of chemicals

Chemicals are used when there is no alternative preparation or pharmaceutical grade material available e.g. betaine is used to treat homocystinuria [10].

Off label drug use

OL use of drugs refers to use outside the conditions of the licence or marketing authorisation. This may be due to:

• Dose

This may need to be lower (or higher) than recommended in the licence, depending on the age and weight of the child e.g. Use of the licenced dose of gentamicin in the preterm neonate results in sub-therapeutic peak levels and potentially toxic trough levels, an OL dose is needed to ensure effective therapy without toxicity [11,12].

• Age

Drugs are often not licenced in children under a certain age, or not at all for children, even if they are commonly used e.g. morphine injection which is not licenced in children.

• Indication

Drugs may be used to treat childhood illnesses not covered by the licence, e.g. diclofenac used in post-operative pain is only licenced to treat juvenile chronic arthritis.

- *Route of administration*

Alternative routes of administration may be necessary, e.g. midazolam intranasally or buccally to treat seizures in children, sodium chloride 30% injection given orally in neonatal patients requiring supplementation in a concentrated form.

- *Contra-indications*

These may need to be disregarded in certain children, e.g. the use of aspirin for its anti-platelet effect in patients with Kawasaki disease. Aspirin is contraindicated in children due to its links with Reye's syndrome.

This situation has existed for many years. The lack of appropriate formulations for children was highlighted in the 1960/70s, when children were termed 'therapeutic orphans' and the 'therapeutic underprivileged' [13,14]. However, in Europe, the issue only really started to be taken seriously in the late 1990s. Following the publication of a joint report by the (then) British Paediatric Association and the Association of the British Pharmaceutical Industry in 1996, the UK government was 'astonished to discover that this situation existed' [15]. They thought that 'the current situation in regard to the testing and licensing of medicines for use by children is unacceptable' [16]. They required that surveillance should investigate the use of UL and OL medicines.

Subsequent years showed a proliferation of such studies from across Europe examining many different areas where children are treated. The following sections review the studies published to date.

1.1.2 PAEDIATRIC INTENSIVE CARE

The first study in Europe was published in 1996 by a group in the UK that prospectively examined UL and OL drug use in paediatric intensive care [17]. In a 4 month period, 166 patients received 862 prescriptions, of which 268 (31%) were UL or OL and 70% of patients received at least one UL or OL drug. Morphine and midazolam injections were the drugs most frequently administered UL or OL.

These authors further reported 44% and 40% rates of combined UL and OL drug use in cardiac PICU and general PICU respectively in a 13 week prospective study [18]. Compared to other wards in the hospital UL and OL drug use was highest in both PICUs and the neonatal intensive care unit (NICU).

In the Netherlands [19,20], different areas of the hospital were studied 1 day a week for 5 weeks. The medium care unit (MCU) admitted 110 patients who received 905 prescriptions; the surgical ICU 33 patients received 308 prescriptions and the PICU 31 patients received 305 prescriptions. 1024 (48%) prescriptions were UL and 390 (18%) OL. In the 435 patient days studied 392 (90%) contained UL drugs or OL prescriptions. The most frequently prescribed UL or OL drug other than in NICU was cisapride.

1.1.3   NEONATAL INTENSIVE CARE

Table 1 shows studies conducted in NICUs. These studies showed that of the drugs most frequently prescribed in NICU, many are used in an UL or OL manner, as no licenced therapeutic alternatives to these drugs are available. The example of gentamicin was described earlier. Similarly, in the Dutch study tobramycin was the fifth most frequently prescribed UL or OL drug for the same reason [19,20]. Furthermore caffeine featured prominently in several studies as a high use UL drug, as did theophylline for the same indication and used OL or UL. Morphine was another example, commonly used with no licenced dose recommendations or appropriate formulations for neonates.

Throughout these sections differences in classification of UL and OL drugs between authors mean that studies are not necessarily directly comparable but give a good overall picture of the issue. Also the Dutch studies tend to show a higher percentage of UL drugs as, unlike in other countries, the hospital pharmacy manufactures many drugs in a 'homemade' ready to administer form on a large scale to overcome the lack of available paediatric formulations.

1.1.4   GENERAL MEDICAL AND SURGICAL PAEDIATRIC IN-PATIENTS

The above studies demonstrate that in critically ill patients the incidence of UL and OL drug use is high, but does the situation apply in more general paediatric patients? Table 2 shows that although the issue is perhaps less pronounced in such areas, it is still significant.

The Australian study highlights the lack of availability of licenced products in one country which are available in other countries [26], e.g. trimethoprim suspension and oxybutynin solution. Differences in the licence from country to country are also highlighted, even for drugs produced by the same pharmaceutical company, e.g. clobazam which is not licenced for children in Australia but is approved in the UK in those over 3 years of age. In Australia, patients and families are financially disadvantaged by the use of UL or OL products since these are not eligible for subsidies under the Pharmaceutical Benefits Scheme which operates there [31].

Even though prescribing habits across Europe vary greatly, around half the children in each of 5 countries were prescribed at least one UL or OL drug, and

Table 1

Studies in neonatal intensive care units

Country & year	Study period	Number of patients	Number of prescriptions	UL	OL *	Patients receiving an UL/OL drug	Most common UL/OL drug	Ref.
UK 1999	13 weeks	70 (49 preterm)	455	10%	55% dose	90%	benzylpenicillin	21
UK 1999	13 weeks	100	323	55% UL+OL				18
France 2000	4 weeks	40 (88% VLBW)	257	10%	63% age		caffeine	22
Israel 2000	4 months (sampled every 2 weeks)	105 (72 preterm)	525	16%	63% dose	93%	OL gentamicin UL theophylline	23
Netherlands 2001	5 weeks (1 day/week)	66	621	62%	14% frequency		caffeine	19,20
Australia 2002	10 weeks	101 (74 preterm)	1442	11%	47% indication	80%	morphine	24

* Reason for highest number of OL prescriptions

VLBW: very low birth weight

Table 2

Studies in general paediatric wards

Country & year	Study period	Number of patients	Number of prescriptions	UL	OL	Patients receiving an UL or OL	Most common UL/OL drug	Ref.
UK 1998	13 weeks	707	2013	7%	18%	36%	diclofenac (surgical) salbutamol (medical)	18,25
Australia 1999	5 weeks	200	735	7%	10%	36%	metoclopramide (surgical) chloral hydrate (medical)	26
UK, Sweden, Germany, Italy, Netherlands 2000	4 weeks	624	2262	7%	39%	67%	cyclizine, salbutamol, budesonide, beclometasone heparin	27
Northern Ireland 2001	8 weeks (1 day/week)	74	237	3%	19%	43%	terbutaline	28
Italy 2002	12 weeks	1461	4265	0.2%	60%	82%	paracetamol	29
Netherlands 2002	20 weeks	293	1017	28%	44%	92%	paracetamol	30

in Italy and the Netherlands, this was even higher [27]. Results were remarkably similar in the UK, Sweden and Germany. Analgesics and bronchodilators were among the five most frequently prescribed OL drugs in 4 of the 5 countries. This high incidence of UL and OL drug use for commonly used drugs is further confirmed in the Italian study where the top 3 classes of prescribed drugs were systemic antibacterials; anti-asthmatics and analgesics/antipyretics, and of these prescriptions 52%; 60% and 59% respectively were OL [29]. Similarly, the Dutch study highlighted that the top 10 UL or OL prescriptions included 7 of the 10 most frequently prescribed drugs [30].

Another British study showed that, within 69 local guidelines for the acute management and elective investigation of children, 86 drugs were recommended of which only 47 (55%) were licenced for use in children [32]. A further 14 were licenced only for children over a certain age or weight, 24 were UL or OL and 1 had unknown status.

1.1.5 OUTPATIENTS AND PRIMARY CARE

Whilst the incidence of UL and OL prescribing in hospitalised patients can be seen to be significant, it is recognised that the majority of children receiving medicines do so either as hospital outpatients or more commonly at home following prescriptions from their general practitioner (GP). Table 3 shows studies exploring the incidence and nature of UL and OL prescribing in primary care settings.

The highest incidence of UL and OL prescribing in primary care is reported to be in neonates and infants [35, 37, 41- 45]. Adolescents were also highlighted as receiving high numbers of UL and OL prescriptions in some studies [42, 43, 45]. Topical treatments for the skin, eye and ear also showed high rates of OL use though this may have been due to the method of classification where if children were not mentioned in the licence, then the product was considered OL. [35, 37, 42, 43].

The study in the Netherlands also reported that the odds for UL or OL drug use increased by 78% when the prescriber was a specialist rather than a GP [42]. Newer drugs were 2-4 times more likely to be used OL than older ones and drugs with a low use in the paediatric population were also more likely to be used UL or OL [42].

Some authors were not concerned about UL and OL use of topical preparations, since most ADRs in children are due to systemic drugs [43]. Others, however, suggest that this is very important due to the increased risk of absorption of substances through the skin due to increased surface area, greater hydration and immaturity in children.

Table 3

Studies in primary care and outpatients

Country & year	Study period and setting	Number of patients	Number of prescriptions	UL	OL *	Patients receiving an UL/OL drug	Most common UL/OL drug	Ref.
UK 2000	1 year (general practice)	1175	3347	0.3%	11% dose		amoxicillin	33
Israel 2000	2 months (outpatients)	132	222	8%	26% dose	42%	ferrous carbonate	34
France 2000	1 day (outpatients)	989	2522	4%	29% age	56%	topical preparations	35
Netherlands 2001	1 year (general practice)	6141	17.452	19%	10%	31%		36
Germany 2002	3 months (outpatients)		1,592.006		13% lack of information		xylometazoline or oxymetazoline	37

* Reason for highest number of OL prescriptions

Table 3 *(continued)*

Country & year	Study period and setting	Number of patients	Number of prescriptions	UL	OL*	Patients receiving an UL/OL drug	Most common UL/OL drug	Ref.
Netherlands 2002	1year (general practice and outpatients)	19,283	68,019	17%	23%		oral contraceptives	38
Netherlands 2002	1year (general practice)	6141	17,453	15%	14% dose	46%	fusidic acid ophthalmological gel	39
France 2002	4 months (outpatients)	1419			19% indication	42%		40
Israel 2002	Seasonal for 1 year. (outpatients)	1802	1925	0.1%	15% dose	16%	amoxicillin	41
Netherlands 2003	1year (general practice and outpatients)	18,943	66,222	17%	21%		ophthamologics & otologics	42
Sweden 2003	1year (outpatients)	357,784	575,526		21% lack of information		dermatologics	43

* Reason for highest number of OL prescriptions

1.1.6 MISCELLANEOUS

Psychotropic agents

Over a 5 year period, 19 British children under 12 years received a prescription for a selective serotonin reuptake inhibitor [46]. The most common was fluoxetine which is not licenced for use in children. The authors postulate that there are likely to be hundreds of children receiving OL antidepressants in general practice. Forty percent of respondents reported that they prescribed OL psychotropic agents to children in an Australian study [47], highlighting that many prescribers, even specialists in their area, are unfamiliar with current labelling of drugs since in fact most prescribing of psychotropics for children is OL.

Newly marketed drugs

A study in primary care in the UK monitored prescriptions for 63 newly marketed drugs in children (2-11 years) and adolescents (12-17 years) [48]. Forty four (70%) of the drugs were prescribed for children, of which only 6 (10%) were licenced in this age group; 58 were prescribed for adolescents, of which 8 were not recommended under 18 years. Most of the drugs had no specific recommendations for adolescents and were therefore assumed to be licenced. Twenty two percent of children and 6% of adolescents received OL drugs. Antiepileptic agents were the most common OL drug, with 30% of children treated for epilepsy receiving OL drugs.

Antidotes

A single study has looked at the licence status of antidotes recommended by an international body for the treatment of human poisonings [49]. Of 77 recommended antidotes only 31 (40%) were licenced in children. Fourteen (18%) were UL and 32 (42%) OL. Age was the main reason for drugs being OL. Some of the UL antidotes, had been used only in an experimental model, whereas others had only been used in small patient numbers in certain poison centres.

Pain management

A 4-week prospective study involving one acute medical and one acute surgical ward showed 235 (33%) of 715 prescriptions for analgesic agents were OL [50]. No medicines were UL. Of the top 6 analgesics administered, 30% paracetamol; 8% ibuprofen; 9% codeine phosphate; 98% diclofenac; 100% pethidine and 79% morphine prescriptions were OL. Paracetamol was OL due to dose, pethidine due to route, codeine and morphine due to age, diclofenac due to indication and ibuprofen OL in terms of the patient's dose or weight.

Systemic anti-infective agents

A population-based cohort study in general practice in the Netherlands demonstrated that, of 2094 children receiving 2855 prescriptions for systemic anti-infective agents in a 1 year study period, 20 prescriptions (0.7%) were UL and 410 (14%) were OL [44]. The most frequently prescribed OL drugs were amoxicillin, azithromycin and trimethoprim. OL doses were the most common reason.

Oncology

One prospective study examined the incidence and nature of UL and OL prescribing in paediatric oncology patients [51]. Inpatient and outpatient prescriptions were analysed for a 4 week period. All patients received at least one UL or OL drug, 19% prescriptions were UL and 26% were OL. UL preparations were used in 40% of prescriptions for cytotoxic agents, due to a lack of commercially available formulations suitable for the paediatric patient. These drugs included mercaptopurine and methotrexate, which have been used in the treatment of paediatric leukaemia for many years, their efficacy and toxicity profiles having been demonstrated by on-going trials. The authors express their disappointment that drugs, which are the mainstay of therapy for paediatric leukaemia and other malignancies, are unavailable in appropriate licenced formulations to facilitate their administration to children.

Gastroenterology

Prescription records for all outpatients and children at discharge under the care of paediatric gastroenterologists were analysed for 6 months in a study in the UK [52]. Of 777 prescriptions received by 308 children, 291 (37%) were OL mainly due to indication, and 93 (12%) were UL. Cisapride was the most common UL medicine (since it is not recommended in children) and domperidone the most frequently prescribed OL drug. These authors examined information regarding the drugs prescribed in 4 commonly used drug formularies. Only *Medicines for Children* [53] contained information about more than half (9/13) of the most commonly prescribed UL or OL drugs. The other formularies poorly reported these medications. The authors highlight the difficulties in finding information on drugs prescribed in specialised paediatric areas. At the same time, general practitioners are often expected to continue to prescribe these drugs following discharge from hospital and access to information is even more scarce in primary care. The authors stress that paediatricians and their sub-specialist colleagues must become aware of the licence status of the medicines they prescribe, and ensure that optimal information is provided to the primary healthcare professionals who are expected to provide continued care of their patients.

1.1.7   RISKS ASSOCIATED WITH UNLICENCED AND OFF-LABEL DRUG USE IN CHILDREN

This summary of studies examining the incidence and nature of UL and OL use of drugs in children shows that the problem is widespread across all areas of paediatric care in countries across Europe and also Australia. However, this is not a new problem and is it really a cause for concern other than for the inconveniences, such as lack of suitable child-friendly formulations, difficulty in finding drug information and problems at the primary/secondary care interface? The little evidence we have as to the real risks of this practice suggest that we are justified in our concerns and that action is required urgently to change things.

In 1995 a British team reported a study conducted in a regional PICU over 28 months looking at the incidence and nature of adverse drug reactions (ADRs) [54]. Thirty-five different drugs were involved in 76 ADRs. Twenty-five were OL in terms of age, dose or indication and one was an UL product. One third of drug prescriptions implicated in an ADR were used outside their product licence.

In 1999 a further study was conducted by Turner and his colleagues to examine the relationship between ADRs and UL and OL prescribing [18]. In a 13 week prospective study, five wards including medical, surgical, neonatal surgical, cardiac ICU and general PICU were studied. The findings in relation to UL and OL drug use have been documented above. ADRs were associated with 112 (3.9%) of 2881 licenced drug prescriptions and 95 (6%) of 1574 UL or OL drug prescriptions. The number of different medications administered to a child was found to be the most statistically significant association with the risk of an ADR, however the percentage of UL and OL drugs was also found to be significantly associated with the risk of an ADR. The authors recommended that further studies were required to confirm this risk.

In a pilot UK regional intensive ADR monitoring scheme, 24 of 95 ADR reports contained at least one suspected drug that was judged to be OL. [55]

An Italian study to investigate the link was published in 2002 [56]. Prospective monitoring of ADRs was carried out in a paediatric unit over 9 months. Of 1619 children admitted to the hospital ward, 29 children (1.8%) experienced an ADR following drugs administered in hospital; 12 children experienced an ADR caused by drugs given in the community. The ADRs were considered serious in 8 cases, and moderate in the rest. Overall, in 16 children (39%), the ADRs were caused by drugs used OL. This included 38% (11/29) of the ADRs that occurred following in hospital drug therapy and 42% (5/12) of the ADRs occurring in community and leading to hospitalisation. Despite the study's limited population size, pharmacological treatment of ADRs was greater and statistically significant for severe ADRs following OL drug use.

A study from France published in 2002 examined prescribing by office based paediatricians and also found that OL drug use was significantly associated with ADRs (relative risk 3.44; 95% CI 1.26, 9.38), particularly when it was due to an indication different to that in the licence[40]. ADR incidence increased from 1.41% overall to 2% with OL drug use.

These studies therefore further suggest that there is a higher risk of ADRs associated with UL and OL prescribing in children and such ADRs may be serious. This confirms that urgent European measures must be taken to change this situation and allow children their basic human rights of equal access to safe, effective and high quality medicines.

REFERENCES

1. Impicciatore, P., Choonara, I., *Br. J. Clin. Pharmacol.*, *48*, 15 (1999).
2. Ceci, A., Felisi, M., Catapano, M., et al., *Eur. J. Clin. Pharmacol.*, *58*, 495 (2002).
3. t'Jong, G. W., Eland, I. A., Stricker, B. H. C., van den Anker, J. N., *Paed. Perinatal Drug Ther.*, *4*, 148 (2001).
4. Anon., *DTB*, *30*, 97 (1992).
5. Brion, F., Nunn, A. J., Rieutord, A., *Acta Paediatr.*, *92*, 486 (2003).
6. Turner, S., *Paed. Perinatal Drug Ther.*, *4*, 24 (2000).
7. Ballantine, N. E., *Pharm. J.*, *265*, 162 (2000).
8. Ballantine, N. E., *Pharm J.*, *265*, 396 (2000)
9. Association of the British Pharmaceutical Industry and Department of Health. *Ensuring best practice in the notification of product discontinuations.* London: Department of Health (2001).
10. Turner, S., Nunn, A. J., Choonara, I., *Paed. Perinatal Drug Ther.,1*, 52 (1997).
11. Conroy, S., *BMJ*, *317*, 204 (1998).
12. Conroy, S., *JPPT*, *8*, 60 (2003).
13. Rylance, G., *Dev. Med. Child. Neurol.*, *21*, 399 (1979).
14. Shirkey, H., *J.Pediatrics,72*, 119 (1968).
15. *Licensing Medicines for Children.* Joint report of the British Paediatric Association and the Association of the British Pharmacutical Industry. (1996).
16. Health Committee. *The specific health needs of children and young people*, House of Commons Health Committee Second report, Volume 1, Session 1996-7 London (1997).
17. Turner, S., Gill, A., Nunn, T., Hewitt, B., Choonara, I., *Lancet*, *347*, 549 (1996).
18. Turner, S., Nunn, A. J., Fielding, K., Choonara, I., *Acta Paediatr.*, *88*, 965 (1999).
19. j'Jong, G. W., Vulto, A. G., de Hoog, M., Schimmel, K. J. M., Tibboel, D., van den Anker, J. N. *N. Engl. J. Med.*, *343*, 1125 (2000).
20. j'Jong, G. W., Vulto, A. G., de Hoog, M., Schimmel, K. J. M., Tibboel, D., van den Anker, J. N., *Pediatrics*, *108*, 1089 (2001).
21. Conroy, S., McIntyre, J., Choonara I., *Arch. Dis. Child. Fetal Neonatal Ed.*, *80*, F142 (1999).

22. Avenel, S., Bomkratz, A., Dassieu, G., Janaud, J. C., Danan C., *Arch. Pediatr., 7,* 143 (2000).
23. Barr, J., Brenner-Zada, G., Heiman, E., et al., *Am. J. Perinatol., 19,* 67 (2002).
24. O'Donnell, C. P. F., Stone, R. J., Morley, C. J., *Pediatrics, 110,* e52 (2002).
25. Turner, S., Longworth, A., Nunn, A. J., Choonara, I. *BMJ, 316,* 343 (1998).
26. Turner, S., *Aust. J. Hosp. Pharm., 29,* 265 (1999).
27. Conroy, S., Choonara, I., Impicciatore, et al., *BMJ, 320,* 79 (2000).
28. Craig, S., Henderson, C. R., Magee, F. A. *IMJ, 94,* 237 (2001).
29. Pandolfini, C., Impicciatore, P., Provasi, D., Rocchi, F., Campi, R., Bonati, M., *Acta Paediatr., 91,* 339 (2002).
30. t'Jong, G. W., van der Linden, P. D., Bakker, E. M., et al., *Eur. J. Clin. Pharmacol., 58,* 293 (2002).
31. Turner, S., *Paed. Perinatal Drug Ther., 4,* 24 (2000).
32. Riordan, F. A. I., *BMJ, 320,* 1210 (2000).
33. McIntyre, J., Conroy, S., Avery, A., Corns, H., Choonara, I., *Arch. Dis. Child., 83,* 498 (2000).
34. Gavrilov, V., Lifshitz, M., Levy, J., Gorodischer, R., *IMAJ, 2,* 595 (2000).
35. Chalumeau, M., Treluyer, J. M., Salanave, B., et al., *Arch. Dis. Child., 83,* 502 (2000).
36. t'Jong, G. W., Eland, I. A., van den Anker, J. N., Sturkenboom, M. C. J. M., Stricker, B. H. C., *Pharmacoepidemiol Drug Safety., 10,* S49 (2001).
37. Bucheler, R., Schwab, M., Morike, K., et al., *BMJ, 324,* 1311 (2002).
38. Schirm, E., Tobi, H., de Jong-van den Berg, L. T. W., *BMJ, 324,* 1312 (2002).
39. t'Jong, G.W., Eland, I. A., Sturkenboom, M. C. J. M., van den Anker, J. N., Stricker, B. H. C., *BMJ, 324,* 1313 (2002).
40. Horen, B., Montastruc, J.-L., Lapeyre-Mestre, M., *Br. J. Clin. Pharmacol., 54,* 665 (2002).
41. Lifshitz, M., Gavrilov, V., Grossman, Z., et al., *Curr. Ther. Res., 63,* 830 (2002).
42. Schirm, E., Tobi, H., de Jong-van den Berg, L. T. W., *Pediatrics, 111,* 291 (2003).
43. Ufer, M., Rane, A., Karlsson, A., Kimland, E., Bergman, U., *Eur. J. Clin .Pharmacol., 58,* 779 (2003).
44. t'Jong, G. W., Eland, I. A., de Hoog, M., Sturkenboom, M. C. J. M., van den Anker, J. N., Stricker, B. H. C., *Paed. Perinatal Drug Ther., 5,* 71 (2002).
45. t'Jong, G. W., Eland, I. A., Sturkenboom, M. C. J. M., van den Anker, J. N., Stricker, B. H. C., *Eur. J. Clin. Pharmacol., 58,* 701 (2003).
46. Martin, R. M., Wilton, L. V., Mann, R. D., Steventon, P., Hilton, S. R., *BMJ, 317,* 204 (1998).
47. Efron, D., Hiscock, H., Sewell, J. R., et al. *Pediatrics., 111, 372* (2003).
48. Wilton, L. V., Pearce, G., Mann, R. D., *Pharmacoepidemiol. Drug Safety., 8,* S37 (1999).
49. Lifshitz, M., Gavrilov, V., Gorodischer, R., *Eur. J. Clin. Pharmacol., 56,* 839 (2001).
50. Conroy, S., Peden, V., *Paediatr Anaesth., 11,* 431 (2001).
51. Conroy, S., Newman, C., Gudka, S., *Annals of Oncology., 14,* 42 (2003).
52. Dick, A., Keady, S., Mohamed, F., et al., *Aliment. Pharmacol. Ther., 7,* 571 (2003).
53. Royal College of Paediatrics and Child Health and Neonatal and Paediatric Pharmacists Group. *Medicines for Children,* London, RCPCH Publications, (1999).
54. Gill, A. M., Leach, H. J., Hughes, J., Barker, C., Nunn, A. J., Choonara I., *Acta Paediatr., 84,* 438 (1995).
55. Clarkson, A., Ingleby, E., Choonara, I., Bryan, P., Arlett, P., *Arch. Dis. Child., 84,* 337 (2001).
56. Impicciatore, P., Mohn, A., Chiarelli, F., Pandolfini, C., Bonati, M., *Paed. Perinatal Drug Ther., 5,* 19 (2002).

1.2 European regulation and the EMEA

Agnès Saint Raymond[1], Kalle Hoppu[2]

[1]*Head of Sector Scientific Advice and Orphan Drugs, EMEA, 7 Westferry Circus, Canary Wharf, London E14 4HB, UK*

[2]*Medical Director, Poison Information Centre, Helsinki University Central Hospital, P.O. Box 340 (Haartmaninkatu 4), 00029 HUS (Helsinki), Finland*

1.2.1 INTRODUCTION

In the Treaties establishing the European Union, the organisation and delivery of health services and medical care are principally the responsibility of the Member States (25 in May 2004). The Community has the responsibility to set a high level of health and to co-ordinate public health issues. To complement national policies directed towards improving public health, Community actions are undertaken under the responsibility of DG Health and Consumer protection, (DG Sanco, one of the Directorates General of the European Commission). The other major aspect of health services and medical care falling under shared competence is marketing authorisation and monitoring of medicinal products. For Community authorisations, it is under the responsibility of DG Enterprise, reflecting the aim to provide not only the European citizens with medicinal products fulfilling the criteria of quality, safety and efficacy, but also the European pharmaceutical industry with faster access to the market.

1.2.2 REGULATORY FRAMEWORK FOR MEDICINAL PRODUCTS IN THE EUROPEAN UNION

The harmonisation of the regulatory framework for medicines in Europe was initiated in 1965. The adoption of Regulation (EEC) 2309/93 [1] in 1993, which was reviewed in 2003, provided the foundation for Community marketing authorisation of medicinal products within the EU. The current regulatory framework provides for a centralised Community procedure (so-called centralised procedure) and, as an alternative, a national procedure of marketing authorisation in a Member State, followed by mutual recognition of the first authorisation by some or all of the other Member States (so-called decentralised, or mutual recognition procedure). The centralised procedure is mandatory for marketing authorisations of all biotechnology products, and optional for innovative products.

1.2.3 THE EUROPEAN AGENCY FOR THE EVALUATION OF MEDICINAL PRODUCTS (EMEA)

The creation of the European Agency for the Evaluation of Medicinal Products Agency in 1994, set up by Regulation (EEC) 2309/93 [1], was a major step further in the process of harmonisation of the regulatory framework for medicines in Europe. The EMEA is a decentralised agency of the European Commission, which co-ordinates activities related to medicinal products for human and veterinary use at the Community level. The Agency is based in London (United Kingdom).

Scientific Committees of the EMEA

The Agency comprises several scientific committees with delegates from each of the Member States, and a secretariat providing scientific, technical and regulatory assistance. Working parties and ad hoc groups may assist the scientific committees. The Agency co-ordinates the network of national agencies and European experts and use their expertise for assessment. Two committees, the Committee for Proprietary Medicinal Products (CPMP), and the Committee for Veterinary Medicinal Products (CVMP) are in charge of providing scientific opinions for marketing authorisations (centralised procedures) to the EU Commission, on the basis of the demonstration of quality, safety and efficacy. The CPMP and CVMP also manage pharmacovigilance and Community arbitration procedures for medicines for human use and veterinary medicinal products, respectively.

Community marketing authorisations

The European Commission issues the administrative decision of marketing authorisation on the basis of the Committees' scientific opinions. A Community authorisation is valid in all Member States; it includes a single tradename and identical product information in all official languages of the Community. In addition, a detailed summary of the evaluation is made publicly available under the format of the European Public Assessment Report (website of the EMEA: www.emea.eu.int)

Since 1995, about 250 products have been authorised through the centralised procedure, about a third being derived from biotechnology. Thirty percent of the products are authorised for use in some of the age groups of children, but 44% more have no paediatric indication, or appropriate formulation, when the indication actually covers a condition affecting children as well as adults[1].

Orphan medicinal products

A third scientific committee, the Committee for Orphan Medicinal Products (COMP) was created more recently in 2000, with the implementation of the European Orphan Regulation (EC) 141/2000 [2], offering Community incentives for the development of medicines intended for rare diseases. The COMP is in charge of providing opinions on designations as orphan products, providing advice to the Commission on policy for orphan products and liaising at international level on such issues. To be designated as orphan, a medicinal product must be intended for serious diseases affecting no more than 5 per 10,000 citizens in the EU; alternatively, for more frequent diseases, the product must demonstrate that its development would not provide the necessary return on investment. In all cases, if satisfactory methods of treatment already exist for the orphan condition, the product must be of significant benefit over existing therapies for the patients affected.

Since 2000, more than 160 products have been designated as orphans and ten marketing authorisations have already been granted for designated orphan products. Of the 160 designated products, 12% percent of orphan products are intended for conditions affecting only children and 55% more for conditions affecting both adults and children[1].

Other responsibilities of the EMEA

In addition to assessing applications, the EMEA and its committees provide scientific advice or protocol assistance (for orphan medicines) to sponsors developing medicinal products. The EMEA also handles quality control of pharmaceutical and clinical data provided in the applications in the form of inspections (Good Manufacturing Practice and Good Clinical Practice) in conjunction with the national inspectorates.

1.2.4 PAEDIATRIC MEDICINES AND THE EMEA

The granting of marketing authorisation for paediatric medicinal products should follow the same demonstration of quality, safety and efficacy, as reques-

---

[1]Figures for August 2003

ted by the regulations. Specific paediatric European experts participate in the assessment as necessary. To facilitate the development of paediatric medicinal products, the CPMP adopted in 1997 a CPMP Efficacy Working Party "Note for guidance on clinical investigation of medicinal products in children" (CPMP/EWP/462/95). This guidance was amended by subsequent international harmonisation; within the ICH Topic E 11, on Clinical investigation of medicinal products in the paediatric population. The guideline was revised, and adopted in 2000 as the "Note for guidance on clinical investigation of medicinal products in the paediatric population" (CPMP/ICH/2711/99), it is operational since January 2001 [3].

1.2.5 THE PAEDIATRIC EXPERT GROUP (PEG)

Taking into consideration the needs, the Agency decided to devote specific resources to paediatric issues, despite the lack of existing legal framework. An *ad-hoc* expert group of the CPMP was created in June 2001 to deal with paediatric issues, even before the adoption of the new European regulation on paediatric medicinal products (*see Chapter 1.3*). The Paediatric Expert Group meets 3-5 times a year and is presently chaired by the CPMP chairman, himself a paediatrician. The group comprises experts in different domains of interest for the use of medicines in children, in particular a pharmacist for issues on formulations suitable for children, toxicologists, neonatologists, pharmacovigilance experts, and various clinical specialists in paediatrics.

The group participates in drafting guidelines for paediatric issues and identifying needs for new ones. In anticipation of the future Regulation, the group is currently reviewing already existing medicinal products that would need studies to fulfil paediatric needs at European level. This work is done with consultation of the appropriate learned societies in the various Member States. (http://www.emea.eu.int/htms/human/peg/pegexpert.htm).

1.2.6 THE US DRUG REGULATION AND THE FOOD AND DRUG ADMINISTRATION

The US drug regulations and procedures are federal, and thus procedures may correspond broadly to the centralised EU procedures. However, the number of countries involved in the EU and different backgrounds of drug regulation still constitute a major difference between the two systems. Although the ICH process has led to the harmonisation of many drug development requirements (*see Chapter 1.4*), significant differences still exist, like the existence of special regulations for paediatric medicinal products at FDA level since 1997 (*see Chapter 1.3*). In the US a new medicinal product must enter the regulatory arena earlier in the form of an Investigational New Drug application (IND), and can be developed in a more binding collaboration between the sponsor and the regulatory authorities. It is outside the scope of this review to go to any detail, for more information see: http://www.fda.gov/cder/index.html.

In the US, the Food and Drug Administration, more specifically its Center for Drug Evaluation and Research (CDER) and Center for Biologics Evaluation and Research (CBER) handle all marketing authorisations. In 1994 a Pediatric Advisory Sub-Committee was created in CDER within the Anti-Infective Drugs Advisory Committee, which contributed to the development of the US paediatric regulations. This Advisory Sub-Committee has original responsibilities, in particular for paediatric labelling of drugs. Subsequent to FDAMA, the 2001 Best Pharmaceuticals for Children Act established within the FDA, in the Office of the Center Director, an Office of Pediatric Therapeutics. Its current functions are ethics and adverse event reporting. There is also, within CDER, a Division of Pediatric Drug Development (DPDD) of the Office of Counter terrorism and Pediatric Drug Development (http://www.fda.gov/cder/pediatric/index.htm). The DPDD is the FDA paediatric resource to foster paediatric drug development throughout the FDA, to provide consultation service to all centers within the FDA, and also partnership with the NIH/NICHD (National Institutes of Health, National Institutes of Child Health and Human Development) to identify and obtain paediatric information on off-patent drugs in the framework of the 2001 Best Pharmaceuticals for Children Act. However, specific paediatric evaluation continues to be performed by the therapeutic divisions of CDER or CBER.

1.2.7 CONCLUSION

The European and the US systems differ in many respects, but on both sides major changes are affecting medicines for use in children. Increased co-operation between the two regions should help synergistic approaches to drug development and trials especially for paediatrics. The awareness of paediatric issues is now well established and it is hoped that the future European Paediatric Regulation will fulfil the therapeutic needs of the European citizens.

REFERENCES

1   Regulation (EEC) 2309/93:   http://pharmacos.eudra.org/F2/eudralex/vol-1/new_v1/2309-93%20_2codified_.pdf
2   Regulation (EC) 141/2000: http://pharmacos.eudra.org/F2/eudralex/vol-1/new_v1/Reg141-2000_en.pdf
3   Note for guidance on clinical investigation of medicinal products in the paediatric population (CPMP/ICH/2711/99): http://www.emea.eu.int/pdfs/human/ich/271199EN.pdf

1.3 A comparison between European and American legislation

Joséphine Géraci Buiche

Department of European and Legal Affairs, AFSSAPS, 93285, Saint-Denis, France

The USA has taken the initiative in adopting a legislative program with the goal of stimulating research in the area of paediatric medication. The different stages of this program are presented in this report. The basic situation is similar in Europe [1]. European-wide regulations are now in the process of being defined for promoting and setting the basic framework for the development of these specialties. This chapter presents and compares the different stages of the legislative process in both the USA and in Europe.

1.3.1 THE USA AND ITS LEGISLATION RELATED TO PAEDIATRIC MEDICATION

Since 1979 [2], the Food and Drug Administration (FDA) has launched a number of initiatives in order to stimulate the research and sale of medication for paediatric use and to expand the therapeutic arsenal available to doctors. This has been done by encouraging specific trials on the usage of commercialised substances for which the safety and effectiveness had only been verified for use in adults.

After several failed attempts, the FDA put a double set of regulations into place, termed the "carrot and the stick." This regulation anticipated both temporary measures aimed at compensating the pharmaceutical industry for research, with an extension of exclusive use, along with obligatory measures without compensation.

Raising awareness and attempts at regulation

The text referred to as the Pediatric Final Rule (or "Specific requirements on content and format of labeling for human prescription drugs") proposed to create a subsection related to the paediatric use of a drug in its printed information, i.e., a section that would be specifically destined for the use of the drug for children [3]. At this time, the FDA identified 5 paediatric sub-populations: the premature newborn, the newborn, the infant, the child and the adolescent.

In 1994, the Center for Drug Evaluation and Research (CDER) and the Center for Biologics Evaluation and Research (CBER) of the FDA put into place the Pediatric Plan [4], intended to encourage laboratories to voluntarily supply paediatric data both during the research phase and after the product's commercialisation.

Another text entitled "Guidance for Industry: the Content and Format for Pediatric Use Supplement" was presented by the CDER and the CBER in May 1996 [5]. The goal was to provide a guide that companies could use to provide all necessary information for paediatric drug use. This series of documents were made available to industry in a pedagogical spirit; they defined the broad lines of what was expected as drug companies carried out studies into paediatric use [6]. Through the publication of these guides, the FDA clearly expressed the will to aide companies at every stage of product development for the paediatric market.

The FDA also published a list of the 10 medicines most often used for children and for which there was little or no information concerning paediatric usage [7]. The criteria behind the identification of these medicines however were not clearly explained, and the list, incomplete as it was, appeared without any real impact.

Despite all these measures based mostly on voluntary action, the number of medicines distributed with information for paediatric use barely increased. In fact, there is no legislation that imposes an obligation on the pharmaceutical industry to develop medical treatments specifically for children. Out of this situation grew the current tendency to introduce legislation that combines aspects of encouragement and constraints.

The concept of paediatric exclusivity is part of the Food and Drug Administration Modernization Act [8]. This text, published in 1998, had a period of validity of three years. It granted an extension of 6 months to the patent of a medicine for all indications when paediatric clinical tests were carried out, valid only upon the first request of the authorisation to market (AMM) for the active substance, which included all adult indications in any pharmaceutical variety. The additional period of exclusivity was added to the patent.

The type of information required was defined by the FDA, based on the public health requirements, and specifically took the form of the written request. This was in fact a contract between the drug company and the FDA; the FDA defined the clinical trials that needed to be performed and the deadlines for the study. Once submitted, the results of the study were examined by the office of paediatric exclusivity of the FDA, to determine if the terms of the contract had been respected. If this was the case, the additional exclusivity was granted.

The Pediatric Rule required that laboratories conduct paediatric trials for all new substances [9]. The rule was expected to be applied without any time limit. The goal of the Pediatric Rule was to ensure that new medicines that might be extensively used on children, or that offer real benefits compared to existing treatments, were released with the appropriate information for all claimed indications. This information had to be supplied at the time when the AMM was filed or rapidly thereafter (within 15 months). The main goal was to include paediatric study in the drug development process as early and systematically as possible, as well as to obtain data simultaneously for both adults and children. In practice, the intention of the FDA has not been to slow access to medicines for adults, but rather to encourage simultaneous development of medicines for children and adults.

In addition, the Pediatric Rule proposed other arrangements, such as the creation of a group of paediatric experts in charge of the list of the initial priorities; the writing of guidelines; and the establishment of protocols for assistance between the FDA and pharmaceutical laboratories that come into effect upon the granting of an AMM. This text also allowed improvement concerning the information that is available for products already on the market.

Assessment and analysis of the regulations

The process was thus legally well defined. A number of advantages and disadvantages have been observed. The six months of exclusivity for sales granted by this rule was a significant incentive because it related to the active drug regardless of the specific form that the product is delivered in. The number of studies that have been carried out since the paediatric exclusivity ruling, is in fact quite remarkable.

The most frequently heard criticism concerning paediatric exclusivity is the excessive encouragement; many feel that the additional period of 6 months of exclusivity delayed the arrival of generic medications to the market. For this reason, it has been suggested to limit the applicability of the rule to paediatric indications. This would, however, reduce significantly the strong effect of encouragement that was originally intended for these measures.

Experience has shown also that the rule has generated the most requests for exclusivity for those drugs that sell best. For medicines with only modest or weak market shares, the paediatric exclusivity approach is not very effective.

It is also important to point out that paediatric exclusivity does not aim to improve the situation for medications without patents. And there is the problem of confidentiality of the data supplied by the pharmaceutical industry because no arrangements have been made to publish the information obtained via the written requests.

The Pediatric Exclusivity Act does not oblige a laboratory to have a program of paediatric development. However, the FDA has the parallel regulatory power, based on the Pediatric Rule, to request that a certain company carry out clinical tests for drugs intended for child use.

At the request of various lobbies, a court of law ruled in 2002 that the FDA stop its practice of appealing to the Pediatric Rule in such cases. Reacting to this development, the American government decided that the needs for public health should overrule the claims of the pharmaceutical lobbies, and that, in order to give the Pediatric Rule the required applicability, it should have the force of law. The American congress thus adapted a document entitled the Pediatric Research Equity Act [10] as an amendment to the Federal Food, Drug and Cosmetic Act. This authorises the FDA to require that certain research be carried out for any medicine that is intended for children.

The system that has been put into place in the USA would appear to offer an interesting and effective means for obtaining complete information about paediatric medicines. Nevertheless, it was necessary to amend certain aspects. The Best Pharmaceutical Act for Children (BPAC) [11] came into force on January 4, 2002 as an extension of paediatric exclusivity and it has a period of validity of 5 years.

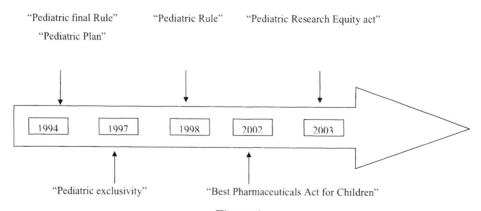

Figure 1.

The BPAC has maintained the concept of an additional period of six months of exclusivity. The incentives are renewed for those companies that conduct paediatric clinical trials. In addition, the text provides an annual list of priorities and describes the arrangements related to the financial support of the clinical trials, including the aspects of labeling information, the publication of the data, as well as the establishment of the specific expertise for paediatric use. The BPAC is thus more flexible than the Pediatric Rule.

1.3.2 EUROPE AND LEGISLATION CONCERNING PAEDIATRIC MEDICINES

The European Union has recognised that these legislative efforts have proven effective in the USA and has decided to propose legislation with the specific goal of stimulating the development of paediatric medicines. However, European legislation on paediatric medicines will certainly require several years before it can be adopted, as was the case with orphan drugs.

The measures that have been taken to prepare a proposal for regulation are:

The Memorandum of July 2000: This describes the need for clinical studies related to paediatric medications, underscoring the need for appropriate treatments in paediatric practice. This text provided the inspiration for the resolution that followed in December 2000.

The Resolution of the Council of 14 December 2000: This document makes a clear statement about the current situation [12], namely that many medicines used in paediatric care have never been evaluated specifically for this use, and they therefore present risks in terms of quality, safety, and effectiveness that are already addressed in the texts related for adult medications.

The creation of the Paediatric Expert Group (PEG): This group was formed within the Committee of Specialty Pharmaceutics of the European Agency for the Evaluation of Medicinal Products Agency (EMEA), with whom the French Agency for Health Products (AFSSAPS, Agence Française de Sécurité Sanitaire des Produits de Santé) already works closely. In fact, its working methods are largely inspired from those of the Paediatric Orientation Committee (COP) of the AFSSAPS.

The publication of the document Better Medicines for Children: This document [13] was made public by the DG III Industry close to the European Commission in February 2002. This document makes the distinction between two situations:

- If the AMM is requested for a new product, or if the product is protected by a patent, the Commission will consider the possibility of extending the initial period of intellectual protection for the product by 6 to 12 months, with the sole condition that the laboratory that holds the patent commits itself to conducting tests based on paediatric use. This extension will follow an agreement between the laboratory and the EMEA, following the example of the system put into place by American legislation and corresponding to the written request.
- If the product enters into the public domain (older products that are still marketed), the Commission proposes a specific status. In this case, a paediatric AMM may be obtained, but only the paediatric use of the medicine thus described will be protected. For this case, the Commission plans the creation of public funds dedicated to research in paediatric medication.

The experiences of the USA have clearly shown that incentives based on intellectual property are effective. However, this proposal is only concerned with those products that are still protected by a patent.

1.3.3 THE EXAMPLE OF RARE DISEASES

Regulations concerning orphan drugs

Medication intended to treat rare diseases were practically absent from the market until the 1990s. Patients, health professionals and the governmental authorities combined forces to obtain a legal framework for these orphan drugs, as well as for the incentives for their development.

The European Parliament and the Council adopted regulations concerning orphan medications on 16 December 1999 [14]. This text gives details about the designation, development and the commercialisation of orphan drugs.

The principles are the following:

- A drug may not be designated as being in the orphan category until this has been approved by the Committee of Orphan Drugs, which is part of the EMEA.
- An exclusivity period of ten years is accorded upon the successful request for AMM for the therapeutic indication defined by the procedure of designation.
- National-level incentives should be initiated in order to encourage research, development, and commercialisation of orphan drugs.
- The financial charges are negotiated on a case by case basis; in this manner, the laboratory may obtain a reduction or exoneration of their charges.

Thus, the regulations adopted by the European Union are inspired to a large degree by the regulations already in place in the United States by 1995 [15].

The "European lag time", referring to the fact that the Europeans adopted regulations in this area only in 1999, can probably be explained by the political complexity within the European territory, with the accompanying diversity in questions of public health.

Points in common between the situations of paediatric and orphan drugs include:

1. The nature of the problem of paediatric medications is similar to that of orphan drugs: the size of the populations is limited to a number that represents a market of limited commercial value.
2. It is critical that paediatric patients, just as patients suffering from an orphan disease, should be able to benefit from drugs that have undergone a perfectly rigorous evaluation, and it is therefore necessary to encourage the pharmaceutical industry to promote the research, development and commercialisation of adequate treatments.
3. The regulatory measures of these two areas need to be considered on the European level rather than on the national level.

As far as for the differences between the two, for orphan drugs, it is relevant to stress that the financial encouragements are given without accompanying constraints. Also, the need for encouragements, not only for new products but also for compounds already on the market, appears to be a requirement in the case of paediatric medications.

Thus, outside certain highly specialized groups of experts (notably paediatricians and paediatric pharmacists), awareness of the shortcomings of paediatric medications has only appeared recently in Europe. The situation of paediatric medications probably became a subject of importance following the Convention relative to the rights of children (OMS) [16], upon the adoption of the Directive 2001/20 of 4 April 2001 [17] and via the connection between orphan drugs and paediatric drugs. In fact, three quarters of orphan drugs are intended for paediatric use.

The adoption and execution of the regulations concerning orphan drugs has allowed professionals and the public health authorities to request a set of regulations concerning paediatric medications in Europe. Another explanation for the timing is the adoption in the USA of regulations for paediatric medications.

1.3.4 A COMPARISON BETWEEN THE UNITED STATES AND EUROPE

The United States have already taken measures that tie together financial encouragement with the obligation to submit data, and thereby to develop new possibilities of medication for the child. Similar measures are indispensable in

Europe. However, it seems reasonable to insist that this is carried out within the specific European context.

The general context

In the first instance, it is important to note that there is a difference in how the term medication is defined. This definition is broader in the US, meaning that certain products, such as cosmetics or dietary supplements, are considered to be medications in the US but not in Europe.

The fundamental principles, however, remain the same: an industrially prepared drug intended for commercialisation must obtain an AMM and satisfy the three basic criteria of safety, quality and effectiveness. The AMM in both cases has the same content and requires the same set of documents and information.

The Americans have at their disposition a direct source of information via the internet sites of the pharmaceutical laboratories. The USA, through the influence of the FDA, applies a policy of transparency of information for all matters related to health and drugs. However, the ability of interpreting these technical documents is not obvious for those who are not specialists of the field.

Finally, the patient in the US has become a consumer of health products with the appearance of drugs taken through self-medication. The number and role of associations of patients is much better developed in the US. This concept of patient associations has appeared in Europe with the preparation of documents concerning orphan drugs, largely under the influence of certain northern European countries.

The context within the pharmaceutical field

The American market can be considered to be much more aggressive than the European market, in large part because of the importance placed on generic medications. In fact, many medicines available in the USA are generic. Also, patent protection is in general better in the USA than in Europe.

The state of pharmacovigilance also differs. In the USA, drugs are withdrawn from the market in a most radical fashion, while in Europe, recommendations for use and suspensions are published by the various national agencies or the EMEA.

Europe is in the process of full revision of the EU code, as instituted by the Directive 2001/83, dated 6 November 2001 [18]. The European Commission proposes a major overhaul of the existing pharmaceutical legislation within the Union to help patients to more rapidly obtain new medications, and to encourage research innovation by applying directives similar to those used in the United States.

Thus, the introduction of a Bolar type amendment [19] within European legislation would permit the formal request for a generic drug ahead of time, such that generics may be commercialised the same day that the patent expires and the drug falls into the public domain. It is important to recall that in 1984, the United States decided to introduce the Waxman-Hatch Act of 24 September 1984 which contained a text known as the Bolar amendment, which explains that it is not counterfeiting to manufacture, use or sell that which is solely and reasonably related to the development and submission of a file with the federal agencies with the authority to regulate the manufacture, the use and the sale of the drug.

It appears clear that there are significant regulatory and cultural differences between the USA and Europe. However, these differences now tend to be blurring since the publication of ICH E11 (see next section).

REFERENCES

1. Turner, S., Gill, A., Nunn, T., Hewitt, B., Choonara, I., *Lancet, 347,* 549 (1996).
2. 44 Federal Register 37, 462, http://www.fda.gov/cder/pediatric/21cfr20157.pdf (1979).
3. Bostein, M. D., Center for Drug and Evaluation and Research, Pediatric Plan, *Drug Information J., 30,* 1125 (1994).
4. "Final Rule", 63 Federal Register 66, 632, (1998).
5. FDA, Department of Health and Human Services, "Specific requirements on content and format of labelling for human prescription drugs; revision of "pediatric use" subsection in labelling," Federal Register (May 1996).
6. "Guidance for Industry: Qualifying for Pediatric Exclusivity Under Section 505A of the Federal Food, Drug, and Cosmetic Act," http://www.fda.gov/cder/guidance/2891fnl.pdf (1999).
7. FDA, "List of approved drugs for which additional pediatric information may produce health benefits in the pediatric population." Docket # 98 N0056
8. Final Rule, 63 Federal Register 66, 632, (1998).
9. Food, Drugs, and Cosmetic Act 505A, 21 U.S.C '355a (1997).
10. http://www.fda.gov, http://www.fda.gov/bbs/topics/NEWS/2003/NEW00979.html (2003).
11. FDA modernization 2002-2005 : Best Pharmaceuticals for Children Act, 4 January 2002 (Public Law No. 107-109). http://www.fda.gov/cder/pediatric/PL107-109.pdf (2002).
12. Résolution du Conseil du 14 décembre 2000 relative aux médicaments pédiatriques, JOCE C17 of 19 January 2001, page 1. (http://europa.eu.int/eur-lex/pri/fr/oj/dat/2001/c_017/c_017 20010119fr00010001.pdf).
13. Consulting Document: Better Medicines for Children, 28 February 2002 (http://pharmacos.eudra.org/F2/pharmacos/docs/Doc2002/feb/cd_pediatrics_en.pdf)
14. Rule 141/2000 of the European Parliament and of the Council of 16 December 1999 concerning orphan medications, JOCE No. L018 of 22 January 2000, page 0001-0005, http://europa.eu.int/eurlex/pri/fr/oj/dat/2000/l_018/l_01820000122fr00010005.pdf (2000).
15. "The US Orphan Drug Act" consists of an amendment to the Federal Food, Drug, and Cosmetic Act (Sections 525, 526, 527, and 528). (http://www.fda.gov/opacom/laws/fdcact/fdcact5b.htm) It was made law on 4 January 1983, Final Rule 57 Federal Register 62076 (1992).
16. Convention relative to the rights of children, of June 2001, CRC/C/28/Add.17 (2001).

17. Directive 2001/20 of 4 April 2001. JOCE, L121 of 1 May 2001 page 34-44. (http://europa.eu.int/eur-lex/pri/fr/oj/dat/2001/l_121/l_12120010501fr00340044.pdf).
18. ICH on 20 July 2000 by the ICH Steering Committee of 20 July 2000. (http://www.nihs.go.jp/dig/ich/efficacy/e11/e11-e.pdf).
19. Directive 2001/83/CEE of the 6 November 2001, instituting the Community code related to human medications, JOCE No. L 311 of 28 November 2001, pages 67 – 128 (http://europa.eu.int/eur-lex/pri/fr/oj/dat/2001/l_311/l_31120011128fr00670128.pdf).

1.4 Clinical trials in children: guidance ICH E11

Natalie Hoog-Labouret

Department of Paediatrics, Hospital Robert Debré, 48 Bd Sérurier, 75019 Paris, and AFSSAPS, France

1.4.1 ICH E11, INTRODUCTORY NOTES

Within the framework of the International Conference on Harmonisation (ICH), a group of recommendations for the development of paediatric medicines was elaborated in July 2000: the ICH E11. (ICH: International Conference of Technical Requirements of Pharmaceuticals for Human Use).

The ICH brings together the authorities in charge of the pharmaceutical industry and their regulation for each of the three large international regions (Europe, USA and Japan).

Their objective is to align the procedures for clinical trials required to evaluate the safety, quality and the effectiveness of medications. The partners are:

- The European Commission and by extension the European Medicines Agency (EMEA), the Committee of Specialty Pharmaceuticals, and the European Federation of Pharmaceutical Industry Associations
- The Food and Drug Administration (FDA) and the Pharmaceutical Research and Manufacturers of America (PhRMA)
- The Japanese Ministry of Health and Welfare and the Japanese Association of Pharmaceutical Manufacturers.

These three groups of partners are often accompanied by observers, such as the World Health Organisation, the Canadian Health Protection Branch, and the European Association for Free Trade.

The creation of the guideline ICH E11 took a period of 4 years. This document describes the procedures that professionals must take in order to put medications on the paediatric market.

This document replaces the American, European and Japanese recommendations that existed previously. It insists on a number of specific points:

1. The definition of the five age classes of paediatrics
2. The need to evaluate medications for their specific use in paediatrics
3. The need to have specific pharmaceutical forms adapted to the different classes of age and to take into account the potential toxicity of the excipients.
4. The ethical aspects of clinical trials with children

1.4.2 THE CLASSES OF AGE

The definition of the five classes of age allows the use of a common language and the harmonisation of information:

1. The premature (less than 36 weeks gestation)
2. The newborn (between 0 and 27 days)
3. The infant (between 28 days and 23 months)
4. The child (between 2 and 11 years)
5. The adolescent (between 12 and 16-18 years, according to the legislation in various countries).

Age should of course be considered as a continuum, but the classes of age serve to provide points of reference for the study of the AMMs filed and allow harmonising of the Summary of Product Characteristics.

For each age class, this text discusses the specificities linked to questions of growth and maturation that could have implications for the development of medicines. For example, the ICH E11 insists strongly on the aspect of the rapid growth and maturation of the central nervous system and of the immune system in the infant. For the adolescent, it stresses the sexual growth and its hormonal consequences on his or her neuro-cognitive development.

1.4.3 EVALUATION

Three types of drugs are specified, corresponding to the different developmental profiles:

- The medications for diseases that are predominantly or exclusively paediatric in nature, for which all the stages of drug development must be carried out with children, with the exception of the initial safety data.
- Medications that are used to treat serious or life-threatening diseases that affect adults as well as children, and for which there are few treatments available. In this case, studies with children should be begun very early.

- Medications for treating other situations. In this case, paediatric development is less urgent and can start after the development of the product for adults. However, if the drug represents a major therapeutic breakthrough, these studies should begin as soon as possible.

One chapter deals with extrapolation of data on the efficacy of the drug in children. When indications are the same in children as adults and the disease process is similar, then only pharmacokinetic data are required for the different age groups. The situation is similar for the extrapolation of data from older children to younger children.

This approach, which is based on the pharmacokinetic – pharmacodynamic relationship, should help reduce the number of unnecessary clinical trials. However, when the relationship between blood plasma concentration and effect cannot be established, or if this relation is obviously different for children compared to adults, then measurements of the drug's activity might be required. These methods are facilitated by the knowledge of the paths of elimination that are implied as a function of the age of the child.

The document stresses the protection of the patient. Practical considerations are given in order to minimise the invasive methods of pharmacokinetic study; the use of microassays in order to keep the volume of blood needed for each sample analysed to a minimum, the use of laboratories that are experienced in the handling of small blood samples, the use of local anesthesia, attempts to coordinate the taking of samples for pharmacokinetic studies so that they coincide with those taken for routine biological examinations, the use of population pharmacokinetic data, development of less invasive methods (saliva or urine testing, etc.). In the case of studies of efficacy, the document insists on the establishment of judgement criteria specific to each age group.

The need for a *long-term follow-up for medications* authorised in paediatrics is underscored in the document. The profile of undesirable side effects might develop differently in the adult. The dynamic process of growth and development (cognitive, sexual, osseous, immunological, cardiac, renal, etc.) might not lead to an immediate expression of an undesirable side effect but rather that these may occur at a later stage.

1.4.4 PHARMACEUTICAL FORMS ADAPTED FOR THE CHILD

Several points need to be considered:

- For oral administration: aromas, colouring agents, formulations as a function of age (suspensions, chewable tablets, micro-granules, etc.). The instructions for administration must be appropriate and safe. Certain technical difficulties (stability, bitterness of certain compounds, etc) are mentioned.

• For parenteral drugs, the document stresses the need for establishing properly adapted concentrations of the active agent, permitting a precise and safe administration of the prescribed dosage. For medications in single-use ampoules, one must make sure that the packaging is adapted to the dose.

• The toxicity of certain excipients can vary as a function of age and between child and adult (e.g., benzoic acid). Depending on the nature of the active principle and the excipients, the use of the medication in newborns may require additional detailed information (e.g., dilution), and perhaps even a new formulation.

1.4.5 CLINICAL TRIALS WITH CHILDREN

The last part of the text is concerned with the ethical problems associated with clinical trials with children. These points include the necessary conditions to insure the protection of children participating in such clinical trials, the role of ethics committees, the means of recruiting candidates, informing, obtaining consent, rigorous clinical and pre-clinical justification for the clinical trial with a solid evaluation of the risk-benefit relationship and selection of methodology that will allow the study to take place with a minimum number of patients and with a minimum amount of discomfort, etc.

1.4.6 THE DIRECTIVE 2001/20/CEE OF 4 APRIL 2001

This directive is concerned with bringing together the various legislative, regulatory and administrative dispositions concerning the application of good practice in the management of clinical trials for medications destined for human use. The directive must be transposed to national law for each member state and came into application in May 2004.

Its objective is to coordinate clinical trials over the entire territory of the European Union and to reinforce the ethical frame within which these tests must be conducted. It recognises that "it is indispensable to carry out clinical tests that include children so that treatments might be improved". Clinical trials with children are therefore possible, but certain specific arrangements must be put into place to ensure a maximum of security. These are included in article 4 of the Directive:

• It is necessary to ensure that "the clear consent of the parents or of the legal guardian must be obtained" and that "the minor apparently agrees." This consent can be cancelled at any moment. The notion of legal representation is handled on a national level according to the laws in effect.

- The minor must have "received information, adapted to his capacity to understand, from competent personnel about the clinical trial, especially its risks and benefits."
- The desire of a minor, capable of forming an opinion and of evaluating this information, to refuse to participate in a clinical trial or to be removed from the trial at any point in time is to be examined by the investigator.
- No incentives or financial advantages are to be given, except for compensation.
- Certain direct advantages that might result from the clinical trial are obtained for the group of patients, and only in the case where this research is essential for the validation of collected data in clinical trials on persons capable of giving their clear consent or by other methods of research; in addition, this research must be linked directly to a clinical condition for which the minor in question suffers, or be such that the tests cannot be conducted except on minors.
- The specific corresponding guidelines of the Agency must be followed (and thus the importance of the directives such as ICH E11).
- Clinical trials have been designed to minimise the pain, the discomfort, the fear and all foreseeable risks related to the disease and to the level of development; the threshold of risk and the degree of harm reached must be specifically defined and constantly re-examined.
- The protocol must be accepted by an ethics committee with competence in paediatrics or after consultation of the clinical, ethical, psychosocial problems related to paediatrics.
- The interests of the patient have priority over those of science or of society.

Article 11 of the directive treats the matter of information exchange and discusses the creation of a European database for clinical trials. This sharing of information should allow researchers to avoid duplication of unneeded clinical trials (*see chapter 1.8*).

1.5 Ethical issues

John McIntyre

Academic Division of Child Health, University of Nottingham, Derbyshire Children's Hospital, Uttoxeter Road, Derby, DE22 3NE, UK

1.5.1 INTRODUCTION

Research, by evaluating the effectiveness of medicines and interventions, has contributed to dramatic improvements in health care. At the same time, an ethical framework has emerged that both protects patients participating in research and guides researchers. For children, a desire to protect them from being experimented on, has been the overriding principle for many years. However, there is also a wish to ensure any treatment given has been rigorously evaluated and this means children need to be included in medical research. Some of the ethical issues relevant in this balance are considered here:

- The regulatory framework
- General guidelines
- Specific guidelines
- Problem areas
- The way forward

1.5.2 REGULATORY FRAMEWORK FOR NEW MEDICINES

It would be unethical if children received poor medicines. Therapeutic misadventures in the past have resulted in legislation aimed at ensuring medicines reaching the patient have information on safety, quality and efficacy. The principles in different countries are broadly similar to the UK. In the UK, the Medicines Control Agency (MCA) ensures medicinal products have marketing authorisation or product licenses before being placed on the market. To acquire this, there must be data, published or unpublished, available for scrutiny to satisfy the MCA on safety, quality and efficacy. A pharmaceutical company may also obtain marketing authorisation throughout the EU via the European Centralised Licensing process. The principles are the same: the company must demonstrate safety, quality and efficacy for the disease(s) and patient group(s) for which they wish to market the drug. Unfortunately children still receive medicines without the benefit of this regulatory framework a situation often referred to as 'therapeutic orphan' [1].

1.5.3   ETHICAL ISSUES IN RESEARCH: GENERAL GUIDELINES

Data that underpins the regulatory framework comes through research. Research must adhere to ethical principles. There are a number of general statements that guide medical research in human subjects notably the Nuremberg Code [2], The Declaration of Helsinki [3], The Guideline for Good Clinical Practice [4] and the European Union Clinical Trials Directive [5]. The high standards demanded by these guidelines apply to research involving children. However the specific needs of children receive relatively little direct mention.

1.5.4   ETHICAL ISSUES IN RESEARCH: SPECIFIC GUIDELINES

A strong lead for including children in ethically sound research came from the US National Institutes of Health (NIH). In 1998 they published the guidelines 'NIH policy and guidelines on the inclusion of children as participants in research regarding human subjects' [6]. This policy states that 'children must be included in all human subjects research conducted or supported by the NIH, unless there are scientific and ethical reasons not to include them'.

The desire for children to no longer be denied the benefits of research is explicit in guidelines produced by The Royal College of Paediatrics and Child Health (RCPCH) Ethics Advisory Committee [7]. It states that 'children are likely to experience either the most lasting benefits or harms from their involvement in research. The attempt to protect children absolutely from the potential harms of research denies any of them the potential benefits'. Six underlying principles form the basis for these guidelines:

1.   Research involving children is important for the benefit of all children and should be supported, encouraged and conducted in an ethical manner.
2.   Children are not small adults, they have an additional unique set of interests.

3. Research should only be done on children if comparable research on adults could not answer the same questions.
4. A research procedure which is not intended directly to benefit the child subject is not necessarily unethical or illegal.
5. All proposals involving medical research on children should be submitted to a research ethics committee.
6. Valid consent should be obtained from the child, parent or guardian as appropriate. When parental consent is obtained, the agreement of school age children who take part in research should also be addressed by the researchers.

The RCPCH also made a position statement in December 1999 'Safeguarding informed parental involvement in clinical research involving newborn babies and infants'[8]. It stated that it would be unethical not to do important clinical research on newborn babies and infants. To fail to do research would lead to stagnation of current practice and a continuation of medical management using untried or unproved remedies on the basis of belief rather than best evidence.

1.5.5 PROBLEM AREAS

When trying to follow the principles set out in guidelines some recurring ethical dilemmas arise:

- Is it an ethical study?
- Risk and benefit
- Consent/assent
- Emergency situations

An ethical study

It is the responsibility of Research Ethics Committees to decide whether a research proposal is ethical. This may be at a local level or, for multicentre trials at a national level. A number of key questions should be considered [5, 9] in reaching a decision and include:

- Is it a relevant question worth asking?
- Has it been done before or can the answer be obtained by study of an adult?
- Have the risks and benefits for the child been evaluated?
- Will the protocol design answer the question?
- Are the researchers competent and the facilities adequate?
- Is information adequate?
- What is the provision for indemnity or compensation?
- What are the arrangements for recruitment?

Risk and benefit

Risk may be considered as the probability of harm and encompasses physical, emotional, social or economic risk. In therapeutic research, the child may directly benefit. This type of research is generally recognised as justifiable and associated risks may be accepted. In non-therapeutic research the patient does not directly benefit, although future patients may do so. An extreme view would be that involving children in such research, whatever the risk, is not justified. The RCPCH view is 'A research procedure which is not intended directly to benefit the child subject is not necessarily unethical or illegal'[7]. However, the degree of risk acceptable is likely be lower.

Allowing or not allowing research therefore requires the thresholds of risk to be stated. In their guidelines the RCPCH categorised risks as minimal, low or high.

Minimal risk, the least possible risk, covered procedures such as questioning, observing or measuring children provided that this was done in a sensitive way. Collection of a single urine sample or using blood from a sample taken as part of treatment are considered to fall in this category.

Low risk covered procedures that might cause brief pain or tenderness and it is acknowledged that for many children injections and venepunctures would constitute low rather than minimal risk.

High risk procedures were thought to be unjustified for research alone and should be carried out only when the research was combined with diagnosis or treatment intended to benefit the child. Examples given included biopsy, arterial puncture and cardio-catheterisation.

The context is also important. An example is blood sampling. In the sick, newborn infant the handling required to obtain a sample and the volume required may constitute more than minimal risk. However, for older children who can understand the reasons, and where the procedure can be skilfully performed with steps to reduce the amount of pain, e.g. local anaesthetic creams, the risk may be minimal.

Consent and assent

Consent and assent are processes. They are not one-off events confirmed by a signed form. Consent comes from the exercise of choice and informed consent from the ability to evaluate options in the exercise of choice. The preconditions for consent to be legal are:

• To be competent of mind
• To be fully informed
• To understand the information
• To decide voluntarily and give authorisation

Competence

The legal and ethical framework governing consent for research is generally regarded to follow the more established precedents set when considering consent for treatment. An ethical problem in England is that 'competence' to give legal consent for treatment is reached at 16 years of age. However, the guiding principle should be 'Gillick competence', i.e., if a child has sufficient understanding and intelligence to understand what is proposed, it is their consent that is required [10]. It is clear that the threshold at which competence is reached cannot be defined by a specific age.

For the young infant and child, it is usually the parents that give permission for them to be included in research. The term consent is often used to encompass this process, even though it cannot meet the usual criteria of fully informed consent. Obtaining consent in this setting will inevitably fall short of the ideal. There are undoubtedly ethical inconsistencies in this area [11] and some have argued an opt out approach may be more appropriate [12]. However, despite the shortcomings, it is clear parents still value involvement and their consent should not be bypassed [13].

Informed

For consent to be informed the following should be discussed with families.

- The purpose of the research
- Whether the child stands to benefit directly and if so how.
- The meaning of relevant research terms.
- The nature of each procedure.
- The potential benefits and harms.
- The name of a researcher who they can contact.
- The name of the doctor responsible for the child.
- How children can withdraw from the project.

Researchers should explain and answer questions throughout the project, ensure other staff know about the research, give clearly written leaflets for families to keep and report the results of the research to families involved.

Voluntary

Legally valid consent should be both freely given. Researchers should approach families and children aware of the implications this carries. The RCPCH Guidelines [4] indicate researchers, in seeking freely given consent, should:

- Offer families no financial inducement.
- Exert no pressure on families.
- Give them as much time as possible.
- Encourage families to discuss the project with others.
- Tell them they may refuse to take part or withdraw at any time.
- Say they need not give a reason for withdrawing.
- Assure them that the child's treatment will not be prejudiced by withdrawal.
- Encourage parents to stay with the children during procedures.
- Respond to families' questions throughout the study.

Emergency situation

In the emergency care of the sick newborn infant and ill child, research has a potential to make significant improvements. However, obtaining informed consent using the comprehensive definition in the emergency situation is problematic.

The RCPCH position statement suggests that, in this emergency stage, parents should be asked to assent to the trial and that the formal consent should be obtained over the next hours or days [8]. While such safeguards are desirable, there has been much debate on the practical realities that such an approach creates [11]. It has been argued that an acceptable alternative in the emergency situation could be one of presumed consent with an opt-out option [12]. Nevertheless, an underlying principle of current research codes places a duty on researchers to obtain informed consent from participants.

1.5.6   THE WAY FORWARD

Researching with children has unique ethical dimensions. It is essential to find ways of ensuring these ethical needs are met rather than abandoning children to continuing poor practices.

In generating research proposals, health professionals with extensive background in working with children are well placed to understand these issues. Their early involvement in study design is crucial if the ethical issues are to be spotted and addressed [14].

Ethics committees throughout Europe need common principles to ensure childrens rights and needs are respected; child specific ethics committees may be an option to enforce the guidelines and recommendations that exist [15, 16].

To generate evidence based treatments requires a culture where open debate about the role of research with children takes place. Children, parents and researchers should work in partnership to overcome many of the misunderstandings that arise from poor communication. Instead of being just recruits in a study, children and families could be involved in all stages of the process – study design, generating information to support the consent process, disseminating the results.

1.5.7 CONCLUSION

The Declaration of Human Rights accepted by all European countries should entitle children to full human rights. Children should benefit from and expect the same safeguards as adults in their health interventions. Achieving this requires research in an ethically sound framework. Although there are unique ethical issues of working with children it is possible to advance the evidence base for them in an ethically sound way.

REFERENCES

1. *The Therapeutic Orphan – 30 Years Later*. Proceedings of a joint conference of the Paediatric Pharmacology Research Unit Network, the European Society of Developmental Pharmacology, and the National Institute of Child Health and Human Development, Washington DC, USA, May 2, 1997, *Pediatrics, 104*, 581 (1999).
2. The Nuremberg Code (1947) *BMJ, 313*, 1448 (1996).
3. World Medical Association. *The Declaration of Helsinki*, Ferney-Voltaire, France, (1964).
4. International Conference on Harmonisation of Technical Requirements for Registration of Pharmaceuticals for Human Use, ICH Tripartite Guidelines, Geneva, (1997).
5. European Union clinical trials directive. *Bull. Med. Ethics, 169*, 13 (2001).
6. NIH policy and guidelines on the inclusion of children as participants in research involving human subjects, Bethesda, Maryland, 1998.
7. McIntosh, N., Bates, P., Brykczynska, et al., *Arch. Dis. Child., 82*, 177 (2000).
8. *Safeguarding informed parental involvement in clinical research involving newborn babies and infants-a postition statement*, Royal College of Paediatrics and Child Health, (1999).
9. Morton, N. S., *Paed. Perinatal Drug Ther., 4*, 135 (2001).
10. Gillick-v-West, Norfolk AHA, *3 All Er 402*, 423-4 (1985).
11. Modi, N., *Semin. Neonatol., 3*, 303 (1998).
12. Manning, D. J., *J. Med. Ethics, 26*, 249 (2000).
13. Mason, S. A., *Lancet, 356*, 2045 (2000).
14. Conroy, S., McIntyre, J., Choonara, I., Stephenson, T., *Br. J. Clin, Pharmacol., 49*, 93 (2000).
15. Chambers, T. L., Kurz, R., *Eur. J. Pediatr., 158*, 537 (1999).
16. Kurz, R., *Good Clin. Prac. J., 7*, 22 (2000).

1.6 Practical issues in relation to clinical trials

Imti Choonara

Academic Division of Child Health, University of Nottingham, Derbyshire Children's Hospital, Derby DE22 3DT, UK

Paediatric drug research needs to involve the pharmaceutical industry working in partnership with paediatric health professionals (doctors, pharmacists and nurses with paediatric expertise). Paediatric clinical pharmacologists, who have both the expertise of other paediatric health professionals and an understanding of clinical pharmacology, should ideally be involved, especially in relation to the design of the clinical trials.

1.6.1 LOCATION

It is accepted that sick children need to be treated by paediatric health professionals. Similarly for clinical trials involving children, paediatric health care professionals need to be involved, ideally within a paediatric unit. One, therefore, cannot perform a paediatric clinical trial in an adult clinical trials unit. Clinical trials and other aspects of paediatric drug research can be performed in district general hospitals [1]. For general paediatric conditions, these units are probably preferable to tertiary centres where one is more likely to see a highly selective patient group that is not representative of children throughout the community.

For those clinical trials that involve children as outpatients, it is important that the children are assessed in a child friendly location, i.e. safe with toys and play therapists available. It may also be appropriate to assess the child at home.

1.6.2 WHICH MEDICINES?

The success of any clinical trial is related to the clinical need for the medicine. Investigators, parents and children are all more likely to participate in a study of a medicine which is likely to result in significant clinical benefit to children than one where there is already satisfactory treatment. Ethics committees are more likely to recognise that a clinical trial is appropriate in children if there is no current treatment available. This does not mean, however, that the clinical trial will automatically be approved, as the design of the study may be inappropriate.

The International Conference on Harmonisation, which includes representatives from the EMEA, the FDA and Japan have issued ICH E11, Clinical Investigation of Medicinal Products in the Paediatric Population [2]. This guidance categorises medicinal products and their value in children into three categories. These are shown in Table 1. The aim is to encourage the study of medicines in the first two groups where there is the greatest potential clinical benefit.

Table 1

ICH classification of medicinal products for children

- Medicinal products for diseases predominantly or exclusively affecting paediatric patients
- Medicinal products intended to treat serious or life threatening diseases occurring in both adults and paediatric patients for which there are currently no or limited therapeutic options
- Medicinal products intended to treat other diseases and conditions

---

1.6.3 TIMING OF STUDIES

The timing of studies in children is dependent upon several factors. These include whether one is dealing with a serious or life-threatening disease for which there is currently no or limited treatment available. In this situation, early paediatric studies are essential. However, where existing treatment is available, then clinical trials in children should only be conducted after initial safety data has been established in adults.

Table 2

ICH classification of children by age

Category	Age
preterm neonates	<36 weeks gestation, 0 – 27 days
full-term neonates	0 – 27 days
infants and toddlers	28 days – 23 months
children	2 – 11 years
adolescents	12 – 17 years

1.6.4   WHICH PAEDIATRIC PATIENTS?

The clinical nature of the medicine will determine whether it needs to be studied in neonates, infants or children. It is important that investigators are aware of the ICH Guidance in relation to the classification of the five different age groups in relation to paediatric patients. These are shown in Table 2.

1.6.5   STUDY DESIGN

A poorly designed study will fail to attract investigators, obtain ethical approval and recruit children. The study design is therefore crucial. Investigators need to ask the following questions:

- Which paediatric age group is most likely to benefit from the medicine?
- How many patients are needed?
- Is more than one centre needed to recruit all the patients?
- Should a placebo be included in the trial design? (Placebo is appropriate if there is no existing treatment for the condition. If however, effective therapy is available, then the use of a placebo is neither appropriate nor ethical).
- How will the pharmacodynamic effect be studied in the particular age group of the study?
- Which pharmacokinetic parameters, if any, need to be determined?
- What is the likelihood of significant drug toxicity?

Regulatory authorities are more likely to raise questions about clinical trials in children than in adults [3]. The duration of clinical trials in children is usually longer than that in adults. Two-thirds of paediatric clinical trials in Finland were completed within 12 months [3].

1.6.6   PHARMACOKINETICS

The minimum number of samples, involving the smallest amount of blood possible, needs to be collected from each patient. Microassays may need to be developed to measure drug concentrations in small volumes of blood. Information regarding the metabolic pathway and pharmacokinetic parameters in adults is essential before commencing pharmacokinetic studies in paediatric patients.

The use of population pharmacokinetics whereby a larger number of children are involved, but fewer samples are collected from each patient, should be considered [4]. Pharmacokinetic studies are usually carried out in a subgroup of the children recruited for the clinical trial. It should not be made a precondition for entry into the clinical trial, as this may result in the loss of a significant number of children from the study.

1.6.7 NON-INVASIVE METHODS

Non-invasive methods, such as the caffeine breath test, should be used if possible. The caffeine breath test has been used as a probe for CYP1A2 enzyme activity [5]. It has been used to study drug interactions (induction and inhibition) and also the effect of disease on drug metabolism [5]. It involves the use of a stable isotope of caffeine and the collection of breath samples for two hours after administration of the caffeine.

1.6.8 PHARMACODYNAMICS

It is often difficult to study pharmacodynamic effect in younger patients. For certain conditions, measuring the pharmacodynamic effect is not difficult, e.g. seizures in patients with epilepsy [6], mortality in children with leukaemia. For other conditions, however, it is more difficult to assess pharmacodynamic effect, e.g. bronchodilators in infants under the age of 18 months, pain relief in pre-verbal children and neonates. There are validated pain scales appropriate for use in paediatric patients of different ages [7]. It is, therefore, essential that one uses a validated pain scale if one is studying an analgesic drug.

1.6.9 RECRUITMENT

Recruitment of children is a crucial issue in relation to ensuring successful completion of a trial. There have been very few studies looking at the issues behind recruiting children to clinical trials. Two studies have highlighted the altruism of parents and children in that they are happy to participate in a clinical trial on the basis that it may improve the management of children in the future [8, 9]. In a prospective study of patient recruitment of a new local anaesthetic agent, the biggest reason for parents and children declining to enter a clinical trial was the time involved [8]. Other reasons related to the wish not to be involved in a clinical trial and the child requesting a local anaesthetic cream they had previously used. Other studies have involved sending a questionnaire to parents of children who had previously been approached to enter a clinical trial [10, 11]. These groups found that safety for the child was the main reason for parents declining to enter a clinical trial. Lack of time was also a major reason for parents not wishing to participate in clinical trials involving their children [10, 11]. Investigators in Europe feel that recruitment of children to clinical trials is less of a problem in Europe than in the USA [12]. There is, however, little documented evidence comparing the viewpoints of parents and children in Europe in comparison with North America.

1.6.10 PHARMACOVIGILANCE

Drug toxicity in children is different to that in adults. This may be due to impaired drug metabolism or altered protein binding, but may also be idiosyncratic. As the child is developing, they may be prone to different toxicities to adults. The principles of pharmacovigilance in children are similar to that in adults. Consideration should be given to setting up an Independent Safety Monitoring Board if there is the potential for significant toxicity.

1.6.11 CONCLUSIONS

Paediatric clinical trials are usually more difficult than similar studies in adults. Paediatric drug research involves patients whereas many adult studies involve volunteers. It is up to paediatric health professionals and the pharmaceutical industry to work together to ensure that we study the right medicines with an appropriate design to ensure that children receive medicines that are fully evaluated scientifically. Such an approach will increase efficacy and hopefully reduce toxicity.

REFERENCES

1. Sammons, H. M., McIntyre, J., Choonara, I., *Arch. Dis. Child.*, *89*, 408 (2004).
2. Spielberg, S. P., *Paed. Perinatal Drug Ther.*, *4*, 71 (2000).
3. Keinonen, T., Miettinen, P., Saano, V., Ylitalo, P., *Paed. Perinatal Drug Ther.*, *5*, 175 (2003).
4. Jacqz-Aigrain, E., Debillon, T., Daoud, D., et al., *Paed. Perinatal Drug Ther.*, *5*, 190 (2003).
5. Webster, E., McIntyre, J., Choonara, I., Preston, T., *Paed. Perinatal Drug Ther.*, *5*, 28 (2002).
6. Appleton, R., Sweeney, A., Choonara, I., Robson, J., Molyneux, E., *Dev. Med. Child. Neurol.*, *37*, 682 (1995).
7. Beyer, J. E., *Paed. Perinatal Drug Ther.*, *2*, 3 (1998).
8. Peden, V., Choonara, I., Gennery, B., Done, H., *Paed. Perinatal Drug Ther.*, *4*, 75 (2000).
9. van Stuijvenberg, M., Suur, M. H., de Vos, S., et al., *Arch. Dis. Child.*, *79*, 120 (1998).
10. Harth, S. C., Thong, Y. H., *Br. Med. J.*, *300*, 1372 (1990).
11. Tait, A., Voepel-Lewis, T., Siewert, M., Malviya, S., *Anesth. Analg.*, *86*, 50 (1998).
12. Hoppu, K., *Pediatrics*, *104*, 623 (1999).

1.7 Specific issues in relation to clinical trials

Corinne Alberti[1], Bruno Giraudeau[2]

[1]*Clinical Epidemiology Unit, Hôpital Robert Debré, 48, Boulevard Sérurier, 75019 Paris, France*

[2]*INSERM CIC 202, Faculty of Medicine, 2bis Bd Tonnellé, 37032 Tours cedex, France*

1.7.1 INTRODUCTION

Randomised clinical trials (RCTs) are the standard method to assess clinical effectiveness [1] and are, with meta-analyses, the standard of excellence for comparisons of treatment effects over time. But clinical questions are most easily answered when a disease is fairly common and the outcome of interest has a high risk of occurring. On the contrary, when diseases are rare, or recruitment of patients is difficult, with modest benefits, clinical trials may not provide a "definite answer". Indeed, they cannot be expected to reach conventional levels of power and statistical precision. Thus, investigators who wish to test new treatments for rare diseases, tend to conduct either single arm studies or comparative studies using historical controls. Alternatively, investigators may attempt to conduct small, underpowered RCTs. These give rise to estimates of outcome that have unacceptably large confidence intervals and this fails to provide clear answers.

In paediatrics, clinical trials are often difficult to conduct. The pharmaceutical industry may be reluctant to study medicines in children because the market for the sale of many drugs for children is smaller than for adults, and therefore investment in paediatric drugs is less attractive financially. Other reasons include ethical difficulties, problems with blood sampling and difficulties in recruiting sufficient numbers of children. Moreover, the incidences of diseases are lower than in adults. Nevertheless, research in children is necessary and presents scientific, ethical and practical challenges. As children grow, their body size and composition, physiology, and cognitive and motor function change. The metabolism and toxicity of medications can vary substantially in children of different ages. Children do not react like scaled-down versions of adults. Some secondary effects may be specific in children. Society wants to spare children from the potential risks involved in research. However, children may be harmed if they are given medications that have been inadequately studied. Researchers in paediatrics are therefore often faced with small sample sizes, which cannot assure an adequately power-controlled trial.

Trial data are often accumulated over a lengthy period of time which means that the emerging information can be used to influence its conduct. As a result, inappropriate use of emerging data can seriously bias the analysis of the clinical trial, which means that investigators have taken the scientifically safest course of action, which is to look at no data until the study is complete [2]. However, such an approach can be in serious conflict with the ethics of clinical research, neglecting opportunities for saving unnecessary resources, particularly in children. In this context, sequential designs are attractive.

Finally, a necessary companion to a well-designed clinical trial is an appropriate statistical analysis of the data from that trial. If a clinical trial produces data revealing differences in effect between two or more interventions, statistical analyses are used to determine whether such differences are real or due to chance. Certain types of analyses, such as sequential analysis or Bayesian analysis, are more amenable to trials with small sample sizes.

In this section, we will introduce three examples of design, namely "N of 1" designs, sequential designs and adaptive designs, which are particularly interesting in the context of paediatrics. We will also discuss the Bayesian philosophy, which can provide useful information for clinicians in the particular context of small clinical trials.

1.7.2 "N OF 1" TRIAL

The "N of 1" method is a powerful and objective technique for assessing an individual's unique response to a given treatment. Its objective is to find the best treatment for a particular patient. Indeed, the establishment of population efficacy does not guarantee that the results can be generalised to a patient with his or her own biological makeup, educational status, and socio-economic status and who is cared for by a particular physician at a different geographical location. "N of 1" trials are useful in answering questions of efficacy of treatment when the physician or the patient doubts that it is present, in studies of rare or unusual disorders that do not lend to large randomised clinical trials, in rapid assessment of outcomes and in situations where little or no carry-over between treatment periods is present. Four critical components are necessary: chronic disease, randomisation (usually cross-over), blinding of physician to treatment assignment (reducing operator bias), defining and quantifying outcomes and explicit criteria for evaluating efficacy of treatment.

Guyatt and colleagues [3] describe one method of conducting an RCT with an "N of 1" trial: a clinician and a patient agree to test a therapy for its ability to reduce or control symptoms of the patient's disease. The patient then undergoes treatment for a pair of periods; during one period of each pair, the experimental therapy is applied, and during the other period, either an alternative treatment or a placebo is applied. The orders of the two periods within each pair is randomised by a method that ensures that each period has an equal chance of applying the experimental or the alternative therapy. Whenever possible, both the clinician and the patient are blind to the treatment being given during either period. The treatment targets are monitored to document the effect of the treatment being applied. Pairs of treatment periods are replicated until the clinician and the patient are convinced that the experimental therapy is effective, is harmful, or has no effect on the treatment targets. This process can be repeated in other patients and a global analysis can then be performed.

This design was employed in the evaluation of methylphenidate in routine practice for children with attention deficit hyperactivity disorder [4], but also in evaluation of the fetal acoustic stimulation as an adjunct to external cephalic version [5]. "N of 1" trials were encouraged at the workshop entitled "Optimal strategies for developing and implementing psychopharmacological studies in preschool children", held in 2000 at the annual meeting of the American Academy of Child and Adolescent Psychiatry. It was suggested that "N of 1" trials would make best use of the preschool children whose severity of psychiatric illness and impairment make multiple interventions necessary and likely to occur when they are treated in the community [6]. The limitations of "N of 1" trials are: (i) they cannot be applied to a disease that can be cured in a short period of time and (ii) they cannot be assessed with the hard clinical endpoints, such as deaths or other irreversible condition indicators. Nonetheless, the "N of 1" study design allows meaningful interpretation of data where there is a small sample size and the condition has a low incidence, because the patient becomes his or her own control.

1.7.3   SEQUENTIAL DESIGN

Usually, data from clinical trial are accumulated gradually over a period of time, which can extend to months or years. Results from patients recruited early in the study are available for interpretation, while patients used later in the study are still being recruited and allocated treatment. It may be considered desirable to stop, if a clear treatment difference is apparent, thereby avoiding the allocation of further patients to the less successful therapy. Investigators may also wish to stop a study, which no longer has much chance of demonstrating treatment difference. Successive examination of accumulating data and stopping rules are based on what is found to be natural consequences of the gradual building of results from clinical trials.

A class of designs which involve repeated inspection of the trial data have been developed [7]. In a study with a sequential design, participants are sequentially enrolled on the study and are assigned a treatment (usually at random). The sequential design may lead the investigator to change the probabilities that participants will be assigned to any particular treatment on the basis of previous results, as they become available (cf. infra). The object is to improve the efficiency, safety or efficacy of the experiment, as it is in progress, by changing the rules by which one determines how participants are allocated to the various treatments. Otherwise, some authors reserve the term of "sequential design" for studies which are inspected after every individual response, corresponding usually to dose-response designs. The term "group sequential design" is used for studies which are inspected after every k responses ($k > 1$). Sequential group designs are useful for the monitoring and accumulation of study data, while they preserve the type I error probability at desired significance level, despite the repeated application of significance tests [8]. Parallel-groups are studied until a clear benefit is seen or it is determined that no difference in treatments exists [8]. The sequential group design allows results to be monitored at specific time intervals throughout the trial, so that the trial may be stopped early if there is clear evidence of efficacy. Safety monitoring can also be done, and trials can be stopped early, if unacceptable adverse effects occur, or if it is determined that the chance of showing a clinically valuable benefit is futile. The triangular test appears to be the most popular group sequential method.

The triangular test

The triangular test has convergent boundaries that define a closed sequential plan and thus a maximum sample size [8]. The triangular test uses a sequential plan defined by two perpendicular axes (Figure 1).

The horizontal axis corresponds to the first statistics, V, which represents the quantity of information accumulated since the beginning of the trial. The vertical axis corresponds to a second statistic, Z, which represents the benefit with experimental treatment compared with the control. Two straight lines, called the boundaries of the test, delineate a continuation region from the regions of non-rejection of the inefficacy hypothesis (below the bottom line) and of rejection of the inefficacy hypothesis (above the top line). The equations of these straight

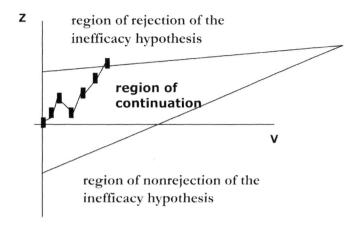

Figure 1. The triangular test: graphical display of successive analyses. Each black rectangle refers to an analysis. V is a statistic which expresses the quantity of information while Z refers to the difference between groups. The trial is stopped as soon as the path between successive analyses crosses one of the two boundaries.

lines depend on the values of the benefit to detect, of its standard deviation, and of type I and II error rates, as well as on the frequency of the analyses defined in terms of the number of patients included between two analyses. At each analysis, the two statistics V and Z are calculated from all data collected since the beginning of the study and define a point on the sequential plan. As long as the sample path stays within the two boundaries, the study is continued and new patients are included. When the sample path crosses one the boundaries, the study is stopped and the conclusion is obtained: crossing the bottom boundary causes the inefficacy hypothesis not to be rejected, whereas crossing the top boundary causes the inefficacy hypothesis to be rejected. Once the trial is stopped, a conventional statistical analysis is performed [8,9]. An application is given by Bellissant and colleagues, showing the interest of the triangular test to assess the efficacy of metoclopramide on gastroesophageal reflux [9]. It was a placebo-controlled, randomised double-blind study aimed at assessing the efficacy of metoclopramide on gastroesophageal reflux in infancy. Two 24-hour oesophageal pH recordings were performed. The study was stopped after the seventh analysis evaluating 39 infants (20 placebos and 19 metoclopramide), because the expected benefit was not reached. The triangular test allowed stopping the study with a 15% reduction in sample size [9].

1.7.4   ADAPTIVE DESIGN: PLAY THE WINNER

Adaptive designs have been suggested as a way to overcome the ethical dilemmas that arise when the early results from an RCT clearly begin to favour one intervention over another. An adaptive design seeks to skew assignment probabilities to favour the better performing treatment in a trial that is under way. It is a data-dependent treatment allocation rule, that sequentially uses accumulating information about the treatment difference during the study to change the probabilities of allocating incoming study patients to the two groups [10]. A particular adaptive design is the randomised "play the winner" rule. This design can be implemented by starting with a container of several balls marked A and B for two treatments. When a patient is available for an assignment, a ball is drawn at random and replaced. If it is type A, the treatment A is assigned to this patient, otherwise it is treatment B. The patient is then followed for a response, which can either be a success or failure. If the response is a success on treatment A or a failure on treatment B, x balls of type A are added to the container. Otherwise, y balls of type B are added. So the container gradually accumulates more balls representing the treatment A or B, depending on which is performing better, in terms of more successes or fewer failures, thus far in the trial [10,11]. Thus, this type of design tends to put more study patients on the better treatment. Despite its attractive design, few clinical trials have been implemented based on this technology [10], because of the need for further statistical development with regard to secondary endpoints, censored data, subgroup analyses and adjustment on co-variates [12].

1.7.5   THE BAYESIAN ANALYSIS

Two philosophies of statistical analysis co-exist: Frequentist and Bayesian [13]. In the frequentist approach, results are expressed in terms of significance levels, point estimates and confidence intervals. The Bayesian approach is subjective and is best envisaged in terms of the changing opinions of one investigator. It can give probabilities that the clinical effect lies in a particular range and also the size of the most likely effect. This is in stark contrast to the often misinterpreted P-value produced by the usual frequentist approach. The frequentist P-value is the probability of the observations occurring when the null hypothesis is true. The Bayesian approach, by contrast, provides probabilities of the treatment effect that apply directly to the next patient who is similar to those treated in any completed or ongoing trial. Whereas conventional (frequentist) confidence limits are unlikely to exclude a null hypothesis, even when the treatment differs substantially, Bayesian methods utilise all available data to calculate probabilities that may be extrapolated directly to clinical practice.

Bayesian statistical approaches involve quantifying the information available about the outcome of interest, in the form of a *prior probability* at the design stage. The investigator acquires all available information about the activities of both the experimental and the control treatments. A prior distribution can then be formulated from the experiences and knowledge gained by others. After the data are collected, these will influence and result in modification of the probability. This is called the *posterior probability*. The combination of observed data and prior opinion is governed by Baye's theorem, which provides an automatic update of the investigator's subjective opinion. The attraction of the Bayesian approach lies in the simplicity of its concept and the directness of its conclusions [14]. Its flexibility for interim inspections is especially valuable in sequential clinical trials. Because the posterior probability can be updated repeatedly, using each successive posterior probability as the prior probability for the next update, it is the natural paradigm for the sequential decision making. Moreover, based on the posterior probability, early stopping criteria can be calculated to allow early termination of the trial, if the estimated effect is unacceptably different to the expected rate, or if the expected gain from further inclusions in reliability of estimating the effect rate is low. Finally, it enables one to predict the probability of the next patient outcomes as an additional tool in decision-making [15].

Tan proposed a scoring system for pooling evidence from small randomised controlled trials, in order to combine with hypothetical scenarios assisting in the design of, and justification for, a small randomised trial [16]. Combination of previous data with trial data by Bayesian techniques may help overcome the problem of recruitment, when few patients can be recruited in a study. An example is given in childhood supratentorial primitive neuroectodermal tumours with the aim to evaluate new therapies for the treatment of such rare cancers [16].

Among other approaches, one can find meta-analysis referring to a set of statistical procedures used to summarise empirical research in the literature. In general, meta-analysis serves as useful tool to answer questions to which single trials were underpowered or not designated to address.

The aim of this section was not to cover all possible design and statistical analyses in small clinical trials, but rather to give examples of what is feasible in this context. Further information may be found in the book *Small Clinical Trials: Issues and Challenges* available at http://books.nap.edu/books/0309073332/html/index.html.

REFERENCES

1. Sacks, H., Chalmers, T. C., Smith, H., Jr., *Am. J. Med.*, *72*, 233 (1982).
2. Clark, W. F., Garg, A. X., Blake, P. G., Rock, G. A., Heidenheim, A. P., Sackett, D. L., *JAMA*, *290*, 1351 (2003).
3. Guyatt, G., Sackett, D., Adachi, J., et al., *Cmaj*, *139*, 497 (1988).
4. Kent, M. A., Camfield, C. S., Camfield, P. R., *Arch. Pediatr. Adolesc. Med.*, *153*, 1292 (1999).
5. Johnson, R. L., Elliott, J. P., *Am. J. Obstet. Gynecol.*, *173*, 1369 (1995).
6. Greenhill, L. L., Jensen, P. S., Abikoff, H., et al., *J. Am. Acad. Child Adolesc. Psychiatry*, *42*, 406 (2003).
7. Whitehead, J., *The design and analysis of sequential clinical trials*, Barnett, V., Chichester (1997).
8. Whitehead, J., *Stat. Med.*, 18, 2271 (1999).
9. Bellissant, E., Duhamel, J. F., Guillot, M., Pariente-Khayat, A., Olive, G., Pons, G., *Clin. Pharmacol. Ther.*, *61*, 377 (1997).
10. Yao, Q., Wei, L. J., *Stat. Med.*, *15*, 2413 (1996).
11. Coad, D. S., Rosenberger, W. F., *Stat. Med.*, *18*, 761 (1999).
12. Rosenberger, W. F., Lachin, J. M., *Control Clin. Trials*, *14*, 471 (1993).
13. Bland, J. M., Altman, D. G., *BMJ*, *317*, 1151 (1998).
14. Spiegelhalter, D. J., Myles, J. P., Jones, D. R., Abrams, K. R., *BMJ*, *319*, 508 (1999).
15. Farge, D., Marolleau, J. P., Zohar, S., et al., *Br. J. Haematol.*, *119*, 726 (2002).
16. Tan, S. B., Dear, K. B., Bruzzi, P., Machin, D., *BMJ*, *327*, 47 (2003).

1.8   A register of clinical trials in children

Chiara Pandolfini, Maurizio Bonati

Laboratory for Mother and Child Health, "Mario Negri" Institute for Pharmacological Research, Milan, Italy

The issue of the underprivileged position of children with respect to optimal drug therapies was raised years ago, but the situation remains inadequate. Children are routinely given drugs that lack specific paediatric information as a result of a combination of factors. The most important is that too few paediatric studies, necessary for obtaining more rational answers to the therapeutic needs of children, are carried out [1-3].The reasons for the scarcity of paediatric clinical studies are well known [4-6] and range from ethical considerations (obtaining an "informed" consent from young patients is difficult and must often involve the parents instead of the patient) to practical issues (the significant differences in pharmacokinetic and pharmacodynamic properties in the different age ranges require series of trials to be performed in various age groups), to economic considerations (the paediatric drug market is also commercially difficult, due to the relatively limited use of most drugs in children). Another important factor is the lack of harmonisation between the available evidence and the product information provided [7]. The consequence is the frequent off-label and unlicensed use of drugs in children, a phenomenon which is documented in numerous studies involving different paediatric settings [8]. This type of drug use carries with it risks of lack of efficacy, adverse effects, calculation errors in adjusting adult formulations to paediatric ones, ineffective treatment through under-dosing, and non-availability of therapeutic advances [9-11].

Action has been taken to improve the situation in different countries, especially through legal changes in the licensing process, which offers the best guarantee for safe and effective medicines. In Europe, increased attention was drawn to the issue in 1997, when the Committee for Proprietary Medicinal Products (CPMP), part of the European Agency for the Evaluation of Medicinal Products (EMEA), created the "Note for guidance on clinical investigation of medicinal products in children" [12], describing how and when drugs should be tested in children. A few studies carried out between 1999 and 2002, however, found no significant improvement in the situation, with only a slight increase in the number of drugs with paediatric indications approved by the EMEA [13-15]. A "Directive on Good Clinical Practice" [16] was adopted in April 2001, which also addressed issues concerning the involvement of children in clinical trials and describes criteria for their protection. Another achievement was reached in 2001, when the International Conference on Harmonisation's (ICH) "Guideline on the clinical investigation of medicinal products in the paediatric population" went into effect in the United States, Europe, and Japan. In 2002, the European Commission took a different approach, with the publication of a consultation document entitled "Better medicines for children" [17], which proposed new regulatory actions to address the lack of suitably adapted paediatric medicinal products. In 2004, the European Commission presented the "Regulation of the European Parliament and Council on medicinal products for paediatric use" [18]. This legislation requires studies in paediatric populations, as part of the licensing dossier for any drugs of possible use to children. More specifically, the consultation document aims to improve the health of children in Europe by: increasing the development of medicines for use in children, ensuring that medicines used to treat children are subject to high quality, ethical research and are appropriately authorised for use in children, and by improving the information available on the use of medicines in the various paediatric populations. In order to achieve this, other strategies were also considered. These included the creation of a Paediatric Board to assist with most aspects of paediatric drug development, the improvement of pharmacovigilance systems and the creation of paediatric clinical trial registers. The strategies taken by the EC so far have shown increasing dedication to improving the status of drug therapies for children.

In the United States, the first formal regulations on the use of drugs in children were published by the FDA in 1979 and involved the introduction of a paragraph on the use of drugs in children in the package insert. These regulations drew attention to the need to generate data from studies in children. In 1994, the FDA introduced the Pediatric Rule [19], simplifying the type of information needed to demonstrate the safety and effectiveness of drugs in children by allowing efficacy data from adult studies to be used in some cases to label products for safe and effective use in children. Although this should have encouraged drug manufacturers to submit paediatric data for review voluntarily, the results were not sufficient. The FDA then created the Pediatric Exclusivity Provision in the

1997 Modernization Act (FDAMA) [20], which granted a six-month patent extension or market exclusivity to drug manufacturers who performed drug studies in children. This action led to more visible results. The 1998 Pediatric Final Rule [21] followed, requiring paediatric studies to be performed for products that would provide greater benefit with respect to existing treatments, that could pose significant risks for paediatric patients in the absence of adequate labelling, and that were likely to be used in a substantial number of children. In 2002, the Best Pharmaceuticals for Children Act [22] renewed FDAMA's Pediatric Exclusivity Provision and permitted the FDA to contract for needed paediatric studies of generic drugs. The incentives put into place were effective and motivated the pharmaceutical industry to generate paediatric data, despite all the difficulties of increasing pharmaceutical research in the United States. Although the results of these provisions were positive, the rule was challenged on the grounds that the FDA exceeded its authority in imposing these requirements on drug companies. Fortunately, after a legal battle involving the FDA, the American Academy of Pediatrics, the pharmaceutical industry, and a number of consumers, the legislation was approved in December 2003 as the Pediatric Research Equity Act of 2003 [23], giving the FDA the power to require drug makers to conduct tests to assess safety and efficacy in drugs for children.

As mentioned previously, despite the efforts carried out so far, the situation remains inadequate and calls for additional strategies aimed at the promotion of the rational use of drugs in children. Clinical trials have a fundamental role in achieving this, since they are the key to distinguishing useful from harmful treatments. However, it is difficult to identify the already insufficient paediatric studies carried out and to thus implement the knowledge derived from them. Most trials become public knowledge only once the results are published. Sometimes, however, studies that are stopped prematurely or with insignificant or negative results remain unpublished [24,25], leading to a distortion of evidence for patients and researchers, duplication of efforts by unknowing researchers and to concealment of potentially significant risks related to the use of certain substances [26]. Another issue related to the inadequacy of paediatric drug information is the waste of resources associated with carrying out clinical studies on drugs and conditions on which ample knowledge already exists, instead of addressing those areas that lack information. One study [27] searched for paediatric RCTs in scientific literature in order to identify the specific therapeutic areas covered by clinical research. It was found that the only things addressed were a few diseases or clinical problems, for which ample knowledge on physiological and therapeutic variables already exists. The 10 RCTs retrieved on acute otitis media, for example, compared the efficacy of 11 compounds. Four of these were multicentre studies co-ordinated by the same investigators, in the same country,

Table 1

Randomised controlled trials on acute otitis media found in the literature [27]

Authors	Nation	Publication year	Study year	Subject's age (months)	Experimental drug	Control
A	F	1997	1993/94	< 48	cefpodoxime proxetil (5 days)	co-amoxiclav (8 days)
A	F	1998	1995/96	< 24	co-amoxiclav (10 days)	co-amoxiclav (5 days)
A	F	1999	1995/96	< 36	ceftriaxone (single dose)	co-amoxiclav (10 days)
A	F	2000	1996/97	< 24	cefpodoxime-proxetil (10 days)	cefpodoxime-proxetil (5 days)
B	I	1996	-	6-60	azithromycin	amoxicillin
C	SF	1996	-	24-96	xylitol	sucrose
D	GR	1997	-	5-60	clarithromycin	cefuroxime
E	SF	1999	-	< 48	prednisolone	placebo
F	GB	2000	-	6-144	cefdinidir	amoxicillin / clavulanate
G	SF	2000	-	-	ne-1530	placebo

over 4 consecutive years, testing 3 antimicrobial agents (Table 1). These findings confirm the fact that the research community still needs to work to avoid unnecessary duplication of studies (and the related waste of resources and unethical involvement of patients).

The most important tools available to assist researchers in designing appropriate, necessary studies are clinical trial registers. Many well-known, established registers already exist; some focus on specific diseases, while others are representative of research carried out in specific geographical areas. However, despite the remarkable efforts made to create numerous, diverse databases addressing the needs of the biomedical research community in the past few years, none have involved a register of drug therapy trials specific to the paediatric field [28]. Furthermore, paediatric trial information from existing registers is generally difficult to access and, despite the well-documented, inadequate situation of drug therapies in children, is not given enough importance. Even the more well-known, freely-available online registers, for example, do not provide accurate search tools for recalling paediatric studies (Table 2). A sample search was repeated for each of three registers to briefly evaluate how precise their results were. The search, which attempted to imitate a query that a lay person would make, involved RCTs in children in general and then specifically on otitis. The results were not accurate; in many cases, they included irrelevant studies, in others, they excluded relevant studies. The effort behind these registers is enormous and commendable, but more importance needs to be given to information concerning children. Registers that take paediatric drug research into consideration should do so in a way, as is done for adults, that would allow the ensuing knowledge to be exploited in the most efficient manner possible to improve children's lives.

A register of completed and ongoing clinical trials in children is therefore a valuable notion. Such a register can be a useful resource for:

- planning new studies,
- promoting communication and collaboration among researchers,
- preventing trial duplication and inappropriate funding (while encouraging appropriate replication),
- identifying the therapeutic needs of children that remain neglected [29].

Such a register can also allow for active monitoring of new or evolved knowledge on drug therapies and facilitate patient access and recruitment into trials. This last potential should not be underestimated, because patients are increasingly searching the internet for health information and because they are a part of research, i.e. they help fund it, they participate in it, and they use the results to improve their health [30,31]. On the part of researchers, the desire to participate in a register by providing trial data comes from the understanding

Table 2

The major clinical trial registers present on the internet and the search options they provide relevant to a sample search: RCTs in children on otitis.

Country	Number of trials included/source	Name	Search options available	Search attempts, number of results, and comments			
				Paed. clin. trials	Comments	Otitis	Comments
UK	110373 (365 "clinical trials")/ Database of research projects funded by, or of interest to, the United Kingdom's National Health Service.	National Research Register (www.update-software.com/national/)	• MeSH* • free text (limitable to certain fields, e.g. "methodology" field)	• MeSH: "child" AND "Clinical trials" =37 • free text:"child and clinical trial" =1466	most results were relevant to children, but it was difficult to separate out clinical trials from other types of research (e.g. systematic reviews, surveys). since the register includes all types of research, not only clinical trials. Many relevant studies were excluded with the MeSH search, even though this should be the most reliable way to search databases	• MeSH: "child" AND "Clinical trials" AND "otitis" =0 • free text:"child and clinical trial and otitis" =1	a total of 67 RCT on otitis were present in the register, but these did not result from either of the combined searches for RCTs on otitis
USA	10,100/Trials sponsored by the National Institutes of Health, other federal agencies, and private industry. Conducted primarily in all 50 States and in over 90 countries.	ClinicalTrials.gov (www.clinicaltrials.gov)	• age group (child/adult/senior) • disease • free text	• limit to child age group (birth-17) 1977 • free text:"child" = 2027	either search option gives many irrelevant results (Parkinson's in patients >40, Alzheimer's in patients >40)	• limit to child age group and Disease: otitis = 2 • free text:"otitis" = 2	only 2 studies on otitis were present in the register, and both resulted from the searches
Internat.	16113/Trials from 27 different registers held by public, charitable and commercial sponsors of trials.	metaRegister (http://controlled-trials.com/mrct/)	• free text	• free text:"child%" (truncation search) = 2051	gives irrelevant studies (e.g. Patient age:"Female of child-bearing potential"	• free text: "child%" AND "otitis" = 15	some results were irrelevant (e.g. Exclusion criteria: "Prophylaxis for otitis media")

* MeSH (Medical Subject Headings): MeSH is the National Library of Medicine's controlled vocabulary thesaurus. It consists of sets of terms naming descriptors in a hierarchical structure that permits searching at various levels of specificity.

that, by doing so, information on the trials is brought to a wider audience, a service is being provided to science and to the public, and their action means taking part in a world-wide collaboration and a growing movement in favour of greater openness about ongoing research.

An additional, considerable legislative measure for any country would be to require registration of trials in an established register, before granting approval by an ethics committee or institutional review board [31]. In Europe, with the creation of the EudraCT database, a first step has been made, although access to its contents will be limited, at least for now. The United States also seem to be moving in this positive direction [32]. In the meantime, a register of clinical trials in Europe dedicated to children is being set up [33].

At the conclusion of the Fifth Framework Programme in 2002, the European Community decided to fund the development of a European register of clinical trials in children, following through with the commitment shown in its consultation document "Better medicines for children" [17]. The project, called DEC-net (EC contract number QLRT-2001-01054) [34], began in January 2003 and involves the creation of an online register able to handle essential data on paediatric clinical trials that can be expanded to include data from all European member states. Its main objective is to aid the dissemination of clinical trial results to ensure that drugs used in children are evaluated properly with respect to safety and efficacy. More specifically, the register aims to become a stepping-stone for co-ordinating paediatric drug research throughout Europe, promoting collaboration among researchers, preventing trial duplication and inappropriate funding, facilitating patient recruitment to trials by providing easy access to information, and identifying the unmet therapeutic needs of children. DEC-net is the first international register dedicated to paediatric trials. The project is being carried out by a network consisting of members from France, Italy, Spain, and United Kingdom with different, complementary clinical backgrounds and research experiences (Figure 1).

Information on all planned or on-going drug therapy trials involving children is being collected through contacts with ethical review boards, national paediatrician associations, pharmaceutical companies, etc., in each of the four countries. Among the trial information collected are the title, the drug under investigation, the disease, the type of study, the study organiser, and the inclusion and exclusion criteria. Some of the essential information is also provided in the original language for the lay public to consult. Information on the outcome of the studies, i.e. the results, conclusions, and publications, will be added periodically. The co-ordinators and participating members of the DEC-net project have access to more detailed trial information and can modify and update the input data. The register is freely accessible from the website (www.dec-net.org), which was created both to host the register and to provide users with useful, updated information on the

Development of the European register of clinical trials on medicines for children

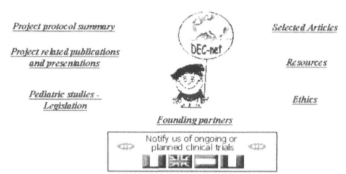

DEC-net
Drug Evaluation in Children
Clinical trials in Europe

Project protocol summary

Project related publications
and presentations

Pediatric studies -
Legislation

Selected Articles

Resources

Ethics

Founding partners

Notify us of ongoing or
planned clinical trials

The DEC-net Register is supported under the European Union's Fifth Framework Programme, Thematic Programme "Quality of Life" (contract QLG4-CT-2002-01054)

Figure 1. DEC-net register homepage (www.dec-net.org) and participating members: Maurizio Bonati, Chiara Pandolfini, Valentina Rossi, Eugenio Santoro -"Mario Negri" Institute for Pharmacological Research, Milan, Italy; Jose Maria Arnau de Bolós, I. Danés Carreras, I. Fuentes Camps - Fundacio Institut Catala de Farmacologia, Barcelona, Spain; Evelyne Jacqz-Aigrain, Sétareh Zarrabian - Hôpital Robert Debré, Paris, France; Imti Choonara, Helen Sammons - Derbyshire Children's Hospital, University of Nottingham, Derby, United Kingdom.

project and links to selected abstracts or articles on clinical trials and related national and international legislation. Links to other existing registers have also been provided; one of these is the "*meta*Register of controlled clinical trials", with which the DEC-net group hopes to actively collaborate in the future.

Many registries on clinical trials in humans, both general and field-specific, have been set up and made available on the Web [35], but none focus specifically on the paediatric population. This initiative will hopefully aid in making therapies more suitable for children.

The goal of drug therapy is the achievement of defined therapeutic outcomes that improve a patient's quality of life while minimising patient risk. If we believe in this, then all possible means to achieve optimal drug therapy must also be given priority for the paediatric population. Legislation and clinical trial registries together have the capacity to make knowledge of paediatric drug therapies reach its maximum potential throughout the world. The two strategies complement each other and the information obtained as a result of both can be compiled. While legislation works slowly, but thoroughly, to improve the future situation, a clinical trial register has the potential to render data available immediately. A register of clinical trials in children, characterised by participation and transparency, would be a unique tool with the capability to increase information on clinical research and, therefore, also increase consent to participation, starting from research *with* and *for* children (and their parents).

REFERENCES

1. Smyth, R. L., Weindling, A. M., *Lancet, 354 (suppl II)*, 2124 (1999).
2. Conroy, S., *Pediatr. Drugs, 4*, 353 (2002).
3. Steinbrook, R., *N. Engl. J. Med., 347*, 1462 (2002).
4. Koren, G., Kearns, G. L., Reed, M., Pons, G., *Clin. Pharmacol. Ther., 73*, 147 (2003).
5. Kodish, E., *J. Pediatr., 142*, 89 (2003).
6. Burns, J. P., *Crit. Care Med., 31 (Suppl. 3)*, S131 (2003).
7. Roberts, R., Rodriguez, W., Murphy, D., Crescenzi, T., *JAMA, 290*, 905 (2003).
8. Pandolfini, C., Bonati, M., *Pharm. Dev. Regul., 1(2)*, 133 (2003).
9. Impicciatore, P., Choonara, I., Clarkson, A., Provasi, D., Pandolfini, C., Bonati, M., *Br. J. Clin. Pharmacol., 52*, 77 (2001).
10. Choonara, I., Conroy, S., *Drug Saf., 25*, 1 (2002).
11. Nunn, A. J., *Arch. Dis. Child., 88*, 369 (2003).
12. Note for guidance on clinical investigation of medicinal products in children, Medicines Control Agency, London, 1997.
13. Impicciatore, P., Choonara, I., *Br. J. Clin. Pharmacol., 48*, 15 (1999).
14. 'T Jong G. W., Stricker, B. H., Choonara, I., Van Den Anker, J. N., *Acta Paediatr., 91*, 1233 (2002).
15. Ceci, A., Felisi, M., Catapano, et al., *Eur. J. Clin. Pharmacol., 58*, 495 (2002).
16. Directive 2001/20/EC of the European Parliament and of the Council of 4 April 2001 on the approximation of the laws, regulations and administrative provisions of the Member States relating to the implementation of good clinical practice in the conduct of clinical trials on medicinal products for human use.
17. European Commission. Better Medicines for Children. Proposed regulatory actions on pae- diatric medicinal products. Brussels, February, 2002. Available at: pharmacos.eudra.org/F2/ pharmacos/docs/Doc2002/feb/cd_pediatrics_en.pdf. (2002).
18. European Commission. Regulation of the European Parliament and Council on medicinal products for paediatric use. Draft for public consultation. March, 2004. Available from: http: //pharmacos.eudra.org/F2/home.html [Cited 2004 March 24].
19. Specific requirements on content and format of labeling for human prescription drugs, revi- sion of the "pediatric use' subsection in the labeling. *Federal Register, 59*, 64240 (1994).
20. Food and Drug Administration Modernization Act. *Pub. L. No. 105-115*. 1997.
21. Regulations requiring manufacturers to assess the safety and effectiveness of new drugs and biological products in pediatric patients. *Federal Register, 63*, 66632 (1998).
22. Best Pharmaceuticals for Children Act, *Pub. L. 107-109*, 107th Cong. (Jan. 4, 2002).
23. Pediatric Research Equity Act of 2003, *Pub. L. No. 108155 (S. 650)* (Dec. 3, 2003).
24. Stern, J. M., *Br. Med. J., 315*, 640 (1997).
25. Antes, G., Chalmers, I., *Lancet, 361*, 978 (2003).
26. Dickersin, K., Rennie, D., *JAMA, 290*, 516 (2003).
27. Clavenna, A., Pandolfini, C., Bonati, M., *Curr. Ther. Res., 63*, 707 (2002).

28. Bonati, M., Impicciatore, P., Pandolfini, C., *BMJ, 320,* 1339 (2000).
29. Horton, R., Smith, R., *BMJ, 319,* 865 (1999).
30. McCray, A. T., *Ann. Intern. Med., 133,* 609 (2000).
31. Tonks, A., *BMJ, 319,* 1565 (1999).
32. Steinbrook, R., *N. Engl. J. Med., 351,* 315-317 (2004)
33. Bonati, M., Pandolfini, C., Choonara, I., Jacqz-Aigrain, E., Arnau, J., *BMJ [e-letter]* (2002) [cited 2004 April 24]. Available from: bmj.com/cgi/eletters/325/7376/1314#28245.
34. Development of the European register of clinical trials on medicines for children *(DEC-net)*. (2003) [cited 2004 April 24]. Available from: www.dec-net.org.
35. Meta-Register of Controlled Trials (mRCT) [cited 2004 May 24]. Available from: http://controlled-trials.com/mrct.

1.9 Epidemiological aspects of drug use

Antonio Clavenna, Maurizio Bonati

Laboratory for Mother and Child Health, "Mario Negri" Institute for Pharmacological Research, Milan, Italy

1.9.1 INTRODUCTION

The importance of obtaining evidence for a more rational use of medicines for children, infants and neonates is widely recognised among professionals and lay people (parents, patients, and politicians). However, there is still a dearth of information regarding safety and efficacy in childhood [1,2]. Many aspects can contribute to this underprivileged position of children with respect to optimal drug therapies: different pharmacokinetic and pharmacodynamic profiles during growth and development; ethical and financial reasons; resources and research capabilities; regulatory guidelines and constraints [3]. Thus, more efforts are needed to guarantee a rational drug use: that children receive medications appropriate (safe and effective) to their clinical conditions, in doses and formulations suitable to their personal requirements, for adequate periods of time, and at the lowest cost to their families and their communities [4]. In such a context, pharmacoepidemiology is a health imperative [5] that, with the appropriate methodologies, can improve the effectiveness and efficiency of health care interventions [6]. Patterns of drug use and prescribing in developed and developing countries have been studied extensively [7], but, once again, the profile in children to date appears to be more a rarity than a routine approach [8].

1.9.2 DRUG UTILISATION PROFILE

In an attempt to define the emerging picture in this field at the beginning of the new millennium, a review of the epidemiological inter-country profiles of paediatric drug exposure in general practice performed from 1997, involving large population sizes was conducted. The general data concerning the selected surveys are summarised in Table 1 [9-15]. The publications dated back to 1999 and referred to data collected from 1997. Six of the seven studies were from Europe, documenting a long-standing tradition in the field [16], especially in Denmark [9,12] and Italy [13,15]. The studies involved 1807children, belonging to all age groups, from neonates to late teens. Only one study [13] specified the diagnoses or symptoms related to given drugs. Age and gender distribution patterns were analysed in six of the seven studies [9,11-15]. The imbalance in the population sizes and age groups considered reflects differences in the criteria used for data collection or inclusion of children. Although every study had some major limitations, due to the fact that important covariables were occasionally considered and that national drug policies differed widely, the rate of treated children (50-65%) and the number of prescribed drugs per child (1.4-3.9) were more similar than previously reported [8]. For both boys and girls, the prescription rate for overall use of drugs (number of prescriptions/number of treated children) was higher for younger children (<2 years), decreased with age until the early teens, and increased thereafter, in particular for girls who took oral contraceptives.

According to the International Anatomic-Therapeutic-Chemical Classification system (ATC), systemic antibiotics were the most prescribed drugs in all studies, followed by anti-asthmatics, whereas wide inter-study (inter-country) differences were seen between the other drug classes. Broad-spectrum penicillins were generally the most used antibiotics, in agreement with indications of their use as first-line therapy for a large number of childhood conditions including pharyngotonsillitis, acute otitis media and sinusitis, which were the most common diseases in the studies [13]. Amoxicillin, which is available in a number of formulations to make it easier to administer to children of all ages, was the most common drug. Cefaclor and clarithromycin were also widely used, but with different rates between countries in which a long list of different compounds for common diseases are used, ignoring choice criteria, and with an increased risk of resistance, andallergic and adverse reactions, and greater economic expenditures [17,18].

For the other therapeutic classes, the drug use profiles differed to a greater extent, suggesting the need for a more rational use of drugs in children as part of an improvement of clinical effectiveness, cost effectiveness and to reduce the potential risk of side effects. Moreover, if the fact that the 20 most prescribed drugs could cover about 70% prescriptions is taken into consideration [13], such wide differences within and between countries imply the need for harmonisation [19]. In such a context, two examples (anti-asthmatic and psychotropic drugs) from the most recent population-based study [15] can highlight the needs in the area [20], as well as the usefulness of drug utilisation monitoring, analysis and comparison [21].

Table 1

Population based studies of drug use in children

Ref.	Study period	Country	Age (years)	Number of children	Treated children (%)	Prescriptions/ treated children	Rank of the most prescribed ATC subgroup				
							J01	R03	S01	R06	D07
9	1 Jan – 31 Dec 1997	Denmark	0-15	95,189	50.6	3.2	1	2	3	5	4
10	1 Jan – 31 Dec 1997	UK	0-12	1807	65.0	2.8	1	2	-	9	6
11	1 Jan –31 Dec 1998	Netherlands	0-16	25,020	60.0	-	1	5	9	4	3
12	1 Jan – 31 Dec 1998	Denmark	0-18	104,897	52.6	-	1	2	3	5	4
13	1 Apr – 30 June 1998	Italy	0-12	9917	64.7	1.4	1	2	10	5	11
14	1 Jan – 31 Dec 1999	Canada	0-17	1,031,731*	n.a.	3.9	1	2	-	-	-
15	1 Jan – 31 Dec 2000	Italy	0-13	520,273	61.1	2.9	1	2	4	3	7

* number of active claimants

JOI = Systemtic antibiotics; RO3 = Anti-asthmatics; SO1 = Ophthalmologics; RO6 = Systemic antihistamines; DO7 = Topical corticosteroids.

1.9.3   ANTI-ASTHMATIC DRUGS

In an Italian study, over 60% of children received at least one drug prescription, and 22% received anti-asthmatic drugs [22]. The prevalence rate of anti-asthmatic drug prescriptions decreased significantly from 34% in children under 1 year to 11% in those over 12 years of age. In children receiving anti-asthmatic drugs, 59% received monotherapy, 30% two drugs, and 11% three or more drugs. 25% of children under 1 year received at least one prescription for beclomethasone. The prevalence of this drug decreased with increasing age, reaching 6% in children 12 to 13 years. The prevalence rates of salbutamol and flunisolide prescriptions also decreased with increasing age (from 12 to 3% and from 7 to 1%), while the prevalence rate of fluticasone prescriptions increased from 1 to 1.4%. 98% of prescriptions for beclomethasone and nearly all prescriptions for flunisolide were for nebulised suspensions, as opposed to inhalers.

In this example, anti-asthmatic drug prescriptions were neither a reliable indicator of quality of care, nor an indicator of asthma epidemiology. The prevalence rate of prescriptions (22%) was higher than both the prevalence of asthma in Italy and the estimated rate in other paediatric contexts [23-25]. Moreover, it was of concern that inhaled steroids were prescribed more often than bronchodilators, with a corticosteroid to bronchodilator ratio of 2. Epidemiological data indicate that more than 50% of asthmatic children suffer from an intermittent kind of disease [26] and do not require long-term prophylactic treatment. The fact that beclomethasone and flunisolide were prescribed more frequently in children under 5 years of age raises some concerns. Most episodes of wheezing that appear in pre-school children disappear with age [27] and it is therefore more difficult to diagnose asthma in children under 6 years of age [28]. Beclomethasone and flunisolide were prescribed as nebulised suspensions, but these formulations should be given to children under 2 years of age, while older children should use a metered dose inhaler with a spacer and a face mask because these devices allow for more appropriate drug delivery [29]. Nebulised suspensions of beclomethasone and flunisolide are licensed in Italy, but not in the US, UK, France and Germany. In these four countries, budesonide is the only corticosteroid licensed for nebuliser administration. While in the other countries nebulised suspensions are licensed only for prophylactic treatment of asthma and for croup, in Italy flunisolide is also licensed for rhinitis, and beclomethasone for rhinitis and inflammatory diseases of the naso-pharyngeal tract. This "extended" label, not supported by evidence, could partially explain the use of these steroids for the treatment of upper respiratory tract infections.

1.9.4 PSYCHOTROPIC MEDICATIONS

The prevalence of psycotropic drug prescriptions in 600'000 italian youths under 18 years of age was 3.4‰ [30]. Antidepressants were prescribed to 1600 youths (2.8‰), antipsychotics to 448 (0.8‰) and lithium to 33 (0.1‰) [30]. The prevalence rate of psychotropic drug prescriptions increased with increasing age and was higher in females. Antidepressants were prescribed mainly to girls, while the prescription prevalence of antipsychotics was higher in boys (Table 2). Co-prescription was rare: 93% of the treated youths received only one class of psychotropic medication, 122 children (6%) received drugs from two classes (mainly antidepressants and antipsychotics) and only 6 children received antidepressants, antipsychotics and lithium. Among youths receiving antidepressant prescriptions, 1138 (71%) were given SSRIs only, 236 (15%) tricyclic antidepressants only, 61 (4%) tricyclics (TCAs) and SSRIs, and 165 (10%) other antidepressants (mainly trazodone or venlafaxine). Paroxetine (0.8‰), sertraline (0.5‰) and citalopram (0.5‰) were the most prescribed antidepressants. Risperidone (0.2‰), amilsulpride (0.1‰) and olanzapine (0.1‰) were the most prescribed antipsychotics.

The psychotropic drug prescription prevalence in Italy was substantially lower than that observed in United States (1 to 2%) and the Netherlands (0.4%) (Table 3). The gender differences in prescription prevalence appear consistent with the epidemiology of psychiatric disorders: antipsychotic drugs were more frequently prescribed to males and antidepressant medications to females. However, school-aged boys received antidepressants more frequently than girls, suggesting that antidepressants were probably also administered for obsessive-compulsive disorder or Attention Deficit Hyperactivity Disorder (ADHD).

The prescription pattern of tricyclic antidepressants appeared different from that of SSRIs. Almost half of the youths receiving tricyclics had only one prescription, whereas one third of children receiving SSRIs had at least 5 prescriptions during 2002. It is likely that tricyclics were prescribed mainly for "acute" conditions (not only psychiatric disorders, but also migraine or enuresis), SSRIs for depression, obsessive-compulsive disorder or ADHD. However, half of the youths did not receive any additional drug prescriptions in the following 6 months, and the therapies prescribed could barely have covered the 8-week trial period needed for evaluating the effects of the treatment [31,32].

The use of antidepressants in children is increasing, although the evidence for the effectiveness and safety of pharmacological treatment of depressive disorder in children and adolescents is scant [33] and widely debated, particularly for SSRIs [34]. To date, the data available on safety and efficacy in the paediatric population are limited. SSRIs are more effective than placebo for the treatment

Table 2

Antidepressant and antipsychotic prescription prevalence by gender and age group per 1,000 Italian children [30]

Age group (years)	Gender	Antidepressants	Antipsychotics	Lithium	Any psychotropics
0-5	M	0.71	0.15	0.01	0.88
	F	0.69	0.20	0.02	0.87
	M/F (95% C.I.)	1.04 (0.74-1.46)	0.79 (0.40-1.57)	0.47 (0.04-5.22)	1.02 (0.75-1.38)
6-13	M	1.92	0.93	0.08	2.77
	F	1.55	0.42	0.02	1.91
	M/F (95% C.I.)	1.24 (1.03-1.49)*	2.21 (1.60-3.06)†	4.71 (1.03-21.51)	1.45 (1.23-1.71)†
14-17	M	5.90	2.40	0.16	7.51
	F	10.60	1.56	0.13	11.49
	M/F (95% C.I.)	0.56 (0.49-0.63)†	1.45 (1.12-1.88)‡	1.18 (0.47-2.99)	0.65 (0.58-0.73)†
0-17	M	2.40	0.97	0.07	3.19
	F	3.25	0.60	0.04	3.69
	M/F (95% C.I.)	0.74 (0.67-0.81)†	1.62 (1.34-1.96)†	1.65 (0.81-3.36)	0.87 (0.79-0.95)§

* p=0.003; † p<0.00001; ‡ p=0.005; § p=0.002.

Table 3

Summary of the characteristics and results of epidemiological studies concerning psychotropic drug prescriptions in the paediatric population

Ref.	Characteristics of the studies				Prevalence (per 1,000 youths)				
	Country	Year	Age (y)	Number of children	Overall	Stimulants	Antidepressants	Neuroleptics	Lithium
55	United States	1996	0-18	6,490	39.0	24.0	10.0	2.0	-
56*	United States	1996	0-20	897,694	62.0	35.6	19.9	5.1	1.8
14	Canada	1999	0-17	1,031,000	-	32.9	16.2	3.8	-
57	Netherlands	1999	0-19	37,670	-	7.4	4.4	3.4	0.2
58*	United States	1994	2-19	900,289	-	-	18.0	-	-
59	Italy	2000	0-15	25,926	-	-	0.3	-	-
60	United States	2002	0-18	380,331	-	-	23.7	-	-
61	Italy	1991	5-19	50,000	-	-	-	1.7	-
62*	United States	1998	0-14	463,377	-	57.3	$1.3^{\dagger}$	-	-
63	United States	1998	0-5	38,664	3.3	1.5	0.4	-	-
64*	United States	1995	2-4	222,967	-	10.9	2.6	0.7	-

* Modified from original paper: aggregated data are presented
† Only SSRI prescriptions were evaluated.

of obsessive-compulsive disorder [35] and the American Academy of Child and Adolescent Psychiatry stated that SSRIs could be considered first-line drugs on the basis of their fewer side effects and lower toxicity profile with respect to clomipramine [31]. Concerning the treatment of major depression, a systematic review of the literature did not find a statistically significant difference between tricyclic antidepressants and placebo [36], while a few randomised control-led trials showed a better response with SSRIs compared to placebo [37-40]. However, solid evidence of the efficacy of psychotropic medications in treating depressed youths is still lacking [41]. Furthermore, the effectiveness of these drugs could have been overestimated due to publication bias [42].

Although fluoxetine was the first drug to be approved for treatment of depression in children [43], the data available concerning its safety, especially in the long-term period, are also scant and little is known about the eventual effects on neurological and behavioural development. Some case-reports suggest that SSRI use could be associated with a growth reduction [44]. Moreover, a 22% incidence of psychiatric adverse events in children and adolescents treated with SSRIs was reported [45], and a possible increase in risk of suicide attempts or ideation is still being debated [46]. According to the statements of several national drug regula-tory agencies, this risk seems to be real, at least for paroxetine and venlafaxine [47-51]. Paroxetine was the most prescribed antidepressant in paediatric patients and was administered to about 1 in 4 of the children treated with antidepressants. Moreover, of the antidepressant users in this sample, 75% received drugs with a benefit-to-risk balance that is considered unfavourable [52]. Furthermore, nearly 56% of antidepressant prescriptions (corresponding to 69% antidepressant users) were unlicensed or off-label: twice the rate generally observed in general practice [53], but consistent with findings observed in another study [54].

1.9.5 CONCLUSIONS

Drugs in children are often over-used and prescribed in an inappropriate manner (i.e., anti-asthmatics). Moreover, the risk of a similar utilisation of other drugs is also evident, with low overall prescription rates, but with high safety con-cerns (i.e., psychotropics). Many children, therefore, receive therapies for which there is no evidence of safety and efficacy, or for which these are questionable. The use of a wide range of drugs, the diverse intra- and inter-country prescribing profiles related to a limited number of very common childhood diseases, and the lack of compliance with therapeutic guidelines, suggest the need for continuous audit and educational programs (also and especially) in paediatric general prac-tice [65-67]. Prescribing physicians and children's parents should therefore make an effort to achieve a more rational drug utilisation. Appropriate, independently funded studies should be planned to guarantee safe and effective evidence based therapeutic approaches for children, adolescents, and their families.

REFERENCES

1. Choonara, I., Nunn, A. J., Kearns, G. (ed.), *Introduction to Paediatric and Perinatal Drug Therapy*, Nottingham Univerity Press, Nottingham 2003.
2. Sanz, E. J., *BMJ, 327,* 858 (2003).
3. Wong, I., Sweis, D., Cope, J., Florence, A., *Drug Safety, 26,* 529 (2003).
4. WHO/DAP. Operational research projects in DAP: an inventory, WHO/DAP, 1995.
5. MacLeod, S. M., *J. Clin. Epidemiol., 44,* 1285 (1991).
6. The U.S. Pharmacopeia Drug Utilization Review Advisory Panel, *J. Am. Pharm. Assoc. (Wash). 40,* 538 (2000).
7. Le Grand, A., Hogerzeil, H. V., Haaijer-Ruskamp, F. M., *Health Policy and Planning, 14,* 89 (1999).
8. Bonati, M., *J. Clin. Pharmacol., 34,* 300 (1994).
9. Thrane, N., Sørensen, H. T., *Acta Paediatr., 88.* 1131 (1999).
10. McIntyre, J., Conroy, S., Avery, A., Corns, H., Choonara, I., *Arch. Dis. Child., 83,* 498 (2000).
11. Schirm, E., van den Berg, P., Gebben, H., Sauer, P., de Jong-van den Berg, L., *Br. J. Clin. Pharmacol., 50,* 473 (2000).
12. Madsen, H., Anderson, M., Halls, J., *Eur. J. Clin. Pharmacol., 57,* 159 (2001).
13. Cazzato, T., Pandolfini, C., Campi, R., Bonati, M., *Eur. J. Clin. Pharmacol., 57,* 611 (2001).
14. Abi Khaled, L., Ahmad, F., Brogan, T., et al., *Paediatrics & Child Health, 8 (Suppl. A)* (2003).
15. Progetto, ARNO. Osservatorio sulla prescrizione farmaceutica pediatrica. CINECA Casalecchio di Reno, Bologna 2002.
16. Dukes, M. N. G. (ed.) Drug utilization studies, WHO Regional Publications European Series n. 45. Copenhagen (1993).
17. Alpern, E. R., Louie, J. P., *Pediatr. Case Rev., 2,* 69 (2002).
18. Woodhead, M., Fleming, D., Wise, R., *BMJ, 328,* 1270 (2004).
19. Bonati, M., Pandolfini, C., *BMJ, 328,* 227 (2004).
20. Clavenna, A., Pandolfini, C., Bonati, M., *Curr. Ther. Clin. Exp., 63,* 707 (2002).
21. Strom, B. L. (ed.) *Pharmacoepidemiology*, 3rd Edition, John Wiley & Sons, 2000.
22. Clavenna, A., Rossi, E., Berti, A., et al., *Eur. J. Clin. Pharmacol., 59,* 565 (2003).
23. Straand, J., Rokstad, K., Heggedal, U., *Acta Paediatr., 87,* 218 (1998).
24. Goodman, D. C., Lozano, P., Stukel, T. A., Chang, C., Hecht, J., *Pediatrics, 104,* 187 (1999).
25. Madsen, H., Andersen, M., Hallas, J., *Eur. J. Clin. Pharmacol., 57,* 159 (2001).
26. Rabe, K. F., Vermeire, P. A., Soriano, J. B., Maier, W. C., *Eur. Respir. J., 16,* 802 (2000).
27. Martinez, F. D., Wright, A. L., Taussig, L. M., Holberg, C. J., Halonen, M., Morgan, W. J., *N. Engl. J. Med., 332,* 133 (1995).
28. Barry, P. W., O'Callaghan, C., *Thorax, 52 (S2),* S78 (1997).
29. National Institute of Health. *Guidelines for the Diagnosis and Management of Asthma.* NIH Publication 97-4051, Betsheda, MD, 1997, p. 68.
30. Clavenna, A., Bonati, M., Rossi, E., De Rosa, M., *BMJ, 328,* 711 (2004).
31. American Academy of Child and Adolescent Psychiatry, *J. Am. Acad. Child. Adolesc. Psychiatry, 37,* 27S (1998).
32. Quitkin, F. M., Petkova, E., McGrath, P. J, et al., *Am. J. Psychiatry, 160,* 734 (2003).
33. Ramchandani, P., *BMJ, 328,* 3 (2004).

34. Jureidini, J. N., Doecke, C. J., Mansfield, P. R., Haby, M. M., Menkes, D. B., Tonkin, A. L., *BMJ*, 328, 879 (2004).
35. Geller, D. A., Biederman, J., Stewart, S. E., et al., *Am. J. Psychiatry*, *160*, 1919 (2003).
36. Hazell, P., O'Connell, D., Heathcote, D., Henry, D *Tricyclic drugs for depression in children and adolescents (Cochrane Review)*, in The Cochrane Library Issue 4, Chichester, UK, John Wiley & Sons, Ltd., (2003).
37. Emslie, G. J., Rush, A. J., Weinberg, W. A., et al., *Arch. Gen. Psychiatry*, *54*, 1031 (1997).
38. Emslie, G. J., Heiligenstein, J. H., Wagner, K. D., et al., *J. Am. Acad. Child. Adolesc. Psychiatry*, *41*, 1205 (2002).
39. Keller, M. B., Ryan, N. D., Strober, M., et al., *J. Am. Acad. Child. Adolesc. Psychiatry*, *40*, 762 (2001).
40. Wagner, K. D., Ambrosini, P., Rynn, M., et al., *JAMA*, *290*, 1033 (2003).
41. Hazell, P., *Clin Evid.*, *10*, 375 (2003).
42. Whittington, C. J., Kendall, T., Fonagy, P., Cottrell, D., Cotgrove, A., Boddington, E., *Lancet*, *363*, 1341 (2004).
43. FDA. Talk Paper FDA Approves Prozac for Pediatric Use to Treat Depression and OCD, available at http://www.fda.gov/bbs/topics/ANSWERS/2003/ANS01187.html (accessed June 2004).
44. Weintrob, N., Cohen, D., Klipper-Aurbach, Y., Zadik, Z., Dickerman, Z., *Arch. Pediatr. Adolesc. Med.*, 696 (2002).
45. Wilens, T. E., Biederman, J., Kwon, A. *et al.*, *J. Child. Adolesc. Psychopharmacol.*, *13*, 143 (2003).
46. FDA Public Health Advisory. *Worsening depression and suicidality in patients being treated with antidepressant medications*, URL http://www.fda.gov/cder/drug/antidepressants/ AntidepressanstPHA.htm (accessed June 2004).
47. Duff, G. *Safety of seroxat (paroxetine) in children and adolescents under 18 years: contraindication in the treatment of depressive illness*, June 10, 2003, URL http://medicines.mhra.gov.uk/aboutagency/regframework/csm/csmhome.htm (accessed June 2004).
48. Duff, G. *Safety of venlafaxine in children and adolescents under 18 years in the treatment of depressive illness*. Semptember 19, 2003, URL http://medicines.mhra.gov.uk/aboutagency/ regframework/csm/csmhome.htm (accessed June 2004).
49. Ministero della Salute, *Bollettino di Informazione sui Farmaci 207*, 5 (2003): .
50. Important drug warning: *Until further information is available, Paxil® (paroxetine hydro-chloride) should not be used in children and adolescents under 18 years of age*, Available at URL http://www.hc-sc.gc.ca/hpfb-dgpsa/tpd-dpt/paxil_e.html (accessed June 2004).
51. Important safety information regarding the use of Effexor® (venlafaxine HCl) tablets and Effexor® XR (venlafaxine HCl) capsules in children and adolescents. Available at URL http://www.hc-sc.gc.ca/hpfb-dgpsa/tpd-dpt/effexor_prof_e.html (accessed June 2004).
52. Committee on Safety of Medicines. *Selective Serotonin Reuptake Inhibitors – Use in Children and Adolescents with Major Depressive Disorder*, December 10th 2003 available at URL http://www.mca.gov.uk/ourwork/monitorsafequalmed/safetymessages/cemssri_101203.pdf (accessed June 2004).
53. Pandolfini, C, Bonati, M., *Pharm. Dev. Regul.*, *1*, 133 (2003).
54. Schirm, E., Tobi, H., de Jong-van den Berg, L. T. W., *Pediatrics*, *111*, 291 (2003).
55. Olfson, M., Marcus, S. C., Weissman, M. M., Jensen, P. S., *J. Am. Acad. Child. Adolesc. Psychiatry*, *41*, 514 (2002).
56. Magno Zito, J., Safer, D. J., dos Reis, S. et al., *Arch. Pediatr. Adolesc. Med.*, *157*, 17 (2003).
57. Schirm, E., Tobi, H., Magno Zito, J., de-Jong-van den Berg, L. T. W., *Pediatrics, 108*, e25 (2001). URL http://www.pediatrics.org/cgi/content/full/108/2/e25 (accessed June 2004).

58. Magno Zito, J., Safer, D. J., dos Reis, S. *et al.*, *Pediatrics, 109*, 721 (2002).
59. Pietraru, C., Barbui, C., Poggio, L., Tognoni, G., *Eur. J. Clin. Pharmacol., 57*, 605 (2001).
60. Delate, T., Gelenberg, A. J., Simmons, V. A., Motheral, B. R., *Psychiatric Services, 55*, 387 (2004).
61. Traversa, G., Spila-Alegiani, S., Arpino, C., Ferrara, M., *J. Child. Adolesc. Psychopharmacol., 8*, 175 (1998).
62. Rushton, J. L., Withmire, J. T., *Arch. Pediatr. Adolesc. Med., 155*, 560 (2001).
63. DeBar, L. L., Lynch, F., Powell, J., Gale, J., *Arch. Ped. Adolesc. Med., 157*, 150 (2003).
64. Magno Zito, J., Safer, D. J., dos Reis, S., Gardner, J. F., Boles, M., Lynch, F., *JAMA, 283*, 1025 (2000).
65. Wallace, P., Drage, S., Jackson, *BMJ, 316*, 323 (1998).
66. Smyth, R. L., *BMJ, 322*, 1377 (2001).
67. Sammons, H. M., McIntyre, J., Choonara, I., *Arch. Dis. Child., 89*, 408 (2004).

Paediatric clinical pharmacology

2.1 The effect of age on drug metabolism

Saskia de Wildt[1], Trevor N. Johnson[2], Imti Choonara[3]

[1]Dept. of Paediatrics, Erasmus University & Sophia Childrens Hospital, Rotterdam, The Netherlands

[2]Academic Unit of Molecular Pharmacology & Pharmacogenetics, University of Sheffield and Pharmacy Department, Children's Hospital, Sheffield, UK

[3]Academic Division of Child Health, University of Nottingham, Derbyshire Children's Hospital, Uttoxeter Road, Derby, UK

2.1.1 INTRODUCTION

There are many factors that affect drug metabolism. There is currently considerable interest in the field of pharmacogenetics, i.e. the effect of genetic make up in relation to the capacity to metabolise different drugs. We wish to give a brief overview of the effect of age on drug metabolism from birth through childhood to adolescence.

The major site of drug metabolism is within the liver. The gastrointestinal tract, blood cells and other organs are also involved in drug metabolism. The biological purpose of drug metabolism is to convert lipophilic (fat soluble) compounds into more polar and thus more water soluble substances that are more readily excreted into bile or urine. The enzymes involved in drug metabolism are not only involved in the breakdown of medicines but also the numerous other chemicals that humans ingest or inhale either deliberately or unwittingly.

The major pathways involved in drug metabolism are divided into phase 1 and phase 2 reactions. Phase 1 involves oxidation, reduction, hydrolysis and hydration reactions. The major pathway is oxidation which involves the cytochrome P450 dependent (CYP) enzymes. The major CYP enzymes are CYP1A2, CYP2B6, CYP2C8-10, CYP2C19, CYP2D6, CYP2E1 and CYP3A4 and 5. The major pathways for phase 2 involve glucuronidation, sulphation, methylation, acetylation and glutathione conjugation.

Reprinted from *Paed. Perinatal Drug Ther.*, 5, 101 (2003).

We plan to highlight the changes that have been previously described in relation to some of the major metabolic pathways and review enzyme activity in paediatric patients of different ages. Specifically we will review the development of CYP1A2 and CYP3A4 as examples of phase 1 and glucuronidation and sulphation as examples of phase 2 metabolism.

We have used the age classification accepted by both the European Medicines Evaluation Agency and also the recent International Conference on Harmonisation [1]. This classification divides paediatric patients into 5 age groups; preterm neonates, full-term neonates, infants from 1 month up to 24 months of age, children between the ages of 2 and 11 years and adolescents from 12 to 17 years.

2.1.2 GENERAL TRENDS IN THE DEVELOPMENT OF PHASE 1 DRUG METABOLISM IN CHILDREN

Total cytochrome P450 content in the fetal liver is between 30 and 60% of adult values and approaches adult values by 10 years of age [2]. Different developmental patterns have been identified for CYPs involving activity in the fetal liver (CYP3A7), minimal activity in the fetal liver but with rapid increase hours after birth (CYP2D6 & CYP2E1) and development in infancy (CYP1A2) [3-6].

For many CYP drug substrates, weight corrected clearance values are often low at birth, but then increase rapidly reaching a maximum by 2 years of age. Hepatically metabolised drugs that exhibit a higher systemic weight normalised clearance in children compared to adults include theophylline [7], diclofenac [8], teniposide [9], phenytoin [10], carbamazepine [11] and omeprazole [12]. Possible reasons for increased hepatic clearance in children include an increased liver volume normalised to body weight [13,14] or a higher concentration of catalytically active CYPs. A recent study failed to detect increased intrinsic CYPs 1A2, 2C8, 2E1 and 3A4/5 activity in paediatric livers in comparison to adult livers [15].

2.1.3 CYP3A4

The CYP3A subfamily is the most abundantly expressed CYP subfamily in the human adult and newborn liver. Moreover, this subfamily is involved in the metabolism of more than half of all drugs, including cyclosporin, tacrolimus, cisapride, midazolam, fentanyl, lidocaine, nifedipine, indinavir and verapamil.

The CYP3A subfamily consists of at least 4 enzymes: CYP3A4, CYP3A5, CYP3A7 and CYP3A43. CYP3A4/CYP3A5 account for 30 to 40% of total CYP content in the adult liver and intestine. CYP3A4 and CYP3A5 are differentially expressed, but have largely overlapping substrate specificity. CYP3A7 is the main CYP isoform in the human fetal and newborn liver. From the few studies available, it appears that the substrate specificity of CYP3A7 is different from CYP3A4.

In vitro studies have shown that CYP3A7 activity is high directly after birth, while CYP3A4 activity is very low [3]. During the first days after birth, a transition occurs from mainly CYP3A7 activity to CYP3A4 activity. Finally, adult levels of CYP3A4 are reached during the first years of life. This developmental pattern of CYP3A4 is reflected by the change in clearance rate of midazolam [16-19] at different ages (Table 1). Midazolam clearance is reduced in infants under the age of 2 years. Although the median plasma clearance reaches adult levels in children over the age of 2 years, it is important to recognise the considerable inter-individual variation [17] (up to 100 fold in one study [18]).

The exact developmental pattern of CYP3A4 activity during infancy remains to be elucidated. Studies of midazolam show a lower clearance (corrected for body weight) for CYP3A substrates in the first two years of life [18]. In contrast, the clearance of both cyclosporin and tacrolimus is higher in infants than older children and adults [20,21]. Young children, therefore, require higher cyclosporin and tacrolimus dosages (in relation to bodyweight) than adults [22,23].

Interestingly, CYP3A is not only localised in the liver, but also in the intestine. Therefore, intestinal drug metabolism also contributes to presystemic clearance of CYP3A substrates. A recent study showed *in vitro* that the ontogeny of intestinal CYP3A activity mirrors that of hepatic CYP3A activity [24]. Therefore, the

Table 1

Age and midazolam clearance

Age group	Number of patients	Mean or median plasma clearance (ml/min/kg)	Range	Ref.
preterm neonates	24	1.8	0.7 – 6.7	16
	?	1.2	?	17
term neonates	?	1.8	?	17
infants 1-24 months	25	3.0	0.5 – 25.8	18*
children 2-11 years	12	9.2	0.5 - 66.7	18*
adolescents 12-17 years	20	10.0	?	19

*Data recalculated for modified age groups

oral bioavailability of CYP3A substrates may be increased consequent to reduced presystemic clearance in newborn infants. This assumption is supported by the finding that midazolam oral bioavailability is higher in preterm infants when compared to adults [25].

2.1.4 CYP1A2

CYP1A2 accounts for approximately 13% of the total cytochrome P450 enzyme expression in the livers of healthy adult humans [26]. Caffeine is a recognised probe to study the activity of CYP1A2 both *in vitro* and *in vivo* [27] (Figure 1).

In vitro caffeine metabolism

In vitro studies have shown that the rates of caffeine N1, N3 and N7 demethylation are significantly lower in the fetus, neonate and infant than the adult [28]. C-8 hydroxylation to 1,3,7 trimethyluric acid was not significantly different between age groups. The production of total dimethylxanthines increased significantly with age up to 300 days. Differences in the maturational profile of each pathway suggest that different CYP isozymes are involved with a delay in maturation of N1 demethylation in comparison with N3 and N7 demethylation.

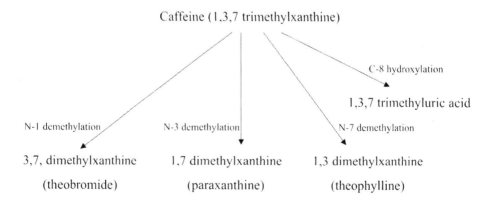

Figure 1. The major pathways of caffeine metabolism.

In vivo caffeine metabolism

In one study, total caffeine demethylation and N3 and N7 demethylation increase exponentially with postnatal age and reach a plateau by 120 days [29]. The maturation of N1 demethylation is delayed and does not occur until after 19 months of age. 8-hydroxylation is mature by as early as one month and may be higher in infants compared to adults. Because the N3 and N7-demethylation pathways account for 88% of the metabolism of caffeine in humans [30], caffeine clearance should give a reasonable estimate of *in vivo* CYP1A2 activity.

Table 2

Age and caffeine clearance

Age group	Number of patients	Mean or median plasma clearance (ml/min/kg)	Range	Ref.
preterm neonates	89	4.9	3 – 17	31
term neonates	1	20.0	-	33
infants 1-24 months	12	72.4	22 – 153	34
children 2-11 years	9	81.6	21 – 132	35
adolescents 12-17 years	–	–	–	–
adults	?	94.0	21- 270	36

Studies in preterm neonates [31,32] have shown reduced clearance of caffeine in comparison with term babies and infants [33,34]. The overall changes in caffeine clearance with age are shown in Table 2. Pons and co-workers have demonstrated significant impairment of the N3 demethylation of caffeine by CYP1A2 in infants under the age of 6 months, thereafter the activity remains fairly constant [35,36].

One study demonstrated an inverse relationship between weight adjusted body clearance and the molar fraction of caffeine excreted unchanged in the urine in neonates and infants aged 3 days to 9 months [37]. There have been relatively few studies of the pharmacokinetics of caffeine in children after the first year of life. In general, most *in vivo* clearance results for caffeine in children mirror the *in vitro* development of CYP1A2 in the human liver [6].

2.1.5 GLUCURONIDATION AND SULPHATION

Many medicines undergo glucuronide conjugation after oxidation. Other medicines undergo direct conjugation with glucuronic acid as a primary metabolic pathway. An important group of phase 2 metabolising enzymes are the UDP-glucuronyltransferases (UGTs); to date, at least 10 different UGTs have been identified. Several drugs are glucuronidated, e.g. morphine, paracetamol, codeine, lorazepam, naloxone, propofol and chloramphenicol [38]. As these drugs are metabolised by one or more different UGT isoforms, and some can also be sulphated, the effect of ontogeny on the pharmacokinetics of these drugs is not uniform.

Sulphation is the other major phase 2 metabolic pathway and results in formation of water soluble metabolites that can be excreted renally [39,40]. Sulfotransferases are the enzymes involved in sulphation. The total number of sulfotransferase enzymes are unknown but are divided into two groups, catechol sulfotransferases and phenol sulfotransferases. The ontogeny of sulphation is different for different drugs. The developmental profiles of glucuronidation and sulphation are illustrated by the examples of morphine and paracetamol, which are markedly different.

Morphine

Morphine undergoes conjugation (UGT2B7) to both morphine 3-glucuronide and morphine 6-glucuronide. The former is the major metabolite and is inactive, whereas morphine 6-glucuronide has considerable analgesic activity. Studies have shown that the sulphation of morphine is a minor metabolic pathway that

Table 3

Age and morphine clearance

Age group	Number of patients	Mean or median plasma clearance (ml/min/kg)	Range	Ref.
preterm neonates	72	3.5	0.5 – 9.6	42
term neonates	44	6.3	0.6 – 39	42
infants 1-24 months	11	13.9	8.3 – 24.1	43
children 2-11 years	18	37.4	20.1 – 48.5	44
adolescents 12-17 years	6	25.4	9 – 53.4	45

does not contribute to the overall clearance [41]. The changes in plasma clearance, therefore, reflect the development of glucuronidation. This is markedly reduced in prematurity and, after infancy, reaches adult levels [42-45] (Table 3). Inter-individual variation is greatest in the neonatal period [42].

Paracetamol

Paracetamol undergoes metabolism by glucuronidation (UGT1A6 and to a lesser extent UGT1A9) and sulphation [46]. Glucuronidation is reduced but there is, however, compensatory sulphation which has a significant impact on the clearance of paracetamol in prepubertal children. There have been relatively few studies looking at the clearance of paracetamol and we have therefore summarised the studies that have determined the plasma half-life [47-51] (Table 4). This is increased in neonates and especially in preterm neonates [47,48]. In infancy and childhood the half-life is the same as in adolescents and adults [49-51]. The ratio of glucuronidation to sulphation, however, changes with development [52]. In contrast to morphine, the sulphation of paracetamol plays a significant role in drug metabolism.

Table 4

Age and paracetamol metabolism

Age group	Number of patients	Mean or median plasma half life	Range	Ref.
preterm neonates	21	5.7	3.5 – 25.2	47
term neonates	12	3.5	2.2 – 5.0	48
infants 1-24 months	15	1.6	0.8 – 2.4	49
children 2-11 years	18	1.7	?	50
adolescents 12-17 years	10	1.5	0.8 – 1.9	51

The relative contributions of glucuronidation and sulphation to the metabolism of morphine [41] and paracetamol [46] are shown in Table 5. In contrast to paracetamol, sulphation is not always available as an alternative metabolic pathway when glucuronidation activity is developmentally low. This is illustrated by the well-known toxicity of chloramphenicol in neonates, which is attribu-

Table 5

Glucuronide to sulphate ratios

Age group	Morphine	Paracetamol
preterm neonates	36	–
term neonates	–	0.34
infants 1-24 months	–	–
children 2-11 years	629	0.75
adolescents 12-17 years	–	1.61

ted to accumulation of drug due to low glucuronidation activity and the lack of alternative metabolic pathways [53].

In summary, due to differences in the ontogeny of the individual UGT isoforms, the different substrate specificity of the individual UGT isoforms and the variations in availability of alternative metabolic pathways, a single ontogenic pattern for glucuronidation is not available. Therefore, up to now, age-related adjustments in dosing of UGT substrates can only be done per individual drug.

2.1.6 CONCLUSIONS

We have used the data for four drugs that are extensively used in children and hence there is considerable information regarding the pharmacokinetics. These four drugs involve both phase 1 (CYP1A2 and CYP3A4) and phase 2 (glucuronidation and sulphation) pathways.

We have demonstrated that for all four drugs, plasma clearance is reduced in the neonatal period (paracetamol: only prolonged plasma half-life has been documented). This reduction in clearance appears to be greater in preterm infants. Adult clearance values appear to be reached within the first two years of life for caffeine, midazolam and morphine. It is important not to extrapolate these findings to all medicines and to all metabolic pathways. For instance studies of cyclosporin and tacrolimus, which undergo metabolism by CYP3A4, show increased clearance in infants. Further studies are required in relation to drug metabolism in infants between the ages of one and twenty four months.

REFERENCES

1. Gennery, B., *Paed. Perinatal Drug Ther.*, *4*, 67 (2000).
2. Shimada, T., Yamazaki, H., Mimura, M., et al., *Drug Metab. Dispos.*, *24*, 515 (1996).
3. Lacroix D, Sonnier M, Moncion A, Cheron G, Cresteil T., *Eur J Biochem.*, *247*, 625 (1997).
4. Treluyer, J. M., Jacqz-Aigrain, E., Alvarez, F., Cresteil, T., *Eur. J. Biochem.*, *202*, 583 (1991).
5. Vieira, I., Sonnier, M., Cresteil, T., *Eur. J. Biochem.*, *238*, 476 (1996).
6. Sonnier, M., Cresteil, T., *Eur. J. Biochem,* *251*, 893 (1998).
7. Ellis, E. F., Koysooko, R., Levy, G., *Pediatrics,* *58*, 542 (1976).
8. Korpela, R., Olkkola, K.T., *Eur. J. Clin. Pharmacol.,* *38*, 293 (1990).
9. Evans, W. E., Sinkule, J. A., Crom, W. R., et al., *Cancer Chemother. Pharmacol.,* *7*, 147 (1982).
10. Suzuki, Y., Mimaki, T., Cox, S., et al., *Ther. Drug Monit.,* *16*, 145 (1994).
11. Summers, B., Summers, R. S., *Clin. Pharmacokinet.,* *17*, 208 (1989).
12. Andersson, T., Hassall, E., Lundborg, P., et al., *Am. J. Gastroenterol.,* *95*, 3101 (2000).
13. Murray, D. J., Crom, W. R., Reddick, W. E., Bhargava, R. Evans, W. E., *Drug Metab. Dispos.,* *23*, 1110 (1995).
14. Noda, T., Todini, T., Watanabe, Y., Yamamoto, S., *Pediat. Radiol.,* *27*, 250 (1997).
15. Blanco, J. G., Harrison, P. L., Evans, W. E., Relling, M. V., *Drug Metab. Dispos.,* *28*, 379 (1999).
16. de Wildt, S. N., Kearns, G. L., Hop, W. C. J., et al., *Clin. Pharmacol. Ther.,* *70*, 525 (2001).
17. Burtin, P., Jacqz-Aigrain, E., Girard, P., et al., *Clin. Pharmacol. Ther.,* *56*, 615 (1994).
18. Hughes, J., Gill, A. M., Mulhearn, H. Powell, E. Choonara, I., *Ann. Pharmacother.,* *30*, 27 (1996).
19. Tolia, V., Brennan, S., Aravind, M. K., Kauffman, R., *J. Pediatr.,* *119*, 467 (1991).
20. Cooney, G. F., Habucky, K., Hoppu, K., *Clin .Pharmacokinet.,* *32*, 481 (1997).
21. Yasuhara, M., Hashida, T., Toraguchi, M., et al., *Transpl. Proc.,* *27*, 1108 (1995).
22. Masri, M. A., Dhawan, V. S., Hayes, K., et al., *Transpl. Proc.,* *24*, 1718 (1992).
23. Schwartz, M., Holst, B., Facklam, D., et al., *Transpl. Proc.,* *27*, 1107 (1995).
24. Johnson, T. N., Tanner, M. S., Taylor, C. J., et al., *Br. J. Clin. Pharmacol.,* *51*, 451 (2001).
25 de Wildt S. N., Kearns, G. L., Hop, W. C., et al., *Br. J. Clin. Pharmacol.,* *53*, 390 (2002).
26. Shimada, T., Yamazaki, H., Mimura, M., Inui, Y., Guengerich, F. P., *J. Pharmacol. Exp. Ther.,* *270*, 414 (1994).
27. Tucker, G. T., Houston, J. B., Huang, S. M., *Clin. Pharmacol. Ther.,* *70*, 103 (2001).
28. Cazeneuve, C., Pons, G., Rey, E., et al., *Br. J. Clin. Pharmacol.,* *37*, 405 (1994).
29. Carrier, O., Pons, G., Rey, E., et al., *Clin. Pharmacol. Ther.,* *44*, 145 (1988).
30. Notarianni, L. J., Oliver, S. E., Dobrocky, P., Bennett, P. N., Silverman, B. W., *Br. J. Clin. Pharmacol.,* *39*, 65 (1995).
31. Lee, T. C., Charles, B., Steer, P., Flenady, V., Shearman, A., *Clin. Pharmacol. Ther.,* *61*, 628 (1997).
32. Thomson, A. H., Kerr, S., Wright, S., *Ther. Drug Monit.,* *18*, 245 (1996).
33. Pons, G., Carrier, O., Richard, M. O., et al., *Dev. Pharmacol. Ther.,* *11*, 258 (1988).
34 Pons, G., Blais, J. C., Rey, E., et al., *Pediatr. Res.,* *23*, 632 (1988).
35 El-Yazigi, A., Shabib, S., Al-Rawithi, S., et al., *J. Clin. Pharmacol.,* *39*, 366 (1999).
36. Aranda, J. V., Collinge, J. M., Zinman, R., Watters, G., *Arch. Dis. Child.,* *54*, 946 (1979).
37. Aldridge, A., Aranda, J. V., Neims, A. H., *Clin. Pharmacol. Ther.,* *25*, 447 (1979).
38. de Wildt, S. N., Kearns, G. L., Leeder, J. S., et al., *Clin. Pharmacokinet.,* *36*, 439 (1999).
39. Alcorn, J., McNamara, P. J., *Clin. Pharmacokinet.,* *41*, 959 (2002).
40. McCarver, D. G., Hines, R. N., *J. Pharmacol. Exp. Ther.,* *300*, 361 (2002).
41. Choonara, I., Ekbom, Y., Lindstrom, B., Rare, A., *Br. J. Clin. Pharmacol.,* *30*, 897 (1990).

42. Choonara, I., *Semin. Neonatal.*, *3*, 137 (1998).
43. McRorie, T. I., Lynn, A. M., Nespeca, M. K., Kent, E., Slattery, J. T., *Am. J. Dis. Child.*, *146*, 972 (1992).
44. Hunt, A., Joel, S., Dick, G., Goldman, A., *J. Pediatr.*, *135*, 47 (1999).
45. Nahata, M. C., Miser, A. W., Miser, J. S., Reuning, R. H., *Dev. Pharmacol. Ther.*, *8*, 182 (1985).
46. Miller, R. P., Roberts, R. J., Fischer, L. J., *Clin. Pharmacol. Ther.*, *19*, 284 (1976).
47. van Lingen, R. A., Deinum, J. T., Quak, J. M. E., et al., *Arch. Dis. Child. Fetal Neonatal Ed.*, *80*, 59 (1999).
48. Levy, G., Khanna, N. N., Soda, D. M., Tsuzuki, O., Stern, L., *Pediatrics*, *55*, 818 (1975).
49. Hopkins, C. S., Underhill, S., Booker, P. D., *Arch. Dis. Child.*, *65*, 971 (1990).
50. Wilson, J. T., Brown, R. D., Bocchini, J. A., Kearns, G. L., *Ther. Drug Monit.*, *4*, 147 (1982).
51. Romsing, J., Ostergaard, D., Sennderovitz, D., Sonne, J., Ravn, G., *Paed. Anaesth.*, *11*, 205 (2001).
52. Alam, S. N., Roberts, R. J., Fischer, L. J., *J. Pediatr.*, *80*, 130 (1977).
53. Weiss, C. F., Glazko, A. J., Weston, J. K., *N. Engl. J. Med.*, *262*, 787 (1960).

2.2 Approaches to studying the development of drug metabolism in children

Trevor N. Johnson
Academic Unit of Molecular Pharmacology and Pharmacogenetics, Division of Clinical Sciences, University of Sheffield, L Floor, Royal Hallamshire Hospital, Sheffield S10 2JF, UK

2.2.1 INTRODUCTION

From birth onwards changes in the pharmacokinetics and pharmacodynamics of administered drugs occur as a consequence of changes in body composition and maturation of organs and enzymes. Accordingly, effective and safe drug therapy in neonates, infants, children and adolescents requires a thorough understanding of human developmental biology and the ontogeny of the processes that govern the absorption, distribution, metabolism, excretion and action of drugs [1].

A number of factors contribute to the developmental changes in metabolic drug clearance in children. These include relative liver size, liver blood flow, extent of protein binding and the maturation of drug metabolising enzymes.

Altered drug metabolism can lead to the development of adverse effects in neonates and small infants that are not generally seen in the adult population. The altered metabolism of sodium valproate in children under 3 years of age is thought to be responsible for a higher incidence of hepatotoxicity [2]. The impaired metabolism of chloramphenicol in neonates can result in the grey baby syndrome (cyanosis and respiratory failure) [3]. Metabolic acidosis following the use of propofol in the critically ill child [4,5] may be due to altered drug metabolism.

This review will focus principally on approaches and problems in determining the development of drug metabolism in humans, with particular reference to phase 1 reactions mediated by cytochromes P450 (CYP) and phase 2 glucuronidation.

Reprinted from *Paed. Perinatal Drug Ther.*, 5, 75 (2002).

2.2.2 ENZYMES INVOLVED IN DRUG METABOLISM

Biotransformation generally converts lipophilic pharmacologically active drug molecules into polar metabolites that are then eliminated by the kidneys or other organs.

The liver is the major site of drug metabolism, with intestinal drug extraction playing a secondary, yet potentially important, role in the pre-systemic elimination of orally administered drugs.

The process of drug metabolism is normally divided into two phases, phase 1 or functionalisation reactions and phase 2 conjugation reactions. Phase 1 reactions introduce or expose a chemically reactive group on which the phase 2 reactions can occur. The phase 2 reaction is usually the true detoxification pathway, resulting in products that are generally water soluble and easily excreted. CYPs catalysing the mixed-function oxidation of xenobiotics comprise the most important group of enzymes involved in phase 1 metabolism. Others include the flavin-containing mono-oxygenases (FMO). Common phase 2 or conjugation reactions include glucuronidation, glycosidation, sulphation, methylation, acetylation and glutathione conjugation [6].

Cytochrome P450

The CYP isoenzymes are a superfamily of haemoproteins that are the terminal oxidases of the mixed function oxidase system found in the endoplasmic reticulum. CYPs play two important roles in the living organism. Firstly, in the biosynthesis or biodegradation of endogenous compounds (steroids, fatty acids, prostaglandins) and, secondly in the oxidative biotransformation of exogenous compounds (drugs, environmental pollutants, toxins).

Approximately 500 CYP enzymes have so far been isolated, spread across different species and grouped according to their amino acid sequences [7]. An amino acid homology greater than 40% defines a family, and greater than 55% defines a sub-family. The symbol CYP is used to denote both human and rat cytochrome P450.

CYP is followed by an Arabic numeral denoting the family (CYP2), a capital letter designating the subfamily (CYP2C) and then another Arabic numeral representing the individual gene or enzyme (CYP2C9). The important human isoforms of the enzyme include CYPs 1A1 and 2, 2B6, 2C8 – 10, 2C19, 2D6, 2E1, 3A4 and 5. The CYP3A isoenzymes are the most abundant enzymes in human liver and small intestine and are involved in the metabolism of around 50% of drugs. The principal substrates, inhibitors and inducers of the major human forms of this enzyme are summarised in Table 1.

Glucuronidation

Glucuronidation is an important detoxification pathway in humans. Many therapeutic drugs and their metabolites are substrates for UDP-glucuronyltransferases (UGT) leading to the formation of usually inactive glucuronides (a notable exception is morphine 6 glucuronide which has analgesic activity [9]) which are then eliminated via the bile or urine [10].

Table 1

Substrates, inducers and inhibitors of each of the major human CYP isozymes [8]

Enzyme	Substrate	Inducer	Inhibitor
CYP1A1 (Extra-hepatic)	chlorinated benzenes	polycyclic hydrocarbons	propofol
CYP1A2	caffeine erythromycin haloperidol paracetamol R-warfarin theophylline tricyclic antidepressants	hydrocarbons omeprazole phenobarbitone phenytoin polycyclic hydrocarbons	quinolone antibiotics
CYP2B6	cyclophosphamide mianserin testosterone	none known	orphenadrine
CYP2C8	carbamazepine diazepam tricyclic antidepressants	phenobarbitone rifampicin	cimetidine (not selective)
CYP2C9/10	diclofenac phenytoin S-warfarin tolbutamide	rifampicin	sulfaphenazole
CYP2C19	diazepam mephenytoin tricyclic antidepressants	phenobarbitone rifampicin	fluconazole omeprazole warfarin
CYP2D6	amitriptyline β-blockers clozepine codeine debrisoquine SSRIs tricyclic antidepressants	none known	amiodarone flecainide fluoxetine quinidine trifluperidol
CYP2E1	caffeine chlorzoxazone isoflurane paracetamol	benzene ethanol isoniazid	dimethyl-sulphoxide disulfiram

Table 1 (continued)

Substrates, inducers and inhibitors of each of the major human CYP isozymes [8]

Enzyme	Substrate	Inducer	Inhibitor
CYP3A4	carbamazepine cisapride clonazepam ciclosporin diazepam erythromycin ethosuximide etoposide fentanyl ketoconazole lignocaine midazolam nifedipine ondansetron rifampicin ritonavir	carbamazepine glucocorticoids omeprazole phenobarbitone phenytoin rifampicin	grapefruit juice propofol naringenin triazole antifungals troleandomycin
CYP3A5	caffeine cyclosporin diltiazem testosterone	dexamethasone	troleandomycin
CYP3A7 (fetal form)	midazolam testosterone		

Individual UGT enzymes are defined according to the family (1 or 2 in humans), subfamily (A or B) and an arabic numeral representing the individual gene product. The liver represents the major site of glucuronidation in humans where five UGT1A and five UGT2B genes are expressed and define capacity [11,12]. A number of UGT enzymes are expressed in extrahepatic sites [12,13]. The important human hepatic enzymes are indicated in Table 2 .

2.2.3 APPROACHES TO STUDYING THE DEVELOPMENT OF DRUG METABOLISM IN HUMANS

Human in vivo approaches

Probe substrates for specific CYP enzymes

A number of *in vivo* substrate probes for specific CYPS have been used and are summarised in Table 3.

One of the main problems with many of the tests relate to difficulties in their application for use in paediatric patients. Many of the tests involve the adminis-

Table 2

Endogenous and exogenous substrates of the major human UGT isozymes
(de Wildt *et al.* [14] with modifications)

UGT isoform	Endogenous substrate	Exogenous substrate	Comment
1A1*	bilirubin	ethinyloestradiol	↓ in Gilberts syndrome
1A3	estrone	norbuprenorphine	
1A4	androstanediol	amitriptyline imipramine	
1A6*		1-napthol 2-napthol naftazone naproxen paracetamol	
1A9	estrone	paracetamol phenol propofol	
1A10		mycophenolic acid	
2B			delayed onset associated with grey baby syndrome
2B4	hyodeoxyicholic		
2B7	androsterone epitestosterone	buprenorphine codeine ibuprofen morphine naproxen propranolol valproic acid	
2B15*	androgen steroids	eugenol 4-hydroxy-biphenyl	
2B17	androsterone dihydrotestosterone testosterone		

* Polymorphic expression

tration of radioactive substrates or intravenous drugs, followed by the collection
of multiple blood samples. This makes it more difficult to gain ethical approval
for their use in children. Even when this is not the case, such as with the caffeine
breath test and urinary 6β hydroxycortisol:cortisol ratio, there is usually a lack of

validation of the tests for use in children. A summary of the consensus view as to the most appropriate *in vivo* probe for a specific CYP are shown in Table 3 with particular reference to those applicable to a paediatric population.

Caffeine breath test

The application of this test to assessing CYP1A2 activity in children has recently been reviewed [32]. Briefly, the caffeine breath test involves the oral administration of a non-radioactive stable isotope of caffeine (^{13}C on the 3-methyl group). The exhaled labelled $^{13}CO_2$ correlates with CYP1A2 activity [33]. The test is relatively non-invasive and suitable for use in young children

Table 3

Recommended *in vivo* probe for CYPs (Tucker, et al. [31] with modifications)

CYP	Probe substrate(s)	Comments	Reference (paediatric studies)
1A2	caffeine	caffeine breath performed in children; results only apply to a small number of drugs used in children	15-20
2B6	bupropion	more validation required; No studies performed in children	
2C8	unclear - paclitaxel?	cannot be given to healthy subjects	
2C9	tolbutamide	no studies performed in children using tolbutamide some evidence from placental transfer and prolonged half-life during first 2 days of postnatal life	22
	diclofenac		21
2C19	omeprazole	potential contamination from 3A4 pathway	23,24
	mephenytoin	availability? no studies performed in children	
2D6	debrisoquine	availability and suitability for use in children?	
	dextromethorphan	alternative has been used in children but potential contamination from 3A4 pathway and urine pH-dependent renal excretion	25-27

Table 3 (continued)

Recommended *in vivo* probe for CYPs (Tucker et al. [31] with modifications)

CYP	Probe substrate(s)	Comments	Reference (paediatric studies)
2E1	Chlorzoxazone		
3A4	Midazolam (oral)	Not selective for 3A4 vs 3A5	28
	Midazolam (oral and iv)	Separates liver vs gut contributions; need for stable isotope labelling for concurrent oral and iv administration; staggered oral and iv dosing may avoid use of labelled drug	28
	Erythromycin	Erythromycin breath test marks 3A4 preferentially to 3A5 in liver only. Results may be influenced by activity of P-glycoprotein transporter. Use of radioactive compound may be an issue especially in children	
	Midazolam (Oral) + Erythromycin (iv)	Suggested as a combined measure of both intestinal and hepatic CYP3A activity	
	6ß hydroxycortosol : cortisol	Has been used in children. Non invasive but results influenced by renal CYP3A5	29,30

[15,20]. A number of investigators have used this test in children to study the effects of disease states and drug interactions [15-18]. Pons *et al.* [19] have demonstrated no detectable changes in expired CO_2 from basal levels in neonates and young infants, whereas changes were measurable in all infants older than 33 days postnatal age and 45 weeks post-conceptional age, in agreement with *in vitro* findings. Lambert *et al.* [20] have demonstrated an increased 2 hour expired CO_2 in pre-pubertal children compared to adults, which is contrary to the *in vitro* findings for the development of CYP1A2 in children [34]. There may be a number of potential problems with the caffeine breath test in the *in vivo* assessment of CYP1A2 activity. Children should be kept still during the test to avoid increased CO_2 production, which may have a dilutional effect on the test [17].

The N3- and N7-demethylation of caffeine are the two major pathways of caffeine metabolism in young infants and maturation of N1-demethylation occurs after 19 months [35]. In the fetus the N3 and N7-reactions appear to be mediated by CYP3A rather than CYP1A2 [36] with a switch over to CYP1A2 in the neonatal-infant stage. Full caffeine N3-demethylation is not observed until between 4 to 6 months of age [37]. Thus, the changes in the pattern of metabolite formation and enzyme involvement in caffeine metabolism with age have to be considered when using this test for the determination of *in vivo* CYP1A2 development in children. Parker *et al.* [18] suggest that the caffeine breath test is unsuitable for use in infants less than 6 months of age. The stable isotope is expensive.

Diclofenac

Diclofenac is metabolised by CYP2C9 to 4-hydroxydiclofenac. Diclofenac is routinely used for analgesia in children. In a pharmacokinetic study of 500 mcg/kg diclofenac administered intravenously, the plasma weight-corrected clearance was higher in children aged 4 to 6 years (7.7 ml/min/kg) compared to adults (3.3 ml/min/kg) [21].

Omeprazole

Omeprazole is metabolised to 5-hydroxyomeprazole, an index of CYP2C19 activity and to omeprazole sulphone, an index of CYP3A4 activity *in vivo* [38]. Few studies have been undertaken in children. However, the clearance of intravenous omeprazole has been shown to be similar to adults in paediatric patients aged 4 months to 19 years [23]. The oral clearance of omeprazole has been shown to be increased in children age 1 to 6 years compared to adults [24].

Mephenytoin

Mephenytoin has been used as a probe substrate to determine CYP2C19 phenotype in humans. (S)-(R)-mephenytoin undergoes rapid hydroxylation to form (S)-4 hydroxymephenytoin (SOHP) with the slower formation of (R)-ethylhydantoin (PEH). The ratio of SOHP to PEH is used as a measure of CYP2C19 activity [39]. The test in adults involves the administration of 100 mg mephenytoin followed by the collection of urine for 8 hours post dose [40,41]. Although mephenytion has been administered to children [42], this test has so far not been applied to a paediatric population, possibly because of worries about toxicity [42,43]. Otherwise, the test is relatively non-invasive. Stability issues have been raised regarding the assessment of CYP2C19 activity based on the SOHP to PEH ratio [44].

Dextromethorphan

Dextromethorphan is metabolised by CYP2D6 to dextrorphan and by CYP3A to 3-methoxymorphinan [45,46]. Dextromethorphan O-demethylation is well established as a measure of CYP2D6 *in vivo* and is highly correlated with CYP2D6 expression in human liver microsomes [47]. The urinary ratio of dextromethorphan to dextrophan is most commonly used as an index of CYP2D6 activity and previous studies have shown a good correlation with other CYP2D6 probe substrates such as desipramine [48] and debrisoquine [49]. In adults, the test involves the administration of 25 to 30 mg of dextromethorphan and the collection of urine samples for 8 to 12 hours post dose [41,50,51].

The dextromethorphan test is relatively non-invasive, involves a relatively innocuous substrate and has been used to determine CYP2D6 polymorphism in children. Evans *et al.* [27] administered a dose of 30 mg to 26 children aged 3 to 21 years and collected urine for 4 hours post dose to determine debrisoquine oxidation phenotype.

The test has been applied to determining dextromethorphan phenotypes in paediatric patients with autoimmune hepatitis [26]. In a recent study 300 mcg/kg dextromethorphan was administered to a group of neonates through the first year of life [25]. Urine was collected into a nappy overnight. CYP2D6 phenotype consistent with genotype was achieved by 14.5 days postnatal age. The maturation of the CYP3A pathway (N-demethylation) was found to be delayed relative to CYP2D6. The use of this substrate in children is potentially problematic, especially in neonates and infants, because dextrorphan formation as a measure of CYP2D6 is dependent on renal function [52], which changes with developmental age. The ratio of dextromethorphan to dextrorphan also depends on urinary pH.

Erythromycin breath test (ERBT)

The ERBT is one of the most extensively validated measures of *in vivo* CYP3A4 activity in the liver in adults and was considered by some to be the gold standard [53,54]. To date, the test has not been applied to the paediatric population. The test is based on the CYP3A4 mediated N-demethylation of erythromycin, producing formaldehyde, which subsequently appears in the breath as CO_2 [55]. The test involves the intravenous administration of a small quantity of ^{14}C labelled erythromycin, thereafter, at timed intervals, the subject blows into scintillation vials containing a CO_2 binding agent. The rate of production of radiolabel is then measured and expressed as the percentage of administered radiolabel exhaled during the first hour after injection [55].

The ERBT has been shown to predict the steady state trough blood levels of ciclosporin [56] and to predict its oral clearance [57]. Lown *et al.* [54] showed the positive correlation of the ERBT with the weight adjusted clearance of midazolam (another *in vivo* probe of CYP3A) and McCrea *et al.* [58] have suggested the use of oral midazolam and the ERBT as a combined probe for hepatic and enterocytic CYP3A activity.

More recently, doubt has been expressed as to whether the ERBT is a useful predictor of the clearance of some CYP3A drug substrates. In contrast to Lown *et al.* [54], Kinirons *et al.* [59] have failed to show correlation between midazolam clearance and the ERBT, likewise a lack of correlation between alfentanyl clearance and the ERBT and has been observed by Krivoruk *et al.* [60]. A possible reason for the discrepancy may be the fact that erythromycin [61], but not midazolam or alfentanyl [62], is also a substrate for P-glycoprotein. This would result in lower intrahepatic concentrations of radiolabelled erythromycin and ultimately lower $^{14}CO_2$ exhalation [63]. Another reason why the ERBT may not predict midazolam clearance is that erythromycin is a substrate for CYP3A4 but not CYP3A5 [64] whereas midazolam is a substrate for both [65]. Hence the ERBT will underestimate overall CYP3A activity in the 20 to 30% of patients with significant CYP3A5 hepatic levels [66].

Other disadvantages of the ERBT include: (i) the use of an intravenous radiolabelled substrate which will preclude its use, especially in the paediatric population, (ii) the test does not measure intestinal CYP3A4 activity, (iii) only the rate of demethylation during the first hour is measured, interindividual differences in protein binding or volume of distribution could significantly affect the test result. Overall, the ERBT may have limited application in the prediction of the CYP3A- mediated metabolism of another substrate, but may be useful for exploring the time course of induction or inhibition associated with drug interactions [67]. There has been some recent debate as to whether the CYP3A N-demethylation of erythromycin is the rate-limiting step in $^{14}CO_2$ production [68].

Midazolam

Midazolam is chiefly metabolised to 1' and 4'-hydroxymidazolam by CYP3A4 and CYP3A5 [65]. To perform the test, an intravenous injection of 15 mcg/kg midazolam is administered and multiple blood samples collected up to 6 hours post dose. Drug and metabolites are measured by GC-MS or HPLC-MS.

Midazolam appears to have some advantages over the ERBT: it is metabolised by both CYP3A4 and CYP3A5 and is not a P-gp substrate. Furthermore administration of the drug intravenously and then, orally may facilitate an overall assessment of intestinal and hepatic CYP3A [69]. Further studies are required on the optimal use of midazolam as an oral CYP3A probe. The main disadvantages of the test are the administration of a clinically sedating dose of midazolam and the need for multiple blood samples. Rather than formally applying this test to children, studies built around the clinical use of midazolam have been applied to assessing the development of CYP3A activity.

The weight-corrected plasma clearance of i.v. midazolam was significantly reduced in neonates compared with adults, 110 to 130 ml/h/kg and 380 to 660 ml/h/kg respectively [70,71], and was even lower in preterm infants younger than 39 weeks gestation, 72 to 96 ml/h/kg [70,72]. Because midazolam is only metabolised by CYP3A7 to a small extent [65], reduced clearance may be explained by low CYP3A4 activity following birth. The surge in CYP3A4 expression after birth is not mirrored by a large increase in midazolam clearance at this stage [73,74]. Another study by Hughes *et al.* [75] showed lower weight-corrected clearance in younger children up to 2 years of age (140 to 180 ml/h/kg) compared with children aged 3 to 13 years (78 ml/h/kg) who, in turn, had higher weight corrected clearance compared with adults (380 to 660 ml/h/kg). In a recent study both the i.v. and oral clearance of midazolam were found to be increased (3000 and 680 ml/h/kg respectively) in children age 6 months to <2 years compared with children age 2 to <12 years (2500 and 600 ml/h/kg respectively) and adolescents 12 to <16 years (1500 and 560 L/h/kg respectively) [28].

Urinary 6β-hydroxy cortisol

This is a simple non-invasive test that is easily performed, even in a neonatal population. The ratio of 6β–hydroxy cortisol to free cortisol is measured in a spot urine sample.

This ratio has been proposed as a measure of hepatic mixed function oxidase activity [76] and subsequently, CYP3A activity [77]. However, there is little evidence to support this. There is lack of correlation between the ERBT and 6βOHC:C ratio [78,79]. Cortisol is also a substrate for P-gp, which may in addition to the influence of renal CYP3A5 [80], explain the lack of correlation between the 6β–OHC:C urinary ratio and midazolam clearance. At best, the 6β–OHC:C ratio may be useful for looking at enzyme induction [81-83]. A number of studies have applied this test to children. The 6β–OHC:C ratio is higher in term compared to premature neonates [30]. In infants 1 to 12 months of age, the mean 6β–OHC:C ratio was lower compared with that of neonates and adults [29].

Further studies are required on the use of probe substrates in determining *in vivo* CYP3A activity in man, especially with respect to their oral administration and with regard to the influence of P-gp. The tests require further validation for use in children.

Drugs in use

Phase 1

The pharmacokinetics of a number of drugs in routine clinical use have been applied to assessing the *in vivo* development of drug metabolism in children. This method is useful where the drug in question is predominantly metabolised by a single CYP enzyme. Ethical approval for such studies is generally easier to obtain compared with administering a probe substrate, especially where i.v. access is already available and the number of blood samples to be collected is small. A number of approaches have been used to minimise the number of blood samples needed for such studies. These include sparse data pharmacokinetics, (small number of blood samples from a larger group of patients) [84,85] and the use of steady state pharmacokinetics [75]. A disadvantage with this approach is that it is opportunistic and thus more limited in its application.

A number of drugs routinely used in paediatric clinical practice and adopted as probes for specific CYP enzymes have already been mentioned including midazolam, diclofenac and omeprazole. Other drugs used in routine clinical practice that have been used to assess age-related changes in CYP activity include: for CYP3A: ciclosporin, tacrolimus, carbamazepine (also CYP2C8), etoposide, nifedipine, and cisapride; and for CYP2C9: phenytoin.

Phase 2

Morphine is largely metabolised by UGT2B7 to morphine-6-glucuronide and morphine-3-glucuronide [9], and has been suggested as a probe substrate for this UGT isoform [86]. A number of *in vivo* studies have already been described on the development changes in morphine clearance in children [87-90].

Paracetamol is mainly metabolised by UGT1A6 (and to a lesser extent by UGT1A9). The rate of paracetamol glucuronidation is negligible in the fetus [91], low after birth and does not reach adult values before 10 years of age. Paracetamol is not the ideal probe for assessing UGT1A6 activity in children because of the compensatory higher sulphotransferase activity of the younger ages [92].

Propofol has been proposed as an alternative probe for UGT1A9 activity [86]. However, propofol is a high extraction drug with clearance being dependent mainly on blood flow rather than enzyme activity, thus, limiting its application as a probe substrate [93].

Human in vitro approaches

Several *in vitro* liver preparations are used in studying hepatic drug metabolism. These include liver slices, intact hepatocytes, S9 fractions and microsomes.

Liver slices and hepatocytes are useful for studying sequential oxidative and conjugative biotransformation in the same system, as well as xenobiotic mediated induction [94]. Many *in vitro* studies are done using human liver microsomes. Microsomes are closed vesicles of fragments of the endoplasmic reticular membrane and contain several drug metabolising enzymes including CYP, FMO and UDP-glucuronyl-transferases.

A number of investigators have used microsomes prepared from the livers of children at various developmental ages and then measured the expression and/or the activity of specific CYP and other enzymes involved in drug metabolism.

The collection of liver biopsies is considered high risk and not justified for research purposes alone, but may be acceptable when the primary purpose is for diagnosis or treatment [95]. In practice, paediatric liver biopsy material is not only difficult to obtain for research purposes, but when available, it is likely to have been collected from a patient with liver disease. Thus, many of the livers used to prepare microsomes have been obtained from aborted fetuses and post mortems in the first few hours after death [96,97]. A recent study of the developmental expression and activity of intestinal CYP3A4 was performed using duodenal and jejunal biopsies harvested at the same time as those to be used for clinical and diagnostic purposes [98]. All human tissue used for *in vitro* studies should be characterised histologically for the presence of disease. *Post mortem* enzyme degradation may occur and tissue should be frozen at -80°C as soon as possible.

The expression of genes coding for specific CYP enzymes is often detected in fresh or frozen tissue, by amplifying the gene in question using a reverse transcriptase polymerase chain reaction (RT-PCR) with appropriate primers.

Enzyme expression is usually measured by a technique, such as Western blotting, which requires separation of proteins by electrophoresis and the detection of the enzyme using specific antibodies.

Enzyme activity is determined by measuring the rate of metabolite appearance from a specific probe substrate. A number of model *in vitro* substrates have been defined for probing specific CYP enzyme activity (Table 4). Preferably, more than one substrate should be used to investigate the ontogeny of CYP3A because of potential differences in the enzyme-substrate interaction [99,100].

To illustrate the application of *in vitro* techniques to investigate the ontogeny of specific CYP enzymes in the human liver, Sonnier and Cresteil [34] utilised liver tissue from stillborn or aborted fetuses (14 to 40 weeks), children aged 1 day to 9 years and from adult donors for kidney transplantation. Microsomes were prepared and CYP1A2 protein was measured by Western blotting, using a rat polyclonal CYP1A1 antibody. Enzyme activity was determined by co-incubating the microsomes with methoxyresorufin. CYP1A2 expression and activity

Table 4

Recommended *in vitro* probe substrates [31]

CYP	Substrates	
	Preferred	Acceptable
1A2	ethoxyresorufin phenacetic	caffeine theophylline
2A6	coumarin	
2B6	S-mephenytoin	bupropion (availability of metabolite standards?)
2C8	paclitaxel	
2C9	diclofenac S-warfarin	tolbutamide (low turnover)
2C19	omeprazole S-mephenytoin	
2D6	bufuralol dextromethorphan	codeine debrisoquine metoprolol
2E1	chlorzoxazone	4-nitrophenol lauric acid
3A4	midazolam testosterone	ciclosporin erythromycin*

* Strongly recommend using at least two structurally unrelated substrates.

was absent in microsomes prepared from fetal and neonatal livers and increased in infants aged 1 to 3 months to attain 50% of the adult value at one year.

Problems with the human *in vitro* approach include the availability of suitable tissue samples, ethical constraints, especially since the Redfern report (The Report of The Royal Liverpool Children's Inquiry) [101], the availability of isoform specific antibodies to assess expression by immunoblotting, the size of the tissue samples available, the non-physiological nature of the preparations used, and the availability of sufficiently sensitive assay systems to measure metabolite formation. The main advantage of this approach is that changes in true human enzymology with age can be ascertained.

Developmental factors may influence the suitability of an *in vitro* probe substrate. For instance, caffeine is metabolised by CYP3A in fetal liver [36], whereas in adults, it is predominantly metabolised by CYP1A2 [102].

Studies on the developmental expression of polymorphic enzymes such as CYP2D6 and CYP2C19 need to account for the numbers of poor metabolisers within each age group. This is itself potentially difficult, as there is little information about the appearance of the different phenotypes during development [103]. Peng *et al.* [104] studied the *in vitro* acetylation of 7-amino-clonazepam in human fetal and adult liver preparations. The adults, but not the fetuses, could be classified into fast and slow acetylators.

In general, enzyme expression and activity correlate closely in adult humans [105]. However, during development this is not always the case. Strassberg *et al.* [106] have demonstrated an early phase in UGT development with the appearance of gene transcripts by 6 months, followed by a later phase characterised by up regulation of UGT protein expression to levels found in adults by 2 years. Even by 2 years, the activity of many UGT enzymes was significantly lower than in adults. The different stages of ontogeny are characterised by dynamic changes in gene expression. Thus, *in vitro* studies drawing conclusions based on a small number of tissue samples, representing a narrow time window, must be viewed with caution [107].

Investigations utilising microsomal and other subcellular liver and intestinal fractions in the study of the development of the major enzymes involved in drug metabolism are shown in Tables 5 and 6. Two reviews are available describing the *in vitro* ontogeny of human phase 1 [107] and phase 2 [108] drug metabolising enzymes.

Animal models

The relative ease of obtaining animals, usually rats, at the relevant stage of development and then harvesting the organs of interest, make the use of animal models attractive for studying the ontogeny of drug metabolism systems.

The development of rat hepatic CYP3A has been shown to be both age- and sex-dependent [130,131]. Enzyme expression and activity increases up to day 25 and then continues to rise in males but falls away in female animals. The expression and inducibility of P450 enzymes during liver ontogeny has been reviewed by Rich and Boobis [132]. In contrast to rat hepatic CYP3A, small bowel CYP3A expression and activity is virtually absent until weaning (day 20) and then surges, reaching a plateau by 40 days sustained into adulthood [131].

Although these data are of interest to toxicologists, there are a number of inherent problems in applying it to humans. There are differences in the expressed enzymes. In man, the predominant CYP3A enzymes are CYP3A7 (fetal form), CYP3A4 and CYP3A5, whereas in rats they are CYP3A1, CYP3A2, CYP3A9 and CYP3A18 [133]. Although they are all involved in the metabolism of certain

Table 5

In vitro study of developing phase 1 enzymes in humans

Tissue and source	*In vitro* preparation	Enzyme CYP	Probe Substrate	Other methods	Relative % expression / activity			Ref.
					Adult	Child (age)	Fetus	
L, AF, SS	microsomes	1A		Western blotting	detected (+++)		not detected	109
AF, SS	microsomes	1A1	ethoxyresorufin –o-demethylase		0		100 (7-9wk)	110
L, AF, OD. B	mRNA and cDNA extract	1A1		RT-PCR	detected (++)		not detected	111
L, AF, SS.OD	microsomes	1A2	methoxyresorufin demethylation	Western blotting	100	50 (1 y)	0	34
L, AF, OD. B	mRNA and cDNA extract	1A2		RT-PCR	detected (+++)		not detected	111
L, AF, PM. SS	microsomes / RNA/section	1A2		immunostaining, Western blotting. Northern blotting	100	53 (5y)	not detected	112
L, SS, B	microsomes	2A6		Western blotting	100	<100 (1y)		113
L, SS, B	microsomes	2B6		Western blotting	100	10 (1y)	-	113
L, AF, OD. B	mRNA and cDNA extract	2B6		RT-PCR	detected (+++)		not detected	111

Table 5 (*continued*)

In vitro study of developing phase 1 enzymes in humans

Tissue and source	*In vitro* preparation	Enzyme CYP	Probe substrate	Other methods	Relative % expression / activity			Ref.
					Adult	Child (age)	Fetus	
L, AF, PM, SS	microsomes / RNA/section	2C		immunostaining, Western blotting, Northern blotting	100	47 (5d) 91 (5y)	<10	112
L, OD, AF, PM	microsomes	2C		RT-PCR	100	5 (1d) 20 (1 m)	1 – 2	114, 115
		2C8 2C9	tolbutamide		100	65 (7d)	50	
		2C10 2C18			100 100	6 (7d) 25 (1m) 85 (1d) 120 (1y)	0** 40	
L, AF, SS	microsomes	2C		Western blotting	detected (+++)		detected (+)	109
L, AF, OD, B	mRNA and cDNA extract	2C8-19*		RT-PCR	detected (++++)		detected (++)	111
L, AF, OD, B	mRNA and cDNA extract.	2D6*		RT-PCR	detected (+++)		detected (+)	111
L, AF, OD	microsomes	2D6*	dextromethorphan		100	detected	0 RNA(++)	116
L, AF, OD	microsomes	2D6*	dextromethorphan	Western blotting, RNA analysis	100	15 (7 d) 50 (1 m)	2 – 5	117

Table 5 (*continued*)

In vitro study of developing phase 1 enzymes in humans

Tissue and source	In vitro preparation	Enzyme CYP	Probe substrate	Other methods	Relative % expression / activity			Ref.
					Adult	Child (age)	Fetus	
L, AF, OD	microsomes	2D6*	dextromethorphan		100		< 5	118
L, AF, OD, B	mRNA and cDNA extract	2E1		RT-PCR	detected (++++)		not detected	111
L, AF, OD, SS	microsomes	2E1	ethanol oxidation	Western blotting, RT-PCR	100		12 - 27	119
L, AF, PM	microsomes	2E1	chlorzoxazone hydroxylation	Northern blotting	100	15 (1 m) 80 (1 y)	0	120
L, AF		2E1		RT-PCR	100		10 - 30	121
L, AF, OD, SS and olfactory tissue	microsomes	2J2		Western blotting	100		near 100	122
L, AF, PM	microsomes	3A4/5*	DHEA hydroxylation, testosterone hydroxylation	Western blotting	100	50 (6 m) 120 (1y)	<10	96
L, AF, OD, B	mRNA and cDNA extract	3A4/5*		RT-PCR	detected (++++)		detected (++)	111
L, AF, PM	microsomes mRNA and cDNA extract	3A7	DHEA	RT-PCR	<10 detected (++)	150 (1d) 50 (8d)	100 detected (++++)	96

Table 5 *(continued)*

In vitro study of developing phase 1 enzymes in humans

Tissue and source	*In vitro* preparation	Enzyme CYP	Probe substrate	Other methods	Relative % expression / activity			Ref.
					Adult	Child (age)	Fetus	
duodenum / jejunum, B, AF, NS	microsomes	3A4	testosterone hydroxylation dextromethorphan	Western blotting	100	50 (1m) 75 (2 y)	5	98
L, AF, PM, SS	microsomes	3A4	cisapride		56	100	3.5	123

* Polymorphism

** Approximately 5% detection of CYP2C9 RNA in fetus compared to adult

Tissue sources: AF: Aborted fetus, L: Liver, SS: Surgical section, OD: Organ donor, PM: Post mortem, B: Biopsy.

DHEA: Dehydroepiandrosterone

Table 6

In vitro studies with hepatic microsomes on the development of phase 2
glucuronidation enzymes in humans

Source	Enzyme UGT	Probe substrate/ other method	Relative % expression / activity			Ref.
			Adult	Child (age)	Fetus	
L, AF, OD	1A1*	bilirubin	100	1 (1 m)	0.1	124
L, AF, SS, PM	1A1 *	bilirubin	100		<14	125
L, AF, PM	1A1 *	bilirubin/ Western blotting	100	<14 (term)	<14	126
L, AF, SS, PM	1A3		100		30	14
L, SS	1A4	amitriptyline	100	5 -10 (18 m)	<5	106
L, AF, SS, PM	1A6*	1-naphthol	100		<14	125
L, AF,OD	1A6*	2-naphthol	100		1	127
L, AF, SS, PM	1A9	phenol	100		<14	125
L, AF, SS, PM	1A9	4-butylphenol	100		5	106
L, AF, SS, PM	2B7*	androsterone	100		<14	125
L, AF, OD	2B7*	morphine	100		10-20	128
L, OD,SS	2B7*	buprenorphine	100		10	106
L, AF, SS, PM	2B17	testosterone	100	<14		125
L, AF	2B17	Western blotting	detected (+++)	detected (+)		129

* Polymorphism

Tissue sources: AF: Aborted fetus, L: Liver, SS: Surgical section, OD: Organ donor, PM:Post mortem.

substrates, e.g. testosterone, major differences exist. For instance, mephenytoin is metabolised by CYP3A in the rat but by CYP2C19 in humans [134,135]. Many enzymes in the rodent have a sex-specific ontogeny (CYP2C11, CYP2C12, CYP2C13, CYP3A2), whereas this is not the case in humans. Animal models also often ignore wide genetic polymorphisms which occur in humans such as CYP2D6 and CYP2C29 [136].

Coupled with the contrasts in enzymology, there are often cross-species developmental miss-matches occurring on a more physiological level. This is especially so when investigating the development of drug metabolism in the small bowel. In rodents, the bowel continues to develop after birth and only becomes functional by the time of weaning [137]. In contrast, the function of human bowel is ahead of need at birth [138]. Neonates have around 50% of the small intestinal CYP3A expression and activity compared with adults [98]. This contrasts with the newborn rat where the enzyme is absent.

Allometric scaling (AS) using *in vivo* data obtained in animal species has been used to attempt to predict pharmacokinetic (PK) parameters in humans [139]. AS assumes that many physiological processes are a function of the size of the animal and, thus, PK parameters should be a function of total body weight [140]. AS is therefore applied in a retrospective manner to compare the PK of a drug across several species and predict the parameters in humans [136]. There are many problems in using AS to predict PK parameters in adult humans due to its empirical nature [141], including interspecies differences in metabolism. Applying the technique to studying the metabolism of drugs in the developing human from PK data derived from developing animals would introduce yet more confounding factors and further reduce the possibility of useful predictions.

Overall, although animal models are of some relevance to acute toxicity testing of drugs during development, they are generally unreliable at predicting the ontogeny of drug metabolism in humans.

In silico approaches

Computer simulation is being employed increasingly within the pharmaceutical industry. The prediction of absorption, distribution, metabolism and excretion parameters for a specific drug are becoming possible from its chemical structure alone [142,143]. An area of rapid development is the prediction of *in vivo* pharmacokinetics parameters from *in vitro* data. Such *in vitro-in vivo* extrapolation (IVIVE) depends on the availability of good quality *in vitro* enzyme kinetic data, such as maximum enzyme velocity (Vmax), Michaelis-Menten constant (Km) for the drug substrate and inhibitor constant (Ki) and inhibitor concentration (I) for the inhibitor.

Computer simulation using SIMCYP a programme developed at the University of Sheffield and incorporating population variability has been successfully applied in the prediction of the variability of a number of drug-drug interactions including methadone-ritonavir, ketoconazole-midazolam, fluconazole-midazolam and sildenafil-ritonavir [144,145].

One future development will be the *in silico* prediction of the likely pharmacokinetic parameters of drugs in children. The *in vitro* developmental patterns of many of the CYP enzymes in children are documented (see earlier), as are other key physiological parameters such as liver size, renal function and intestinal development. The first stage will be to simulate the changes and variability in clearance of a drug, such as midazolam with age, and compare the results with those available in the literature. Additional *in vitro* data will be required to build a truly paediatric version of SIMCYP, but once validated, it will form a valuable tool both for the pharmaceutical industry and in the clinical setting.

2.2.4 CONCLUSIONS

An array of approaches exist for studying the development of drug metabolism in children, each with its own advantages and disadvantages. Future studies are required not only to extend the current data on the development of specific enzyme systems with age, but also to further validate methods for use in children. Drug metabolism is an important determinant of both drug dose and frequency of administration. Thus, it is important that information on the changes in drug metabolism with age is accounted for in the clinical use of medicines for children. To this end, the development of validated *in silico* approaches may facilitate the reliable prediction of pharmacokinetic parameters for drugs to be used in children. This, in turn, may lead to a more scientific approach to dosage recommendation and the design of clinical studies in children.

An area that is currently generating research interest, is the relationship between the ontogeny of drug metabolising enzymes and the pathogenesis of adverse drug reactions (ADRs) in children. Delayed maturation of enzymes systems over the first 2 to 4 weeks of life may contribute to the development of type A (dose-dependent) ADRs. Conversely, it is postulated that the increased metabolism, and thus, circulating metabolites of possibly reactive metabolites in young children may predispose to certain type B (idiosyncratic, non-dose-dependent) ADRs. Future research in this area must not only focus on the ontogeny of the drug metabolising enzymes and transporter proteins, but also other systems involved in the generation of ADRs, such as the development of the immune system. Only by understanding the underlying mechanisms of ADRs in children, can we ultimately begin to prevent them from occurring.

REFERENCES

1. Morselli, P. L., Franco-Morselli, R., Bossi, L., *Clin. Pharmacokinet., 5*, 485 (1980).
2. Fisher, E., Siemes, H., Pund, R., Wittfoht, W., Nau, H., *Epilepsia, 33,* 165 (1992).
3. Weiss, C. F., Glasko, A. J., Weston, J. K., *N. Engl. J. Med., 262*, 787 (1960).
4. Parke, T. J., Stevens, J. E., Rice, A. S., et al., *BMJ, 305,* 613 (1992).
5. Strickland, R. A., Murray, M. J., *Crit. Care Med., 23,* 405 (1995).
6. Gibson, G. G., Skett, P. *Introduction to drug metabolism*, 2nd Edition, Blackie Academic and Professional, London, 1997.
7. Nelson, D. R., Koymans, L., Kamataki, T., et al., *Pharmacogenetics, 6,* 1 (1996).
8. Chang, G. M. W., Kam, P. C. A., *Anaesthesia, 54*, 42 (1999).
9. Coffman, B. L., Rios, G. R., King, C. D., et al., *Drug Metab. Dispos., 25*, 1 (1997).
10. Dutton, G. J. *Glucuronidation of drugs and other compounds*, Boca Raton, Florida, CRC, Boca Raton Press, 1980.
11. Tukey, R. H., Strassburg, C. P., *Ann. Rev. Pharmacol. Toxicol., 40,* 581 (2000).
12. Strassburg, C. P., Nguyen, N. Manns, M. P., et al., *Gastroenterology, 116,* 149 (1999).
13. Strassburg, C. P., Kneip, S., Topp, J., *J Biol Chem., 275*, 36164 (2000).
14. de Wildt, S. N., Kearns, G. L., Leeder, J. S., van den Anker, J. N., *Clinical Pharmacokinet., 36*, 439 (1999).
15. Parker, A. C., Preston, T., Heaf, D., Kitteringham, N. R., Choonara, I., *Br. J. Clin. Pharmacol., 38*, 573 (1994).
16. Parker, A. C., Pritchard, P., Preston, T., Dalzell, A. M., Choonara, I., *Br. J. Clin. Pharmacol., 43*, 467 (1997).
17. Parker, A. C., Pritchard, P., Preston, T., Smyth, R. L., Choonara I., *Arch. Dis. Child., 77,* 239 (1997).
18. Parker, A. C., Pritchard, P., Preston, T., Choonara, I., *Br. J. Clin. Pharmacol., 45,* 176 (1998).
19. Pons, G., Blais, J. C., Rey, E., et al., *Pediatric Res., 23,* 632 (1988).
20. Lambert, G. H., Schoeller, D. A., Kotake, A. N., Flores, C., Hay, D., *Dev. Pharmacol. Ther., 9,* 375 (1986).
21. Korpela, R., Olkkola, K. T., *Eur. J. Clin. Pharmacol., 38,* 293 (1990).
22. Christesen, H. B., Melander, A., *Eur. J. Endocrinol., 138,* 698 (1998).
23. Jacqz-Aigrain, E., Bellaich, M., Faure, C., e al., *Eur J Clin Pharmacol., 47,* 181 (1994).
24. Andersson, T., Hassall, E., Lundborg, P., et al., *Am. J. Gastroenterol., 95,* 3101 (2000).
25. Leeder, J. S., Gotschall, R. R., Gaedigk, A., Adcock, K., Wilson, J. T., Kearns, G. L., *Drug Metab Rev., 32(Suppl. 2)*, 281(abstract #289) (2000).
26. Jacqz-Aigrain, E., Laurent, J., Alvarez F., *Br. J. Clin. Pharmacol., 30,* 153 (1990).
27. Evans, W. E., Relling, M. V., Petros, W. P., Meyer, W. H., Mirro, J., Crom, W. R., *Clin. Pharmacol. Ther., 45,* 568 (1989).
28. Reed, M. D., Rodarte, A., Blumer, J. L., et al., *J. Clin. Pharmacol.,41,*1359 (2001).
29. Nakamura, H., Hirai, M., Ohmori, S., et al., *Eur. J. Clin. Pharmacol., 53,* 343 (1998).
30. Vauzelle-Kervroedan, F., Rey, E., Pariente-Khayat, A., et al., *Eur. J. Clin. Pharmacol., 51,* 69 (1996).
31. Tucker, G. T., Houston, J. B., Huang, S. M., *Clin Pharmacol Ther., 70,* 103 (2001).
32. Webster, E., McIntyre, J., Choonara, I., Preston, T., *Paed Perinatal Drug Ther., 5*, 28 (2002).
33. Wietholtz, M., Voegelin, M., Arnaud, M. J., Bircher, J., Preisig, R., *Eur. J. Clin. Pharmacol., 21*, 53 (1981).
34. Sonnier, M., Cresteil, T., *Eur J. Biochem., 251,* 893 (1998).
35. Carrier, O., Pons, G., Rey, E., et al., *Clin. Pharmacol. Ther., 44,* 145 (1988).
36. Cazeneuve, C., Pons, G., Rey, E. , *Br. J. Clin. Pharmacol., 37* 405 (1994).

37. Aranda, J. V., Collinge, J. M., Zinman, R., Watters, G., *Arch. Dis. Childhood*, *54*, 946 (1979).
38. Tassaneeyakul, W., Vannaprasaht, S., Yamazoe, Y., *Br. J. Clin. Pharmacol.*, *49*, 139 (2000).
39. Wilkinson, G. R., Guengerich, F. P., Branch, R. A., *Pharmacol. Ther.*, *43*, 53 (1989).
40. Eap, C. B., Bondolfi, G., Zullino, D., et al., *Ther. Drug Monit.*, *23*, 228 (2001).
41. Eap, C. B., Guentert, T. W., Schaublin-Loidl, M., et al., *Clin. Pharmacol. Ther.*, *59*, 322 (1996).
42. Ilyes, I., Tornai, A., Kirilina, S., Gyorgy, I., *Acta Paediatrica Hungarica*, *26*, 307 (1985).
43. Troupin, A. S., Ojemann, L. M., Dodrill, C. B., *Epilepsia*, *17*, 403 (1976).
44. Zhang, Y., Blouin, R. A., McNamara, P. J., Steinmetz, J., Wedlund, P. J., *Br. J. Clin. Pharmacol.*, *31*, 350 (1991).
45. Barnhart, J. W., *Toxicol. Appl. Pharmacol.*, *55*, 43 (1980).
46. Kerry, N. L., Somogli, A. A., Bochner, F., Mikus, G., *Br. J. Clin. Pharmacol.*, *38*, 243 (1994).
47. Rodrigues, A. D., Kukulka, M. J., Surber, B. W., et al., *Anal. Biochem.*, *219*, 309 (1994).
48. Spina, E., Gitto, C., Avenoso, A., Campo, G. M., Caputi, A. P., Perucca, E., *Eur. J. Clin. Pharmacol.*, *51*, 395 (1997).
49. Kupfer, A., Schmid, B., Preisig, R., Pfaff, G., *Lancet*, *2*, 517 (1984).
50. Jones, D. R., Gorski, J. C., Haehner, B. D., O'Mara, E. M. Jr., Hall, S. D., *Clin. Pharmacol. Ther.*, *60*, 374 (1996).
51. Streetman, D. S., Bleakley, J. F., Kim, J. S., et al., *Clin. Pharmacol. Ther.*, *68*, 375 (2000).
52. Rostami-Hodjegan, A., Kroemer, H. K., Tucker, G. T., *9*, 277 (1999).
53. Kinirons, M. T., O'Shea, D., Downing, T. E., et al., *Clin. Pharmacol. Ther.*, *54*, 621 (1993).
54. Lown, K. S., Thummel, K. E., Benedict, P. E., et al., *Clin. Pharmacol. Ther.*, *57*, 16 (1995).
55. Watkins, P. B., Murray, S. A., Winkelman, L. G., Heuman, D. M., Wrighton, S. A., Guzelian, P. S., *J. Clin. Invest.*, *83*, 688 (1989).
56. Watkins, P. B., Hamilton, T. A., Annesley, T. M., Ellis, C. N., Kolars, J. C., Voorhees, J. J., *Clin. Pharmacol. Ther.*, *48*, 120 (1990).
57. Turgeon, D. K., Normolle, D. P., Leichtman, A. B., Annesley, T. M., Smith, D. E., Watkins, P. B., *Clin. Pharmacol. Ther.*, *52*, 471 (1992).
58. McCrea, J., Prueksaritanont, T., Gertz, B. J., *J. Clin. Pharmacol.*, *39*, 1212 (1999).
59. Kinirons, M. T., O'Shea, D., Kim, R. B., et al., *Clin. Pharmacol. Ther.*, *66*, 224 (1999).
60. Krivoruk, Y., Kinirons, M. T., Wood, A. J. J., Wood, M., *Clin. Pharmacol. Ther.*, *56*, 608 (1994).
61. Schuetz, E. G., Yasuda, K., Arimori, K., Schuetz, J. D., *Arch. Biochem. Biophys.*, *350*, 340 (1998).
62. Kim, R. B., Fromm, M. F., Wandel, C., et al., *J. Clin. Invest.*, *101*, 289 (1998).
63. Kinirons, M. T., O'Shea, D., Kim, R. B., Wood, A. J. J., Wilkinson, G. R., *Clin. Pharmacol. Ther.*, *67*, 577 (2000).
64. Wrighton, S. A., Brian, W. R., Sari, M. A., et al., *Mol. Pharmacol.*, *38*, 207 (1990).
65. Gorski, J. C., Hall, S. D., Jones, D. R., Vandenbranden, M., Wrighton, S. A., *Biochem. Pharmacol.*, *47*, 1643 (1994).
66. Wrighton, S. A., Ring, B. J., Watkins, P. B., Watkins, P. B., Vanderbranden, M., *Mol. Pharmacol.*, *36*, 97 (1989).
67. Cheng, C. L., Smith, E., Carver, P. L., et al., *Clin. Pharmacol. Ther.*, *61*, 531 (1997).
68. Chiou, W. L., Jeong, H. Y., Wu, T. C., Ma, C., *Clin. Pharmacol. Ther.*, *70*, 305 (2001).
69. Thummel, K. E., O'Shea, D., Paine, M. F., et al., *Clin. Pharmacol, Ther.*, *59*, 491 (1996).
70. Burtin, P., Jacqz-Aigrain, E., Girard, P., *Clin. Pharmacol. Ther.*, *56*, 615 (1994).
71. Reves, J. G., Fragen, R. J., Vinik, H. R., Greenblatt, D. J., *Anaesthesiology*, *62*, 310 (1985).
72. Harte, G. J., Gray, P. H., Casteel, H. B., Steer, P. A., Charles, B. G., *J. Paediatr. Child. Health*, *33*, 335 (1997).
73. Wells, T. G., Ellis, E. N., Casteel, H. B., *Clin. Pharmacol. Ther.*, *49*, 160 (1991).

74. Jacqz-Aigrain, E., Daoud, P., Burtin, P., Maherzi, S., Beaufils, F., *Eur. J. Clin. Pharmacol.*, *42*, 329 (1992).
75. Hughes, J., Gill, A. M., Mulhearn, H., Powell, E., Choonara, I., *Ann. Pharmacother.*, *30*, 27 (1996).
76. Park, B. K., *Br. J. Clin. Pharmacol.*, *12*, 97 (1981).
77. Waxman, D. J., Attisano, C., Guengerich, F. P., Lapenson, D. P., *Arch. Biochem. Biophys.*, *263*, 424 (1988).
78. Watkins, P. B., Turgeon, D. K., Saenger, P., et al., *Clin. Pharmacol. Ther.*, *52*, 265 (1992).
79. Hunt, C. M., Watkins, P. B., Saenger, P., et al., *Clin. Pharmacol. Ther.*, *51*, 18 (1992).
80. Schuetz, E. G., Schuetz, J. D., Grogan, W. M., et al., *Arch. Biochem. Biophys.*, *294*, 206 (1992).
81. Ohnhaus, E. E., Breckenridge, A. M., Park, B. K., *Eur. J. Clin. Pharmacol.*, *36*, 39 (1989).
82. Kovacs, S. J., Martin, D. E., Everitt, D. E., Patterson, S. D., Jorkasky, D. K., *Clin. Pharmacol. Ther.*, *63*, 617 (1998).
83. Tran, J. Q., Kovacs, S. J., McIntosh, T. S., Davis, H. M., Martin, D. E., *J. Clin. Pharmacol.*, *39*, 487 (1999).
84. Mahmood, I., *Ther. Drug Monit.*, *22*, 532 (2000).
85. Sun, H., Pelsor, F., Lazor, J., Selen, A., Lesko, L., *Clin. Pharmacol. Ther.*, *67*, 163 (Plll-86) (2000).
86. Burchell, B., Coughtrie, M. W., *Environ. Health Perspect.*, *105 Suppl 4*, 739 (1997).
87. Choonara, I. A., McKay, P., Hain, R., Rane, A., *Br. J. Clin. Pharmacol.*, *28*, 599 (1989).
88. Anderson, B. J., McKee, A. D., Holford, N. H., *Clin. Pharmacokinet.*, *33*, 313 (1997).
89. Koren, G., Butt, W., Chinyanga, H., Soldin, S., Tan, Y. K., Pape, K., *J. Pediatrics, 107*, 963 (1985).
90. McRorie, T. I., Lynn, A. M., Nespeca, M. K., Opheim, K. E., Slattery, J. T., *Am. J. Dis. Child.*, *146*, 972 (1992).
91. Rollins, D. E., Bahr, C. V., Moldeus, P., Rane, A., *Science, 205*, 1414 (1979).
92. Levy, G., Khanna, N. N., Soda, D. M., Tsuzuki, O., Stern, L., *Pediatrics, 55*, 818 (1975).
93. Kataria, B. K., Ved, S. A., Nicodemus, H. F., et al., *Anaesthesiology, 80*, 104 (1994).
94. Ventatakrishnan, K., von Moltke, L. L., Greenblatt, D. J., *J. Clin. Pharmacol.*, *41*, 1149 (2001).
95. Royal College of Paediatrics and Child Health: Ethics Advisory Committee. Guidelines for the ethical conduct of medical research involving children, *Arch. Dis. Child.*, *82*, 177 (2000).
96. Lacroix, D., Sonnier, M., Moncion, A., Cheron, G., Cresteil, T., *Eur. J. Biochem.*, *247*, 625 (1997).
97. Treluyer, J. M., Cheron, G., Sonnier, M., Cresteil, T., *Biochem. Pharmacol.*, *52*, 497 (1996).
98. Johnson, T. N., Tanner, M. S., Taylor, C. T., Tucker, G. T., *Br. J. Clin. Pharmacol.*, *51*, 451 (2001).
99. Ueng, Y. F., Kuwabara, T., Chun, Y. J., Guengerich, F. P., *Biochemistry, 36*, 370 (1997).
100. Kenworthy, K. E., Bloomers, J. C., Clarke, K. E., Houston, J. B., *Br. J. Clin. Pharmacol.*, *48*, 716 (1999).
101. Redfern, M., Keeling, J., Powell, *The report of the Royal Liverpool Children's inquiry, HMSO 2000.* Also available at www.rclinquiry.org.uk/
102. Berthou, F., Flinois, J. P., Ratanasavanh, D., Beaune, P., Riche, C., Guillouzo, A., *Drug Metab. Dispos.*, *19*, 561 (1991).
103. Rane, A., *Pediatrics, 104*, 640 (1999).
104. Peng, D., Birgersson, C., von Bahr, C., Rane, A., *Pediatr. Pharmacol.*, *4*, 155 (1984).
105. Murry, G. I., Barnes, T. S., Sewell, H. F., Ewan, S. W. B., Melvin, W. T., Burke, M. D., *Br. J. Clin. Pharmacol.*, *25*, 465 (1988).
106. Strassburg, C. P., Strassburg, A., Kneip, S. et al., *Gut, 50*, 259 (2001).
107. Hines, R. N., McCarver, D. G., *J. Pharmacol. Exp. Ther.*, *300*, 355 (2002).

108. McCarver, D. G., Hines, R. N., J. Pharmacol. Exp. Ther., 300, 361 (2002).
109. Cresteil, T., Beaune, P., Kremers, P., Celier, C., Guengerich, F. P., Leroux, J. P., *Eur. J. Biochem.*, *151*, 345 (1985).
110. Yang, H.-Y. L., Namkung, M. J., Juchau, M. R., *Biochem. Pharmacol.*, *49*, 717 (1995).
111. Hakkola, J., Pasanen, M., Purkunen, R., et al., *Biochem. Pharmacol.*, *48*, 59 (1994).
112. Ratanasavanh, D., Beaune, P., Morel, F., Flinois, J. P., Guengerich, F. P., Guillouzo, A., *Hepatology*, *13*, 1142 (1991).
113. Tateishi, T., Nakura, H., Asoh, M., et al., *Life Sci.*, *61*, 2567 (1997).
114. Treluyer, J. M., Gueret, G., Cheron, G., Sonnier, M., Cresteil, T., *Pharmacogenetics*, *7*, 441 (1997).
115. Treuyer, J. M., Benech, H., Colin, I., Pruvost, A., Cheron, G., Cresteil, T., *Pediatric Res.*, *47*, 677 (2000).
116. Jacqz-Aigrain, E., Cresteil, T., *Dev. Pharmacol. Ther.*, *18*, 161 (1992).
117. Treluyer, J. M., Jacqz-Aigrain, E., Alvarez, F., Cresteil, T., *Eur. J. Biochem.*, *202*, 583 (1991).
118. Jacqz-Aigrain, E., Funck-Brentano, C., Cresteil, T., *Pharmacogenetics*, *3*, 197 (1993).
119. Carpenter, S. P., Lasker, J. M., Raucy, J. L., *Mol. Pharmacol.*, *49*, 260 (1996).
120. Vieira, I., Sonnier, M., Cresteil, T., *Eur. J. Biochem.*, *238*, 476 (1996).
121. Boutelet-Bochan, H., Huang, Y., Juchau, M. R., *Biochem. Biophys. Res. Commun.*, *238*, 443 (1997).
122. Gu, J., Su, T., Chen, Y., Zhang, Q.-Y., Ding, X., *Toxicol. Appl. Pharmacol.*, *165*, 158 (2000).
123. Treluyer, J. M., Rey, E., Pons, G., Cresteil, T., *Br. J. Clin. Pharmacol.*, *52*, 419 (2001).
124. Kawade, N., Onishi, S., *Biochem. J.*, *196*, 257 (1981).
125. Leakey, J. E., Hume, R., Burchell, B., *Biochem. J.*, *243*, 859 (1987).
126. Burchell, B., Coughtrie, M., Jackson, M., et al., *Dev. Pharmacol. Ther.*, *13*, 70 (1989).
127. Pacifici, G. M., Franchi, M., Giuliani, L., Rane, A., *Dev. Pharmacol. Ther.*, *14*, 108 (1989).
128. Pacifici, G. M., Sawe, J., Kager, L., Rane, A., *Eur. J. Clin .Pharmacol.*, *22*, 553 (1982).
129. Coughtrie, M. W., Burchell, B., Leakey, J. E., Hume, R., *Mol. Pharmacol.*, *34*, 729 (1988).
130. Wright, M. C., Edwards, R. J., Pimenta, M., Ribeiro, V., Ratra, G. S., *Biochem. Pharmacol.*, *54*, 841 (1997).
131. Johnson, T. N., Tanner, M. S., Tucker, G. G., *Biochem. Pharmacol.*, *60*, 1601 (2000).
132. Rich, K. J., Boobis, A. R., *Micros. Res. Tech.*, *39*, 424 (1997).
133. Mahnke, A., Strotkamp, D., Ross, P. H., Hanstein, W. G., Charbot, G. C., Neff, P., *Arch. Biochem. Biophys.*, *337*, 62 (1997).
134. Shimada, T., Guengerich, F. P., *Mol. Pharmacol.*, *28*, 215 (1985).
135. Shimada, T., Misono, K. S., Guengerich, F P., *J. Biol. Chem.*, *261*, 909 (1986).
136. Bonate, P. L., Howard, D., *J. Clin. Pharmacol.*, *40*, 335 (2000).
137. Cummings, A. G., Steele, T. W., Labrooy, J. T., Shearman, D. J. C., *Gut*, *29*, 1672 (1988).
138. Menard, D., *Acta Paediatr.*, *405*, S1 (1994).
139. Zeugge, J., Schneider, G., Coassolo, P., Lave, T., *Clin. Pharmacokinet.*, *40*, 553 (2001).
140. Chappell, W. R., Mordenti, J., *Adv. Drug Res.*, *20*, 2 (1991).
141. Lave, T., Dupin, S., Scmitt, C., Chou, R. C., Jaeck, D., Coassolo, P., *J. Pharm. Sci.*, *86*, 584 (1997).
142. Ekins, S., Ring, B., Grace, J., McRobie-Bell, D. J., Wrighton, S. A., *J. Pharmacol. Toxicol. Methods*, *44*, 313 (2000).
143. Ekins, S., Waller, C. L., Swaan, P. W., Cruciani, G., Wrighton, S. A., Wikel, J. H., *J. Pharmacol. Toxicol. Methods*, *44*, 251 (2000).
144. Tooley, A., Rostami-Hodjegan, A., Lennard, M. S., Tucker, G. T., *Clin. Pharmacol. Ther.*, *68*, 216 (2000).
145. Yang, J. S., Rostami-Hodjegan, A., Tucker, G. T., *Br. J. Clin. Pharmacol.*, 484P (2001).

2.3 Drug disposition

Gérard Pons, Jean-Marc Treluyer, Elisabeth Rey

Paediatric Clinical Pharmacology, Hospital Saint Vincent de Paul,
University René Descartes, 82 Av. Denfert-Rochereau, 75674 Paris Cedex 14, France

2.3.1 INTRODUCTION

An understanding of drug disposition in neonates, infants and children is crucial for drug evaluation in children and for developing specific dose regimens. The maturation of physiological processes has an important influence on the fate of drugs in the body.

2.3.2 ABSORPTION OF DRUGS

Oral

Drug absorption is assessed by the rate of absorption (time to maximum drug concentration) and bioavailability. Bioavailability is estimated from the area under the concentration-time curve of the drug, the calculation of which requires several blood samples. In most instances this is considered too invasive to be ethically acceptable. As a consequence, data on drug bioavailability and absorption are scanty in children particularly in neonates and infants. There are several documented physiological maturational changes that allow one to speculate on the consequences on the absorption of drugs: changes in the intraluminal pH of the gastro-intestinal tract that can directly affect the stability and the degree of ionisation of a drug, particularly intra-gastric pH changes [1], gastric emptying and intestinal motility [2-4], the transport of biliary salts into the intestinal lumen [5,6], and intestinal microflora [7]. Frequent feeding in neonates may have a large influence on the absorption of drugs. Nevertheless, there are very few available data on the rate and the extent of drug absorption in children.

Neonates are thought to have a decrease in the rate and extent of drug absorption. Processes of both passive and active transport through the intestinal wall are considered fully mature by approximately four months of age [4].

Developmental differences in the activity of intestinal drug-metabolising enzymes and efflux transporters, that can markedly alter the bioavailability of drugs as part of the first-pass effect, are incompletely characterised [8]. The intestinal activity of CYP 1A1 appears to increase with age [9].

Rectal

Rectal administration seems to be more influenced by the formulation than by intestinal maturation. Absorption of drugs from suppositories is often erratic and delayed.

Intramuscular

Absorption of drugs via the intramuscular route is variable and depends on changes in muscular blood flow. Changes in muscular blood flow occur during maturation and during acute ilness. Intramuscular administration of drugs may lead to side effects, such as sciatic nerve injury, local necrosis of the muscular tissue or of the subcutaneous fat. The absorption of drug via intramuscular administration may be unpredictable in acute situations. Due to the pain related to intramuscular injection, this route is unpopular among children and parents.

Transcutaneous

Transcutaneous absorption of drugs after local administration is often ignored despite of the fact that there is systemic exposure that can lead to side effects. Skin absorption does not seem to be reliable for systemic therapeutic purposes [10]. Transcutaneous absorption of drugs is greater in neonates and infants than in adults. This phenomenon is mostly related to the higher skin surface area relative to body weight and to the increase of the hydration of the stratum corneum (the limiting factor of cutaneous absorption of drugs), for medications applied on the surface of the buttocks where the diapers may play the role of an occlusive dressing.

2.3.3 DRUG DISTRIBUTION

Drug distribution varies as a function of age. These changes are related to the maturational evolution of the size of the body compartments and protein binding. The body water represents about 75% of body weight in the neonate

(85% in the premature neonate). It decreases with age to reach a value close to that of an adult (60%) at one year of age. Extracellular water represents about 45% of body weight in the neonate. It decreases with age to reach around 25% at one year of age and 15 to 20% at puberty [11]. The fat compartment is smaller in neonates (15% at term and 1% in the premature neonate) than in adults. It increases to around 25% in the infant, decreases later on, and increases again before puberty.

Protein binding

For most drugs known to be highly bound to plasma proteins in adults, the unbound fraction is more important in the neonate than in infants during the first year of life [12]. This is related in part to the lower plasma concentration of binding proteins (albumin for weak acids and α-1-glycoprotein for weak bases). The volume of distribution of most drugs, expressed per unit of body weight, is higher in neonates and infants than in adults. Afterwards it decreases with age.

Transporters

Transporters are important in relation to the distribution of a drug. P-glycoprotein, a member of the ATP-binding family of transporters that functions as an efflux transporter capable of extruding selected toxins and xenobiotics is one such example. The expression and localisation of P-glycoprotein in specific tissues facilitate its ability to limit cellular uptake of xenobiotic substrates to these sites (e.g., the blood-brain barrier, hepatocytes, renal tubular cells, and enterocytes [13]. There are limited data on the ontogeny of the expression of P-glycoprotein in humans. They suggest a pattern of localisation in neonates similar to that in adults, but with lower level of expression.

2.3.4 DRUG METABOLISM

Phase 1 reactions (such as demethylation, hydroxylation) and phase 2 reactions (glucuronidation, acetylation) are usually lower in neonates than in adults [14-16] explaining lower clearances and longer half-lives [17]. The consequences may be severe as illustrated by the historical examples of the life-threatening "grey baby syndrome" in neonates treated with chloramphenicol and more recently the example of cisapride with QT prolongation, "torsades de pointes" [18]. Some metabolic pathways are, however, almost mature at the time of birth, e.g. sulphation, methylation of theophyline to caffeine, methylation of thiopurine [19]. This explains the different metabolic patterns observed in neonates compared to adults.

After birth, metabolic activity and clearance sometimes dramatically increase [17]. The maturational profile of the metabolic pathways is different from one metabolic pathway to another and adult profile is reached at various ages according to the metabolic pathways involved. Demethylation and acetylation of caffeine mature rapidly in infancy, for example [14-16]. These different profiles are dependant on the ontogeny of the enzymes involved. CYP3A7 is the major hepatic isoform in the fetus while CYP3A4/5 is the major one in adults. Shortly after birth, CYP3A7 progressively decreases and is replaced by CYP3A4/5 within the first weeks of life [20]. CYP2D6 [21] and CYP2E1 [22] appear shortly after birth. The rate of demethylation of caffeine in adolescent girls appears to decline to levels seen in adults once girls reach Tanner stage 2, whereas it occurs at Tanner stage 4 or 5 in adolescent boys [23], thus demonstrating an apparent sex-based difference in the ontogeny of CYP1A2.

Maturation has an influence on the expression of the genotype into the phenotype of genetically determined metabolic pathways. During the first few weeks of life all patients appear to be slow metabolisers. As age increases, the percentage of fast metabolisers increases up to the percentage observed in adults. For acetylation of caffeine, this ratio is reached during the third year of life and for isoniazid, the percentage seems to be reached at 4 years [16,24]. Most metabolic pathways appear more active in infants and young children than in adults [23,15]. The maturational profile can therefore be represented with an asymetrical bell-shaped curve with low clearance in neonates, high clearance in infants and young children compared to adults [25]. The maturational profile of various metabolic pathways can be modified by *in utero* induction, due to xenobiotics such as anticonvulsants passively transfered through the placenta [26].

2.3.5   DRUG ELIMINATION

Glomerular filtration, measured using creatinine clearance, is low in the neonates. At birth, maturation is proportional to gestational age. It increases rapidly during the first 2 to 3 days of life and reaches adult values during the third week of life [27]. The clearance of gentamicin increases linearly as a function of gestational age and exponentially as a function of postnatal age [28].

Tubular secretion is diminished in neonates and reaches adult values after the second month of life. These changes explain similar changes in the clearance of drugs secreted through the tubule, such as penicillins (penicillin G, ampicillin, tircacillin) [29,30]. Tubular reabsorption is passive. The half-life of penicillins decreases as a function of gestational age and postnatal age. Drug reabsorption is mostly passive. Tubular reabsorption increases when liposoluble drugs reach the lumen of the tubule unchanged. Consequently, the lower clearance in neonates requires a lower maintenance dose and the longer half-life allows a larger dose interval.

2.3.6   PRACTICAL CONSEQUENCES

Dosing is most often calculated per unit of body weight. For some drugs (anticancer agents, steroids), the dose is calculated as a function of body surface area. Maturational profile, as a function of age, requires an individual drug-dose adjustment, particularly for drugs with a narrow therapeutic interval. Individual dose adjustment requires appropriate drug formulation.

REFERENCES

1. Grand, R. J., Watkins, J. B., Torti, F. M., *Gastroenterology, 70*, 790 (1976).
2. Signer, E., *Acta. Paediatr. Scand., 64*, 525 (1975).
3. Gupta, M., Brans, Y. W., *Pediatrics, 62*, 26 (1978).
4. Heimann, G., *Eur. J. Clin. Pharmacol., 18*, 43 (1980).
5. Watkins, J. B., Ingall, D., Szczepanik, P., et al., *N. Engl. J. Med., 288*, 431 (1973).
6. Melhorn, D. K., Gross, S., *J. Pediatr., 79*, 581 (1971).
7. Long, S. S., Swensson, R. M., *J. Pediatr., 91*, 298 (1977).
8. Hall, S. D., Thummel, K. E., Watkins, P. B., et al., *Drug Metab Dispos, 27*, 161 (1999).
9. Stahlberg, M. R., Hietanen, E., Maki, M., *Gut, 29*, 1058 (1988).
10. Evans, N. J., Rutter, N., Hadgraft, J., et al., *J. Pediatr., 107*, 307 (1985).
11. Friis-Hansen, N. B., *Pediatrics, 28*, 169 (1961).
12. Pacifici, G. M., Viani, A., Taddeucci-Brunelli, G., *Dev. Pharmacol. Ther., 10*, 413 (1987).
13. Lin, J. H., Yamazaki, M., *Clin. Pharmacokinetics., 42*, 59 (2003).
14. Carrier, O., Pons, G., Rey, E., et al., *Clin. Pharmacol. Ther., 44*, 145 (1988).
15. Pons, G., Blais, J. C., Rey, E., et al., *Pediatr. Res., 23*, 632 (1988).
16. Pons, G., Rey, E., Carrier, O., et al., *Fund. Clin. Pharmacol., 3*, 589 (1989).
17. Pons, G., Carrier, O., Richard, M. O., et al., *Dev. Pharmacol. Ther., 11*, 258 (1988).
18. Treluyer, J. M., Rey, E., Sonnier, M., et al., *Br. J. Clin. Pharmacol., 52*, 419 (2001).
19. Ganiere-Monteil, C., Medard, Y., Lejus, C., et al., Eur. *J. Clin. Pharmacol., 60*, 89 (2004).
20. Lacroix, D., Sonnier, M., Moncion, A., et al., *Eur. J. Biochem., 247*, 625 (1997).
21. Treluyer, J. M., Jacqz-Aigrain, E., Alvarez, F., et al., *Eur. J. Biochem., 202*, 583 (1991).
22. Viera, I., Sonnier, M., Cresteil, T., et al., *Eur. J. Biochem., 238*, 476 (1996).
23. Lambert, G. H., Schoeller, D. A., Kotake, A. N., et al.,*Dev. Pharmacol. Ther., 9*, 375 (1986).
24. Pariente-Khayat, Pons, G., Rey, E., et al., *Pediatr Res., 29*, 492 (1991).
25. Yee, G. C., Lennon, T. P., Gmur, D. J., et al., *Clin. Pharmacol. Ther., 40*, 438 (1986).
26. Rating, D., Jager-Roman, E., Nau, H., et al., *Pediatr. Pharmacol., 3*, 209 (1983).
27. Aperia, A., Broberger, O., Elinder, G., et al., *Acta Paediatr. Scand., 70*, 183 (1981).
28. Pons, G., d'Athis, P., Rey, E., et al., *Ther. Drug Monit., 10*, 421 (1988).
29. McCracken, G. H., Jr., Ginsberg, C., Chrane, D. F., et al., *J. Pediatr., 82*, 692 (1973).
30. Kaplan, J. M., McCracken, G. H., Jr., Horton, L. J., et al., *J. Pediatr., 84*, 571 (1974).

2.4 Pharmacokinetics and pharmacodynamics

Noel E. Cranswick

Royal Children's Hospital, Melbourne, Australia

2.4.1 INTRODUCTION

Pharmacokinetics (PK) is the study of the effects of the body upon the drug. This usually describes the measured concentration of drug in the vascular space (plasma or serum concentrations). An understanding of the PK of a drug is fundamental to the safe and efficacious use of medicines. We often use mathematical relationship (or models) to try and predict the amount of drug in the body for a given dose of drug. PK is often made confusing to the clinician by complex equations. Fortunately, only a brief reference to mathematical formulae is necessary for basic understanding of pharmacokinetics.

Pharmacodynamics (PD) is the study of the action of drugs upon the body. As with PK, PD attempts to predict this action upon the basis of mathematical models. This describes the effect of the drug at a particular concentration.

It is now well recognised that children do not act as scaled down adults in terms of their pharmacology. Drugs have appeared to have unpredictable or unexpected effects when first used in the paediatric population. PK and PD concepts have helped significantly advance our understanding of why children are different and we are now discovering how to scale drug doses appropriately for children.

2.4.2 PHARMACOKINETICS

Pharmacokinetics usually describes the amount of drug in the body at any time in terms of constants, often called parameters and variables. Typical parameters include volume of distribution (Vd) and clearance (CL). The most common independent variable is time (t), but can include dose (D) or dosing interval (τ). This approach is known as the parametric approach to pharmacokinetics and is the approach that will be taken through the remainder of this chapter. Parameters are usually fully described by the mean value and a variance; typically using a normal or log-normal distribution.

An alternative to this is the nonparametric approach, which does not make assumptions about the shape of the distribution and predictions determined solely by the doses and measured drug concentrations in the population. The distribution is identified from the raw data and a weighting scheme with a maximum likelihood function. The nonparametric approach will not be discussed further in the chapter and specialist references should be consulted.

2.4.3 DRUG TRANSFER

Non-compartmental analysis

The non-compartmental analysis is the most straightforward model in which parameters are estimated. The typical parameters estimated are area under the concentration/time curve (AUC), maximum serum concentration (C_{max}) and half-life ($t_{1/2}$). This method makes no assumptions about how many compartments are present. There is elimination from a central compartment (the blood or serum) by a first order process (Figure 1a). Vd and CL are usually not estimated but it is possible to calculate them.

The weakness of this approach is that it does not allow for drug concentration prediction for future dosing or when subjects, different to those in the original study, receive the drug. It is, however, the most common type of analysis requested by regulatory agencies, as it makes relatively few assumptions about the data.

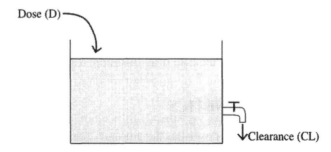

Figure 1a. A diagrammatic representation of a single compartment model with first order elimination

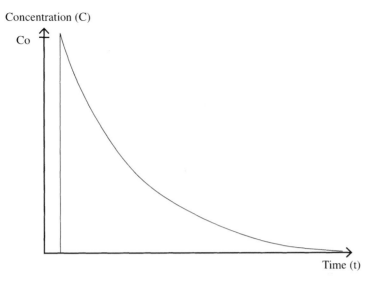

Figure 1b. A concentration time curve of an intravenous (IV) bolus drug with first order (linear) elimination using a linear scale.

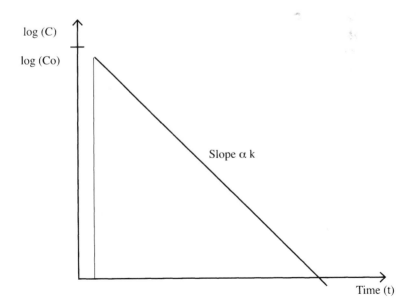

Figure 1c. A concentration time curve of an intravenous (IV) bolus drug with first order (linear) elimination on a logarithmic scale.

The AUC is usually calculated by adding up the trapezoidal areas between each of the concentration/time points. If a straight line is plotted between consecutive points, the calculation is said to be linear, while, if it is plotted along a first order elimination (log) line, it is said to be log linear.

The C_{max} is simply read off the graph as the highest concentration recorded following a given dose. The $t_{1/2}$ is calculated as the time taken for a given concentration to halve, once the distribution phase is complete. It is most easily estimated on a log/linear plot where the elimination phase is a straight line (Figure 1c). If there is more than one elimination phase, then the dominant phase is usually recorded; this is usually, but not necessarily, the terminal phase of elimination.

First order kinetics

Almost all drugs, whether cleared through the kidneys or metabolised by the liver, can have their elimination described by a first order process, as described above. That is, the amount of drug eliminated at a particular time is proportional to the amount of drug present and, as a result of this, $t_{1/2}$ can be calculated. The time taken to completely eliminate the drug from the body is independent of the initial amount of drug present and depends only upon the $t_{1/2}$. The elimination phase commences once the drug is distributed through the body, although actual elimination begins immediately after dosing. A doubling of dose results in a doubling in serum drug concentrations, so called linear kinetics.

Zero order kinetics

Only a few drugs are eliminated by zero-order kinetics. This means that the amount of drug cleared at any time is constant, independent of the amount of drug present. Drugs that can be modeled by a zero-order process include alcohol, phenytoin and high dose aspirin. A consequence of this is that the greater the dose of drug, the longer it takes to eliminate the drug (i.e., the more alcohol you consume, the longer that you remain drunk). A doubling of dose will result in a greater than doubling in the serum concentration; so called nonlinear kinetics. This is commonly recognized with phenytoin dosing. There may be a transition phase from first order to zero order kinetics. This can be mathematically modeled using the Michaelis-Menten equation, further described under pharmacodynamics.

$$E(C) = E_0 + \frac{(E_{max} \times C^n)}{((EC_{50})^n + C^n)}$$

E(C) is the observed effect at concentration C; E_0 is the baseline effect value; Emax is the maximal effect and EC_{50} is the concentration at half maximal effect and n is a constant.

One compartment model

The one compartment model assumes that the body is made of a single container (compartment) of water and, after dosing, the drug is uniformly distributed throughout the compartment (See Figure 1a). The volume of this compartment is Vd and usually measured in litres (L). In the real world, there may be drug outside the compartment and so, the estimated size of Vd may be greater than the total body volume, hence Vd is often referred to as the apparent volume of distribution. For drugs that are mostly distributed in extra-cellular water, such at aminoglycosides, Vd approximated that of extra-cellular water volume.

Infants and children change their water body composition as they grow. Premature infants consist of up to 90% water. This amount alters through childhood until, as adolescents, they approach approximately 60% body water found in adults. This means that water-soluble drugs have higher Vd in younger children, and so need a greater amount of drug (per kilogram) to achieve the same drug concentration compared to adults.

The drug is eliminated by first order kinetics, as represented by the tap in Figure 1a. The rate at which water runs out of the compartment is analogous to clearance. The slope of the elimination curve (plotted on semi-log graph) can estimate the elimination rate constant, K (or k_e or k_{elim}). The relationship between k, Vd and CL is shown below.

$$K = \frac{0.693}{t_{1/2}} = \frac{CL}{Vd}$$

Two compartment model

Some drugs do not behave as if they are distributed in a single compartment and a multi-compartment model may be required to predict drug concentrations. A diagrammatic representation of a two-compartment model is shown in Figure 2a. The drug usually has both its input and output into the "central" compartment (A), while there is exchange with the peripheral compartment (B). The exchange between the two compartments can be represented by either inter-compartmental rate constants or clearances. A two-compartment model is usually recognised in the data when the K appears to change with time (Figure 2b).

Multi-compartment models

Some drugs require more complicated models to represent their kinetic profiles, such as multi-compartment models. These may either be linear or mamillary,

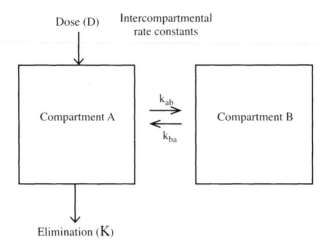

Figure 2a. A diagrammatic representation of a two compartment model with first order elimination

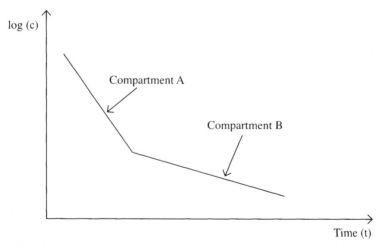

Figure 2b. A concentration time curve for a two compartment model, with first order elimination, on a logarithmic scale.

the mamillary model named, as its structure is similar to a mammary gland. The mathematics of these models is complicated and it is usually preferable to use a simpler model, if it fits the data as well as a more complicated model.

2.4.4 SYSTEMIC AVAILABILITY

Calculation of drug bioavailability

Up to this point, we have assumed that the entire drug administered has entered the body. This is usually true for intravenous injections. However, for most drugs administered by other routes, only a percentage is actually absorbed into the central compartment. The fraction absorbed is called the bioavailability (F). F is calculated as the amount absorbed by a given route compared to an IV dose. For example, with an oral drug, the F_{oral} calculation is shown below. It should be noted that F is a parameter.

$$F = \frac{AUC_{oral}}{AUC_{IV}}$$

Factors affecting availability

The most important influence upon F is the route of administration, although the drug formulation can also influence the amount of drug absorbed. Most drugs are administered orally, although alternative routes are sometimes employed to increase F.

Oral bioavailability

Any drug has several barriers to overcome when administered orally, including being spat out or vomited. It must survive the acidity of the stomach and then the alkaline environment of the small intestine where the majority of drugs are absorbed. Once absorbed through the bowel wall, drugs pass via the portal circulation to the liver where they may undergo first-pass metabolism. Some drugs also undergo first-pass metabolism at the bowel wall. Individuals with low levels of hepatic enzyme activity may have increased bioavailability, due to a lower first pass effect. Co-administration of enzyme inhibitors can have a similar effect. The fraction of drug removed from the blood as it passes through the liver is known as the extraction ratio.

The co-administration of grapefruit juice increases the bioavailability of drugs usually metabolised by CPY 3A4. Grapefruit juice inhibits this enzyme, both at the brush border and the liver [1]. The inhibition at the brush border appears to be highly significant. Drugs affected include cisapride and ciclosporin. The effect may be exacerbated by the concomitant use of other CYP 3A4 inhibitors.

P-glycoproteins

P-glycoproteins are a family of membrane bound transporters that move xenobiotics including drugs. In the intestine, they are located in the luminal brush border and move drugs back into the bowel lumen. P-glycoproteins are also thought to be an important mechanism of drug resistance in tumours. Inhibitors of P-glycoproteins may be useful therapeutically to increase the efficacy of some cytotoxic agents used in oncology.

Rectal

In children, the rectal route is often used to administer medicines. There are several reasons for this. The lack of a first pass effect for drugs absorbed from the rectal mucosa is commonly quoted, although rarely important in children. Some medicines actually have better bioavailability orally than rectally. More commonly, the reason is that children find it more difficult to refuse or expel medicines administered rectally. However, recent studies suggest that rectal absorption may not be as reliable as first thought [2]. For example, paracetamol is erratically absorbed from the rectum in children and this route should generally be avoided.

Other routes

Drugs are also administered by intramuscular injection, subcutaneously and via the transdermal route. Young infants may demonstrate altered PK profiles due to their smaller muscle mass and thin skin. For example, topical medication, which would not be significantly absorbed in adults, may demonstrate a high bioavailability in premature neonates resulting in unwanted and serious adverse effects. However, transdermal administration may be useful in neonates where other routes are impractical.

2.4.5 VOLUME OF DISTRIBUTION

Vd is clearly an important parameter to estimate in pharmacokinetic analysis. The calculation of volume of distribution is usually based upon the concentration of drug at the time of administration, before any has been eliminated. While the actual concentration at t=0 is usually not known, the data are extrapolated back to that time (Figure 3) and Vd is calculated as shown below;

$$Vd = \frac{D}{C_0}$$ where D is the dose and C_0 is the plasma concentration at the time of administration

Factors affecting distribution

The most important factor that influences Vd is size. A small infant clearly has a lower volume of distribution for a drug compared to a large adult. Usually, it

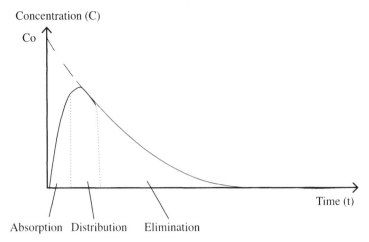

Figure 3. A concentration time curve for an oral dose in a single compartment model, with first order elimination, on a linear scale. The absorption, distribution and elimination phases are indicated. Also the extrapolation of the concentration time curve to estimate C_0 is shown.

is thought that Vd scales linearly with weight. However, changes in body composition (e.g. water and fat), may alter this relationship. Children often display a disproportionately large Vd for their weight for water-soluble drugs. Also, extremes of weight may alter this relationship. Individuals who are over 120% of their ideal body weight may not have a Vd proportional to their mass.

Other factors, such as protein binding, may be important for individual drugs distribution. However, alterations in protein binding rarely affects free drug concentration, as increases in free drug are compensated for by concurrent alterations in drug clearance.

2.4.6 CLEARANCE

Clearance is calculated from the concentration time curve. There are several equations for estimating clearance. The ratio of dose to AUC is the easiest to visualise.

$$CL = \frac{D}{AUC}$$

Total body clearance is a sum of the clearances through all pathways, so total clearance can be calculated by the addition of the individual organ clearances, if these are known.

$$CL_{Total} = CL_{Renal} + CL_{Hepatic}$$

For drugs eliminated by a single pathway such as the kidneys, total clearance is equivalent to renal clearance. At steady state, drug concentration is determined by clearance.

Factors affecting clearance

The most important determinants of clearance are drug dependent. Clearance is usually renal or hepatic or a combination of both.

Renal function varies throughout life and in general scales to body mass. Larger individuals have greater renal blood flow and glomerular filtration rate (GFR) and hence, greater clearance. However, the relationship is not simply linear and children have higher renal clearance than adults. Furthermore, neonates have a lower renal clearance for their weight compared to adults. There are major changes in renal function and body fluids in the first few days after birth, which further complicates dosing in the first few days of life. Prematurity further complicates this, as a premature infant will have a lower GFR than a more mature neonate at the same weight. Drugs that are cleared renally often require dose adjustment for gestational age as well as weight.

Individuals with renal impairment will require adjustments based upon the degree of dysfunction. This is often estimated using formulae based upon measures of serum creatinine; however direct measures of renal function, such as creatinine or inulin clearance, are more accurate, especially in children where muscle mass may vary. In some circumstances, calculated drug clearance can be used to estimate renal dysfunction.

Hepatic clearance is often more difficult to predict than renal clearance. Hepatic capacity depends both upon liver blood flow and intrinsic enzyme activity. There are currently no easily measurable physiological parameters that can be used to estimate total hepatic function. Serum liver enzyme measurements indicate liver damage rather than function, while measures of synthetic function only alter in severe liver dysfunction and are not useful in normal physiological circumstances.

Some drugs have high extraction from the liver and are limited by hepatic blood flow. Changes to blood flow, as occurs in severe hepatic disease, may lower clearance, but changes in enzyme activity are less likely to impact upon drug elimination.

Many drugs have low liver extraction and clearance is limited by enzyme capacity rather than blood flow. Alterations in enzyme capacity can significantly influence clearance levels. Examples include individuals with intrinsically low enzyme capacity (genetically determined) or when exposed to enzyme inhibitors.

In childhood, drug metabolising hepatic enzyme activity changes with age and development [3]. This is covered by the term ontogeny. In general, neonates have very low intrinsic enzyme activity. The activity increases to levels above that found in adults in the toddler years. Most enzymes have decreased to adult levels by the teenage years. This has implications for the dosing of children throughout childhood and may explain why many doses in children cannot be simply scaled down linearly for adults.

Allometric scaling

In recent years, it has been realised that many physiological processes do not scale linearly in animals [4]. Empirically, it has been found that metabolic functions such as heart rate and renal function are better described by the allometric model.

$$\text{Renal Function } \alpha \text{ } Wt^{3/4}$$

The allometric model scales over the full range of animal size from ants to elephants. In developing pharmacokinetic models for children, both renal and hepatic function is better described by the allometric weight than by a simple linear model.

In recent times, body surface area (BSA) calculations have successfully calculated drug doses in children, yet very few drugs are actually excreted through the skin. The allometric model may explain the discrepancy. It is probably not body surface area per se, but rather that the formula used to calculate BSA approximates an allometric model.

2.4.6   WHICH PRIMARY PARAMETERS

Under some circumstances, other primary parameters are required to model drug action. Some of these are used to explain variation in drug absorption, such as absorption rate constants and lag times. Others are used to explain transfer between compartments. In general, it is preferable to use physiological parameters when possible. Parameters, such as clearance and Vd, have much more meaning in the real world than rate constants (e.g., k). Multi-compartment parameters describe the volumes of each compartment and the transfer rates between compartments.

2.4.7   SECONDARY PARAMETERS

Half-life $(t_{1/2})$

The half-life, $t_{1/2}$, is the time taken for the amount of drug (and hence, concentration) to halve in the body. Its calculation assumes that absorption and distribution are complete. In a single compartment model, there is a single $t_{1/2}$ while multiple compartment models have several $t_{1/2}$ associated with the elimination phase.

As the time to halve the drug amount in the body is constant, the time to essentially complete drug elimination is constant, usually five half-lives, and independent of the starting amount of drug. In overdose, however, drug elimination may be prolonged if the elimination route is saturated.

Area under the curve (AUC)

The AUC is a measure of the total drug exposure. It is dependent upon the dose of drug, the Vd and drug elimination. For some drugs, the AUC is an important determinant of efficacy and toxicity. For example, the efficacy of the immunosuppressant, ciclosporin, is determined by its AUC rather than peak (C_{max}) or trough (C_{min}) concentrations. Likewise, the ototoxicity related to aminoglycosides appears to be related to the total drug exposure over a course of treatment.

2.4.8 THERAPEUTIC RANGE

The therapeutic range is the optimal concentration range for a drug for which the drug is efficacious yet not toxic. It is usually defined as between the minimum effective concentration and the minimum toxic concentration. As can be seen from Figure 4a, concentrations above the range will result in toxicity while low concentrations mean loss of efficacy. This can be especially important in drugs with narrow therapeutic ranges, such as phenytoin and theophylline. Drugs with narrow therapeutic ranges may benefit from serum drug concentration monitoring [5].

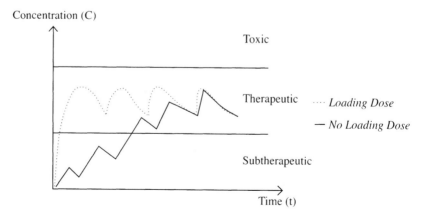

Figure 4a. Multiple dosing to achieve concentrations in the therapeutic range, with and without a loading dose.

2.4.9 APPLICATION OF PHARMACOKINETIC PRINCIPLES TO MULTIPLE DOSE REGIMENS

Multiple dose regimens are defined by the attainment of steady-state; that is when the rate of input is equal to the elimination rate. It takes five half-lives to reach steady state. Dosing regimens aim to maintain the drug concentration within the therapeutic range once at steady-state. For drugs with a narrow therapeutic range, more frequent dosing will result in less variation between C_{max} and C_{min}, reducing the likelihood of toxicity or sub-therapeutic levels. The total daily dose is derived from the Dose (D) / Dosing interval (τ).

Children may require a more frequent dosing regimen when compared to adults, as they often have more rapid drug elimination.

In order to attain a therapeutic regimen, a loading dose may be employed (Figure 4a). This will be determined by the volume of distribution and the target concentration. In general, for a drug that is dosed at its half-life (i.e., $\tau=t_{1/2}$; a common practice), the steady state concentration will be approximately double that attained with a single dose and, in such circumstances, the loading dose would often be about double the steady-state dose. However, multi-compartment kinetics may make the calculations more complicated. Steady-state concentration usually refers to the mean drug concentration, although C_{max} and C_{min} can also be defined at steady state. Some drugs will have steady-state concentrations that are different to what would be predicted from a single dose. An example would be lamotrigine, which induces its own metabolism and hence, its clearance increases with time.

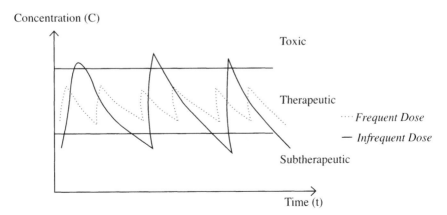

Figure 4b. A concentration time curve at steady state for a multiple dosage schedule, demonstrating that reducing the dosing interval (with no change in total daily dose) will improve the time that the concentration remains within the therapeutic range.

2.4.10 PHARMACODYNAMICS

Pharmacodynamics is the measurement and description of the effect of the drug on the body. Drugs will interact at a specific site, usually a receptor or enzyme, resulting in an alteration of the function. Pharmacodynamic models attempt to describe this interaction and explain the size of effect in terms of drug concentration or dose. The model may need to include an effect compartment, which is separate from the central drug concentration compartment. Most drug-effect relationships are not linear and are best described by a saturable model. This is often described by the Michaelis-Menten equation.

$$E(C) = E_0 + \frac{(E_{max} \times C^n)}{((EC_{50})^n + C^n)}$$

At low concentrations, there is an approximately linear relationship between the drug and effect. However, as the concentration increases, the effect plateaus and approaches the maximum effect, E_{max}. The concentration at which 50% of the maximal effect is reached, is defined as the EC_{50} and gives a measure of the potency of a drug. A potent drug has a lower EC_{50}. This is different to efficacy, which is defined by the drug's E_{max} (Figure 5a and 5b).

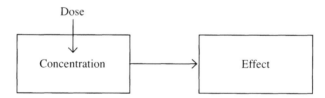

Figure 5a. A simple pharmacodynamic model.

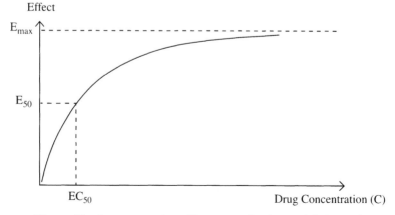

Figure 5b. A concentration effect curve for the model shown in 5a.

The effect compartment may be separated from the central compartment where the drug concentration is measured and result in a seemingly variable effect at a particular drug concentration. Alternatively, the drug may not act directly upon a receptor but have an indirect or delayed effect. Specific models have been developed to better understand these phenomena. Figure 6a and 6b are examples of a dose-response curve when there is a delay in the drug passing from the central compartment to the active site. Specific texts should be consulted for further explanations of pharmacodynamic models.

2.4.11 COMPUTER PROGRAMMES

The use of computer programmes has become standard for pharmacokinetic and dynamic analysis. Many programmes allow for both PK and PD analysis and some programmes are suitable for both dense (standard) and sparse data analysis. Some useful programmes are listed on the next page.

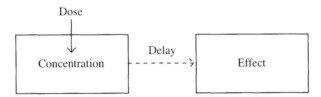

Figure 6a. A simple pharmacodynamic model with a time lag between concentration and effect.

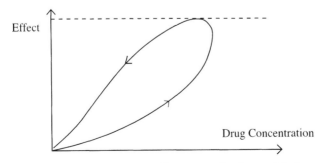

Figure 6b. A concentration effect curve for the model shown in 6a.

Standard Kinetic/Dynamic Programmes

Programme	Contact
ADAPT II	http://bmsr.usc.edu/Software/Adapt/adptmenu.html
EZ-Fit	http://www.jlc.net/~fperrell/
Kinetica	http://www.innaphase.com/
PK Solutions	http://www.summitpk.com/pksolutions/
RxKinetics	http://www.rxkinetics.com/
SAAM II	http://www.saam.com/
USC*PACK	http://www.usc.edu/hsc/lab_apk/software.html
Winnonlin	http://www.pharsight.com/

Population Programmes and add-ons

Programme	Contact
Kinetica	http://www.innaphase.com/
NONMEM	http://www.globomax.net/products/nonmem.cfm
S-plus	http://www.insightful.com/products/default.asp
Trial Simulator	http://www.pharsight.com/
WinBUGS	http://www.mrc-bsu.cam.ac.uk/bugs/welcome.shtml
Winnonmix	http://www.pharsight.com/
Xpose	http://xpose.sourceforge.net/

ABBREVIATIONS

τ - Dosing interval

$C(t)$ - Concentration at time T

CL – Clearance

$CL_{Hepatic}$ – Hepatic clearance

CL_{Renal} – Renal clearance

CL_{Total} - Total clearance

C_{max} - Maximum concentration

C_{min} - Minimum concentration

Css - Steady state concentration

D - Dose

F - Bioavailability

k - Elimination rate constant (also k_e or k_{elim})
LD - Loading dose
$t_{1/2}$ - Half-life
Vd - Volume of distribution
E(C) - Observed effect at concentration C
E_0 - Baseline effect value
E_{max} - Maximal effect
EC_{50} - Concentration at half maximal effect
n - Constant

REFERENCES

1. Goho, C., *Pediatr. Dent. 23*, 4, 365 (2001).
2. Anderson, B. J., N. H. Holford, *Anesthesiology 88*, 4, 1131 (1998).
3. Reed, M. D., J. B. Besunder, *Pediatr. Clin. North Amer.*, *36*, 5, 1053 (1989).
4. Anderson, B. J., N. H. Holford, et al., *Anaesth. Intensive Care 25*, 5, 497 (1997).
5. Boreus, L. O., *Clin Pharmacokinet. 17, Suppl. 1*, 4 (1989).

SUGGESTED AND FURTHER READING

Bauer, L., *Applied Clinical Pharmacokinetics*, McGraw-Hill/Appleton & Lange, New York (2001).

Begg, E., *Instant Clinical Pharmacology*, Blackwell Publishing, Oxford (2003).

Birkett, D. J., *Pharmacokinetics made easy*, McGraw Hill, Sydney (1998).

Gibaldi. M., Perrier, D., *Pharmacokinetics* (2nd ed.), Marcel Dekker, New York (1982).

Pecile, A., Rescigno, A., *Pharmacokinetics: Mathematical and Statistical Approaches to Metabolism and Distribution of Chemicals and Drugs*, Plenum Press, New York (1988).

Rowland, M., Tozer T. M., *Clinical Pharmacokinetics: Concepts and Applications* (3rd ed.), Lea & Febiger, Philadelphia (1995).

Shargel, L., Yu, A. B. C., *Applied Biopharmaceutics and Pharmacokinetics* (4th ed.), McGraw-Hill/Appleton & Lange, New York (1999)

2.5 Population pharmacokinetics and pharmacodynamics

Alison H. Thomson

Pharmacy Department, Western Infirmary and Division of Cardiovascular and Medical Sciences, University of Glasgow, Western Infirmary, Glasgow G11 6NT, UK

2.5.1 INTRODUCTION

When designing drug dosage regimens for clinical use, one of the most important factors to be defined during drug development is how drugs are handled by the body, i.e. the pharmacokinetic parameters need to be characterised. Initial studies are typically performed in healthy subjects then, in later phases of drug development, studies are undertaken using patients with the underlying disease. Historically, there were limited opportunities for evaluating drug handling in infants and children and, consequently, paediatric drug dosage regimens were developed empirically from adult regimens by adjusting for weight or body surface area. Unfortunately, this approach fails to recognise age-related differences, both in drug handling and drug response, that might influence drug dosage requirements independently of "size". The "population approach" allows drug handling to be studied in paediatric patients with a technique that is less invasive and, consequently, has fewer ethical and practical limitations than conventional methodologies.

The concept of "population pharmacokinetics" was first introduced into clinical pharmacology in the late 1970s. Initial studies largely utilised data collected from routine therapeutic drug monitoring in "current patients" to develop dosage guidelines that could be used for "future" patients. This methodology differs from traditional pharmacokinetic data analysis procedures in which "rich" data from small groups of subjects (e.g. 10-20 blood samples from each of 10-20 patients) are analysed to obtain individual pharmacokinetic profiles and parameters. Although population techniques can utilise rich data, they are more commonly associated with "sparse" data, i.e. 2-3 blood samples collected from each of, say, 100 patients. This creates the possibility of undertaking studies in patients who would otherwise be excluded from research protocols, but might still receive a drug for therapeutic purposes. The population approach is therefore particularly applicable to the paediatric population, in which conventional studies are difficult to perform. This section will briefly describe the methodology used to undertake a population pharmacokinetic study, with illustrations from paediatric applications that have utilised this approach.

2.5.2 PHARMACOKINETICS

The relationship between the drug dosage regimen ("how much, how often") and the magnitude of the clinical response is fundamental to the successful treatment of disease. The study of pharmacokinetics focuses on the processes of absorption, distribution and elimination and provides a link between the profile of drug concentration over time and the prescribed dosage regimen. The principal pharmacokinetic parameters that determine these relationships are bioavailability (the fraction of the administered dose that reaches the systemic circulation), rate of absorption, volume of distribution and clearance.

Many differences in pharmacokinetic parameters between paediatric and adult patients have been identified: these differences are most extreme in neonates and infants but they are not always predictable. For example, the bioavailability of topically administered drugs, such as corticosteroids, may be enhanced in young infants and lead to an increased potential for toxicity. In contrast, the bioavailability of orally administered drugs is more variable, especially in newborn infants in whom gastrointestinal flora is different, intestinal and liver enzymes responsible for first pass drug metabolism are immature and both gastric emptying and gut motility are erratic. Differences in body composition, characterised by a higher proportion of body water to fat in young infants relative to adults, result in higher relative volumes of distribution of water-soluble drugs and, consequently, lower peak concentrations if the same weight-related dose is used. Concentration-time profiles are also influenced by the efficiency of drug removal from the body. Immaturity in metabolic capacity and age-related deficiencies in some metabolising enzymes can reduce clearance by hepatic metabolism. Similarly, immaturity of renal function, especially in premature neonates, leads to a low clearance of drugs eliminated by excretion in the urine.

In summary, it is often difficult to predict pharmacokinetic differences in paediatric patients and population studies may provide some of the answers.

2.5.3 PHARMACODYNAMICS

The variability in pharmacodynamics (drug response) between individuals is often as wide as or wider than the variability in pharmacokinetics. However, pharmacodynamic studies are often more difficult to conduct than pharmacokinetic studies. "Response" can take many forms: ranging from all or none, such as death or survival; through a graded response, such as partial, moderate or complete resolution of symptoms; to a continuous measurable response, such as respiratory peak flow rate or heart rate. A further complication is that medicines often elicit several "responses" – some are desirable, others are not.

Pharmacodynamic analysis aims to investigate the relationship between the drug dosage regimen and the response, both therapeutic and toxic. Concentration, rather than dose-effect relationships, may be examined and a combined pharmacokinetic-pharmacodynamic analysis may clarify the nature of such relationships. The dose or concentration-response relationship may be described by a simple linear relationship between the concentration of the drug in the body and the response or by a nonlinear relationship that includes a maximum response at higher concentrations. Furthermore, there may be a delay (or lag) between the maximum concentration and the maximum effect; this has sometimes erroneously led to the conclusion that no relationship exists between drug concentration and response. For example, response may be more closely related to the total exposure of the patient to the drug, as indicated by the area under the concentration-time curve, or to the time that the drug concentration remains above a threshold. Consequently, the dosage schedule itself influences the response to some drugs.

Concentration-effect relationships may differ in paediatric patients from those observed in adults. This may be due to differences in pharmacokinetic factors (such as lower protein binding) or altered responses (such as due to differences in receptor number and function). A further challenge is the actual measurement of response in infants and young children: often they are unable to articulate their symptoms, and, additionally, their clinical conditions can change very rapidly.

2.5.4 THE POPULATION APPROACH

Background

Differences in drug handling (pharmacokinetics) and pharmacological response (pharmacodynamics) offer powerful support for clinical pharmacological studies to be conducted in paediatric patients but, until recently, practical and ethical concerns have limited the involvement of paediatric patients in clinical trials. In addition, practical difficulties in taking multiple blood samples from a child and limitations in the volume of blood that can be withdrawn from an infant or neonate, have made traditional pharmacokinetic studies difficult to perform. However, increasing awareness of the need for paediatric clinical research has been prompted by the recognition by the FDA in the United States that medicines are routinely administered to children "off label", without clinical pharmacological data to support the dosage regimens used [1]. Since then, other regulatory authorities have focussed on paediatric patients [2] and have recommended the use of population approaches to help design drug dosage regimens for this group of patients [3].

Data requirements

The essence of population methodology is that data are obtained from a large number of patients. Of itself, the data collection may be "rich" (many samples per individual), or sparse (1-3 samples per individual), or a combination of rich and sparse sampling. Whatever the source, the data are analysed simultaneously, taking account of the correlations between samples collected within the same individual.

For pharmacokinetic studies, both single and multiple dose data are ideal, since this creates an opportunity to examine drug accumulation. In routine therapeutic drug monitoring and in observational studies, single or multiple "trough" concentrations of drugs are often measured. However, this information is only useful to a limited extent because troughs contain little information about drug absorption and distribution. The ideal protocol for a population study incorporating only sparse data, would be that samples have been obtained within different time "windows" that cover the full dosage interval. Random allocation of different sampling times to different patients is an alternative strategy.

The appropriate number of patients for a population study is difficult to determine and many "population" studies containing less than 30 subjects have been reported in the literature. These are not true "population" studies, but reflect the use of population methodology to analyse the data. The traditional view requires at least 50 subjects in the "population", but much higher numbers if the patients' characteristics are highly variable. For example, a study that included patients ranging from neonates to adolescents would need large numbers to ensure sufficient patients to cover the full age range, whereas smaller numbers may be adequate if the "population" only comprised neonates.

Although "routine" clinical data have been utilised in many population analyses, this approach has additional limitations because dosage and sampling details may not be recorded accurately. If dosage and sampling times are "assumed", this will add to the variability and may adversely influence the results. Additional data that also need to be collected include clinical information likely to influence drug handling, such as age, gestational age (for neonates), creatinine concentration, potentially interacting drugs, etc. A clear protocol, detailing the factors to be considered in the analysis, is fundamental to achieve successful identification of the clinical characteristics that influence dosage requirements.

Methodology

Development of a basic model

The aims of population analysis are to define the typical pharmacokinetic and / or pharmacodynamic parameters in the population of patients under investigation, along with the clinical factors that influence these parameters. Estimation of variabilities, both between patients and within patients, are also major goals of this approach.

The first phase of a population pharmacokinetic or pharmacodynamic study usually involves determining the most appropriate structural model to describe the data. Published literature, or the results of small, "rich" data studies, may provide starting estimates of the parameters and the likely models (one-compartment, two-compartment, linear, nonlinear, etc.). The sampling protocol will also influence the model, because sparse data rarely support complex models. However, parameters, such as absorption rate, can be fixed to previously determined values if appropriate.

These preliminary analyses may also try to define the best model for residual variability i.e. the error between the measured and the model-predicted concentrations. This residual variability includes assay error over the concentration range (which is described separately with some population pharmacokinetic packages), errors in sampling time, model misspecification and variability within an individual between different sampling times. Sometimes "inter-occasion variability" is also estimated, especially when more than one course of therapy has been administered. High inter-occasion variability would suggest that knowledge of how a patient handles a drug during one course of therapy, would not help to predict how they would handle the drug during a second course.

Covariate analysis

After the preliminary analyses are complete and the basic pharmacokinetic and error models have been identified, the next stage is to try to reduce interindividual variability by examining the influence of clinical factors ("covariates"), such as age, weight, renal function, hepatic function, other drug therapy and other disease states, on the pharmacokinetic parameters (drug clearance, volume of distribution, etc.). Current population software packages usually also provide individual estimates of these parameters for each patient. Potentially influential factors can then be identified by means of plots, multiple regression analysis or generalised additive modelling [4]. In paediatric patients, plots are often very useful, since the range of ages and weights can be very wide. For example, Figure 1 shows the relationship between individual estimates of vancomycin clearance and weight in a group of neonates, while Figure 2 illustrates the relationship between clearance and creatinine concentration [5].

Potentially relevant factors are then added to the model in a sequential fashion. Statistical comparisons of models, reduction in intra-individual variability and improvement in the prediction of measured concentrations are used to determine the best model. For example, in the vancomycin study, inter-individual variability in clearance fell from 67 to 22% when weight and serum creatinine were included in the model. A similar approach would be used if the data were dose and response or dose, concentration and response, but the structural models would be different.

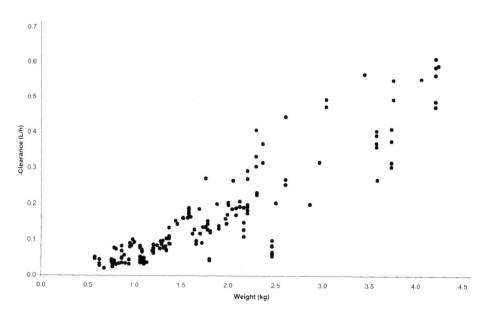

Figure 1. Plot of vancomycin clearance versus weight from a population analysis of vancomycin pharmacokinetics in neonates. See Grimsley, C., Thomson, A.H., *Arch. Dis. Child.*, *81*, F221 (1999).

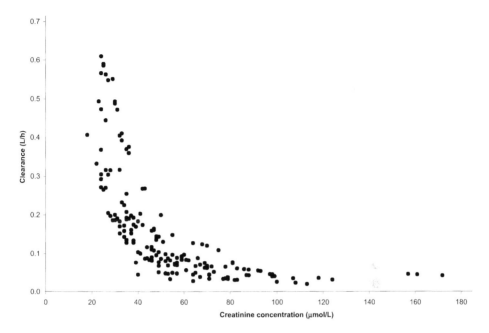

Figure 2. Plot of vancomycin clearance versus creatinine concentration from a population analysis of vancomycin pharmacokinetics in neonates. See Grimsley, C., Thomson, A.H., *Arch. Dis. Child., 81*, F221 (1999).

Paediatric patients: specific issues

Due to the high correlation between age and weight in paediatric patients, it may be difficult to distinguish between differences in drug handling that are simply related to size from those related to maturity. However, allometric approaches, which are used to predict drug handling in man from animal data, indicate that clearance is related to weight to the power 0.75, whereas volume of distribution is linearly related to size [6]. This indicates that, if the population under investigation has a wide age (and weight) range, as is often the case with paediatric studies, the modelling approach should consider both linear and non-linear relationships between clearance and weight.

Difficulties often arise in determining renal function in infants and young children and a number of equations have been published that allow estimation of glomerular filtration rate (GFR) from clinical characteristics. Two recent studies have used the population approach to determine a model for GFR in patients with renal disease [7] and patients with cancer [8]. Estimates of GFR are often used as covariates in population models for drugs that have a significant renal component to their elimination.

Software for population pharmacokinetic analysis

There is a range of software that has been used to perform population analyses, but some are more relevant to academic research and require sophisticated statistical and programming skills. Programs that have been more widely used include NONMEM [9], which has been used in the majority of published population studies, NPEM [10] and WinNonMix [11].

Outcomes of population analysis

The results of population analyses can be used in a number of ways. For example, to devise dosage regimens that achieve target concentration-time profiles for patients with a range of clinical characteristics; to describe the variability in concentrations (or response) that would be expected in response to a particular dosage regimen; to derive initial estimates for MAP Bayesian analysis in the area of therapeutic drug monitoring; or to help elucidate limited sampling protocols for future studies with sparse data [12]. Validation of results is an important consideration and is ideally performed by utilising the dosage regimens that emerge from the population analysis and then determining whether the expected concentrations or clinical outcomes are achieved (external validation). However, this is not always feasible and sometimes "internal" validation procedures have to be performed instead. A comprehensive review of population analysis and validation methodology is available at the FDA website [1].

2.5.5 PAEDIATRIC APPLICATIONS OF POPULATION PHARMACOKINETIC ANALYSIS

Introduction

The application of population analysis to paediatric research expanded rapidly during the 1990s, as the pharmaceutical industry began to develop paediatric research programmes. In many cases, the use of the population approach in paediatric patients has mirrored its application in adults. Early studies utilised data collected in the course of routine drug monitoring, while more recent studies have examined data collected during drug development or clinical research. The following are illustrative examples of some of the results to date.

Antimicrobial agents

As data are readily available through routine therapeutic drug monitoring, population studies on the aminoglycoside antibiotics have proved popular. Dosage recommendations arising from these studies have changed over the years on account of the increasing survival of very premature neonates and the greater emphasis on high peak and low trough concentrations. Although the majority of studies have focussed on neonates, more recent research, especially with the newer aminoglycosides, has included wider age ranges.

The quinolone derivative, ciprofloxacin, has also been subjected to population analysis. However, two studies conducted in paediatric patients came to slightly different conclusions [12, 13]. Both groups identified an influence of age on clearance, but one found that weight was more important [13]. Similarly, one group found that a diagnosis of cystic fibrosis increased clearance [12], whereas the other only found an influence of cystic fibrosis on absorption rate [13]. These results illustrate that the results of a population analysis depend on the data available and the modelling approach used.

Paediatric use of antimalarials is high in certain countries and population analyses have been performed with several drugs including quinine, mefloquine, artemisinin and atovaquone. Of interest is one study that examined proguanil pharmacokinetics in a large group of patients with a range of ethnic backgrounds. Patients were categorised as "poor" or "extensive" metabolisers and, not surprisingly, this factor proved to be a major determinant of clearance [14]. This study illustrates the value of having data on phenotype (or possibly genotype) for drugs whose metabolism is genetically controlled.

Antiviral agents

With the success of antiretroviral therapy in adults, population approaches have been used to help determine the appropriate dosage regimens for the treatment of HIV infections in infants and children. Some of these studies have also examined pharmacokinetic-pharmacodynamic relationships. For example, an attempt to link enfuvirtide exposure to viral clearance proved unsuccessful [15], whereas a direct link between drug concentrations and haematological toxicity has been identified for zidovudine [16].

Anticonvulsants

As therapeutic drug monitoring has been used for many years to help optimise anticonvulsant therapy, there is extensive literature on the pharmacokinetics and dose requirements of most anticonvulsants, especially phenytoin, carbamazepine, phenobarbitone and valproic acid, in paediatric patients [17,18]. Some studies have employed population methodology to analyse routine clinical data, while others used conventional techniques and more formal research protocols.

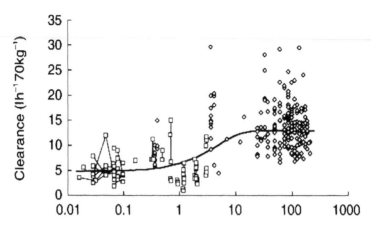

Figure 3. Plot of paracetamol clearance versus Ln age from a population analysis of paracetamol pharmacokinetics in infants and children. See Anderson, B.J., Woollard, G.A., Holford, N.H., *Br. J. Clin. Pharmacol., 50*, 125 (2000).

Anaesthetic agents and analgesics

Midazolam concentrations, generated from administration by infusion, have been examined in population pharmacokinetic studies conducted in critically ill neonates and young children [19,20]. These studies found relationships between midazolam clearance and weight, age, renal function, hepatic function and other drug therapy. Other anaesthetic agents that have been investigated include fentanyl, alfentanyl and propofol and, in some cases, the results have been used to devise models for use in computer controlled infusion devices.

Figure 3 illustrates the changes in paracetamol clearance with age that were identified when an allometric approach was used to analyse data from patients who ranged in age from neonates to young adults [21].

A further extension of this work related analgesic response to paracetamol concentrations and identified a dosage regimen that was associated with a satisfactory clinical response in patients undergoing tonsillectomy [22]. A similar approach has also been used to examine morphine pharmacokinetics and to devise a dosage regimen that would achieve concentrations associated with pain control in patients with cancer [23].

Treatment of respiratory disease

Theophylline has been used for many years to treat bronchospasm in children and to stimulate respiration in premature neonates with apnoea. Routine monitoring of theophylline concentrations has generated data suitable for population analysis and many studies have been performed in a range of ethnic groups. Not just the clinical characteristics of the patients themselves may be important, but also the interventions used, as illustrated by a study that examined theophylline handling in patients receiving extracorporeal membrane oxygenation (ECMO). A lower theophylline clearance and higher volume of distribution was identified in neonates and children who received this treatment [24]. More recently, caffeine has replaced theophylline for the treatment of premature apnoea; population studies have proved useful in confirming optimal dosage regimens of this drug.

The application of population analysis to drug development has been demonstrated in a study which found that a 4 mg dose of montelukast given to asthmatic children aged 2-5 years achieved an area under the concentration-time curve consistent with that obtained by a 10 mg dose in adults [25].

Drugs used in cancer chemotherapy

Deciding on the dose of a chemotherapeutic agent is particularly difficult, since both underdosing and overdosing can have life-threatening consequences. Accordingly, this area has received particular attention for paediatric population research. Initial focus has been on drugs that are easily measured, such as methotrexate, and other drugs for which individualisation of dosage improves outcome, such as carboplatin and busulphan. However, more recent studies have examined a wide range of drugs, including daunorubicin, prednisolone, etoposide, ifosfamide, cytosine arabinoside and temozolomide. In contrast to what has been observed in adults, body surface area may be more descriptive of clearance in a paediatric population and has proved superior to weight for a number of drugs (although non-linear weight models may not always have been considered). It is likely that this is an area that will expand considerably in the future.

Gastrointestinal disease

A range of drugs used in gastrointestinal disease, including ranitidine, famotidine, ondansetron and cisapride have been examined using population approaches in paediatric patients. Both pharmacokinetic and pharmacodynamic data have been analysed, as illustrated by a study that described the nonlinear pharmacodynamic relationship between serum concentrations of famotidine and gastric pH [26].

Other drugs

Some drugs, such as indomethacin, imipramine and methylphenidate, have indications that are specific to children, whereas for others, such as sotolol, the concentration-effect relationship may differ from than observed in adults. Population analysis has proved valuable for these drugs in quantifying variability and helping to elucidate the relationships between drug concentration and effect.

2.5.6 CONCLUSIONS

The population approach is an applied research technique that can be used to define the typical pharmacokinetic parameters that control the relationships between the drug dosage regimen and the concentration-time profile and also the pharmacodynamic parameters that relate drug concentration measurements to response. It aims to quantify variability in these parameters and to identify factors that influence these relationships. This methodology is particularly applicable to the paediatric population because it can handle both sparse and rich data, thereby reducing some of the practical and ethical limitations of paediatric research.

ACKNOWLEDGEMENTS

The author would like to thank Dr Henry Elliott for helpful comments on this chapter.

REFERENCES

1. http://www.fda.gov/cder/guidance.
2. http://www.emea.eu.int.
3. Tett, S. E., Holford, N. H. G., McLachlan, A. J., *Drug Inf. J., 32*, 693 (1998).
4. Mandema, J. W., Verotta, D., Sheiner L. B., *J. Pharmacokinet. Biopharm. 20*, 511 (1992).
5. Grimsley, C., Thomson, A. H., *Arch. Dis Child. Fetal Neonatal Ed., 81*, F221 (1999).
6. Holford, N. H. G., *Clin. Pharmacokinet., 30*, 329 (1996).
7. Léger, F., Bouissou, F., Coulais, Y., Tafani, M., Chatelut, E., *Pediatr. Nephrol., 17*, 903 (2002).
8. Cole, M., Price, L., Parry, A., et al., *Br. J. Cancer, 90*, 60 (2004).
9. Beal, S. L., Sheiner, L. B., NONMEM (User's Guide), parts 1-VII. Technical Report, University of California, San Francisco (1992).
10. USC*PACK PC Pharmacokinetic Programs, USC Laboratory of Applied Pharmacokinetics, Los Angeles, California, http://www.lapk.org.
11. WinNonMix, Pharsight Corporation, Mountain View, California USA (http://www.pharsight.com).
12. Payen, S., Serreau, R., Munck, A., et al., *Antimicrob. Agents Chemother., 47*, 3170 (2003).
13. Rajagopalan, P., Gastonguay, M. R., *J. Clin. Pharmacol., 43*, 698 (2003).
14. Hussain, Z., Eaves, C. J., Hutchinson, D. B., Canfield, C. J., *Br. J. Clin. Pharmacol., 42*, 589 (1996).
15. Soy, D., Aweeka, F. T., Church, J. A., et al., *Clin. Pharmacol. Ther., 74*, 569 (2003).
16. Capparelli, E. V., Englund, J. A., Connor, J. D., et al., *J. Clin. Pharmacol., 43*, 133 (2003).
17. Battino D., Estienne, M., Avanzini, G., *Clin. Pharmacokinet., 29*, 257 (1995).

18. Battino, D., Estienne, M., Avanzini, G., *Clin. Pharmacokinet.*, *29*, 341 (1995).
19. Burtin, P., Jacqz-Aigrain, E., Girard, P., et al., *Clin. Pharmacol. Ther, 56*, 615 (1994).
20. de Wildt, S. N., de Hoog, M, Vinks A. A., van der Giesen, E., van den Anker, J. N. *Crit. Care Med., 31*, 1952 (2003).
21. Anderson, B. J., Woollard, G. A., Holford, N. H. G., *Br. J. Clin. Pharmacol., 50*, 125 (2000).
22. Anderson, B. J., Woollard, G. A., Holford, N. H. G., *Eur. J. Clin. Pharmacol., 57*, 559 (2001).
23. Hunt, A., Joel, S., Dick, G., Goldman, A., *J. Pediatr., 135*, 47 (1999).
24. Mulla, H., Nabi, F., Nichani, S., Lawson, G., Firmin, R.K., Upton, D.R., *Br. J. Clin. Pharmacol. 55*, 23 (2003)
25. Knorr, B., Nguyen, H. H., Kearns, G. L., et al., *J. Clin. Pharmacol., 41*, 612 (2001).
26. James, L. P., Marshall, J. D., Heulitt, M. J., Wells, T. G., Letzig, L., Kearns, G. L., *J. Clin. Pharmacol., 36*, 48 (1996).

2.6 Pharmacogenetics

Evelyne Jacqz-Aigrain[1], Anders Rane[2]

[1]*Department of Paediatric Pharmacology and Pharmacogenetics,
Robert Debré Hospital, Paris, France*

[2]*Department of Laboratory Medicine, Division of Clinical Pharmacology,
Karolinska Institute, Karolinska University Hospital, Stockholm, Sweden*

Intra- and inter-individual variations in drug metabolism and disposition are central to differences in the therapeutic response to a standard dosage regimen. In adults, variability is the result of complex interactions between genetic determinants and the environment. This explains why drug dosage must be adjusted to each patient in order to obtain the required therapeutic effect, while reducing the incidence of adverse effects. Adjusting dosage is particularly important when drugs have a narrow therapeutic range and metabolic variability is high. In children, the variability of drug disposition is even more complex, as the expression of genetic and environmental factors is modified by the impact of the physiological development /maturation [1].

2.6.1 DEFINITION

The history of pharmacogenetics started in the 1930s (Table 1). Pharmacogenetics has been defined as "…a monogenic trait caused by the presence in the same population of more than one allele at the same gene locus and more than one phenotype regarding drug interaction with the organism, with the frequency of the less common allele being greater than 1%." [2]. In recent years, the term pharmacogenomics has been used interchangeably with pharmacogenetics.

Such polymorphisms generally remain undetected in the absence of drug intake. They may be characterised by the genotype, referring to an individual's genetic constitution and/or the phenotype referring to the impact of environmental factors on the expression of an individual's genotype. In the field of drug metabolism, different phenotypic groups are commonly denoted as slow or fast metabolisers.

Table 1

The early history of pharmacogenetics

- 1932 : First inherited difference in an inherited response to a chemical : inability to taste phenylthiourea [71]

- 1957 : "Inheritance might explain many individual differences in the efficacy of drugs and in the occurrence of adverse drug reactions"

- World War II : Haemolysis in African –American soldiers treated with primaquine

- 1959 : "Pharmacogenetics : the role of genetics in drug response" [72]

- 1960 : Genetic influence on isoniazid blood concentrations [31]

- 1964 : Genetic variation in ethanol metabolism

- 1977 : CYP2D6 polymorphism [5-8]

More recently, polymorphisms of ABC transporters and various target molecules, such as receptors, has become an emerging research field within pharmacogenetics.

2.6.2 POLYMORPHISMS OF DRUG METABOLISING ENZYMES

Drug biotransformation

Most drugs are lipophilic and metabolised into hydrophilic compounds which are eliminated through the kidneys. Biotransformation reactions are usually classified into phase 1 and phase 2 reactions, in various combinations depending on the drug concerned.

Phase 1 reactions generally introduce a functional group (by oxidation, reduction, or hydrolysis) to the drug. They depend primarily on cytochromes P450, a superfamily of haem-containing enzyme proteins. The cytochromes P450 are mostly located in the liver and the intestine, but are also present in many other organs. At least fifty different cytochromes P450 have been characterised in humans. The accepted nomenclature based on amino acid sequence homology, identifies the members of the human cytochromes P450 superfamilies as numbers 1, 2, 3, 4, etc. added to the letters CYP (corresponding to the CYP families involved in drug metabolism), a letter (A to F) identifying the subfamily, and an Arabic number designating the isoenzyme. These cytochromes P450 have a high, but overlapping, substrate specificity. In adults, the predominant hepatic cytochrome is CYP3A4. All (except CYP2D6) are inducible, induction being dose dependent and reversible. Phase 2 enzymes catalyse conjugations with endogenous substrates (e.g. acetate, sulphate, glucuronic acid) which increases the polarity and facilitates renal excretion.

Many enzymes operating in the metabolism of drugs are subject to genetic polymorphisms. For most of them, the molecular basis responsible for the allelic variants with compromised function is known, so that it is possible to determine individual genotypes. By measuring the *in vivo* metabolism of drug substrates specific to the polymorphic enzyme activity, it is possible to identify different phenotypes, e.g., fast and slow metabolisers.

The therapeutic consequences depend on the importance of the deficient enzyme for the overall metabolism of the drug and/or the influence on the metabolism pattern. In poor metabolisers, the accumulation of the drug may be associated with excessive therapeutic or toxic effects. In fast metabolisers the therapeutic outcome will depend on the possible formation of active and/or toxic metabolites. In fast metabolisers, excessive metabolism is a result causing risk of undertreatment and insufficient clinical efficacy. Disequilibrium between phase 1 and phase 2 reactions may account for a high risk of adverse consequences including induction of cancer or immunotoxicity.

The pattern of maturation of various metabolic activities differs depending on the enzyme [3], with central impact of age on the phenotypic expression of the genotype. This is the reason why the mutual relation between different phenotypes in children may differ from that described in adults [4].

Pharmacogenetics of drug metabolising enzymes

Cytochromes P450

CYP2D6. This polymorphism has been thoroughly explored and was discovered following therapeutic accidents occurring with the use of debrisoquine for hypertensive adults [5]. More than 50 mutations and 70 different "poor metaboliser" alleles have been described, with large ethnic differences in frequency [6]: 7 to 10% of Caucasians and 1 to 2% of Asians are poor metabolisers, while 2 to 3 % are ultrarapid metabolisers, due to gene duplications or gene multiplications [7].

This polymorphism is inherited through an autosomal recessive gene. Homozygous individuals are characterised by negligible or no metabolism of a variety of drugs [8]. Several β-adrenoceptor-blocking agents, antidepressants, and antiarrythmic drugs belong to this group. Using dextromethorphan as a marker for CYP2D6 activity, the ontogenesis of the enzyme activity was examined in hepatic tissue samples from human fetuses, newborns and children [9]. CYP2D6 activity was not detectable in the fetus, increased rapidly during the postnatal period, independently of gestational age, but remained low during the first month of life (about 20% of adults levels). These data are consistent with the lack of CYP2D6 mediated O-demethylation despite catalytic CYP3A mediated N-demethylation of codeine [3] in human fetal liver. Phenotypic studies conducted in small populations of children, have shown that the adult phenotypic distribution pattern was attained at 3 years of age [10].

CYP2D6 is implicated in the metabolism of more than 50 drugs, including antiarrhythmic drugs, beta-blockers, antidepressants and antipsychotic drugs. In most cases, the clinical consequence of poor CYP2D6 activity is drug accumulation and toxicity. In children, the number of drugs dependent on CYP2D6 is limited. However, codeine and tramadol are metabolised to active compounds (morphine and O-desmethyl tramadol, respectively) and the analgesic efficacy of these drugs is low in patients with the slow metaboliser (developmental or pharmacogenetic) phenotype [11].

CYP2C19. This polymorphism is inherited through an autosomal recessive gene. Homozygous individuals are characterised by negligible or no metabolism of a variety of drugs, such as S-mephenytoin, citalopram, diazepam, omeprazol, lanzoprazole, etc. Three to 5% of Caucasians and more than 20% of Orientals are deficient in CYP2C19 and mephenytoin hydroxylase activity and they are unable to metabolize S-mephenytoin to 4-hydroxy mephenytoin [12]. At the genotype level, more than five defective alleles with varying ethnic distribution have been identified. As a consequence of poor CYP2C19 metabolism, substrate accumulation may occur. Although the polymorphism may be of clinical importance in certain situations in paediatric pharmacotherapy, very little is known about the clinical implications in children [13].

CYP2C9. The gene of CYP2C9 has at least three important allelic variants denoted as CYP2C9*1 (which corresponds to the reference gene), CYP2C9*2, and CYP2C9*3. CYP2C9 is responsible for the oxidative metabolism of widely used drugs, such as anticoagulants (warfarin), non-steroidal antiinflammatory drugs (indomethacin, diclofenac, celecoxib etc.), phenytoin, tolbutamide, angiotensin receptor antagonists, etc. Therefore, polymorphic metabolism catalysed by this enzyme may have clinical implications, predominantly in the treatment of cardiovascular disease and epilepsy.

CYP2C9 is not present in fetal liver. Adult activity is reached by 6 months of age. By 3 to 4 years of age, the enzyme activity generally exceeds that of adults.

CYP3A subfamily [14]. The CYP3A subfamily is the predominant cytochrome P450 subfamily, accounting for 30% of the cytochromes in the liver

[15] and 70% of the cytochromes in the small intestine [15,16]. The members of this subfamily are involved in the metabolism of more than 70% of currently used drugs. The CYP3A subfamily comprises three major isoforms: CYP3A4, CYP3A5, CYP3A7 and a recently identified CYP3A43. They are very closely related, as they share at least 85% amino acid sequence.

The CYP3A4 enzyme is the predominant form in adults, identified both in the liver and the gut [17]. It is variably expressed [16,18,19], but none of the identified genetic polymorphisms seem to have any functional correlate. CYP3A4 enzymes are important for the metabolism of a large number of commonly used drugs within the groups of antiepileptics, immunosuppressants, cytotoxics, antibiotics etc. As the enzyme is frequently subject to induction and inhibition, there are possible clinical implications [20]. CYP3A5 is also present in the liver, but is predominantly an extrahepatic form of the CYP3A enzymes. CYP3A5 mRNA is present in the intestine and can be detected using sensitive methods such as real time quantitative RT-PCR [21,22].The variability of CYP3A5 expression is under genetic control. Individuals with two CYP3A5*1 alleles have high expression and up to 10 fold increase in CYP3A5 protein. This is observed in 10 to 30% of livers in Caucasians [23,24]. The contribution of CYP3A5 to drug metabolism is important, as CYP3A5 when it is expressed can contribute to over 50% of the total CYP3A content.

CYP3A7 is the major CYP isoform detected in embryonic, fetal and newborn liver [25]. It may be expressed at a low level in the adult liver [26] but seems to be absent in the adult intestine. A fourth isoform, CYP3A43, has been detected at very low levels, i.e. below 0.1% of the total CYP3A content in the adult liver. This isoform seems to have only negligible or no metabolic activity.

CYP1A2. The CYP1A2 phenotype was initially studied using caffeine as a probe. More recently, it was demonstrated that CYP1A2 protein is absent in microsomes prepared from fetal and neonatal livers. Its level increases in infants 1-3 months to attain 50% of adult values at one year of age. Methoxyresorufine demethylase and imipramine demethylation are partially metabolised by CYP1A2 and follow the same profile of ontogenic development.

Additional phase 1 enzymes are listed in Table 2.

Conjugation reactions

Gluthathione S-tranferases (GST). GSTs constitute a large family of enzymes implicated in the detoxification of reactive chemical moieties by conjugation with reduced gluthatione. GSTs are highly polymorphic and this variability results from GST duplications, deletions and/or promotor/coding gene mutations. A complete deletion of GST-M1 and GST-T1 genes is evidenced in 50%

Table 2

Major phase-1 and phase-2 enzymes (adapted from Leeder 1997)

Enzyme	Substrate examples	CYP inhibitors	CYP inducers
CYP1A1	polycyclic hydrocarbons		
CYP1A2	caffeine carbamazepine phenacetin theophylline	fluvoxamine	cigarette smoke charcoal broiled-meat
CYP2A6	paracetamol nicotine		barbiturates
CYP2B6	cyclophosphamide ifosfamide	qunidine fluoxetine	
CYP2C8	diazepam diclofenac tolbutamide		
CYP2C9*	S-warfarin phenytoin losartan tolbutamide ibuprofen naproxen		rifampicin
CYP2C19*	omeprazole diazepam S-mephenytoin		rifampicin
CYP2D6*	codeine ethylmorphine tricyclic antidepressants haloperidol antiarrhythmic drugs perphenazine perhexiline serotonin reuptake inhibitors zuclopenthixol S-mianserine tolterodine	quinine quinidine	
CYP2E1	ethanol paracetamol	disulfiram	ethanol isoniazid

Table 2 (continued)

Enzyme	Substrate examples	CYP inhibitors	CYP inducers
CYP3A4	ciclosporin clomipramine cortisol erythromycin nifedipine tacrolimus	erythromycin ketoconazole grape fruit juice	carbamazepine corticosteroids phenobarbital phenytoin rifampicin
CYP3A7	dehydroepiandro-sterone ethinylestradiol		carbamazepine rifampicin
NAT	caffeine dapsone isoniazid sulfametoxazole		
TPMT	6-mercaptopurine 6-thioguanine		
UDPGTs	chloramphenicol diclofenac ketoprofen morphine paracetamol		cigarette smoke phenobarbital

CYP *: polymorphic enzyme

and 25% of Caucasians, respectively [27,28]. Additional GST polymorphisms affecting GST-P1 and GST A have been reported. GST polymorphisms are implicated in the risk of *de novo* cancer and cancers secondary to chemotherapy, in decreased drug efficacy, and in relapse of malignant conditions [29,30].

N-acetyltransferase type 2 (NAT2). NAT2 activity displays genetic polymorphism mediated by as an autosomal recessive trait [31]. Its distribution is extremely variable from one ethnic group to another. In Caucasian populations, 50 to 70% of individuals are slow acetylators, whereas the percentage is only 5% in Eskimo populations, and more than 90% in Egyptians [32].

Ontogeny of NAT2 has been studied *in vivo* and *in vitro*. During fetal life, NAT2 activity becomes detectable at 16 weeks of pregnancy. After birth, studies using caffeine have demonstrated that all children are slow metabolisers up to about the age of 2 months, after which the proportion of fast metaboliser phenotype is successively increased. At about 3 years of age, phenotypic distribution is similar to that found in adults [33].

Although data may be contradictory, several studies have suggested an association between the appearance of hypersensitive reactions to a combination of trimethoprim-sulfamethoxazole and the individual slow metaboliser genotype, in both adults and children [34,35].

Thiopurine methyltransferase (TPMT). TPMT is a cytosolic enzyme metabolising thiopurine drugs. The phenotypic distribution of TPMT activity may be measured in red blood and is trimodal in healthy adult volunteers: 89% have high activity, 11% have an intermediate activity, and approximately 1 individual in 300 has severely compromised enzyme activity. This polymorphism is inherited as an autosomal recessive trait [36]. Three major point mutations in the TPMT gene (localised in chromosome 6p22.3) are responsible for the low TPMT activity and identified in the genotyping process.

In vivo, the thiopurine pro-drug 6-mercaptopurine (6-MP), is metabolised by three competing pathways. The bioactivation pathway is catalysed by hypoxanthine guanine phosphoribosyl transferase into active 6-thiopurine nucleotides, mainly 6-thioguanine nucleotides (6-TGN), which can damage DNA. The first inactivation pathway is oxidation by xanthine oxidase. The second inactivation pathway for 6-MP and 6-thiopurine nucleotides is S-methylation catalysed by TPMT [37].

The TPMT polymorphism is of great importance in the treatment of acute childhood lymphoblastic leukaemia [38,39] and in certain inflammatory diseases [40]. Patients deficient in TPMT activity are at very high risk of developing myelosuppression when treated with standard doses of 6-mercaptopurine [41]. In children with ALL having detectable TPMT activity, 6-TGN concentrations were negatively correlated with TPMT activity and were predictive of better outcome [38].

The data related to the developmental aspects of TPMT activity are very limited. TPMT seems to be present in fetal livers at one third the level in adults. In newborn infants, results of TPMT activity has been reported to be either at least 50% higher [42] or similar to adults [43].

5'-Uridine-diphosphate-glucuronosyl transferases (UDPGTs). The UDPGTs constitute a large superfamily of at least 10 different enzymes that have the ability to conjugate their substrates with glucuronic acid. UDPGTs are involved in the metabolism of endogenous substrates (bilirubin and ethinyl estradiol) and xenobiotics, including drugs such as morphine and paracetamol. The UDPGT1 gene is important in the glucuronidation of bilirubin and decreased activity has been demonstrated in patients with Crigler-Naijar or Gilbert's syndrome [44]. Additional pharmacogenetic polymorphisms have been reported for the UDPGTs [14]. Serious adverse reactions caused by immature glucuronidation capacity were described in neonates receiving chloramphenicol. However, data regarding the substrate specificity and the developmental aspects of the different UDPGT isoforms are sparse [14].

2.6.3   POLYMORPHISM OF ABC TRANSPORTERS

The ATP binding cassette (ABC) family comprises at least eight multi-drug resistance-associated proteins. They have a central role in the absorption, distribution and elimination of many drugs. P-glycoprotein (P-gp) is a transmembrane ATP-dependent efflux pump. It is encoded by the *MDR1* (ABCB1) gene [45] and present in many tissues such as liver, kidney and blood brain barrier. In the gastrointestinal tract, its location in the brush border of the apical surface of mature enterocytes in the small intestine contributes to the chemical protection of the organism. Its expression is genetically controlled and *MDR1* expression is correlated with a mutation in exon 26 (C3435CT) of the *MDR1* gene. Substrates of P-glycoprotein include glucocorticoids, anticancer drugs, immunosuppressants, HIV1 protease inhibitors and many other drugs. As an example, digoxin pharmacokinetics is affected by the C3435T polymorphism in exon 26 [46], the T variant rendering digoxin a higher bioavailability due to deficient pump function. Polymorphisms of additional transport proteins (MRP1 to MRP5) have been reported, but their clinical relevance remains under evaluation [46].

2.6.4   POLYMORPHISM OF TARGET PROTEINS

Target proteins can be either a receptor, an enzyme, or another type of protein. Genetic polymorphisms affecting these drug targets can contribute to the pathogenesis of the disease and modify the response to specific medications in children. As an example, polymorphisms of the β2-adrenergic receptor (ADRB2) have been implicated in the response to β2-agonists in patients with asthma [47]. Polymorphisms in the promotor region (variable number of tandem repeats) affecting ALOX5 gene expression have been associated with the response to inhibitors of ALOX5 [48]. A number of additional polymorphisms of potential clinical importance were described in adults (for review see [49]) (Table 3).

2.6.5   METHODS IN PHARMACOGENETICS

At the gene level, pharmacogenetic alterations leading to a poor metaboliser allele may include: 1) partial or total gene deletion, insertion or duplication (as described for the CYP2D6 alleles), 2) SNPs (Single Nucleotide Polymorphisms) affecting the coding region, the perigenic region, or the non-coding region. The larger the gene, the larger the number of expected SNPs. On average, about four

Table 3

Drug target genetic polymorphisms with potential implications for drug effects in children

Drug target	Therapeutic consequences
β-adrenergic receptors	sensitivity to β-agonists in asthma
angiotensin converting enzyme	sensitivity to angiotensin converting enzyme inhibitors
angiotensin II type 1 receptor	vascular response to phenylephrine
5-hydroxytryptamine receptor	response to neuroleptics
sulfonylurea receptor	response to sulfonylurea hypoglycemic agents
dopamine D2 receptor	drug induced tardive dyskinesia

SNPs of functional importance are expected per gene. Such SNPs are generally located in the coding region and associated with an amino acid change, i.e. non-synonymous [50].

The functional significance of all the pharmacogenetic variations identified through rapid-throughput sequencing requires demonstration of a correlation between genotype and phenotype, either functionally or by powerful statistical approaches [51,52]. Indeed, the majority of the SNPs do not have any clinical importance, depending primarily on the role of the polymorphic trait and the therapeutic index.

2.6.6   IMPLICATIONS OF PHARMACOGENETICS IN THERAPEUTICS

Impact on pharmacokinetics

The disposition of drugs entering the body may be divided into absorption, distribution, metabolism, and elimination. Most of the pharmacogenetic polymorphisms identified so far affect drug metabolism, as reflected by lower metabolic clearance and prolonged elimination half-life of many drugs in deficient metaboliser phenotypes.

Pharmacogenetic differences in absorption are less commonly described. Vitamine B12 malabsorption may be classified as a pharmacogenetic difference in uptake. More recently, genetic variation in P-gp expression in the intestine was described. In addition, the CYP3A/P-gp complex in the enterocytes share common substrates (immunosuppressant drugs such as ciclosporine or FK506, anticancer drugs such as etoposide, vincristine, antiretroviral drugs) and their coordinated action may results in a reduced bioavailability of drugs administered orally by local metabolism and extrusion into the intestinal lumen, respectively [46].

Pharmacogenetics of protein binding and excretion have been sparsely studied, although aminoacidurias (cystinurias) might be regarded as pharmacogenetic differences in excretion [29].

Pharmacogenetics and cancer therapy in children

As many genetic determinants affect the patient's response and survival, pharmacogenetics gives promise for optimisation of cancer therapy. Recent studies have identified correlations between response/toxicity to anticancer drugs and pharmacogenetic traits (Table 4) [53,54]. In addition, progress in understanding the molecular mechanisns of oncogenesis and in targeting the genetic alterations of cancer cells will allow the discovery of new cancer drugs and help to identify patients who will benefit from individualised therapy [55].

Table 4

Pharmacogenetics of anticancer drugs in children

Polymorphic enzyme	Anticancer drugs	Therapeutic consequences	References
TPMT thiopurine methyl transferase	6-mercaptopurine	myelotoxicity in deficient individuals; 6-MP dosage individualisation; secondary cancer	38 41
DPD dihydropyridine dehydrogenase	5-fluorouracil	myelosuppression; neurotoxicity	73 74
UDPGT 1A1	irinotecan	myelosuppression; diarrhea	75
MTHFR methylene tetrahydrofolate reductase	methotrexate	risk of mucositis	76
TS thymidylate synthase	cyclophosphamide doxorubicin	conflicting results regarding relapse rate in ALL children	77
GST gluthathione S-transferase	busulfan	outcome of chemotherapy	78 79

Table 5

Pharmacogenetics and adverse drug reactions

Enzyme	Drug	Adverse effect	References
CYP1A2	antipsychotics	dyskinesia	80
CYP2C9	warfarin tolbutamide	haemorrhage; hypoglycaemia	81
CYP2C19	diazepam	prolonged sedation	13
CYP2D6	propafenone perhexiline	higher risk of ADRs in UDPGT1A1 deficiency; hepatotoxicity	82,83
DHPD	5-fluorouracil	myelosuppression	74
UDPGT1A1	irinotecan	diarrhoea; myelosuppression	75

Pharmacogenetics and adverse drug reactions, including teratogenic effects

Adverse drug reactions are a significant cause of morbidity and mortality in children. According to recent reviews,
- 9% of hospitalised children experience an ADR;
- 2% of paediatric admissions are caused by one or several ADRs [56];
- 25% of prescriptions involved in ADRs are off-label prescriptions [57];

In a prospective pharmacovigilance survey involving 1419 children, off-label prescriptions were at high risk of being associated with ADRs (relative risk 3.44, CI: 1.26-9.38) [58]. In a recent review by Philips *et al* [59], 59% of drugs in the ADR studies were metabolised by at least one polymorphic enzyme in comparison to 22% of randomly selected drugs.

ADRs may be secondary to:
- accumulation of the parent drug ;
- formation of toxic metabolites;
- Immune-mediated ADR, most often associated with the HLA (human leukocyte antigen) [60].

Numerous examples of genetic predisposition to ADRs exist in the literature, some of which are presented in Table 5 [61,62].

Similarly, polymorphisms of a wide variety of genes have been implicated in the susceptibility to obstetrical diseases (pregnancy loss, pre-eclampsia) and fetal malformations [63-65].

2.6.7 PROSPECTS IN PHARMACOGENETICS IN DRUG TREATMENT OF CHILDREN

From genotype to phenotype

The human genome is now known to contain about 30 000 identified genes and some 3.5 billion base pairs [66,68].

There are great expectations about the impact of this new knowledge on the possibilities to identify and treat diseases. Proteomics, the science about the pattern and function of gene products i.e. an enzyme, a receptor, a transporter protein etc., will be very important in the future. However, the *in silico* prediction of the function is not in itself sufficient, and complementary *in vitro* tests are mandatory. Ultimately, the function of these targets must be tested *in vivo*, in intact cells, or in the intact organism. For drugs for paediatric use, this scenario must be preceded by development of drugs for adult patients, unless the disease is unique for the paediatric age group.

The drug industry has already invested in the expectations to find new drug targets that are directly or indirectly related to disease or disease treatment. However, so far, no examples of drugs exist to describe the usefulness of this approach.

From phenotype to genotype

The identification of outliers in pharmacological studies of drug response or adverse reactions will remain a useful approach to discover genes of clinical importance, e.g. drug metabolising enzymes, receptor systems etc. [49]. Such observations have often led to discoveries of new pharmacological strategies, principles and polymorphisms [5,69]. This emphasises the need for reproducible and reliable clinical endpoints that can be related to findings in the genes of relevance for the symptoms observed.

Studies of monogenic traits, for example in the field of drug metabolism, are relatively easy, since the drug dispositional parameters serve as functional correlates rather than a symptom or a disease. In general, the outcome of therapy and many disorders are influenced by several genes, as well as by ill-understood effects of the environment. The relative rarity of paediatric diseases makes it difficult to collect sufficient number of patients for data analysis. This puts in a claim for new approaches, e.g. by IT, to capture large patient materials in a limitless and paperless way. Only then, will it be possible to find outliers and to look for associations between the deviating phenotypes and specific SNPs or haplotypes. Obviously, the clinical endpoints, whether they be related to drug response, non-response, ADRs etc, need to be robust and well defined. This is a critical, often neglected, but limiting factor in functional pharmacogenomics on a clinical level. It is even more important in paediatric drug therapy.

Advanced and efficient technology for screening of large numbers of SNPs is now available and includes DNA microarrays, pyrosequencing techniques, etc. It is mandatory to involve bioinformatics, which is a discipline that has evolved in response to the need of handling and interpretation of the massive amount of data that is generated.

Safety is one of the mainstays in drug therapy. The dimension of the problem of ADRs in drug therapy has recently been highlighted [70]. In the United States, ADRs constitute the fifth to sixth most important cause of death (>100 000 in 1994). There are reasons to believe that this is also a significant problem in drug treatment of children and infants. The aim of the scientific and medical community is to identify patients at risk of adverse reactions to drug treatment. Functional pharmacogenomics, at the clinical level, will be important to identify known genetic variants of relevance for safety. New drugs with undesirable or unacceptable ADR profiles are seldom accepted, no matter how effective they are.

2.6.8 CONCLUSIONS

The majority of hitherto characterised pharmacogenetic polymorphisms affect drug metabolism, although new areas are now being opened with the identification of pharmacogenetic polymorphisms in transporter proteins and receptor functions. The physiological development and maturation of the child may interact with the expression of genotypic variants, in a way that may be different from the expression in adult patients. This is important to consider in the use, as well as in the development of drugs for children. In other words, the relationship between phenotype and genotype may be different from what it is seen in adults. As a result, variability in pharmacokinetics and response to standard dosage regimen is greater in children than in adults. Carrying out pharmacological studies as a function of age is therefore important for the definition of dosage regimens suitable for children and for limiting the risk of side effects and toxicity [60].

REFERENCES

1. Rane, A., *Drug metabolism and disposition in infants and children*, in *Neonatal and Paediatric Pharmacology* (3rd ed.), Aranda, J.. Yaffe, S. (eds.), Lippincott Williams & Wilkins, (2004).
2. Meyer, U. A., *Ann. Rev. Pharmacol. Toxicol., 1,* 66 (1991).
3. Ladona, M. G., Lindström, B., Thyr, C., Rane, A., *Br. J. Clin. Pharmacol., 32,* 295 (1991).
4. Leeder, J. S., Kearns, G. L., *Ped. Clin. North Am., 44,* 55 (1997).
5. Mahgoub, A., Idle, J. R., Dring, L. G., Lancaster, R., Smith, R. L., *Lancet, 2,* 584 (1997).
6. http://www.imm.ki.se/CYPalleles
7. Johansson, I., Johansson, I., Lundqvist, E., et al., *Proc. Nat. Acad. Sci. USA , 15,* 11825 (1993).
8. Eichelbaum, M. and Gross, A.S., *Pharmacol. Ther. 46,* 377 (1990).
9. Jacqz-Aigrain E., Cresteil, T., *Dev. Pharmacol. Ther., 18,* 161 (1992).
10. Evans, W. E., Relling M. V., Petros W. P. et al., *Clin. Pharmacol. Ther., 45,* 568 (1989).
11. Poulsen, L., Brosen, K., Arendt-Nielsen L. et al., *Eur. J. Clin. Pharmol., 51,* 289 (1989).
12. Kupfer, A., Desmodm P. V., Schenkerm, S. et al., *Pharmacologist, 21,* 173 (1979).
13. Desta, Z., Zhao, X., Shin, J. G., Flockhart, D. A., *Clin. Pharmacokinet.,* 41, 913 (2002).
14. de Wildt, S. N., Kearns G. L., Leeder, S., van den Anker, J. N., *Clin. Pharmacokinet., 37,* 485 (1999).
15. Paine, M. F., Khalighi, M., Fischer, J. M., et al., *J. Pharm. Exper. Ther., 283,* 1552 (1997).
16. Wacher, V. J., Silverman, J. A., Zhang,Y. C., Benet, L. Z., *J. Pharm. Sci., 87,* 1322, (1998).
17. Koch, I., Weil, R., Wolbold, R., et al., *Drug Metab. Disp., 30,* 1108 (2002).
18. Lown, K. S., Kolars, J. C., Thummel, et al., *Drug Metab. Disp., 2,* 947 (1994).
19. Lown, K. S., Mayo, R. R., Leichtman, A. B., et al., *Clin Pharm. Ther., 62,* 248 (1997).
20. Park, B. K., et al., *Pharmacol. Ther., 68,* 385 (1995) .
21. Fakhoury, M., Litalien, C., Medard, Y., Peuchemaur, M., Cave, H., Jacqz-Aigrain, E., Submitted, (2004).
22. Thörn, M., Finnström, N., Lööf, L., Rane, A., Submitted, (2004).
23. Hustert, E., Haberl, M., Burk, O., et al., *Pharmacogenetics, 11,* 773 (2001).
24. Paulussen, A., Lavrijsen, K., Bohets, H., et al., *Pharmacogenetics, 10,* 415 (2000).
25. Lacroix, D., Sonnier, M., Moncion, A., Cheron, G. and Cresteil, T., *Eur. J. Biochem., 247,* 625 (1997).
26. Burk, O., Tegude, H., Koch, I., et al., *The American Society for Biochemistry and Molecular Biology, 277,* 24280 (2002).
27. Board, P.G., *Am. J. Hum. Genet., 33,* 36 (1981),
28. Pemble, S., Schroeder, K.R., Spencer et al., *Biochem J., 300,* 271 (1994).
29. Nebert, D. W., *Clin. Genet., 56, 247* (1999)
30. Woo, M. H., et al., *Leukemia, 14,* 226 (2000).
31. Evans, D. A., Manley, K. A., McKusick, V. A., *BMJ, 5197,* 485 (1960).
32. Relling, M.V., Lin, J. S., Ayers, G. D., Evans, W. E., *Clin. Pharmacol. Ther., 52,* 643 (1992).
33. Pariente-Khayat, A., Pons, G., Rey, E., et al., *Pediatr. Res., 29,* 492 (1991).
34. Delomenie, C., Grant, D. M., Mathelier-Fusade, P., et al., *Br. J. Clin. Pharmacol., 38,* 581 (1994).
35. Zielinska, E., Niewiarowski, W., Bodalski, J., *Eur. J. Clin. Pharmacol., 54,* 779 (1998).
36. Krynetski, E. Y., Tai, H. L., Yates, C. R., et al., *Pharmacogenetics, 6,* 279 (1996).
37. Weinshilboum, R. M., Sladek S. L., *Am. J. Hum. Genet. 32,* 651 (1980).
38. Lennard, L., Lilleyman, J. S., Van Loon, J., Weinshilboum, R. M., *Lancet, 336,* 225 (1990).
39. Dervieux, T., Medard, Y., Verpillat, P., et al., *Leukemia, 15,* 1706 (2001).
40. Colombel, J. F., Ferrari, N., Debuysere, H., et al., *Gastroenterology, 118,* 1025 (2000).

41. Relling, M. V., Hancock, M. L., Rivera, G. K., et al., *J. Natl. Cancer Inst.*, *91*, 2001, (1999).
42. McLeod, H. L., Krynetski, E. Y., Wilimas, J. A., Evans, W. E., *Pharmacogenetics*, *5*, 281 (1995).
43. Ganiere-Monteil, C., Medard, Y., Lejus, C., et al., *Eur. J. Clin Pharmacol.*, *60*, 89 (2004) .
44. Mackensie, P. I., Miners, J. O., McKinnin, R. A., *Clin. Chem. Lab. Med.*, *38*, 889 (2000)
45. Tanigawara, Y., *Ther. Drug Monit.*, *22*, 137 (2000).
46. Hoffmeyer, S., et al., *Proc. Natl. Acad. Sci.*, *97*, 3473 (2000).
47. Dysdala, C. M., et al., *Proc. Nat. Acad. Sci.*, *97*, 10483 (2000).
48. Palmer, L. J., Silverman, E. J., Weiss, S. T., Drazan, J. M., *Am. J. Respir. Crit. Care. Med.*, *4*, 861 (2002).
49. Evans, W. E., Relling, M. V., *Science*, *286*, 487 (1999).
50. Daley, G. Q., Lander, E. S., *Nat. Genet.*, *22*, 213 (1999).
51. Mc Kusick, V. A., *Genomics*, *45*, 244 (1997).
52. Cargill, M., Altshuler, D., Ireland, J., et al., *Nat. Genet.*, *22*, 231 (1999).
53. Nebert, D. W., McKinnon, R. A., Puga, A., *DNA Cell Biol.*, *5*, 273 (1996).
54. Relling, M. V., Dervieux, T., *Nature Rev. Cancer*, *1*, 99 (2001)
55. Goetz, M. P., Mayo, *Clin. Proc.*, *79*, 376 (2004); MacLeod H. L., Jinsheng Y., *Cancer Invest.*, *21*, 630 (2003).
56. Impicciatore, P., Choonara, I., Clarkson, A., Provasi, D., Pandolfini, C., Bonati, M., *Br. J. Clin. Pharmacol.*, *52*, 77 (2001).
57. Clarkson, A., Ingleby, E., Choonara, I., Bryan, P., Arlett, P., *Arch. Dis. Child.*, *84*, 337 (2001).
58. Horen, B., Montastruc, J. L., Lapeyre-Mestre, M., *Br. J. Clin. Pharmacol.*, *54*, 665 (2002).
59. Phillips, K. A., Veenstra, D. I., Oren, E., Lee, J. K., Sadee, W., *JAMA*, *286*, 2270 (2002).
60. Park, B.K., Pirmohamed, M., Kitteringham, N. R., *Br. J. Clin. Pharmacol.*, *34*, 377 (1992).
61. Kearns, G. L., *Curr. Opin. Pediatr.*, *7*, 220 (1995).
62. Meyer, U. A., *Lancet* 356, 1667-1671
63. Spielberg, S. P., *N. Engl. J. Med.*, 1981, (2000)
64. Van Dyke, D. C., Ellingrod, V. L., Berg, M. J., Niebyl, J. R., Sherbondy, A. L., Trembath, D. G., *Ann. Pharmacother.*, *34*, 639 (2000).
65. Tempfer, C. B., Schneeberger, C., Huber, J. C., *Pharmacogenomics*, *5*, 57 (2004)
66. International Human Genome Sequencing Consortium, *Nature*, *409*, 860 (2001).
67. Venter, J. C., Adams, M. D., Myers, E. W., et al., *Science*, *291*, 1304 (2001).
68. Goldstein, D.B., Tate, S.K., Sisodiya, S., *Nature Reveiws*, *4*, 937 (2003).
69. Eichelbaum, Spannbrucker, N., Steincke, B., et al., *Eur. J. Clin. Pharmacol.*, *16*, 183 (1979).
70. Lazarou, J., Pomeranz, B. H., Corey, P. N., *JAMA*, *279*, 1200 (1998).
71. Snyder, L. H., *Ohio J. Sci.*, *32*, 436 (1932)
72. Vogel F., *Ergeb. Inn. Med. Kinderheilkd.*, *12*, 52 (1959).
73. Diasio, R. B., Beavers, T. L., Carpenter, J. T., *J. Clin. Invest.*, *81*, 47 (1988).
74. Gonzales, F. J., Fernandez-Salguero, P., *TiPS*, *16*, 325 (1995).
75. Lyer, L., King, C.D., Whitington, P. F., et al., *J. Clin. Invest.* *101*, 847 (1998).
76. Ulrich, C.M., Yasui, Y., Storb, R., et al., *Blood*, *98*, 231 (2001).
77. Krajinovic, M., Costea, I., Chiasson, S., *Lancet*, *359*, 1033 (2002).
78. Stanulla, M., Schrappe, M., Brechlin, A. A. M., Zimmermann, M., Welte, K., *Blood*, *95*, 1222 (2000).
79. Davies, S. M., Robinson, L. L., Buckley, J. D., et al., *J. Clin. Oncol.* 19, 1279 (2001).
80. Basile, V. S., Ozdemir. V., Masellis, M., et al., *Mol. Psychiatry* 5, 410 (2000).
81. Aithal, G. P., Day, C. P., Kesteven, P. J., Daly, A. K., *Lancet*, *353*, 717 (1999)
82. Dilger, K., Meisel, P., Hofmann, U., Eichelbaum M., *Ther. Drug Monit.*, *22*, 366 (2000).
83. Farrell, G. C., *Semin. Liver. Dis.*, *22*, 185 (2002).

2.7 Therapeutic drug monitoring

Rafael Gorodischer

Department of Paediatrics, Soroka University Medical Center and Faculty of Health Sciences, Ben- Gurion University of the Negev, Beer-Sheva, Israel 84101

2.7.1 INTRODUCTION

Clinicians recognise that not all patients respond similarly to the same medication, even when doses are normalised for body weight or surface area. This may be attributed to lack of compliance with the therapeutic regimen, but the main reason is the existence of pharmacokinetic differences between individuals. Individual variations in drug biotransformation may result from genetic factors, age, concomitant drug therapy, other environmental factors (foods, toxins, etc.) and disease. Genetically determined cytochrome P450 polymorphism can cause several-fold differences in the metabolic rate between rapid and slow metabolisers [1]. Some conjugation pathways, such as glucuronidation, are reduced in the newborn and accumulation of chloramphenicol was associated with the "grey baby syndrome" [2]. Elevated serum concentrations of theophylline are seen when administered concomitantly with erythromycin, which interferes with cytochrome P450-mediated metabolism [3]. Ciclosporin blood concentrations increase by concurrent administration of grapefruit juice, a CYP3A4 inhibitor [4]. Children with cystic fibrosis or burns have more rapid clearance of several drugs as compared with unaffected individuals [5,6].

2.7.2 INDIVIDUALISATION OF DRUG THERAPY

The usually recommended doses are designed for the "average" patient. The inter-patient variability in drug pharmacokinetics is not reflected in those recommended dosage regimens. In principle, the most appropriate way of adjusting drug dosage to a particular patient would be by assessing the clinical effect of the drug. This is possible when the response is easily measurable (i.e. blood pressure, urine output). But the effects of most medications are difficult to evaluate directly in the clinical situation.

Individualisation of therapy for many medications is done by measuring their concentrations in blood samples. Drug concentration in biological fluids often relates better to clinical response than drug dosage. Also, drug concentration in blood provides information about the patient's pharmacokinetic profile. By determining drug blood concentrations at steady state (C_{ss}), the ratio between drug clearance and the amount of the drug that reaches the systemic circulation (Cl/F) may be estimated for the specific patient ($Cl / F = dosing\ rate / C_{ss}$). Combined with the clinical observation, this measurement guides the physician on whether or not adjustment of the dosage regimen is necessary in order to obtain maximal efficacy with a minimum of unacceptable adverse effects.

That means that proper use of a therapeutic drug monitoring (TDM) service entails more than just drug measurement in blood samples: it includes evaluation of the clinical indication for testing, the laboratory analysis, as well as the appropriate interpretation of the results conveyed to the clinician for therapeutic action.

Furthermore, optimal interpretation of drug concentrations may be accomplished by application of probability inference, that is, by calculating the probability that an event (toxicity *vs.* non-toxicity, effectiveness *vs.* non-effectiveness) will occur from the frequency with which it occurred in a prior patient representative sample. Despite its advantages, this Bayesian forecasting is not sufficiently applied in TDM services as yet [7].

In addition to the measurement of drug and active metabolites in blood, individualisation of therapy for some drugs (such as mercaptopurine, thioguanine, azathioprine) is done on the basis of the individual patient's genotype. It is expected that, with advancements in the field of pharmacogenomics, characterisation of the individual's genome will become the preferred approach for individualisation of drug therapy.

Which drugs to measure in TDM?

Table 1 summarises the requirements for a drug that may be considered for TDM. The issue of drug analysis, and particularly the accuracy and precision, are of major significance. Some immunoassays lack specificity and cross react with drug metabolites or with endogenous substances. Examples are cross reactivity in immunoassays between carbamazepine and its active metabolite 10, 11

Table 1

Characteristics of drugs that are good candidates for TDM

A quantitative relationship exists between blood concentration and clinical effect
A poor correlation exists between dose and serum concentration
The pharmacokinetics of the drug is well known
The analytical technique is specific, precise and accurate
The therapeutic index of the drug is small
Significant consequences are associated with therapeutic failure

epoxide [7], and between digoxin and digoxin-like substances [8]. HPLC and a microparticle enzyme technique reportedly minimise digoxin-like substances cross reactivity [9].

Indications and interpretation of drug blood concentrations

Usual indications for ordering blood concentrations are assessment of patient compliance, lack of response, adverse effects and drug interactions. While TDM contributes considerably to individualisation of drug therapy and patient care, adjustment of dosage must be based on clinical grounds, with TDM as a support element.

The so called "therapeutic range" of a plasma drug concentration is actually a target concentration, and does not always result in the desired clinical response. The lowest limit is approximately the concentration that produces half of the maximal possible therapeutic effect. Some individuals may benefit from concentrations below this lowest limit. The upper limit is determined by toxicity, and it is the concentration at which no more than 5 to 10% of patients will develop adverse effects. This means that some patients will experience toxicity at lower concentrations than the upper limit of the "therapeutic" range, while others may benefit from concentrations above that limit.

Timing of sampling

Timing of blood sampling for drug analysis is of paramount importance. Generally, drugs are administered in repetitive doses at constant time intervals, or as a continuous infusion, in order to maintain a steady state concentration of the drug in plasma within a therapeutic range. At steady state, the rate of drug elimination equals the rate of drug availability. After a constant dosage regimen is initiated (or modified) and no loading dose is given, for drugs with linear metabolism, 94% of the steady state is reached at approximately four half-lives of the drug. If a sample is obtained before the steady state is reached, the value will not reflect the drug's clearance. In practice, in the case of small therapeutic index drugs, the first sample is usually taken at two half-lives in order to prevent toxicity, and repeated after 4 to 5 half-lives for the steady state value. This concept does not apply to drugs with saturable metabolism (dose dependent or non-linear kinetics) in the usual dosage range, such as phenytoin, theophylline, and salicylate, among others. In this situation, as metabolism is saturated, drug clearance decreases, the plasma concentration increases disproportionately as compared to the increase in the rate of drug administration, and the steady state is achieved slowly.

While, during continuous infusion, no fluctuations are expected in plasma concentration at steady state, during intermittent dosage the concentration of the drug rises and falls in each inter-dose interval. Drug measurements in samples obtained too soon after dosing during an intermittent regime are of no value and even misleading, as they reflect mainly the rates of absorption and of distribution, and the size of the volume of distribution, but not drug clearance. Due to slow distribution, plasma concentrations of digoxin, for instance, peak well before maximal effect and a value of >2 ng/ mL (potentially toxic when taken 6 to 8 hours after the dose) must be expected when sampling too soon after dosing. In the usual clinical situation, the sample is taken just before the next planned dose, although the obtained value will represent the minimal concentration, and not the average concentration at steady state.

2.7.3   ADVANTAGES AND LIMITATIONS OF TDM

Is TDM cost-effective?

Studies comparing TDM with no TDM indicate that, when properly used, TDM is an intervention that improves cost-effectiveness of therapy with several medications [10,11]. Individualisation of drug therapy by an adequate TDM service should improve the quality of care and reduce costs.

Limitations of TDM

A serious limitation of TDM is that, for most medications monitored, target concentrations have not been determined following statistical analysis of sufficiently large patient populations. In many instances, target concentrations are based on data obtained in adults and not in children. Thus, potential age, genetic, environmental and disease differences in target concentrations have not been well studied. In addition, there are limitations related to the practice of TDM, as in many settings it is not optimally used. Over-monitoring unnecessarily increases the costs of care, and under-monitoring has potential adverse consequences for the patient. Inadequate time sampling and inappropriate indication are common in a high proportion of TDM assays [12,13].

2.7.4 PAEDIATRIC AND NEONATAL TDM

Paediatric and neonatal TDM have unique features. Placental or breast milk passage of medications administered to the mother may interfere with infant's TDM. Body weight, clearance and protein binding of many drugs increase rapidly in young infants. For instance, the clearance of caffeine and of theophylline (among other medications) exhibit considerable inter-individual differences, are slow in the neonate, and undergo rapid maturation in the first weeks and months of life [14-16]. Phenytoin Vmax is greater in infants above one month of age than in neonates and older children and adults [17-19]. The therapeutic concentration range of theophylline for apnoea in neonates is lower than for the prevention of asthma in older children (5-15 *vs.* 10-20 mcg/ml). Conversion of theophylline to caffeine occurs in the neonate, and both are active molecules in the prevention of apnea of prematurity. Endogenous digoxin-like substances are present in neonates and in young infants and interfere with digoxin immunoassays [9]. The timing of blood sampling is problematic in small patients, as a drug dose given through an intravenous line may take considerable time to completely reach the patient's circulation [20]. A significant unique issue in neonatal intensive care units is the need to analyse drug concentrations in very small sample volumes, as the vascular compartment is small in premature infants, multiple drugs are administered, and the possibility of drug interactions is an important consideration. Paediatric TDM is also useful in the detection of medication errors. Sometimes, an excessive dose is erroneously administered because of a mistake in the dilution of concentrated stock formulations intended for adults [21].

The use of saliva, instead of blood, has been suggested in therapeutic drug monitoring in children. Although saliva sampling is a painless, non-invasive and less costly procedure [22], no studies are available comparing in which biological fluid, saliva or plasma, drug concentration correlates better with the drug therapeutic and toxic effects.

Table 2

TDM for anti-infective agents[*]

Medication	Plasma $t_{1/2}$ (h)	Target concentration (µg/ml)	Comments	Ref.
chloramphenicol	5-10 (neonate 24)	20-30 (peak) 5-10 (trough)	interacts with phenytoin	29
flucytosine	5	25-100	bone marrow suppression with concentrations >125 µg/mL	30
gentamicin	2	6-10 (peak) <2 (trough)	once-daily dosing schedule modifies target concentrations	31
tobramycin	2	6-10 (peak) <2 (trough)	once-daily dosing schedule modifies target concentrations	32
amikacin	2	20-30 (peak) <10 (trough)	once-daily dosing schedule modifies target concentrations	33
vancomycin	5	25-40 (peak) <10 (trough)	see text	24

[*] Plasma $t_{1/2}$ is expressed as representative approximate means and reflects order of magnitudes; there is marked inter-individual variation for some drugs.

2.7.5 SPECIFIC DRUGS: SPECIAL CONSIDERATIONS

Anti-infective agents (Table 2)

Aminoglycosides

The traditional target serum concentration ranges for *gentamicin*, *tobramycin* and *amikacin* must be reviewed following the present practice of once-daily dosing schedule. Peak concentrations are greater, and trough concentrations (at 24 h) of

0.5 to 2 μg/mL may indicate toxicity during a once-daily dose schedule. A target AUC (area under the curve) with a Bayesian approach has been suggested [23].

Glycopeptides

Little data that correlate *vancomycin* serum concentrations with efficacy are available. As a consequence, there is poor consensus regarding post-dose sampling times, target ranges and toxic vancomycin concentrations [24]. Studies in adult populations have shown that patients with higher trough concentrations have a greater incidence of renal damage [11,25,26] and that TDM for vancomycin is cost-effective [11]. Measurement of peak concentrations may not be needed in neonates [27]. TDM may be useful in individualisation of *teicoplanin* therapy in patients with high clearance [28].

Antiretroviral agents

The limited available data suggest that TDM of *non-nucleoside reverse transcriptase inhibitors* and *protease inhibitors* (but not *nucleoside reverse transcriptase inhibitors* [34]) contribute to individualisation of therapy in patients with HIV infection. It is unclear however whether AUC, trough concentrations, or a ratio of trough concentrations to the IC_{50} of the drug is the best measurement [35,36]. The difficulty in monitoring several anti-HIV drugs in a single subject is being solved by the use of tandem mass spectrometry [37]. TDM of anti-HIV drugs (as for antituberculous medications) is also important to assess compliance, as poor compliance may result in emergence of resistant strains [38].

Antiepileptic medications (Table 3)

Routine measurements of antiepileptic drugs in all patients is not recommended. A randomised controlled trial of TDM for all adult patients with newly diagnosed epilepsy showed no improvement in overall therapeutic outcome and the majority of patients could be satisfactorily treated by adjusting dose on clinical grounds [39]. TDM is of greatest value in relation to antiepileptic drugs in selected patients and in special situations. *Phenytoin* exhibits non-linear elimination kinetics, and this is a pharmacokinetic reason to monitor its plasma concentrations. The variable ratio of *carbamazepine* to its active metabolite *10,11 epoxide* complicates the interpretation of plasma carbamazepine concentrations, and for this reason some TDM laboratories determine both the parent drug and this metabolite. Due to the large fluctuation in plasma valproic acid concentrations, many laboratories no longer offer routine TDM of this anticonvulsant. There is general agreement about the target serum concentrations for phenytoin, carba-

Table 3

TDM for antiepileptic medications*

Medication	Plasma $t_{1/2}$ (h)	Target concentration	Comments	Ref.
carbamazepine	15	4-12 µg/ml	active metabolite 10,11 epoxide may also be measured	41
clonazepam	23	20-80 ng/ml	higher $t_{1/2}$ values in first days of life	42
ethosuximide	40	40-100 µg/ml	seizures may decrease with lower concentrations	43
phenobarbital	20-130	15-40 µg/ml	stimulates metabolism of its own and of many drugs	44
phenytoin[†]	6-24	10-20 µg/ml (total) 1-2 µg/ml (free)	$t_{1/2}$ dependent on plasma concentration	19
valproic acid	14	50-100 µg/ml	multiple drug interactions	45

*Plasma $t_{1/2}$ is expressed as representative approximate means and reflects order of magnitudes; there is marked inter-individual variation for some drugs.
[†]Phenytoin $t_{1/2}$ is concentration dependent. Adjustment of plasma concentration is required in hypoalbuminaemia

mazepine and phenobarbitone. It is important, however, to recognise that some patients can be controlled well, despite a sub-therapeutic serum concentration. This has been demonstrated in an elegant clinical trial by Woo and colleagues [40], who studied adult patients who were well controlled on monotherapy with either phenytoin or phenobarbitone, despite sub-therapeutic serum concentrations. They found no significant difference in efficacy between the group where the levels were kept in the sub-therapeutic range and those where the dosage was increased until it reached the therapeutic range. The former group had less side effects and therefore one needs to treat the patient and not a serum concentration. No generally accepted target ranges are available for the newer antiepileptic drugs, and overlap between drug concentrations related to toxicity and to non-response has been reported.

Cardiac medications (Table 4)

A metanalysis showed that TDM is beneficial in patients treated with *digoxin* [46]. Data indicate that TDM is useful in the management of patients with Class

Table 4

TDM for cardiac medications*

Medication	Plasma $t_{1/2}$	Target concentration	Comments	Ref.
amiodarone	25 days	0.5-2.5 µg/ml	serum conc. does not distinguish responders and non-responders; toxicity correlates with amiodarone conc. >2.5 and desethylamiodarone > 3 µg/mL	48-50
digoxin	36 h; (neonate 70 h; infant 45 h)	0.8-2 ng/ml	EDLS interferes with interpretation of serum digoxin in young infants	51
flecainide	10 h	0.2-1 µg/ml	milk blocks flecainide absorption	49,52,53
lidocaine	2 h	2-5 µg/ml	beta blockers and cimetidine increase serum concentrations	54
mexiletine	9 h	0.5-2 µg/ml	rifampin, phenobarbital and phenytoin increase clearance	49,55
nitroprusside	<10 min; thiocyanate: 4 days	toxic conc: thiocyanate 35-100 µg/ml; cyanide >2 µg/ml	nitroprusside metabolism releases cyanide, which is metabolised to thiocyanate	56

* Plasma $t_{1/2}$ is expressed as representative approximate means and reflects order of magnitudes; there is marked inter-individual variation for some drugs.

I antiarrhythmic agents (such as *lidocaine, mexiletine, flecainide*) and with *amiodarone* [47].

Cytotoxic medications

Various anticancer medications exhibit a clear relationship between systemic exposure and therapeutic or toxic effects. Mercaptopurine (MP) plasma AUC correlates with relapse-free survival in children with acute lymphocytic leukaemia (ALL) [57]. As MP is a prodrug, measurement of the active intracellular metabolite thioguanine nucleotide (TGN) may be superior. A greater erythrocyte TGN concentration correlates with a better event-free survival in children with

ALL [58]. MP activation to TGN competes with its enzymatic inactivation by thiopurine S-methyltransferase (TPMT). Erythrocyte TGN concentration correlates negatively with TPMT activity. About 0.3% of the population is deficient in TPMT, and 10% has intermediate TPMT activity. These patients are at risk of haematological toxicity when treated with MP or azathioprine, and may be best treated by measuring TPMT activity, determining the TPMT genotype, or monitoring erythrocyte TGN [59]. Individualisation of the methotrexate dose to the patient's methotrexate clearance improves the outcome in children with ALL [60,61].

Table 5

TDM for miscellaneous medications[*]

Medication	Plasma $t_{1/2}$ (h)	Target concentration	Comments	Ref.
caffeine	premature: 40-230	neonatal apnea: 5-20 µg/ml	serum measurements once every 1-2 weeks suffices	70
ciclosporin	20	100-400 ng/ml (blood, trough)	target concentrations not well defined; concentrations are method dependent	62
lithium	20	0.6-1.2 mEq/l; seizures may occur at conc. >2.5 mEq/l	NSAIDs and thiazides may decrease lithium excretion	71
salicylate [†]	2.5-19	antiinflammatory: 150-300 µg/ml	aspirin is converted to salicylic acid during absorption; salicylate $t_{1/2}$ is dose-dependent	72
tacrolimus	4-40	5-20 ng/ml	monitor renal function	73
theophylline	premature: 12-64 children: 2-6	neonatal apnea: 5-15 µg/ml asthma: 10-20 µg/ml	monitor both theophylline and caffeine in newborns receiving theophylline; clearance is markedly age-dependent	74, 75

[*] Plasma $t_{1/2}$ is expressed as representative approximate means and reflects order of magnitudes; there is marked inter-individual variation for some drugs.

[†] Ingested aspirin is rapidly hydrolysed to salicylic acid in the gastrointestinal mucosa, plasma, liver and erythrocytes. Salicylate $t_{1/2}$ is concentration dependent.

Immunosuppressant medications (Table 5)

An inverse relationship has been demonstrated between the maximum *ciclosporin* blood concentration and the rate of rejection of liver transplants [62]. Targeting ciclosporin concentration at 2 hours post-dose ("absorption profiling") is beneficial [63,64]. High erythrocyte TPMT activity was found to be linked to poor outcome in paediatric renal transplant patients treated with *azathioprine* [65]. A relationship has been shown between clinical effect and *tacrolimus* blood [66] and *mycophenolic acid* plasma [67,68] concentrations. However, the role of TDM in therapy with mycophenolic acid and *sirolimus* in children has yet to be defined [69].

2.7.6　TDM IN DEVELOPING COUNTRIES

Due to the limited resources allocated for health care, TDM is rarely practiced in developing countries. *Chloramphenicol* is an antibiotic widely used in the developing world, for the management of typhoid fever and of bacterial meningitis; in view of its variable pharmacokinetics, monitoring chloramphenicol therapy has been recommended [29,76]. Also, TDM of *antituberculous drugs* has been advocated as a measure of compliance [37,77].

2.7.7　FUTURE DIRECTIONS

The rapidly advancing field of pharmacogenomics bears the potential to considerably improve the task of individualising drug therapy. It is likely that, in the future, genotyping techniques will become routine diagnostic tools in the detection of sequence variants in genes encoding for drug metabolising enzymes, drug transporters and drug receptors. That molecular information, in conjunction with other factors affecting drug response (age, disease, concomitant drug therapy, food...), will improve our ability to select the best drug and the right dose for each patient [78].

ABBREVIATIONS

TDM	therapeutic drug monitoring
AUC	area under the concentration-time curve
MP	mercaptopurine
TGN	thioguanine nucleotide
TPMT	thiopurine-S-methyltransferase
ALL	acute lymphocytic leukemia
HIV	human immunodeficiency virus
EDLS	endogenous digoxin-like substances

REFERENCES

1. Weinshilboum, R., *N. Engl. J. Med., 348*, 529 (2003).
2. Weiss, C. F., Glazko, A. J., Weston, J. K., *N. Engl. J. Med., 262*, 787 (1960).
3. Periti, P., Mazzei, T., Mini, E., Novelli, A., *Clin. Pharmacokinet., 23*, 106 (1992).
4. Brunner, L. J., Pai, K. S., Munar, M. Y, et al., *Pediatr. Transplant., 4*, 313 (2000).
5. Rey, E., Treluyer, J. M., Pons, G., *Clin. Pharmacokinet., 35*, 313 (1988).
6. Weinbren, M. J., *J. Antimicrob. Chemother., 44*, 319 (1999).
7. Shen, S., Elin, R. J., Soldin, S. J., *Clin. Biochem., 34*, 157 (2001).
8. Dasgupta, A., *Am. J. Clin. Pathol., 118*, 132 (2002).
9. Chicella, M., Branim, B., Lee, K. R., Phelps, S. J., *Ther. Drug Monit., 20*, 347 (1998).
10. Schumacher, G. E., Barr, J. T., *Ther. Drug Monit., 20*, 539 (1998).
11. Fernandez-de Gatta, M. M., Calvo, M. V., Hernandez, J. M., Caballero, D., San Miguel, J. F., Dominguez-Gil, A., *Clin. Pharmacol. Ther., 60*, 332 (1996).
12. Gross, A. S., *Br. J. Clin. Pharmacol., 46*, 95 (1998).
13. Bates, D. W., Soldin, S. J., Rainey, P. M., Micelli, J. N., *Clin. Chem., 44*, 401 (1998).
14. Aranda, J. V., Collinge, J. M., Zinman, R., Watters, G., *Arch. Dis. Child., 54*, 946 (1979).
15. Aranda, J. V., Sitar, D. S., Parsons, W. E., et al., *N. Engl. J. Med., 295*, 413 (1976).
16. Simons, F. E. R., Simons, K. J., *J. Clin. Pharmacol., 18*, 472 (1978).
17. Loughnan, P. M., Greenwald, A., Purton, W. W., et al., *Arch. Dis. Child., 52*, 302 (1977).
18. Bauer, L. A., Blouin, R. A., *Clin. Pharmacokinet., 8*, 545 (1983).
19. Levine, M., Chang, T., *Clin. Pharmacokinet., 19*, 341 (1990).
20. Gould, T., Roberts, R. J., *J. Pediatr., 95*, 465 (1979).
21. Koren, G., *J. Clin. Pharmacol., 42*, 707 (2002).
22. Gorodischer, R., Burtin, P., Hwang, P., Levine, M., Koren, G., *Ther. Drug Monit., 16*, 437 (1994).
23. Duffull, S. B., Kirkpatrick, C. M. J., Begg, E. J., *Br. J. Clin. Pharmacol., 43*, 125 (1997).
24. Tobin, C. M., Darville, J. M., Thomson, A. H., Sweeney, G, Wilson, J. F., Mac-Gowan, A. P., White, L. O., *J. Antimicrob. Chemother., 50*, 713 (2002).
25. Welty, T. E., Copa, A. K., *Ann. Pharmacother., 28*, 1335 (1994).
26. Zimmerman, A. E., Katona, B. G., Plaisance, K. I., *Pharmacotherapy, 15*, 85 (1995).
27. Tan, W. H., Brown, N., Kelsall, A. W., McClure, R. J., *Arch. Dis. Child Fetal Neonatal Ed., 87*, F214 (2002].
28. Pea, F., Brollo, L., Viale, P., Pavan, F., Furlanut, M., *J. Antimicrob. Chemother., 51*, 971 (2003).
29. Ekblad, H., Ruuskanen, O., Lindberg, R., Iisalo, E., *J. Antimicrob. Chemother., 15*, 489 (1985).
30. Cutler, R. E., Blais, A. D., Kelly, M. R., *Clin. Pharmacol. Ther., 24*, 333 (1978).
31. Evans, W. E., Feldman, S., Ossi, M., et al., *J. Pediatr., 94*, 139 (1979).
32. Aarons, L., Vozeh, S., Wenk, M., et al., *Br. J. Clin. Pharmacol., 28*, 305 (1989).
33. Bauer, L. A., Blouin, R. A., *Eur. J. Clin. Phamacol., 24*, 639 (1983).
34. Aarnoutse, R., Schapiro, J., Boucher, C., Hekster, Y., Burger, D., *Drugs, 63*, 741 (2003).
35. Acosta, E. P., Gerber, J. G., *AIDS Res. Human Retroviruses, 18*, 825 (2002).
36. Back, D. J., Khoo, S. H., Gibbons, S. E., Merry, C., *Br. J. Clin. Pharmacol., 52*, 89S (2001).
37. Volosov, A., Soldin, S. J., *Ther. Drug. Monit., 23*, 485 (2001).
38. Walson, P. D., *Clin. Chem., 44*, 415 (1998).
39. Jannuzzi, G., Cian, P., Fattore, C., et al., *Epilepsia, 41*, 222 (2000).
40. Woo, E., Chan, Y. M., Yu, Y. L., Chan, Y. W., Huang, C. Y., *Epilepsia, 29*, 129 (1988).
41. Bertilsson, L., Tomson, T., *Clin. Pharmacokinet., 11*, 177 (1986).
42. Dreifuss, F. F., Penry, J. K., Rose, S. W., et al., *Neurology, 25*, 255 (1975).

43. Bauer, L. A., Harris, C., Wilensky, A. J., et al., *Clin. Pharmacol. Ther., 31*, 741 (1982).
44. Browne, T. R., Evans, J. E., Szabo, G. K., et al., *J. Clin. Pharmacol., 25*, 51 (1985).
45. Zaccara, G., Messori, A., Moroni, F., *Clin. Pharmacokinet., 15*, 367 (1988).
46. Ried, L. D., Horn, J. R., MaKenna, D. A., *Ther. Drug Monitor., 46*, 945 (1989).
47. Campbell, T. J., Williams, K. M., *Br. J. Clin. Pharmacol., 52*, 21S (2001).
48. Rotmentsch, H. H., Belhassen, B., Swanson, B. N., et al., *Ann. Int. Med., 101*, 462 (1984).
49. Campbell, N. P. S., Pantridge, J. F., Adgey, A. A. J., *Br. Heart J., 40*, 796 (1978).
50. Kannan, R., Yabek, S. M., Garson, A., et al., *Am. Heart J., 114*, 283 (1987).
51. Steinberg, C., Notterman, D. A., *Clin Pharmacokinet., 27*, 345 (1994).
52. Russell, G. A. B., Martin, R. P., *Arch. Dis. Child., 64*, 860 (1989).
53. Perry, J. C., McQuinn, R. L., Smith, R. T., et al., *J. Am. Coll. Cardiol., 14*, 185 (1989).
54. Nattel, S., Gagne, G., Pineau, M., *Clin. Pharmacokinet., 13*, 293 (1987).
55. Labbe, L., Turgeon, J., *Clin. Pharmacokinet., 37*, 361 (1999).
56. Beekman, R. H., Rocchini, A. P., Rosenthal, A., *Circulation, 64*, 553 (1981).
57. Koren, G., Ferrazini, G., Sulh, H., Langevin, A. M., et al., *N. Engl. J. Med., 323*, 17 (1990).
58. Lennard, L., Lilleyman, J. S., Van Loon, J., Weinshilboum, R. M., *Lancet, 336*, 225 (1990).
59. Yates, C. R., Krynetski, E. Y., Loennechen, T., Fessing, M. Y., et al., *Ann. Int. Med., 126*, 608 (1997).
60. Evans, W. E., Pratt, C. B., Taylor, H., Barker, L. F., Cromm, W. R., *Cancer Chemother. Parmacol., 3*, 161 (1979).
61. Evans, W. E., Relling, M. V., Rodman, J. H., et al., *N. Engl. J. Med., 338*, 499 (1998).
62. Keown, P., Kahan, B. D., Johnston, A., et al., *Transpl. Proc., 30*: 1645 (1998).
63. Belitski, P., Dunn, S., Johnston, A., Levy, G., *Clin. Pharmacokin., 39*, 117 (2000).
64. Trompeter, R., Fitzpatrick, M., Hutchinson, C., Johnston, A., *Pediatr. Transplant., 7*, 282 (2003).
65. Dervieux, T., Medard, Y., Baudouin, V., et al., *Br. J. Clin. Pharmacol., 48*, 793 (1999).
66. Laskow, D. A., Vincenti, F., Neylan, J. F., Mendez, R., Matas, A. J., *Transplantation, 62*, 900 (1996).
67. Van Gelder, T., Hilbrans, L. B., Vanrenterghem, Y., et al., *Transplantation 68*, 261 (1999).
68. Jacqz-Aigrain, E., Khan-Shaghaghi, E., Baudouin, V., et al., *Pediatr. Nephrol., 14*, 95 (2000).
69. Del Mar Fernandez De Gatta, M., Santos-Buelga, D., Dominguez-Gil, A., Garcia, M. J., *Clin. Pharmacokinet., 41*, 115 (2002).
70. Gorodischer, R., Karplus M., *Eur J. Clin. Pharmacol., 22*, 47 (1982).
71. Sproule, B., *Clin. Pharmacokinet., 41*, 639 (2002).
72. Roberts, M. S., Rumble, R. H., Wanwimolrut S., et al., *Eur. J. Clin. Phamacol., 25*, 253 (1983).
73. Jusko, W. J., Piekoszewski, W., Lintmalm, G. B., et al., *Clin. Pharmacol. Ther., 57*, 281 (1995).
74. Muttitt, S. C., Tierney, A. J., Finer, N. N., *J. Pediatr., 112*, 115 (1988)
75. Milavetz, G., Vaughn, L. M., Weinberger, M. M., Hendeles, L., *J. Pediatr., 109*, 351 (1986).
76. Ismail, R., The, L. K., Choo, E. K., *Ann. Trop. Paediatr., 18*, 123 (1998).
77. Yew, W. W., *Clin. Chim. Acta, 313*, 31 (2001).
78. Evans, W. E., McLeod, H. L., *N. Engl. J. Med., 348*, 538 (2003).

2.8 Drug toxicity and pharmacovigilance

Imti Choonara

Academic Division of Child Health, University of Nottingham, Derbyshire Children's Hospital, Uttoxeter Road, Derby, DE22 3NE, UK

2.8.1 INTRODUCTION

Adverse drug reactions (ADRs) are common clinical problems in both paediatric and adult medicine. Over 9% of hospitalised children experience an ADR and 2% of paediatric admissions to hospital are the consequence of ADRs [1]. The toxicity of many medicines in children is different to that seen in adults.

It can be difficult to evaluate drug toxicity in children. Studies of drug-induced toxicity are usually retrospective reviews and isolated case reports. It is difficult to be certain about the association between drug exposure and possible toxicity in the majority of cases. Thus, it is essential to maintain a high index of suspicion for the occurrence of drug toxicity in infants and children. The identification of new patterns of ADRs is dependent upon health professionals being alert and aware of the possibility of drug toxicity. Some of the major cases of ADRs that have occurred in both neonatal and paediatric patients are described below (Table 1).

2.8.2 DRUG TOXICITY IN THE NEONATE

The newborn infant historically has suffered some of the most severe ADRs reported in paediatric patients. It is important to learn from previous cases of drug toxicity to try and prevent future episodes [2].

Percutaneous drug toxicity

One of the earliest reported ADRs was methaemoglobinaemia due to aniline dye, which had been used to stamp the name of the institution on diapers. As reported in the British Medical Journal over a century ago, seventeen newborn infants developed cyanosis after absorbing the dye percutaneously [3]. Subsequently, there have been other cases of infants developing cyanosis, due to percutaneous absorption of aniline dye.

This particular ADR illustrates the possibility of enhanced drug toxicity through percutaneous absorption. In general, absorption of compounds is enhanced by issues such as the prolonged contact of a wet diaper with the perineum. Newborn infants have a higher surface area to weight ratio than both children and adults. Therefore, medicines that are administered topically result in greater relative exposure in newborn infants. Permeability through the epidermis is greatest in the preterm infant. Other examples of percutaneous toxicity in newborn infants include neurotoxicity due to hexachlorophane and hypothyroidism due to iodine [2].

Sulphonamides and protein binding

Newborn infants who received a combination of penicillin and sulphisoxazole were found to have a significantly higher mortality that those who received

Table 1

Major adverse drug reactions in neonatal and paediatric patients

Year	Drug/Compound	Age group	ADR	Mechanism
1886	aniline dye	neonates	methaemoglobinaemia	percutaneous absorption
1956	sulphisoxazole	neonates	kernicterus	protein displacing effect on bilirubin
1959	chloramphenicol	neonates	grey baby syndrome	impaired metabolism
1979	sodium valproate	young children (< 3 years)	hepatic failure	abnormal metabolism?
1980	salicylate	children	Reye's syndrome	unknown
1990	propofol	children	metabolic acidosis	unknown; dose related?

oxytetracycline [4]. In particular, the newborn infants who received the sulpho-namide had a higher incidence of kernicterus. These infants developed seizures and on post mortem examination, had yellow staining of the brain. Although this higher mortality was described in 1956, it was almost a decade later that studies showed that sulphonamides have a higher binding affinity than bilirubin for albumin [5]. This results in a marked increase in the free fraction of bilirubin in the plasma. The conjugation of bilirubin is minimal, due to reduced activity of glucuronosyltransferase in the neonate (Table 1).

Chloramphenicol and impaired metabolism

A few years after the reported mortality of sulphonamides in sick neonates, the grey baby syndrome was reported in association with the use of the antibiotic chloramphenicol [6]. The affected infants developed abdominal distension, vomi-ting, cyanosis, cardiovascular collapse, irregular respiration and subsequent death shortly after therapy with chloramphenicol was started. Pharmacokinetic studies in the neonate showed accumulation of chloramphenicol in plasma, which subse-quently impaired cellular oxidative metabolism, leading to cardiovascular insta-bility and collapse [7]. A reduction in the total daily dosage from 100 mg per kg to 50 mg per kg prevented the development of the grey baby syndrome [7]. This illustrates the importance of understanding the impact of developmental changes in drug disposition and metabolism, especially in the neonatal period, and using this information to design age-appropriate drug dosing regimens. It should be noted that had appropriate clinical trials been conducted so as to characterise the impact of age on chloramphenicol disposition, the tragedy associated with the "grey baby syndrome" could well have been averted.

2.8.3 DRUG TOXICITY IN CHILDREN

Different patterns of drug toxicity between adult and paediatric patients is not confined to the neonate. There are numerous examples of both different ADRs and different incidences of ADRs in children as opposed to adults [2]. The mechanism behind the increased incidence of certain ADRs in children is unknown. In contrast, there are some ADRs that are more common in adults than children. Hepatotoxicity following paracetamol overdose is a significant pro-blem in adults, but is infrequent in pre-pubertal children. This difference appears to be related to an enhanced capacity for sulphation in children.

Hepatotoxicity and sodium valproate

Hepatotoxicity was first reported in 1979 and there have been more than 100 deaths. Hepatotoxicity in association with sodium valproate is much more frequent in children [8]. A retrospective study showed that children under the age of 3 years were a high risk group along with patients on polypharmacy and those with developmental delay [8].

The mechanism of sodium valproate hepatotoxicity is thought to be related to abnormal drug metabolism, either a reduction in fatty acid beta-oxidation (the major detoxification pathway for valproic acid) and/or age-associated increases in the activity of specific cytochromes P450 responsible for the generation of putative hepatoxic metabolic intermediates. Enzyme induction by other anticonvulsants is thought to result in increased production of these toxic metabolites.

Reye's syndrome and salicylates

Reye's syndrome was originally described in 1963. Children usually had a preceding viral infection and subsequent drowsiness which led to coma, hypoglycaemia, seizures and liver failure. The association between Reye's syndrome and salicylates had been raised as a possibility in 1965 by an observant physician who noted that at least 15 of the 31 cases reported had received aspirin prior to admission [9]. In 1980 the association between Reye's syndrome and the use of salicylates during a preceding viral infection was confirmed [10]. During an outbreak of influenza A, seven children were admitted to hospital with Reye's syndrome. These seven children were compared with 16 controls from the same class who did not have signs or symptoms of Reye's syndrome. All of the children with Reye's syndrome took salicylates whereas only 8 of the 16 controls took salicylates. The patients took larger doses of salicylates than the control group and the level of salicylate consumption was associated with the severity of the Reye's syndrome.

The restriction of the use of salicylates for general antipyresis and analgesia in children aged 12 and under has resulted in a dramatic reduction in the incidence of Reye's syndrome [11]. Several cases have subsequently been reported of salicylates associated with Reye's syndrome in children between the ages of 12 and 16 [12, 13]. In the UK, it is now recommended that salicylates are avoided in children under the age of 16 years with a febrile illness. The mechanism of the toxicity is unknown.

Metabolic acidosis and propofol

Propofol is an ultra-short acting, parenteral anaesthetic agent that has gained wide acceptance for the induction and maintenance of anaesthesia in both adults and children. It's short elimination half-life and rapid distribution kinetics allows patients to recover rapidly upon discontinuation of infusion with much fewer adverse effects than have been observed with other parenteral anesthestic agents (e.g., barbiturates, ketamine). Propofol has also been used as a sedative to facilitate prolonged mechanical ventilation in critically ill children. To date, there have been at least 12 children dying following the use of propofol as a sedative in the UK [13-15]. These children developed severe metabolic acidosis and lipaemia. Many of the children were originally admitted to hospital with upper airway obstruction and thus, were not expected to suffer multi-organ failure. The mechanism(s) associated with the production of metabolic acidosis by propofol in children without concomitant hypoxemia is not known. It is, however, noted that the dose of propofol used in the children who died was considerably higher than has been recommended by other groups, thus implying an enhanced risk associated with the extent of drug exposure (i.e., dose) [15].

2.8.4 FORMULATION ERRORS

The introduction of sulphonamides for the treatment of infection in 1935 was a major advance in medical care. However, sulphonamides are relatively insoluble in water and, consequently, there was a problem in preparing a paediatric formulation. In 1937 the use of diethylene glycol as a solvent to prepare elixir of sulphanilamide - done without appreciation that the vehicle was a potent toxin - was responsible for the deaths of at least 76 American children and adults [16]. Unfortunately, this historical tragedy has been repeated numerous times. Diethylene glycol has been used as a solvent for paracetamol, resulting in the death of 47 children in Nigeria in 1992, 51 children in Bangladesh in 1995 and 85 children in Haiti in 1998 [2]. It is important to remember that medicines contain not only the desired active compound but also numerous other chemicals which are added to make the drug more palatable, soluble, stable or for a variety of other reasons (e.g., to add colouring, to enhance drug suspension). Thus, it is important to consider every component of a drug formulation as a substance with the potential of producing an ADR in the paediatric patient.

Benzyl alcohol is used for its antibacterial properties in ampoules of sodium chloride and water that are intended for intravenous administration. Metabolic acidosis, hepatic and renal failure and cardiovascular collapse (i.e., gasping syndrome) have been described among 10 premature newborn infants who were receiving multiple injections of sodium chloride for flushing catheters and bacteriostatic water in association with medicines that have been reconstituted [17]. Both solutions contained 0.9% benzyl alcohol, which was postulated as the causative agent in neonates and young infants with a reduced capacity to metabolise and excrete this toxic "inactive ingredient" in the doses that were unwittingly administered.

Table 2

Major formulation errors in neonatal and paediatric patients

Year	Drug	Error	Deaths	Age group	Country
1937	sulphanilamide elixir	diethylene glycol used as solvent	76	children and adults	USA
1972	talc baby powder	contained 6.3% hexachlorophene	36	infants and young children	France
1982	sodium chloride, water	benzyl alcohol concentrations high	10	neonates	USA
1984	vitamin E	emulsifiers toxic?	38	neonates	USA
1992	paracetamol	diethylene glycol used as solvent	47	children	Nigeria
1995	paracetamol	diethylene glycol used as solvent	51	children	Bangladesh
1998	paracetamol	diethylene glycol used as solvent	85	children	Haiti

In 1984 an intravenous formulation of vitamin E was withdrawn from the market following the death of 38 neonates. It was postulated that the emulsifiers used to make the vitamin E water miscible for intravenous use may have been responsible for the deaths [18].

Inappropriate formulation of a baby powder containing talc in France resulted in the death of 36 infants and young children. The baby powder contained 6.3% hexachlorophene, which is a known neurotoxin. The affected children developed an encephalopathy; in total 204 children became ill and 36 died [19]. The major formulation errors are highlighted in Table 2.

2.8.5 PHARMACOVIGILANCE

The first studies looking for adverse drug reactions (ADRs) occurring in children in hospital were carried out in the USA and UK in the 1970s [1]. A detailed surveillance programme was established in Boston Children's Hospital. Subsequently, there have been further studies of paediatric inpatients, outpatients

Table 3

Incidence of ADRs in different groups of neonatal and paediatric patients

Patients	%	Reference
paediatric inpatients	9.5 (CI 6.8, 12.3)	1
paediatric outpatients	1.5 (CI 0.7, 3.0)	1
paediatric admissions	2.1 (CI 1.0, 3.8)	1
neonatal inpatients	11-30	21
paediatric oncology/haematology inpatients	56	22
paediatric epilepsy outpatients	53	23

and admissions to hospital secondary to ADRs in numerous countries around the world [1]. A systematic review and meta-analysis of 17 prospective studies reported an overall incidence of 9.5% in paediatric inpatients and 1.5% in paediatric outpatients [1] (Table 3). Subsequent studies have confirmed these figures. Two per cent of paediatric admissions to hospital are due to ADRs.

The incidence of ADRs increases dramatically with an increase in the number of medicines prescribed [20]. The disease process and the medications used affect the incidence of ADRs with a high incidence in neonates [21], oncology [22] and epilepsy patients [23] (Table 3).

Targeted pharmacovigilance

Studies that have looked at a particular drug/group of drugs in relation to a specific ADR have been useful and aided the construction of guidelines to minimise such ADRs in the future. Examples include avoiding the use of sodium valproate in children under the age of 3 years [8] and studies of opiate induced respiratory depression, which is infrequent in neonates and children [24]. Respiratory depression following the use of diazepam in children with acute seizures, however, is not uncommon, especially in comparison with adults [25].

Novel methods

The regulatory authorities in many countries have relied on spontaneous reporting systems and have found these useful in adults and also in relation to certain ADRs occurring in children, for example visual field defects and vigabatrin. It is recognised that there is significant under-reporting of ADRs in spontaneous reporting schemes and this is likely to be more marked in paediatric patients. In order to raise awareness and stimulate reporting of ADRs, a Paediatric Regional Monitoring Centre was established in the Trent region of the UK [26]. The three year project was funded by the Medicines Control Agency and Trent NHS. The scheme resulted in a significant increase in the reports of ADRs (both severe and mild). The scheme also identified fatal suspected ADRs, which were unlikely to have been reported had the scheme not been in existence [27].

Another study, carried out in collaboration with the regulatory authorities in the UK, looked at spontaneous reports of suspected ADRs that were associated with deaths in children [13]. This study described over 300 deaths and a wide range of suspected ADRs illustrating that children may suffer as wide a range of ADRs as adults. Hepatic failure was the most frequently reported ADR associated with a fatality and anticonvulsants were the group of medicines most frequently described. Sodium valproate was the single medicine most frequently associated with a fatal suspected ADR. The study also highlighted the failure of health professionals to follow guidelines in relation to minimising ADRs.

2.8.6 CONCLUSION

Drug therapy has been a major advance in the medical care of children but, unfortunately, has resulted in significant toxicity. Some of the major ADRs that have occurred in children over the last 120 years have been highlighted. It is important that health professionals learn from these tragedies in order that we can minimise future cases of drug toxicity by understanding the differences in drug metabolism and distribution in paediatric patients. Novel methods of pharmacovigilance are required to improve recognition of existing ADRs and also to detect previously unrecognised ADRs. There are cost implications associated with novel methods of paediatric pharmacovigilance. The prevention of both mortality and morbidity in children however would justify such expenditure.

REFERENCES

1. Impicciatore, P., Choonara, I, Clarkson, A., Provasi, D., Pandolfini, C., Bonati, M., *Br. J. Clin. Pharmacol., 52*, 77 (2001).
2. Choonara, I., Rieder, M., *Paed. Perinat. Drug. Ther., 5*, 12 (2002).
3. Rayner, W., *B.M.J., 1*, 294 (1886).
4. Silverman, W. A., Anderson, H. D., Blanc, W. A., Crozier, D. N., *Pediatrics, 18*, 614 (1956).
5. Dunn, P. M., *J. Obstet. Gynaecol. Br. Commonwealth, 71*, 128 (1964).
6. Sutherland, J. M., *Am. J. Dis. Child., 97*, 761 (1959).
7. Weiss, C. F., Glazko, A. J., Weston, J. K., *N.E.J.M., 262*, 787 (1960).
8. Dreifuss, F. E., Santilli, N., Langer, D. H., Sweeney, K. P., Moline, K. A., Menander, K. B., *Neurology, 37*, 379 (1987).

9.   Giles, H. McC., *Lancet, 1*, 1075 (1965).
10. Starko, K. M., Ray, G. C., Dominguez, L. B., Stromberg, W. L., Woodall, D. F., *Pediatrics, 66*, 859 (1980).
11. Belay, E. D., Bresee, J. S., Holman, R. C., Khan, A. S., Shahriari, A. B., Schonberger, L. B., *N.E.J.M., 340*, 1377 (1999).
12. McGovern, M. C., Glasgow, J. F. T., Stewart, M. C., *BMJ, 322*, 1591 (2001).
13. Clarkson, A., Choonara, I., *Arch. Dis., Child., 87*, 462 (2002).
14. Parke, T. J., Stevens, J.E., Rice, A. S. C., et al., *Paed. Anaesthesia, 8*, 491 (1998).
15. Bray, R. J., *Paed. Anaesthesia, 8*, 491 (1998).
16. Geiling, E. M. K., Cannon, P. R. *JAMA, 111*, 919 (1938).
17. Gershanik, J., Boecler, B., Ensley, H., McCloskey, S., George, W., *N.E.J.M., 307*, 1384 (1982).
18. Phelps, D. L. *Pediatr., 74*, 1114 (1984).
19. Martin-Bouyer, G., Lebreton, R., Toga, M., Stolley, P. D., Lockhart, J., *Lancet, 1*, 91 (1982).
20. Turner, S., Nunn, A. J., Fielding, K., Choonara, I., *Acta Paed., 88*, 965 (1999).
21. Bonati, M., Marchetti, F., Zullini, M. T., Pistotti, V., Tognoni, G., *Adverse Drug React. Acute Poisoning Rev., 9*, 103 (1990).
22. Collins, G. E., Clay, M. M., Falletta, J. M., *Am. J. Hosp. Pharm., 31*, 968 (1974).
23. Choonara, I., *Br. J. Clin. Pract., 42*, 21 (1988).
24. Gill, A. M., Cousins, A., Nunn, A. J., Choonara, I., *Ann Pharmacother, 30*, 125 (1996).
25. Norris, E., Marzouk, O., Nunn, A. J., McIntyre, J., Choonara, I., *Dev. Med. Child. Neurol., 41*, 340 (1999).
26. Clarkson, A., Ingleby, E., Choonara, I., Bryan, P., Arlett, P., *Arch. Dis. Child., 84*, 337 (2001).
27. Clarkson, A., Choonara, I., Martin, P., *Paed. Anaesthesia, 11*, 631 (2001).

2.9 Poisoning in childhood

Michael Riordan[1], George Rylance[2]

[1] *Department of Paediatrics, Yale University, New Haven, Connecticut, USA*
[2] *School of Clinical Medical Sciences (Child Health),*
University of Newcastle upon Tyne, UK

2.9.1 ASSESSMENT

Poisoning is common in the paediatric population. Serious consequences, however, are rare and the vast majority of children will not require specific treatment.

History and clinical examination

The nature and quantity of the substance ingested should be identified as precisely as possible. In very young children the poison is often easily identifiable, but the dosage is difficult to ascertain. In these cases, the maximum amount of toxin which could have been ingested should be identified from the packaging. Care must be taken not to overlook the involvement of other children in a poisoning incident.

Poison identification

The possibility of poisoning should be considered in children presenting with unexplained, sudden illness. The presence of unusual patterns of presentation, symptoms, signs or routine investigations should increase suspicion.

Prominent activity of the autonomic nervous system can be associated with a broad range of toxins. Physical manifestations might include skin changes such as flushing, sweating, hot dry skin; pupillary changes; hypertension and tachycardia; muscle fasciculation, weakness or paralysis. Clear demarcations between sympathetic and parasympathetic signs may not be present.

Severe metabolic abnormalities or unusual combinations of derangement should raise the possibility of poisoning. Severe acidosis is associated with methanol, ethylene glycol and salicylate poisoning. Non ketotic hypoglycaemia is seen in ethanol poisoning. Hyperglycaemia is seen in organophosphate poisoning, acetone and theophylline toxicity.

Is it poisonous?

A careful history may avoid the need to treat many common paediatric ingestions. If it is possible to establish the ingested dose of a toxin as below the treatment level, then no further treatment is required. Many commonly ingested substances are not toxic – oral contraceptive pills cause mild gastro-intestinal disturbance but are not otherwise toxic. Mercury absorbed by inhalation or through the skin is extremely toxic, however mercury swallowed as a consequence of thermometer breakage is not toxic as it is poorly absorbed from the gastrointestinal tract.

2.9.2 MECHANISMS OF TOXICITY

Toxins exert effects via a diversity of molecular mechanisms. The mechanism of action of poisons can be broadly classified according to their effects on cellular structure and function:

Impaired cellular energy production

A variety of toxins act by inhibiting components of the respiratory chain, resulting in impaired cellular energy production. Cyanide binds to ferric iron in the cytochrome a-a3 complex, inhibiting its action and blocking the final step in oxidative phosphorylation. Arsenic reacts with adenosine diphosphate, uncoupling oxidative phosphorylation and resulting in ATP depletion. Iron disrupts energy production by direct disruption of mitochondria.

Receptor agonists and antagonists

Many toxins exert effects by binding to cellular receptors, acting as either agonists or antagonists to perturb homeostasis. The toxic effects of prescription drugs are often predictable as an extension of their pharmacological action. Anti hypertensive β blockers are highly toxic in overdose, binding to β adrenoreceptors, blocking sympathetic tone and producing profound bradycardia and hypotension.

Disruption of electromechanical function

Poisons can disrupt electromechanical function by altering membrane permeability or polarisation state. The antidiabetic drug, sulphonylurea, increases insulin release by depolarising pancreatic beta cells. Ingestion of a single tablet can produce symptomatic hypoglycaemia in children.

Depletion of intermediate metabolites

Some poisons act indirectly, depleting cellular stores of important intermediate metabolites. The anti-tuberculosis drug, isoniazid, reacts with pyridoxine to form a compound which is rapidly excreted in the urine. Pyridoxine is necessary for the production of the inhibitory neurotransmitter gamma-aminobutyric acid (GABA). Isoniazid poisoning leads to a deficiency of GABA and results in intractable seizures.

Generation of toxic metabolites

The generation of toxic metabolites underlies the toxicity of a number of compounds and enhances the toxicity of others. Ethylene glycol, most commonly encountered in antifreeze, brake fluid or windscreen wash, is metabolised by alcohol dehydrogenase to a variety of toxic metabolites, including glycolic and oxalic acid, which cause convulsions, coma, acidosis, hypocalcaemia and renal failure.

Paracetamol is metabolised in the liver to a variety of products, the majority of which are not toxic. N-acetyl-p-benzoquinone imine (NAPQI) is a highly reactive product, produced by paracetamol oxidation, which is normally rendered inert within the liver by combination with glutathione. In paracetamol overdose, NAPQI production overwhelms cellular reserves of glutathione and toxic levels result in severe liver damage. Paracetamol hepatotoxicity is common in adolescents, but rare in pre-pubertal children who have greater reserves of glutathione and enhanced sulphation.

Cell membrane disruption

Alcohols and other organic solvents exert their toxic effects by disruption of the cell membrane.

Inhibition of protein function or synthesis

The inhibition of protein function or synthesis can have devastating consequences. Carbon monoxide (CO) binds to haem-proteins disrupting oxygen transport and cellular metabolism. The plant toxin ricin irreversibly inhibits protein synthesis by damaging ribosomes.

Disruption of homeostasis

Some toxins exert their principle effect by the disruption of homeostasis. Aspirin induces hyperventilation by a direct action on the respiratory centre. In addition, aspirin uncouples oxidative phosphorylation in skeletal muscle, leading to pyrexia and metabolic acidosis.

2.9.3 PHARMACOKINETICS

Absorption of toxins

The majority of toxins encountered by children are taken by mouth and absorbed via the gastrointestinal tract. Notable exceptions include mercury, which is only toxic when absorbed by inhalation or via the skin; and carbon monoxide inhalation.

Factors affecting the absorption of toxins in the gastrointestinal tract include the solubility and formulation of the toxin and any effect toxin may have on gastrointestinal transit time.

In general, liquid preparations are subject to faster absorption than tablets – a disadvantage offset by the fact that such preparations are often of a lower dosage or concentration as they are intended for children. Many toxins induce diarrhoea or vomiting following ingestion, a response which may help to reduce absorption. Anti-histamines and certain anti-motility drugs, prescribed most frequently for the treatment of diarrhoea, decrease gut motility. The resulting prolongation of gastro-intestinal transit time results in delayed onset of symptoms and enhanced absorption of toxin.

The risks of ingestion of slow release preparations can be overlooked if allowance is not made for a delay in the peak plasma concentration of toxin. Standard preparations of iron and aspirin tablets can aggregate in the stomach – decreasing the surface area available for absorption and effectively producing a sustained release of toxin.

The bio-availability of toxins is also determined by the extent to which they are metabolised prior to entering the systemic circulation. Insulin is not toxic when swallowed, due to complete breakdown within the gut.

Distribution of toxins

Toxins entering the systemic circulation undergo distribution within the body as determined by the physiochemical properties of the poison. The pattern of distribution between compartments will depend on the ability of the toxin to: (a) cross plasma membranes; (b) bind to plasma or tissue components; (c) exist in an ionised form; and (e) dissolve in fat.

Values for the apparent volume of distribution, V_d, the notional volume of fluid required to contain a total dose, D, of drug in the body at the same concentration as that present in the plasma, C_p [3], are available for many drugs and can be used to calculate the approximate amount of toxin consumed, based on measurements of plasma concentration. For example, a 2-year-old, weighing 12 kg has a four hour plasma paracetamol concentration of 250 mg/L. The V_d of paracetamol is 1.0 L/ kg [4], therefore the minimum dose taken to give the current level $D = V_d \times weight \times C_p$; so D is 3000 mg (3 g) of paracetamol.

Calculating the dose of drugs taken, on the basis of V_d, can be useful from a medico-legal perspective. The final value for D is an approximation, as V_d measurements are not possible in humans using toxic doses of drugs. In the example given, D will be underestimated, as V_d is expected to increase with high doses of paracetamol due to increased protein binding. In addition, this method of analysis makes no allowance for metabolism or excretion of drug occurring prior to the blood test.

Metabolism of toxins

Enzymatic modification of toxins occurs principally in the liver. Under normal conditions the processes of oxidation, reduction or hydrolysis are used to add reactive groups to many molecules, prior to inactivation by conjugation. Metabolic pathways can become swamped in the presence of large quantities of toxin and reactive intermediates accumulate producing cell damage. The rate at which toxins are metabolised depends on the enzyme system involved and the concentration of drug present.

Liver enzymes are sensitive to induction by a number of drugs and illnesses. The toxicity of drugs metabolised by the liver can be enhanced in patients taking these agents, as toxic intermediates accumulate at a faster rate. For example, ethanol increases the activity of the P450 mixed function oxidase in the liver, resulting in enhanced paracetamol toxicity.

Metabolism of drugs, such as ethanol, phenytoin or salicylate, is limited by the availability of modifying enzymes. Relatively small increases in dosage, or reductions in the dosage interval, can produce dramatic changes in plasma concentration and symptoms of toxicity. This effect is termed *saturation* or *zero order kinetics*.

Elimination of toxins

Removal of toxic metabolites, and those drugs largely excreted unchanged from the body, is predominantly via the kidney. Speed of renal clearance will depend on the rate of delivery of toxin to the kidney, the efficiency with which the toxin enters the urine, and whether the toxin is subject to active tubular secretion or reabsorption.

Biliary excretion is uncommon but important in that drugs eliminated via this route may be reabsorbed – this effect is termed enterohepatic circulation and can exacerbate poisoning with agents such as carbamazepine, phenobarbital, theophylline or digoxin. The use of repeated doses of activated charcoal can reduce enterohepatic circulation of such toxins.

Awareness of the routes of elimination of toxins is important in managing poisoning, as many toxins impair the function of the organ responsible for their detoxification and removal. For example, ethylene glycol can be excreted unchanged by the kidney; however, products of oxalate metabolism can cause acute renal failure.

2.9.4 INVESTIGATION

Taking a careful history may obviate the need for blood tests, and particular attention should be paid to safe ingestion levels. Specific tests can be performed to assess the concentration of a number of toxins within the plasma; examples include paracetamol, salicylate, iron, digoxin, ethanol, carbon monoxide and lithium. Threshold levels are useful in the management of severe poisoning episodes; however, treatment should not be delayed pending results if action is indicated by clinical condition.

Routine measurement of plasma paracetamol and salicylate is advisable in older children presenting with deliberate ingestion of drugs or alcohol, as the cost of missing serious ingestion outweighs the price of the test [5].

Screening techniques are available to identify drugs, particularly those of abuse, from samples of blood and urine. These tests are expensive and rarely alter clinical management [6]. However, samples should be obtained and stored acutely if medico-legal or social consequences are anticipated. Calculation of the anion gap can be a useful predictor of the presence of organic acids in the blood (Figure 1).

2.9.5 TREATMENT

Preventing absorbance

There is no place for the use of emetics, such as ipecac syrup [7], in the treatment of poisoning. There is no evidence that emetics improve outcome in poisoning and their use may decrease the efficacy of subsequent specific treatment.

Gastric lavage has a very limited role in the treatment of a small number of specific poisonings, where the procedure can be performed within 1 hour

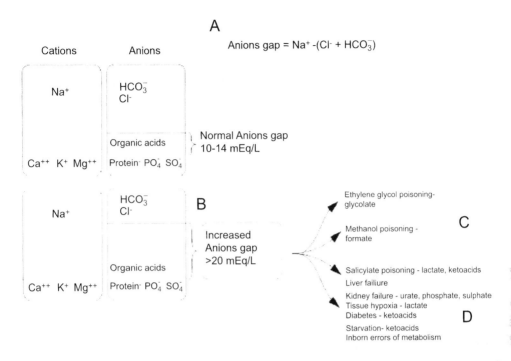

Figure 1. The number of anions in the plasma must equal the number of cations to ensure electrical neutrality. The sum of differences between the measured cations (sodium) and measured anions (chloride and bicarbonate) is termed the anion gap (A). In patients with a metabolic acidosis the presence of a significant elevation in the anion gap (B) suggests the presence of unmeasured exogenous (C) or endogenous (D) organic acids within the blood. Significant elevation of the anion gap in the presence of metabolic acidosis should raise the possibility of poisoning. Other variables can effect the anion gap, e.g. serum albumin concentration, and this calculation is not a substitute for specific investigation where indicated.

of ingestion [8]. There is no evidence that lavage improves outcome and the treatment itself carries significant risks. Gastric lavage is contraindicated if a corrosive substance or volatile hydrocarbon has been ingested. Adequate airway protection is essential in the presence of an altered level of consciousness.

Activated charcoal is effective in the treatment of poisoning with a number of toxins; however, routine use in the treatment of poisoning is inappropriate [9]. Charcoal is activated by heating in a stream of gas at high temperature. Activation generates small particles with a highly developed internal pore structure, increasing the surface area available for absorption of toxins and generating reactive carbon moieties able to bind a variety of toxins. Activated charcoal is most efficacious when administered within 1 hour of toxin ingestion. Substances for which treatment with activated charcoal is ineffective are listed in Table 1.

Enhancing excretion

Active elimination techniques have a limited role in the management of poisoning. Their use should be restricted to situations where prolonged exposure to high concentrations of toxin is predictably deleterious. Examples of such situations would include haemodynamic instability despite supportive measures, intractable seizures or organ failure.

Excretion of toxins undergoing enterohepatic circulation can be enhanced by the administration of repeated doses of activated charcoal [10]. Important examples include carbamazepine, barbiturates, dapsone, quinine and theophylline. Careful monitoring of bowel sounds is essential.

Alkalinisation of the urine can be used to enhance the excretion of weakly acidic drugs, such as salicylate (Figure 2). The unionised form of the drug is filtered and reabsorbed in the kidney. Urinary alkalinisation increases the proportion of ionised drug in the tubule, preventing its reabsorption.

Whole bowel irrigation [11] can be used to physically eliminate highly toxic substances that are not absorbed by activated charcoal and have a long gastrointestinal transit time. Treatment is based on the enteral administration of large quantities (30 ml/kg/hr) of osmotically balanced polyethylene glycol electrolyte solution to induce a liquid stool. Substances for which this technique may prove useful include iron and sustained-release or enteric-coated preparations.

Dialysis, haemoperfusion and haemofiltration have all been used to actively enhance toxin excretion (Table 2). Whilst many case reports exist in the literature, the efficiency of such methods is very difficult to assess clinically. These

Table 1

Agents which are not amenable to treatment with activated charcoal.

alcohols (e.g., ethanol, isopropanol)

essential oils

petrochemicals

iron

lithium

bleach

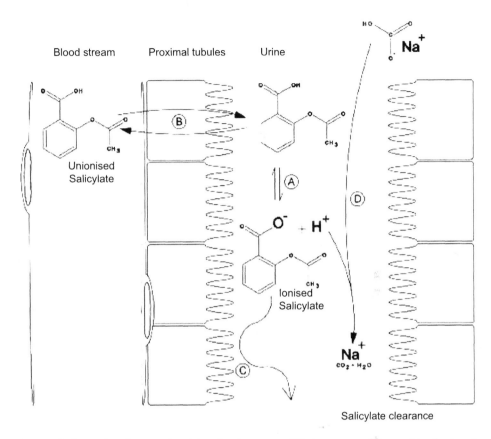

Figure 2. Salicylate is a weak acid and exists at equilibrium as a mixture of ionised and unionised forms (A). Unionised salicylate can cross between the blood stream and the urine (B); ionised drug is unable to cross out of the tubule and is excreted in the urine (C). Treatment with intravenous sodium bicarbonate (D) results in alkalinisation of the urine, neutralises hydrogen ions within the proximal tubule and increases the proportion of drug existing in the ionised form.

methods are ineffective for benzodiazepines, tricyclic compounds [21], phenothiazines, chlordiazepoxide and dextropropoxyphene.

2.9.6 SUPPORTIVE TREATMENT

In the acute setting, supportive treatment does not differ from that generally applied to any child presenting with sudden illness. Initial resuscitation should focus on assessment of airway, breathing and circulation. Subsequent management should address specific problems. Special circumstances encountered in the poisoned child can include:

Table 2

Drugs where enhanced elimination may be appropriate following poisoning

Technique	Drug	Reference
Dialysis	salicylate	
	methanol	
	ethylene glycol	10
	vancomycin	11
	lithium	12
	isopropanol	13
Haemoperfusion	carbamazepine,	
	barbiturates	
	theophylline	14
Haemofiltration	aminoglycoside	15
	theophylline	16
	iron	17
	lithium overdose	18

Dehydration

Many toxins induce vomiting, diarrhoea or diuresis, all of which may contribute to dehydration. Fluid resuscitation, possibly guided by invasive monitoring, may be necessary. Electrolyte disturbances are not uncommon and should be considered in any child requiring intravenous fluids.

Hypotension

Hypotension should be treated with intravenous fluids. Patients failing to respond to volume replacement require treatment with inotropes, dopamine or dobutamine are the agents most commonly used. The inotropic effect of glucagon has been used in the management of ß-blocker and tricyclic induced hypotension.

Acidosis

Metabolic acidosis is a common consequence of poisoning. Mild acidosis does not require correction and, in some cases, may improve toxin clearance.

Arrhythmias

Arrhythmias in the poisoned child do not necessarily indicate direct cardiac drug toxicity [22] and may be relatively benign in otherwise healthy children [23]. Initial treatment should aim to address abnormalities of electrolyte and acid-base balance, hypoxia or hypercarbia. Only if supportive measures prove inadequate should specific therapy, aimed at correcting an arrhythmia, be considered. Treatment with sodium bicarbonate is a safe first line therapy.

Hypoglycaemia

Hypoglycaemia occurs as a consequence of poisoning with a variety of toxins, including ethanol, β-blockers, ethylene glycol and sulphonylurea compounds, and should be corrected using intravenous boluses of 10% dextrose. Resistant hypoglycaemia may respond to treatment with octreotide, a somatostatin analogue which inhibits pancreatic insulin release [24].

Gastro-intestinal upset

Nausea, vomiting and diarrhoea are common following toxin ingestion and, in severe cases, may require intravenous fluids. Treatment with anti-emetic or anti-diarrhoeal drugs is contraindicated as toxin clearance will decrease and drug interactions can occur.

2.9.7 ANTIDOTES

A number of antidotes have been developed to reduce the toxicity of specific poisons. A variety of antidote strategies exist.

Inhibition of toxin metabolism

The harm caused by poisons, which must undergo metabolic processing to exert any toxic effect, can be modulated by preventing their metabolism. Treatment of ethylene glycol poisoning is based on inhibition of the enzyme alcohol dehydrogenase. Historically, treatment involved administering ethanol-utilising competitive inhibition of alcohol dehydrogenase to prevent significant metabolism of the ethylene glycol. Although elegant from a pharmacological perspective (Figure 3), ethanol can produce profound hypoglycaemia in young children and this antidote strategy was not without its risks. Fomepizole, a specific inhibitor of alcohol dehydrogenase, is now the treatment of choice (Figure 3).

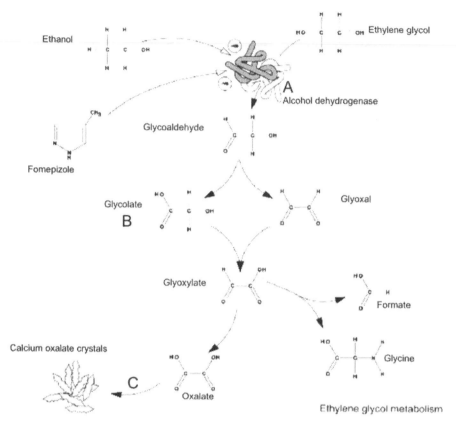

Figure 3. Alcohol dehydrogenase (A) within the liver converts ethylene glycol to gly-coaldehyde, which is subsequently converted to glycolate and glyoxal. Glycolate interfe-res with cellular metabolism and is responsible for the profound acidosis associated with ethylene glycol poisoning (B) [1]. Intermediate metabolites are converted to glyoxylate, the majority of which is converted to oxalate. Oxalate combines with calcium and precipitates (C) in the kidney resulting in acute renal failure. Ethanol is an alternative substrate for alcohol dehydrogenase and will competitively inhibit production of gly-coaldehyde. Fomepizole is a specific inhibitor of alcohol dehydrogenase which lacks the side effect profile of ethanol.

Replacing depleted metabolic stores

In some instances, the effect of poisons which act indirectly, depleting cellular stores of important intermediate metabolites, can be modulated by the administration of supplemental metabolites. Seizures associated with isoniazid poisoning can be treated with intravenous supplementation of pyridoxine (Figure 4). The liver damage associated with paracetamol poisoning can be prevented by glutathione supplementation using N-acetylcysteine (Figure 5).

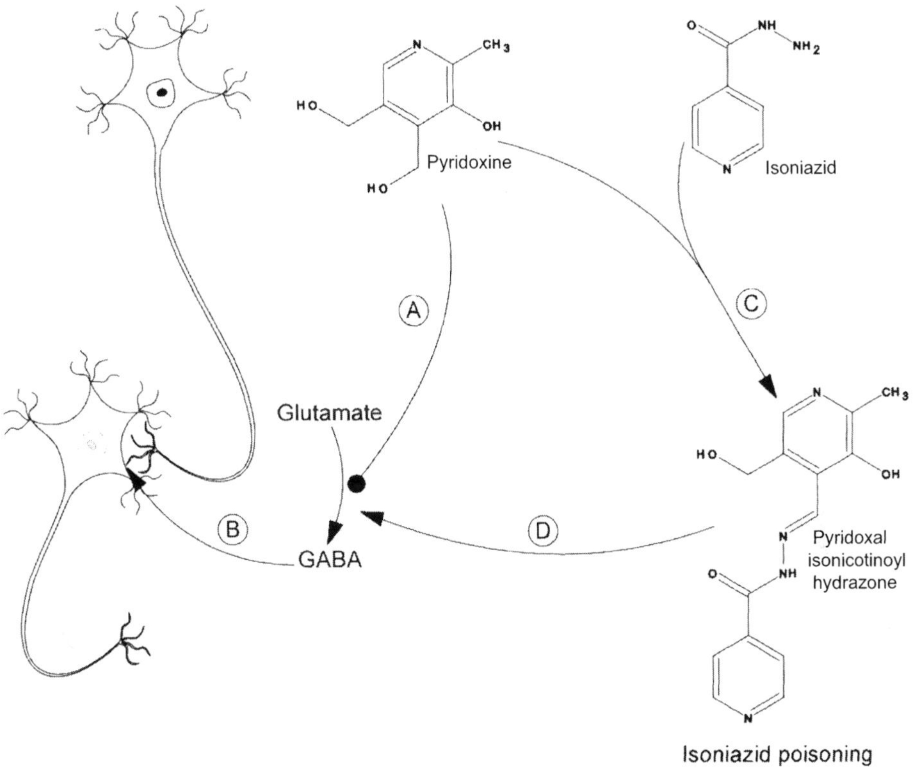

Isoniazid poisoning

Figure 4. Pyridoxine (A) is an essential co-factor for glutamic acid decarboxylase, the enzyme responsible for the synthesis of gamma amino butyric acid (GABA). GABA functions as an inhibitory transmitter in many different central nervous system (CNS) pathways (B). In overdose isoniazid reacts with pyridoxine (C) to form pyridoxal iso-nicotinoyl hydrazone which is excreted in the urine, depleting pyridoxine supplies and reducing GABA synthesis. In addition, pyridoxal isonicotinoyl hydrazone directly inhibits glutamic acid decarboxylase (D). Deficiency of GABA results in seizures and can be reversed by the administration of pyridoxine. Benzodiazipines, a class of drug commonly used to treat seizures, act by potentiating the effects of GABA within the CNS and are therefore ineffective in the treatment of seizures induced by isoniazid poisoning.

Antagonism

Direct antagonism of toxins can be undertaken by the administration of drugs having an opposing action to the poison. High dose glucagon can be used as an antidote to treat profound hypotension associated with severe beta blocker poisoning. β-blockers competitively antagonise the binding of catecholamines to ß-receptors, glucagon stimulates myocardial adenylate cyclase directly, bypassing beta receptors and increasing cardiac contractility.

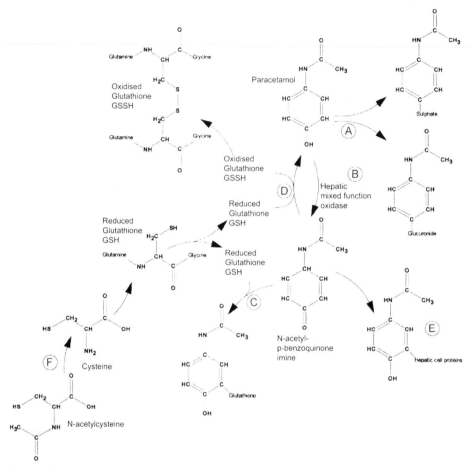

Figure 5. Paracetamol metabolism occurs in the liver. In pharmacological doses paracetamol is inactivated by conjugation with sulphate or glucuronide and the resulting compounds excreted in the urine (A). In overdose a significant proportion of paracetamol undergoes metabolism by the mixed function oxidase pathway (B) leading to the production of a variety of toxic metabolites, principle amongst whish is N-acetyl-p-benzoquinone imine (NAPQI). NAPQI can be neutralized within the liver by conjugation (C) with glutathione or converted back to paracetamol by reduction (D) coupled to the oxidation of glutathione[2]. Hepatotoxicity occurs when glutathione stores are depleted; NAPQI accumulates and begins to react non-specifically with hepatic cellular proteins resulting in cell death (E). The antidote N-acetylcysteine is converted to cysteine (F) which links glutamine and glycine to form glutathione.

Atropine can be used to block the excessive activity of acetylcholine associated with organophosphate poisoning. Organophosphates are found in insecticides, and produce irreversible acetyl cholinesterase inhibition, leading to accumulation of acetylcholine and stimulation of muscarinic receptors at parasympathetic postganglionic synapses. Patients not responding to treatment with atropine can be treated with pralidoxime, which reactivates inactivated acetyl cholinesterase.

Preventing absorption or enhancing excretion

In general, measures to prevent toxin absorbance and enhance excretion focus on poisons present in the gastro-intestinal tract. Treatment strategies also exist to prevent cellular absorbance of certain toxins from the plasma and to enhance excretion by the liver or kidneys. The iron chelating agent, desferrioxamine, can be used intravenously to prevent cellular uptake of toxic levels of iron, following ingestion of iron tablets or multivitamins containing iron. In a similar way, antibodies directed against digoxin can be administered to prevent its absorption and enhance excretion. Bound toxins are unable to enter cells and can be safely excreted.

2.9.8 PREVENTION

Strategies to reduce the incidence of childhood poisoning include public awareness campaigns, warning labels and statutory changes to improve the child resistance of packaging. Greatest care is necessary when children themselves are prescribed relatively toxic medication - in such circumstances any taboo on taking medication is absent and the toxicity of the medication needs to be stressed to the parents and suitable arrangements for storage actively discussed.

REFERENCES

1. Walder, A. D., Tyler, C. K., *Anaesthesia, 49*, 964 (1994).
2. Corcoran, G. B., Wong, B. K., *J. Pharmacol. Exp. Ther., 238*, 54 (1986).
3. Rang, H. P., Dale, M. M., *Pharmacology*, Churchill Livingstone (1987).
4. Anderson, B. J., Woollard, G. A., Holford, N. H. G., *Br. J. Clin. Pharmacol., 50*, 125 (2000).
5. Ashbourne, J. F., Olson, K. R., Khayam-Bashi, H., *Ann. Emerg. Med., 18*, 1035 (1989).
6. Belson, M. G., Simon, H. K., Sullivan, K., Geller, R. J., *Pediatr. Emerg. Care, 15*, 383 (1999).
7. Krenzelok, E. P., McGuigan, M., Lheur, P., *J. Toxicol. Clin. Toxicol., 35*, 699 (1997).
8. Vale, J. A., *J. Toxicol. Clin. Toxicol., 35*, 711 (1997).
9. Chyka, P. A., Seger, D., *J. Toxicol. Clin. Toxicol., 35*, 721 (1997).
10. Vale, J. A., Krenzelok, E. P., Barceloux, G. D., *J. Toxicol. Clin. Toxicol., 37*, 731 (1999).
11. Tenenbein, M., *J. Toxicol. Clin. Toxicol., 35*, 753 (1997).
12. Vale, A., Meredith, T., Buckley, B., *BMJ, 289*, 366 (1984).
13. Bunchman, T. E., Valentini, R. P., Gardner, J., Mottes, T., Kudelka, T., Maxvold, N. J., *Pediatr. Nephrol., 13*, 773 (1999).
14. Okusa, M. D., Crystal, L. J., *Am. J. Med., 97*, 383 (1994).

15. Simon, N. M., Krumlovsky, F. A., *Ration. Drug Ther.*, *5*, 1 (1971).
16. Pond, S. M., *Emerg. Med. Clin. North Am.*, *2*, 29 (1984).
17. Pond, S. M., *Med. J. Aust.*, *154*, 617 (1991).
18. Higgins, R. M., Hearing, S., Goldsmith, D. J., Keevil, B., Venning, M. C., Ackrill, P., *Postgrad. Med. J.*, *71*, 224 (1995).
19. Banner, W., Jr., Vernon, D. D., Ward, R. M., Sweeley, J. C., Dean, J. M., *Crit. Care Med.*, *17*, 1187 (1989).
20. Ayuso, G. A., Leon R. M. A., Mestre. S. J., Diaz B. R. M., Sirvent C. J. M., Nolla, P. M., *Rev. Clin. Esp.*, *185*, 195 (1989).
21. Henry, J. A., *Hum. Toxicol.*, *1*, 359 (1982).
22. Commerford, P. J., Lloyd, E. A., *Med. Clin. North Am.*, *68*, 1051 (1984).
23. Mofenson, H. C., Caraccio, T. R., Schauben, J., *Pediatr. Clin. North Am.*, *33*, 723 (1986).
24. McLaughlin, S. A., Crandall, C. S., McKinney, P. E., *Ann. Emerg. Med*, *36*, 133 (2000).

The administration of medicines to children

3.1 Routes of administration and formulations

Anthony J. Nunn

Department of Pharmacy, Royal Liverpool Children's NHS Trust, Liverpool L12 2AP, UK

3.1.1 INTRODUCTION

Clinical pharmacology and paediatric medicine usually focus on the drug subs-tance (active ingredient) and less frequently on the drug preparation and its other ingredients. However, pure drug substances are rarely administered to patients but are presented as dosage forms suitable for the intended route of administra-tion, for example tablets, capsules and liquids for oral administration; injection solutions or powders to be reconstituted for injection; pressurised inhalers and nebuliser solutions for administration into the lung. The route of administration may be determined by the physical and chemical characteristics of the drug; the characteristics of the patient and of their illness. Thus, intravenous injections will be preferred for treatment of life-threatening infections or for drugs poorly absorbed from the gut whilst the convenience of oral administration will be pre-ferred for management of a long term illness such as epilepsy.

Drug substances must be turned into suitable preparations by formulation scientists, taking into account acceptability to the patient or carer; the need for physical, chemical and microbial stability to provide an adequate expiry period during distribution, storage and use; interactions with packaging and administra-tion materials and the problems of handling materials on a manufacturing scale.

This section will not consider formulation and manufacturing science in detail but will consider those inactive ingredients or excipients added to the prepa-ration of the drug substance and their clinical effects. It will also consider the routes of drug administration linked to appropriate formulations and the need to manipulate dosage forms designed for adults when appropriate authorised drugs or preparations are not available for children.

3.1.2 EXCIPIENTS

To produce drug preparations appropriate for different routes of adminis-tration and to ensure their acceptability and long term stability requires the addition of certain other chemicals or substances called excipients. Excipients should be pharmacologically inactive but many have some effect and can pro-duce adverse reactions in some patients. Those that are absorbed from the gut, skin and mucus membranes or injected, may have systemic effects and may be metabolised and eliminated by the liver and kidney. Developmental phar-macology may be as important for the handling of some excipients as it is for 'active' substances. Drug regulatory agencies determine which excipients are acceptable and they usually set exposure limits and labelling requirements for the finished product [1].

Examples of different types of excipients and their functions are shown in Table 1 [2].

Table 1
Common excipients and their function

Function	Examples
antioxidant	ascorbic acid, vitamin E
antimicrobial preservative	chloroform, ethyl alcohol, hydroxybenzoates
solvents and co-solvents	ethyl and benzyl alcohols, propylene glycol, oils
chelating agents	EDTA
emulsifiers and suspending agents	lecithin, tragacanth, keltrol
stabilisers	povidone
pH adjusters	hydrochloric acid, sodium hydroxide
aids to powder flow and compression	magnesium stearate
aids to dissolution of tablets	starch
capsule materials	gelatin
sweeteners and flavours	sucrose, aspartame, essential oils
colouring agents	azo dyes (E102 tartrazine; E110 sunset yellow)
coatings	polymethacrylates

Ethyl alcohol

Alcohol may be used as a co-solvent or antimicrobial preservative and may be included at a high concentration in some preparations such as phenobarbital elixir (38%) and ritonavir oral solution (43%). Administration of large quantities may elevate blood alcohol levels [3] and the taste may be unacceptable to some children. Chronic administration may induce liver enzymes. The presence of alcohol should be apparent from the patient information leaflet for products authorised in the EU [4].

Benzyl alcohol and benzoate salts

Benzyl alcohol and benzoate salts may be used as co-solvents or preservatives and may not be tolerated by neonates when metabolism is immature. 'Gasping syndrome' and death have been described in neonates treated with intravenous saline preserved with benzyl alcohol [5,6]. Preparations such as lorazepam injection and amiodarone injection contain benzyl alcohol and have been contraindicated for children up to 3 years of age. Use of such excipients is potentially denying useful drugs to this age group.

Sweeteners and aspartame

Sucrose is a common and useful sweetener used to disguise poor taste. It can cause dental caries in long term use if dental hygiene is poor [7-9]. Aspartame is commonly used as an alternative but is a source of phenylalanine unsuitable for some patients with phenylketonuria [10]. Sorbitol and mannitol may induce diarrhoea in large amounts [11] whilst lactose (a common bulking agent) may not be tolerated in lactase deficiency, for example following severe gastroenteritis.

Colourings and preservatives

Tartrazine and other azo dyes may produce a variety of hypersensitivity reactions particularly in patients sensitised to aspirin or with asthma. Hydroxybenzoate preservatives may produce allergy and bronchospasm [12].

Parents will sometimes describe hyperactivity reactions to colourings and preservatives in food and medicines. Clinical trials have not demonstrated problems but parents who have rechallenged their children are often convinced of a problem [12]. If individuals have shown hypersensitivity or hyperactivity reactions, their drug preparations should be changed with great caution and knowledge of the excipients in the new preparation [13].

Some other problems

Propylene glycol is a solvent that may cause hyperosmolality and lactic acidosis when infused intravenously, particularly if there are renal problems [14,15]. Inadvertent or deliberate substitution of diethylene glycol in oral preparations of sulphanilamide and paracetamol syrups has resulted in many deaths, some in the last decade [16].

Changes in the excipients may lead to altered bioavailability. This was well demonstrated when calcium was removed from oral phenytoin preparations with a significant increase in bioavailability of phenytoin [17].

3.1.3   GENERAL CONSIDERATIONS

The route of administration and formulations should be acceptable to the child patient and their carer. Many carers and older children in UK do not like the rectal route of administration, but it may be acceptable for the unconscious child in hospital or to control fits if the oral or buccal route is not available. Whilst the intranasal route may be uncomfortable, it may be preferred by some children to injection. Many medicines are administered orally to children so taste, texture and smell are very important for acceptability. The age at which children are able to swallow tablets is variable and may depend upon the size of the tablet and taste of the alternative liquid [18]!

Most medicines are administered by the child's carer. Administration devices should allow accurate measurement and controlled delivery. Syringes or pipettes designed for oral use are generally more accurate and convenient than spoons. Medicines should not be added to the total contents of a babies milk feed in case the whole feed is not taken.

Because childhood and adolescence span a variety of ages, weights, preferences and abilities, a range of preparations may be required to allow accurate and convenient dose administration to all paediatric age groups [19]. Such ranges of preparations may be limited because of commercial constraints, but where the market is large, the pharmaceutical industry has demonstrated that it can be innovative and meet the needs of the diverse paediatric population. Thus, ondansetron is available as tablets in two strengths, rapidly dissolving buccal 'melts', oral liquid, injections in two strengths and suppositories; paracetamol is available as oral liquid and suppositories in several strengths, soluble tablets, rapidly dissolving buccal melts and injection [20].

For administration by the oral and inhalation routes in particular, some co-operation and co-ordination is required. Skill and experience are needed to determine which will be the most suitable preparation for individual patients. Whilst some pre-school children will swallow small tablets, most prefer a palatable liquid medicine. For many children, tablets must be crushed or capsules opened and the contents administered with food or liquid. The stability and bioavailability of preparations may be altered and, if such manoeuvres are intended, the characteristics of the modified preparations should be studied and information made available. Devices for administration by inhalation in the treatment of asthma are well studied and have been adapted to infants, children and young adults of all ages and abilities [21,22].

Because many medicines for children are used 'off-label', the commercial drug preparation has been designed for adults and is often presented as an adult single dose. Administration to the child may require manipulation, such as cutting of tablets or dispersing them in a known volume of water, so that a proportion can be administered. Segmenting tablets may not deliver accurate doses [23, 24] and many tablets may disperse but not dissolve, such that proportions cannot be taken accurately, e.g., aspirin and amlodipine.

Injections intended for adults may contain many times the dose needed for a child, such that a dose calculation error may not be spotted, because the dose can be administered from a single ampoule. For example, morphine 10 mg in 1 ml ampoules contains 100 times the dose for a neonate. If the dose is calculated accurately, measurement of the very small volume or dilutions to obtain measurable volumes may lead to error. Thus 100 microgram of morphine for a neonate is contained in 0.01 ml of the 'adult' preparation.

Formulations should be adapted for the age and ability of the child. If suitable formulations of off-label medicines are not commercially available, they may be prepared extemporaneously or manufactured on a small scale by pharmacists [25]. Pharmacists may also prepare novel (unlicensed) preparations. Importation of suitable authorised formulations from countries with robust regulatory arrangements (such as Europe, USA, Australasia) should be considered as a lower risk alternative to extemporaneous preparation but professional and patient information should be translated into the local language [26].

3.1.4   ORAL ROUTE

Because of the well known problems administering tablets and capsules to young children, liquid medicines are preferred [27]. However, it may be an unreliable route if the child is uncooperative. Taste, smell and colour of the liquid medicines will all be important in determining acceptability and will be major factors in compliance and concordance. The chemical form of the active drug may be altered to improve acceptability e.g. metronidazole benzoate, but it may be necessary to add a variety of excipients such as sweeteners and flavours in an attempt to improve acceptability. In principle, the minimum number of excipients should be used and selected from those least likely to cause adverse effects [28,29]. Taste perception changes during childhood and may be quite different to that of adults who are often used as taste testers for medicines [30, 31]. It can be ethically acceptable to use child patients to test taste [32,33] and to use child volunteers for testing the taste of low risk medicines [34-36]. *In vitro* methods of testing are being developed which will allow experimentation with taste masking agents in preparations considered too toxic to use with human volunteers e.g. cytotoxic agents [37].

Oral liquid medicines are best measured and administered from an oral dosing syringe (e.g. Baxa Oral Dispensing System, http://www.baxa.co.uk [Accessed 6 May 2004]) which will not connect to intravenous apparatus and avoids the possibility of inadvertent IV administration if injection syringes are used. Concentrated liquids with the dose administered in drops (approximately 0.05 ml) are attractive because of their low volume, but the accuracy of drop delivery has been questioned [38].

Osmolality of oral liquid medicines is important and high osmolality of infant feeds and pharmaceuticals has been associated with necrotising enterocolitis in the newborn, although a causal relationship is not established [39-41]. Oral liquids with high osmolality may irritate the stomach and produce nausea or vomiting; they should be diluted prior to administration. Solutions intended for injection are sometimes administered orally if no other liquid preparation is readily available e.g. labetalol, dexamethasone. However, their formulation should be investigated and those with high osmolality, extremes of pH or unsuitable excipients avoided.

Buccal administration

Buccal administration of small volume liquid medicines (e.g. midazolam for seizures) or solid dose fast-dissolving tablets or wafers (ibuprofen, ondansetron, paracetamol, piroxicam) can be useful for young and older children. 'Melt' technology uses freeze-dried active drug with flavours, which dissolve in the saliva and are difficult to spit out. Orodispersible tablets may dissolve in the mouth or be dissolved in a small amount of liquid before administration. Taste of these preparations and of chewable tablets should be acceptable [42]. Melting wafers may be difficult to divide to provide a dose appropriate to age or weight. Chewable tablets are considered safe to administer to young children [43]. Depending on their physicochemical properties, drugs administered by the buccal route may be absorbed locally (fentanyl) or swallowed and absorbed in the stomach and intestine (ondansetron, piroxicam).

Modified release preparations

Preparation of modified release liquid dosage forms is a considerable challenge for the pharmaceutical industry and few are available for young children. However, some products containing modified release granules presented in capsules e.g. theophylline, may be opened and administered with food. Drugs with a long elimination half life may be preferred to avoid repeated doses e.g. amlodipine for nifedipine. Modified release preparations or drugs with a long half life are particularly useful to avoid administration whilst the child is at school e.g. methylphenidate, atomoxetine. Polymethacrylate resins used for positioned release of some high-dose pancreatic enzymes have been associated with colonic strictures [44].

3.1.5 INTRAVENOUS ROUTE

For sick infants and children, intravenous administration may be the only appropriate route for the administration of antibiotics and supportive drugs. Accessing small veins in babies and young children may be difficult and peripheral venous access may need to be re-established frequently as cannula material, infusion solutions and drugs produce mechanical, chemical or osmolar irritation to the vein [45]. Use of topical local anaesthetic creams can reduce the pain and stress of cannulation and peripherally sited cannulae may permit drug administration over several days if attention is given to adequate dilution and speed of injection. Guaranteed venous access may be required for the most seriously ill and will require insertion of a catheter into the central veins. Dilution of drugs administered into the central veins is less critical than into the periphery and this may allow higher concentrations to be administered when fluids are restricted. Attention must still be given to an appropriate rate of administration.

Many preparations for intravenous administration will require dilution prior to administration or flushing into the circulation and the volumes and fluids used can be significant and important for fluid and electrolyte balance. Ensuring that drugs have reached the systemic circulation is essential when monitoring plasma drug levels e.g. for aminoglycosides, if inappropriate dosage adjustment is to be avoided [46, 47]. Most intravenous drugs should be administered slowly to avoid 'speed shock' or cardio-respiratory collapse. Using the central venous route of administration may reduce the need for drug dilution but does not remove the need for slow administration.

Many intravenous drugs are presented as freeze-dried powders to be reconstituted prior to administration. A proportion of the reconstituted volume of liquid is then measured to provide the calculated dose. When a powder dissolves, it occupies a volume (the 'displacement volume') which must be taken into account. Manufacturers' directions often assume that the total contents of the injection vial are to be administered such that their recommended volume of diluent produces a drug concentration which is difficult for calculating proportions for paediatric doses. Many hospitals produce their own tables of reconstitution directions to ensure that simple calculations can be made and to give advice on the diluents, concentration and stability for infusions and the rate of administration [48,49].

3.1.6 OTHER INJECTIONS

Intramuscular injections are often painful and may cause necrosis or nerve damage if muscle mass is small and may not be absorbed effectively in neonates [50]. Most vaccines are administered intramuscularly or subcutaneously. If regular insulin or growth hormone is required they are usually administered subcutaneously. Many antibiotics and analgesics can be administered intramuscularly if necessary. Children will tolerate these routes of administration if well trained and psychologically prepared. However, most children prefer not to receive repeat intramuscular injections of antibiotics and analgesics which should be administered intravenously or enterally if possible. Using topical local anaesthesia before venepuncture can establish a route of access which may remain available for several days, allowing many injections with only one uncomfortable procedure. If subcutaneous or intramuscular injection is necessary, adequate preparation of the child, parental holding, distraction and local application of ice or cooling may all reduce pain perception [51,52]. Needle size and length and technique are important for correct administration and the correct site must be used to avoid nerve damage [53,54]. Buffering lidocaine hydrochloride [55] with sodium bicarbonate reduces pain of local infiltration anaesthesia without reduction in effect whilst use of lidocaine as a diluent for some cephalosporins [56] may reduce pain if the intramuscular route cannot be avoided. 'Pen' injectors with very fine needles or automated needle insertion may be useful for regular self-administration and needle-free transcutaneous injection using air pressure may be useful for drugs such as growth hormone [57,58].

3.1.7 PULMONARY ROUTE

The administration of metered-dose aerosol inhalers can pose particular problems for young children because of difficulties with breath co-ordination. The inspiratory effort required may be too great to activate dry powder devices or breath-activated inhalers, especially during an asthma attack. Thus, choice of inhaler device is critical to the success of therapy but may also present life style issues for school attenders and adolescents. Spacer devices, with masks for infants, will be required for young children and also for older children during the acute attack. The convenience of a small metered dose inhaler or powder inhaler may be more appropriate for delivering 'preventer' therapy to older children and adolescents [59].

Many practitioners now recommend multiple doses of beta-2 agonist therapy delivered via a spacer device in preference to nebulisers during an acute asthma attack [60]. Nebulisers that deliver drug only during efficient inspiration are being developed and may be particularly useful for administering antibiotics to children with cystic fibrosis. Lung deposition may be improved and less antibiotic aerosol discharged into the atmosphere but further development is required to provide the optimum system [61].

3.1.8   OTHER ROUTES OF ADMINISTRATION

There are many other routes of administration for systemic effect that may be considered for specific drugs including nasal, rectal and transdermal. Children will vary in the age at which they consider them acceptable or convenient. There may be particular issues with the rectal route in countries such as UK where concerns over potential accusations of child abuse have been expressed. Topical administration for local effect has not been considered in this chapter, but it should be remembered that drugs may be absorbed systemically and cause adverse effects when only local effect was intended [62]. For example, application of corticosteroids under occlusive dressings has caused Cushingoid features and adrenal suppression.

Transdermal route

Transdermal administration is limited by the physicochemical properties of drugs, by the maturation of the skin [63] and, ultimately, by the availability of suitable commercial preparations. Transdermal 'patches' can provide controlled release of medication into the systemic circulation over several days e.g. fentanyl, hyoscine, lidocaine. However, the size and release characteristics of the patches are usually intended for adults. The amount released has been modified for children by cutting the patch into halves or quarters or by covering part of the patch to prevent release e.g. hyoscine hydrobromide as Scopaderm TSS® for excessive salivation. The pharmacist should provide advice on the suitability of different patches for this type of modification, since inappropriate modification could lead to 'dumping' of the contents of the patch onto the skin.

Intranasal administration

Several drugs are administered intranasally for local effect, such as corticosteroids for allergic rhinitis, but absorption across the mucous membrane may be useful for systemic effect for drugs such as desmopressin for diabetes insipidus or nocturnal enuresis and sumatriptan for migraine. Bioavailability may be low and local irritation may limit usefulness. Midazolam hydrochloride injection has been administered intranasally for sedation and to control convulsions, but may be painful, although onset of action is rapid [64]. Diamorphine may be useful for rapid analgesia [65].

Rectal administration

Traditional suppositories and rectal enemas may be available commercially or prepared extemporaneously. Enemas usually provide faster absorption than suppositories if they are retained. Midazolam hydrochloride injection is authorised for administration by the rectal route for premedication and sedation, but is often used as an anticonvulsant. Several oral anticonvulsants may be administered rectally if necessary [66]. When these preparations are not presented in containers designed for rectal administration, a syringe and smooth plastic applicator should be used.

Administration via nasogastric or gastrostomy feeding tubes

Most liquid medicines can be administered through feeding tubes, but should be flushed with warm water (sterile water for the newborn) to prevent blockage. Many tablets may be crushed or capsules opened and contents mixed with water, but blockage of smaller tubes is a risk if the powders do not dissolve. Further advice should be sought from manufacturers or experienced practitioners [67].

Enteric coated or modified release tablets should not be crushed. It may be possible to add drugs to enteral feeds if compatibility is known [68]. Phenytoin presents particular problems and requires administration separated from feeds and adequate flushing of tubes, if bioavailability is to be maintained [69].

3.1.9 EXTEMPORANEOUS PREPARATION

Extemporaneous preparation describes the manipulation by pharmacists of various drugs and chemicals using traditional compounding techniques to produce suitable medicines when no commercial form is available. The technique is widespread in paediatric practice and may use commercial dosage forms (e.g. tablets, capsules, injections) as starting materials or pure chemical ingredients.

Formulations may be published in national reference works and journals or may have been constructed 'in house'. The physical, chemical and microbiological shelf life of the products may have been established with appropriate tests or may have been assigned arbitrarily. In a recent unpublished UK survey, it was noted that 54% of 112 *paediatric extemporaneous formulations* had inadequate data on shelf life. Products for individual patients may have little quality assurance whilst those manufactured on a larger scale should at least be checked for chemical composition. Rarely are bioavailability studies performed to demonstrate that extemporaneous preparations have the same absorption characteristics as commercial preparations. This may be especially important if different salts are used e.g. midazolam hydrochloride injection given by the buccal route compared to midazolam maleate buccal solution. Prescribers should be made aware when extemporaneous preparation is necessary and pharmacists should take steps to assure the quality of their products. (http://www.npqa.org. Home/Advisory Documents/Standards Manual [Accessed 3 May 2004]).

There are several reference works (see Table 2) containing formulations for children's medicines and providing information on the ingredients; method of preparation; stability and shelf life. It can be difficult to keep up to date with developments such that Intranet-based publications (http://www.pharminfotech.co.nz; http://www.npqa.org [Accessed 3 May 2004]) or specific journals such as International Journal of Pharmaceutical Compounding (http://www.ijpc.com [Accessed 3 May 2004]) should be consulted.

Oral liquids are comparatively quick to prepare extemporaneously. They allow flexibility in dosage from a single strength preparation, if the volume required is accurately measured using a syringe or pipette designed for oral administration.

However, oral liquids may be difficult to formulate to ensure palatability;

Table 2
Useful sources of information about paediatric formulation

Grassby, P. F., *UK formulary of extemporaneous preparations*, Penarth, Paul F Grassby, 1995. ISBN 0 9524880 0 0, www.npqa.org

---

Woods, D. J., *Formulation in pharmacy practice* (2nd ed.) www.pharminfotech.co.nz

---

Nahata, M. C., Hipple, T. F., *Pediatric drug formulations* (4th ed.) Cincinnati: Harvey Whitney Books Co., 2000. ISBN 0-929375-23-8

Trissel, L. A., *Stability of compounded formulations*, Washington: American Pharmaceutical Association, 2000. ISBN 1-58212-007-2

physical, chemical and microbial stability. The formulations may require excipients (especially preservatives) that produce adverse reactions in some babies or children.

Capsules are hard gelatin shells which can be filled with powder manually or semi-automatically. Because of difficulty with swallowing capsules, for many younger children, they are simply used as containers for powder since the capsules are often opened before administration and the contents given with liquid or food. Powders can also be individually weighed and presented in powder papers or individual containers of plastic or glass. The powder is administered in liquid or food. In general, if stored under suitable conditions away from moisture, oral capsules and powders should have greater stability than oral liquids but are more time consuming to prepare. They are fixed dosage forms so many different strengths may be required to satisfy the varying dosage requirement of children of different ages.

Segments are tablets cut into halves, quarters or smaller pieces (e.g. aspirin, captopril). They are often crushed before administration in liquid or with food. Segments from tablets are quick to cut, probably have similar stability to the original tablet but cannot be cut with great accuracy of dose [23,24] . It is likely that nurses, attempting to give the correct paediatric dose, segment tablets at ward level. Segmenting tablets and dispersing in liquid is probably more accurate than dispersing the whole tablet in liquid and attempting to take a proportion [70].

A recent survey of hospital pharmacists in Europe [71] showed that there are considerable differences between countries in the types of preparation made extemporaneously. For example, whilst pharmacists in UK, Ireland, Norway and Sweden tend to prepare oral liquids; France and Spain tend to prepare capsules; Finland and Italy tend to prepare powders.

Whilst most extemporaneous preparations are for oral, rectal or topical administration, injections may also be prepared if appropriate facilities are available. Centralised preparation of injections for individual patients may reduce risk by presenting the appropriate dose in a suitable volume, but may also achieve economies by reducing wastage [72].

Time, expertise and facilities in hospital pharmacies limit the type of preparations that can be prepared extemporaneously. Modern dosage forms using, for example, 'melt' technology and sustained release technology are rarely available.

3.1.10 CONCLUSION

The availability of licensed formulations of medicines specifically designed for children is far from optimal. Improvements in licensing regulations may provide the incentives to produce a variety of modern dosage forms that help improve compliance and concordance and exert minimum effect on life style. Until this happens, it is important that carers and pharmacists have sufficient information to adapt 'adult' dosage forms to the needs of children.

REFERENCES

1. 3AQ9A – Excipients in the dossier for application for Marketing Authorisation of a medicinal product. In *Eudralex Collection Vol. 3, The Rules Governing Medicinal Products in the European Union*, Brussels, European Commission, 1994.   http://pharmacos.eudra.org /F2/eudralex/vol-3/pdfs-en/3aq9aen.pdf.   [Accessed on 8 May 2004].
2. Rowe, R. C., Sheskey, P. J., Weller, P. (eds.). *Handbook of pharmaceutical excipients*. 4th ed. London, Pharmaceutical Press, 2003.
3. American Academy of Pediatrics Committee on Drugs: Ethanol in liquid preparations intended for children, *Pediatrics, 73*, 405 (1984).
4. 3BC7A – "Excipients in the Label and Package leaflet of Medicinal Products for Human Use." in *Eudralex Collection Vol. 3, The Rules Governing Medicinal Products in the European Union*, Brussels, European Commission, 2003. http://pharmacos.eudra.org /F2/eudralex/vol-3/pdfs-en/3bc7aen.pdf. [Accessed on 8 May 2004].
5. Gershanik, J., Boecler, B., Ensley, H., McCloskey, S., George, W., *N. Engl. J. Med., 307*, 1384 (1982).
6. Brown, W. J., Buist, N. R M., Gipson, H. T. C., Huston, R. K., Kennaway, N. G., *Lancet, 1*, 1250 (1982).
7. Roberts, I. F., Roberts, G. J., *BMJ, 2*, 14 (1979).
8. Manley, M. C., Calnan, M., Sheiham, A., *Soc. Sci. Med., 39*, 833 (1994).
9. Maguire, A., Rugg-Gunn, A. J., Butler, T. J., *Caries Res., 30*, 16 (1996).
10. Butchko, H. H., Stargel, W. W., Comer, C. P., et al., *Regul. Toxicol. Pharmacol., 35(2 Part 2)*, S1 (2002).
11. Pawar, S., Kumar, A., *Paediatr. Drugs, 4*, 371 (2002).
12. American Academy of Pediatrics Committee on Drugs, 'Inactive' ingredients in pharmaceutical products: update, *Pediatrics, 99*, 273 (1997).
13. Nunn, A. J., Hunter, R., Ryan, S., *Pharm. J., 272*, 414 (2004).
14. Cate, J. C., Hedrick, R.S., *N. Engl. J. Med., 303*,1237 (1980).
15. Chicella, M., Jansen, P., Parthiban, A., et al., *Crit. Care Med., 30*, 2752 (2002).
16. Wax, P. M., *Ann. Intern. Med., 122*, 456 (1995).
17. Tyrer, J. H., Eadie, M. J., Sutherland, J. M. et al., *BMJ, 4*, 271 (1970).
18. Schirm, E., Tobi, H., de Vries, T. W., Choonara, I., De Jong-van den Berg, L. T., *Acta Paediatr., 92*, 1486 (2003).
19. Breitkreutz, J., Wessel, T., Boos, J. "Dosage forms for oral administration to children," in Choonara, I., Nunn, A. J., Kearns, G., (eds.) *Introduction to Paediatric and Perinatal Drug Therapy*, Nottingham, Nottingham University Press, 2003, pp. 189-205.
20. Mehta, D. K. (ed.). *British National Formulary* (47th ed.), London, British Medical Association and Royal Pharmaceutical Society of Great Britain, 2004.
21. *Guidance on the use of inhaler systems (devices) in children under the age of 5 years with chronic asthma*, London: National Institute of Clinical Excellence, 2000.
22. *Asthma inhaler devices for older children* (Technology Appraisal No 38), London, National Institute of Clinical Excellence, 2002.
23. Sedrati, M., Arnaud, P., Fontan, J.-E., Brion, F., *Am. J. Health Systems Pharmacy, 13*, 205 (1994).
24. Horn, L. W., Kuhn, R. J., Kanga, J. F., *J. Pediatric Pharm. Practice, 4*, 38 (1999).
25. Nahata, M. C., *Pediatrics, 104*, 607 (1999).

26. Nunn, A. J., *Arch. Dis. Child., 88*, 369 (2003).
27. Steffensen, G. K., Pachai, A., Pedersen, S. E., *Ugeskr Laeger, 160*, 2249 (1998).
28. Weiner, M., Bernstein, L. I., *Adverse reactions to drug formulation agents: a handbook of excipients*, New York, Marcel Dekker (1989).
29. Weiner, M. L., Kotkoskie, L. A. (eds), Excipient toxicity and safety, New York, Marcel Dekker, 1999.
30. Liem, D. G., Mennella, J. A., *Chem. Senses, 28*, 173 (2003).
31. Mennella, J. A., Jagnow, C. P., Beauchamp, G. K., *Pediatrics, 107*, E88 (2001).
32. Marshall, J., Rodarte, A., Blumer, J., Khoo, K. C., Akbari, B., Kearns, G., *J. Clin. Pharmacol., 40*, 578 (2000).
33. Stevens, R., Votan, B., Lane, R., et al., *Pediatr. Hematol. Oncol., 13*, 199 (1996).
34. Angelilli, M. L., Toscani, M., Matsui, D. M., Rieder, M. J., *Arch. Pediatr. Adolesc. Med., 154*, 267 (2000).
35. Dagnone, D., Matsui, D., Rieder, M. J., *Pediatr. Emerg. Care, 18*, 19 (2002).
36. Tolia, V., Johnston, G., Stolle, J., Lee, C., *Paediatr. Drugs, 6*, 127 (2004).
37. Uchida, T., Tanigake, A., Miyanaga, Y., et al., *J. Pharm. Pharmacol., 55*, 1479 (2003).
38. Ansermot, N., Griffiths, W., Bonnabry, P., *Pharm. Hosp., 37*, 233 (2002).
39. Ernst, J. A., Williams, J. M., Glick, M. R., Lemons, J. A., *Pediatrics, 72*, 347 (1983).
40. Atakent, Y., Ferrara, A., Bhogal, M., Klupsteen, M., *Clin. Pediatr., 23*, 487 (1984).
41. Tuladhar,, R., Daftary, A., Patole, S. K., Whitehall, J. S., *Int. J. Clin. Pract., 53*, 565 (1999).
42. Sugimoto, M., Matsubara, K., Koida, Y., Kobayashi, M., *Pharm. Dev. Technol., 6*, 487 (2001).
43. Michele, T. M., Knorr, B., Vadas, E. B., Reiss, T. F., *J. Asthma, 5*, 391 (2002).
44. Smyth, R. L., *Arch. Dis. Child., 74*, 464 (1996).
45. Phelps, S. J., Helms, R. A., *J. Pediatr., 111*, 384 (1987).
46. Nazeravich, D. R., Otten, N. H., *Am. J. Hosp. Pharm., 40*, 1961 (1983).
47. Leff, R. D., Johnson, G. F., Erenberg, A., Roberts, R. J., *Am. J. Hosp. Pharm., 42*, 1358 (1985).
48. Phelps, S. J (ed.), *Teddy Bear Book, Pediatric Injectable Drugs* (6th ed.), Bethesda, USA, American Society of Health-Systems Pharmacists, 2002.
49. *Paediatric Injectable Therapy Guidelines*, Liverpool, UK, Royal Liverpool Children's NHS Trust, 2000.
50. Bergeson, P. S., Singer, S. A., Kaplan, A. M., *Pediatrics, 70*, 944 (1982).
51. Reis, E. C., Roth, E. K., Syphan, J. L., Tarbell, S. E., Holbkov, R., *Arch. Pediatr. Adolesc. Med., 157*, 1115 (2003).
52. Reis, E. C., Holubkov, R., *Pediatrics, 100*, E5 (1997).
53. Losec, J. D., Gyuro, J., *Pediatr. Emerg. Care, 8*, 79 (1992).
54. Pope, B. B., *Nursing, 32*, 50 (2002).
55. Palmon, S. C., Lloyd, A. T., Kirsch, J. R., *Anesth. Analg., 86*, 379 (1998).
56. Foster, T. S., Shrewsbury, R. P., Coonrod, J. D., *J. Clin. Pharmacol., 20*, 526 (1980).
57. Polillio, A. M., Kiley, J., *Pediatr. Nurs., 23*, 46 (1997).
58. Houdijk, E. C. A. M., Herdes, E., Delemarre-Van de Waal, H. A., *Acta Paediatr., 86*, 1301 (1997).
59. British Thoracic Society and Scottish Intercollegiate Guidelines Network, British guideline on the management of asthma, *Thorax , 58 (suppl I)*, i1 (2003).
60. Delgado, A., Chou, K. J., Silver, E. J., Crain, E. F., *Arch. Pediatr. Adolesc. Med., 157*, 76 (2003).
61. Byrne, N. M., Keavey, P. M., Perry, J. D., Gould, F. K., Spencer, D. A., *Arch. Dis. Child., 88(8)*, 715 (2003).
62. Rudy, S. J., Parham-Vetter, P. C., *Dermatol. Nurs., 15*, 150 (2003).
63. Bronaugh, R. L., Maibach, H. I. *Percutaneous absorption*, 2nd ed., New York, Marcel Dekker (1989).

64. Fisgin, T., Gurer, Y., Tezic, T., et al., *J. Child Neurol.*, *17*, 123 (2002).
65. Kendall, J. M., Reeves, B. C., Latter, V. S., *BMJ*, *322*, 261 (2001).
66. Smith, S., Sharkey, I., Campbell, D., *Paed. Perinatal Drug Ther.*, *4*, 140 (2001).
67. Rosemont Pharmaceuticals Ltd. *Swallowing difficulties protocol: achieving best practice in medication administration.* http//:www.rosemontpharma.com/Education/Protocol.htm. [Accessed on 9 May 2004].
68. Pickering, K., *Nurs. Times*, *99*, 46 (2003).
69. Au Yeung, S. C., Enson, M. H., *Ann. Pharmacother.*, *34*, 896 (2000).
70. Woods, D. J., *Extemporaneous formulations – problems and solutions*, in Choonara, I., Nunn, A. J., Kearns, G. (eds.), *Introduction to Paediatric and Perinatal Drug Therapy*, Nottingham, Nottingham University Press, pp. 177-87 (2003).
71. Brion, F., Nunn, A. J., Rieutord, A., *Acta Paediatr.*, *92*, 486 (2003).
72. Nunn, A. J., Fairclough, S. *A centralised intravenous additive (CIVA) service for children*, *VFM Update 15 (May)*, London, Department of Health, pp.18-19, (1995).

3.2 Novel approaches to the route of administration

Jörg Breitkreutz

*Institute for Pharmaceutical Technology and Biopharmaceutics,
Westphalian Wilhelms-University Münster, Corrensstraße 1, 48149 Münster, Germany*

3.2.1 INTRODUCTION

An appropriate drug formulation is the basis for efficient drug therapy in children. Paediatric formulations should allow administering medicines to children accurately and without error. If children refuse to take the medicine or if the formulation concept fails due to a paediatric particularity, the efficacy of the therapy is at risk and medication errors are probable. However, the paediatric population comprises a wide range of ages, body weights and abilities. Rapid changes in development, especially in newborns and infants, mean that the dose and the formulation of a drug must be adjusted frequently. The pharmaceutical companies are not able to provide a huge number of formulations for a single drug substance, which comprises all ages, development stages and particularities of children. This is due to the fact that a broad product portfolio is rather expensive in development and maintenance. As a consequence, gaps in paediatric drug formulations [1], as well as a widespread off-label or unlicensed drug use in children have been identified recently [2-4]. In January 2003, the Committee for Proprietary Medicinal Products (CPMP) of the European Agency for the Evaluation of Medicinal Products (EMEA) published a concept paper on formulations of choice for the paediatric population [5]. The committee stresses the need to collect all data that is available and to identify the gaps in our knowledge that require further research. The evaluation should result in a set of guidelines on the development of paediatric formulations. The state of the art of paediatric drug formulation and some future perspectives are the subject of this chapter.

3.2.2 PARENTERAL ADMINISTRATION

If a child is hospitalised due to a serious illness or if the required drug to treat an illness is unavailable for oral use, the medication may be administered parenterally by either intravenous (IV), intramuscular (IM), subcutaneous (SC) or intraosseous (IO) route.

The intravenous administration of infusions or injections is the preferred parenteral pathway. If an IV-line cannot be placed, the IO route may be an alternative as the marrow sinusoids of the long bones drain into the systemic venous circulation. Formulations based on nanoparticles (e.g. SLN®) and liposomes show advantages for some drugs. IM and SC injections may be painful and drug absorption may be erratic as it depends on a number of factors such as drug solubility and distribution, local blood flow, and tissue fluid concentration. The drug reaches the systemic circulation more slowly than by IV application. Sometimes, the prolongation of drug absorption is part of the therapeutic principle; examples include diazepam, phenytoin or insulin. The compliance of the child is usually low in parenteral administration and needle phobia can develop, especially if bad experiences from complications like nerve injuries, blood vessel damages or abscesses have occurred. Needle-free injectors have been developed which are spring-powered (Injex®) or gas driven (Intraject®, PowderJect®). A few products, for example ZomaJet®2 containing somatropin for paediatric use, have recently entered the market, and some are still under development. The initial enthusiasm about needle-free injections has vanished for a number of reasons. The major issues are that the application of the injectors is not always pain-free, the handling is more difficult than for a classical syringe, the costs are significantly higher, and targeting the drug safely into the tissue of each individual patient, e.g. by calculating "skin factors" is difficult. However, as the novel drugs developed by pharmaceutical biotechnology are mainly peptides, proteins or nucleotides and oral formulations of these compounds are not at hand, the market for needle-free injectors may increase in the near future.

3.2.3 ORAL ADMINISTRATION

The oral route is the predominant route for drug administration to children. However, there are some obstacles in paediatric use that need special developments. As mentioned above, the pharmaceutical companies are not able to provide a huge number of different formulations. Therefore, some recent research has focused on formulations that are suitable for adults and children as well.

Liquid formulations

Liquid drug formulations for oral administration are available as drops, solutions, syrups, emulsions or suspensions. Homogeneous liquids like drops, solutions and syrups show some advantages, as they ensure uniform drug concentration. The dose can be reduced to any extent if a suitable dispensing item is provided. The application and measurement of the dose of liquid drugs must be ensured and thoroughly communicated. Over 7,000 dispensing cup errors are reported to the American Association of Poison Control Centers annually [6]. Dosage cups and dosing spoons are convenient for children who can drink from cups or spoons without spilling the contents. For younger children, dosage droppers or oral syringes are preferred. As most pharmaceutical companies provide only one dosing instrument per package, a market for medical companies has been established with a number of stand-alone dosing instruments, including comic paintings and figures that may improve compliance of children.

There are three main disadvantages of liquid drug formulations for paediatric use:

• Many drug substances are insoluble or unstable in aqueous solutions. The drug needs excipients to improve drug solubility or to ensure stability, but various excipients are of major concern in paediatric use.
• Many drug substances have an unpleasant taste. After recognising a bad taste, most children refuse further doses of the drug.
• Liquid drug formulations show low feasibility of modifying the drug release.

Antimicrobial excipients, such as propylene glycol, benzyl alcohol and benzoic acid that are widely used in medications for adults, should be avoided due to their lethal potential for newborns and young infants. Polyethylene glycol, ethanol and other organic solvents should be carefully administered. In the case of parabens, Cremophor EL and sorbic acid, allergies may develop. Hence, the applicability of excipients, even if regarded as generally safe in adults, is rather limited for paediatric drug formulations. To stabilise antibiotic drugs, the pharmaceutical companies have developed dry powder formulations that are mixed with tap water by the parents or nurses just before the first application. Drug solutions or suspensions are obtained. Usually they are stable for one or two weeks after preparation. Most of them have to be stored in the refrigerator. A number of new entities and generic products with different drugs, flavours and colours have entered the market in such formulations. Some fundamentals of these preparations have been questioned recently. For example, many parents do not prepare the liquid formulations correctly [7]. Classical quality control of the finished product, known from the pharmaceutical practice, is missing in these

a)   b)

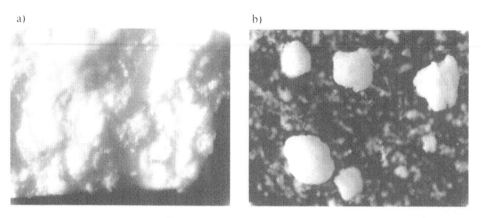

Figure 1. Amagesan Solutab®, a tablet formulation with taste-masked drug particles, a) cross-section of a Solutab® tablet; b) suspended taste-masked amoxicillin particles after addition of tap water.

extemporaneous preparations. Often detergents are added to the dry powder to improve solubility of the drug or reconstitution of the suspension. Foam builds up when the bottle is shaken. This complicates the correct dosing of water and of the medicine. If suspensions are obtained, dose uniformity is sometimes deficient [8], which has been attributed both to individual shaking behavior, as well as to insufficient stability of the prepared formulations [9].

As many drugs exhibit an unpleasant bitter taste, many children will not take oral medicines. Children are rather sensitive to a bitter taste. This may have been an advantage in evolution as many toxic plant ingredients have a bitter taste, but in drug therapy of children this is a key issue. Whereas the salty taste sensation of liquids can be usually masked by non-toxic excipients, the bitter taste remains in the form of an unpleasant aftertaste. Recently found molecules that block bitter taste sensation receptors must still pass toxicological tests. New developments in pharmaceutical technology allow the preparation of small-sized coated drug particles [10]. Recently, novel taste-masked products were introduced, containing film-coated particles in dry powder or in tablet formulations intended to prepare an antibiotic suspension. If biting on the particles is avoided, the child does not recognise the unpleasant taste for minutes (Figure 1).

Solid formulations

Granulates, powders, capsules and tablets are classical solid drug formulations. Compared with liquid formulations, solid preparations exhibit higher dosage form stability, and owing to the absence of water, higher drug stability, as well as high content uniformity of the single dose. Solid drug formulations can usually be produced with non-toxic excipients and offer a variety of measurements to control the release of the drug. Depending on the absorption properties of the drug substance, solid formulations can have the same, a lower or, in particular cases, a higher bioavailability compared with a drug solution. Granules and powders for paediatric use are usually designed for the extemporaneous preparation of liquids or suspensions, or for mixing with food by the medical staff or by parents. Attention should be drawn to the drug stability after mixing. Drugs can degrade rapidly on contact with food ingredients and enzymes. Monolithic formulations like tablets and capsules are difficult to administer to infants. As known from toxicology reports, infants are able to swallow large objects, but they dislike it and therefore may refuse the oral administration of tablets or capsules. Rapid-dissolving drug formulations [11] are an interesting new approach for paediatric purposes if the drug has an acceptable taste. Their most important property is the instantaneous disintegration in the saliva.The complex processing, in most cases based on a lyophilisation procedure, results in sponge-like structures with a large surface that enables rapid disintegration (Zydis®, Expidet®, Rapid® formulations). Natural milk products, breast milk and formula milk can also serve as sponge-forming liquids with very low toxicity [12] (Figure 2).

Disadvantages of rapid-dissolving drug formulations are the high manufacturing costs, the water sensitivity and the poor form stability. Only a few drugs

Figure 2. Scanning electron micrograph of a lyophilised breast milk formulation.

are currently available in rapid-dissolving formulations. Recently developed self-emulsifying drug delivery systems [13] contain relatively high amounts of detergents that may limit their use in paediatrics. Chewable tablets (Singulair®), extremely thin tablets (FDTAB®) or drug-loaded films (Rapifilm®) may attract children but usually the use of these formulations is restricted to drugs with a pleasant taste.

In recent pharmaceutical-technological research, various strategies have been developed to reduce the size of the classical solid dosage forms. Minitablets [14] and minicapsules [15] have been developed with sizes of less than 5 mm. As for paediatric use various dosages are required, drug formulations with multiple monolithic particles offer the best opportunities. Dose adaptation of multiple-unit systems is easy, comfortable and more exact than splitting tablets into pieces. Modification of drug release, e.g. by film coating, can be achieved for each unit. The units are offered in multiple-dose containers or a single dose is contained in a capsule or a tablet. Minitablets exhibit an excellent dose and shape uniformity. As they can pass the pylorus of the stomach in fastened and full state, they can minimise variations in bioavailability and accelerate the onset of drug action. Although minitabletting is more expensive than classical methods, one can assume that the best solid drug formulation for paediatric use at the present time is the minitablet. A single formulation for both children and adults is achieved when the appropriate amount of minitablets for an adult dose is filled into a cap-sule shell. When re-opening the capsule, paediatric doses with accurate content uniformity can be withdrawn. A special dosing instrument (Fig. 3b) can assist in safely dosing from multiple-dosage containers. The first products with minitablet technology, containing pancreatin with an enteric coating and valproic acid with a sustained-release coating, have already entered the market.

3.2.4 RECTAL ADMINISTRATION

Rectal drug administration using suppositories or rectal capsules is indicated when nausea or vomiting occurs, or if the child refuses oral medication, or if there are other reasons limiting the oral administration. Liquid or semi-solid formulations are used for local treatments or as laxatives. The acceptance of rectal drug administration varies depending on the country, being widely used in Western Europe but limited use in North America. Rectal bioavailability is poor for most drug substances (Table 1).

Antibiotic therapy with suppositories is limited because the bioavailability of most antibiotics is poor and the required systemic blood levels cannot be achie-ved. Even drugs that are often rectally administered, such as paracetamol, show less than 70% rectal bioavailability compared to an oral formulation. Recently,

a)   b)

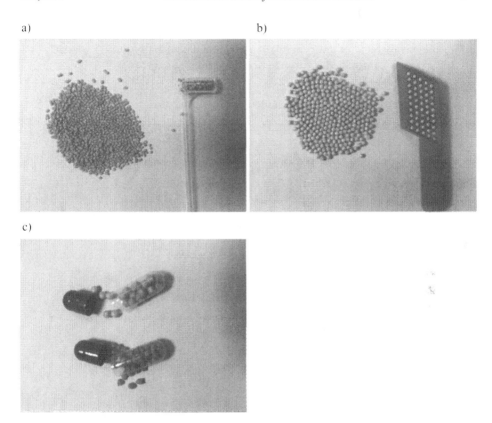

c)

Figure 3. Solid drug formulations of pancreatin: a) Enteric-coated extrusion pellets from multiple-dose container with dosing spoon; b) Enteric-coated minitablets from multiple-dose container with dosing spoon; and c) Enteric-coated minitablets and granulates in hard gelatin capsules

some efforts have been made to improve bioavailability by using novel suppository matrices including detergents and emulsifiers. In the future, new chemical entities with protein or nucleotide backbone may be formulated for rectal administration to avoid degradation in the intestine.

3.2.5   TRANSDERMAL ADMINISTRATION

The skin is a drug transport barrier perfectly constructed by nature. There are only a few drugs which can permeate the skin to a therapeutic extent. These are glycerol trinitrate, oestrogens, testosterone and nicotine. In the paediatric population, transdermal application of drugs is difficult because skin thickness and hydration may differ significantly, depending on the developmental stage of

Table 1

Rectal bioavailability of drugs in relation to oral bioavailability ($AUC_{rect.}/AUC_{p.o.}$)
(Modified and condensed table from Sakran [16])

Drug substance	Oral formulation	Suppository base	Bioavailability [%]
levodopa	capsule	cocoa butter	0
phenytoin	suspension	hard fat (triglyceride)	0
allopurinol	tablet	cocoa butter	6
tamoxifen	tablet	hard fat (triglyceride)	28
phenoxymethyl penicillin	tablet	hard fat (triglyceride)	33
aspirin (acetylsalicylic acid)	tablet	cocoa butter	63
paracetamol	tablet	cocoa butter	68
theophylline	tablet	hard fat (triglyceride)	79
naproxen	tablet	hard fat (triglyceride)	97

the child, the application site and the intactness of the stratum corneum. Some drugs, such as scopolamine, are toxic if transdermally applied to infants. Fatal toxic effects have occurred from accidental ingestion of new or used transdermal patches [17]. Most novel transdermal products have been insufficiently tested in infants. For example, fentanyl is licensed for patients above 12 years and buprenorphine above 18 years, although individual therapeutic trials have been reported.

3.2.6 ALTERNATIVE ROUTES OF ADMINISTRATION

Nasal, otic, ophthalmic or pulmonary administration is usually restricted to local drug therapy in paediatrics. As the compliance and the abilities of children to absorb medication vary, other administration routes should be preferred for systemic drug therapy. However, some progress has been made in nasal drug formulation. Today, various preservative-free solutions are available. Sterility is guaranteed by the packaging process or by special drug delivery systems, like the 3K®- or COMOD®-system for nasal solutions. As preservatives may reduce the mucus transport and their toxicity in children is often unclear, preservative-free delivery systems are required in modern paediatric drug therapy.

The application of inhalers is often limited by poor coordination and by the low inspiratory flow of the child. Spacer devices with or without face masks, e.g. Aerochamber® and Babyhaler®, and nebulisers, e.g. Pari®, are predominantly used for infants [18]. Pressurised metered dose inhalers (pMDI) require good coordination of actuation and inhalation that cannot be expected from most children. Breath actuated pMDI (Autohaler®, Easi-Breath®) significantly facilitate the coordination and have become increasingly popular. Still, these inhalers need an appropriate inspiratory flow to actuate the drug delivery. Various dry powder inhalers (DPI) are available, and most of them are breath actuated. They contain single doses, e.g. Spinhaler® and Inhalator® Ingelheim, or multiple doses that are factory-metered, e.g. Diskus/Accuhaler™, or device-metered, e.g. Turbohaler®, Twisthaler®, Easyhaler® and Novolizer® [19]. An interesting new approach represents the MAGhaler® or Jethaler®. The single dose of the drug is accurately shaved from a tablet. The required inspiratory flow of 30 l/min is significantly lower compared to other DPI. Power-assisted inhalers, e.g. the Nektar® Inhaler containing insulin, are currently being tested in clinical trials. As the lung represents a good application site for proteins and nucleotides, a number of novel inhalation products are being developed in the pharmaceutical industry and can be expected soon. Hopefully, these new products will enrich the therapeutic options in paediatrics.

REFERENCES

1. Breitkreutz, J., Wessel, T., Boos, J., *Paed. Perinat. Drug Ther., 3*, 25 (1999).
2. Conroy, S., Choonara, I., Impicciatore, P., et al., *BMJ, 320*, 79 (2000).
3. Ufer, M., Rane, A., Karlsson, A., et al., *Eur. J. Clin. Pharmacol., 58*, 779 (2003).
4. Lifshitz, M., Gavrilov, V., Gorodischer, R., *Eur. J. Clin. Pharmacol., 56*, 839 (2001).
5. EMEA, Quality Working Party (QWP), *CPMP/QWP/415/03.*
6. Litovitz, T., *Ann. Pharmacother., 26*, 917 (1992).
7. Young, S. L., *J. Pharm. Pract., 9*, 3 (1996).
8. Deicke, A., Süverküp, R., *Eur. J. Pharm. Biopharm., 49*, 73 (2000).
9. Deicke, A., Süverküp, R., *Eur. J. Pharm. Biopharm., 48*, 225 (1999).
10. Sugao, H., Yamazaki, S., Shizowa, H., Yano, K., *J. Pharm. Sci., 87*, 96 (1998).
11. Seager, H., *J. Pharm. Pharmacol., 50*, 375 (1998).
12. Yener, G., Topaloglu, Y., Breitkreutz, J., *S.T.P. Pharma., 10*, 401 (2000).
13. Pouton, C. W., *Adv. Drug Deliv. Rev., 25*, 47 (1997).
14. Lennartz, P., Mielck, J. B., *Int. J. Pharm., 173*, 75 (1998).
15. Breitkreutz, J., Kleinebudde, P., Boos, J., *Pharm. Ztg., 147*, 3210 (2002).
16. Sakran, W. S. A., *Ph.D. Thesis*, Cairo (1994).
17. American Academy of Paediatrics, Committee on Drugs, *Pediatrics, 100* (1997).
18. O'Callaghan, C., Barry, P., *Paed. Perinat. Drug Ther., 1*, 59 (1997).
19. Ashurst, I., Malton, A., Prime, D., Sumby, B., *PSTT, 3*, 246 (2000).

3.3 Medication errors

David H. Cousins

National Patient Safety Agency, 4 – 8 Maple Street, London, W1T 5HD, UK

3.3.1 PATIENT SAFETY, MEDICAL ERRORS AND ADVERSE EVENTS

The World Health Organisation identified risks in the health care system in the form of adverse events as a major risk to world health alongside malnutrition, tobacco and alcohol consumption, high blood pressure and unsafe sex etc [1]. There is a growing understanding that as well as substantial benefits; health care practices may be a source of disease and death. The World Health Organisation passed Resolution 55.18 in 2002 that recognised the need to promote patient safety as a fundamental principle of all health care systems [2]. The resolution also urged all Member States to pay the closest possible attention to the problem of patient safety and to establish and strengthen science-based systems, necessary for improving patients' safety and the quality of healthcare, including the monitoring of drugs, medical equipment and technology.

In the Institute of Medicine Report (USA) 'To Err is Human' [3], patient safety is defined as freedom from accidental injury due to medical care (medical errors). The World Health Profession Alliance (comprising the International Council of Nurses, The International Pharmaceutical Federation and The World Health Organisation) published a fact sheet on patient safety [4]. They defined an adverse event as harm or injury caused by the management of a patients' disease or condition by health care professionals rather than by the underlying disease or condition itself.

In recent years, there have been a number of major international reports indicating the high level of adverse events in healthcare systems worldwide.

The Quality in Australian Health Care study (QAHCS) published in 1995, drew attention to adverse events and iatrogenic injury in Australian hospitals. The study identified that 16.6% of patients whose hospital charts were reviewed suffered an adverse event [5].

The Institute of Medicine Report (USA) in 1999 stated that 44,000 to 98,000 deaths each year result from medical mistakes [3]. The Department of Health in England in 2000 published a report 'An Organisation with a Memory' identifying the important impact of adverse events in the NHS [6]. Adverse events were reporting to be occurring in around 10% of admissions to hospitals in the UK. This was estimated as 850,000 adverse events a year. The cost of these events was approximately £2 billion a year in additional hospital stays alone and the National Health Service was paying out every around £400 million settlement of clinical negligence claims each year, and these costs were rising.

Errors in any industry have been defined, using a systems approach, as either active or latent errors. Active errors can be observed, and are the incorrect and potentially deleterious activities directly arising when an individual uses an incorrect plan or performs a task incorrectly. Examples of these types of error are slips and lapses, which result from some failure in the execution (slips) or memory storage (lapse) of a action sequence, regardless of whether or not the plan which guided them was adequate to achieve its objective. Latent errors are design flaws in the system that remain dormant, but given a particularly set of circumstances, increase the likelihood of an active error [7].

3.3.2 MEDICATION ERRORS, ADVERSE DRUG EVENTS AND ADVERSE DRUG REACTIONS

The Harvard Medical Practice Study found that adverse events involving medicines are the single largest group of medical errors, comprising over 19% of detected events [8].

There is a plethora of terms and definitions surrounding the topic of medication errors in the literature. One popular definition was developed by the National Co-ordinating Committee for Medication Error Reporting Programmes (NCCMERP) in the USA [9]. A medication error is defined as "any preventable event that may cause or lead to inappropriate medication use or patient harm, while the medication is in the control of the health care professional, patient, or consumer. Such events may be related to professional practice, health care products, procedures, and systems including: prescribing; order communication; product labelling, packaging, and nomenclature; compounding; dispensing; distribution; administration; education; monitoring; and use."

Table 1

Definition of patient safety incidents involving medicines [10].

No.	Class name of category	Definition
1.0	Medication Error	any error in the process of prescribing, dispensing, preparing, administering, monitoring drug therapy regardless of whether an injury occurred or the potential for an injury was present.
2.0	Adverse Drug Event (ADE)	when an injury occurred as a result of the drug therapy.
2.1	preventable ADE	where an injury resulted due to an error taking place in some part of the medication process.
2.2	non-preventable ADE	where an injury occurred with no error having taken place in the medication process. For example a patient experiences a known side effect to a drug for the first time, and there was no way that the prescriber could have prospectively predicted or prevented the side effect for occurring.
2.3	ameliorable ADE	where the severity of the drug induced injury could have been substantially reduced if different actions had been taken. This usually involves events where there has been an error in the drug monitoring process.
3.0	Potential ADE	incidents that have the potential for injury but no harm was experienced by the patient.
3.1	potential ADE preventable	where an error occurred that had the potential to cause injury but was intercepted by someone within the medication process, preventing the incorrect medicine reaching the patient.
3.2	potential ADE non-intercepted	medication errors that have the potential to cause injury but failed to do so after the medication reached the patient.

This definition is too imprecise and liable to misinterpretation for some authors. The term 'inappropriate medication use' from the above definition can be interpreted as any deviation from accepted standards or procedures no matter how trivial. Bates *et al.* 1995 has developed a set of definitions to help better define patient safety incidents involving medicines (Table 1) [10].

From these definitions, Bates indicates that not all medication errors are adverse drug events (ADEs) or potential ADEs. Many medication errors are excluded from being classified as 'events' if the error is considered unlikely to cause harm, e.g., when a small dose of paracetamol is administered rectally instead if orally.

The interrelationship between these definitions is shown in Figure 1. These definitions help to illustrate that only a small percentage of medication errors have the potential for patient harm or actually cause patient harm. The diagram also helps clarify that there are some events traditionally called adverse drug reactions that can be more precisely defined as non preventable ADEs, when no medication error has occurred. However, where a patient has a known adverse drug reaction with the same or similar drug, then, in these circumstances, the incident can be classified as a preventable ADE.

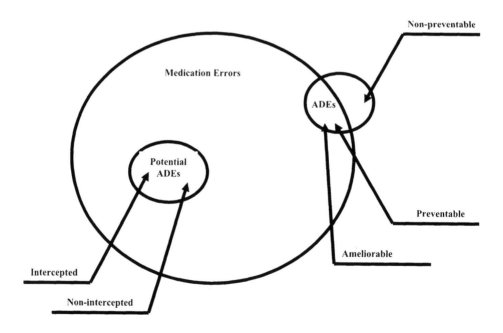

Figure 1. Diagram illustrating the interrelationship of medication errors and types of adverse drug events [10]

Bates *et al.* in 1993 reported a medication error rate of 5% [11]. Of these, 7 in 100 medication errors were classified as potential ADEs and 1 in 100 resulted in an ADE. However, this study, and the majority of similar studies, have centred on medication errors in adults.

An initiative to establish a National Reporting and Learning System for all adverse events, including those involving medicines, is underway in the United Kingdom. The National Patient Safety Agency (NPSA) has been set up to collect and analyse information on adverse events in the National Health service, assimilate other safety related information from within the UK and worldwide, learn lessons and ensure that they are fed back into practice, where risks are identified - produce solutions to prevent harm, specify national goals, establish mechanisms to track progress (see www.npsa.nhs.uk).

Following consultation with patients and healthcare staff over the best term to use for the reporting system, the NPSA uses the term Patient Safety Incident for any unintended or unexpected incident which could have or did lead to harm for one or more patients receiving NHS funded healthcare. This term is used for incident reports involving potential harm or near misses. (See www.npsa.nhs.uk for more information).

3.3.3   THE INCIDENCE OF PAEDIATRIC MEDICATION ERRORS

There is a wide range (0.09-18.2%) of medication error rates in children in published studies [12-21]. In some studies the term medication error is used, and in other studies preventable adverse drug events (actual and potential) are reported. Different measurement methods are used in these reports, ranging from voluntary incident reporting [12-17], retrospective medical chart review [18-20] or prospective observational studies of drug administration [21].

What these studies tell us is that the issue of ADEs in children is an important one. It is not possible however, to extrapolate published error rates from different healthcare settings, in different countries, using different data collection methods to your own healthcare setting. Rather, it is important to begin to record and measure these events in your own environment and take action to prevent and reduce these events.

There is recent evidence that potential ADEs may be more common in the paediatric population in the USA, suggesting that the epidemiologic characteristics of ADEs may be different between children and adults [13]. The picture is the same in England, where it has been extrapolated that at least 1675 preventable ADEs occur each year in paediatric inpatients, of which 85 could be classified as moderate/severe reactions [22].

The paediatric population is a high risk group because the number of potential ADEs is likely to be higher than that found in hospitalised adult patients [23]. Factors placing paediatric patients at increased risk of ADEs are as follows:

- Different and changing pharmacokinetic parameters between patients at various ages and stages.
- Need for calculation of individualised doses based on the patient's age, weight (mg/kg), body surface area (mg/m^2) and clinical condition.
- Lack of medicinal products with licensed indications for use in children, leading to the use of unlicensed or off-label products with inadequate information on dosing, pharmacokinetics, safety, efficacy and clinical use of the medicine in children.
- Lack of medicinal products with dosage forms and concentrations appropriate for administration to neonates, infants and children. This leads to the preparation of dosage formulations from adult designed dose strengths and forms requiring complex calculations and dilutions and the potential for 10, 100, and 1000 fold errors from a single dose unit and lack stability, compatibility, and bioavailability data.
- Problems in the continuity of supply of specially prepared dosage forms and concentrations.
- Need for precise dose measurement and appropriate drug delivery systems.

3.3.4   TYPES OF PAEDIATRIC MEDICATION ERROR

In the USA, there is a national medication error reporting programme operated by the US Pharmacopoeia [24]. In 2001, 368 hospitals/facilities released 105,603 medication incident reports. These reports included 14 deaths. There were 3,361 (3.5%) reports involving children and there were 2 child deaths. Only 5.6% of the reports involving children can be classified as an ADE. However, the percentage of reports describing temporary harm was higher in children vs adults, i.e. 0.9% for children and 0.4% adults. The percentage of reports of errors leading to deaths was higher in children i.e. 0.1% for children and 0.02% for adults. Table 2 indicates the stage in the medication process linked with the medication error report. The largest category was drug administration errors. Table 3 indicates the most frequent type of errors in children; wrong dose or quantity is the largest category.

In other countries, national systems for reporting incidents have not been established. In the absence of such a reporting system, one group of authors reviewed ADEs reported in local and national newspapers. The results of this review of ADE newspaper reports between 1993 – 2000 in the UK are presented in Table 4 [25]. Over an eight year period, there were 83 ADEs reported

Table 2

Steps in the medication process for paediatric
medication safety incident reports [24]

Medication process step	%
prescribing	8.6
dispensing	22.6
administration	51.7
documentation	15.5
monitoring	1.5

Table 3

Types of paediatric medication safety incident reports
(US Pharmacopoeia 2002)

Error type	%
wrong dose/quantity	29
omission	24
wrong time	15
unauthorised dose	11
extra dose	7
wrong drug preparation	6
prescribing error	5
wrong patient	3
wrong administration	2
wrong dosage form	2
wrong route	1

(Incident reports may contain one or more error types).

involving at least 1,146 children. There were at least 29 deaths, seven of which
involved premature neonates.

Results from individual studies [12-21] follow a similar pattern to those
seen in the national incident reporting data seen in the USA and UK. The most
common type of error reported was dosing errors and often 10 times the actual

Table 4

Types of paediatric medication incidents reported
in the local and national UK press 1993-2000 [25]

Type of error	No.	%	Fatalities	%
wrong dose	32	40	12	42
wrong drug	16	19	5	18
wrong route	3	4	3	10
incorrect container	3	4	2	8
incorrect rate of administration	2	3	2	7
omitted in error	4	5	1	4
wrong strength	3	4	1	4
wrong patient	4	5	-	0
duplicate dose	3	4	-	0
expired drug	3	4	-	0
incorrect label	2	3	-	0
miscellaneous	5	5	2	7
total	80	100	29	100

dose required.

Others included the wrong drug, route of administration, medication adminis-tration record transcription or documentation error, frequency of administration or wrong time, missed dose and wrong patient. Drugs given to patients with known allergies, drug interaction, drug-drug incompatibilities, and rate of intra-venous drug administration are also frequently recorded.

Table 5 shows the drugs included in ADE reports. Many of the drugs involved with an error have a narrow therapeutic index. The intravenous route is the most common route that is associated with medication errors in children.

Reports and studies concerning medication errors in adults and children are usually from hospitals. There are far fewer studies relating medication errors for patients at home in the community. In a six month prospective study carried out by sixteen poison centres in France, 1108 medication errors were analysed [26]. The mean age of the children involved was 3.2 years and 30% were under 1 year of age. In 87% of cases, a member of the patients family committed the error in

Table 5

Drug or drug groups involved in
medication errors in children [25, 29-35]

Type of error	Drug/drug group involved
wrong dose	adrenaline
	asparaginase
	benzylpenicillin
	BCG vaccine
	diamorphine
	digoxin
	IV fluids
	magnesium sulphate
	methylphenidate
	morphine
	phenytoin
	sedatives
	sodium nitroprusside
	tacrolimus
wrong formulation	penicillin
wrong route	vincristine
	nystatin oral suspension
wrong strength	chloroform water double strength
	magnesium sulphate

(Incident reports could be allocated to one or more error types)

medication use. In 31.5% of cases, the medication was administered to the child by the parents without medical consultation or the advice of a pharmacist. In 30% of cases, there were errors in execution of the prescription by the parents, i.e. errors in preparing and administering the medicine. There was injury in 186 patients (17%) and 161 patients (15%) were hospitalised, usually for observation only. Two patients died. The drugs most commonly misused included cough suppressants containing morphine analogues, salicylates and ear, dose and throat drops.

3.3.5 CASE REPORTS

There are benefits in publishing case reports concerning rare events, describing harms from medicines [27,28]. Although these reports have focused on describing non-preventable ADEs, it can be argued that there is even greater benefit to be gained from publishing reports of preventable ADEs (actual and potential). Such reports help organisations and individual practitioners examine current practice and determine whether current safeguards would have prevented similar errors from occurring in their own practice. Eight case reports are presented in this section that help to provide qualitative information concerning these events and help to place these events in their clinical context. Risk management methods used to eliminate or minimise the recurrence of these types of incidents are described below in Section 3.3.6.

Unclear prescribing

Case study 1: An unclear prescription for morphine (tenfold error)

A six week old baby boy was admitted to hospital for a hernia repair. While in the recovery room, a nurse misread a prescription and this resulted in the baby having 4 mg of morphine administered instead of 0.4 mg (400 micrograms). The patient's mother noticed him going blue after his return to the ward. The baby went on to have fits and stopped breathing. He was successfully resuscitated and transferred to the intensive care unit. He was kept in the intensive care unit overnight before going home a few days later [29].

Case study 2: Unclear chemotherapy prescription (tenfold error)

A five year old boy being treated for lymphoblastic leukaemia was given 4.8 ml of asparaginase, instead of the prescribed 0.48 ml by a senior nurse. A day after the treatment was administered, the patient's father was telephoned by hospital medical staff, who told him that his son must be taken immediately from his school to the hospital isolation ward for tests. The boy was tested and monitored for several days and did not exhibit any symptoms as a result of the overdose. [29].

Dose calculation errors

Case study 3: Miscalculation of a morphine dose 1

A female baby, born seven weeks prematurely, died at 28 hours old when a junior doctor miscalculated a dose of intravenous morphine resulting in the administration of a 100 times overdose. The doctor is reported to have worked out the dose on a piece of paper and then checked it on a calculator, but the decimal point was inserted in the wrong place and 15 instead of 0.15 mg was prescribed. The dose was then prepared and handed to the senior registrar who administered it without double checking the calculation and, despite treatment with naloxone, the baby died 55 minutes later [30].

Case Study 4: Miscalculation of a morphine dose 2

A four month old boy was given morphine overdoses on three separate occasions. The baby boy had been treated in a hospital special baby care unit since being born three months prematurely weighing less than 1 kg. The baby had been receiving regular intravenous doses of morphine, as his mother was a registered heroin addict when she gave birth. The baby was given 10 times the correct dose of morphine on three separate occasions, as the nurse administering the injection had misread the decimal point on the prescription. The boy subsequently died. At the inquest, the pathologist said that the three morphine overdoses had played no part in the baby's death: the death was attributed to complications arising from the premature birth [29]

Case Study 5: Miscalculation of a nitroprusside infusion

A three month old boy died after receiving an overdose of sodium nitroprusside infusion following a major heart operation. In the operating theatre a junior anaesthetist working under the direction of a consultant anaesthetist had prepared a syringe with four times the child's dose of the drug. A senior nurse in the paediatric intensive care unit noticed that the boy's blood pressure was erratic three hours after the operation and that his face was blue and swollen. She checked the dose of the drug and discovered the error. The poisons unit was contacted for advice and they recommended that sodium thiosulphate be used as a cyanide antidote. The hospital pharmacy only stocked disodium editate, which the medical staff refused to use after receiving expert advice from the poisons unit. A supply of sodium thiosulphate injection therefore had to be hurriedly sent from a neighbouring hospital and was eventually administered to the patient. Unfortunately the boy died two days later [31].

Wrong formulation error

Case study 6:　Use of the wrong formulation of penicillin　by the intrathecal route.

A seventeen month old boy was admitted to hospital with suspected meningitis. A junior doctor prepared and administered 300mg benzylpenicillin by the intrathecal route. The correct dose of benzylpenicillin for a 9 kg child of his age is less than 1 mg, although intrathecal treatment is generally discouraged. The boy started to have seizures just minutes after the injection and died a short time later, despite full resuscitation procedures.

The junior doctor admitted that she had not prepared an intrathecal injection previously, and said that she had seen thousands of bottles of penicillin but had not noticed a label warning of paralysis and death if administered into the spinal column.

The circumstances relating to this error are very tragic. During a medical ward round, two weeks before the boy was admitted, a consultant paediatrician had had a casual discussion with the junior doctor about the possible benefits of benzylpenicillin via the intrathecal route when children were extremely ill with meningitis. However, the consultant had made no mention of the dangers of using penicillin by this route of administration or of the appropriate dosage and later said that he had not foreseen any situation where the junior doctor was going to use this route to administer benzylpenicillin [32].

Wrong route

Case study 7: Intravenous administration of oral nystatin suspension.

Oral medication may be administered by the IV route, due to the use of inappropriate syringes. In a recent case, a child was accidentally given nystatin oral suspension through an IV cannula.

The suspension had been drawn up into a 1ml syringe and left with the mother for her to administer – involvement of mothers with oral medicines being part of the paediatric department policy as preparation for patient discharge. The mother was unhappy to give the medicine herself and told a student nurse who said she would give it. The medicine was then administered intravenously; the student immediately realised the error and informed a staff nurse. The paediatric registrar was contacted and he removed the cannula, which was full of the suspension. The child was discharged from hospital two days later without any apparent adverse effect [33].

Communicating and interpreting paediatric medicine information

Case study 8: Errors of interpreting dosing information for itraconazole

A three month old baby boy with chronic granulomatous disease was admitted to the paediatric ward for IV antibiotic treatment. His drugs on admission were co-trimoxazole and itraconazole, both as ongoing prophylaxis. The co-trimoxazole was correctly prescribed as 120 mg daily, but the itraconazole was prescribed as 200 mg daily. The only information found in the paediatric formularies available in the pharmacy department, was 5 mg/kg/day for oncology patients; the patient weighed 7.2 kg, which would give a daily dose of about 35 mg.

The patient apparently had supplies of his own medicine, brought to the hospital with him. The bottle of itraconazole liquid 10 mg/ml was examined by the pharmacist. It carried the dosage instructions. Take four 5 ml daily, and the senior house officer confirmed that she had copied the dosage from this. The nurse clerking the patient recorded the itraconazole dose as 1.2 ml once daily, and it is likely that this is the dose actually given by the child's mother and grandmother.

There was no information available in the patient's medical notes, as he was under the care of a specialist hospital for his disease. No shared-care arrangements were in place, and there were no consultant letters or discharge summaries. The patient's GP had a letter from the specialist hospital, advising that the patient had been started on itraconazole and would no longer require nystatin, but there was no mention of the itraconazole dose. The pharmacist contacted the pharmacy department in the specialist hospital, and another pharmacist provided the dosage information from the patient's previous dispensing records which was 12 mg once daily. Based on the information from the patient's carer and the information from the specialist hospital pharmacy, the paediatrician was happy to change the prescribed dose from 200 mg to 12 mg [34].

3.3.6 METHODS TO PREVENT MEDICATION ERRORS ON CHILDREN

Developing a fair, open and non punitive culture where medical errors are reported, discussed, reviewed and where appropriate system changes introduced to prevent further errors is a very important means of reducing medical errors including medication errors [14,35-37].

It is interesting to see how industries from outside healthcare have used incident reporting to reduce the number of serious incidents. Figure 2 illustrates an increase in the number of air transport incident reports in the UK that provided systems improvements that led to the reduction on the percentage (and actual number of) serious incident reports [38].

Other methods developed in industries outside of healthcare can be used to identify and develop initiatives to reduce serious medication error risks. The use of prospective risk assessment methods to identify current risks and safety controls and determine whether these are adequate or whether new improved safeguards need to be introduced and then re-audited [39].

Koren reported on the outcomes of initiatives to reduce medication errors in the Hospital for Sick Children in Toronto over a nine-year period [12]. Compared to a baseline, there was a steady and a statistically significant decrease in medication errors. Total errors (actual and potential) decreased for nurses and doctors by half and for pharmacists by 75%. The initiatives used to produce this decrease in errors are listed below.

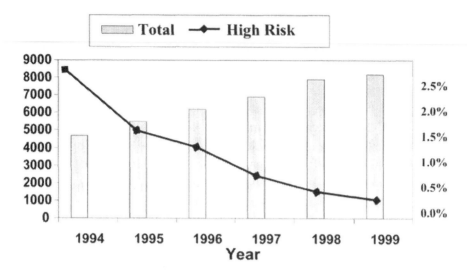

Figure 2. Air safety reports (UK), volume and risk

- Introduction of an electronic prescribing system (without decision support) to eliminate handwritten prescriptions and the requirement to transcribe prescriptions.
- The publication of a list of hazardous drugs that can kill a child e.g., digoxin, heparin, insulin, IV potassium. Implementation of a double check system to calculate hazardous drugs.
- All hazardous drugs not required for immediate use, removed from ward stock.
- All medical trainees participate in an in-service test on the calculation of drug doses.
- A hospital committee investigates all cases of major medication error to ensure immediate remedial action and steps to prevent further errors, and these steps and action are followed by an audit to determine implementation and overall success.
- A computer system calculates the drug dosage, and volume of medicines to be administered to children based on a body weight and age.

Specific initiatives listed here have been supported by other studies. Electronic prescribing either as part of an electronic medical record or as a stand alone system [40-43], more clinical involvement of pharmacists [40,44], training programmes [45].

The introduction of electronic prescribing will help clarify prescriptions and will be an important risk management method that will help to minimise errors involving unclear and incomplete prescriptions, as in case studies 1 and 2 described above.

Case studies 3 and 4 involve *dose calculation errors*. There are several reports and studies to indicate that healthcare staff have difficulties correctly performing dosage calculations for patients [46-50] and drug dosage errors are the single largest category of medication error seen in paediatrics.

Recommendations to prevent tenfold medication dose errors [51]:

- Implement and enforce rules regarding zeroes in all medication documentation, labelling and communication including typed documents.
- Always place a zero before the decimal for numbers <1 (e.g., 0.1 not .1)
- Never place a trailing zero following a decimal point for numbers >1 (e.g., 1 not 1.0)
- No trailing zeros (e.g., 1.25 not 1.250)
- Require patient weight, dosing equation and final calculated dose in that order when using dose equations.
- Require independent dose check or dose calculations by a second caregiver before administration.
- Estimate calculated dose mentally, then recalculate all doses twice with a calculator.
- Check dose calculations against a pre-calculated chart of medication doses.
- Avoid calculations whenever possible by using standardised doses and dosing tables.
- Avoid the use of decimal points when not necessary, e.g., prescribe amikacin 11 mg instead of 10.8 mg.
- Never use the abbreviation u for the word unit.
- Provide dose in multiple unit when converting e.g., levothyroxine 0.025 mg = 25 micrograms.
- Spell out numbers with multiple zero's e.g., penicillin 150,000 units = one hundred and fifty thousand units.
- Require a pharmacist to have documented patient weight prior to dispensing medication for patients.
- Require a pharmacist to review all medication orders, check dose equations and recalculate all doses.
- Require all doses of dangerous mediations to be double checked by experienced staff.
- Establish standardised reference for dosing and calculating medication with a standardised format and unit expression.
- Create individualised weight-based emergency drug dosing charts for each patient on admission to the hospital and place at the bedside.

- Establish maximum dose ranges for dangerous medications.
- Limit the availability of and access to medications on patient care units, have pharmacy prepare medications.
- Obtain accurate patient medication histories through structured interviews.
- Remove lines from carbon and NCR copies of orders.
- Increase staff awareness of the risk of tenfold errors.
- Involve patients and family in medication safety and encourage questioning of anything unusual.

Case studies 5-7 describe incidents involving the selection of the wrong formulation and route and help to illustrate the importance of safety controls that requires double checking of the medicines doses and routes of administration by a pharmacist and by a second practitioner prior to administration. It is also essential that established standardised reference sources, such as 'Medicines for Children' in the UK [52], are available and used in practice to confirm indications for use, doses and methods of administration for children.

Guidelines for preventing medication errors in paediatrics were developed and published by the Institute for Safe Medication Practices and The Pediatric Pharmacy Advocacy Group in the USA in 2001 [23]. The guidelines included recommendations to prevent prescribing errors, drug administration errors, and recommendations to improve corporate medicines management systems to reduce latent factors that can contribute to medication errors. These guidelines are listed on the next page.

Recommendations to prevent medication errors in prescribing medicines for children [23]

- Patient's name in full.
- Patient's age (date of birth) and current weight.
- Information regarding diagnosis and other patient specific data appropriate to the circumstances should be included.
- Any known allergies should be included. (If there are no known allergies – this information must be included).
- Drug name, dosage form and drug strength. If the medication is rarely used, the name should be printed. Concentrations should be expressed in metric units.
- Number or amount to be dispensed. If appropriate, quantity to be dispensed should be expressed in metric units.
- Include calculations, or at least mg/kg/day dosing, so calculations can be independently double checked; i.e. amoxicillin 40 mg oral every 8 hours (40 mg/kg/day).

- The prescriber's name and pager or telephone number should be included.
- Complete instructions for the patient including indication (i.e. purpose of the drug), directions for use including dose, frequency of dosing, route of administration, intended duration of therapy, and number of authorised refills.

Recommendations for the safe administration of medicines to children following discharge from hospital

- Written information concerning the medicines to be used outside the hospital should be provided to the carer, sometimes school teacher, the hospital doctor in secondary care (where the child has been seen in tertiary care), general practitioner and, where appropriate, the community pharmacist. This is particularly important if the medicine supplied is not available commercially and needs to be extemporaneously prepared.
- During discharge counselling, patients and/or caregivers should be asked to demonstrate the measurement of a dose if the medication is dispensed in a liquid form.
- The patient and/or caregiver should be asked to demonstrate any manipulation required in the preparation and administration of a dose, especially if special techniques are required, e.g. dilution of an oral medicine, injection, nebuliser or inhaler.
- Appropriate measuring devices should be provided or recommended.
- Where possible the measuring device should be demonstrated.
- Written education material should provide relevant information in a clear and easily understood format.
- The use of household teaspoons and tablespoons should be discouraged because of their variability and resulting in accuracies.

Case 8 describes an incident caused by poor communication between healthcare professionals and emphasises the importance of complete and detailed communication concerning medicines used in children.

Recommendations for corporate medicines management systems to reduce latent factors that can contribute to medication errors [23]

- Ensure that medicinal products are appropriate for use in paediatric populations (e.g., concentrations, dosage form, active ingredient).
- Ensure that all devices (e.g. pumps, administration sets, burette chambers) selected for use are appropriate for the paediatric population.
- Ensure that oral syringes that are not compatible with intravenous devices are used to measure and administer oral liquid medicines.
- Ensure that appropriate measuring devices are available for oral administra-

tion and standards are set to address the minimum measurable volumes for these devices (i.e. measuring 0.2 ml in a 10 ml syringe).

• Ensure that, at the time a new medicines is being reviewed for inclusion in the hospital formulary, the committee should evaluate potential errors: 1) Sound alike and look-alike confusion, 2) labelling and packaging that appears similar to other drugs on the formulary, 3) the requirement of monitoring tests to assure safety, 4) associated with clearly defined dosage information and finally, 5) allows for standardised dosing to reduce the need for complex calculations.

National Guidelines for the Prevention of Medication Errors in Pediatric Inpatient Setting, incorporating many of the safety controls described previously, were published by the Amercian Academy of Pediatrics Committee's on Drugs and Hospital Care in 2003 [53]. In England in 2004, the Department of Health published a national report 'Improving Safe Medication Practice' that included a section on preventing medication incidents in children [54]. The report reviews the frequency, nature and causes of medication errors, the risk factors inherent in current medication processes, and specific risk factors for some medicines and patient groups.

Other methods developed in industries outside of healthcare can be used to identify and develop initiatives to reduce serious medication error risks. The use of prospective risk assessment methods to identify current risks and safety controls and determine whether these are adequate or whether new improved safeguards need to be introduced and then re-audited [39].

3.3.7 CONCLUSION

In the introduction it was mentioned that the World Health Organisation asked the world-wide healthcare community to pay the closest possible attention to the problem of patient safety and to establish and strengthen science-based systems, necessary for improving patients' safety. From this review we have seen that medication errors in children are a serious problem leading to mortality and morbidity. However, there is lack of clarity and standardisation over definitions and methods used to study medication errors and difficulty in generalising medication error rates from published studies.

It is to be hoped that over the next few years, there will be initiatives to help standardise these definitions and methods and provide more generalisable data indicating evidence based methods that are proven to minimise serious medication errors (preventable ADEs) in children.

In addition to these initiatives and research studies, individual practitioners and organisations should report and review medication error incidents. A systems approach should be adopted to ensure that lessons are learnt and preventative action taken from retrospective incident reports. Prospective risk assessment should be undertaken to review medication systems and products used with children.

Finally, it is important that all healthcare staff receive training concerning patient safety, and the need to function as a team using each other's resources to achieve optimal patient care.

REFERENCES

1. World Health Organisation. World Health Report 2002, *Reducing Risks and Promoting Health Life*.http://www.who.int/whr/en/. (2002).
2. World Health Assembly. Fifty-Fifth World Health Assembly Resolution 55.18, *Quality of care: patient safety*: www.who.int/gb/EB_WHA/PDF/WHA55/ewha5518.pdf (2002).
3. Institute of Medicine. *To Err is Human. Building a Safer Health System*, Washington, National Academy Press: http://www.nap.edu/books/0309068371/html/, (1999).
4. World Health Professions Alliance. *Fact Sheet on Patient Safety*, www.whpa.org/factptsafety.htm, (2002).
5. Wilson, R. M., Runciman, W. B., Gibberd, R. W., Harrison, B. T., Newby, L., Hamilton, J. D. *Med. J. Australia, 163*, 458 (1995).
6. Department of Health [England]. *An Organisation with a Memory*, London, http://www.doh.gov.uk/orgmemreport/index.htm (2000).
7. Reason J. *Human Error*. Cambridge University Press, Cambridge, (1990).
8. Leape, L. L., Brennan, T. A., Laird, N., *N. Engl. J. Med., 324*, 377 (1991).
9. National Co-ordinating Council for medication error reporting programmes. *Council defines terms and sets goals for medication error reporting and preventing*www.nccmerp.org/press/press1995-10-16.html, (1998).
10. Bates, D. W., Cullen, D. J., Laird, N., et al., *J. Am. Med. Assoc., 274*, 29 (1995).
11. Bates, D. W., Leape, L. L., Petrycki, S. J., *Gen. Intern. Med., 8*, 289 (1993).
12. Koren, G. J., *Clin. Pharmacol., 42*, 707 (2002).
13. Kaushal, R., Bates, D.W., Landrigan, C., et al., *J. Am. Med. Assoc., 285*, 2114 (2001).
14. Ross, L. M., Wallace, J., Paton, J. Y., *Arch. Dis. Child., 83*, 492 (2000).
15. Selbst, S. M., Fein, J. A., Osterhoudt, K., Ho, W., *Ped. Emerg. Care., 15*, 1 (1999).
16. Wilson, D. G., McArtney, R. G., Newcombe, R. G., et al., *Eur. J. Ped., 157*, 769 (1998).
17. Paton, J., Wallace, J., *Lancet, 349*, 956 (1997).
18. Kozer, E., Scolnik, D., Macpherson, A., et al., *Pediatrics, 110*, 737 (2002).
19. Marino, B. L., Reinhardt, K., Eichelberger, W. J., Steingard, R. *Outcomes management for Nursing Practice., 4*, 129 (2000).
20. Bordun, L. A., Butt, W. J. *Paed. Child. Health., 28*, 309 (1992).
21. Schneider, M.-P., Cotting, J., Pannatier, A., *Pharmacy World and Science., 20*, 178 (1998).
22. Stephenson, T., *Arch. Dis. Child., 83*, 497 (2000).
23. Levine, S.R., Cohen, M. R., Blanchard, N. R., et al., *Pediatr. Pharmacol Ther., 6*, 426 (2001).
24. United States Pharmacopeia. *Summary of information submitted to Medmarx in the year 2001*, www.usp.org/medmarx, (2002).
25. Cousins, D. H., Clarkson, A., Conroy, S., Choonara, I., *Paed. Perinat. Drug Ther., 5*, 52 (2002).
26. Jonville, A. P., Autret, E., Bertrand, P. P., Barbier, P., Gauchez, A. S. *DICP, 10*, 1113 (1991).
27. Aronson, J. K., *BMJ, 326*, 1346 (2003).
28. Vandenbroucke, J. P., *Ann. Intern. Med., 134*, 330 (2001).

29. Cousins, D. H., Upton, D. R., *Pharmacy in Practice, 7*, 220 (1997).
30. Cousins, D. H., Upton, D. R., *Pharmacy in Practice, 7*, 368 (1997).
31. Cousins, D. H., Upton, D. R., *Pharmacy in Practice, 6*, 172 (1996).
32. Cousins, D. H., Upton, D. R., *Pharmacy in Practice, 5*, 28 (1995).
33. Cousins, D. H., Upton, D. R., *Pharmacy in Practice, 9*, 18 (1999).
34. Cousins, D. H., Upton, D. R., *Pharmacy in Practice, 9*, 334 (1999).
35. Department of Health (England). *Building a safer NHS for patients*, http://www.doh.gov.uk/ buildsafenhs, (2001).
36. Gill, A., Hooton, M., Fairclough, S. *Journal of the Guild of Healthcare Pharmacists [UK], 3*, 26 (2002).
37. Frey, B., Buettiker, V., Hug, M. I., et al., *Eur. J. Pediatr., 161*, 594 (2002).
38. British Air Transport Industry. Personal communication to the National Patient Safety Agency, (2002).
39. Health and Safety Executive (UK). *Five steps to risk assessment*. http://www.hse.gov.uk/ pubns/indg163.pdf (2003).
40. Fortescue, E. B., Kaushal, R., Landrigan, C. P., et al., *Pediatrics., 111*, 722 (2003).
41. Mullett, C. J., Evans, R. S., Christenson, J. C., Dean, J. M., *Pediatrics., 108*, E75 (2001).
42. Balaguer-Santamaria, J. A., Fernandez-Ballart, J. D., Escribano, J. S., *An. Esp. Pediatr., 55*, 541 (2001).
43. Bates, D. W., Teich, J. M., Lee, J., et al., *Am. Med. Inform. Assoc., 6*, 31 (1999).
44. Leape, L. L., Cullen, D. J., Clapp, M. D., et al., *J. Am. Med. Association., 282*, 267 (1999).
45. Munoz-Labian, M., Pallas-Alonso, C., de La Cruz-Bertolo, J., et al., *An. Esp. Pediatr., 55*, 535 (2001).
46. Rowe, C., Koren, T., Koren, G., *Arch. Dis. Child., 79*, 56 (1998).
47. Potts, M. J., Phelan, K. W., *Arch. Pediatr. Adolesc. Med., 150*, 738 (1996).
48. Baldwin, L., *BMJ, 310*, 1154 (1995).
49. Rolfe, S., Harper, N. J. N., *BMJ, 310*, 1173 (1995).
50. Lesar, T. S., *Ann. Pharmacother., 36*, 1833.
51. Royal College of Paediatrics and Child Health and the Neonatal and Paediatric Pharmacists Group. *Medicines for Children*, (2003).
52. Stucky, E. R., Amercian Academy of Pediatrics Committee on Drugs. *American Academy of Pediatrics Committee on Hospital Care, Pediatrics, Pediatrics III*, 431 (2003)
53. Department of Health (England). *Building a Safer NHS for Patients, Improving Safe Medication Practice*, January 2004. www.dh.gov.uk.

3.4 Compliance or concordance in children?

Sharon Conroy[1], Tony Nunn[2], Steve Tomlin[3]

[1]*Academic Division of Child Health, University of Nottingham, Derbyshire Children's Hospital, Uttoxeter Road, Derby DE22 3DT, UK*

[2]*Royal Liverpool Children's NHS Trust, Eaton Road, Liverpool L12 2AP, UK*

[3]*Guy's & St Thomas' NHS Trust, London SE1 9RT, UK*

3.4.1 COMPLIANCE OR CONCORDANCE IN CHILDREN?

Concordance is a term that seems to be misunderstood by many. For some, it is merely the latest in a series of terms used to describe compliance or adherence. For others, it is a radical shift in the way patients taking medicines are approached [1]. Concordance in its true sense, is fundamentally different from either compliance or adherence. It focuses on the consultation processes between health care professional and patient rather than on the specific behaviour of whether or not the patient takes his medicine as prescribed. It has an underlying principle of a shared approach to decision-making [1]. Concordance may be described as an agreement, following negotiation between patient and healthcare professional, respecting the beliefs and wishes of the patient to determine whether, when and how medicines will be taken [2]. It depends on the patient (or customer) being fully engaged during a consultation, whether it is in a doctor's surgery or over the counter at a pharmacy, so that they understand what the condition is, and why particular medicines are being recommended and prescribed [2]. By being fully informed, patients may be more likely to take the medicines prescribed for them and stick to the regimen. Alternatively, they may wish to be told very little and prefer a totally didactic approach. However, a concordant consultation may result in the patient refusing treatment, which may not please the healthcare professional, but leaves the patient in control.

Compliance refers to a specific patient behaviour: did the patient take the medicine in accordance with the wishes of the health care professional? Otherwise, were the instructions simply followed? It is therefore possible to have a non-compliant (or non-adherent) patient, but it is not possible to have a non-concordant patient. Only a consultation or a discussion between the parties concerned can be non-concordant [1].

Concordance is usually established between two people, doctor and patient, but the use of drugs in children usually involves a third party, the parent/carers. Where the patient is a child, the pathway to concordance raises several dilemmas [3]. Do we agree that children should have a more active role in making decisions about their health and what medicines they take than they currently do? Should healthcare professionals continue to tell parent/carers about their child's illness and its treatment, or should more time be spent on direct communication with the child? How old should children be, to be addressed directly about their illness? What do they want to know? Do children and parent/carers have the same ideas about taking medicines?

There is a great lack of research in children in this area and the above questions cannot be satisfactorily answered. Many factors have been identified as contributory to non-compliance or failure to reach concordance of patients, but relatively few have focussed on children and their medicines especially in those under 7 years of age [3]. In 1997, the Royal Pharmaceutical Society of Great Britain (RPSGB) introduced to many, and advocated to all, the concept of 'concordance' – yet failed to mention the word 'child' [4]. This chapter will therefore focus principally on compliance, due to the lack of information on concordance in children.

3.4.2 WHAT DO PARENT/CARERS THINK?

As part of the development of a Children's National Service Framework, parent/carers of children with long term medication needs were consulted about their views on medicines [5]. Key messages included the need for clear, understandable, up to date information in a variety of media and formats to allow parent/carers and children to make fully informed choices about their medication. Healthcare professionals need to be trained to listen and communicate effectively. Parent/carers and young patients wanted to have the final say in decisions and choices about medication. Licensing of medicines was not of great concern, providing that the medicines were shown to be safe and effective, and that adequate information was provided. Parent/carers' top priority was choosing medicines which are clinically effective and reviewed regularly to ensure continuing effectiveness.

3.4.3 CONSEQUENCES OF NON-COMPLIANCE

Most research relates to rates of compliance which in children have been reported to range from 25 to 82% [6]. Worrying consequences of non-compliance have been documented in children prescribed antibiotics for prophylaxis of recurrent urinary tract infections [6]; patients with epilepsy [7] and in children on 'maintenance' chemotherapy for common acute lymphoblastic leukaemia, where lack of compliance may explain a substantial proportion of late relapses [8,9]. Non-compliance may be a greater problem several months post-diagnosis in such patients [8]. Patients with a better understanding of instructions and doses may be more likely to comply with therapy. Similarly, non-compliance is a major cause of mortality in paediatric liver transplant recipients with 3 of 30 children presenting in acute rejection being inadequately immunosuppressed due to non-compliance [10].

Asthma is the most common chronic childhood illness. Mortality and morbidity rates associated with asthma are increasing in children and poor compliance with maintenance treatment may be responsible [11]. Undertreatment of asthma has been associated with increased morbidity and healthcare utilisation, including number of symptomatic days or nights, emergency visits, hospitalisation and physician visits [12]. This is reported to be a particular problem in inner city children in the US as measured by caregiver reports, physician reports, pharmacy refill data and insurance claims data [12].

3.4.4 FACTORS INCREASING NON-COMPLIANCE

Age

Several studies have suggested that adolescents are less compliant than younger children [13]. Compliance of infants and toddlers is largely determined by the ability and willingness of the parent/carer to understand and follow the treatment plan. As age increases, children develop the ability to treat themselves but continue to need parent/carer supervision. School age children gradually learn to regulate their own behaviours, spending less time at home with their parent/carers and becoming increasingly influenced by their peers and social environment. Adolescents, though capable of greater autonomy in following treatment recommendations, may struggle with other issues. Difficulties surround medicine taking in adolescence particularly in chronic illness such as cancer, diabetes, arthritis, renal disease, epilepsy and asthma [8,-10,14] with many contributory reasons. These include:

- disagreement with parent/carers over responsibility and drive for independence from parent/carers;
- poor understanding of illness, less perceived vulnerability to disease;
- patients and families not accepting the seriousness of the disease;
- misunderstanding and uncertainty about the benefits of medication;

- forgetting;
- embarrassment e.g. of using inhalers;
- complex regimen, frequent dosing, frustrations with length of treatment;
- fear of side effects;
- desire to have the same lifestyle as their peers;
- physical, emotional and social changes accompanying adolescence;
- feeling dependent and powerless;
- unjustified or inappropriate criticism of non-compliance.

Support from others, especially parent/carers and doctors, seems to be very important with good compliance being encouraged by perceived acceptance of the patient's autonomy.

Palatability

In infants, compliance rates less than 50% may be due to difficulties of administering unpalatable medicines [15]. Young children are unlikely to understand or be motivated to take a poor tasting medicine. In older children also, concordance may be encouraged if the medicine is acceptable to them. Different medicines and brands have different flavours and acceptability to individual children [16]. Assessment of palatability should be a key part in the development of a medicine particularly where administration to children is likely. This should involve children as good or bad taste, as determined in adult taste test studies, do not correlate with compliance of infants and young children taking oral antibiotic suspensions [17]. Mothers report 'tricks' are needed to disguise taste in up to 40% of cases such as mixing products with jam, syrup or ice cream [17].

Practitioners should be familiar with the taste of liquid medicines and obtain feedback on their acceptability from patients and carers [18]. More research in children is needed to establish acceptable tastes, textures, volumes and colours of medicines that would encourage concordance.

Duration of therapy and frequency of dosing

Shorter lengths of course and less frequent daily dosing of antibiotic courses have been shown to be more acceptable to children [15,19]. Once or twice daily compared to more frequent daily dosing and limiting courses to 7 days or less encourages compliance. Similarly, in paediatric epilepsy and psychiatry studies, compliance rates drop with increasing number of daily doses [14,20].

Administering medicines in schools is another issue. There are problems in who will give the medicine, where to store them and who takes responsibility for ensuring that the child takes the dose. Advice is given to schools by the health and education departments [21] but, whilst many have developed their own policies enabling appropriate access to medicines for child pupils, the approach is inconsistent. Pharmacists may help by dispensing school medicines separately, labelled with appropriate directions to avoid midday doses being taken to school in unlabelled bottles or envelopes, however, teachers are not trained in medicines administration.

Once or twice daily dosing avoids this problem. In some conditions, preparations with an enhanced duration of activity are necessary to facilitate extended dosing intervals. Children with Attention Deficit Hyperactivity Disorder requiring methylphenidate may be bullied at school when they take the drug that allows them to cope with afternoon lessons and after school activities. A once daily formulation is available which can be administered at home in the morning and avoids these problems [22]. It is, however, more expensive than other formulations, and many health authorities struggle to justify the expense involved.

Children up to 11 years old have been reported to find it difficult to understand the concept of long-term preventative drugs and struggle to comprehend why someone should need to take a drug when they are not actually ill [3]. It has also, however been suggested that children with asthma are quite different in this perspective and their degree of autonomy in the use of drugs is often considerable.

Patient/parent/carer beliefs

Parent/carers and doctors have been shown to be significantly more likely to agree about the medication prescribed for a child with asthma when the parent/carer scored highly in an 'Asthma Belief Scale' which reflected attitudes to preventative therapy, confidence in managing asthma attacks and concerns over side effects of medication [12]. The study identified an important intermediate step in the pathway to optimal treatment of paediatric asthma which was the caregiver's acknowledgment and acceptance of medication prescribed. Parent/carers specifically not believing in the appropriateness of daily medication use (including symptom free days) and unaddressed concerns about side effects were associated with discordance.

Cultural influences are also very important. Different ethnic groups have different attitudes, values and beliefs about health and illness that could affect compliance. An example is in races who only believe there is illness if there is pain. Such beliefs greatly affect compliance with preventative medicines and medicines for asymptomatic conditions [10]. Other factors include social position, experience of stigma, access to health care, language difficulties in understanding and reading information, including medicine instructions and the illness and risks and benefits of treatment [10].

3.4.5 WHAT CAN WE DO TO HELP?

Choice

Allowing asthmatic children a choice in their inhaler device, and to adminis-ter steroids once-daily at a time of their choosing may reduce missed doses by increasing patient satisfaction and allowing treatment delivery which interferes with daily life as little as possible [23,24]. Prescribing of inhalers, which fit a single spacer device, will help so that children do not need two different devices.

Allowing children with eczema to pick their favourite emollient from a tray of all those available has been shown to enhance compliance by empowering the child to use the product of their choice, thus avoiding those which they find unpleasant and are not willing to use [25]. An audit of emollient use in children with eczema revealed a large proportion of children not using the aqueous cream which had been prescribed for them, due to their suffering immediate cutaneous reactions to it [25].

The best choice of drug in terms of side effects is very important. For exam-ple, weight gain, hair loss or change in structure, gingival hyperplasia, hirsutism, exacerbation of behavioural problems with subsequent consequences on peer relations and socialisation are very important in children with epilepsy. Matching the medicine to the patient is extremely important, especially in teenagers [20].

Excessive body hair, trembling hands and weight gain were reasons for poor compliance in adolescent patients post-transplant resulting in non-compliance rates of 40-60% in 14-21 year olds [10].

Communication

Patient/parent/carer/healthcare professional communication regarding medi-cation can influence health outcomes. This may include satisfaction with care, recall and understanding of medical information, quality of life and even health status [12]. Poor parent/carer-physician communication and information sharing during an asthma care encounter were shown to be important contributing fac-tors in observed disagreements regarding prescribing of a preventor medication [12]. Good communication ensuring that medical recommendations are fully understood and discussed, together with the provision of the type and amount of information desired by post-transplant adolescents has been shown to improve compliance [10].

Education

Education of parent/carers and patients has been suggested to be central to concordance [26]. The effect of an intervention involving education and demonstration of therapies by specialist dermatology nurses, based on best practice, was assessed in 51 patients with eczema referred to the clinic as their eczema had been impossible to control in the community. Demonstration of the treatments resulted in an 800% increase in the use of emollients, an 89% reduction in eczema severity score and no overall increase in the use of topical steroids [26]. Twenty-two percent of patients achieved control with emollients alone by their third clinic visit.

- Children should be prescribed
  - the shortest length of effective treatment course;
  - the lowest number of daily doses possible;
  - the most palatable medicine available (as decided by the child);
  - provision of good patient and parent/carer information explaining reasons for, and details of how to take the medicine, in verbal and written forms.
- Healthcare professionals should be aware that special efforts should be made with children generally, and especially those less than three years old and adolescents.
- Giving children choice in their therapy, e.g. inhaler device, topical therapy, form and flavour of medicine, time and number of daily doses is likely to enhance concordance and likelihood of successful treatment.

There are however, major obstacles to these straightforward suggestions.

Lack of suitable formulations

Many drugs are not authorised for use in children. Licensed drugs are prescribed outside the terms of the product license (off label) in relation to age, indication, dose or frequency, route of administration, or formulation. Unlicensed drugs are also used which have not been subjected to the licensing process at all. Two thirds of children in European hospitals receive drugs prescribed in an unlicensed or off label manner [27].

Consequently, many medicines are only commercially available as a tablet or capsule in adult strengths. Suitable products for children such as oral liquids, powders or capsules are extemporaneously produced in large volumes on a regular basis by hospital pharmacy departments across Europe [28]. There is often little information to support the bioavailability or stability of the preparation. The result may be unpleasant to take with a short shelf life causing unneeded hassle for the family to organise further supplies. Standards of extemporaneous dispensing procedures and quality of ingredients used are inconsistent and unregulated. Mistakes have happened with devastating consequences [29].

For 75% of extemporaneously dispensed liquids and capsules, a suitable licensed alternative is available in another country [28]. Importation is an alternative to extemporaneous preparation, but there are problems around free movement of medicines between countries and the process can be complicated.

Other means must sometimes be used in order to deliver a child's dose. Examples include [30]:

- using injections orally;
- crushing tablets or opening capsules and using the powder to make an oral liquid or repackage in smaller strength powder sachets or capsules;
- using concentrated preparations designed for adults;
- halving or quartering tablets;
- dissolving or dispersing tablets in a specified volume and administering an aliquot using an oral syringe;
- preparing medicines from chemicals.

These methods have their own associated problems and are far from ideal [30]. Such complicated methods of drug delivery are only likely to impair the chances of achieving concordance.

Patient information and education

It has been established that patient information leaflets (PILs) facilitate concordance [4]. However, these are not usually available for patients taking unlicensed medicines. For off label medicines, manufacturer's leaflets may be confusing and cause doubt, they must however legally be provided with dispensed medicines. They may actually reduce the chance of achieving concordance.

The readability of some PILs is poor. Of 10 PILs for common paediatric medicines, 9 could not have been read by 13 year olds or 9% of adults [31]. It is suggested that PILs should have a reading age of 10-11 years or be equivalent to tabloid newspapers [31].

Advances in drug delivery

Exciting formulation developments are becoming increasingly available for adults. Transcutaneous delivery systems, fast dissolving drug formulations and multiple unit dose systems [32] would resolve many problems increasing non-compliance. Children however, tend not to be considered when these innovative products are being developed.

The European Commission offers hope in its consultation document 'Better Medicines for Children' which aims to increase the availability of suitable licensed medicines for children [33]. It is hoped that action will urgently follow this consultation in order to provide opportunities and means of improving the chances of concordant relationships with paediatric patients.

REFERENCES

1. Britten, N., Weiss, M., *Pharm. J., 271*, 493 (2003).
2. Editorial. *Pharm. J., 271*, 480 (2003).
3. Sanz, E. J., *BMJ, 327*, 858 (2003).
4. Marinker, M., Blenkinsopp, A., Bond, C., *et al.*, (eds.). *From compliance to concordance: achieving shared goals in medicine taking,* London, Royal Pharmaceutical Society of Great Britain, 1997.
5. Shortis, M., *Listening and responding to parent/carers and young people: Report of a National Service Framework Consultation event relating to children's medicines (May 2003).* http://www.doh.gov.uk/nsf/pdfs/childrens-med-consult.pdf (accessed 30 Dec 2003).
6. Sclar, D. A., Tartaglione, T. A., Fine, M. J., *Infectious Agents Dis., 3*, 266 (1994).
7. Lisk, D. R., Greene, S. H., *Postgrad. Med. J., 61*, 401 (1985).
8. Tebbi, C. K., Cummings, M., Zevon, M. A., Smith, L., Richards, M., Mallon, J., *Cancer, 58*, 1179 (1986).
9. Lilleyman, J. S., Lennard, L., *BMJ, 313*, 1219 (1996).
10. Carter, S., Taylor, D., Levenson, R., *A question of choice – compliance in medicines taking, a preliminary review,* Medicines Partnership, London 2003.
11. Milgrom, H., Bender, B., Ackerson, L., Bowry, P., Smith, B., Rand, C., *J. Allergy Clin. Immunol., 98*, 1051 (1996).
12. Riekert, K. A., Butz, A. M., Eggleston, P. A., Huss, K., Winkelstein, M., Rand, C. S., *Pediatrics, 111*, e214 (2003).
13. World Health Organisation. *Adherence to long-term therapies, evidence for action.* Geneva 2003. (www.who.int).
14. Hack, S., Chow, B., *J. Child. Adolesc. Psychopharmacol., 11*, 59 (2001).
15. Hoppe, J. E., Wahrenberger, C., *Clin. Ther., 21*, 1193 (1999).
16. El-Chaar, G. M., Mardy, G., Wehlou, K., Rubin, L. G., *Pediatr. Infect. Dis. J., 15*, 18 (1996).
17. Higa, S. K., Chan, D. S., Bass, J. W., Alfaro, P. J., *J. Ped. Pharm. Prac., 2*, 265 (1997).
18. Bell, A. E., *J. Ped. Pharm. Prac., 4*, 43 (1999).
19. Cohen, R., Levy, C., Doit, C., et al., *Pediatr. Infect. Dis. J., 15*, 678 (1996).
20. Nordli, D. R., *Epilepsia, 42(Suppl. 8)*, 10 (2001).
21. Department for Education and Skills and Department of Health, Supporting Children with Medical Needs: A Good Practice Guide, (1996).
22. Pelham, W. E., Gnagy, E. M., Burrows-Maclean, L., et al., *Pediatrics, 107*, E105 (2001).
23. Hyland, M. E., *Drugs, 58 (Suppl. 4)*, 1 (1999).
24. Müller, C., *Drugs, 58 (Suppl. 4)*, 35 (1999).
25. Cork, M. J., Timmins, J., Holden, C., et al., *Pharm. J., 271*, 747 (2003)
26. Cork, M. J., Britton, J., Butler, L., Young, S., Murphy, R., Keohane, S. G., *Br. J. Dermatol., 149*, 582 (2003).
27. Conroy, S., Choonara, I., Impicciatore, et al., *BMJ, 320*, 79 (2000).
28. Brion, F., Nunn, A. J., Rieutord, A., *Acta Paediatr., 92*, 486 (2003).
29. Anon., *Pharm. J., 264*, 390 (2000).
30. Nunn, A. J., *Arch. Dis. Child., 88*, 369 (2003).
31. Gabriel, V., Stephenson, T. J., *J. Ped. Pharm. Prac., 3*, 29 (1998).
32. Breitkreutz, J., Wesse, T., Boos, J., *Paed. Perinatal Drug Ther., 3*, 25 (1999).
33. European Commission. Better medicines for children, Consultation document, Brussels, February 2002.

Drugs in pregnancy and lactation

4.1 Epidemiology of drug use in pregnancy

Gideon Koren

The Motherisk Program, University of Toronto, The Ivey Chair in Molecular Toxicology, University of Western Ontario, Canada

4.1.1 INTRODUCTION

More than half of all pregnancies are unplanned, leading millions of women every year to expose their unborn babies to xenobiotics to which they have been exposed. This reality is well reflected in the prevalent drugs we are asked to counsel about at the Motherisk program in Toronto: analgesics, antihistamines and decongestants are, by far, the most prevalent among medicinal compounds [1]. Alcohol (ethanol) is consumed by over 50% of women in developing countries. Not surprisingly, then, this is the most prevalent non-medicinal xenobiotic consumed by pregnant women, followed by cigarettes and marijuana.

There are few prospective studies on the prescription of drugs during pregnancy, conducted in Europe [2], Denmark [3], Italy [4], Hungary [5] Sweden [6] and Norway [7]. In the study by Lacroix [8], based on the records of the French Health Insurance Service, of drug prescription during pregnancy in 1000 women in southeast France, 99% of women received a prescription for at least one drug during pregnancy with a mean of 13.6 medications per women.

Because embryogenesis occurs mostly during the first trimester of pregnancy, a large number of women expose their fetuses unintentionally during the potentially most sensitive of times. This reality, *per se*, leads to very high levels of anxiety among women, their families and health professionals, even when there is large body evidence on fetal safety [1].

4.1.2 PREGNANCY-INDUCED CONDITIONS

Large numbers of pregnant women experience gestation-related morbidity, such as nausea and vomiting of pregnancy (NVP), bacteruria and gestational diabetes, often necessitating drug therapy. NVP ("morning sickness") may serve as a useful paradigm for perception, misperceptions and practice of drug therapy for such conditions.

Starting in the late 1950s, Bendectin® was the major agent to treat NVP worldwide. This controlled-release combination of doxylamine and pyridoxine (originally also with dicyclomine) was used by 40% of pregnant American women in the late 1970s.

Because of increasing numbers of legal cases, the company removed the drug from the world market in 1983, despite compelling epidemiological evidence of its fetal safety [9], resulting in a 3-fold increase in hospitalisation rate for severe forms of NVP (hyperemesis gravidarum). In contrast, in Canada Diclectin®, a generic form of Bendectin, has been increasingly used, leading to significant decrease in hospitalisation rates of Canadian women for morning sickness [10].

The fears of women taking apparently safe drugs for morning sickness have led the Motherisk Program to dedicate a special Helpline to advise and counsel women on safe and effective means to control these debilitating symptoms.

4.1.3 CHRONIC CONDITIONS

A critically important group of women are those who need to continue drug therapy during pregnancy to ensure their health. These include a large number of conditions from organ transplant to depression, or from diabetes to asthma. In general, there is compelling evidence that women are discouraged from continuing needed therapy in pregnancy, and those receiving medication are often prescribed low and ineffective doses [1].

Using depression as an example, women are commonly advised to stop "cold turkey" their antidepressants, often resulting in suicide ideation, suicide attempts, hospitalisation, and replacement of antidepressants with alcohol and drugs of abuse [11].

This state of affairs is appalling when one considers that the selective serotonin reuptake inhibitors (SSRIs) have been shown to be apparently safe, both in terms of dysmorphology, as well as neurodevelopment of offspring [11].

There is large body of evidence that uncontrolled or suboptimally controlled maternal conditions affect fetal well-being. A simplistic approach, therefore, of discontinuing medications due to perceived (and often unproven) risk, while ignoring the risk of uncontrolled maternal condition, is unwarranted and often dangerous.

In reality, less than 30 medicinal molecules have been proven to be genuine human teratogens, and yet physicians and pregnant women abstain from using hundreds of medications [1].

Interesting areas in the epidemiology of drug use in pregnancy are instances where the agent has been linked to teratogenicity, but there is evidence that such data are not relevant for humans. For example, the new quinolones have been shown to cause arthropathy in weight-bearing joints in dog puppies. Consequently, the labels of these medications warn clinicians from their use in pregnancy. In reality, no such effects have been documented in neonates exposed *in utero* [12]. Yet women tend to terminate such pregnancies more commonly than among unexposed women [12].

A similar situation has been documented with the rubella vaccine. While the rubella virus is highly teratogenic, there is no such evidence for the attenuated live virus used for the vaccine. Yet women who have inadvertently received the vaccine in pregnancy tend to abort their pregnancies due to such misinformation.

4.1.4 STUDYING THE SAFETY OF DRUGS IN PREGNANCY

Ethically, it is impossible to subject pregnant women to medications for the sake of studying fetal safety. However, because more than half of all pregnancies are unplanned, millions of fetuses are exposed every year to medicinal drugs. Capturing this human experience is the major source of information on safety/risk of drugs in pregnancy.

Retrospective collection of such data, (*i.e.,* after the child was born), has been shown to be comfounded by disease state and many other variables not collected. Importantly, it has been shown that retrospective ascertainment overly represents adverse fetal outcome by a factor of 3- to 4-fold, as parents of healthy children are less likely to call drug companies to report on pregnancy outcome [13].

Prospective collection of exposed cases (*i.e.,* during pregnancy, and before pregnancy outcome is known), allows collection of potential confounders that may affect pregnancy outcome. In the Motherisk program, we compare babies exposed to the drug in question to an unexposed group born to healthy women, termed "non-teratogenic control" as well as to an unexposed group born to mothers and suffering from the same condition. These, then, are called "disease matched controls". The main challenge in these studies is in their limited power. One needs, for example, over 800 exposed and control cases to rule out a 2-fold increase in major malformation rates. This issue can be partially resolved by meta-analysis of studies measuring the same endpoint.

Case control studies offer much higher sensitivity, by comparing the exposure rates to a specific drug in mothers who give birth to children with a specific malformation *vs.* those who give birth to healthy children. The teratogenicity of prostaglandins in causing the Moebius syndrome in Brazil [14], was confirmed with this methodology.

4.1.5 CONCLUSION

In conclusion, the issues of surrounding drug exposure in pregnancy are presently often controlled by emotion-based medicine rather than evidence-based medicine. A major change in approach is needed to stop the orphaning of women from the rapid advances in drug therapy.

ACKNOWLEDGMENT

Supported by grants from from the Canadian Institutes for Health Research, and Duchesney Inc., Canada.

REFERENCES

1. Koren, G. *Maternal-Fetal Toxicology* (3rd Ed.), Marcel Dekker, NY, (2001).
2. Bonati, M., Bortolus, R., Marchetti, F., Romero, M., Tognoni, G., *Eur. J. Clin. Pharmacol.*, *38*, 325-328 (1999).
3. Olesen, C., Hald Steffenssen, F., Lauge Nielsen, G., et al., *Eur. J. Clin. Pharmacol.*, *55*, 139 (1999).
4. Danati, S., Baglia, G., Spinelli, A., Grandolfo, M. E., *Eur. J. Clin. Pharmacol.*, *56*, 323 (2000).
5. Creizel, A. E., Racz, J., *Teratology*, *42*, 505 (1990).
6. Boethius, G., *Eur. J. Clin. Pharmacol.*, *12*, 37 (1977).
7. Nordeng, H., Eskild, A., Nesheim, B. I., Aursnes, I., Jacobsen, G., *Eur. J. Clin. Pharmacol.*, *57*, 259 (2001).
8. Lacroix, I., Damase-Michel, C., Lapeyre-Mestre, M., Montastruc, J. L., *Lancet*, *356*, 1735 (2000).
9. Einarson, T., Leeder, J. S., Koren, G., *Drug Intell. Clin. Pharm.*, *22*, 813 (1988).
10. Neutel, C. I. "Variation in rates of hospitalisation for excessive vomiting in pregnancy by Bendectin/Diclectin® use in Canada," in *Nausea and vomiting of pregnancy, State of the Art 2000*, Koren, G., Bishai, G.R. (eds.)., Motherisk Program, Toronto (2000).
11. Bonari, L., Bennett, H., Einarson, A., Koren, G., *Can. Fam. Physician*, *50*, 37 (2004).
12. Loebstein, R., et al., *Chemother.*, *42*, 1336 (1998).
13. Bar Oz, B., Moretti, M. E., Moreels, G., Van, T., Helboom, T., Koren, G., *Lancet*, *354*, 1700 (1999).
14. Pastuszak, A. P., et al., *N. Engl. J. Med.*, *338*, 1881 (1998).

4.2 Placental transfer of drugs

Raphaël Serreau, Evelyne Jacqz-Aigrain

Department of Paediatric Pharmacology, Robert Debré Hospital, Paris, France

The quantity of drugs reaching the fetus depends on the physico-chemical characteristics of molecules and on maternal pharmacokinetic parameters and placental factors, which vary according to the term of pregnancy.

Drugs can be divided into three groups depending on their degree of placental transfer: limited, high, or in-excess. The degree of transfer governs the quantity of drugs that reaches the fetus and, therefore, therapeutic options and the possible toxic risks of the various molecules.

4.2.1 PLACENTAL TRANSFER MECHANISMS AND THEIR QUANTIFICATION

Simple passive diffusion

The placental transfer of drugs in most cases takes place through simple passive diffusion. Rapidity of transfer depends on several parameters.

- **K**: *diffusion constant*. This is dependent on the physico-chemical characteristics of molecules: molecular weight, liposolubility, and degree of ionisation.
- **A/X**: *area-to-thickness ratio of the exchange membranes*. The ratio varies with term of pregnancy.
- **Cm - Cf**: *concentration gradient between maternal and fetal circulation*. Only the free non-ionised fraction of the drug passes through the membranes [1]. It is defined by the following equation: $V = K \cdot A/X \cdot (C_m - C_f)$.

Involvement of other membrane passage mechanisms in the placental transfer of drugs is very limited

Sugars are mainly involved in facilitated diffusion, as has been reported for several drugs such as pancuronium, with which a reduction of placental passage has been observed in the case of association with succinylcholine chloride [2]. Active transport occurs for a few drugs (methyldopa, 5-fluorouracil) and particularly vitamins, amino acids, and some ions (sodium and potassium). Endocytosis is active transport achieved through vacuolisation, arising in immunoglobulin transfer. Passage through membrane pores is probably not involved in placental drug passage [3].

Factors influencing placental transfer (Table 1) [3]

Physico-chemical characteristics of the drug

Molecular weight. Most drugs have a molecular weight of less than 1000 and pass through the placenta. Drugs of greater molecular weight such as heparin (molecular weight 6,000 to 20,000) cannot pass through the placenta.

Table 1

Placental transfer parameters of drugs by passive diffusion

Molecular characteristics
molecular weight dissociation (pKa) liposolubility
Placental factors
surface area thickness vascularisation metabolism
Maternal and fetal factors
protein binding blood pH vascularisation metabolism

Lipid solubility, measured by the organic phase/water partition coefficient.

Dissociation. The degree of dissociation depends on the pKa of the molecule (defined by the pH value for which 50% of the molecule is in a non-dissociation form and 50% in the ionised form) and the pH of its environment.

Placental factors

Exchange surface area and thickness. Exchange surface area increases gradually during pregnancy whereas membrane thickness diminishes. At term, a normal human placenta is a disc about 20 cm in diameter, 3 cm thick, weighing about 500 g. Its surface area is evaluated at 1.5 m² at three months into pregnancy and 15 m² at term [4]. Membrane thickness is 25 mm in early pregnancy and 3.5 mm in late pregnancy.

Blood flow. Uterine blood flow increases during pregnancy. At term, flow is 500 ml/min [5]. Variations modify the fast or "flow-dependent" passage of non-ionised and very lipid-soluble molecules.

Biotransformation. Placental biotransformations are numerous. Enzymatic activity in phase 1 (functionalisation) and in phase 2 (conjugation) studied *in vitro* using homogenates or subcellular preparations, is weak in comparison with activity in the liver. Amid cytochromic P450 activities [6], the activity of P450 19A1 (aromatase), responsible for placental conversion of androgens into oestrogens, is significant [7]. Cytochrome P4501A1, which metabolises aromatic hydrocarbons, is present and particularly marked in women who smoke during pregnancy [8]. Finally, a cytochrome P450 3A7 has been identified but its role is unknown as yet [9]. So far, data is scant on phase 2 reactions. Glutathion-S-transferase activity has recently been detected [10].

Maternal and fetal factors

Protein binding. Weakly acidic drugs mainly bind to albumin. Basic drugs also bind to acid α1-glycoproteins. Binding percentage varies with plasma proteins (their nature, concentration, affinity constant) and with the kind of drug and its concentration.

Binding protein concentrations differ in the maternal and fetal compartments. The mother-to-fetus ratio of plasma albumin concentration is 0.28 at 12 to 15 weeks of pregnancy and approximately 1 at term, whereas the mother-to-fetus ratio of α–1-acid glycoproteins is 0.1 at 12 to 15 weeks, and close to 0.4 at term. There are also apparent differences in affinity of proteins, connected to the presence of endogenous ligands such as bilirubin or free fatty acids [3].

Blood pH. At the end of pregnancy, fetal blood pH is about 0.1 to 0.15 units less than the blood pH of the mother. This difference is increased in the case of fetal acidosis (fetal distress for instance), which facilitates an accumulation of basic drugs in the fetus.

Maternal pharmacokinetics. Maternal physiological modifications bring about considerable pharmacokinetic differences. Absorption modifications vary and are connected to the increase of the rate at which the stomach empties and to the decline of intestinal motility. Distribution is modified by changes in body weight (and its water and fat contents), cardiac output, and plasma proteins. Thus, a plasma volume increase of about 50%, to which is associated a decrease in plasma protein content (albumin in particular), increases the distribution volume of many drugs, such as ampicillin and caffeine. In contrast, drug renal clearance is gradually modified by a large increase of renal blood flow and glomerular filtration. Because of this, many drugs for which renal elimination is predominant, such as digoxin, will be eliminated more rapidly [11].

Evaluation of placental transfer of drugs

Animal models. The study of placental transfer of drugs in animals is of limited interest. Similarity to the human placenta is only to be found in monkeys and rodents [12]. The influence of morphological differences is probably considerable.

Methods with pregnant women. An evaluation of placental transfer of drugs in humans uses various methods. By using the perfused human cotyledon method or the perfused human placenta method, it has been possible to examine *in vitro* the placental transfer of many molecules (for review [13,14]). Because such models are limited, confrontation with *in vivo* data is important.

In vivo methods used with pregnant women are limited for obvious ethical reasons. More often than not, the fetal-maternal drug ratio is calculated on the basis of maternal blood concentrations and those of the cord at birth. This ratio, measured after a single or repeated administrations, depends on the drug regimen (dosage and route of administration), on the term at the time of birth, and time elapsed between the last drug administration and birth. Some authors have succeeded in pooling cord concentration data and thus have calculated fetal elimination half-life, as has been the case with tobramycin [15]. Modelling techniques can be used to evaluate pharmacokinetic parameters in maternal and fetal compartments and their evolution during pregnancy [11].

4.2.2 PLACENTAL TRANSFER OF DRUGS–
A CLASSIFICATION

Drugs can be defined according to the degree of placental transfer. Three broad categories exist, of which the first two can accept most molecules.

High placental transfer. The drug crosses the placenta rapidly, and at equilibrium, fetal concentrations are close to maternal plasma concentrations.

Limited placental transfer. Fetal plasma concentrations are lower than maternal concentrations.

Excess placental transfer. Fetal concentrations are higher than maternal concentrations, probably because of limited retropassage of the drug to maternal circulation.

Certain endogenous compounds and drugs do not cross the placenta, or only to a very limited extent [16]. Examples are the hypoglycaemic drugs, pituitary hormones, insulin, TSH, heparin, and low molecular weight heparins.

Drugs are presented in therapeutic categories, on the basis of abundant data available in the literature. Only those medications which are most used during pregnancy are addressed. In most cases, placental transfer is quantified at the end of pregnancy, on the basis of fetal concentration measured in the cord blood. Numbers of patients may vary from a few individuals to several hundreds.

Antibiotics, antiparasitics, and antiviral drugs (Table 2)

Studies of ampicillin have included several hundred pregnant women and reveal high placental transfer. On the contrary, cephalosporins mostly have a limited placental transfer rate and this is also the case of aminoglycosides [17-19].

Table 2

Antibiotics and antiviral drugs

Placental transfer	High	Limited
penicillins	ampicillin amoxicillin methicillin	piperacillin
cephalosporins		cefotaxime cefuroxime
aminoglycosides		amikacin gentamicin kanamycin
miscellaneous antibiotics	fosfomycin thiamphenicol	rifampicin sulfasalazine
antiparasitic agents	pyrimethamine sulfadiazine	spiramycin
antiviral agents	aciclovir ganciclovir	

In the case of toxoplasmosis seroconversion during pregnancy, spiramycin which concentrates in the placenta and crosses the placental barrier, is used [20]. When the fetus is infected, a combination of pyrimethamine and sulfadiazine is the reference treatment [21].

Acyclovir crosses the placenta by simple diffusion and accumulates in the amniotic fluid [22]. Ganciclovir, studied in a perfused human placenta system, also has high placental transfer activity [23]. Protease inhibitors do not cross the placenta to an appreciable extent. Limited transfer may result from their high degree of plasma protein binding and their backwards transport through P-glycoprotein, largely expressed in the placenta. [24]

Drugs acting on the central nervous system (Table 3)

There is a great deal of data for anticonvulsants [25-27]. All the benzodiazepines have a high placental transfer rate. In particular, diazepam attains fetal plasma concentrations greater than maternal concentrations and concentrates in the central nervous system and the liver. The elimination half-life of this drug is much longer for neonates than for adults, and it may provoke a withdrawal syndrome [28,29].

Fluoxetine is an anti-depressant inhibiting the recapture of serotonine. Its elimination half-life is short, but it is metabolised as norfluoxetine with a much longer half-life, 7 days on average. Adverse effects observed in the newborn [30-32] are evidence of the placental transfer of the drug, and of its metabolite and their slow elimination in the newborn.

Table 3

Drugs acting on the central nervous system

Placental transfer	In excess	High	Limited
anticonvulsants	valproic acid	phenobarbital phenytoin	carbamazepine
benzodiazepines	diazepam	chlordiazepoxide lorazepam midazolam nitrazepam	
psychotropic drugs		lithium fluoxetine	

Cardiovascular agents (Table 4)

Placental transfer of digoxin is thought to be high and several studies report a mother-to-fetus concentration ratio at birth close to one. This transfer is probably overestimated because of the presence at birth of digoxin-like substances which exaggerate fetal concentrations [33]. Placental transfer may be reduced, particularly in the case of an hydropic fetus [34].

Table 4

Cardiovascular agents

Placental transfer	High	Limited
digitalis	digoxin	
beta-blockers	betaxolol sotalol	propranolol
antiarrhythmic drugs	flecaine	amiodarone
antihypertensive drugs	clonidine methyldopa	
diuretics		furosemide hydrochlorothiazide

Anaesthetics, analgesics, curares (Table 5)

Anaesthetics are mostly used at the end of pregnancy. An emergency Caesarean section under general anaesthesia may be necessary when there is acute fetal distress or for maternal reasons. Peridural anaesthesia is more frequent, involving bupivicaine, lidocaine, or mepivacaine for which placental passage is limited [35].

Miscellaneous drugs (Table 6)

Placental perfusion techniques have shown that the human placenta can inactivate on average 50% of prednisolone, but only approximately 5% of dexamethasone and betamethasone [36,37]. Metabolism and placental passage of prednisone and prednisolone were studied at equilibrium the day before delivery by Caesarean section for 10 women, after infusion of (H^3) labelled prednisone and

Table 5

Anaesthetics, analgesics and muscle relaxants

Placental transfer	In excess	High	Limited
general anaesthetics	ketamine	secobarbital thiopenthal	methoxyflurane propofol
local anaesthetics			bupivacaine lidocaine mepivacaine
analgesics		morphine	alfentanil pentazocine pethidine
curares			atracurium pancuronium d-tubocurarine vecuronium

prednisolone [38]. Pregnancy does not modify the maternal prednisone–prednisolone balance. The placenta inactivates a significant fraction of prednisolone, the mother-fetus gradient of prednisolone is 10:1, and for prednisone, the gradient is 1:1. A similar study at the time of delivery showed that dexamethasone crosses the placental barrier with a gradient of about 1 [39].

Table 6

Miscellaneous drugs

Placental transfer	High	Limited
corticosteroids		dexamethasone betamethasone
nonsteroidal anti-inflammatory drugs	salicylic acid indomethacin	
analgesics	paracetamol	
antiulcer drugs	ranitidine	cimetidine metoclopramide omeprazole
xanthines	aminophylline caffeine	theophylline

4.2.3   CONCLUSION

Placental passage of drugs never ceases to vary during pregnancy. Quantification is difficult in pregnant women and may be approximated by using *in vitro* and/or pharmacokinetic models. Pharmacological data in some cases may help to select therapy used during pregnancy to limit placental passage and fetal exposure or, on the contrary, to gain higher placental passage and treat the fetus.

Placental passage of drugs can cause major fetal teratogenic and/or toxic effects and neonatal complications. These accidents follow a complex process and multiple factors are involved such as dosage, treatment indications, and they also probably involve genetic metabolic variability.

REFERENCES

1. Pacifici, G. M., Nottoli, R., *Clin. Pharmacokinet.*, 28, 235 (1995).
2. Abouleish, E. Jr., Wingard, L. B., de la Vega, S., *Br. J. Anaesth.*, 52, 531 (1980).
3. Nau, H., Plonait, S. L., "Physicochemical and structural properties regulating placental transfert of drugs," in *Fetal and neonatal physiology* (2nd ed.), Polin and Fox (eds.), W. B. Sanders Company, pp.146-160 (1988).
4. Kaufmann, P., Schellen, I. "Placental development," in *Fetal and neonatal physiology*, Polin, R. A., Fox,W. W. (eds.), pp. 47-56 (2002).
5. Metcalfe, J., Bakke, O. M., Johannessen, K. H., Lund, T., *J. Clin. Invest.*, 34, 1632 (1955).
6. Lewandowski, M., Hodgson, E., *Int. J. Biochem.*, 17, 1149 (1985).
7. Kellis, J. T., Vickery, L. E., *J. Biol. Chem.*, 262, 4413 (1987).
8. Manchester, D. K., Parker, N. B., Bowman, C. M., *Pediatr. Res.*, 18, 1071 (1984).
9. Schuetz, J. D., Kauma, S., Guzelian, P. S., *J. Clin. Invest.*, 92, 1018 (1993).
10. McRobbie, D. J., Glover, D. D., Tracy, T. S., *Gynecol. Obstet. Invest.*, 42, 145 (1996).
11. Mattison, D. R., Malek, A., Cistola, C., "Physiologic adaptations to pregnancy: impact on pharmacokinetics," in *Pediatric Pharmacology – Therapeutic Principles in Practice* (2nd ed.), Yaffe-Aranda, W. B. (ed.), Saunders Company, pp. 81-96 (1993).
12. Leiser, R., Kaufmann, P., *Exp. Clin. Endocrinol.*, 102, 122 (1994).
13. Bourget, P., Pons, J. C., Delouis, C., Fermont, L., Frydman, R., *Ann. Pharmacother.*, 28, 1031 (1994).
14. Sastry, B. V., *Advanced Drug Delivery Reviews, 38,* 17 (1999).
15. Bourget, P., Fernandez, H., Demirdjian, S., *Arch. Fr. Pédiatr.*, 48, 543 (1991).
16. Garcia-Bournissen, F., Feig, D. S., Koren, G., *Clin. Pharmacokinet.*, 42, 303 (2003).
17. Good, R. G., Johnson, G. H., *Obstet. Gynecol.*, 38, 60 (1971).
18. Stewart, K. S., Shafi, M., Andrews, J., Williams, J. D., *J. Obstet. Gynecol.*, 80, 902 (1973).
19. Yoshioka, H., Monma, T., Matsuda, S., *Pediatr. Pharmacol.*, 80, 121 (1972).
20. Forestier, F., Daffos, F., Rainaut, M., Desnottes, J. F., Gaschard, J. C., *Arch. Fr. Pediatr.*, 44, 539 (1987).
21. Dorangeon, P. H.,., Ray, R., Max-Chemla, C., et al., *Presse Med.*, 19, 2036 (1990).
22. Fletcher, C. V., *J. Lab. Clin. Med.*, 120, 821 (1992).
23. Gilstrap, L. C., Bawdon, R. E., Roberts, S. W., Sobhi, S., *Am. J. Obstet. Gynecol.*, 170, 967 (1994).
24. Marzolini, C., Rudin, C., Decosterd, L. A., et al., *AIDS*, 12, 889 (2002).
25. Ishizaki, T., Yokochi, K., Chiba, K., Tabuchi, T., Wagatsuma, T., *Pediatr. Pharmacol.*, 1, 291 (1981).

26. Nau, H., Kuhnz, W., Egger, H. J., Rating, D., Helge, H., *Clin. Pharmacokinet.*, *7*, 508 (1982).
27. Myllynen, P. K., Pienimaki, P. K., Vahakangas, K. H., *Eur. J. Clin. Pharmacol.*, *58*, 677 (2003).
28. Mandelli, M., Morselli, P. L., Nordio, S., et al., *Clin. Pharmacol. Ther.*, *17*, 564 (1975).
29. Cooper, J., Jauniaux, E., Gulbis, B., Bromley, L., *Reprod. Biomed. Online*, *2(3)*, 165 (2001).
30. Spencer, M. J., *Pediatrics*, *92*, 721 (1993).
31. Mijanna, M. J., Bennett, J. B., Izatt, S. D., *Pediatrics*, *100(1)*, 158 (1997).
32. Hendrick, V., Stowe, Z. N., Altshuler, L. L., Hwang, S., Lee, E., Haynes, D., *Am. J. Psychiatry*, *160(5)*, 993 (2003 May).
33. Lupoglazoff, J. M., Jacqz-Aigrain, E., Guyot, B., Chappey, O., Blot, P., *Br. J. Clin. Pharmacol.*, *35*, 251 (1993).
34. Younis, J. S., Granat, M., *Am. J. Obstet. Gynecol.*, *157*, 1268 (1987).
35. Yurth, D. A., *Clin. Perinat.*, *9*, 13 (1982).
36. Levitz, M., Jansen, V., Dancis, J., *Am. J. Obstet. Gynecol.*, *132*, 363 (1978).
37. Blanford, A. T., Murphy, B. E. P., *Am. J. Obstet. Gynecol.*, *127*, 264 (1977).
38. Beitins, I. Z., Bayard, F., Ances, I. G., Kowarski, A., Migeon, C. J., *J. Pediatr.*, *81*, 936 (1972).
39. Osathanondh, R., Tulchinsky, Y. D., Kamali, H., Fencl, M. M., Taeusch, H. W., *J. Pediatr.*, *90*, 617 (1977).

4.3 Drug administration during pregnancy–evaluation of fetal toxicity: animal studies

Pierre Guittin

Drug Safety Evaluation, Aventis, 94400 Vitry sur Seine, France

4.3.1 INTRODUCTION: THE NEED FOR ANIMAL TERATOLOGY STUDIES

Teratological investigations on mammalian fetuses really began in the 1930s. Before the thalidomide tragedy, there was no compelling reason to develop tests for pregnant animals, because the medical profession thought them unnecessary [1]. Since 1994, reproductive toxicology studies must be performed according to harmonised ICH (International Conference on Harmonisation) guidelines, which are the reference for new compound registration [2]. The primary objective should be to detect any indication of embryo-fetal toxicity in laboratory animals and then to conduct a human risk assessment.

In addition, evaluation of fetal toxicity is critical for paediatric risk assessment because embryo-fetal findings can also be predictive of paediatric impairment.

Due to the complexity of the developmental processes and the dynamic interchange between the dam and the embryo/fetus, *in vitro* tests currently available cannot provide suitable assurance of an absence of effect, nor provide good information for human risk assessment [2]. Thus, there is a clear need for relevant animal teratology studies. Due both to animal ethics and the scientific viewpoint, it is not acceptable to conduct developmental toxicity studies with dams that are suffering or with dams in poor general health [3].

4.3.2   STUDY PLAN ENDPOINTS

Requirement for two relevant species

Safety evaluation programs include two relevant species in which the compound is pharmacologically active for detection of the embryo-fetal toxicity studies; one rodent and one non-rodent species are requested. Due to the placental similarities between humans and rodents or rabbits, the majority of the regulatory embryo-fetal toxicity studies are routinely performed in the rat and rabbit. The biological activity with species and/or tissue specificity of compounds should be considered. Some are biologically active in conventional toxicity species, while others have species-specific biological activity. Non-human primates are rarely used as alternative test species, since historical data are poor and the rate of spontaneous anomalies is often high. However, for biotechnology products when only one species can be identified or where the biological activity of the product is well understood, one relevant species may suffice. Toxicity studies in non-relevant species may be misleading and are discouraged.

Intra- or inter-species concordance of adverse effects deserves some special consideration in this risk integration process. Not all species or individual animals within a species are equally sensitive to developmental insult. Therefore, intra-species concordance of effects may be demonstrated by the occurrence of related adverse effects across dose groups. Concordance between the test species should be analysed with respect to the metabolic and drug disposition profiles, and the general toxicity profiles of the test species compared to humans.

Pertinent treatment durations

The dam treatment duration of the embryo-fetal toxicity studies is focused on the critical period of the gestation where organogenesis takes place. For example, the administrations are performed to pregnant rats, mice or rabbits once daily on gestation days 6 through 17, 15 or 18, respectively. This approach allows maximisation of the developmental impact in animal studies.

The drug administration effects during the fetal period are not investigated in the embryo-fetal toxicity studies, but during the pre- and postnatal toxicity study. In this study, the dam administration lasts from implantation to the end of lactation, covering organogenesis, the fetal period and the postnatal period up to weaning. Such a design allows the detection of delayed and postnatal outcome of fetal toxicity.

Critical dose selection

Dose selection is one of the most critical issues in the design of developmental toxicity studies. Doses are designed to incorporate as the high dose, an exposure level that induces slight maternal toxicity, whatever the expected embryo-fetal toxicity. The determination of this high dose depends upon several criteria including the feasibility of the technique such as route of administration, the presumed therapeutic doses, and the demonstration of slight maternal toxicity. In order to define the criteria of a slight maternal toxicity, range-finding studies are recommended with the intent of demonstrating maternal toxicity. The mid dose is chosen with the intent of establishing a dose-response relationship. For approaching the human risk assessment and for determining the safety margin, the low dose should demonstrate a no observed effect level.

Concern is increased for compounds that provoke an increase in the severity of effects with an increase in dose, an increase in the incidence of animals affected with an increase in dose, or a high incidence of effects across all dosed groups.

Detailed fetal examination

In rodent embryo-fetal toxicity studies, all fetuses are examined externally and approximately one half of the fetuses are examined for visceral alterations. The remaining fetuses in each litter are examined for skeletal alterations. Fetuses are alternately assigned to skeletal and visceral examination within each litter; this method does not jeopardise the quality of the fetal assessment. Due to the lower number of fetuses per litter in rabbits, all fetuses are examined first for external and visceral alterations and then for skeletal examination (e.g., stained with Alizarin Red [4]).

With these 3 sets of examinations, all the organs are examined in detail. If any finding is doubtful or if an unusual finding is suspected, special studies may be indicated. For example, defects such as those of the great vessels or the heart septum need adequate dissection of the fetal heart. It is uncommon that an anomaly will be observed isolated and/or in only one set of examination types. Thus, as the fetuses are examined at least twice, a drug related fetal finding is often confirmed by concomitant findings. Because fetus anomalies are rare findings, it is important to examine all the doses including the mid- and low-dose fetuses in order to detect a dose-effect relationship.

The quality of the fetal examination is ensured by people particularly well trained and on relevant historical fetal data from the laboratory. To improve the quality of the fetal investigation, blind fetal examination is often considered as a method for minimising operator bias, especially for subjective findings.

Quality of the ICH studies

Following the ICH guidelines, mated females are divided into at least 4 (1 control and 3 treated) groups of approximately 20 to 26. Toxicokinetic evaluation should be decided on a case-by-case basis. If toxicokinetics are done, satellite groups are added. In most of the cases, it is preferable to perform separate toxicokinetic studies designed according the fetal findings observed in the previous embryo-fetal toxicity studies.

The selected route is the intended clinical route or an acceptable alternative. Due to the stressful constraint of inhalation chambers in pregnant animals, a parenteral route is used for most of the human inhalation compounds.

Embryo-fetal toxicity studies and pre/postnatal toxicity studies should be adapted to detect and investigate any hazard of the novel drug. Factors such as species specificity, immunogenicity, pharmacodynamics, altered physiological state of the dam, and physical properties of the agent must be carefully considered in designing appropriate study plans for evaluating developmental safety. Therefore, developmental toxicity studies should not be performed just to fulfill a regulatory requirement. These animal studies should be designed, based on the available technology, to provide information for the regulatory scientist to make an informed assessment of safety. Preclinical fetal data for each novel therapy is based on knowledge of its biology and the intended clinical use. This case-by-case approach has provided the necessary flexibility to appropriately evaluate constantly changing product data. The flexible approach designed by the ICH guidelines provided a suitable quality of the data and contributed largely to enhancing the teratological risk assessment.

4.3.3 ANALYSIS OF RESULTS

When studies are well conducted, the interpretation of these results for human risk assessment is possible and relevant [5]. Four key points should be analysed carefully to improve the human risk assessment of the fetal toxicity.

Maternal toxicity risk

The purpose of embryo-fetal toxicity studies and pre/postnatal toxicity studies is to determine if developmental toxicity appears at a lower, similar, or higher dose than does maternal toxicity. Minimal toxicity should be observed in dams at the highest dose level tested. In most cases, a 10-15% decrease in body weight gain in rodents is considered to be a good indication of minimal maternal toxicity. In rabbits, a prolonged decrease in food consumption should be considered a better indicator of maternal toxicity, as it is more closely and more rapidly related to poor health conditions than body weight changes [6]. High levels of maternal toxicity can induce bias in the interpretation of the study results, because such toxicity can result in nonspecific developmental toxicity. Thus, when there is evidence of severe maternal toxicity, these experimental groups should be discounted and the interpretation should be focused on a lower highest dose.

Maternal toxicity does not always induce developmental toxicity. Thus, a concern arises when the maternal "no observed adverse effect level" (NOAEL) is lower than the developmental NOAEL, since a toxic effect on the embryo or the fetus could be missed. On the other hand, developmental effects observed in the presence of maternal toxicity should not be systematically viewed as secondary to the maternal toxicity. If the maternal toxicity can by itself lead to developmental toxicity, the presence of maternal toxicity, even slight, is not sufficient evidence to discount potential embryo-fetal adverse effects of a drug. Whether or not there is maternal toxicity, the occurrence of malformations constitutes a warning signal and should never be discounted based simply on maternal status [6].

Based on the severity of the maternal toxicity, we assume three different interactions with embryo-fetal effects. Firstly, when no maternal toxicity is detected, the embryo-fetal toxicity should be dose-related in order to consider that the effects are drug-related. Secondly, when there is severe maternal toxicity, embryo-fetal toxicity is often assumed to be related to maternal toxicity. Finally, when maternal toxicity is minimal, the causes of embryo-fetal toxicity are not obvious and additional investigations should be performed, such as using higher dose levels for shorter treatment periods [6].

Harmonised fetal terminology

Fetal anomalies are complex and difficult to understand by non-teratologists. Thus, the International Federation of Teratology Societies worked to provide a harmonised laboratory animal glossary [7]. Such Harmonisation is a key enhancement for providing a better extrapolation of the animal fetal findings to humans.

Moreover, the rate and the type of fetal anomalies provide critical data to evaluate a possible incidence increase and the severity of the findings. Thus, they should be compared with historical data for each species studied. For example, based on the common incidence of cleft palate in some susceptible strains of mouse, this finding was not considered to be a predicting interference with human embryo development.

In toxicity studies, the malformation is defined as "a rare and irreversible morphologic alteration affecting organ, body part, or the whole body, and resulting from an adverse event occurring during intrauterine development". The definition provided by Chahoud *et al.* [8] is similar, i.e. "a permanent structural change that is likely to adversely affect the survival or health of the species under investigation".

Variations also need to be accurately defined in order not to be considered malformations. In toxicity studies, the term variation is defined as "a frequent, often reversible insult resulting from intrauterine injury and unlikely to adversely affect survival or health". A variation is an event occurring at a level higher than a threshold based on historical control data in the laboratory and in the animal strain. Variations might also be placed in the context of possible delayed development. Thus, only the delayed fetal development would be taken into account, and not all its correlated effects. However, significantly increased incidences of variations should not be ignored, as they are signs of embryo-fetal toxicity.

Various fetal toxicities

The four types of embryo-fetal toxicity are intra-uterine death, dysmorphogenesis, alterations to growth, and functional toxicities. Embryotoxicity and fetotoxicity were well defined in the ICH guidelines as "any adverse effect on the conceptus resulting from prenatal exposure, including structural or functional abnormalities or postnatal manifestations of such effects" [2]. These effects include death, malformation, variation, delayed ossification, body weight reduction, and also perinatal functional impairment.

Toxicities causing death to the developing conceptus may be evident at any time from early conception to weaning. Thus, a positive signal may appear as peri-implantation loss, early or late resorption, abortion, stillbirth or neonatal death. In rodents and rabbits, embryo-fetal death is often a sign of maternal toxicity, but in some cases, it can occur without any maternal toxicity and is related to the compound.

Alterations to growth are generally defined by growth retardation, although excessive growth or early maturation may also be considered an alteration to growth. The most common metric for growth is body weight, and delayed ossification is often related to the fetal weight. Delayed ossification corresponds to the association of several ossification sites that are less developed than the normal expected ossification. Compound-associated delayed ossification should be considered if it is dose-related and when deviations from normal values are well defined. Delayed ossification should be evaluated as a sign of delayed embryo-fetal growth, but should also be analysed in relation to possible maternal toxicity. As a consequence, when delayed ossification is associated with fetal weight reduction and maternal toxicity, these observations may be minimised. When not associated with fetal weight decrease or maternal toxicity, delayed ossification can be considered to be a sign of compound-related embryo-fetal toxicity.

Relevant risk assessment

Intra- and inter-species concordance of observed effects should be considered in the estimation of human risk of adverse developmental outcomes. If a specific type of embryo-fetal toxicity has been demonstrated in the two animal species, it may logically be assumed that a similar effect represents the most likely adverse event to be seen in humans treated with the drug. In the event that dissimilar but related adverse effects within the two major categories of developmental effects are detected in multiple species (for example alterations to growth in one species and developmental death in another), it may be assumed that some level of risk of the related endpoints within the category may be demonstrated in human development [9].

Integration of the impact of maternal toxicity and hierarchisation of the animal injuries are two key points for providing a suitable understanding of animal data. Thus, these animal studies are expected to provide reliable results, and an acceptable prediction of human risk when no human data are yet available. Drug-related malformation observed in animals, whatever the degree of severity, should be regarded as a signal of potential hazard for a clinician who is contemplating the use of an agent in the treatment of a pregnant woman [6]. When an animal species exhibits a developmental effect, potential adverse outcome in human pregnancy, i.e. relevance for human extrapolation, should be established according to several approaches (Table 1).

When variations are observed but without malformation, the human risk extrapolation can be questioned. Experience has shown that variations do not have significant or permanent clinical consequences in most cases, particularly if they occur at much higher than the therapeutic doses. The importance of the safety margin has direct implications for patient counseling. For instance, interruption of the pregnancy is not justified in the case of an exposure far below that causing adverse effect in experimental animals.

Table 1

Relevant animal study criteria for risk assessment

interspecies reproducibility of the fetal effects

dose relatedness of effects

type and severity of birth defects

comparative interspecies kinetics in order to be compared with human kinetics

knowledge of the mechanism of action of the compound

effects of other agents in the chemical class

4.3.4 CONCLUSION

Embryo-fetal toxicity studies and pre/postnatal toxicity studies are well designed, based on ICH guidelines. Such animal studies using a high level of quality data became relevant models for the human risk assessment of drug administration during pregnancy. Differences in response, if they exist, between the animal species should be evaluated and further investigation should guide human extrapolation.

Among the key points, embryo-fetal effects rather than maternal effects are to be considered as the primary endpoint. If embryo-fetal effects occur with maternal toxicity, the dam influence should be clearly assessed, and might be used to provide a safety margin for the tested drug. Determination of the lowest developmental toxic dose, apart from any maternal toxicity, should be performed when possible, as maternal toxicity is not always the cause of harmful fetal effects.

In the case of positive animal embryo-fetal findings, the interpretation should take into account various parameters, including the incidence and the severity of the embryo-fetal insults, interspecies reproducibility, level of exposure, metabolic pathways and possible mechanistic investigations. If animal findings are doubtful, further investigations are necessary to estimate more effectively the human risk.

In summary, recommendations to clinicians can take into account two key criteria; first the drug developmental effect such as malformations or embryo-fetal toxicity. Secondly, one should take into account the occurrence of experimental embryo-fetal effects at human therapeutic exposure levels or with a wide safety margin established from the animal exposure at the NOAEL. The ultimate goal of using the optimal animal model and/or strategy in preclinical developmental assessment is to enhance the predictive value and to provide the best information to the prescriber and patient via product labelling.

REFERENCES

1. Dally, A., *Lancet, 351*, 1197 (1998).
2. ICH (International Conference on Harmonisation of Technical requirements for the Registration of Pharmaceuticals for Human Use). Detection of toxicity to reproduction for medicinal products. *Fed. Reg, 59*, 48746 (1994).
3. Guittin, P., Decelle, T., *ILAR Journal, 43*, S80 (2002).
4. Dawson, A. B., *Stain. Tech, 1*, 123 (1926).
5. Guittin, P., Soubrié, C., Autret, E., et al., *Therapie, 53*, 355 (1998).
6. Guittin, P., Eléfant, E., Saint-Salvi, B., *Reprod. Tox., 14*, 369 (2000).
7. Wise, D. L., Beck S. L., Beltrame, D., et al., *Teratology, 55*, 249 (1997).
8. Chahoud, I., Buschmann, J., Clark, R., et al., *Reprod. Toxicol., 13*, 77 (1999).
9. Guittin, P., *Arch. Pediatr., 7*, 401 (2000).

4.4 Evaluation of fetal toxicity: human data

Anders Rane

Department of Laboratory Medicine, Division of Clinical Pharmacology, Karolinska Institute, Karolinska University Hospital, Huddinge, SE-141 86 Stockholm, Sweden

4.4.1 INTRODUCTION

Drug testing for fetal toxicity and teratogenicity is not only a test of the mother compound as there are circulating metabolites that may also reach the fetal compartment. Thus, we also test the metabolites of the drug. If the drug is a racemate, about twice as many chemical moieties are tested compared to an enantiomeric or non-stereospecific drug.

Once entered into the fetal compartment, the drug may be metabolised in the fetal tissue. Fetal metabolism is an important issue, as biotransformation is not only an inactivation process but may also be a bioactivation process. There are hundreds of clinically used drugs that are bioactivated to pharmacologically active, or even toxic metabolites.

Drug metabolic reactions may be classified into oxidations (which are catalysed by the members of the cytochrome P450 enzyme system), conjugations (with different endogenous moieties such as glucuronic acid, acetyl groups and glutathione), reductions and various kinds of hydrolytic reactions. In this context, oxidations are the most interesting, because such reactions often lead to the formation of reactive metabolites that may bind to macromolecules in the tissues and cause adverse manifestations.

A large body of knowledge has evolved from the past decades' research concerning the cytochrome P450 system (CYP). A logical classification system identifies the specific enzyme gene as a figure indicating the gene family, followed by a letter for the gene subfamily and another figure for the actual gene, e.g. CYP1A1. The CYP gene superfamily has a long evolutionary history, from the aquatic and amphibian organisms billions of years ago to today's animal fauna. There is a large diversity of the cytochrome P450 system. There is also an increasing amount of data on the different members of the subfamilies, their chromosomal locations, amino acid sequences, and tissue specificity, as well as their genetic polymorphisms.

4.4.2 ONTOGENIC DEVELOPMENT OF DRUG METABOLI-SING ENZYMES IN LABORATORY ANIMALS

How do the CYP enzymes develop in the fetus, when, and by what mechanisms? In the late 1950s, Fouts and Adamson [1] were among the first to demonstrate that fetuses of most laboratory animals are unable to catalyse drug metabolic reactions. Their results were confirmed by several investigators in other laboratories. In our laboratory, we showed that fetuses of both rabbits and rats had no, or negligible, CYP activity [2,3]. Postnatally, there is a surge of the enzyme activity up to the weaning period. Thereafter, it decreases to the level of the adult animals.

There are two types of developmental pattern of these enzymes. In the first, the prenatal enzyme activity is deficient. In this situation the fetus is dependent on the maternal metabolism for its elimination of the drug and the metabolites. Second, the drug may be metabolised in the fetus. In both situations the developing fetus is exposed to metabolites of various kinds that have been generated in the maternal and/or fetal compartment and that may have other fetotoxic properties than the mother compound. Ideally, preclinical toxicological screening should aim at testing *all* chemical drug moieties; the ones administered and those formed in the organism. If possible, it should be performed at concentrations/doses close to those achieved in treated patients.

This story is not complete without briefly mentioning the postnatal situation. In herbivore mammals, to which group most of our common laboratory animals belong, the enzyme activity is low in the prenatal period. This may be due to repressing factors of maternal origin and/or lack of "activators". After birth, the enzyme activity increases. The mechanisms of the postnatal surge are poorly understood. Endocrine "activators", enzyme stimulating compounds in the milk, influence of growth hormone (GH), or elimination of maternal "repressors" have been proposed. After weaning, a decrease in enzyme activity is often observed, possibly due to dietary changes. For some isozymes, however, the postnatal activity does not increase until after the weaning period.

Gonzalez and collaborators [4] have studied the CYP2E and CYP2C families, which are constitutive enzymes. These isozymes are controlled by a tissue-specific, developmentally programmed and sex-specific transcription process. The CYP2E1 isozyme is developed immediately after birth, probably due to a DNA binding protein that triggers the transcription and enzyme synthesis a few hours after birth. On the other hand, there are isozymes, e.g. CYP2C6, which do not develop until a few weeks after birth.

The situation is different higher up on the evolution ladder, e.g. in subhuman primates. Data have shown that measurable drug oxidative enzyme activity is present near term in the stumptailed monkey (*Macaca arctoides*) and marmoset (*Callithrix jacchus*), albeit at a low levels [5,6]. In this respect, the subhuman primates are different from our common laboratory animals and they are more similar to human.

4.4.3   ONTOGENESIS OF DRUG METABOLISM IN HUMANS

The human fetal liver has an unequivocal capacity to metabolise and catalyse oxidations of various drugs, which is in contrast to the situation in laboratory animals. Studies from different research groups have demonstrated oxidation rates between a few percent (as for benzo(a) pyrene hydroxylation) to 70% (as for ethylmorphine N-demethylation) of adult values. It is known that benzo(a)pyrene hydroxylation is catalysed by CYP1A, which develops only very late in gestation. The half-lives of caffeine and theophylline in newborn are 95 and 24-36 hours compared to 4 and 3-9 hours respectively, in adults. This late development of the metabolic clearance of caffeine and theophylline, which are substrates of CYP1A enzymes, is consistent with the slow postnatal maturation of this enzyme.

A series of opiates, like morphine, codeine, ethylmorphine and dextromethorphan, have been used as tools in studies of drug metabolism during development. They are subject to various oxidative reactions, e.g. N- and O-dealkylation reactions, although their major metabolic pathway includes glucuronidation. To our surprise, we found that the N-demethylation of ethylmorphine to norethylmorphine was as high in the human fetal as in the human adult liver [7]. This pathway is believed to be catalysed by an isozyme with putative importance for the metabolism of certain steroid hormones. By use of a library of antibodies against different isozymes, we were able to show that the human fetal liver contains an isozyme identified by a monoclonal antibody against a CYP isozyme induced by the antiglucocorticoid pregnenolone-16α-carbonitrile (PCN). This isozyme is active in some steroid metabolic pathways [8] and was later denoted as CYP3A7. The intensity of the immunoidentified protein band in Western blot was linearly related to the rate of N-demethylation of codeine, ethylmorphine and dextromethorphan [9]. In contrast, the quantitatively insignificant CYP2D6 which catalyses the O-dealkylation, was not measurable with the corresponding antibody in the human fetal liver specimens.

One can conclude that there is a clear difference in the development of the isozymes catalysing the N-demethylation and the O-dealkylation of these drugs. Our results were corroborated by Treluyer *et al.* [10] who studied the CYP2D6 by immunoblotting of hepatic microsomes from fetuses and neonates. The bands were quantitated as negative, intermediate, or positive. The proportion of positive results increased during the gestational period. At term and in neonates younger than 24 hours, about 50% of the liver samples were positive, but in neonates older than 24 hours, approximately 85% of the liver biopsy samples were positive. Thus, it is apparent that the isozyme responsible for the O-dealkylation of opiates develops much later than the CYP3A isozyme [9].

As mentioned earlier, the CYP3A isozyme is essential in the metabolism of endogenous steroids. In view of its importance for the fetal development, it is not surprising that it is expressed early in gestation. The N-demethylating activity of ethylmorphine is inhibited by various steroids (progesterone, dehydroepiandrosterone, etc.), which yields indirect evidence that the drug and the steroid substrates utilise the same catalyst [9].

4.4.5   DRUG DISPOSITION IN PREGNANCY

Many drugs are disposed of differently in pregnant women compared with non-pregnant women. An increased drug metabolism and/or excretion is often observed. Phenytoin, when given to pregnant epileptic women, yields lower plasma concentrations per dose unit than in non-pregnant women. Metoprolol is another such example and the difference in apparent oral clearance between non-pregnant and pregnant women is 3-4 fold [11]. Drugs excreted via the kidneys often have a higher renal clearance in pregnancy as a result of the increase in glomerular filtration and creatinine clearance. This is the case for ampicillin and many other antibiotics. If possible, the drug therapy should be controlled by therapeutic drug monitoring in plasma. However, the observed changes in drug disposition in pregnant women do not automatically authorise the doctor to adjust the dose since there may be concomitant changes in sensitivity to the drugs.

4.4.6   PASSAGE OF DRUGS ACROSS THE PLACENTA

Most drugs pass the placenta and other biological membrane through passive diffusion of the unbound moiety in blood. Lipid soluble drugs, such as psychoactive agents, generally pass at a higher rate than more water soluble drugs like antibiotics. Most drugs are administered intermittently through i.v. or oral administration. Figure 1 demonstrates that "pseudo-equilibrium" occurs in the moment when the ratio between maternal and fetal blood concentration is 1.0.

The rate of transfer across the placenta, estimated as time to pseudo-equilibrium, may be estimated through repeated analyses of the maternal/fetal concentration ratio in samples collected concomitantly from the mother and the umbilical cord during legal abortion or parturition.

Information about time to pseudo-equilibrium and the maternal/fetal ratio may be used for estimation when a drug should be given in relation to the predicted parturition or Caesarean section in order to minimise the effects on the fetus. This is shown in Figure 1. An anaesthetic agent with a rapid transplacental passage (Fig. 1, left) is compared with a drug with a slow rate of transfer (Fig. 1b, right). Both agents are supposed to give optimal anaesthesia to the mother during the time marked by the shadowed area. If the fetus is born at time A, it is to be noted that the *relative* fetal concentration of the drug is lower for the least lipid soluble drug (Fig. 1, right). This drug would be preferable if the agents are equal in other respects. On a later time point (such as time B), the fetal concentration may be higher than the maternal concentration (Fig 1, right). Yet, the *actual* drug

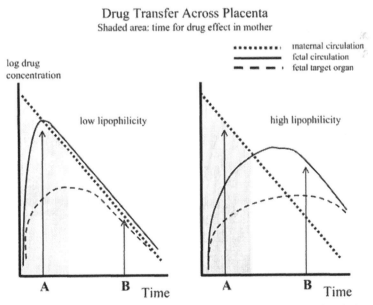

Figure 1. Concentration versus time profiles in the maternal and fetal circulation after an i.v. dose to the mother. On the left this is shown for a lipophilic drug with a rapid transfer, and on the right for a drug with a relatively high lipophilicity.

concentration in the newborn is below the anaesthetic levels indicated by the shadowed area. Thus, a maternal/fetal concentration ratio of <1 does not necessarily implicate a pharmacological effect on the newborn, since the concentration at that time may be below the pharmacological concentration range.

4.4.7 EFFECTS ON THE FETUS

If a fetotoxic effect arises, the type and degree of injury is dependent of the stage of development and maturation of the conceptus at the time of drug exposure. It is also dependent on the total drug load, i.e. the dose, duration of treatment and the drug clearance capacity in the maternal and fetal compartments. The drug may affect the conceptus/fertilised zygote already in the first week after ovulation, during its passage down to the uterus and development to morula and blastocyst. The organ systems are formed during the first weeks after implantation, i.e during the *pre-embryonic* period. In the *embryonic* period, between the third and tenth week after conception, most organs are developed. In this period, the fetus is most sensitive to teratogenic agents. During the ensuing maturation phase, the *fetal* period, morphological malfor-

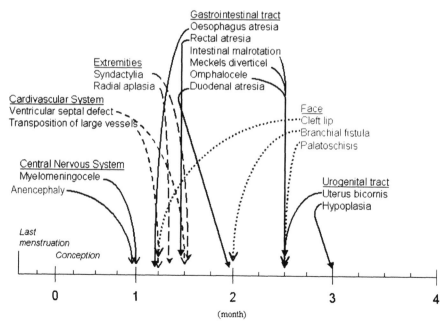

Figure 2. Estimated times for concluded organ formation in the human fetus.

Table 1

Drugs with risk of morphological teratogenic effect in humans

Drug	Reported effects	Reference
antiepileptics	multiple malformations	12,13
barbiturates	skeletal malformations	14,15
phenytoin	clef lip, cardiac malformations	15,16
carbamazepine	reduced head circumference	17,18
parametadion trimetadion	mental retardation, palatal and dental malformations	19-21
valproic acid	spina bifida	22,23
diethylstilbestrol	vaginal adenosis, vaginal adenocarcinoma	24,25
folic acid antagonists, cytostatics, etc.	multiple malformations, intrauterine growth, retardation and fetal death	26,27
gestagens – androgens	virilisation of female fetuses, cardiac malformations, urogenital malformations	28,29
iodine 131, 125	fetal thyroid destruction	30
lithium	congenital cardiac malformations	31,32
retinoids	multiple malformations (in e.g. CNS, heart)	33
thalidomide	defects in extremities, ocular, ear and visceral malformations	34
warfarin	chondrodystrophia calcificans, nose hypoplasia, ocular malformations, and mental retardation	35
oestrogens	feminisation of male fetuses	36,37

mations are generally not seen since the organs are already formed. On the other hand, drugs may cause many pharmacological and toxic reversible or irreversible effects. Many such effects are known and may affect the development of the autonomic nervous system.

Knowing the time points when different organ systems have matured it is possible to rule out a malformation of a specific organ system with a high degree of certainty. Drug treatment after these time points will in all probability not result in increased risk of morphological injuries (Fig. 2). The rate of spontaneous malformations is about 2-3%. The fraction of all malformations caused by drugs is very small, as is the number of drugs with confirmed or plausible teratogenic effects (Table 1).

Behavioural teratogenic effects have been studied to a comparatively limited degree (Table 2). These "invisible " drug effects are difficult to study in animals and man. Such effects may be caused by psycho-active agents through interaction with the autonomic nervous system and perturbation of the development of signal systems in the CNS. Endocrine adverse effects may also arise from hormonal treatment or from drugs that interfere with endocrine mechanisms. Such drugs should not be used in pregnant women. Alternative possible remedies should be sought.

Table 2

Drugs suspected of giving higher risk of behavioural disorders when administered to
fetuses or newborns; documentation is mostly based on animal studies

Drug	Plausible mechanism of action	Effects	References
neuroleptics haloperidol phenothiazines	inhibitory effect on development of dopamine pathways in CNS	motoric movement dysfunctions and learning disturbances	38-40
tricyclic antidepressants shown for imipramine, clomipramine	affect β-adrenoreceptors	disturbance in psychomotor development	41
analgesics morphine and derivatives morphine antagonists	affect opiate receptors	sexual behavioural disturbances	42-45
diethylstilbestrol	hormone endocrine effect	sexual behavioural disturbances	46
cimetidine	hormone endocrine effect	sexual behavioural disturbances	47

4.4.8 SUMMARY

In summary, some CYP isozymes are developed early in human gestation
(CYP3A), other enzymes are developed at birth (CYP2C, CYP2D), while still
others are developed later on (CYP1A). Many reaction steps are supposedly sites
of metabolic interactions between drugs and steroids.

Information required for better interpretation of preclinical drug toxicity/
teratogenicity data include routes of drug metabolism in the test species *and* in
man, including metabolism pattern in pregnant women, and in the human fetus.
In the absence of such data one may conclude that teratogenicity tests in animals
have limited predictive value for the potential of the drug to cause malformations
in human.

REFERENCES

1. Dvorchik, B.H., "Hepatic drug metabolism in the fetal stumptailed Monkey (*Macaca arctoi-des*)," in *Role of Pharmacokinetics in Prenatal and Perinatal Toxicology*, Neubert, D., et al., eds., *Georg Thieme Publishers*, 275 (1978).
2. Fouts, J. R., Adamson, R. H., *Science, 129*, 897 (1959).
3. Gonzalez, F. J., Yano, M., Liu, S-Y., *J. Basic Clin. Physiol. Pharmacol., 3 (Suppl.1)* (1992).
4. Högstedt, S., Lindberg, B., Rane, A., *Eur. J. Clin. Pharmacol., 24*, 217 (1983).
5. Ladona, M. G., Park, S. S., Gelboin, H. V., Hammar, L., Rane, A., *Biochem. Pharmacol., 37*, 4735 (1988).

6. Ladona, M. G., Lindström, B., Thyr, C., Peng, D., Rane, A., *Br. J. Clin. Pharmacol.*, *32*, 295 (1991).
7. Neubert, D.; Siddal, R. A.; Tapken, S.; Hiddleston, W. A.; Higgins, J. E., *Georg Thieme Publishers*, 299 (1978).
8. Pacifici, G. M., Park, S. S., Gelboin, H. V., Rane, A., *Pharmacol. Toxicol.*, *62*, 101 (1988).
9. Rane, A., Berggren, M., Yaffe, S. J., Ericsson, J. L. E., *Xenobiotica, 3*, 37 (1973).
10. Treluyer, J-M., Jacqz-Agrain, E., Alvarez, F., Cresteil, T., *Eur. J. Biochem.*, *202*, 583 (1991).
11. Hogstedt, S., Lindberg, B., Rane, A., *Eur. J. Clin. Pharmacol.*, *24*, 217 (1983).
12. Lowe, C. R., *Lancet 1973*, 9 ;
13. Shapiro, S., et al., *Lancet 1976 (2)*, 272.
14. Seip, M., *Acta. Paediatr. Scand. 65*, 617 (1976);
15. Smith, D. W., *Am. J. Dis. Child. 131*, 1337 (1977).
16. Hanson, J. W., Smith, D. W., *J. Pediatr. 87*, 285 (1975); .
17. Niebyl, J. R., et al., *Obstet. Gynecol. 53*, 139 (1979);
18. Hiilesmaa, V. K. et al., *Lancet 1981 (2)*, 165.
19. Zackai, E. H., et al., *J. Pediatr. 87*, 280 (1975),
20. Feldmann, G. L., et al., *Am. J. Dis. Child. 131*,1389 (1977);
21. Rosen, R. C., Lightner, E.S. *J. Pediatr. 92*, 240 (1978).
22. Brown, N. A. et al., *Lancet 1980 (1)*, 660;
23. Bjerkedal, T., et al., *Lancet 1982 (2)*, 1096.
24. Herbst, A. L., et al., *N. Engl. J. Med. 292*, 334 (1975);
25. Fowler, W. C., Edelmann, D. A., *Obstet. Gynecol. 51*,459 (1978),
26. Nicholson, O. H., *J. Obstet. Gynaecol. Brit. Com. 75*, 307 (1968),
27. Hassenstein, E., Riedel, H., *Geburtshilfe Frauenheilkd. 38*, 131 (1978),
28. Wilkins, L., *JAMA 172*, 1028 (1960);
29. Nora, J. J., Nora, A. H., *Lancet 1973 (1)*, 941;
30. Shepard, T. H., *Catalogue of teratogenic agents* (3rd ed.), Johns Hopkins University Press (1980);
31. Weinstein, M. R., Goldfield, M. D., *Am. J. Psychiatry132*, 529(1975);
32. Rane, A. et al. *J. Pediatr. 93*,296 (1978).
33. Rosa, F. W., *Lancet 1983 (2)*, 513;
34. Lentz, W., Knapp, K., *Arch. Environ. Health 5*, 100 (1962);
35. Shaul, W. L., Hall, J. G., *Am. Obstet. Gynecol. 127*, 191 (1977),
36. Gill, W. B. et al., *J. Reprod. Med. 16*, 147 (1976);
37. Bibbo, M., et al., *Obstet. Gynecol. 49*, 1 (1977).
38. Kolata, G. B., *Science 202*, 732 (1978).
39. Lundborg, P., *Brain Res. 44*, 684 (1972).
40. Lundborg, P., Engel, J., *Naunyn Schmiedebergs Arch. Pharmacol. 279*, 31 (1973).
41. Jason, K. M., Cooper, T. B., Friedman, E., *J. Pharmacol. Exp. Ther. 217*, 461 (1981).
42. Meyerson, B. J., *Eur. J. Pharmacol. 69*, 453 (1981).
43. Hetta, J., Terenius, L., *Neurosci. Lett. 16 (3)*, 323 (1980).
44. Vathy, I. U., et al., *Pharmacol. Biochem. Behav. 19*, 777 (1982).
45. Meyerson, B. J., Berg, M., *Neurosci. Lett. 18 (3)*, 323 (1985).
46. Ehrhardt, A. A., et al., *Arch. Sex. Behav. 14*, 1 (1985).
47. Anand, S., van Thiel, D. H., *Science 218*, 493 (1982).

4.5 Fetal pharmacology and therapy

Evelyne Jacqz-Aigrain

Department of Paediatric Pharmacology and Pharmacogenetics, Robert Debré Hospital, Paris, France

4.5.1 INTRODUCTION

Clinicians have to deal with a large variety of situations where the mother and the fetus are exposed to drugs. The evaluation of the risk-to-benefit ratio of drug administration during pregnancy often constitues a challenge [1]. The human teratogenic effects of thalidomide and retinoids have led the medical profession to realise how dangerous medications can be, even without adverse maternal effects [2-4]. Warfarin induces congenital malformations and fetal haemorrhages by competing with vitamin K [5]. The administration of these drugs is contra-indicated during pregnancy. Long term consequences of fetal exposure to drugs is unpredictable. The high incidence of vaginal adenocarcinoma, developing in young women exposed to stillbestrol during pregnancy, is a well-known example of possible delayed side-effects of drugs administered during pregnancy [6].

In addition to these drastic situations, clinicians are frequently faced with a compromise between either a mandatory maternal treatment and its poten-tial fetal effects, or the maternal and fetal side effects when fetal treatment is required. Hypertension, life-threatening maternal arrhythmia or epilepsy are situations where maternal therapy is required, not only for maternal reasons, but also for the possible fetal consequences of the maternal disease. Most physicians are now aware of potential fetal risk, including teratogenic effects, fetal growth retardation and fetal and neonatal distress. Therefore, such treatments are asso-ciated with close monitoring of the fetus.

There are multiple applications of fetal therapy. In this review, we present the recent progress in performing fetal diagnosis, in understanding the pharmacokinetics of drugs during pregnancy, and in evaluating the risk-to-benefit ratio of these drugs in different areas of fetal therapy [7].

4.5.2 PHARMACOKINETICS DURING PREGNANCY

Physiological changes in the mother

The physiological changes that occur during pregnancy in the mother can alter the uptake, distribution, metabolism and elimination of drugs that may be used to treat maternal, fetal or pregnancy associated diseases.

Absorption is modified by decreased intestinal motility and increased gastric emptying time. This may delay the time to peak concentration in the mother, and either decrease or increase drug absorption. Modifications in body weight and fat, total body water and plasma proteins can alter the distribution of drugs and the amount of drug needed for therapeutic effectiveness. Maternal hepatic and extrahepatic metabolism may be affected by hormonal changes and an increase in hepatic blood flow. Elimination of xenobiotics are modified by major changes in renal blood flow and glomerular filtration rate.

Placental physiology

For ethical and procedural reasons, studies of physiology and metabolism of the placenta, as well as factors determining placental transfer, are limited and have been conducted using term human placenta.

The physiology of the placenta changes throughout pregnancy: surface area increases, thickness of tissue layers decreases, and the placenta can carry out oxidations *in vitro*. However, the metabolism of xenobiotics is less important than the biotransformation of endogenous substrates such as steroid hormones.

Placental drug transfer is dependent on the physiochemical characteristics of drugs: most drugs have a molecular weight from 250 to 500 and cross the placenta easily. The rate of transfer is dependent upon lipid solubility, degree of ionisation measured by the pKa value, and protein binding in the maternal and fetal circulation. Only the free and non-ionised drug in the maternal circulation is available for transfer [8,9].

Drug disposition in the fetus [10]

Following their transfer across the placenta, drugs enter the umbilical vein. The umbilical venous blood flow is either diverted into the portal vein to perfuse the hepatic tissue or bypass the liver via the ductus venosus. This relative distribution between the two circulations may be of major importance in determining the quality of drug reaching fetal tissues and, in particular, the central nervous system.

The human fetal liver is able to oxidise numerous substrates. Among cytochromes P450, the CYP3A subfamily is present as early as twenty weeks' gestation. In contrast, the CYP2D family exhibits a perinatal ontogenesis and the CYP1A family a postnatal ontogenesis. Other factors affecting drug disposition in the fetus include excretion by the fetal kidney, and swallowing of amniotic fluid. Some drugs accumulate in the amniotic fluid compartment after urine excretion and exchanges occur across through the fetal skin.

In most cases, drugs exert their effects through interactions with receptor sites. Adrenergic receptor function develops early during fetal life. In addition, receptor interactions exist with digoxin, insulin and glucocorticoids. However, major differences between adults and fetuses have been reported in the number of receptors or in drug affinity for these receptors. Much more information is needed to understand the factors controlling drug disposition and effect in the fetus.

Pharmacokinetics of various drugs during pregnancy

The pharmacokinetics of many drugs are modified during pregnancy [7]. For ampicillin, the effect of increased rate of elimination and volume of distribution doubles the amount of drug needed to obtain the same plasma concentrations.

Caffeine is characterised by an increased volume of distribution and a decreased rate of elimination. Therefore, it is necessary to decrease the dose of caffeine throughout pregnancy to obtain the same plasma concentrations as in the non-pregnant state.

Digoxin is mainly excreted unchanged by the kidney. When administered intravenously to pregnant women to treat fetal arrhythmia, the systemic clearance of digoxin was increased, probably due to increased renal blood flow and glomerular filtration rate. As a consequence, elimination half-life was shortened.

4.5.3 EXAMPLES OF FETAL THERAPY

Major progress has been made in the recent years in investigating fetal disorders. Echography now allows precise diagnosis of congenital malformations (such as digestive, urinary tract or cardiac abnormalities). Diagnosis of abnormal chromosomal pattern is possible with chorion biopsy, amniocentesis or fetal blood sampling [11]. Such procedures have their own complications but their frequency is reduced in experienced hands. The therapeutic decision of whether to carry on pregnancy and to treat fetal diseases or to induce abortion is obviously a matter of individual cases and differences in legislation in various countries. These investigational procedures, particularly the routine use of cord taps, have increased the diagnostic and treatment options for life-threatening fetal disorders.

An historical example of fetal therapy is the maternal administration of penicillin for the treatment of fetal syphilis [12]. Other examples include the use of antenatal glucocorticoid therapy to reduce both the incidence of hyaline membrane disease and perinatal mortality [13] and phenobarbital for its enzyme inducing properties [14].

The situations requiring fetal drug therapy are numerous and examples we present below can be classified as: 1) high risk fetal diseases, 2) alterations in hormonal milieu, and 3) treatment of materno-fetal diseases of pregnancy.

Treatment of high risk fetal diseases

Treatment of fetal supraventricular tachyarrhythmias

Fetal supraventricular tachyarrythmias with normal cardiac anatomy are estimated to complicate the course of 1/10 000 pregnancies. The diagnosis, the mechanism and the tolerance of the arrhythmia can be evaluated by M. Mode and Doppler echocardiography. The prognosis is dependent on the cardiac function and risk of cardiac failure, the mechanism of the arrhythmia and the response to treatment.

Antiarrhythmic drugs are usually administered to the mother for fetal treatment. The selection of the drug is based on antiarrhythmic efficacy, pharmacokinetics, potential maternal and fetal toxic effects [15]. The drug administered as first choice is usually digoxin and the control of the arrhythmia is obtained in only 30 to 40% of cases. When return to sinus rhythm is not obtained, various other drugs are additionally prescribed, including amiodarone [15,16]. Sotalol has excellent placental transfer and is effective in the treatment of fetal arrhythmia [17].

Flecainide is a promising antiarrythmic drug for this indication. However, data on its use during pregnancy are limited, particularly the evaluation of maternal and fetal toxic effects. The overall efficacy of treatment is over 80% in the few series available in the literature [15,18,19]. However, a large discrepancy is noted in terms of drugs and doses administered.

Treatment of fetal toxoplasmosis

Congenital toxoplasma infection is defined by the transmission of the parasite to the fetus and is dependent on the time of acquisition of maternal infection during pregnancy. The later maternal infection is acquired, the more frequently parasites are transmitted to the fetus, reaching 90% of transmission during the last weeks before delivery [20]. Congenital toxoplasmosis is defined by the clinical manifestations of the disease in the newborn. It is subclinical, mild or severe and depends on the age of the fetus at the time of transmission. The period of highest risk is between 10 and 24 weeks. It is known that the majority of infants with congenital toxoplasmosis have major sequelae. In the early 1950s, at a time when treatment was not undertaken, the major sequelae among 105 children less than five years of age, were mental retardation (96%), convulsions (84%), impaired vision (61%) and deafness (15%).

Epidemiological surveys of the prevalence of specific antibodies in women in the childbearing age were conducted in 1960 and 1985 and demonstrate that the frequency of high-risk negative women varied widely between countries. The observed decrease in prevalence in France (87% in 1960-1970 and 73% in 1985) is probably related to changes in eating habits [20,21].

Antenatal diagnosis of fetal toxoplasmosis is performed by fetal blood sampling, ultrasonography and amniocentesis. Specific diagnosis tests include demonstration of fetal toxoplasma IgM and presence of *Toxoplasma Gondii* in fetal blood amniotic fluid or both, by inoculation into mice or tissue cultures. PCR reaction tests on amniotic fluid are recent techniques used to detect toxoplasma DNA [22].

The treatment of congenital toxoplasma infection is based on the administration of spiramycin and, if the fetal infection is confirmed, an additional treatment with pyrimethamine and sulfadiazine is administered in 3-week courses alternating with 3 weeks of daily spiramycin.

Spiramycin is a macrolide antibiotic active against toxoplasma. The incidence of fetal toxoplasmosis is reduced from 60% to 23% by spiramycin administration. This effect may reflect a limitation of placental infection and transplacental passage to the fetus [23].

Pyrimethamine is a folic acid antagonist and sulfadiazine is a sulfonamide. They act synergistically against toxoplasma, as demonstrated by their combined activity which is eight times what would be expected if their effects were additive [24].

Pyrimethamine and sulfadiazine are both potentially toxic. Sulfonamides may induce hypersensitivity and pyrimethamine exhibits dose-related toxicity inducing platelet depression and leukopenia. In addition, teratogenic effects are reported in rats, but their frequency seems to be reduced by the co-administration of folic acid. Very little is known of the pharmacology of the combination of these drugs during pregnancy [25].

Future tools in the fetal treatment of congenital toxoplasmosis include the improvement of the reliability of prenatal diagnosis, particularly using PCR, and conducting large trials of treated patients to analyse maternal and fetal tolerance of treatment.

The use of antiretroviral drugs during pregnancy [26].

Antiretroviral therapy is widely prescribed in pregnancy to reduce the risk of vertical transmission. In recent years, therapeutic schedules have been either concurrent therapy (nucleoside analogue reverse transcriptase inhibitors (NRTIs) with or without non-nucleoside transcriptase inhibitor), or even triple therapy with nucleoside analogues.

Among NRTIs, zidovudine was demonstrated to significantly reduce mother-to-child transmission of HIV-1 and it has been widely used in pregnancy, both as monotherapy and combination therapy. Animal studies have shown teratogenic effects in rodents and vaginal tumours during adult life in mice exposed to high doses *in utero*. In monkeys, mitochondrial damages has been reported in cerebrum, cardiac and skeletal muscles with persistent neuro-mitochondrial diseases in children exposed either to zidovudine alone or in association with other antiretroviral drugs [27,28]. Additional data from long-term follow up of children exposed *in utero* to antiretroviral drugs are required to confirm these findings. The other NRTIs are mostly used in combination therapy and to determine their safety is a challenge because the number of pregnant women exposed to each of them is relatively small.

Protease inhibitors are used in combination therapy during pregnancy. Experimental animal data have evaluated the teratogenicity and developmental toxicity of these drugs. In the absence of maternal toxic effects associated with high doses, animal studies were negative for arazanavir, lopinavir, nelfinavir, ritonavir and saquinavir, but positive for amprenavir (abortions in rabbits and ossifications defects in rabbits and rodents) and indinavir (cervical ribs).

Safety animal data were negative for nevirapine, while teratogenicity was associated with exposure to delavirdine in rats (ventricular septal defects), and efavirenz in cymomolgus monkeys (anencephaly, microphthalmy, cleft palate).

Alterations in fetal hormonal milieu

Fetal treatment of congenital adrenal hyperplasia (CAH)

The sexual differentiation occurs very early in male fetuses under the influence of testosterone. In contrast, sexual differentiation occurs later and is not stimulated by any hormones in female fetuses. As early as 7 weeks' gestation, circulating concentrations of testosterone are higher in male than in female fetuses but the peak concentration of testosterone occurs at 17 weeks' gestation.

The congenital adrenal hyperplasia (CAH) is related in most cases (>90%) to a deficit in steroid 21 hydroxylase, leading to overproduction of D4 epiandrosterone and testosterone. 21 hydroxylase deficiency is transmitted as an autosomal recessive trait. The "classic form" causes ambiguous external genitalia in females with or without salt wasting. But there is also a "wild-late, form", causing symptoms of masculinisation at puberty and infertility [29].

The prenatal treatment of congenital adrenal hyperplasia due to 21 hydroxylase deficiency is instituted with the aim of preventing the *in utero* virilisation of CAH-affected female fetuses, which is caused by overproduction of adrenal androgens.

Fetal adrenal suppression is based on maternal administrated glucocorticoids that induce suppression of the increased adenocorticotropin (ACTH) levels by the fetal pituitary adrenal axis. It must be started early, before 7 weeks' gestation, and therefore, before the possible antenatal diagnosis of CAH in a high-risk mother, and either continued until term in case of a fetal CAH female or discontinued in the other cases.

Prenatal diagnosis was initially made in females after amniocentesis at 16 to 18 weeks' pregnancy. It was based on data provided by an index case, using HLA typing. Prenatal diagnosis is now made earlier during pregnancy, by direct molecular studies of the 21 hydroxylase gene using chorionic villus biopsies [30]. If the baby is affected, treatment is maintained thoughout pregnancy. Concentrations of several steroids, including 17 hydroxyprogesterone and adrenal androgens, are determined in order to verify fetal adrenal gland suppression [31]. It is now clear that the first choice drug for the treatment of CAH is dexamethasone. The drug is not inactivated by the placenta, but crosses the placenta with a fetal-to-maternal ratio of near unity. It is not linked with transcortine, and has a long half-life of 4 to 6 hours. The administered dose was initially 1 mg/day but was increased to 1.5 mg/day in 2 to 3 divided doses. The analysis of failures of antenatal treatment suggests that they may be related to low doses, sometimes administered once daily, late initiation or interruption of treatment and individual variations in dexamethasone disposition during pregnancy. The maternal side effects of treatment are very limited and the benefits of treatment clearly outweigh the risks [32].

Treatment of fetal dysthyroidism

The development of the fetal thyroid function is initiated during the tenth week of gestation [33]. Fetal thyroid function during pregnancy can be evaluated by ultrasound examination and measurement of thyroid hormones in cord blood.

Hyperthyroidism during pregnancy is usually attributable to Grave's disease, an autoimmune disease caused by the presence of thyroid-stimulating immunoglobulin (TSI). TSI crosses the placenta and may cause intrauterine hyperthyroidism. The treatment of hyperthyroidism during pregnancy is based on maternal administration of antithyroid drugs that easily cross the placenta. Propylthiouracil is the drug of choice, but carbimazole and thiamiazole are also prescribed during pregnancy. Careful titration of the dose should be continued throughout pregnancy [34].

Fetal hypothyroidism with goitre was successfully diagnosed *in utero* by obstetrical ultrasound revealing a goitre. Congenital hypothyroidism occurs with a incidence of 1/4000 and 1/5000 and only 10 to 20% may be attributed to dyshormonogenesis. Intra-amniotic administration of levothyroxine sodium (T4) every 2 weeks increased fetal concentrations of T4 and decreased fetal concentrations of TSH. The usefulness of such treatment needs to be evaluated. However, it may reduce the central nervous system deficits and the specific defects in hearing and speech associated with late post-natal treatment [35-37].

Treatment of materno-fetal diseases

Feto-maternal alloimmune thrombocytopenia and rhesus immunisation

Feto-maternal alloimmune thrombocytopenia is caused by feto-maternal incompatibility for platelet antigens. Treatment options include maternal administration of corticosteroids and immunoglobulins or repeated intrauterine platelet transfusions. A recent European study demonstrated that the maternal administration of intravenous gamma globulins increases the fetal platelet count in fetuses with alloimmune thrombocytopenia and reduces the risk of intracranial hemorrhage [38].

Fetal blood transfusions [39] are performed for the treatment of severe rhesus immunisation (erythroblastosis fetalis). In the past, transfusions were decided upon after amniocentesis, but Doppler velocity assessment in the middle-fetal cerebral artery now allows accurate timing of intrauterine transfusions [39].

The use of antiinflammatory drugs during pregnancy

During pregnancy, prostanoids play a major role in the regulation of maternal vascular tone, placental and fetal circulation (placenta, umbilical blood flow, ductus arteriosus), uterine contractility and induction of labor [40].

Non-steroidal antiinflammatory agents are inhibitors of prostanoid synthesis by acting on both COX-1 and COX-2 enzymes. They are primarily administered during pregnancy and the perinatal period for the following reasons: 1) acetyl salicylic acid is currently administered to prevent idiopathic fetal growth retardation of vascular origin [41], and 2) indomethacin is prescribed to try to stop premature labor and to reduce polyhydramnios [42-44].

Aspirin is known to irreversibly acetylate and therefore inactivate cyclo-oxygenase, while indomethacin is a competitive inhibitor of the enzyme. Aspirin is extensively protein bound and, during pregnancy, crosses the placenta. The fetal-to-maternal ratio of aspirin was reported to be over 1.5 at birth and is related to the binding characteristics of fetal serum. Indomethacin is highly absorbed and protein-bound. It is metabolised by the liver and excreted by the kidney. Its elimination half-life is longer in neonates than in adults.

Aspirin and idiopathic fetal growth retardation

Idiopathic fetal growth retardation is associated with reduced uteroplacental perfusion, abnormal platelet behavior and local thrombosis, resulting in placental infarction.

Many studies have been conducted since 1978 to treat pre-eclampsia and other chronic placental insufficiency syndromes using low doses of aspirin [150 mg/ day) [45]. The inclusion criteria were initially a past history of pregnancy complicated with severe fetal growth retardation, placental infarction, idiopathic fetal growth retardation, fetal or neonatal death, or pregnancy-induced hypertension. In all the published studies, the incidence of intra-uterine growth retardation was reduced, the birth weight was higher and the incidence of fetal death, eclampsia and pre-eclampsia, reduced even if it did not reach statistical significance in some studies. The Epreda trial is a European double-blind randomised controlled trial, including 319 women with either 1 (220 patients) or 2 (103 patients) cases of previous fetal growth retardation or hypertension induced complication. This study demonstrated that low dose aspirin has a preventive effect on recurrent fetal growth retardation or fetal death in high-risk mothers [45].

Indomethacin administration to inhibit preterm labour

Indomethacin remains currently prescribed to inhibit preterm labour. Short-term administration is known to reduce renal blood flow and glomerular filtration rate, both in animals and in humans. Long-term prenatal exposure may lead to renal dysfunction complicated with neonatal anuria and death. (46,47). In neonates, the impact of indomethacin on drug disposition and pharmacokinetics of drugs primarily eliminated by renal excretion, was clearly demonstrated with ceftazidine, demonstrating that materno-fetal treatment may affect neonatal pharmacokinetics and require dosage adjustment in neonates [48].

4.5.4   MATERNAL-FETAL SURGERY

Maternal-fetal surgery for treatment of progressive congenital anomalies has evolved dramatically in the recent years. The first human trials were undertaken to evaluate the intrauterine repair of congenital diaphragmatic hernia [49]. Surgery was undertaken in many additional fetal anomalies, including lower urinary tract obstruction, spina bifida, and fetal gastroschisis.

4.5.5 CONCLUSION

The diagnosis of a fetal anomaly is important to allow appropriate counselling and specialised neonatal care. There are many indications for fetal treatment. In most cases, major questions remain to be elucidated and fetal drug therapy will benefit from investigations conducted in the following areas: investigations of normal and pathological fetal development using non-invasive procedures, development of pharmacokinetic modelling for fetal drug therapy, evaluation of non-invasive *versus* invasive drug administration, evaluation of the immediate benefit/risk ratio and long term risks of treatment by large clinical trials, validation of animal models for toxicity and teratogenicity. In addition, the possible implications of pharmacogenetics in the fetal side-effects of drugs, including teratogenic effects remain to be investigated [50,51].

REFERENCES

1.   Koren, G., Pastuszak, A., Ito, S., *N. Engl. J. Med., 338*, 1128 (1998).
2.   McBride, W. G., *Lancet, 2*, 1358 (1961).
3.   Lammer, E. J., Chen, D. T., Hoar, R. M. et al., *N. Engl. J. Med., 313*, 837 (1985).
4.   Rosa, F. W. "Retinoid embryopathy in humans", in G. Koren (ed.), *Retinoids in clinical pratice. The risk-benefit ratio*, M. Dekker, New York, 1993, pp. 77-109.
5.   Holzgreve, W., Carey, J. O., Hall, D. B., *Lancet, 2*, 794 (1976).
6.   Herbst, A. L., Ulfelder, H., Poskanzer, C. D., *N. Engl. J. Med., 284*, 878 (1971).
7.   Ward, R. M., *Clin. Pharmacokinet., 28*, 343 (1995).
8.   Pacifici, G. M., Nottoli, R., *Clin. Pharmacokinet., 28*, 235 (1995).
9.   Jacqz-Aigrain, E., *Paed. Perinat. Drug Ther., 2*, 36 (1999).
10.   Levy, G., *Obstet. Gynecol., 58*, 9S (1981).
11.   Kumar, S., O'Brien, *Br. Med. J., 328*, 1002 (2004).
12.   Wendel, Jr., J. G., Sheffield, J. S., Hollier, L. M., Hill, J. B., Ramsey, P. S., Sanchez, P. J., *Clin. Infec. Dis., 35*, S200 (2002).
13.   Bolt, R. J., van Weissenbruch, M. M., Lafeber, H. N., Delemarre-van de Waal, H. A., *Pediatr. Pulmonol., 32*, 76 (2001).
14.   Grauel, L., Almanza, M., Alvarez, R., et al., *J. Perinat. Med., 16*, 431 (1988).
15.   Azancot Benisty, A., Jacqz-Aigrain, E., Guirgis, N. M., De Crepy, A., Oury, J. F., Blot, P., *J. Pediatr., 121*, 608 (1992).
16.   Strasburger, J. F., Cuneo, B. F., Michon, M. M., et al., *Circulation, 109*, 375 (2004).
17.   Oudijk, M. A., Ruskamp, J. M., Ververs, F. F., et al., *J. Am. Coll. Cardiol., 42*, 765 (2003).
18.   Maxwell, D. J., Crawford, D. C., Curry, P. V. M., Tynan, M. J., Allan, L. D., *Br. Med. J., 297*, 107 (1988).
19.   Van Engelen, A. D., Weijtens, O., Brenner, J. I., et al., *J. Am. Coll. Cardiol., 24*, 1371 (1994).
20.   Dupouy Camet, J., Gavinet, M. F., Paugam, A., Tourte-Schadfer, C., *Med. Mal. Inf., 23*, 139 (1993).
21.   Wong, S. Y., Remington, J. S., *Clin. Infect. Dis., 18*, 853 (1994).
22.   Hohfeld, P., Daffos, F., Costa, J. M., Thulliez, P., Forestier, F., Vidaud, M., *N. Engl. J. Med., 331*, 695 (1994).
23.   Couvreur, J., Desmonts, G., Thulliez, P., *J. Antimicrob. Chemother., 22*, 193 (1998).

24. Hohfeld, P., Daffos, F., Thulliez, P., Aufrant, C., Couvreur, J., MacAleese, J., Descombey, D., Forestier, F., *J. Pediatr., 115*, 765 (1989).
25. Switala, I., Dufour, P., Ducloy, A.S., et al., *J. Gynecol. Obstet. Biol. Reprod. (Paris), 22*, 513 (1993).
26. Safety and toxicity of antiretroviral agents in pregnancy. http:aidsinfi.nih.gov/guidelines/ perinatal – 2003.
27. Blanche, S., Tardieu, M., Rustin, P., et al., *Lancet, 354*, 1084 (1999).
28. Barret, B., Tardieu, M., Rustin, P., et al., *AIDS, 15*, 1769 (2003).
29. New, M. I., *Ann. Rev. Med., 49*, 311 (1998).
30. Morel, Y., Miller, W. L., *Adv. Hum Genet., 20*, 1 (1991).
31. Forest, M. G., David, M., Morel, Y., *J. Steroid Biochem. Mol. Biol., 45*, 75 (1993).
32. Forest, M. G., Dörr, H. G., *Pediatr. Res., 33*, S3 (1993).
33. Burrow, G. N., Fisher, D. A., *N. Engl. J. Med., 331*, 1072 (1994).
34. Atkins, P., Cohen, S. B., *Drug Safety, 23*, 229 (2000).
35. Perelman, A. H., Johnson, R. L., Clemons, R. D., Finberg, H. J., Clewell, W. H., Trujillo, L., *J. Clin. Endocrinol. Metab., 71*, 618 (1990).
36. Morine, M., Takeda, T., Minekawa, R., et al., *Ultrasound Obstet. Gynecol., 19*, 506 (2002).
37. Matsumoto, T., Miyakoshi, K., Kasai, K., et al., *Fetal Diagn. Ther., 18*, 459 (2003).
38. Birchall, J. E., Murphy, M. F., Kaplan, C., Kroll, H., *Br. J. Haematol., 122*, 275 (2003).
39. Mari, G., Deter, R. L., Carpenter, R. L., et al., *N. Engl. J. Med., 342*, 9 (2000).
40. Friedman, S. A., *Obstet. Gynecol., 71*, 1 (1998).
41. Schiff, E., Peleg, E., *N. Engl. J. Med., 321*, 351 (1989).
42. Higby, K., Suiter, C. R., *Drug Safety, 121*, 35 (1999).
43. Vermillion, S. T., Landen, C. N., *Semin. Perinatol., 25*, 256 (2001).
44. Kramer, W. B., Van den Veyver, I. B., Kirshon, B., *Clin. Perinatol., 21*, 615 (1994).
45. Uzan, S., Beaufils, M., *Am. J. Obstet. Gynécol., 160*, 763 (1989).
46. Gloor, J. M., Muchant, D. G., Norling, L. L. *J. Perinatol, 13*, 425-427 (1993)
47. Van der Heijen, B. J, Carlus, C., Narcy, F., Bavoux, F, Delezoide, A. L., Gubler, M. C. Am. *J. Obstet. Gynecol., 171*, 617-623 (1994)
48. Van den Anker, J. N., Schoemaker, R. C., Hope, W. C., *Clin. Pharmacol. Ther., 58*, 650 (1995).
49. Harrison, M. R., Langer, J. C., Adzick, N. S., *et al., J. Pediatr. Surg., 25*, 47 (1990).
50. Van Dyke, D. C., Ellingrod, V. L., Berg, M. J., Niebyl, J. R., Sherbondy, A. L., Trembath, D. G., *Ann. Pharmacother., 34*, 639 (2000).
51. Tempfer, C. B., Schneeberger, C., Huber, J. C., *Pharmacogenomics, 5*, 57 (2004).

4.6 Drug toxicity during brain development

Pierre Gressens, Ignacio Sfaello[†] and Philippe Evrard

INSERM E 9935 and Paediatric Neurology Service, Robert Debré Hospital, 48 Blvd. Sérurier, F-75019 Paris, France

4.6.1 INTRODUCTION

Genes from neural cells interact with the environment to modulate brain development at the successive ontogenic steps during pre- and postnatal life [1-2]. Schematically, environmental factors potentially interfering with the *in utero* brain development can be divided into several classes, as summarised in Table 1.

In Western countries, recreational drugs are used widely, even among pregnant women, constituting a major public health problem. Drug-addicted mothers generally combine several risk factors for the brain of their fetus, making animal models critical to understand the role and the pathophysiological mechanisms of each of these toxic factors.

The fetus can also be exposed to therapeutic drugs taken by the mother if they are able to cross the placenta. In this context, drugs which have an effect on the mother's brain are most likely to have effects on the developing brain of the fetus. Indeed, a large number of the molecules (for example, neurotransmitters such as glutamate, GABA, dopamine, serotonin, acetylcholine, opiates, adenosine) used by the adult brain, and which are targets for neuroactive drugs such as antiepileptic, antidepressant, anaesthetic or sleep control drugs, are also used to control or modulate the successive steps of pre- and postnatal brain development.

[†]Present address: CETES, Servicio de Neurología Infanto – Juvenil, Universidad Católica de Córdoba, Argentina.

Table 1

Environmental factors and conditions which can potentially* influence
the developing brain.

therapeutic drugs	retinoic acid anti-thyroid drugs oestroprogestative drugs testosterone and derivatives anti-mitotic drugs lithium and psychotropic drugs benzodiazepines anesthetics anti-epileptic drugs ...
recreational drugs	tobacco caffeine ethanol cocaine heroin L.S.D. marijuana
physical and chemical agents	dioxins heavy metals organic solvents ionising radiations head trauma repeated head shaking
maternal factors and conditions	catecholamines thyroid hormones deficiency diabetes non-treated phenylketonuria coagulopathies hypoxic-ischemic conditions chorioamniotitis hyperthermia
infectious agents (human pathogens)	herpes simplex virus I and II varicella zoster human cytomegalovirus benign lymphocytic choriomeningitis virus rubella virus parvovirus B19 coxsackie virus (B group) pestivirus influenza virus BK and JC viruses toxoplasma gondii listeria monocytogenes treponema pallidum (syphilis)

* For some substances and conditions listed in the present table, the effect on the developing brain is highly probable or proven in animals and/or in humans. In contrast, other agents have been included based on potential, not yet proven, effects; however, based on their described activities, these substances are theoretically capable to interfere with and/or modulate brain development.

4.6.2 ETHANOL

Fetal alcohol syndrome (FAS) was initially described in 1968 [3]. FAS and other congenital deficits linked to ethanol exposure remain frequent (3/1000 live births) and represent one of the most frequent cause of non-genetic mental retardation.

The typical FAS displays three types of clinical signs (Table 2): facial dysmorphic features (small palpebral fissures, long upper lip, flat indistinct philtrum, midface hypoplasia, epicanthal folds, depressed nasal bridge), intra-uterine growth retardation, and brain lesions associated with cerebral dysfunction ranging from moderate cognitive impairment to severe mental retardation. Other malformations, as well as visual and auditive impairments, are frequently present in infants with FAS.

Beyond the complete form of FAS, *in utero* ethanol exposure could lead to severe neurological sequelae in the absence of the typical facial dysmorphic features [4]. Although population-based studies have shown a dose-response relationship between maternal consumption of ethanol and neurological outcome of children [5], it is very difficult to predict. Estimates are based on the amount of ethanol consumed by the pregnant mother, the clinical phenotype of her child, probably due to large variations in the metabolic activities of mothers and fetuses, and genetic susceptibilities in respect to ethanol and metabolites. Similarly, there is no predictable amount of ethanol which could be consumed safely by a pregnant mother without any risk for her fetus. A recent study [6] showed the existence of fetal brain abnormalities in 100% of cases where mothers had a

Table 2

Fetal effects of ethanol

human studies	• dysmorphic features predominating at the level of the face
	• intra-uterine growth retardation and microcephaly
	• abnormalities of brain cytoarchitectonics
	• mental retardation and behavioural problems
animal studies	• intra-uterine growth retardation and microcephaly
	• inhibition of neuronal proliferation
	• neuronal migration disorders
	• increased neuronal cell death
	• disturbances of gliogenesis (too early transformation of radial glial fibers into astrocytes and delayed production of astrocytes destined to the superficial neocortical layers)

heavy ethanol consumption (100 to 500 ml ethanol per day, 4 to 7 days a week), in 80-90% of cases where mothers had a moderate consumption (100 to 200 ml per day, 1 to 4 days a week) and in 30% of cases where mothers had an episodic ethanol consumption (35 to 100 ml, 3 times during the whole pregnancy). Similarly, some children born of mothers consuming moderate amounts of ethanol during pregnancy might suffer from behavioural problems, even in the absence of any dysmorphic or physical feature [7]. It is generally admitted that the same amount of ethanol is more toxic when given acutely when compared to administration over a longer period of time.

In the neonatal period, dysmorphic features and growth retardation (affecting both body and brain size) are the most evident clinical characteristics. In contrast with tobacco exposure, ethanol-induced growth delay generally does not completely disappear with time. Following ethanol exposure, the newborn can already present behavioural problems such as suction difficulties, sleep disturbances or alteration of motor activity, muscle tone, or orientation. However, brain malformations will manifest themselves mostly at pre-school and school ages, with mental retardation (IQ between 50 and 100), motor problems, speech difficulties, precise motor control impairment, and behavioural disturbances including hyperactivity, sleep troubles, stereotyped behaviors, and attention deficit.

Post-mortem exams of infants exposed *in utero* to ethanol can show reduced brain weight and disturbances of the horizontal cortical lamination, neuronal ectopias (sometimes invading the meninges) and/or a reduced thickness of the cortical mantle. Several mechanisms underlying the effects of ethanol on the developing brain have been unravelled in animal models (Table 2):

- i) ethanol stimulates GABA-A receptors and inhibits the glutamatergic N-methyl-D-aspartate receptor, two key receptors in multiple steps of brain development [8,9], acute ethanol administration has also been shown to interfere with other systems of neurotransmitters including serotonin, dopamine and neuropeptide Y,
- ii) ethanol inhibits the proliferation of neuronal precursors, impairs their migration and induces excess neuronal cell death [10-13],
- iii) ethanol also disturbs several critical steps of neocortical gliogenesis [10,12], leading to abnormalities in neuronal function, neuronal survival and cerebral cytoarchitectonics.

4.6.3 OPIATES

It is estimated that, in the United States, one newborn in a thousand has been exposed *in utero* to opiates. Heroin and its substitute methadone are the two most commonly encountered opiates in pregnant women. Although this is true for most illicit drugs, heroin intake is generally accompanied by an important consumption of one or several other toxics such as cocaine, barbiturates, ethanol, or smoking. Furthermore, the low socio-economical level of many women consuming heroin explains the great difficulties to obtain accurate data concerning the toxicity or safety of opiates for the developing human brain. In this context, the only interpretable studies are based on mothers treated with methadone and who are otherwise drug-free. These epidemiological studies show that methadone (and this is probably also true for heroin) does not induce major malformations [14]. However, human newborns exposed *in utero* to methadone frequently display a withdrawal syndrome which can be biphasic (Table 3). Follow-up of these children show that prenatal methadone does not affect cognitive performances measured by an IQ level at four years of age or by more comprehensive cognitive tests performed at six years of age [13]. However, these studies also reveal an increased frequency of attention deficits, hyperactivity and fine motor skill impairments.

Studies performed in animal models of prenatal exposure to methadone confirm the lack of effect on learning capabilities and on brain morphogenesis [13,15]. However, an increasing number of studies suggest that opiates are important modulators of brain development. Excess stimulation of opiate receptors can induce cellular and molecular alterations in cultured neurons. The opioid system is partly responsible for the induction of maternal behaviour, at a limbic level [16,17] and, in rhesus monkies, administration of naloxone reduces maternal affectivity and grooming [18]. Endogenous opioid peptides are also key for the behavior of the offspring and modulate the early responses to stress and novelty [19]. The relevance of these experimental data to human fetuses and infants remain to be confirmed.

Table 3

Withdrawal syndrome in newborns exposed *in utero* to opiates

acute withdrawal syndrome	• occurring within 3-6 postnatal weeks
	• hyperactivity, hyperexcitability, increased response to auditive stimuli
	• sleep disturbances
	• prolonged cry
subacute withdrawal syndrome	• occurring around 4-6 postnatal months
	• hyperactivity, motor agitation
	• sleep disturbances

4.6.4　COCAINE

Implications for the fetus and the newborn of maternal cocaine addiction represent a major problem in North America and is becoming a growing issue in Western Europe. It is estimated that, in some urban areas of the United States, 10 to 45% of pregnant women consume cocaine. Beyond the increased risk of prematurity and of intra-uterine growth retardation, *in utero* cocaine exposure is associated with microcephaly (very common feature), cerebral malformations (mostly affecting the midline), focal hypoxic-ischaemic or haemorrhagic damage, as well as disturbances of cerebral cytoarchitectonics (disorders of neuronal migration and differentiation, abnormalities of vasculo-mesenchymal tissues) [20,21] (Table 4). Experimental studies performed in mice and monkeys have shown that prenatal cocaine exposure induces an inhibition of neuronal production, disorders of neuronal migration and differentiation, anomalies of gliogenesis, as well as long-term anatomical, molecular and biochemical alterations in the aminergic systems [22-24] (Table 4).

Table 4

Fetal effects of cocaine

human studies	• pre-term delivery
	• intra-uterine growth retardation; microcephaly (up to 16 % of cases in some studies)
	• abnormalities of the midline: corpus callosum agenesis, lack of septum pellucidum, septo-optic dysplasia
	• cerebral infarcts, sub-arachnoid, sub-ependymal and intraventricular haemorrhages
	• schizencephalies
	• neuronal heterotopias, neuronal immaturity
	• cerebral dysgeneses mimicking neuropathological aspects of type II lissencephaly
	• acute and subacute behavioural problems (withdrawal syndrome)
	• persistent neuropsychological and behavioural disorders
animal studies	• inhibition of neuronal production
	• abnormalities of gliogenesis: inhibition of radial glial cell (specialised cell guiding migrating neurons) production and impairment of astrocyte production
	• disturbed addressing of neurons into cortical layers
	• abnormal neuronal differentiation
	• disturbed establishment of dopaminergic circuits
	• deficits in memory, learning and behaviour tests

The postnatal syndrome may include sleep disturbances, feeding difficulties, hypertonic tetraparesis and, in some cases, epileptic fits. These clinical findings are often accompanied by abnormal neurophysiological exams and are generally transient, disappearing within the first year of life.

However, based on behavioural studies performed in animal models and on recent clinical data obtained in humans [25-27], it seems clear that some children who had been exposed *in utero* to cocaine will develop long-term neuropsychological disturbances, even though their IQ will generally be within the normal range [28]. This group of children will often display moderate, but significant, deficits which become evident at school age when they exhibit difficulties to concentrate, weak resistance to distracters, aggressive behaviour, or great impulsivity. In addition, they are more likely to develop anxiety or depressive symptoms. Previously described cytoarchitectonic alterations and aminergic neurotransmission impairments could represent anatomical and biochemical basis for these neuropsychological disturbances. As important as these, *in utero* exposure to cocaine seems to be a risk factor for sudden infant death syndrome.

4.6.5   CAFFEINE

Throughout the world, a large proportion of women drink caffeine-containing beverages. In France, the average consumption is 2 cups of coffee per day (equivalent to about 4 mg/kg/day caffeine) [29]. During pregnancy, the half-life of caffeine is tripled and, at least in animal models, caffeine is concentrated in the developing embryonic brain. Caffeine is an inhibitor of adenosine receptors and modulates circulating levels of adenosine [30,31]. Caffeine also stimulates hyaluronic acid production by astrocytes [32], inhibits cholesterol synthesis [29] and produces an early evagination of the telencephalic vesicles [33-35]. Numerous epidemiological reports studying the effects of caffeine on brain development do not take into consideration several potential confounding factors and lead to contradictory results [29,30]. Pertinent data extracted from well designed studies could be summarised as follows (Table 5) [29,30]:

- i) The risk of malformation in infants of mothers consuming moderate amounts of caffeine is very low.
- ii) To our knowledge, there is no solid study evaluating the risk of malformations in cases of heavy maternal caffeine consumption (> 8 cups / day).
- iii) Similarly, no study has investigated the potential moderate alterations of brain development and functioning in the absence of obvious cerebral malformation in the offspring of mothers drinking low to moderate amounts of caffeine. However, recent data obtained in animal models, as well as human data, suggest a potential risk for the developing brain which remains to be con-

Table 5

Effects of caffeine on the human fetus.

Low to moderate consumption of caffeine during pregnancy	- no demonstrated risk for congenital malformation - no available study evaluating the risk for minor defects of brain development - exacerbation of the effects of ethanol, nicotine and other vasoconstrictive molecules
Moderate to heavy caffeine consumption (more than 4 cups per day) during pregnancy	- increased risk of sudden infant death possible
Heavy caffeine consumption (more than 8 cups per day) during pregnancy	- no available study evaluating the potential risk for brain development in this subset of patients

firmed in control studies. Indeed, two cups of coffee taken by a pregnant women does not modify umbilical vein blood flow but induces a significant reduction of placental blood flow, this reduction of placental blood flow could trigger an increased production of catecholamines which could interfere with brain development.

• iv) Caffeine seems to exacerbate deleterious effects of ethanol, smoking, and other vasoconstrictive agents as anti-migraine drugs. A recent study showed that caffeine intake increases the risk of first-trimester spontaneous abortion, even in non-smoking women [36].

• v) Finally, recent epidemiological studies [37] suggested that *in utero* caffeine exposure increases the risk of sudden infant death syndrome, although another study [38] did not confirm this association.

4.6.6   SMOKING AND NICOTINE

Smoking by pregnant women is increasing in spite of the epidemiological and experimental data demonstrating that smoking has direct and indirect effects on the developing brain, which lead to serious complications, including cognitive impairment and sudden infant death syndrome (SIDS).

Among the numerous potentially toxic chemicals present in tobacco, CO and nicotine are the two most studied compounds. Both can act on the developing brain, either directly or indirectly through haemodynamic effects.

Smoking during pregnancy is a risk factor for intrauterine growth retardation (IUGR) and obstetrical complications (retro-placental haematoma, placenta praevia or premature rupture of membranes) which can lead to pre-term delivery [39]. IUGR is associated with a higher risk of moderate cognitive impairment and learning difficulties at school age. Very early pre-term infants are at high risk to develop white matter lesions (periventricular leukomalacia) or severe intra-ventricular haemorrhages, which are significantly associated with cerebral palsy and cognitive impairments.

Neuropathological studies have shown that rat embryos exposed to maternal smoking have a reduced growth of the forebrain and a reduced density of cerebellar Purkinje cells [40]. The level of gliosis of the inferior olivary nucleus in infants who died from SIDS is proportional to the intensity of maternal smoking during pregnancy [41].

A direct toxicity of tobacco has been described in some studies. CO can act as a neurotransmitter and can reduce the catecholamine levels in the brain. Nicotine will bind to cholinergic nicotinic receptors which are present at high concentration in the fetal brain from midgestation on. In animal models, low doses of nicotine, acting on these brain receptors, can induce neural cell death, inhibit neural cell proliferation and induce long-term changes in brain catecholamine and serotonin levels [42,43].

These experimental studies support the epidemiological data showing a deleterious effect of smoking on the fetal brain, with cognitive sequelae. These studies also raise the question of the potential effects on the developing brain from constant exposure to nicotine with nicotine patches used by women who want to stop smoking during pregnancy.

4.6.7 EPILEPSY AND ANTI-EPILEPTIC DRUGS

The established antiepileptic drugs (AEDs) (benzodiazepines, phenytoin, phenobarbital, carbamazepine and valproate) used by pregnant women are associated with a higher risk of malformations including microcephaly, cleft lip and palate, and cardiac defects, growth retardation, minor abnormalities of face and fingers. The incidence of major malformations in children born to mothers treated with AEDs is 4 to 6 % compared with 2 to 4 % for the general population. Neural tube defects occur in 1 to 2 % of exposed to valproate during pregnancy and in 0.5 to 1 % of those exposed to carbamazepine. Fearing that discontinuation of antiepileptic drugs during pregnancy induces dangerous epileptic fits and even status epilepticus, for decades many epileptologists and general textbooks often suggested that many of these defects are due to other reasons, including genetic factors causing epilepsy of the mother and inherited by the fetus. Holmes *et al.* [44] clearly demonstrated that the higher risk of malformations is mainly a teratogenic feature. Recent studies confirm that risks of malformations are highest in fetuses exposed to multiple AEDs and in those exposed to higher doses.

With newer, second generation, AEDs, such as lamotrigine, oxcarbazepine, topiramate, tiagabine, vigabatrin and levetiracetam, there is very little knowledge on teratogenicity. For a number of reasons, these drugs appear to be more favourable than the older ones as treatments for epilepsy in women of childbearing age [45]. As far as the teratogenic potential of the newer antiepileptic drugs is concerned, one asks what is basically known and how should this influence prescribing? They possess a good pharmacokinetic profile that makes them more stable during pregnancy, and they have a low potential for interaction with other drugs. They are also less likely than the older AEDs to be metabolised to compounds that are teratogenic. Furthermore, most of them do not possess antifolate properties. There is, however, a significant lack of information regarding the teratogenic profile of these newer agents in humans. For lamotrigine, the percentage of outcomes with major birth defects after lamotrigine monotherapy does not seem to differ from that reported in the recent literature for women with epilepsy receiving other antiepileptic drug monotherapy (4%) [46]. It should be kept in mind that more than 2000 prospective pregnancies are needed to detect a drug effect that occurs in 4 to 8% of exposed fetuses [47]. The teratogenic potential of AEDs will be defined more effectively as global prospective registries develop to record pregnancy outcomes after AED exposure [48]. In the last decade, pregnancy registries have been activated by collaborative groups of physicians in Europe (EURAP) and North America (NAREP), to enrol a large number of exposed women to be monitored prospectively with standardised methods.

Clear clinical recommendations have been derived from the above data [49]. Neural tube defects are formed by day 28 after conception and other malformations often occur early in gestation. Preconception counselling is thus essential. Modifications of on-going treatment should take place before conception, to allow optimal therapy with the lowest possible dose, preferably on monotherapy, administered at least two or three times daily during pregnancy, to avoid peaks of concentration. Sustained release preparations are also recommended when available. High peak plasma concentrations of valproate have been suggested to be associated with an increase risk of malformations in rodents [50]. Spina bifida being considered a particular concern, systematic supplementation, started prior to conception, with folic acid 2.5-5 mg daily is essential. Folic acid should be administered to all women of childbearing age who have epilepsy and take anticonvulsant therapy [51]. Therapy is needed on a daily basis up to the 12th week after conception.

Pre- and postnatal disturbances of development in cognition by AEDs become a first priority in research about epilepsy. No controlled data is available concerning postnatal neuropsychological consequences of exposure of the human fetal brain to first and second generations AEDs. Animal models already demonstrate how these drugs could impair the development of neurobiological equipment for future cognition. In newborn rats, corresponding to mid gestation in the human, clinically relevant doses of several common antiepileptic drugs (phenytoin, phenobarbital, diazepam, clonazepam, vigabatrin, or valproic acid) and anaesthetics (combination of midazolam, nitrous oxide, and isoflurane) dramatically exacerbate apoptotic neuronal cell death when administered during the period of growth spurt [52].

4.6.8   SELECTIVE SEROTONIN REUPTAKE INHIBITORS (SSRI)

The cerebral cortex is widely innervated by serotonin (5-HT)-containing axons originating from neurons in the raphe nuclei. This system is present at early stages of brain development and different studies have shown that serotonin is important for neuronal proliferation, migration and differentiation, and for apoptotic cell death inhibition [53]. Furthermore, there is a serotonergic regulation of somatosensory cortical development [54]. SSRI, used as antidepressants in humans, will increase 5-HT concentrations in the synaptic cleft. Therefore, if taken by a pregnant woman, SSRI could potentially disturb several steps of fetal brain development. Experimentally, excess of 5-HT, alters the segregation of retinogeniculate and somatosensory projections in monoamine oxidase A knockout newborn mice [55].

REFERENCES

1. Evrard, P., Marret, S., Gressens, P., *Acta Paediatr. Suppl., 422*, 20 (1997).
2. Evrard, P., Nassogne, M. C., Mukendi, R. et al., *Enviromental determinants of prenatal nervous system development, with particular emphasis on hypoxic-ischemic pathology*, in "Brain lesions in the newborn: hypoxic and haemodynamic pathogenesis," Lou, H.C., Greisen, G., Falck-Larsen, J., (eds.), Munksgaard, Copenhagen, 92 (1994).
3. Lemoine, P., Harousseau, H., Borteyru, J. P., *Ouest Médical, 25*, 476 (1968).
4. Mattson, S. N., Riley, E. P., Gramling, L., Delis, D. C., Jones, K. L., *J. Pediatr., 131*, 718 (1997.
5. Streissguth, A. P., Barr, H. M., Sampson, P. D., Bookstein, F. L., *Drug Alcohol Depend., 36*, 89 (1994).
6. Konovalov, H. V., Kovetsky, N. S., Bobryshev, Y. V., Ashwell, K. W., *Early Hum. Dev., 48*, 153 (1997).
7. Larroque, B., Kaminski, M., Dehaene, P., Subtil, D., Delfosse, M. J., Querleu, D., *Am. J. Public Health, 85*, 1654 (1995).
8. Wafford, K. A., Burnett, D. M., Leidenheimer, N. J., et al., *Neuron, 7*, 27 (1991).
9. Lovinger, D. M., White, G., Weight, F. F., *Science, 243*, 1721 (1989).
10. Gressens, P., Lammens, M., Picard, J. J., Evrard, P., *Alcohol., 27*, 219 (1992).
11. Ikonomidou, C., Bittigau, P., Ishimaru, M. J. et al., *Science, 287*, 1056 (2000).
12. Miller, M. W., Robertson, S., *J. Comp. Neurol., 337*, 253 (1993).

13. Miller, M. W., *Development of central nervous system: effects of alcohol and opiates*, Wiley-Liss (New York) (1992).
14. Blinick, G., Wallach, R. C., Jerez, E., Ackerman, B. D., *Am. J. Obstet. Gynecol., 125*, 135 (1976).
15. Nassogne, M. C., Gressens, P., Evrard, P., Courtoy, P. J., *Brain Res. Dev. Brain Res., 110*, 61 (1998).
16. Mann, P. E., Pasternak, G. W., Bridges, R S., *Physiol. Behav., 47*, 133 (1990).
17. Panksepp, J., Nelson, E., Siviy, S., *Acta Paediatr. Suppl., 397*, 40 (1994).
18. Martel, F. L., Nevison, C. M., Rayment, F. D., Simpson, M. J., Keverne, E. B. , *Psychoneuroendocrinology, 18*, 307 (1993).
19. Martel, F. L., Nevison, C. M., Simpson, M. J., Keverne, E. B., *Dev. Psychobiol., 28*, 71 (1995).
20. Volpe, J. J. , *N. Engl. J. Med., 327*, 399 (1992).
21. Kaufman, W. E., *Soc. Neurosci, 16*, 305 (1990).
22. Gressens, P., Kosofsky, B. E., Evrard, P., *Neurosci. Lett., 140*, 113 (1992).
23. Kosofsky, B. E., Wilkins, A. S., Gressens, P., Evrard, P., *J. Child Neurol., 9*, 234 (1994).
24. Lidow, M. S., *Synapse, 21*, 332 (1995).
25. Delaney-Black, V., Covington, C., Ostrea, E. Jr., et al., *Pediatrics, 98*, 735 (1996).
26. Levitt, P., Harvey, J. A., Friedman, E., Simansky, K., Murphy, E. H., *Trends Neurosci., 20*, 269 (1997).
27. Mayes, L. C., Bornstein, M. H., Chawarska, K., Granger, R. H., *Pediatrics, 95*, 539 (1995).
28. Hurt, H., Malmud, E., Betancourt, L., Braitman, L. E., Brodsky, N. L., Giannetta, J., *Arch. Pediatr. Adolesc. Med., 151*, 1237 (1997).
29. Nehlig, A., Debry, G., *Neurotoxicol. Teratol., 16*, 531 (1994).
30. Nehlig, A., Daval, J. L., Debry, G., *Brain Res. Brain Res. Rev., 17*, 139 (1992).
31. Conlay, L. A., Conant, J. A., deBros, F., Wurtman, R., *Nature, 389*, 136 (1997).
32. Marret, S., Delpech, B., Girard, N. et al., *Pediatr. Res., 34*, 716 (1993).
33. Marret, S., Gressens, P., Maele-Fabry, G., Picard, J., Evrard, P., *Brain Res., 773*, 213 (1997).
34. Sahir, N., Bahi, N., Evrard, P., Gressens, P., *Brain Res. Dev. Brain Res., 121*, 213 (2000).
35. Sahir, N., Mas, C., Bourgeois, F., Simonneau, M., Evrard, P., Gressens, P., *Cereb. Cortex, 11*, 343 (2001).
36. Cnattingius, S., Signorello, L. B., Anneren, G. et al., *N. Engl. J. Med., 343*, 1839 (2000).
37. Ford, R. P., Schluter, P. J., Mitchell, E. A., Taylor, B. J., Scragg, R., Stewart, A. W., *Arch. Dis. Child, 78*, 9 (1998).
38. Alm, B., Wennergren, G., Norvenius, G. et al., *Arch. Dis. Child, 81*, 107 (1999).
39. Marpeau, L., Gravier, A., *Réalités Gynécol Obstet, 48*, 1 (2000).
40. Chen, W. J., Parnell, S. E., West, J. R., *Alcohol, 15*, 33 (1998).
41. Storm H, Nylander G, Saugstad OD., *Acta Paediatr., 88*, 13 (1999)
42. Oliff, H. S., Gallardo, K. A., *Front Biosci., 4*, D883 (1999).
43. McFarland, B. J., Seidler, F. J., Slotkin, T. A., *Brain Res. Dev. Brain Res., 58*, 223 (1991).
44. Holmes, L. B., Harvey, E. A., Coull, B. A. et al., *N. Engl. J. Med., 344*, 1132 (2001).
45. Palmieri, C., Canger, R., *Epilepsia, 43*, 1161 (2002).
47. Morrell, M. J., *Am. Fam. Physician, 66*, 1489 (2002).
48. Beghi, E., Annegers, J. F., *Epilepsia, 42*, 1422 (2001).
49. Tettenborn, B., Genton, P., Polson, D., *Epileptic. Disord., 4 Suppl 2*, S23 (2002).
50. Ceylan, S., Duru, S., Ceylan, S., *Neurosurg. Rev., 24*, 31 (2001).
51. Yerby, M. S., *Epilepsia, 44*, 1465 (2003).
52. Bittigau, P., Sifringer, M., Genz, K., et al., *Proc. Natl. Acad. Sci., U.S.A, 99*, 15089 (2002).
53. Vitalis, T., Parnavelas, J. G., *Dev. Neurosci, 25*, 245 (2003).
54. Luo, X., Persico, A. M., Lauder, J. M., *Dev. Neurosci, 25*, 173 (2003).
55. Upton, A. L., Salichon, N., Lebrand, C. et al., *J. Neurosci, 19*, 7007 (1999).

4.7 Neonatal effects of drugs administered during pregnancy

John N. van den Anker

Department of Pediatics, Pharmacology and Physiology, The George Washington University School of Medicine and Health Sciences, Division of Pediatric Clinical Pharmacology, Children's National Medical Center, Washington, D.C., and Department of Paediatrics, Erasmus MC-Sophia, Rotterdam, The Netherlands

4.7.1 INTRODUCTION

Early in the 19th century, it was observed that, when pregnant women were anesthetised with ether before delivery, the smell of ether could not be detected in the cut ends of the umbilical cord of the delivered child [1]. It was therefore concluded that ether did not cross the placenta and the idea that the placenta acted as a barrier to drugs was born. It was not until the thalidomide tragedy of the late 1950s and early 1960s that the concept of the placenta as a barrier for therapeutic agents was reversed [2]. The thalidomide disaster convinced medical practitioners that every drug and chemical was a potential reproductive hazard. Although several compounds have been proven to be teratogenic, the vast majority of medicines administered during pregnancy do not pose a significant reproductive hazard when used in recommended doses.

This chapter will not deal with potential teratogenic effects of drugs administered during pregnancy but will focus on neonatal effects of frequently used medicines during different stages of pregnancy.

4.7.2 ANALGESICS

The condition of a neonate at birth is affected by many factors, such as maternal obstetric and medical problems, therapeutic interventions, and maternal drug use [3]. With the administration of analgesics and/or anesthetics additional factors such as adequacy of maternal oxygenation, haemodynamic stability, as well as the direct and indirect effects of regional versus general anesthesia, are of great importance. There is currently a paucity of data regarding the adverse effects of medicines administered during pregnancy, on preterm versus term neonates. Preterm neonates are commonly viewed as being more prone to serious adverse effects of drugs administered during pregnancy. The preterm neonate has a higher incidence of asphyxia [4]. Furthermore, a poorly developed blood-brain barrier can result in a higher concentration of any drug in the central nervous system of these infants [5]. The preterm neonate has lower plasma protein concentrations with reduced binding sites available for drugs. In addition, the protein binding sites have less affinity for medicines. Preterm neonates are more prone to develop hyperbilirubinaemia and it is well known that bilirubin competes with drugs for available protein binding sites. The net result will be enhanced effects of drugs. These effects are further exaggerated in asphyxiated neonates. Despite the information derived from animal studies [6,7], there have been no large-scale prospective studies conducted to evaluate the effects of anesthetic drugs and opioids in preterm versus term neonates, and acidotic versus non-acidotic preterm neonates to date. Therefore, it is advisable to exercise caution when administering analgesics and anesthetics to the pregnant woman.

Pethidine

Pethidine is widely used during pregnancy, and it is metabolised to norpethidine. The half-lives of pethidine and norpethidine in neonates are extremely prolonged (20 and 60 hours, respectively). The parent drug and its metabolite cause low Apgar scores, prolonged time to sustained respiration, and respiratory acidosis. It is therefore prudent to expect neonatal respiratory depression if the pregnant woman has received any pethidine within a few hours before delivery [8].

Fentanyl/Alfentanil/Remifentanil

These synthetic opioids are frequently used during the last stages of pregnancy. The use of fentanyl (using patient-controlled analgesia) was associated with a 44% incidence of neonatal respiratory depression, with Apgar scores < 6 at 1 minute [9]. Of women whose neonates required naloxone as a rescue medication, total fentanyl administration was significantly higher than in those who did not receive naloxone.

The use of alfentanil is even more worrisome, because alfentanil is highly bound to $\alpha 1$-acid-glycoprotein and preterm neonates have a decreased concentration of $\alpha 1$-acid-glycoprotein which will lead to higher free concentrations in these infants. When used in doses > 10 µg/kg in the pregnant woman, it can cause severe neonatal respiratory depression [10]. Therefore, it is important to realise that, with both fentanyl and alfentanil, the risk of neonatal respiratory depression is ever-present [11]. The newer synthetic opioid remifentanil seems promising, but there is an urgent need for studies that will document the short- and long-term neonatal effects of these drugs.

4.7.3 CORTICOSTEROIDS

Neonatal renal function is crucial for the elimination of not only drugs that are primarily excreted by the kidneys, but also for many active metabolites (e.g., morphine-6-glucuronide, midazolam-1OH-glucuronide, etc.). Therefore the following paragraphs on corticosteroids, angiotensin-1 converting enzyme inhibitors, angiotensin-2 receptor inhibitors, and nonsteroidal anti-inflammatory drugs will focus on the effects of these drugs administered during pregnancy on the renal clearing capacity of the neonate.

Pharmacological acceleration of fetal lung maturation with prenatal corticosteroid treatment was first described in sheep 35 years ago [12], then confirmed by others in sheep [13], rabbits [14], and humans [15]. Comparative studies with different corticosteroids revealed pharmacological advantages for the fluorinated corticosteroids, betamethasone and dexamethasone. Subsequent studies have shown that prenatal corticosteroid treatment bestows beneficial effects on neonates beyond those involving their lungs [16,17].

Bethamethasone is a synthetic glucocorticoid with a potency equivalent to dexamethasone. The drug is prescribed to pregnant women with an increased risk of preterm delivery before the 34th week of gestation. The objective of this treatment is to accelerate maturation of the alveolar epithelium and stimulate synthesis of lipid and protein components of the pulmonary surfactant complex to prevent hyaline membrane disease. Much of the data available regarding renal responses to antenatal glucocorticoid treatment are derived from animal studies. For example, prolonged fetal betamethasone infusions have been shown to increase GFR and urine flow in both near-term fetal and newborn lambs [18]. Although fetal cortisol infusion may increase fetal renal blood flow, betamethasone-induced increases in GFR result primarily from an increase in filtration fraction [18]. Thus, although glucocorticoids increase blood pressure and thus will indirectly alter renal perfusion pressure, glucocorticoid-induced increases in GFR are primarily related to changes in renal vasculature resistance. This phenomenon is of interest because an increase in filtration fraction, rather than total renal blood flow, appears to be the primary mechanism for the marked perinatal increase in GFR observed in term newborn lambs [19]. Antenatal betamethasone treatment significantly increases GFR in preterm newborn lambs supported by mechanical ventilation [20]. The effects of prenatal exposure to betamethasone on the GFR of preterm infants have been studied by several investigators [21, 22-24]. These studies did not show an increase of the GFR during the first week of life after prenatal exposure to glucocorticoids. However, in three of these studies, creatinine clearance was used as a less reliable marker for the GFR in preterm infants and a small number of children was studied [22-24]. This might have prevented the authors from demonstrating an increase in the GFR in the first week of life after prenatal exposure to glucocorticoids. The most recent study was hampered by the fact that most pregnant women who were treated with betamethasone were also treated with indomethacin, thereby minimising the number of women who were treated with betamethasone alone [21]. However, betamethasone reversed indomethacin-induced decreases in GFR [21]. It was hypothesised that an increase in renal plasma flow, due to betamethasone, may overcome intrarenal vasoconstriction secondary to the decreased synthesis of intrarenal prostaglandins by indomethacin.

4.7.4 ANGIOTENSIN CONVERTING ENZYME INHIBITORS

All components of the renin-angiotensin system (RAS) exist within the fetal kidney during the early stages of development and participate as promoting factors for the growth of this organ, more specifically its angiogenesis, and have an important role in controlling intrarenal haemodynamics [25-27]. In the early fetal stage, renin-containing cells are present in the developing intrarenal branches of the renal artery. Renin is also distributed in other vascular parts including the arcuate, interlobar and afferent arterioles. Renin mRNA gene expression markedly increases throughout fetal life to peak in the perinatal period. This gene could be under the influence of adrenergic input, as its expression is abolished with renal denervation [26]. In early life, renin is almost exclusively detected in the juxtaglomerular apparatus. Renin acts on plasma angiotensinogen to form angiotensin I.

Angiotensin converting enzyme (ACE) is a dipeptyl carboxypeptidase that releases the pressor peptide angiotensin II (ANG II) from angiotensin I and inactivates bradykinin as well [28]. ACE is present in both vascular and extravascular tissues (brush border) of the kidney. While the extravascular localisation of ACE is not fully known, this enzyme is found on glomerular endothelial cells at the place where the capillary invades the inferior cleft of the S-shaped body. ACE may participate in the tubular handling process of ANG II, as it has been found on the apical and basolateral membranes already in the early nephron stage. This glomerular distribution in the fetal kidney looks different as compared to the more mature kidney where ACE is essentially found in the peritubular endothelial cells. The switch of ACE from glomerular to peritubular vessel with maturation has been well documented and occurs progressively during infancy. In addition to the renal haemodynamic regulation, the ANG II locally generated in the glomerulus also stimulates angiogenesis through the stimulation of its receptors.

It has been demonstrated that ANG II acts as a growth factor for renal cells and, therefore, plays a crucial role in the development of the kidney through its two receptors: angiotensin II receptor 1 (AT1-R) and angiotensin II receptor 2 (AT2-R) [29]. Both receptors are indeed independently present in mammalian fetal kidney tissues and AT2-R seems to predominate [29-33]. AT2-R mRNA has been found in almost all fetal tissues including the metanephros, and undifferentiated mesenchymal and connective tissues. AT1-R has been found more specifically in the adrenal, liver and kidney. Within the kidney, both AT1-R and AT2-R mRNAs are expressed in the metanephros at 14 days of gestation when branching of the ureter bud has already started. AT1-R expression in the immature glomeruli coincides with mesangial cell differentiation from the pericyte, continues throughout adulthood in the glomeruli and in the tubulointerstitium, whereas AT2-R expression decreases after birth, except in large cortical blood vessels. AT1-R mRNA is expressed in mature glomeruli, in maturing S-shaped bodies and in the proximal and distal tubule as well. Early in the embryologic period, AT2-R mRNA is first expressed in mesenchymal cells adjacent to the stalk of the ureter epithelium. The expression is then extended in the mesenchyme cells of the nephrogenic area and in the collecting ducts. AT2-R is also invariantly expressed in the epithelial cells of the macula densa.

The studies performed in order to determine the localisation of AT1-R and AT2-R will help identify the specific and crucial role of ANG II for kidney development. Via AT1-R, ANG II stimulates proliferation, regulates nitric oxide synthetase (NOS) expression, has growth-promoting effects and acts on glomerular mesangial and tubular cell differentiation during nephrogenesis [34]. It further mediates biological actions like maintenance of circulatory homeostasis and cell proliferation [35]. Furthermore, ANG II participates in the downregulation of AT2-R and renin gene expression. In the growth retarded fetus, AT2-R expression has been down-regulated and it has been postulated that this down-regulation is associated with a higher risk of hypertension in adulthood.

Given the fundamental role of the RAS, either *in utero* for general renal morphogenesis or during the first days of life to promote adequate glomerular filtration, administration of drugs with ACE inhibitory effects or acting as AT1-R or AT2-R inhibitors during pregnancy or during the first days of life are contraindicated.

4.7.5 NONSTEROIDAL ANTI-INFLAMMATORY DRUGS

Two cyclooxygenase (COX) isoforms are known: COX-1, which is expressed constitutively in almost all organs, and COX-2, which is usually absent in most organs but can be induced by various stimuli [36]. These enzymes have a key role in the biosynthesis of prostanoid derivatives [37]. In adults, renal prostaglandin synthesis is thought to counterbalance vasoconstrictive agents (e.g., ANG II), and renal vasodilatory prostanoids are primarily derived from COX-2 [38].

In the fetus, prostaglandins are crucial in the early phase of nephrogenesis, more specifically in the glomerulogenesis and in the differentiation of the nephrons. For example, metabolites of arachidonic acid modulate the activity of $Na^+/K^+/ATPase$ along the nephron and this action is age-dependent [39]. Vasodilator prostanoids counteract the high vascular resistance *in utero* and during the first days of life. PGE_2 and PGI_2 could also act as potent and rapid stimulators of renin secretion through prostaglandin receptors, located on renal juxtamedullar cells as has been demonstrated in more mature animals [40]. Experimental studies show that a constitutive cortical, as well as a medullary, COX-2 are overexpressed in fetal life and during the first days of life and accounts for the high excretion of vasodilator prostaglandin [41,42].

Numerous prostaglandin receptors have been identified and their role in renal development becomes increasingly clear [43]. Four prostaglandin E2 receptor subtypes have been identified in the kidney as well: EP1, EP2, EP3, and EP4. They are localised both on glomerular vessels (EP1, EP2, and EP4) and on different parts of the tubule (EP1, EP3, and EP4). Overexpression of some prostaglandin receptors (EP2 and EP4) has been demonstrated in the glomerular afferent vessels of the developmental kidney, allowing for increased activity of vasodilator prostanoids. The vasodilation of the afferent arteriole via these receptors is the way by which prostaglandins counter-act the high vascular resistance generated by the ANG II mediated vasoconstriction of efferent arterioles. It is the main mechanism for maintaining glomerular filtration in fetal and early postnatal life. Overexpression of the tubular EP3 receptor (located in the distal tubule and collecting duct) is needed for amniotic fluid formation and to excrete water during the first days of life [44-46]. In addition, it has been postulated that the down-regulation of the apical collecting duct water channel AQP2 also results in the excretion of hypotonic urine in utero and during the first days of life [47,48]. Embryonic calcium-sensing receptor expression is another mechanism involved in the blockade of arginine vasopressin action, resulting in hypotonic urine, during antenatal life [49].

NSAIDs inhibit the enzymatic activity of both COX-1 and COX-2, and thereby block the formation of prostanoids [37]. COX-2 selective NSAIDs, as well as the conventional non-selective NSAIDs, such as indomethacin, may cause a reversible decline in GFR and renal perfusion [50]. Experimental data have also shown that endogenous PGE_2 downregulates inducible NOS (iNOS) induction and that the decrease of PGE_2 production by indomethacin COX inhibition results in enhancement of IL-1β-induced steady state iNOS mRNA levels and NO production in mesangial cells [51]. While it has not yet been demonstrated in glomerular vessels, this mechanism highlights a possible feedback mechanism which could exist between prostaglandins and NO, as it has been shown for ANG II.

In the perinatal period, NSAIDs are used for the following reasons: 1) as a tocolytic agent, 2) for closure of a patent ductus arteriosus, and 3) to reduce polyuria in patients with congenital salt losing tubulopathies. Numerous case reports have shown transient fetal/neonatal oliguria following exposure to non-selective NSAIDs [52, 53-55]. In addition, Butler-O'Hara *et al.* reported a significant prolonged rise in plasma creatinine in infants exposed prenatally to indomethacin [56]. Conventional NSAIDs may even cause fatal renal failure in the neonate [55]. The renal pathology associated with antenatal NSAIDs exposure is characterised by small and immature glomeruli and cystic dilations in the renal cortex [52,55]. Whether these functional and histological changes can also be attributed to an imbalance of vasodilatory prostanoids and vasoconstrictive agents still needs to be demonstrated.

The availability of COX-2 selective inhibitors made some investigators believe that the detrimental effects of the non-selective NSAIDs possibly could be related to inhibition of COX-1 and that administration of COX-2 selective inhibitors would not result in perinatal renal impairment [57]. However, following this initial enthusiasm about the fact that COX-2 inhibitors could be renal sparing in the perinatal period, data from recent studies have shown severe fetal oliguria [58] and even fatal renal failure in neonates antenatally exposed to the COX-2 selective inhibitor nimesulide [59,60]. This may indicate that COX-2 is more essential for normal renal development and function than COX-1. This has been demonstrated in rodents [61-64].

Based on the current literature, NSAIDs should not be used during renal development. However, indomethacin is still frequently prescribed to inhibit preterm uterine contractions before the 34th week of gestation. Short-term exposure to indomethacin leads to a reduction of the GFR, whereas conflicting data exist about the effect on the GFR after long-term exposure [54, 65-69]. Animal studies have indicated that the inhibition of prostaglandin synthesis by indomethacin increases renal vascular resistance [70]. This subsequently results in an impaired renal blood flow and a concomitant reduction in the GFR [70]. To investigate the impact of prenatal exposure to betamethasone and indomethacin on the pharmacokinetics of ceftazidime, 136 preterm infants were studied [71]. Twenty-five of these infants were treated with indomethacin alone, and 21 infants were treated with both indomethacin and betamethasone. The results of this study clearly demonstrated that prenatal exposure to indomethacin alone significantly decreases ceftazidime clearance and increases serum half-life of ceftazidime. The co-administration of betamethasone prevented these changes. These results indicate that, after prenatal exposure to indomethacin alone, additional dosage adjustments are indicated [71].

NSAIDs induced reduction of neonatal GFR is an important example that *in utero* exposure to drugs can have a profound effect on the pharmacokinetics of drugs administered to the newborn infant. Prescribing clinicians should be aware of possible fetal drug exposure and its potential consequences on neonatal renal clearing capacity.

REFERENCES

1. Caton, D., *Anesthesiology, 46*, 132 (1977).
2. McBride, W., *Clin. Pharmacokinet., 28*, 235 (1995).
4. Low, J. A., Wood, S. I., Killen, H. L., et al., *Am. J. Obstet. Gynecol., 162*, 378 (1990).
5. Malinow, A. M., Dailey, P. A. "Anesthesia for preterm labor and delivery," in Hughes, S. C., Levinson, G., Rosen, M. A. (eds.), *Anesthesia for obstetrics,* Lippincott Williams and Wilkins (2002).
6. Morishima, H. O., Pedersen, H., Santos, A. C, et al., *Anesthesiology, 71*, 110 (1989).
7. Pedersen, H., Santos, A. C., Morishima, H. O. et al., *Anaesthesiology, 68*, 367 (1988).
8. Belfrage, P., Boreus, L. O., Harvig, P. et al., *Acta Obstet. Gynaecol. Scand., 60*, 43 (1981).
9. Morley-Forster, P. K., Weberpals, J., *Int. J. Obstet. Anaesth., 7*, 105 (1998).
10. Cartwright, D. P., Dann, W. L., Hutchinson, A., *Eur. J. Anaesth., 6*, 103 (1989).
11. Morley-Forster, P., Reid, D. W., Vandeberghe, H., *Can. J. Anesth., 47*, 113 (2000).
12. Liggins, C. C., *J. Endocrinol., 45*, 515 (1969).
13. DeLemos, R. A., Shermeta, D. W., Knelson, J. H. et al., *Am. Rev. Respir. Dis., 102*, 459 (1970).
14. Kotas, R. V., Avery, M. E., *J. Appl. Physiol., 30*, 358 (1971).
15. Liggins, C. C., Howie, R. N., *Pediatrics, 50*, 515 (1972).
16. Van Marter, L. J., Leviton, A., Kuban, K. C. et al., *Pediatrics, 86*, 331 (1990).
17. Halac, E., Halac, J., Begue, E. F. et al., *J. Pediatr., 117*, 132 (1990).
18. Stonestreet, B. S., Hansen, N. B., Laptook, A. R., Oh, W., *Early Hum. Dev., 8*, 331 (1983).
19. Nakamura, K. T., Matherne, G. P., McWeeny, O. J., Smith, B. A., Robillard, J. E., *Pediatr. Res., 21*, 29 (1987).
20. Berry, L. M., Ikegami, M., Woods, E., Ervin M. G., *Reprod. Fertil. Dev., 7*, 491 (1995).
21. Van den Anker, J. N., Hop, W. C., De Groot, R. et al., *Pediatr. Res., 36*, 578 (1994).
22. MacKintosh, D., Baird-Lambert, J., Drage, D., Buchanan, N., *Dev. Pharmacol. Ther., 8*, 107 (1985).
23. Al-Dahhan, J., Stimmler, L., Chantler, C., Haycock, G. B., *Pediatr. Nephrol., 1*, 131 (1987).
24. Zanardo, V., Giacobbo, F., Zambon, P. et al., *J. Perinat. Med., 18*, 283 (1990).
25. Tufro-McReddie, A., Gomez, R. A., *Sem. Nephrol., 13*, 519 (1993).
26. Ito, H., Wang, J., Strandhoy, J. W., Rose, J. C., *J. Soc. Gynecol. Investig., 8*, 327 (2001).
27. Wang, J., Rose, J. C., *Am. J. Physiol., 277*, R1130 (1999).

28. Berecek, K. H., Zhang, L., *Adv. Exp. Med. Biol., 377*, 141 (1995).
29. Wolf, G., *Adv. Exp. Med. Biol., 377*, 225 (1995).
30. Kakuchi, J., Ichiki, T., Kiyama, S. et al., *Kidney Int., 47*, 140 (1995).
31. Shanmugam, S., Llorens-Cortes, C., Clauser, E., Corvol, P., Gasc, J. M., *Am. J. Physiol., 268*, F222 (1995).
32. Robillard, J. E., Page, W. V., Matthews, M. S., Schutte, B. C., Nuyt, A. M., Segar, J. L., *Pediatr. Res., 38*, 896 (1995).
33. Butkus, A., Albiston, A., Alcorn, D. et al., *Kidney Int., 52*, 628 (1997).
34. Fischer, E., Schnermann, J., Briggs, J. P., Kriz, W., Ronco, P. M., Bachmann, S., *Am. J. Physiol., 268*, F1164 (1995).
35. Maric, C., Aldred, G. P., Harris, P. J., Alcorn, D., *Kidney Int., 53*, 92 (1998).
36. Herschman, H. R., *Lipid Metab., 1299*, 125 (1996).
37. Smith, W., *Am. J. Physiol., 263*, F181 (1992).
38. Qi, Z., Hao, C M., Langenbach, R. I. et al., *J. Clin. Invest., 110*, 61 (2002).
39. Li, D., Belusa, R., Nowicki, S., Aperia, A., *Am. J. Physiol., 278*, F823 (2000).
40. Jensen, B L., Schmid, C., Kurz, A., *Am. J. Physiol., 271*, F656 (1996).
41. Zhang, M. Z., Wang, J L., Cheng, H. F., Harris, R. C., McKanna, J. A., *Am. J. Physiol., 273*, F994 (1997).
42. Khan, K. N., Stanfield, K M., Dannenberg, A. et al., *Pediatr. Dev. Pathol., 4*, 461 (2001).
43. Breyer, M. D., Breyer, R. M., *Am. J. Physiol., 279*, F12 (2000).
44. Bonilla-Felix, M., Jiang, W., *Am. J. Physiol., 271*, F30 (1996).
45. Joppich, R., Haberle, D. A., Weber, P. C., *Pediatr. Res., 15*, 278 (1981).
46. Bonilla-Felix, M., John-Phillip, C., *Am. J. Physiol., 267*, F44 (1994).
47. Bonilla-Felix, M., Jiang, W., *J. Am. Soc. Nephrol., 8*, 1502 (1997).
48. Bonilla-Felix, M., Vehaskari, V. M., Hamm, L. L., *Pediatr. Nephrol., 13*, 103 (1999).
49. Chattopadhyay, N., Baum, M., Bai, M., et al., *Am. J. Physiol., 271*, F736 (1996).
50. Reinalter, S., Jeck, N., Brochhausen, C., et al., *Kidney Int., 62*, 253 (2002).
51. Tetsuka, T., Daphna-Iken, D., Srivastava, S. K., Baier, L. D., DuMaine, J., Morrison, A. R., *Proc. Natl. Acad. Sci., 91*, 12168 (1994).
52. Kaplan, B. S., Restaino, I., Raval, D. S. et al., *Pediatr. Nephrol., 8*, 700 (1994).
53. Gloor, J. M., Muchant, D. G., Norling, L. L., *J. Perinatol., 13*, 425 (1993).
54. Vanhaesebrouck, P., Thiery, M., Leroy, J. G, et al., *J. Pediatr., 113*, 738 (1988).
55. Van der Heijden, B. J., Carlus, C., Narcy, F., Bavoux, F., Delezoide, A. L., Gubler, M. C., *Am. J. Obstet. Gynecol., 171*, 617 (1994).
56. Butler-O'Hara, M., D'Angio, C. T., *J. Perinatol., 22*, 541 (2002).
57. Sawdy, R., Slater, D., Fisk, N., Edmonds, D. K., Bennett, P., *Lancet, 350*, 265 (1997).
58. Holmes, R. P., Stone, P. R., *Obstet. Gynecol., 96*, 810 (2000).
59. Peruzzi, L., Gianoglio, B., Porcellini, M. G, et al., *Lancet, 354*, 1615 (1999).
60. Balasubramaniam, J., *Lancet, 355*, 575 (2000).
61. Norwood, V. F., Morham, S. G., Smithies, O., *Kidney Int., 58*, 2291 (2000).
62. Morham, S. G., Langenbach, R., Loftin, C. D. et al., *Cell. 83*, 473 (1995).
63. Komhoff, M., Wang, J. L., Cheng, H. F. et al., *Kidney Int., 57*, 414 (2000).
64. Dinchuk, J. E., Car, B. D., Focht, R. J. et al., *Nature, 378*, 406 (1995).
65. Van der Heijden, A. J., Provoost, A. P., Nauta, J. et al., *Pediatr. Res., 24*, 644 (1988).
66. Wurtzel, D., *Obstet. Gynecol., 76*, 689 (1990).
67. Gerson, A., Abbasi, S., Johnson, A., Kalchbrenner, M., Ashmead, G., Bolognese, R., *Am. J. Perinatol., 7*, 711 (1990).
68. Simeoni, U., Messer, J., Weisburd, P., Haddad, J., Willard, D., *Eur. J. Pediatr., 148*, 371 (1989).
69. Dudley, D. K., Hardie, M. J., *Am. J. Obstet. Gynecol., 151*, 181 (1985).
70. Duarte-Silva, M., Gouyon, J. B., Guignard, J. P., *Kidney Int., 30*, 453 (1986).
71. Van den Anker, J. N., Schoemaker, R. C., Hop, W. C., et al., *Clin. Pharmacol. Ther., 58*, 650 (1995).

4.8 Management of neonates exposed to illicit drugs during pregnancy

Claude Lejeune

Neonatal Department, Louis Mourier Hospital, 92701 Colombes, France

4.8.1 INTRODUCTION

In Western Europe, in contrast with the situation in North America, pregnant women with drug addictions usually take several drugs, one of which is an opiate. The consumption of cocaine is less frequent.

Health care for these women has become more humane with the establishment of centres that specialise in pre- and post-natal care [1]. These centres are characterised by:

- a warm welcome by a multidisciplinary team, well-integrated within a city-hospital network, and therefore aware of pregnancies at risk, allowing for early intervention;
- a risk-reduction policy based on substitution treatment with either methadone or buprenorphine in high doses;
- the open objective of avoiding child placement by actively promoting the establishment of a harmonious bond between mother and child.

4.8.2 NEONATAL ABSTINENCE SYNDROME (NAS)

The neonatal abstinence syndrome (NAS) is the most serious complication arising from the maternal use of opiates (heroin and/or opiates of substitution such as methadone or high-dose buprenorphine), with an incidence of 40 to 90% in neonates born to addicted mothers in this group. The syndrome manifests itself most frequently during the first 24 to 36 hours of life for heroin and high-dose buprenorphine, and appears later for methadone, between the second and seventh day [1,2].

The clinical symptoms of NAS include:

- irritability of the central nervous system: intense agitation, tremors, muscular jolting or less frequently convulsions, continuous shrill crying and sleeping difficulties;
- digestive problems: poor sucking, vomiting, diarrhoea, all of which can lead to a state of dehydration and / or a poor rate of weight gain;
- respiratory or neurovegetative difficulties.

The consumption of other psychoactive substances during pregnancy (alcohol, cocaine, cannabis, barbiturates, and above all benzodiazepines) can aggravate NAS [3,4].

Whether a correlation exists between the posology of the substitution treatment for the mother and the severity of NAS is very controversial. This sort of correlation was found by some for methadone [5,6], but not by others [7-9] and no correlation has been found for high-dose buprenorphine [1,10].

Treatment depends on the severity of the NAS; which can be estimated by clinical scoring systems, such as the Finnigan [11] and Lipsitz [12] scores, the use of which is recommended by the American Academy of Pediatrics [13]. In practice, the score is measured daily from the first day of birth, at intervals of 12 hours, for all newborns who have been exposed to opiates *in utero*.

4.8.3 SYMPTOMATIC TREATMENT

A number of simple measures are effective in minimising the NAS. These include:

- reduction in sound and light stimulation;
- use of a fortified and segmented diet, or continuous-tube or parenteral feeding in the case of severe dehydration and/or undernourishment;
- rocking, caressing, skin-to-skin contact, ideally by the mother.

In our current experience, we find that mothers in a well-balanced state following substitution therapy and having undergone a period of medical, psychological and social support during pregnancy, participate much better in the nursing of their newborn [1,14-17]. These mothers, with a heightened sense of their competence as parents, then follow more closely the therapeutic program that is proposed.

Breastfeeding becomes, in this general perspective, very desirable. There has been debate about the problem of the passage of substitution opiates into breast milk. Recent publications have shown that methadone [18-21] and buprenor-phine [22,23] are not significantly excreted in breast milk. Some authors have concluded that there is no reason not to recommend breastfeeding in this con-text [18, 21-24]. We must take into account the positive role that breastfeeding plays in the development of the bond between mother and child, and remember that there is now an almost general consensus to recommend breastfeeding for women who are seropositive for hepatitis C (even if they are PCR positive). More than 80% of former intravenous substance abusers are seropositive for hepatitis C, and of these, about 80% have a positive PCR. In this context, the only case for which breastfeeding is not indicated is for women with positive HIV serology.

4.8.4 MEDICAL TREATMENT

Several general reviews, meta-analyses and recommendations [13,17,25-27] have shown that there have been very few controlled or randomised studies con-cerning the choice and the appropriateness of medical treatments for NAS. Two studies have highlighted the large range of therapeutic protocols currently used in Europe [28,29]. Most authors support medical treatment when the Finnegan or Lipshitz score exceeds 8, particularly in the case of digestive difficulties and/or convulsions. However, there is far less agreement on the selection of the appro-priate medical treatment.

4.8.5 USE OF OPIATE-DERIVED MEDICATIONS

Most recent recommendations call for the use of opiates as a preferred first approach [13,17,25-27], but the specific choice of medicine is a source of discus-sion. The classical treatment, in both American and European literature, is with *paregoric elixir*. A number of criticisms have been raised about its use with the newborn child [27,30,31]:

- the dosage of opiate varies from country to country (in France, for example, a solution of 0.05% morphine is employed, 0.5 mg/ml, i.e., 10 drops provided 0.1 mg of morphine);
- most importantly, the paregoric elixir is a complex mixture of a variety of substances, some of which (other than opium) are active and potentially harmful. For example, it contains benzoic acid, camphor (a nervous system stimulant) and alcohol (44 to 46%). Its use was however recommended in a recent review [24].

The most recent recommendations call for the use of a *morphine solution*, without addictive agents [13,27,31,32]. In our own practice, we use *morphine chlorohydrate* (analgesic solution containing 10 mg per 10 ml, i.e., 1 ml provides 1 mg), but there are no controlled studies of its application. In our experience, this treatment is effective and without side effects. The initial total daily dose is 500 microg/kg, which is divided into smaller doses administrated 4-6 hourly. The total daily dose may be as high as 1 mg/kg/day, which in most cases results in a spectacular improvement in the Finnegan and Lipsitz scores. The dose can then be progressively reduced in stages lasting 2 to 4 days as determined by the evolution in the score (measured at least twice daily).

Other teams use *morphine sulphate* [28] or the *tincture of opium* (400 micrograms of morphine per ml). For the latter, the initial total dose is 400 microg/kg/day, in 6 to 8 daily doses. The dose can be increased, if the score remains high, by 50% per day or by 10% every 4 hours. When the symptoms of abstinence are under control, the total dose is reduced by 10% every 1 to 3 days until it reaches the level of 200 microg/kg/day, at which point the treatment is stopped [17].

In an overview study [32], it was observed that, when administering morphine solution, it might be useful to increase the number of individual doses so that the interval between treatments is reduced to every 3 to 4 hours, rather than every 6 hours as described in the protocol given above. This practice seems to be correlated with a shorter hospitalisation among the 41 newborns in the study whose mothers who only took methadone during the pregnancy.

Several attempts to use *methadone* have been published. This approach is used, for example, by a European centre [28]. The initial dose is 50 - 100 microg/kg every 6 hours, with an increase of 50 microgram/kg per dose until the symptoms come under control. Methadone can then be given every 12 to 24 hours and then stopped entirely when the total daily dose of 50 microg/kg is reached [13].

4.8.6   USE OF OTHER DRUGS

Phenobarbital is used in cases of convulsions (confirmed by an electroencephalogram) [30] or in the case of extreme agitation, but this substance has no effect on digestive difficulties [13]. An intravenous loading dose of 20 mg/kg, is followed by doses of 4 to 5 mg/kg/day.

A Cochrane review found no difference in the rate of failure comparing opiates to phenobarbital [25]. A small study has suggested that phenobarbital and tincture of opium in combination are more effective than tincture of opium alone[33]. Phenobarbital may be of benefit for NAS due to exposure to other psychoactive substances [13].

The family of *benzodiazepines*, and in particular *diazepam*, in use for quite some time, are generating increasing controversy [13,24,27] because of their very long-term action [30], the complications they cause in sucking [34,35], and the fact that they are less effective than opiates in controlling the risk of convulsions [27,30,34,36,37]. A meta-analysis of two studies has shown that there is a significant reduction in the therapeutic failure rate (RR 0.43; 95% IC 0.23-0.80) when the patient is treated with opiates instead of diazepam for those newborns whose mothers only took opiates or who took opiates with other psychoactive substances [25].

Chlorpromazine is deemed effective particularly for digestive difficulties (500 microg/kg every 6 hours) [31]. A long elimination time (half-life of 3 days) as well as a number of adverse effects limit its applicability with newborns [13].

Finally, the use of *clonidine* has been studied in a single uncontrolled study for the treatment of NAS [38], and a second study, started in 2002, is underway [26].

4.8.7 CONCLUSIONS

The medical, psychological and social care of pregnant drug-addicted women and their newborn results in a reduction in pre- and post-natal complications, especially premature delivery [1,14-16]. In contrast, the specific medical treatment for the Neonatal Abstinence Syndrome remains largely empirical, highly variable among teams, and based on a very limited number of controlled studies.

REFERENCES

1. Lejeune, C., Simmat-Durand, L., et al., *Grossesse et substitution*. Enquête sur les femmes enceintes substituées à la méthadone ou à la buprénorphine haut dosage et caractéristiques de leurs nouveau-nés, Paris, OFDT, pp. 142 (2003).
2. Shaw, N. J., McIvor, L., *Arch. Dis. Child. Fetal Neonatal. Ed., 71*, F203 (1994).
3. Fundaro, C., Solinas, A., Martino, A. M., et al., *Minerva Pediatr. 46*, 83 (1994).
4. Wilbourne, P. L., Dorato, V., Miller, W. R., Curet, L. B., *Am. J. Obstet. Gynecol., 182*, S177 (2000).
5. Dashe, J. S., Sheffield, J. S., Olscher, D. A., Todd, S. J., Jackson, G. L., Wendel, G. D., *Obstet. Gynecol., 100*, 1244 (2002).
6. Doberczak, T. M., Kandall, S. R., Friedman, P., *Obstet. Gynecol., 81*, 936 (1993).
7. Berghella, V., Lim, P. J., Hill, M. K., Cherpes, J., Chennat, J., Kaltenbach, K. A., *Am. J. Obstet. Gynecol., 189*, 312 (2003).
8. Kaltenbach, K. A., *Drug Alcohol Depend., 36*, 83 (1994).
9. Brown, H. L., Britton, K. A., Mahaffey, D. et al., *Am. J. Obstet. Gynecol., 179*, 459 (1998).
10. Marquet, P., Lavignasse, P., Gaulier, J. M., Lachatre, G., "Case study of neonates born to mothers undergoing buprenorphine maintenence treatment," in *Buprenorphine therapy of opiate addicts*, P. Kintz, P., Marquet, P. (eds.), Human Press Ed., pp. 125-35 (2002).
11. Finnegan, L. P., Connaughton, J. F., Kron, R. E., Emich, J. P., *Addict, Dis,. 2*, 141 (1975).
12. Lipsitz, P. J., *Clin. Pediatr., 14*, 592 (1975).
13. Committee on drugs, *Pediatrics, 101*, 1079 (1998).
14. Mazurier, E., Chanal, C., Miaroui, M., et al., *Arch. Pédiatr., 7 (suppl. 2)*, 281 (2000).
15. Ward, J., Hall, W., Mattick, R. P., *Lancet, 353*, 221 (1999).

16. Jernite, M., Viville, B., Escande, B., Brettes, J. P., Messer, J., *Arch. Pédiatr.*, *6*, 1179 (1999).
17. Greene, C. M., Goodman, M. H., *Neonatal Network, 22*, 15 (2003).
18. Wojnar-Horton, R. E., Kristensen, J. H., Yapp, P. et al., *Br. J. Clin. Pharmacol.*, *44*, 543 (1997).
19. Geraghty, B., Graham, E. A., Logan, B., Weiss, E. L., *J. Hum. Lact.*, *13*, 227 (1997).
20. Hoegerman, G., Schnoll, S., *Clin. Perinatol.*, *18*, 51 (1991).
21. MacCarthy, J. J., Posey, B. L., *J. Hum. Lact.*, *16*, 115 (2000).
22. Marquet, P., Chevrel, J., Lavignasse, P., Merle, L., Lachatre, G., *Clin. Pharmacol. Ther.*, *62*, 569 (1997).
23. Jernite, M., Viville, B., Diemuncsh, P., et al., *Arch. Pédiatr.*, *7*, 1014 (2000).
24. Kandall, S. R., *Clin. Perinatol.*, *26*, 231 (1999).
25. Osborn, D. A., Cole, M. J., Jeffery, H. E., *Cochrane Database of systematic reviews, 2*, CD 002059 (2002)
26. Osborn, D. A., Jeffery, H. E., Cole, M. J., *Cochrane Database of systematic reviews, 2*, CD 002053 (2002)
27. Theis, J. G., Selby, P., Ikizler, Y., Koren, G., *Biol. Neonate, 71*, 345 (1997).
28. Micard, S., Brion, F., *Arch. Pédiatr.*, *10*, 199 (2003).
29. Morrison, C. L., Sinet, C., *Eur. J. Pediatr.*, *155*, 323 (1996).
30. Levy, M., Spino, M., *Pharmacotherapy, 13*, 202 (1993).
31. Suresh, S., Anand, K. J. S., *Semin. Perinatol.*, *22*, 425 (1998).
32. Jones, H. C., *Fam. Med., 31*, 327-30 (1999)
33. Coyle, M. G., Ferguson, A., Lagasse, L., et al., *J. Pediatr., 140*, 561-4 (2002)
34. Volpe, J. J., "Teratogenic effects of drugs and passive addiction" in Volpe, J. J. (ed.), *Neurology of the newborn* (2nd ed.), W.B. Saunders Publishers, pp. 664-97 (1987).
35. Kron, R. E., Litt, M., Eng D., et al., *J. Pediatr., 88*, 637 (1976).
36. Herzlinger, R. A., Kandall, S. R., Vaughan, H. G., *J. Pediatr., 91*, 638 (1977).
37. Wijburg, F. A., De Kleine, M. J., Fleury, P., Soepamti, S., *Acta Paediatr. Scand., 80*, 875 (1991).
38. Hoder, E. L., Leckman, J. F., Poulsen, J., et al., *Psychiatry Res., 13*, 243 (1984).

4.9 Medication and breast feeding

E. Rey[1], R. Serreau[2]

[1] Pharmacology Clinic, University René-Descartes, Hospital Cochin Saint Vincent de Paul, Paris, France

[2] Paediatric Pharmacology Unit, Hospital Robert Debré, Paris, France

4.9.1 INTRODUCTION

Breastfeeding is encouraged by the WHO for the physical and psychological well-being of the newborn child. In addition, immunological studies have shown the importance of maternal breastfeeding for periods exceeding 6 months in relation to protecting against infections [1,2]. In Europe, approximately 50% of women breastfeed, but there is considerable variation between countries. The majority of women stop breastfeeding before the third month [3,4].

The use of medication, both prescribed and over-the-counter, is frequent during pregnancy. Many medicines are continued post-partum. In this situation, the possibility of side effects in the newborn needs to be considered [5].

It is possible to choose medications that are compatible with breastfeeding on the basis of available studies and pharmacological criteria [6], in order to find a response to a number of practical questions that are frequently asked:

- If a mother needs treatment and she breastfeeds, should breastfeeding continue?
- In the case of chronic maternal pathology, if the medication has been taken during the entire pregnancy, is this situation compatible with breastfeeding?
- If a newborn/infant has clinical symptoms, could this be related to the use of a medication by the mother?

4.9.2 PHYSIOLOGY OF LACTATION

Milk is produced by cells of the mammary epithelium where it is stored. The milk is a suspension of lipids and proteins in a solution of minerals and sugars. The colostrum, the first form of milk that is produced during the 2 to 3 days following birth, is a yellow-orange liquid composed largely of proteins; it has a composition very similar to that of plasma. In contrast, between days 15 and 20 after birth, the milk assumes its "mature" composition, much richer in lipid and lactose. There is a very large variability in the composition of milk between meals and in the course of a single meal.

4.9.3 PASSAGE OF DRUGS INTO BREAST MILK

There are many factors that influence the drug transfer into breast milk and the dose consumed by the infant.

Drug factors

Most drugs cross from maternal blood to milk by passive diffusion of unionised free drug. This passive diffusion is driven by a concentration gradient, established when there is a difference in the concentration of chemicals on either side of a semi-permeable membrane. The rate of diffusion depends on the physico-chemical characteristics of the drug.

One of these is the molecular weight. Most drugs have a molecular weight smaller than 500D which allows passage through capillary membranes. Drugs with a molecular weight of 6000-20000 D (heparin, insulin), however, do not cross into breast milk.

Ionised or electrically charged molecules cannot passively diffuse through biological membranes. The ionisation of drugs depends on their pKa (pH at which 50% of drug are ionised), therefore weak acids are ionised and weak bases are non-ionised in the maternal plasma. While average plasma pH is 7.4, the mean pH of breast milk is 7.2 which is significantly lower. In breast milk, some weak bases will be ionised preventing diffusion back, with the molecule being trapped in the breast milk. More acidic drugs will concentrate in the maternal plasma.

Drugs highly bound to albumin are unable to diffuse through tissue and only the free fraction will diffuse. This restricts many drugs to the plasma compartment. The binding by the major milk whey proteins (casein, alpha-lactalbumin, lactoferrin, and IgA) is lower than in plasma, due to the lower concentration and affinity of these proteins.

Lipid-soluble drugs cross the cell membrane rapidly and concentrate in milk lipids. The milk-to-plasma concentration ratio of a lipid-soluble drug is dependent on the lipid concentration of the milk [6].

Maternal factors

As most drugs are excreted into milk by passive diffusion, the drug concentration in milk is directly proportional to the corresponding concentration in maternal plasma. The maternal plasma concentration depends on maternal pharmacokinetics of the drug, the dose administered, the route of administration, the frequency of administration, and the timing of a drug dose in relation to breastfeeding.

The physiological changes which occur during pregnancy influence the pharmacokinetic processes of drug absorption, distribution, and elimination. The main variations are seen in the distribution of the drug with an increase in the volume of distribution and a decrease in protein binding, leading to a lower total plasma concentration, but to an increase in the unbound fraction which is pharmacologically active. Furthermore, the glomerular filtration rate is usually increased, which enhances renal drug elimination. Hepatic metabolism is increased, decreased or unchanged. The substantial physiological changes normalise by approximately 3 months postpartum [7]. However, it is possible that maternal drug disposition shows great variations in the immediate postpartum period because of the physiological stress of labour and delivery.

The dose of the medication is another factor which will determine the amount of drug entering the breast milk: the higher the dose, the more drug passes into breast milk, therefore the recommended dose should be the lowest needed.

The interval between administration of the drug and breastfeeding will influence the amount of drug that is transferred to breast milk: higher when the maternal plasma concentration is peaking than when it is at the minimum plasma concentration.

Factors associated with the newborn

For any given newborn, the drug effects will depend on a number of factors, including (1) his or her age, which influences his capacity to absorb and then eliminate the medication; (2) the amount of milk consumed; (3) specific characteristics of the medication; and in particular (4) the data related to the safety of use which may be available during the period of breastfeeding.

For most medications, pharmacokinetic data or safety information has been obtained in animal studies and is not available for the humans. In most cases, this data was obtained during feeding periods of short duration.

4.9.4   AMOUNT OF DRUG EXCRETED INTO BREAST MILK

The actual amount of drug excreted into breast milk depends on the milk yield and the drug concentration in the milk which is dependent on the factors mentioned above. Regarding the potential effect of the drug on the suckling infant, the quantity of any active metabolites present in the milk must also be considered.

The difference between total milk yield and milk transferred to the infant is greatest on day 4 post partum (approximately 140 g). The difference then decreases to about 50 g after 2 months and 20 g after 5 months. The average infant milk intake is 753 ± 89 ml/day between months 3 and 5.

It is unclear how the frequency and duration of lactation affects drug excretion in breast milk. The yield from each breast is not similar, but depends on how well the breast was emptied during the previous feed. Furthermore, there are differences in drug excretion when the amount of drug in each breast is compared.

The infant's suckling pattern and physiological development will influence the amount of drug that will be ingested. The amount of milk consumed during each feeding, the number of feedings each day, as well as the length of each feed will determine the potential exposure of the infant to the concentration of the drug in the milk.

Calculation of the dose received by the nursing infant

Limited information is available about the true amount of drug that crosses into breast milk. The most commonly reported ratios are the M/P point, which is the ratio of milk drug concentration to simultaneous plasma drug concentration at a given time. The accuracy of this ratio is dependent on when, during the dosing interval, drug concentrations are measured, as peak milk drug concentrations do not coincide with peak plasma concentrations: drug concentrations peaking later in milk than in plasma. Therefore, if an M/P ratio is determined at the time of peak plasma concentration, it can be underestimated, conversely if it is measured at the time of peak milk ratio, it can be overestimated. This estimate assumes a parallel decline in milk and plasma drug concentrations throughout the dosage interval, in support of the hypothesis that there is a rapid equilibration of drug between plasma and milk which is probably an important pitfall.

This ratio is more reliable when it comes from studies where the area under the concentration-time profile has been measured over a whole dose interval. A pharmacokinetic study should be specifically designed for this purpose. An approach whereby, during a 24 hour period, milk is extracted by breast pump at regular intervals, to simulate normal feeding, has been favoured. The total drug content is then determined and represents the daily dose received by the infant, assuming that an infant would be able to ingest the total amount of milk collected.

The estimated drug intake by the suckling infant can be expressed on a mg/kg basis, as a percentage of the normal infant dose adjusted to body weight and age, if the information is available. In the absence of such data, the dose in milk may be expressed as the relative dose, i.e., a percentage of the actual dose received by the mother (on a mg/kg basis) that is then received by the infant (on a mg/kg basis).

4.9.5 CLASSIFICATION OF MEDICATIONS BY CATEGORY OF RISK IN THE CASE OF BREASTFEEDING

The following characteristics should be taken into account in relation to the decision to avoid breastfeeding:

- Medications having the following pharmacokinetic or metabolic properties: significant transfer to milk, active metabolites, lengthy elimination (for example psychotropic medications), or narrow therapeutic window (cardiovascular drugs such as digoxin);
- Medications not indicated or not used on newborns;
- Medications that have been reported to cause undesirable side effects in newborns (for example, lithium and aspirin).

Medications have been classified by risk category, from 1 to 5 based on these factors and published data [8]:

1. **safe:** the product has been widely used during breastfeeding without any increase of undesirable effects for the child. Controlled studies with breastfeeding mothers, when available, have shown that there is no risk for the child or that the bioavailability of the drug, when administered orally, is negligible.

2. **reasonably safe:** there are very little available data concerning women who breastfeed, but there has been no evidence of an increase of undesirable side effects for newborn children.

3. **potentially unsafe:** no data is available at all for mothers during breastfeeding, or the risk of side effects for the newborn is real. Animal studies have shown a slight risk. The drug must not be used for the mother unless the expected benefit clearly outweighs the potential risk for the child.

4. **hazardous:** data is available that shows the drug is present in the newborn, or a significant concentration has been detected in the milk. The drug must not be used for the mother unless the expected benefit clearly outweighs the potential risk for the child.

5. **contraindicated:** data from human studies have shown a significant risk of undesirable side effects, which might hold serious consequences for the newborn child. There is no anticipated benefit that can justify the administration of this product during breastfeeding.

Tables 1-4 summarise the existing data [8-13] with regards to the safety of medicines and breastfeeding. The tables show two systems of risk classification: the 1-5 scale devised by Hale [8] described above and that of the Drug Committee of the American Academy of Pediatrics (AAP) [9] that uses the following scheme:

*1: Breastfeeding authorised (updated in September 2001).

*2: Medications that might have an effect on newborns or infants (with an effect that has not yet been reported).

*3: Medications associated with undesirable side effects for newborns or infants.

*4: Cytotoxic medications that interfere with the cellular metabolism of the newborn or infant.

Table 1

Medications compatible with breastfeeding

	AAP [9]	Hale TW [8]
tetracycline	1	2 in the neonate
sulfafurazol	1	2*
quinine	1	2
digoxin	1	2
labetalol	1	2
warfarin	1	2
carbimazole	1	3
ionponic acid	1	2
prednisone	1	2
carbamazepine	1	2
valproic acid	1	2
ofloxacin	1	3

* contraindications in cases of G6PD deficiencies

Table 2

Contraindicated medication during breastfeeding

	AAP [9]	Hale TW [8]
tetracycline	1	4
ciprofloxacin	1	4
	2	5
prednisone	1	4*
lithium carbonate	1	4
ergotamine tartrate	3	4
bromides	1	5
gold salts	1	5
heroine	5	4
cocaine	5	5
ciclosporin	4	3
methotrexate	5	5 (depending on how it is used)
doxorubicine	5	5
atropine	1*	3
ramipril	not reviewed	3, 4
quinapril	not reviewed	2, 4

* high dose / prolonged

4.9.6 CONCLUSION: PRACTICAL RECOMMENDATIONS

(1) If the passage of the medication into milk has been studied [10,11], it is important to estimate the maximum quantity of medication that the child will receive, taking into account the quality of the known data, the dosage for the mother, the characteristics of the breastfeeding, and the clinical parameters of the newborn.

(2) In the absence of this information, possibilities include:
- a temporary cessation of breastfeeding or a delay in the start of breastfeeding. It is possible to feed the child with pooled breast milk while maintaining lactation with the use of a pump.
- substitution of the medication for another that is better characterised

Table 3

Medications to be evaluated on a case-by-case basis

DCI	AAP [9]	Hale TW [8]
sulfamethoxazole	not reviewed	3
levofloxacin	not reviewed	3
norfloxacin	not reviewed	3
mefloquine	not reviewed	2
flecainide acetate	1	4
nicardipine	not reviewed	3
povidone iodide	not reviewed	4
lopamidol	not reviewed	3
metformin	not reviewed	3
glimepiride	not reviewed	4
glipizide	not reviewed	3
meprobamate	not reviewed	3
oxazepam	not reviewed	3
amitriptyline	2	2
clomiparamine	2	2
phenobarbital	3	3
lamotrigine	2	3
mesalasine	3	3
indomethacin	1	3
aspirin	3	3
codeine	1	3

- the selection of a medication that is strongly bound to plasma proteins, having the lowest possible oral bioavailability, a short half-life for elimination, a low milk-plasma ratio, lack of active metabolites, and presenting the least number of undesirable side effects possible.

(3) Administer the medication at times furthest from the concentration peak observed in the milk, in order to reduce the exposure of the newborn.

(4) In the rare cases where the treatment presents risks for the newborn and is therefore not compatible with breastfeeding, either the drug should be stopped or breastfeeding avoided.

Table 4

Drugs that have been associated with significant effects on some nursing infants and should be given to nursing mothers with caution (adapted from [10])

Drug	Reported Effect	Ref.
acebutolol	hypotension; bradycardia; tachypnea	14
5-aminosalicylic acid	diarrhoea (1 case)	15
atenolol	cyanosis; bradycardia	16
bromocriptine	suppresses lactation; may be hazardous to the mother	17
aspirin (salicylates)	metabolic acidosis (1 case)	18
clemastine	drowsiness, irritability, refusal to feed, high-pitched cry, neck stiffness (1 case)	19
ergotamine	vomiting, diarrhoea, convulsions (doses used in migraine medications)	20
lithium	one-third to one-half therapeutic blood concentration in infants	21
phenindione	anticoagulant: increased prothrombin and partial thromboplastin time in 1 infant; not used in United States	22
phenobarbital	sedation; infantile spasms after weaning from milk containing phenobarbital, methaemoglobinemia (1 case)	23
primidone	sedation, feeding problems	23
sulfasalazine	bloody diarrhoea (1 case)	24

REFERENCES

1. Hanson, L. A., Carlsson, B., et al., *Acta Paediatr. Scand., Suppl. 299*, 38 (1982).
2. Hanson, L. A., Bergstrom, S., *Acta Paediatr. Scand., 79*, 481 (1990).
3. Crost, M., Kaminski M., *Arch. Pediatr., 5*, 1316 (1998).
4. Branger, B., Cebron, M., et al., *Arch. Pediatr., 5*, 489 (1998).
5. Ito, S., *N. Engl. J. Med., 343*, 118 (2000).
6. Pons, G., Rey, E., Matheson, I., *Clin. Pharmacokinet., 27(4)*, 270 (1994).
7. Atkinson, H. C., Begg E. J., et al., *Clin. Pharmacokinet., 14*, 217 (1988).
8. Hale, T. W., *Medications and mothers' milk* (11th Edition), Amarillo, Pharmacosoft Publishing, 2004.
9. American Academy of Paediatrics. Committee on Drugs: "The Transfer of Drugs and Other Chemicals Into Human Milk," *Pediatrics, 108*, 776 (2001) and http://www.aap.org/

10. Lawrence, R. A., Howard, C. R., *Pediatr. Clin. North Am., 48*, 517 (2001).
11. Howard, C. R., Lawrence, R. A., *Pediatr. Clin. North Am., 48*, 485 (2001).
12. Briggs, G. G., Freeman, R. K., Yaffe, S. J., *Drugs in Pregnancy and Lactation* (6th ed.), Lippincott Williams and Wilkins (2002).
13. Dossier du CNIMH, *Revue d'évaluation sur le médicament, XIII 5-6*, 1 (1996).
14. Boutroy, M. J., Bianchetti, G., Dubruc, C., Vert, P., Morselli, P. L., *Eur. J. Clin. Pharmacol., 30*, 737 (1986).
15. Klotz, U., Harings-Kaim, A., *Lancet, 342*, 618 (1993).
16. Schimmel, M. S., Eidelman, A. I., Wilschanski, M. A., Shaw, D., Jr., Ogilvie, R. J., Koren, G., *J. Pediatr.,114*, 476 (1989).
17. Katz, M., Kroll, D., Pak, I., Osimoni, A., Hirsch, M., *Obstet. Gynecol., 66*, 822 (1985).
18. Jamali, F., Keshavarz, E., *Int. J. Pharm., 8*, 285 (1981).
19. Kok, T. H., Taitz, L. S., Bennett, M. J., Holt, D. W., *Lancet, 1*, 914 (1982).
20. Fomina, P. I., *Arch. Gynecol., 157*, 275 (1934).
21. Sykes, P. A., Quarrie, J., Alexander, F. W., *Br. Med. J., 2*, 1299 (1976).
22. Eckstein, H. B., Jack, B., *Lancet, 1*, 672 (1970).
23. Kuhnz, W., Koch, S., Helge, H., Nau, H., *Dev. Pharmacol. Ther., 11*, 147 (1988).
24. Branski, D., Kerem, E., Gross-Kieselstein, E., Hurvitz, H., Litt, R., Abrahamov A., *J. Pediatr. Gastroenterol. Nutr. 5*, 316 (1986).

4.10 Clinical drug trials in pregnancy: challenges and solutions

Gideon Koren

The Motherisk Program, University of Toronto and Ivey Chair in Molecular Toxicology, University of Western Ontario, Canada

4.10.1 INTRODUCTION

Due to obvious ethical issues, women and their unborn babies are commonly excluded from drug trials. As a result, these two groups of patients are orphaned from the revolution in drug therapy witnessed in the last generation.

The removal of Bendectin® from the American market despite being safe in pregnancy (see Chapter 4.1) sent a chilling signal to drug companies, essentially discouraging them from studying pregnant patients. Consequently, the advance in therapeutics in pregnancy lags substantially behind the same conditions or drugs in non-pregnant women, or men.

However, this approach is highly unwarranted, as uncontrolled maternal conditions may affect adversely both the mother and her fetus. Hence, a rational approach must also incorporate the risks of the untreated maternal condition.

Because embryogenesis is completed by the end of the first trimester of pregnancy, if a drug is not affecting brain development (which continues throughout gestation), there is no apparent reason not to study it during the second and third trimesters of pregnancy.

Several recent developments may mark important milestones in changing the approach to drug trials in pregnancy.

4.10.2 CHANGES IN DRUG DISPOSITION IN LATE PREGNANCY

During the last few years an increasing body of evidence has suggested that, in late pregnancy, there is a substantial increase in the clearance rate of various drugs, including nicotine, lamotrigine, fluoxetine, citalopram [1-2]. This means that women may need larger doses to achieve therapeutic steady state concentrations. For example, using nicotine replacement therapy has failed to prevent smoking in late pregnancy when compared to placebo, probably because the dose regimen used in late pregnancy was insufficient.

Similarly, many women with depression are not controlled clinically in late pregnancy, probably because doses that were adequate before pregnancy are grossly inappropriate in late pregnancy [3].

Late pregnancy is characterised by major increases in GFR, hepatic blood flow, decrease in protein binding and altered drug compliance. These changes lead, in most instances to lower systemic exposure to medications, both hepatically and renally eliminated.

4.10.3 LEARNING FROM THE GLYBURIDE MILESTONE

Gestational diabetes (GD) affects up to 10 to 15% of late pregnancies. Left untreated, it may adversely affect pregnancy outcome. The hallmark of therapy is dietary control and insulin, as this naturally occurring hormone does not cross the placenta appreciably. However, insulin therapy is expensive, unavailable in many areas, and is associated with low compliance rates.

The use of oral hypoglycaemic drugs has been largely contraindicated, because these drugs cross the placenta, increasing the risk of neonatal hypoglycaemia [4].

In 1994, using the *ex vivo* placental perfusion studies, Elliot and colleagues documented that unlike other "older" sulfanylurea, glyburide did not cross the placental barrier from the mother to the fetus [5].

In 2000, the same group published the results of a randomised controlled trial comparing pregnancy outcome with insulin vs. glyburide. The offspring were not different in any outcome characteristics, including birth weight, rates of hypoglycemia or mortality. Critically, while maternal glyburide levels were in the therapeutic range (50-150 ng/mL), the drug was undetected in any of the umbilical blood samples [6].

The mechanisms preventing a relatively small, non-polar molecules from crossing the placenta has not been elucidated yet. It has been hypothesised that this may be a combination of a relatively short half-life with extremely high protein binding, plus the effect of glyburide being a substrate of P-glycoprotein, and thus possibly pumped from the fetal to the maternal circulations [7].

4.10.4 A PROPOSED FRAMEWORK FOR DRUG TRIALS IN PREGNANCY

The objective of this section is to synthesise for the first time known principles and concepts, and to propose a framework for trials of medicines in pregnancy.

Principles

- Studies should be conducted first in the second and third trimester of pregnancy.
- High priority should be given to studies of agents, which can be expected to address an unmet maternal/fetal risk, or improve maternal or fetal outcomes compared to existing therapy.
- High priority should be given to agents not likely to affect CNS development, which continues throughout gestation.
- A pharmacokinetic study should precede an efficacy-effectiveness study, as in late pregnancy, women may need larger doses due to faster clearance rate.
- Before a study is initiated during the first trimester of pregnancy, human safety data should be available. Such data should be prospective observational data with an unexposed comparison group.
- Participants in studies should consent after being made aware of the available safety data and its limitations, the risk of the untreated condition, as well as the known risks and benefits of available data on alternative therapeutics.

The recent breakthrough in paediatric drug trials in the USA secondary to enacting financial benefits through extension of exclusivity to products studied in children, should logically lead to similar moves for drug trials in pregnancy. However, the legal-ethical equation in pregnancy is much more complex than in childhood, and it seems less likely that the pharmaceutical industry will agree to participate in the very litigious climate of today. Hence, it is critical that national and international research funding agencies consider the pregnant patient as a high priority.

ACKNOWLEDGEMENT

This work is supported by grants from the Canadian Institutes for Health Research.

REFERENCES

1. Dempsey, D., Jacob, P., Benowitz, N. Z., *J. PET, 301*, 594 (2002).
2. Heikkinen, T., Ekblad, U., Palo, P., Laine, K., *Clin. Pharmacol. Ther., 73*, 330 (2003).
3. Bonari, L., Bennett, H., Einarson, A., Koren, G., *Can. Fam. Physician, 50*, 37 (2004).
4. Koren, G., *Maternal-Fetal Toxicology: A Clinician's Guide* (3rd ed.), Marcel Dekker, NY, (2001).
5. Elliott, B. D., Schenker, S., Langer, O., Johnson, R., Prihoda, T., *Am. J. Obstet. Gynecol., 171*, 653 (1994).
6. Langer, O., Conway, D. L., Berkus, M. D., Xenakis, E. M., Gonzales, O., *N. Engl. J. Med., 343*, 1134 (2000).
7. Garcia-Bournissen, F., Feig, D. S., Koren, G., *Clin. Pharmacokinet., 42*, 303 (2003).

Medicines in neonates

5.1 Problems with medicines in the newborn infant

John N. van den Anker

Department of Paediatrics, Pharmacology and Physiology, Division of Paediatric Clinical Pharmacology, George Washington University School of Medicine and Health Sciences, Children's National Medical Center, Washington, D.C., USA; and Department of Paediatrics, Erasmus MC-Sophia Children's Hospital, Rotterdam, The Netherlands

5.1.1 INTRODUCTION

The history of drug therapy shows that newborn infants are more prone to adverse reactions to medicines. In 1956, Silverman et al. [1] reported an excessive mortality rate and an increased incidence of kernicterus among preterm neonates receiving a sulphonamide antibiotic as compared to neonates receiving chlortetracycline. Then, in 1959, Sutherland [2] described a syndrome of cardiovascular collapse in newborns receiving high doses of chloramphenicol for presumed infections. More recently, the therapeutic misadventures experienced by low birth weight infants exposed to a parenteral vitamin E formulation [3] and the "gasping syndrome" by infants who received excessive amount of benzyl alcohol [4,5] underscore the persistence of problems with the use of medicines in neonates. As a result of these experiences, neonatologists and paediatricians have recognised that rational drug therapy for neonates is often confounded by a combination of unpredictable and often poorly understood pharmacokinetic and pharmacodynamic interactions [6-9].

A more specific approach to neonatal therapeutics requires a thorough understanding of human developmental biology, as well as insights regarding the dynamic ontogeny of the processes of drug absorption, drug distribution, drug metabolism, and drug excretion.

This chapter will focus on the pharmacokinetic determinants of neonatal drug therapy that may explain and potentially prevent aforementioned problems with medicines in neonates.

5.1.2   DRUG ABSORPTION

Developmental changes in the gastrointestinal tract can alter the ***extent of drug absorption***. Changes in luminal pH directly impact upon both drug stability and the degree of ionisation, thus influencing the relative amount of drug available for absorption [10,11]. An acid pH favors absorption of acid drugs (low pK_a), because, in such an environment, the drug will be largely in an un-ionised, more lipid-soluble form. In contrast, a relatively high pH (as in state of achlorhydria) will enhance the translocation of basic drugs and retard the absorption of acidic drugs.

At birth, gastric pH is usually between 6 and 8, but it falls rapidly to between 1.5 and 3.0 within several hours [12,13]. This decrease in gastric pH is quite variable but seems to be independent of both birth weight and gestational age (GA). Even preterm infants with GAs of 24-29 weeks are able to produce and maintain an intragastric median pH below 4 with an inverse relationship between gestational age and initial acid production [14]. In addition, Grahnquist *et al.* have described the presence of gastric H,K-ATPase in stomach biopsies from infants 25-42 weeks gestation with it's expression increasing with increasing GA [15]. The subsequent pattern of gastric acid secretion remains controversial. Initial descriptions have suggested that after birth, acid secretion shows a biphasic pattern: the highest acid concentrations occur within the first 10 days of life and the lowest between days 10 and 30. Agunod *et al.*, using betazole stimulation, have confirmed these early findings and found that the volume of gastric juice and acid concentration were dependent on age and that gastric acid secretions approached the lower limit of adult values by 3 months of age [16]. These data clearly demonstrate the large degree of variability in the intragastric pH over the first 30 days of life.

Additionally, the ability to solubilise and subsequently absorb lipophilic drugs or formulations can be influenced by age-dependent changes in biliary function. For example, pleconaril, when formulated in a lipid-base vehicle, shows apparent capacity limited absorption in neonates, which may be the result of inadequate bile salt concentrations necessary to sufficiently solubilise this drug [17].

Development can similarly influence the ***rate of drug absorption***. Gastric emptying rate and intestinal motility are primary determinants of the rate at which drugs are presented to, and dispersed among, the mucosal surface of the small intestine. Gastric emptying rate during the neonatal period is variable and is characterised by irregular and unpredictable peristaltic activity [18-20]. The rate of gastric emptying appears to be directly affected by gestational and postnatal age, as well as the type and composition of feeding used [13, 18-20]. The osmolality of the meal does not appear to be a factor, but slower emptying is seen in feeding with long-chain fatty acids compared with medium-chain triglycerides, an important variable considering that infant formulas differ in their fatty acid content [19]. In addition, age-associated changes in the splanchnic blood flow in the first 2-3 weeks of life [21-23] may influence absorption rates by altering the concentration gradient established across the intestinal mucosa.

Despite their incomplete characterisation [24], developmental differences in activity of intestinal drug metabolising enzymes and active efflux transporters in the small intestine have the potential to markedly impact upon ***oral drug bioavailability***. Although data on developmental expression of P-glycoprotein in humans are absent, we include a summary of P-glycoprotein because of its importance for drug absorption, distribution and elimination in the neonate.

P-glycoprotein (P-gp) is normally found within the cellular membranes of the intestinal tract (duodenum, ileum, jejunum, colon), apical membrane of hepatocytes, renal proximal tubular cells and on the luminal side of the capillary endothelial cells that make up the blood brain barrier [25-27]. This efflux protein has affinity for a broad range of hydrophobic substrates and effectively "pumps" xenobiotics out of cells, influencing the amount of a drug that may be absorbed into systemic circulation, influencing the rate at which a drug may be cleared by the liver or kidney, or influence the amount of drug that enters the central nervous system [25-28]. Moreover, the degree of expression and/or modulation of P-gp activity by inhibitors or inducers are probably the basis for a number of clinically important drug-drug interactions. The variability in intestinal absorption characteristics for many drugs is likely a direct result of the variability of P-gp expression within the intestinal tract as well as the presence or absence of P-gp modulators [25,26,29].

Drug absorption following different routes of administration

The intravenous route for drug delivery is preferred in neonates. However, neonates frequently have poor intravenous access, and extravascular routes then becomes a viable and effective alternative for the administration of many drugs.

Absorption of drugs given intramuscularly

An important factor influencing absorption of drugs from an intramuscular injection site is the blood flow to and from the injection site. This may be compromised in newborns with poor peripheral perfusion from low cardiac output states or the respiratory distress syndrome [30]. The rate and extent of absorption from an intramuscular injection site are also influenced by the total surface area of muscle coming into contact with the injected solution [15]. The ratio of skeletal muscle mass to body mass is less for neonates than for adults [30].

However, the impact of these factors may be offset by the increase in skeletal muscle capillary density observed in infants as compared to older children [31]. Accordingly, evidence supports more efficient intramuscular drug absorption and higher peak plasma concentrations for a number of agents (e.g., antibiotics and anticonvulsants) in neonates and infants as compared to adults [32,33].

A final consideration relative to intramuscular drug absorption is muscle activity, which may affect the rate of absorption and, therefore, affect the peak serum concentration. Sick, immobile neonates or those receiving a paralysing agent to facilitate mechanical ventilation may show reduced absorption rates following intramuscular drug administration.

Percutaneous absorption

The skin represents an often overlooked, but important organ for systemic drug absorption [34]. Enhanced percutaneous absorption during childhood is accounted for by a number of factors [35-39]. Theoretically, if a newborn receives the same percutaneous dose of a compound as an adult, the systemic availability per kilogram of body weight will be approximately 2.7 times greater in the neonate.

There are numerous reports in the literature of neonatal toxicity related to the cutaneous exposure to drugs and chemicals. They include hexachlorophene [40], pentachlorophenol-containing laundry detergents [41], hydrocortisone [42], and aniline-containing disinfectant solution [43].

Finally, if the integrity of the integument is compromised (denuded, burned, or inflamed skin, for example), then percutaneous translocation of compounds into the blood will be enhanced.

Rectal absorption

Rectal bioavailability for several agents is enhanced in the young, consequent to reduced metabolism rather than enhanced mucosal translocation. However, enhanced lower gastrointestinal motility can enhance rectal expulsion of solid dosage forms [44] and thereby, reduce drug absorption (e.g., erythromycin and paracetamol) [45-47]. In addition, physicochemical properties of the drug and/or formulation (e.g., melting temperature and particle size) may affect rectal drug absorption in the neonate.

Knowledge of the venous drainage system for the lower gastrointestinal tract is imperative in understanding the potential bioavailability of drugs administered rectally. The inferior and middle rectal veins, which drain the anus and lower rectum, respectively, drain directly into the systemic circulation by means of the inferior vena cava, whereas the superior rectal vein, which drains the upper part of the rectum, empties into the portal vein by means of the inferior mesenteric vein. Therefore, drugs administered into the superior aspect of the rectum will be subjected to the hepatic first-pass effect because portal blood enters the liver, whereas drugs administered lower into the rectum will initially bypass the liver.

5.1.3 DRUG DISTRIBUTION

The movement of drugs and other compounds from the systemic circulation into various body compartments, tissues, and cells is termed distribution. The distribution of most drugs in the body is influenced by a variety of age-dependent factors, including protein binding, body compartment sizes, haemodynamic factors such as cardiac output and regional blood flow, and membrane permeability [48,49].

The apparent volume of distribution (Vd) describes the relationship between the amount of drug in the body and its plasma concentration. Vd is the volume needed to contain the total body store of drug if the concentration throughout the whole body were the same as in plasma.

Developmental aspect of protein binding

The binding of drugs to plasma protein is dependent on the concentration of available binding proteins, the affinity constant of the protein(s) for the drug, the number of available binding sites, and the presence of pathophysiologic conditions or endogenous compounds that may alter the drug-protein binding interaction [50-52].

The affinity of albumin for acidic drugs increases, as do total plasma protein levels, from birth into early infancy [53]. In addition, although plasma albumin may reach adult levels shortly after birth, the albumin level in blood is directly proportional to gestational age, reflecting both placental transport and fetal synthesis [54,55].

Albumin is not the only plasma protein that binds drugs. Basic drugs are bound by several plasma proteins, including α_1-acid glycoprotein [51]. Piafsky and Mpamugo [56] showed significant reductions in both α_1-acid glycoprotein plasma concentrations and, as a consequence, decreased binding of the basic drugs lidocaine and propranolol to cord blood as compared with adult controls.

Competition for protein binding

There are a number of endogenous molecules, which, like drugs, may bind to plasma proteins [48] and displace drugs from these binding sites. Although most clinicians perceive this "interaction" as a primary mechanism of drug toxicity, the clinical significance of protein displacement drug-drug interactions are usually very limited or none [52]. However, during the neonatal period, this interaction may be of greater importance due to immaturity in clearance organ function.

Nonesterified fatty acids are reversibly bound to albumin [57] and are present at relatively high concentrations in the plasma of newborn infants. Significant reductions in albumin binding of phenytoin have been demonstrated at high serum levels of free fatty acids (FFA), and a linear correlation between unbound plasma phenytoin concentrations and the ratio of serum FFA to albumin concentration in neonates have been described.

Although these values are rarely attained, they have been observed under certain pathophysiological conditions such as gram-negative septicaemia. Interestingly, similar elevations in FFA levels have not been reported with gram positive septicaemia, which is common in newborns.

Bilirubin is noncovalently bound to albumin, and this association is freely reversible [58,59]. The binding affinity of albumin for bilirubin increased with age [60,61]. This increased affinity, at least during the first week of life, is related to gestational age [60]. The lower bilirubin binding affinity of albumin in neonates is believed to be a contributing factor in their susceptibility to kernicterus [61]. However, other factors, such as the effect of hypothermia, acidosis, hypoglycaemia, hypoxaemia, sepsis, birth asphyxia, and hypercapnia on the permeability of the blood brain barrier and on bilirubin-albumin binding must be considered [62]. A number of drugs are thought to be able to compete with and displace bilirubin from binding sites on the albumin molecule, thus increasing the risk for the development of kernicterus [62].

Developmental aspects of body compartment sizes

Alterations in body water compartment sizes will affect the volume of distribution of a drug. In the young fetus, total-body water comprises nearly 92 % body weight, with the extra-cellular fluid volume responsible for 25 % body weight; body fat is less than 1 %[63-65]. At term, total-body water falls to approximately 75 % body weight, and the amount of fat increases to approximately 15 %. By 40 weeks gestation, measurements of extracellular fluid volume range from 350-440 ml/kg of body weight [66] and correlate more closely with body weight than with gestational age. The intracellular fluid volume increases from 25 % body weight in the young fetus to 33 % at birth to approximately 37 % of body weight at 4 months of age.

The clinical relevance of this gradual reduction in the size of body water compartments with age cannot be overemphasised. In order to achieve comparable plasma and tissue concentrations of drugs distributing into the extracellular fluid, higher doses per kilogram of body weight must be given to infants than adults.

5.1.4 DRUG METABOLISM

The primary organ for drug metabolism is the liver, but the kidneys, intestine, lungs and skin are also capable of transformation [67]. At every level, from the ontogenetic changes in hepatic blood flow and portal oxygen tension to the developmental alterations in protein binding and xenobiotic metabolising enzyme activities, there is the real effect of age and various pathophysiologic states on the processes associated with hepatic clearance. Dramatic developmental changes in the physiological and biochemical process that govern drug disposition and effect occur during the first year of life. These important processes and description of the ontogeny of hepatic function, including the ontogeny of CYP450 hepatic enzymes (Phase I) and conjugation pathways (Phase II) have been reviewed [68,69].

Insight into the ontogeny of drug metabolism can be derived from *in vitro* observations or pharmacokinetic studies of drugs primarily metabolised by a single CYP isoform. For instance, when administered intravenously, midazolam clearance reflects hepatic CYP3A activity [70], which increases approximately 5-fold over the first 3 months of life [71]. In addition, phenytoin's apparent half life is prolonged (± 75 h) in preterm infants, but decreases to ± 20 h in term infants less than one week postnatal age and to ± 8 h after two weeks of age [72]. Saturable phenytoin metabolism, a hallmark of CYP2C9 activity, does not appear until approximately 10 days postnatal age, demonstrating the developmental delay in CYP2C9 activity [73].

In contrast to information pertaining to phase I enzymes, the ontogeny of phase II enzymes is less well established. Nevertheless, available data indicate that individual glucuronosyl transferase (UGT) isoforms have unique maturational profiles. For example, glucuronidation of morphine (a UGT2B7 substrate) can be detected in premature infants as young as 24 weeks gestational age [74]. In addition, morphine clearance is closely correlated with post-conceptional age (PCA) and increases approximately four-fold between 27 and 40 weeks PCA, necessitating dosage increases of a similar magnitude to maintain effective analgesia [75].

5.1.5 RENAL EXCRETION

Most drugs and/or their metabolites are excreted from the body by the kidneys. Renal excretion is dependent on glomerular filtration, tubular reabsorption, and tubular secretion.

At birth, anatomical and functional immaturity of the kidney limit glomerular and tubular functional capacity, which results in inefficient drug elimination and a prolonged elimination half-life [76,77]. Rapid improvements in glomerular and tubular functions occur during the postnatal period, greatly enhancing renal drug elimination. The main factors involved in the development of renal function are gestational age and the dramatic sequential haemodynamic changes after birth, in a situation initially dominated by high vascular resistance and extremely low blood flow. At birth, the GFR is 2-4 ml/min in term neonates and it may be as low as 0.6-0.8 ml/min in preterm infants. The increase in GFR after birth is important and usually greater in term than in preterm infants [78-81]. This increase is due to the increase in cardiac output which is associated with specific changes in renal vascular resistances, resulting in an increase in renal blood flow, changes in renal blood flow distribution, and a higher permeability of the glomerular membrane.

In addition to these rapid developmental changes in GFR, a wealth of new information concerning mechanisms of tubular drug transport has become available over the past few years [82]. Transporter protein science is a rapidly evolving field of pharmacology, and new transporter proteins are continuously discovered. However, the clinical implications of many of these new discoveries are not yet established.

Glomerular filtration rate

The newborn kidney's main physiological limitation is its very low GFR, maintained by a delicate balance between vasoconstrictor and vasodilatatory renal forces. These forces recruit maximal attainable filtration pressure in the face of minimal renal blood supply, resulting from a combination of a low mean arterial blood pressure and a high intrarenal vascular resistance. The low GFR of the newborn kidney, although sufficient for growth and development under normal conditions, limits the postnatal renal functional adaptation to endogenous and exogenous stress [83]. Such stress may result from renal hypoperfusion caused by anoxia, sepsis, and/or exposure to nephrotoxic medications. Vasoactive forms of nonsteroidal anti-inflammatory drugs (NSAIDs) such as aspirin, indomethacin, and ibuprofen (nonspecific cyclooxygenase inhibitors) and the new selective cyclooxygenase-2 inhibitors such as rofecoxib can induce renal hypoperfusion, resulting in generally reversible, oliguric acute renal failure (ARF). This adverse renal effect of cyclooxygenase inhibition appears to be specific for the term and particularly the premature newborn [84,85]. The mechanism of action of these drugs abolishes the vasodilatory effect of prostaglandins. When prostaglandin synthesis is inhibited, the vasoconstrictor state of the newborn kidney is unopposed. These observations are of great clinical importance because NSAIDs are prescribed during pregnancy for the management of pre-eclampsia, polyhydramnios, and premature birth. These drugs easily pass the placenta, so the fetus is readily exposed to their toxic effects. Postnatally, recurrent boluses of NSAIDs are administered to promote the pharmacological closure of a haemodynamic significant patent ductus arteriosus. Indomethacin has traditionally been the drug of choice, but recently, ibuprofen has been advocated for its decreased renal toxicity [86]. However, Chamaa et al. [84] showed in the newborn rabbit (a well established model for evaluating developmental changes in neonatal renal function) that ibuprofen is not less nephrotoxic than indomethacin. Because all specific and nonspecific cyclooxygenase inhibitors can cause ARF in the newborn, caution is advised when administering any of these compounds to the very young. Specific and nonspecific cyclooxygenase inhibition *in utero* may lead to renal morphological changes and even end stage renal disease at birth [87,88].

The most important factors that influence the GFR in the neonatal period are gestational age, prenatal drug exposure (i.e., betamethasone, angiotensin-1 converting enzyme inhibitors, angiotensin-2 receptor inhibitors, and nonsteroidal anti-inflammatory drugs), postnatal age, the existence of a patent ductus arteriosus, and postnatal exposure to indomethacin or ibuprofen, dopamine, and furosemide.

REFERENCES

1. Silverman, W. A., Anderson, D. H., Blanc, W. A., et al., *Pediatrics, 18,* 614 (1956).
2. Sutherland, J. M., *Am. J. Dis. Child, 97,* 761 (1959).
3. Lorch, V., Murphy, D., Hoersten, L. R., Harris, E., Fitzgerald, J., Sinha, S. N., *Pediatrics, 75,* 598 (1985).
4. Lovejoy, F. H., *Am. J. Dis. Child, 136,* 974 (1982).
5. Christensen, M. L., Helms, R. A., Chesney, R.W., *Pediatrics, 104, (Suppl),* 593 (1999).
6. Lobstein, R., Koren, G., *Ther. Drug Monit., 24,* 15 (2002).
7. Garland, M., *Obstet. Gynecol. Clin. North Am., 25,* 21 (1998).
8. Berlin, Jr., C. M., *Adv. Pediatr., 44,* 545 (1997).
9. Kearns, G. L., *J. Allergy Clin. Immunol., 106 (3 Suppl),*S128 (2000).
10. Martinez, M. N., Amiden, G. L., *J. Clin. Pharmacol., 42,* 620 (2002).
11. Zhou, H., *J. Clin. Pharmacol., 43,* 211 (2003).
12. Hess, A. F., *Am. J. Dis. Child, 6,* 264 (1913).
13. Grand, R. J., Watkins, J. B., Torti, F. M., *Gastroenterology, 70,* 790 (1976).
14. Kelly, E. J., Newell, S. J., Brownlee, K.G., Primrose, J.N., Dear, P. R. F., *Early Human Dev., 35,* 215 (1993).
15. Grahnquist, L., Ruuska, T., Finkel, Y., *J. Pediatr. Gastroenterol. Nutr., 305,* 533 (2000).
16. Agunod, M., Yomahuchi, N., Lopez, R., Lubby, A. L., Glass, G. B. J., *Am. J. Dis., 14,* 400 (1969).
17. Kearns, G. L., Bradley, J. S., Jacobs, R. F., et al., *Pediatr. Inf. Dis. J., 19,* 833 (2000).
18. Cavell, B. , *Acta Paediatr. Scan., 70,* 639 (1981).
19. Dumont, R. C., Rudolph, C. D., *Ped. Clin. North Am., 23,* 655 (1994).

20. Carlos, M.A., Babyn, P.S., Macron, M. A., Moore, A. M., *J. Pediatr., 130,* 931 (1997).
21. Yanowitz, T. D., Yao, A. C., Pettigrew, K. D., et al., *J. Appl. Physiol., 87,* 370 (1999).
22. Martinussen, M., Brubakk, A. M., Vik, T., Yao, A. C., *Pediatr. Res., 39,* 275 (1996).
23. Martinussen, M., Brubakk, A. M., Linker, D. T., Vik, T., Yao, A. C., *Pediatr. Res., 36,* 334 (1994).
24. Hall, S. D., Thummel, K. E., Watkins, P.B., et al., *Drug Metab. Dispos., 27,* 161 (1999).
25. Kim, R. B., *Drug Metab. Rev., 34,* 47 (2002).
26. Johnson, W. W., *Methods Find Exp. Clin. Pharmacol., 24,* 501 (2002).
27. Anthony, V., Skach, W. R., *Curr. Protein Peptide Sci., 3,* 485 (2002).
28. Watchko, J. F., Daood, M. J., Mahmood, B., Vats, K., Hart, C., Ahdab-Barmada, M., *J. Perinatol., 21,* S43 (2001).
29. Schuetz, E. G., Furuya, K. N., Schuetz, J. D., *J. Pharmacol. Exp. Ther., 275,* 1011 (1995).
30. Radde, I. C., "Mechanisms of drug absorption and their development," in *Textbook of Pediatric Clinical Pharmacology,* MacLeod, S. M., Radde, I.C. (eds.), PSG Publishing Company, pp. 17-43 (1985).
31. Carry, M. R., Ringel, S. P., Starcevich, J. M., *Muscle Nerve, 9,* 445 (1986).
32. Kafetzis, D. A., Sinaniotis, C. A., Papadatos, C. J., Kosmidis, J., *Acta Paediatr. Scand., 68,* 419 (1979).
33. Sheng, K. T., Huang, N. N., Promadhattaveddi, V., *Antimicrob. Agents Chemother.,* 200 (1964).
34. Choonara, I., *Arch. Dis. Child, 71,* F73 (1994).
35. Rutter, N., *Clin. Perinatol., 14,* 911 (1987).
36. Okah, F. A., Wickett, R. R., Pickens, W. L., Hoath, S. B., *Pediatrics, 96,* 688 (1995).
37. Fluhr, J. W., Pfisterer, S., Gloor, M., *Pediatr. Dermatol., 17,* 436 (2000).
38. Radde, I. C., McKercher, H. G., "Transport through membranes and development of membrane transport," in *Textbook of Pediatric Clinical Pharmacology,* MacLeod, S.M., Radde, I. C. (eds.), PSG Publishing Company, pp. 1-16 (1985).
39. Marks, J., Rawlins, M. D., "Skin diseases," in *Avery's Drug Treatment: Principles and Practice of Clinical Pharmacology and Therapeutics* (3rd ed.), Speight, T. M. (ed.), ADIS Press, pp. 439-479 (1987).
40. Tyrala, E. E., Hillman, L. S., Hillman, R. E., Dodson, W. E., *J. Pediatr., 91,* 481 (1977).
41. Armstrong, R. W., Eichner, E. R., Klein, D. E., et al., *J. Pediatr., 75,* 317 (1969).
42. Feinblatt, B. I., Aceto, T., Beckhorn, G., Bruck, E., *Am. J. Dis. Child, 112,* 218 (1966).
43. Fisch, R. O., Berglund, E. B., Bridge, A. G., et al., *JAMA, 185,*760 (1963).
44. Di Lorenzo, C., Flores, A. F., Hyman, P. E., *J. Pediatr., 127,* 593 (1995).
45. Stratchunsky, L. S., Nazarov, A. D., Firsov, A. A., Petrachenkova, N.A., *Eur. J. Drug Metab. Pharmacokinet., 3,* 321 (1991).
46. Van Lingen, R. A., Deinum, J. T., Quak, J. M., et al., *Arch. Dis. Child Fetal Neonatal Ed., 80,* F59 (1999).
47. Coulthard, K. P., Nielson, H. W., Schroder, M., et al., *J. Paediatr. Child Health, 34,* 425 (1998).
48. Radde, I. C., "Drugs and protein binding," in *Textbook of Pediatric Clinical Pharmacology,* MacLeod, S. M., Radde, I.C. (eds.), PSG Publishing Company, pp. 32-43 (1985).
49. Bjorkman, S., *J. Pharm. Pharmacol., 54,* 1237 (2002).
50. Muller, W. E., Wollert, V., *Pharmacology, 19,* 59 (1979).
51. Piakfsky, K. M., *Clin. Pharmacokinet., 5,* 246 (1980).
52. Benet, L. Z., Hoener, B. A., *Clin. Pharmacol. Ther., 71,* 115 (2002).
53. Morselli, P. L., Franco-Morselli, R., Bossi, L., *Clin. Pharmacokinet., 5,* 485 (1980).
54. Gitlin, D., Boseman, M., *J. Clin. Invest., 45,* 1826 (1966).
55. Hyvarinen, M., Zeltzer, P., Oh, W., Stieham , E. R., *J. Pediatr., 82,* 430 (1973).
56. Piafsky, K. M., Mpamugo, L., *Clin. Pharmacol. Ther., 29,* 272 (1981).

57. Thiessen, H., Jacobsen, J., Brodersen, R., *Acta Paediatr. Scand., 61*, 285 (1972).
58. McDonagh, A. F., Lightner, D. A., *Pediatrics, 75*, 443 (1985).
59. Hansen, T. W., *Clin. Perinatol., 29*, 765 (2002).
60. Ebbesen, F., Nyboe, J., *Acta Paediatr. Scand., 72*, 665 (1983).
61. Kapitulnik, J., Horner-Mboshan, R., Blondheim, S. H., Kauffman, N. A., Russell, A., *J. Pediatr., 86*, 442 (1975).
62. Brodersen. R., *J. Pediatr., 96*, 349 (1980).
63. Friis-Hansen, B., *Acta Paediatr. Scan., 305*, (suppl): 7 (1983).
64. Butte, N., Heinz, C., Hopkinson, J., Wong, W., Shypailo, R., Ellis, K., *J. Pediatr. Gastroenterol. Nutr., 29*, 184 (1999).
65. Butte, N. F., Hopkinson, J. M., Wong, W. W., Smith, E. O., Ellis, K. J., *Pediatr. Res., 47*, 578 (2000).
66. Fink, C. W., Cheeck, D.B., *Pediatrics, 26*, 397 (1960).
67. Litterst, C. L., Minnaugh, E. G., Reagan, R. L., *et al.*, *Drug Metab. Disp., 3*, 259 (1975).
68. Alcorn, J., McNamara, J., *Clin. Pharmacokinet. Part I, 41*, 959 (2002), *Part II, 41*, 1077 (2002).
69. Kearns, G. L., Abdel-Rahman, S. M., Alander, S.W., Blowey, D. L., Leeder, J. S., Kauffman, R. E., *New Engl. J. Med., 349*, 1157 (2003).
70. Kinirons, M. T., O'Shea, D., Kim, R. B., et al., *Clin. Pharmacol. Ther., 66*, 224 (1999).
71. Payne, K., Mattheyse, F. J., Liedenberg, D., Dawes, T., *Eur. J. Clin. Pharmacol., 37*, 267 (1989).
72. Loughnan, P. M., Greenwald, A., Purton, W. W., et al., *Arch. Dis. Child, 52*, 302 (1977).
73. Bourgeois, B. F., Dodson, W. E., *Neurology, 33*, 173 (1983).
74. Barrett, D. A., Barker, D. P., Rutter, N., Pawula, M., Shaw, P. N., *Br. J. Clin. Pharmacol., 41*, 531 (1996).
75. Scott, C. S., Riggs, K. W., Ling, E. W., et al., *J. Pediatr., 135*, 423 (1999).
76. Besunder, J. B., Reed, M. D., Blumer, J. L., *Clin. Pharmacokinet., 4*, 189 (1988).
77. Van den Anker, J. N., *Acta Paediatr., 85*, 1393 (1996).
78. Bueva, A., Guignard, J. P., *Pediatr. Res., 36*, 572 (1994).
79. Coulthard, M. G., *Early Hum, Dev., 11*, 281 (1985).
80. Arant, Jr., B. S., *J. Pediatr., 92*, 705 (1978).
81. Drukker, A., Guignard, J. P., *Curr. Opin. Pediatr., 14*, 175 (2002).
82. Burckhardt, B. C., Burckhardt, G., *Rev. Physiol. Biochem. Pharmacol., 146*, 95 (2003).
83. Toth-Heyn, P., Drukker, A., Guignard, J. P., *Pediatr. Nephrol., 14*, 227 (2000).
84. Chamaa, N. S., Mosig, D., Drukker, A., et al., *Pediatr. Res., 48*, 600 (2000).
85. Drukker, A., Mosig, D., Guignard, J. P., *Pediatr. Nephrol., 16*, 713 (2001).
86. Van Overmeire, B., Smets, K., Lecoutere, D., et al., *N. Engl. J. Med., 343*, 674 (2000).
87. Kaplan, B. S., Restaino, I., Raval, D. S., et al., *Pediatr. Nephrol., 8*, 700 (1994).
88. Peruzzi, L., Gianoglio, B., Porcellini, M. G., et al., *Lancet, 345*, 1615 (1999).

5.2 Neonatal RDS–prenatal corticosteroids

Henry L. Halliday

Department of Child Health, Queen's University Belfast, Grosvenor Road, Belfast BT12 6BJ, Northern Ireland

Regional Neonatal Unit, Royal Maternity Hospital, Belfast, Grosvenor Road, Belfast BT12 6BB, Northern Ireland

5.2.1 INTRODUCTION

Respiratory distress syndrome

Respiratory distress syndrome (RDS) is an acute illness, usually of preterm infants, which develops within 4 hours of birth [1]. It presents with rapid respirations, indrawing of the chest, expiratory grunting and cyanosis. In very preterm infants, RDS may present with apnoea at birth, in which case the characteristic clinical features described above will be absent. On chest radiograph, there is reticulogranular mottling and air bronchograms but, in severe cases, the lungs may be completely white, due to fluid retention in the airspaces and widespread atelectasis [1]. RDS is the second commonest cause of neonatal death after congenital malformations.

The underlying cause of RDS is primary surfactant deficiency which leads to alveolar collapse at end-expiration, reduced lung compliance, decreased functional residual capacity, increased work of breathing, hypoxaemia and respiratory acidosis [1]. If not reversed, these changes lead to progressive respiratory and cardiac failure before death. Well recognised complications of RDS include pulmonary air leaks (pneumothorax and pulmonary interstitial emphysema), persistent ductus arteriosus (PDA), intraventricular haemorrhage (IVH) and chronic lung disease (CLD).

Surfactant synthesis and secretion

It is not until about 24 weeks' gestation that surfactant appears in the human lung, coinciding with the onset of the saccular phase of development [2]. However, explants of lungs from 15 to 20 week fetuses in culture rapidly develop mature type II pneumocytes that synthesise surfactant proteins. Extremely early maturation can be further stimulated by glucocorticoids. Although many hormones and growth factors can modulate lung maturation *in vitro,* only glucocorticoids consistently show these effects *in vivo* [2]. However, changes in the surfactant system in fetal sheep require more than 4 days or multiple weekly courses of glucocorticoids [3]. It is now believed that the effects of glucocorticoids to induce maturation are primarily to mature lung structure rather than stimulation of surfactant synthesis or secretion [2].

Effects of glucocorticoids on lung maturation

Observations from transgenic mice show that glucocorticoids function primarily as modulators of development [2]. However, prior to these studies it was believed that glucocorticoids exerted their effect by maturation of the surfactant system [4]. Indeed, in 1968 Liggins, when investigating the physiology of parturition in sheep, found that fetal lung maturation was accelerated after an infusion of glucocorticoids [5], and he proposed that this was due to induction of surfactant synthesis. This observation led to the first randomised controlled trial which demonstrated an approximate 50% reduction in the incidence of RDS in babies of mothers given betamethasone [6].

5.2.2 GLUCOCORTICOIDS USED IN CLINICAL TRIALS

Three glucocorticoids capable of crossing the placental barrier to the fetus have been used in randomised trials: betamethasone, dexamethasone and hydrocortiosone. Some trials used methylprednisolone, but this steroid does not cross the placenta. The evidence from the small trials in which it was used do not support the use of hydrocortisone to reduce the risks of RDS [7].

Betamethasone and dexamethasone differ only in the configuration of the methyl group on position 16 of the sterol ring (Figure 1).

Both drugs have a plasma half-life of 3-5 hours and a long biological half-life of 36-54 hours and cross the placenta in biologically active forms [8]. However, the small difference between the 2 molecules (methyl group on position 16) leads to considerable differences in biological effects. Both drugs modulate the rate of differentiation of numerous fetal organs including the lung, heart, liver, kidney, gut and skin [9]. There is evidence from animal studies that fetal exposure to excess glucocorticoids at critical stages of development can have life long effects [10,11]. However, if betamethasone and dexamethasone are used in recommended doses, the fetus should receive only physiological stress levels

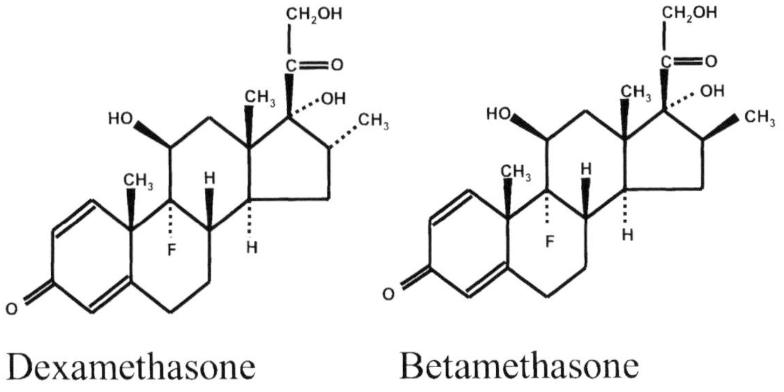

Dexamethasone   Betamethasone

Figure 1. The chemical structures of dexamethasone and betamethasone. The methyl group on the 16 carbon atom in the sterol rings is in the alpha configuration in dexamethasone and in the beta configuration in betamethasone

of glucocorticoid and not pharmacological levels [12]. Recently, however, there have been concerns about adverse effects of repeated courses of prenatal steroids [13] and this will be discussed later.

5.2.3 RANDOMISED CLINICAL TRIALS

Single course of prenatal glucocorticoids

Since the first randomised trial of Liggins and Howie [6] over 30 years ago, many studies have assessed the effects of prenatal glucocorticoids in improving the outcome of preterm birth [7]. Eighteen trials are included in Crowley's systematic review [7], which includes data from over 3700 babies. Prenatal administration of 24 mg of betamethasone, of 24 mg of dexamethasone, or 2 g of hydrocortisone to women expected to be delivered preterm reduces the risk of neonatal mortality, respiratory distress syndrome and intraventricular haemorrhage in their babies (Table 1).

Sub-group analyses show that reduction in the risk of RDS occurs in all preterm babies except those less than 28 weeks' gestation [7] (Table 2) but this may relate to the very small numbers available for analysis at this gestation. There is a trend towards a reduction in RDS in babies born less than 24 hours after treatment, but a significant effect is not seen until 48 hours have elapsed [7] (Table 2). The treatment effect in babies born after 7 days is of borderline significance. Betamethasone and dexamethasone are both associated with a significant reduction in incidence of RDS (Table 2). The evidence from the small trials in which

Table 1

Effects of prenatal glucocorticoid treatment on various outcomes

Outcome	No. trials	No. babies	OR	95% CI
RDS	18	3735	0.53	0.44-0.63
neonatal death	14	3517	0.60	0.48-0.75
IVH	5	596	0.48	0.32-0.72
NEC	4	1154	0.59	0.32-1.09
CLD	3	411	1.57	0.87-2.84
surfactant use	1	121	0.41	0.18-0.89
stillbirth	12	3306	0.83	0.57-1.22
fetal/neonatal infection	15	2675	0.82	0.57-1.19
neurological abnormality	3	778	0.62	0.36-1.08

OR = odds ratio; 95% CI = 95% confidence interval; RDS = respiratory distress syndrome; IVH = intraventricular haemorrhage; NEC = necrotising enterocolitis; CLD = chronic lung disease

hydrocortisone was used does not support its use in preterm birth. The response to prenatal glucocorticoid therapy is not affected by fetal gender. In the small number of babies from twin and triplet pregnancies treated, it is not possible to show a significant benefit from prenatal glucocorticoid treatment (Table 2).

There is no convincing evidence of any adverse effects of prenatal glucocorticoid treatment in any of these trials [7]. The stillbirth rate overall is not affected (Table 1) but, in a subgroup analysis of the first randomised trial [6], there was an increase in fetal death in hypertensive women. No fetal deaths were observed in this subgroup in the other 3 trials for which data are available [7]. However, meta-analysis shows an increased risk of stillbirth in hypertensive pregnancies [7] (Table 3).

Prenatal glucocorticoid treatment does not increase the risk of fetal or neonatal infection overall (Table 1) or in cases of preterm prelabour rupture of the membranes [7,14] (Table 3). Maternal infection is not increased overall, but in women whose membranes were ruptured for more than 24 hours before delivery, infection is increased [7] (Table 3). Follow-up data on growth and neuro-development have been reported up to 12 years in 3 studies [15-17], and there is no evidence of long-term adverse effects of postnatal glucocorticoid treatment (Table 1).

Table 2

Effects of prenatal glucocorticoid treatment on RDS in various sub-groups

Subgroup	No. Trials	No. Babies	OR	95% CI
< 34 wk gestation	7	1048	0.36	0.27-0.48
> 34 wk gestation	8	744	0.65	0.33-1.29
< 30 wk gestation	8	349	0.48	0.30-0.77
< 28 wk gestation	4	48	0.64	0.16-2.50
< 24 h after 1st dose	6	349	0.70	0.43-1.16
24 h to 7 d after 1st dose	4	728	0.38	0.25-0.57
> 7 d after 1st dose	3	265	0.41	0.18-0.98
dexamethasone	5	1400	0.56	0.43-0.73
betamethasone	11	2176	0.49	0.39-0.63
hydrocortisone	2	172	0.69	0.32-1.47
males	3	627	0.43	0.29-0.64
females	3	555	0.36	0.23-0.57
twins/triplets	2	140	0.72	0.31-1.68

Table 3

Potential adverse effects of prenatal glucocorticoid treatment

Potential adverse effect	No. Trials	No. Babies	OR	95% CI
stillbirth – maternal hypertension	4	239	3.75	1.24-11.30
maternal/neonatal infection – PROM > 24 h	2	163	2.31	0.77 - 6.99
maternal infection - overall	11	2109	1.31	0.99 - 1.73
maternal infection – PROM > 24 h	1	42	6.04	1.47-24.70
maternal infection – PROM at entry	3	320	1.26	0.68 - 2.28

PROM = prolonged rupture of the membranes

A recent observational study, comparing prenatal betamethasone and dexa-methasone with no treatment in a cohort of 883 live born infants, found different rates of periventricular leucomalacia (PVL) in the 3 groups [18]. In infants whose mothers had been treated with dexamethasone, the rate of PVL was 11.0% compared to 4.4% for betamethasone and 8.4% in those not treated with gluco-corticoids. The authors speculated that the presence of the preservative, sulphite, in the dexamethasone preparation had a deleterious effect. These observations need to be confirmed in a randomised controlled trial.

Repeated courses of prenatal glucocorticoids

Repeated courses of prenatal corticosteroids became common practice after the Crowley meta-analysis revealed that after 7 days had elapsed following a single course of treatment the beneficial effects appeared to have disappeared [7]. This practice led to concern amongst neonatologists that any benefit might be outwei-ghed by adverse effects on fetal growth and development [19]. Recently, 3 rando-mised trials comparing a single course with repeated weekly courses of prenatal corticosteroids have been reported [20-22]. A systematic review of these 3 trials has been published in the Cochrane Library [23]. Repeated courses of prenatal steroids do not reduce the risk of RDS (RR 0.96; 95% CI 0.72-1.26), but there were fewer babies with severe neonatal lung disease in the repeat course group (RR 0.64; 95% CI 0.44-0.93; NNT 12; 7-50). Also, infants in the repeat course group were less likely to need surfactant treatment (RR 0.65; 95% CI 0.46-0.92; NNT 12, 7-50). However, the authors of the systematic review concluded that, although repeat doses may reduce severity of neonatal lung disease, there was insufficient evidence on benefits and risks to recommend repeated courses of prenatal steroids until further trials had been performed [23].

5.2.4 GUIDELINES FOR PRENATAL GLUCOCORTICOIDS

There are very few contraindications to prenatal glucocorticoid treatment when preterm birth is likely. Diabetic control may be more difficult, hyperten-sion may need additional treatment and infectious indicators in the mother and baby may be obscured, but these are not contraindications to treatment [1]. In 1994, a Consensus Conference organised by the National Institutes for Health in the USA suggested the following recommendations for use of glucocorticoids to prevent or reduce the severity of RDS [24]:

- All fetuses between 24 and 34 weeks' gestation at risk of preterm birth should be considered as candidates for prenatal treatment with glucocorticoids.
- The decision to use prenatal glucocorticoids should not be altered by fetal race or gender or by the availability of surfactant replacement therapy.
- Women eligible for tocolytic therapy should also be eligible for treatment with glucocorticoids.
- Treatment consists of 2 doses of 12 mg betamethasone given intramuscularly

24 hours apart or 4 doses of 6 mg dexamethasone given intramuscularly 12 hours apart. Optimal benefits begin 24 hours after initiation of therapy and last for at least 7 days.

- Treatment with glucocorticoids for less than 24 hours is associated with significant reductions in neonatal mortality, RDS and IVH, thus prenatal glucocorticoids should be given, unless immediate delivery is anticipated.
- In preterm premature rupture of the membranes at less than 30-32 weeks' gestation, in the absence of clinical chorioamnionitis, prenatal glucocorticoid use is recommended because of the high risk of IVH at these early gestations.
- In complicated pregnancies where delivery prior to 34 weeks' gestation is likely, prenatal glucocorticoid use is recommended, unless there is evidence that the drug will have an adverse effect on the mother or delivery is imminent.

A year later in 1996, the Royal College of Obstetricians and Gynaecologists in the UK published similar guidelines for the use of prenatal glucocorticoids, but these extended the upper gestational age limit to 36 weeks [25].

5.2.5 COMBINATION THERAPIES

Prenatal glucocorticoids and postnatal surfactant

There is evidence that the combination of prenatal glucocorticoids and postnatal surfactant leads to improved outcomes compared to either treatment alone [26,27]. Indeed, the contribution of prenatal glucocorticoid therapy and prophylactic surfactant is the most cost effective intervention for RDS in infants of less than 30 weeks' gestation [28].

Prenatal glucocorticoids and prenatal thyrotropin releasing hormone

There have been 11 randomised trials comparing treatment with prenatal thyrotropin releasing hormone (TRH) in addition to glucocorticoids with glucocorticoids alone in women at risk of very preterm birth for prevention of RDS [29]. The conclusion of this systematic review was that combination therapy does not reduce the risk of RDS, chronic oxygen dependence or improve fetal, neonatal or childhood outcomes [29]. Indeed, the data show prenatal TRH has adverse effects for women and their infants. There was an increased risk of neonatal ventilation (RR 1.14; 95% CI 1.03-1.29) and low 5 minute Apgar score (1.80; 1.14-1.92) and poorer neuro-developmental outcomes at 12 months in 1 trial [30]. Administration of TRH to women at risk of preterm delivery cannot be recommended [1].

Prenatal glucocorticoids and tocolytic agents

Tocolytic agents have been used to decrease uterine activity when preterm labour is likely. Various drugs have been used, including beta-mimetics [31], calcium channel blockers [32] and magnesium sulphate [33]. Beta-mimetics (such as ritrodrine) are effective in prolonging pregnancy for up to 48 hours but not longer [31]; this intervention may be beneficial to allow time for postnatal glucocorticoids to have effect [7] or to allow transfer to a tertiary perinatal centre for delivery. Calcium channel blockers appear to be more effective than beta-mimetics and they are effective in combination with prenatal glucocorticoids in reducing RDS (RR 0.63; 95% CI 0.46-0.88) [32]. Magnesium sulphate, however is ineffective and also has the adverse effect of increased neonatal and paediatric mortality [33].

Uterine tocolysis with beta-mimetics or calcium channel blockers in combination with prenatal glucocorticoid therapy is recommended for suspected preterm labour, unless there are contraindications [31].

5.2.6 CONCLUSIONS

Prenatal glucocorticoid therapy is indicated when preterm birth is likely before 36 weeks' gestation. Two doses of 12 mg betamethasone or 4 doses of 6 mg dexamethasone, both given 12 hourly, reduce the risks of RDS, neonatal mortality and intraventricular haemorrhage [7]. Betamethasone may have a better adverse event profile [18], but this has not been proved in a randomised comparative trial. Prenatal glucocorticoid therapy in combination with postnatal surfactant treatment or prenatal tocolytic therapies are evidence-based but use with TRH or magnesium sulphate is contra-indicated.

REFERENCES

1. Halliday, H. L., "Respiratory distress syndrome," in *Neonatal respiratory disorders* (2nd ed.), Arnold, p. 247 (2003).
2. Jobe, A. H., Ikegami, M., *Ann. Rev. Physiol.,* 62, 825 (2000).
3. Ikegami, M., Jobe, A. H., Newnham, J., Polk, D. H., Willet, K. E., Sly, P. *Am. J. Respir. Crit. Care Med.,* 156, 178 (1997).
4. Ballard, P. L., *Endocr. Rev., 10,* 165 (1989).
5. Liggins, G. C., *J. Endocrinol., 42,* 323 (1968).
6. Liggins, G. C., Howie, R. N., *Pediatrics, 50,* 515 (1972).
7. Crowley, P., *The Cochrane Library,* CD000065 (2003).
8. Ozdemir, H., Guvenal, T., Cetin, M., Kaya, T., Cetin, A., *Pediatr. Res, 53,* 98 (2003).
9. Ballard, P. L., Ballard, R. A., *Am. J. Obstet. Gynecol., 173,* 254 (1995)
10. Seckl, J. R., *Clin. Perinatol., 25,* 939 (1998).
11. Edwards, H. E., Burnham, W. M., *Pediatr. Res., 50,* 433 (2001).
12. Ballard, P. L., Liggins, G. C., *J. Pediatr., 101,* 468 (1982).

13. Aghajafari, F., Murphy, K., Willan, A., et al., *Am. J. Obstet., Gynecol., 185,* 1073 (2001).
14. Harding, J..E., *Am. J. Obstet. Gynecol., 194,* 131 (2001).
15. Collaborative Group on Antenatal Steroid Therapy, *J. Pediatr., 104,* 259 (1984).
16. MacArthur, B. A., Howie, R. N., Dezoete, J. A., Elkins, J., *Pediatrics, 70,* 99 (1982).
17. Smolders-de Haas, H., Neuvel, J., Schmand, B., Treffers, P. E., Koppe, J. G., Hoeks, J., *Pediatrics, 86,* 65 (1990).
18. Baud, O., Foix-LíHelias, L., Kaminski, M., *N. Engl. J. Med., 341,* 1190 (1999).
19. Halliday, H. L., *Prenat. Neonat. Med., 5,* 201 (2000).
20. Guinn, D. A., Atkinson, M. W., Sullivan, J., et al., *JAMA., 286,* 1581 (2001).
21. Aghajafari, F., Murphy, K., Ohlsson, A., Amankwah, K., Matthews, S., Hannah, M. E., *J. Obstet. Gynaecol. Can., 24,* 321 (2002).
22. McEvoy, C., Bowling, S., Williamson, K., *Pediatrics, 110,* 280 (2002).
23. Crowther, C. A., *The Cochrane Library,* CD 003035 (2003).
24. NIH Consensus Developmental Panel, *JAMA, 273,* 413 (1995).
25. Royal College of Obstetricians and Gynaecologists, www.rcog.org.uk/guidelines/corticosteroids
26. Farrell, E. E., Silver, R. K., Kimberlin, L. V., Wolf, E. S., Dusik, J. M., *Am. J. Obstet. Gynecol., 161,* 628 (1989).
27. Jobe, A. H., Mitchell, B. R., Gunkel, J. H., *Am. J. Obstet. Gynecol., 168,* 508 (1993).
28. Egberts, J., *Pharmaco-Economics, 8,* 324 (1995).
29. Crowther, C. A., Alfirevic, Z., Haslam, R. R., *The Cochrane Library,* CD000019 (2004).
30. Crowther, C. A., Hiller, J. E., Haslam, R. R., et al., *Pediatrics, 99,* 311 (1997).
31. King, J. F., Grant, A. M., Keirse, M. J., Chalmers, I., *Brit. J. Obstet. Gynaecol., 95,* 211 (1988).
32. King, J. F., Flenady, V. J., Papatsonis, D. N. M., Dekker, G. A., Carbonne, B., *The Cochrane Library,* CD002255 (2003).
33. Crowther, C. A., Hiller, J. E., Doyle, L. W., *The Cochrane Library,* CD001060 (2002).

5.3 Neonatal RDS–surfactant

Henry L. Halliday

Department of Child Health, Queen's University Belfast, Grosvenor Road, Belfast BT12 6BJ, Northern Ireland;

Regional Neonatal Unit, Royal Maternity Hospital, Belfast, Grosvenor Road, Belfast BT12 6BB, Northern Ireland

5.3.1 INTRODUCTION

Pulmonary surfactant is a complex mixture of phospholipids, neutral lipids and specific proteins. It lowers surface tension at the air-liquid interface of the lung to prevent alveolar collapse at end-expiration [1]. The major cause of neonatal respiratory distress syndrome (RDS) is a primary deficiency of surfactant [2], and surfactant replacement therapy has had a major impact in improving the outcome of this disorder in preterm infants [3]. Surfactants were the first drugs designed primarily for use in the newborn, and those licensed for treatment or prevention of RDS fall into 2 broad categories: synthetic and natural. Synthetic surfactants are composed mainly of phospholipids (usually dipalmitoylphosphatidylcholine, DPPC, also known as colfosceril palmitate) but do not contain surfactant proteins. Natural surfactants are derived from animal lungs, and they contain both phospholipids and surfactant proteins B and C [4].

5.3.2 CHEMISTRY AND METABOLISM OF SURFACTANT

Phosphatidylcholine (PC) is the major component of pulmonary surfactant, contributing about 60% of total phospholipids [4]. DPPC is the primary surface tension lowering phospholipid (Figure 1), but it is ineffective on its own because

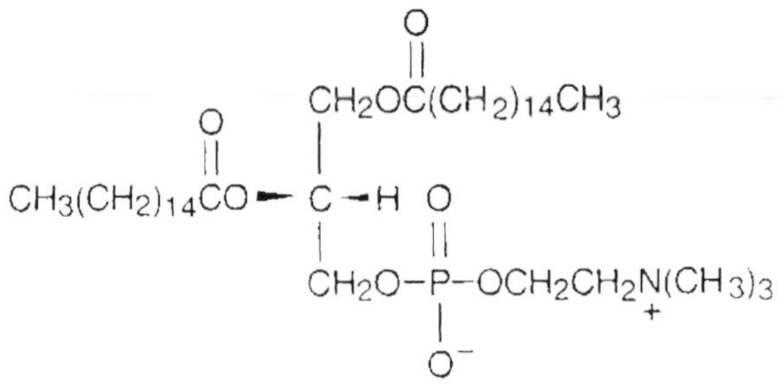

$$
\begin{array}{l}
\qquad\qquad\qquad\quad \overset{\displaystyle O}{\overset{\displaystyle \|}{CH_2OC(CH_2)_{14}CH_3}} \\[2mm]
\qquad\quad \overset{\displaystyle O}{\overset{\displaystyle \|}{}} \quad | \\
CH_3(CH_2)_{14}CO \leftarrow C \rightarrow H \quad O \\
\qquad\qquad\qquad | \qquad\quad \| \\
\qquad\qquad\quad CH_2O-P-OCH_2CH_2N(CH_3)_3 \\
\qquad\qquad\qquad | \qquad\qquad\qquad\quad + \\
\qquad\qquad\qquad O^-
\end{array}
$$

Figure 1. Chemical structure of dipalmitoylphosphatidylcholine, the primary surface-active agent in natural and synthetic surfactant preparations

it spreads and absorbs slowly under physiological conditions. Surfactant proteins and phosphatidylglycerol (PG) are needed for adequate spreading and adsorption *in vitro* and *in vivo*. Four surfactant-associated proteins have been identified SP-A, SP-B, SP-C and SP-D [3]. They are synthesised and secreted in alveolar type II cells and clara cells. SP-A and SP-D are large hydrophilic proteins that play a role in innate immunity in the lung [3]. When natural surfactant preparations are prepared from animal lungs, using chloroform-methanol extraction, these proteins are left behind. The 2 smaller hydrophobic proteins, SP-B and SP-C, which improve spreading and adsorption of phospholipids, are present in natural surfactant preparations [5].

Endogenous surfactant lipids in healthy neonates have a pool size of about 100 mg/kg whereas, in preterm infants with RDS, this is usually 5-10 mg/kg [6]. Surfactant is secreted from type II pneumocytes in the form of lamellar bodies that unravel to form tubular myelin before surface adsorption occurs, allowing a monolayer of lipids and protein to form [6,7]. *De novo* synthesis and secretion of surfactant phospholipids are slow processes. PC is recycled with a turnover time of about 10 h and an efficiency of more than 90% [6]. The biological half life of labelled PC is about 6 days in newborn lambs. Surfactant proteins have metabolic pathways that are quite distinct from those for the lipids [6]. Surfactant treatment does not adversely affect endogenous synthetic and secretory pathways by feedback inhibition and this is true for both PC [6] and SP-A [8]. Surfactant treatment of preterm infants with RDS is beneficial, in part because of the favourable metabolism in the immature lung that can recycle surfactant without catabolising the exogenous surfactant [6].

5.3.3　BIOPHYSICAL AND PHYSIOLOGICAL EFFECTS OF SURFACTANT

In vitro assessment of surface tension is performed using either the Wilhelmy balance or the pulsating bubble surfactometer [9]. Natural surfactant preparations containing SP-B and SP-C reduce minimum surface tension after compression to < 1 mN/m compared to about 30 mN/m for synthetic surfactants [9,10].

The physiological effects of surfactants can be assessed in animal models including the immature rabbit [9], lamb [11] or baboon [12]. Pulmonary compliance increases significantly more in preterm rabbits treated with natural compared to synthetic surfactants (76% *vs*. 29%) [9]. Histological studies of the lungs also show increased alveolar expansion after natural surfactant treatment [9].

5.3.4　SURFACTANT PREPARATIONS

The first surfactants used in clinical trials in the 1960s contained only DPPC given by aerosol and they were ineffective [13,14]. In 1980 a bovine preparation (Surfactant-TA) was shown to cause a striking improvement in gas exchange in preterm babies with RDS (15). This success led to development of a number of a commercial natural surfactant preparations derived from bovine and porcine lungs (Table 1). Although at least 3 synthetic surfactants were used in clinical trials, only 2 were later licensed for treatment of neonatal RDS (Table 2).

5.3.5　CLINICAL TRIALS

Following some pilot trials in the 1980s, many randomised controlled trials were conducted involving all the surfactants listed in Tables 1 and 2. Details of these trials have been summarised elsewhere [5]. There is convincing evidence that severity of RDS can be reduced by either prophylactic surfactant treatment in the delivery room or surfactant replacement of established disease within 12 hours of birth. About 40 randomised controlled trials have been reported and meta-analyses of these trials are published in the Cochrane Library [16-18]. Both natural and synthetic surfactants, given either prophylactically or for treatment of RDS, reduce the risk of pneumothorax and neonatal mortality (Table 3). In general the numbers needed to treat (NNT) for natural surfactants are lower for both pneumothorax and neonatal death than those of synthetic surfactants. Meta-analyses of comparative trials of both types of surfactants confirm that natural surfactants lead to better survival and fewer pneumothoraces than synthetic surfactants [19,20]. Subsequently both these synthetic surfactants have been discontinued by the companies that produce them.

Table 1

Natural surfactant preparations in clinical use to treat RDS

Generic name	Common name	Company	Source	Composition		Dose (mg/kg)	Volume (mL/kg)	Number of doses
				Phospholipids (%)	Proteins (1-2%)			
Surfacten	surfactant – TA	Tokyo Tanabe	bovine mince	84	SP-B, SP-C	120	4	3
Beractant	survanta	Abbott	bovine mince	84	SP-B, SP-C	100	4	4
Calfactant	infasurf	Forest	bovine lavage	95	SP-B, SP-C	100	3	3
Bovine Lipid Surfactant	bLes	BLES	bovine lavage	95	SP-B, SP-C	135	5	4
Bovactant	alveofact	Boehringer	bovine lavage	88	SP-B, SP-C	50	1.2	4
Poractant Alfa	curosurf	Chiesi	porcine mince	99	SP-B, SP-C	100-200	1.25-2.5	3

SP-B = surfactant protein –B; SP-C = surfactant protein-C

Table 2

Synthetic surfactant preparations used to treat RDS

Generic Name	Common Name	Company	Composition	Dose (mg/kg)	Volume (mL/kg)	Number of doses
colfosceril palmitate	Exosurf	Wellcome	DPPC 13.5 mg hexadecanol 1.5 mg tyloxapol 1.0 mg	67.5	5	3
pumactant	ALEC	Britannia	DPPC 7: PG 3	100	1.2	4

Table 3

Effects of prophylaxis or treatment with natural or synthetic surfactants on pneumothorax and neonatal mortality

Surfactant	Pneumothorax				Neonatal mortality			
	RR	95% CI	NNT	95% CI	RR	95% CI	NNT	95% CI
prophylactic synthetic	0.67	0.50-0.90	20	12-67	0.70	0.58-0.85	15	10-31
prophylactic natural	0.35	0.26-0.49	7	5-9	0.60	0.49-0.83	14	9-35
synthetic treatment	0.64	0.55-0.76	12	9-18	0.73	0.61-0.88	22	14-53
natural treatment	0.43	0.35-0.52	6	5-8	0.68	0.57-0.80	11	8-21

From The Cochrane Library, 2003

RR = relative risk   95% CI = 95% confidence interval   NNT = number needed to treat

5.3.6 TREATMENT REGIMENS

Recent studies have shown that natural preparations are superior to synthetic ones [19] (see above). New generation synthetic surfactants are being developed, but there is a paucity of clinical trials. Timing, size and frequency of dosing together with mode of administration differ according to clinical opinion and depending on which surfactant preparation is used (Table 1).

Dose

In trials, surfactants have been given in doses of 25 to 200 mg/kg. For Surfactant-TA 120 mg/kg is more effective than 60 mg/kg [21] and for Curosurf a initial dose of 200 mg/kg gives better acute physiological responses than 100 mg/kg in babies with severe RDS but there is no difference in long term outcome [22]. There is now general agreement that an initial dose of surfactant of at least 100 mg/kg should be used, but timing of treatment in relation to severity of RDS may affect the decision as regards dose [1].

Timing of first dose

For infants of < 31 weeks' gestation, there is good evidence that prophylaxis (within 10-15 minutes of birth) in the delivery room is superior to much later rescue treatment [26,27]. Prophylaxis with natural surfactant preparations leads to a reduction in neonatal mortality, severe RDS, chronic lung disease and intraventricular haemorrhage in babies < 31 weeks gestation compared to later treatment [27]. When these studies were performed, however, use of prenatal corticosteroids to mature the fetal lung and prevent RDS was infrequent. With increased use of prenatal steroids and taking into consideration over-treatment of babies not likely to develop severe RDS, a policy of universal prophylaxis for all infants < 31 weeks gestation is probably not justified. The best option may be to give prophylactic surfactant in the delivery room to all babies < 28 weeks' gestation and to those from 28-30 weeks who need intubation for resuscitation [28]. Infants > 30 weeks' gestation should be given surfactant when they have clear evidence of RDS and need more than 40% oxygen. Whilst 100 mg/kg should be sufficient for prophylaxis, a dose of 200 mg/kg may be needed to treat established severe RDS.

Total dose

The total cumulative dose of surfactant needed to treat RDS has been examined in at least 3 studies [22,29,30]. For Curosurf, a total dose of up to 300 mg/kg gives similar long-term results as up to 600 mg/kg with no differences in mortality or oxygen dependency at 28 days [27]. In two studies of Exosurf, up to 4 doses were not superior to 2 doses for treatment of RDS [29,30].

Mode of administration

Currently, all surfactants are administered intratracheally through an endotracheal tube in one, two or more boluses [1,28]. For Survanta, the infants should be positioned to allow surfactant to reach the dependent lung [31]. A sterile feeding tube is often used to deliver surfactant into the lower trachea. After instillation, the baby is either manually ventilated for about 60 seconds or reconnected to the ventilator to allow the surfactant to distribute rapidly in the lungs. Some surfactants have been administered slowly through an endotracheal tube adaptor with a side port [29], however there is evidence from animal experiments that slow instillation of natural surfactant leads to a blunted acute response or rapid relapse [32].

Curosurf may be administered in one or two boluses but it can also be given to babies who need only continuous positive airway pressure (CPAP) using a short endotracheal intubation before returning to CPAP [33]. Attempts to avoid intubation altogether by giving surfactants by nebulisation have so far been unsuccessful [34], probably because current delivery systems are inefficient.

5.3.7   ACUTE RESPONSES AND VENTILATOR SETTINGS

There are differences in acute physiological responses after administration of natural and synthetic surfactants [9,11,35]. The more rapid responses with natural surfactants are most likely due to presence of surfactant proteins B and C, but speed of instillation and total dose of phospholipids may also be important. The need to adjust the ventilator after surfactant treatment will also vary according to the type of surfactant used. Following a recent UK trial the synthetic surfactant, Pumactant, which was shown to be inferior to Curosurf, [36] has been withdrawn by the company.

When using natural surfactant, it is important to reduce inspired oxygen concentrations and ventilator settings soon after administration to avoid hyperoxaemia and overdistention of the lungs [1,28]. Functional residual capacity increases within 1 hour of surfactant treatment [37] and to prevent over-distension peak inspiratory pressure and inspiratory time should be reduced [1]. No suctioning of the airways should be performed during the first 4 hours after surfactant administration unless it is clinically indicated.

Not all babies respond optimally to surfactant treatment with some showing either a blunted response or early relapse. These infants may have secondary surfactant deficiency associated with pneumonia, asphyxia, pulmonary hypoplasia or ARDS rather than true RDS. In some infants relapse may be due to pulmonary oedema from a PDA. Excessive fluid intake and failure to reduce ventilator settings after treatment are also associated with a poorer outcome [38].

5.3.8   ADVERSE EFFECTS

There are very few adverse effects of surfactant treatment [4]. In babies treated with prophylactic synthetic surfactant there is an increased risk of PDA and pulmonary haemorrhage [16]. With later surfactant treatment unmasking of a PDA can occur because of rapid fall in pulmonary vascular resistance. If this is suspected by a decrease in diastolic blood pressure treatment with indomethacin or ibuprofen should be considered [1,28]. Fall in blood pressure is due to lowering of peripheral vascular resistance [39] associated with activation of NO synthase [40]. Transient suppression of the EEG has been found in some babies [41] but this is not due to reduced cerebral blood flow [42,43] and is not associated with any long term adverse effects [44].

The risk of sensitisation against animal proteins in natural surfactants would appear to be minimal. Circulating anti-surfactant immune complexes have been detected in surfactant treated babies but with similar frequency and levels as in untreated babies who had RDS [45]. This phenomenon probably represents transient leakage of surfactant proteins into the circulation in RDS. Treatment with bovine surfactant may reduce this protein leak and limit immune complex production [46]. Long-term follow-up studies of surfactant treated babies have not shown any difference in neurological or pulmonary outcomes between treated and control infants, despite improved survival of more immature infants in the treated groups [44,45,47].

5.3.9   NEW GENERATION SYNTHETIC SURFACTANTS

New synthetic surfactant preparations are being developed to replace the current animal-derived or natural surfactants. These include KL_4 (Surfaxin), recombinant SP-C (Venticute) and analogues of SP-B and SP-C [48, 49]. Studies in animals have been undertaken and some clinical trials have been planned or are in progress. These surfactants may be more resistant to inactivation by proteins which have leaked into the airways and thus may have an important place in treatment of ARDS where there is secondary surfactant deficiency. However, as they are likely to be easier to produce than existing surfactants, they may also have a role in treatment of RDS.

5.3.10   CONCLUSIONS

Surfactant replacement for neonatal RDS has been one of the major therapeutic developments in paediatrics and it has already had a great impact on improving morbidity, mortality and resource use in preterm infants in the developed world [3, 50].

REFERENCES

1. Speer, C. P., *Curr. Paediatr.*, *4*, 5 (1994).
2. Avery, M. E., Mead. J., *Am. J. Dis. Child.*, *97*, 517 (1959).
3. Curley, A. E., Halliday, H. L., *Early Hum. Devel.*, *61*, 67 (2001).
4. Jobe, A. H., *N. Engl. J. Med.*, *328*, 861 (1993).
5. Halliday, H. L., *Paediatr. Perinat. Drug Therapy*, *1*, 30 (1997).
6. Jobe, A. H., Ikegami, M., *Clin. Perinatol.*, *20*, 683 (1993).
7. Wright, J. R., Clemets, J. A., *Am. Rev. Respir. Dis.*, *135*, 426 (1987).
8. Corcoran, J. D., Sheehan, O., O'Hare, M. M. T., Halliday, H. L., *Appl. Cardiopulm. Pathophysiol.*, *5*, 245 (1995).
9. Corcoran, J. D., Berggren, P., Sun, B., Halliday, H. L., Robertson, B., Curstedt, T., *Arch. Dis. Child.*, *71*, F165 (1994).
10. Hall, S. B., Venkitaraman, A. R., Whitsett, J. A., *Am. Rev. Respir. Dis.*, *145*, 999 (1992).
11. Cummings, J. J., Holm, B. A., Hudak, M. L., Hudak, B. B., Ferguson, W. H., Egan, E. A., *Am. Rev. Respir. Dis.*, *145*, 999 (1992).
12. Maeta, H., Vidyasagar, D., Raju, T. N., Bhat, R., Matsuda, H., *Pediatrics*, 81, 277 (1988).
13. Robillard, E., Alarie, Y., Dagenais-Perusse, P., Baril, E., Guilbeault, A., *Can. Med. Assoc. J.*, *90*, 55 (1964).
14. Chu, J., Clements, J. A., Cotton, E. K., Klaus, M. H., Sweet, A. Y., Tooley, W. H., *Pediatrics*, *40*, 709 (1967).
15. Fujiwara, T., Maeta, H., Chida, S., Morita, T., Watabe, Y., Abe, T., *Lancet*, 1, *55* (1980).
16. Soll, R. F., *The Cochrane Library*, CD 000511 (2003).
17. Soll, R. F., *The Cochrane Library*, CD 001079 (2003).
18. Soll, R. F., *The Cochrane Library*, CD 001149 (2003).
19. Soll, R. F., *The Cochrane Library*, CD 000144 (2003).
20. Halliday, H. L., *Drugs*, *51*, 226 (1996).
21. Konishi, M., Fujiwara, T., Naito, T., et al., *Eur. J. Pediatr.*, *147*, 20 (1988).
22. Halliday, H. L., Tarnow-Mordi, W. O., Corcoran, J. D., Patterson, C. C., *Arch. Dis. Child.*, *66*, 276 (1993).
23. Dunn, M. S., Shennan, A. T., Possmayer, F., *Pediatrics*, *86*, 564 (1990).
24. Speer, C. P., Robertson, B., Curstedt, T., *Pediatrics*, *89*, 13 (1992).
25. Soll, R. F., *The Cochrane Library*, CD 000141 (2003).
26. Soll R. F., *The Cochrane Library*, CD 000510 (2003).
27. Egberts, J., Brand, R., Walti, H., Bevilacqua, G., Bréart, G., Gardini, F., *Pediatrics*, *100*, e4 (1997).
28. Halliday, H. L., "Respiratory distress syndrome," in *Neonatal Respiratory Disorders* (2nd ed.), Arnold, p. 247 (2003).
29. OSIRIS Collaborative Group, *Lancet*, *340*, 1363 (1992).
30. Pramanik, A. K., *Am. J. Perinatol*, *9*, 507 (1992).
31. Zola, E. M., Overbach, A. M., Gunkel, J. H., et al., *J. Pediatr.*, *122*, 453 (1993).
32. Segerer, H., van Gelder, W., Angenent, F. W., *Pediatr. Res.*, *34*, 490 (1993).

33. Verder, H., Albertsen, P., Ebbesen, F., *Pediatrics, 103*, e24 (1999).
34. Berggren, E., Liljedah, M., Winbladh, B., et al. *Acta Paediatr., 89*, 460 (2000).
35. Vermont-Oxford Neonatal Network, *Pediatrics, 97*, 1 (1996).
36. Matthews, J. N. S., Ainsworth, S. B., Beresford, M. W., Milligan, D. W. A., *Lancet, 355*, 1387 (2000).
37. Svenningsen, N. W., Bjorklund, L., Vilstrup, C., Werner, O., *Biol. Neonate, 61 (suppl 1)*, 44 (1992).
38. Hallman, M., Merritt, T. A., Bry, K., Berry, C., *Pediatrics, 91*, 552 (1993).
39. Moen, A., Rootwelt, T., Robertson, B., Curstedt, T., Hall, C., Saugstad, O. D., *Pediatr. Res., 40*, 215 (1996).
40. Yu, X-O., Feet, B., Moen, A., Curstedt, T., Saugstad, O. D., *Pediatr. Res.*, 42, 151 (1997).
41. Hellström-Westas, L., Bell, A. H., Skov, L., Greisen, G., Svenningsen, N. W., *Pediatrics, 89*, 643 (1992).
42. Edwards, A. D., McCormick, D. C., Roth, S. C., *Pediatr. Res., 32*, 532 (1992).
43. Skov, L., *Neuropediatr., 23*, 126 (1992).
44. Halliday, H. L., in *Surfactant in Clinical Practice*. Harwood Academic Publishers, G. Belilacqua, et al. (eds.), p. 149 (1992).
45. Collaborative European Multicentre Study Group, *Eur. J. Pediatr., 151*, 372 (1992).
46. Chida, S., Phelps, D. S., Soll R. F., Taeusch, H. W., *Pediatrics, 88*, 84 (1991).
47. Long, W., *Semin. Perinatol., 17*, 275 (1993).
48. Spragg, R. G., *Biol. Neonate, 81 (suppl 1)*, 20 (2002).
49. Johansson, J., Curstedt T., Robertson, B., *Pediatr. Pathol. Mol. Med., 20*, 501 (2001).
50. Schwartz, R. M., Luby, A. M., Scanlon, J. W., Kellogg, R. J., *N. Engl. J. Med., 330*, 1476 (1994).

5.4 Non-steroidal antiinflammatory drugs (NSAIDs), prostanoids and the developmental kidney: a subtle compromise

J. P. Langhendries

NICU, CHC-Site St Vincent, B-4000 Rocourt-Liège, Belgium

5.4.1 INTRODUCTION

Non-steroidal antiinflammatory drugs (NSAIDs) are used in neonates in different clinical conditions. They selectively inhibit cyclo-oxygenases (COXs) -1 and -2 which synthesise prostanoids [1]. The developmental kidney is highly dependent on prostanoids and this explains the provocative title. Many studies exploring the action of prostaglandins on the kidney have been done in adult animals or in adult tissue (*in vitro* data). Extrapolating the results of these studies to the immature human kidney therefore has to be made with caution. Nevertheless, they help one understand the pharmacokinetic/pharmacodynamic (PK/PD) relationship of NSAIDs, especially regarding toxicity in the neonate. Indomethacin is the main drug belonging to the NSAIDs class, which has been used in premature infants and most of the studies in these neonates refer to this drug, while new experimental data are now available for ibuprofen.

5.4.2 PROSTANOIDS ACTION ON THE KIDNEY THROUGH THEIR SPECIFIC RECEPTORS

The action of prostanoids on the kidney is mediated through prostaglandin receptors belonging to the G-protein family [2]. The role of COXs in renal development appears increasingly evident while, by far, not fully understood [3]. In the fetus, COX-2 is crucial for nephrogenesis in mid-pregnancy [4-6]. Arachidonic acid metabolites modulate some enzyme activities along the nephron with an age-dependent developmental effect [7]. Experimental studies show that cortical as well as medullar COX-2 are over-expressed in fetal life and in the first days of life. This accounts for the high amount of vasodilator prostaglandins being secreted [3]. Both the fetus and the neonate, in the first postnatal days of life, face a high vascular resistance environment. In the renal glomerular vasculature, the efferent arteriole is under the influence of vasoconstrictive agents, such as angiotensin II (ANG II). Maintaining a high vascular resistance at this level during this period is the way by which the immature kidney is able to filtrate. PGE2 and PGI2 are the main vasodilator prostanoids which act on the glomerular afferent arterioles. These prostanoids counteract this high renal vascular resistance *in utero* in the first days of life and maximise the glomerular filtration through the EP2 and EP4 receptors (EP for *E-prostanoid),* and the IP2 receptor (IP for *I-prostanoid).* EP2 prostaglandins appear to play a role in regulating renin release through their receptors on renal juxtamedullar cells, providing a useful feedback mechanism [8].

Experimental studies in adult animals are in favour of a regulatory action of prostanoids on nitric oxide (NO) synthesis and *vice versa* [9], but this has been disputed by others [10]. NO could act more specifically on both inductive and constitutive COXs-2 in a model of renal inflammation [11]. A vasodilator compensatory mechanism, mediated by vasodilator COXs products, has even been suggested in the presence of NO deficiency [12]. It remains questionable if these last data may be transposable as such to the immature kidney. It appears that the action of prostanoids on the kidney is not limited to the glomerular level but also involves the tubular level and is dependent on the sodium load. In the mature kidney, EP1 receptor stimulation inhibits Na^+ and water reabsorption through a Ca^{++}-coupled mechanism at the basolateral level. In the salt-depleted animal, EP2 and EP4-receptor stimulation, both at the basolateral and the luminal tubular level of the collecting tubule, give rise to an increase of cAMP generation, contributing to increased Na^+ reabsorption [2]. The over-expression of the tubular EP3 receptor is one of the main features of the developing kidney [13]. This receptor is located in the distal tubule and collecting duct. Its localisation in the thick ascending limb of Henle's loop and the early distal convoluted tubule is uncertain. By decreasing cyclic cAMP production at this level, prostaglandin E2 interferes with the concentrating mechanisms of the arginine vasopressin hormone (AVP), rendering the immature kidney able to produce hypotonic urine and amniotic liquid [13-15]. This action is facilitated by the under-expression of aquaporin 2 channel (AQP2) [16-18]. The presence of a calcium-sensing receptor (CaR), which is over-expressed in early fetal life, is the other mechanism which potentiates the blockade of AVP [19]. This is an example of the perfectly adapted immature tubule to its role during pregnancy and in the first hours of postnatal life.

5.4.3 NON-STEROIDAL ANTIINFLAMMATORY DRUGS (NSAIDS) AND THE IMMATURE KIDNEY

Non-selective NSAIDs, such as both indomethacin and ibuprofen, have been used in neonates. They reversibly inhibit the enzymatic activity of both COX-1 and COX-2 [1,20]. Nimesulide, a Cox-2 selective inhibitor, is not recommended in the perinatal period owing to its toxicity [21-23]. Keeping in mind the above information on the crucial action of prostanoids on the perinatal kidney, especially those products synthesised from the over-expressed COX-2 at this time, it is evident that all these drugs administered during the last trimester of the pregnancy or during the first postnatal days may effect this organ in different ways, depending on the volume and salt-depleted state. Their use is therefore only recommended in some very restrictive well-balanced clinical conditions in term of risk/benefit ratio.

Clinical conditions

During pregnancy, indomethacin has been proposed as an effective tocolytic agent. The passage across the placenta is, however, evident and fetal concentrations, not far from those observed in the mother, have been found. It explains the adverse effects *in utero* which have been sometimes observed: 1) a profound oliguria which may result in a postnatal renal insufficiency; this effect has been used as a therapeutic goal in treating polyhydramnios or to reduce polyuria in patients affected by congenital hypokalemic salt losing tubulopathies; 2) *in utero* closure of the ductus arteriosus has been described, as well as pulmonary hypertension in the postnatal period; 3) postnatal necrotising enterocolitis and/or intestinal perforation have also been observed. Renal toxicity from the prenatal use of indomethacin is usually reversible but may be associated with fatal renal failure [24].

In neonates, indomethacin and ibuprofen (in clinical trials) have been used in the following clinical conditions: 1) the closure of a symptomatic PDA; 2) prophylactic treatment with indomethacin to prevent cerebral haemorrhage in the preterm neonate. There is currently insufficient evidence to support this use [25]. 3) Indomethacin has been used as a therapeutic agent in neonates to reduce polyuria in patients affected by congenital hypokalemic salt losing tubulopathies. Indomethacin-induced changes in renal blood flow (RBF) velocity waveform have been documented 10 minutes after its administration in preterm babies, at the dose of 100 mcg/kg [26]. The administration of indomethacin in the premature infant may be associated with serious adverse effects. An oliguria – which is sometimes the goal – has been described with a decrease of the glomerular filtration rate (GFR) and a transitory increase in serum creatinine.

NSAIDs may affect other organs as well. A decrease in the cerebral blood flow has been described from the use of indomethacin in premature infants, which explain some benefit in preventing cerebral haemorrhage [25], but not observed while using ibuprofen [27]. Intestinal perforation has also been reported after treatment with indomethacin. The adverse effects, which arise from the use of indomethacin in the neonate, appear to be related to a decrease of prostanoid synthesis [28], which may be transitory in case of prolonged treatment [29]. The last observation underscores some mechanisms of adaptation. Perinatal indomethacin does not seem to affect long term renal growth and function [30].

Pharmacokinetics/pharmacodynamics relationship

The glomerular filtration rate (GFR) is low at birth, due to the high vascular resistance. The postnatal changes in the renal flow distribution explain the progressive increase of the GFR after birth, the magnitude of which will be low until nephrogenesis is achieved (around 34 weeks of gestation). Indomethacin, used as a tocolytic agent before the 34th week of gestation, will influence the clearance of drugs given to premature neonates in the postnatal period as has been demonstrated for ceftazidime [31]. Studies in preterm lambs have shown that PDA itself may affect renal function, due to the poor renal perfusion related to the left to right shunting and the cardiac failure [32]. It is well known that the general fluid overload state associated with PDA in premature infants and the PDA-related poor renal perfusion will also increase the volume of distribution and decrease the clearance of drugs. NSAIDs administered to haemodynamically unstable babies will also affect the pharmacokinetics of the drugs concomitantly given which are eliminated through the kidney, rendering their prescription very difficult. This is especially true for drugs with a narrow therapeutic index such as aminoglycosides. An increase in the dosing interval is recommended alongside monitoring of the serum trough concentration to avoid drug accumulation. NSAIDs are usually administered through the intravenous route in neonates, maximising, in such a way, their bioavailability. NSAIDs normally show a high inter- as well as intra-individual variation in pharmacokinetics, most often related to the differences in gestational ages and/or in clinical states (Table 1) [33,34].

Indomethacin is highly protein-bound and the theoretical risk of bilirubin displacement has to be taken into account. In adults, indomethacin is eliminated by the liver through a demethylation process followed by deacylation, which precedes conjugation with glucuronide. Enterohepatic recirculation of indomethacin is probable and could account, at least in part, for the prolonged half-life in neonates

Table 1

NSAIDs pharmacokinetics in premature infants (ranges)

NSAIDs	Vd (L/kg)	T1/2β (h)	Cl (mL/kg/h)
indomethacin	0.21 - 0.45	11 - 20	4 - 25
ibuprofen	0.17 - 0.53	4 - 50	2.5 - 20

[33]. In premature babies, the indomethacin clearance is 10 to 20 times slower.

Controversy still exists regarding the association of indomethacin and furosemide. The latter drug has been thought to increase the incidence of PDA through a prostaglandin-mediated mechanism [35]. As a consequence, some authors have hypothesised that concomitant use of furosemide, alongside indomethacin for PDA closure, may decrease the nephrotoxic effect of the latter drug [36]. Another study, however, showed no benefit with this combination [37]. In a recent meta-analysis, Brion and Campbell concluded that there was no evidence to support the concomitant administration of furosemide and indomethacin in PDA-treated patients and that furosemide was contra-indicated in the presence of dehydration [38]. Seri *et al.* found a benefit from the use of dopamine in preventing the indomethacin-induced renal side effects in PDA-treated preterms [39], but this was not confirmed by others [40].

REFERENCES

1. Smith, W., *Am. J. Physiol., 263*, F18 (1992).
2. Breyer, M. D., Breyer, R. M., *Am. J. Physiol., 279*, F12 (2000).
3. Komhoff, M., Grone, H. J., Klein, T., Seyberth, H. W., Nusing, R. M., *Am. J .Physiol., 272,* F460 (1997).
4. Zhang, M. Z., Wang, J. L., Cheng, H. F., Harris, R. C., McKanna, J. A., *Am. J. Physiol., 273,* F994 (1997).
5. Komhoff, M., Wang, J. L., Cheng, H. F., Langenbach, R., McKanna, J. A., Harris, R. C., Breyer, M. D., *Kidney Int., 57,* 414 (2000).
6. Dinchuk, J. E., Car, B. D., Focht, R. J., et al., *Nature, 378,* 406 (1995).
7. Li, D., Belusa, R, Nowicki, S., Aperia, A., *Am. J. Physiol., 278,* F823 (2000).
8. Jensen, B. L., Schmid, C., Kurtz, A., *Am. J. Physiol., 271,* F659 (1996).
9. Luckhoff, A., Mulsch, A., Busse, R., *Am. J. Physiol., 258,* H960 (1990).
10. Gryglewski, R. J., *Sem. Thromb. Hemostasis, 19,* 158 (1993).
11. Salvemini, D., Seibert, S., Masferrer, J. L., Misko, T. P., Currie, M. G., Needleman, P., *J. Clin. Invest., 93,* 1940 (1994).
12. Baylis, C., Slangen, B., Hussain, S., Weaver, C., *Am. J. Physiol., 271,* R1327 (1996).
13. Bonilla-Felix, M., Jiang, W., *Am. J. Physiol., 271,* F30 (1996).

14. Joppich, R., Haberle, D. A., Weber, P. C., *Pediatr. Res., 15,* 278 (1981).
15. Bonilla-Felix, M., John-Phillip, C., *Am J Physiol., 267,* F44 (1994).
16. Bonilla-Felix, M., Jiang, W., *J. Am. Soc. Nephrol., 8,* 1502 (1997).
17. Bonilla-Felix, M., Vehaskari, V. M., Hamm, L. L., *Pediatr. Nephrol., 13,* 103 (1999).
18. Kim, Y. H., Earm, J. H., Ma, T., et al., *J. Am. Soc. Nephrol., 12,* 1795 (2001).
19. Chattopadhyay, N., Baum, M., Bai, M., et al., *Am. J. Physiol., 271,* F736 (1996).
20. Herschman, H. R., *Bba-Lipid Lipid Metab., 1299,* 125 (1996).
21. Holmes, R. P., Stone, P. R., *Obstet. Gynecol.., 96,* 810 (2000).
22. Balasubramaniam, J., *Lancet, 355,* 575 (2000).
23. Komhoff, M., Wang, J. L., Cheng, H. F., et al., *Kidney Int., 57,* 414 (2000).
24. van der Heijden, B. J., Carlus, C., Narcy, F., Bavoux, F., Delezoide, A. L., Gubler, M. C., *Am. J. Obstet. Gynecol., 171,* 617 (1994).
25. Schmidt, B., Davis, P., Moddemann, D., et al., *N. Engl. J. Med., 344,* 1996 (2001).
26. Van Bel, F., Guit, G. L., Schipper, J., Van De Bor, M., Baan, J., *J. Pediatr., 118,* 621 (1991).
27. Patel, J., Roberts, I., Azzopardi, D., Hamilton, P., Edwards, D., *Pediatr. Res., 47,* 36 (2000).
28. Winther, J. B., Hoskins, E., Printz, M. P., Mendoza, S. A., Kirkpatrick, S. E., Friedman, W. F., *Biol. Neonate, 38,* 76 (1980).
29. Seyberth, H. W., Rascher, W., Hackenthal, R., Wille, L., *J. Pediatr., 103,* 979 (1983).
30. Ojala, R., Ala-Houhala, M., Ahonen, S., et al., *Arch. Dis. Child. Fetal Neonatal Ed., 84,* F28 (2001).
31. Van den Anker, J. N., Schoemaker, R. C., Hop, W. C., et al., *Clin. Pharmacol. Ther., 58,* 650 (1995).
32. Gleason, C. A., Clyman, R. I., Heymann, M. A., Mauray, F., Leake, R., Roman, C., *Am. J. Physiol., 254,* F38 (1988).
33. Thalji, A. A., Carr, I., Yeh, T. F., Raval, D., Luken, J. A., Pildes, R. S., *J. Pediatr., 97,* 995 (1980).
34. Van Overmeire, B., Tow, D., Schepens, P. J. C., Kearns, G. L., van den Anker, J. N., *Clin. Pharmacol. Ther., 70* 336 (2001).
35. Green, T. P., Thompson, T. R., Johnson, D. E., Lock, J. E., *N. Engl. J. Med., 308,* 743 (1983).
36. Yeh, T. H., Wilks, A., Singh, J., Betkerur, M., Lilien, L., Pildes, R. S., *J. Pediatr., 101,* 433 (1982).
37. Romagnoli, C., Zecca, E., Papacci, P., et al., *Clin. Pharmacol. Ther., 62,* 181 (1997).
38. Brion, L., Campbell, D. E., *Pediatr. Nephrol., 13,* 212 (1999).
39. Seri, I., Tulassay, T., Kiszel, J., Csomor, S., *Int. J. Pediatr. Nephrol., 5,* 209 (1984).
40. Fajardo, C. A., Whyte, R. K., Steele, B. T., *J. Pediatr., 121,* 771 (1992).

5.5 Pharmacology of analgesics in neonates

John N. van den Anker[1] and Dick Tibboel[2]

[1]Department of Paediatrics, Pharmacology and Physiology, Division of Paediatric Clinical Pharmacology, George Washington University School of Medicine and Health Sciences, Children's National Medical Center, Washington, D.C., USA; and Department of Paediatrics, Erasmus MC-Sophia Children's Hospital, Rotterdam, The Netherlands

[2]Department of Paediatric Surgery, Erasmus MC-Sophia Children's Hospital, Rotterdam, The Netherlands

5.5.1 INTRODUCTION

Neonates were believed to be unable to experience pain until 1987 [1]. Since that moment, the use of analgesic agents in this specific patient population has increased exponentially. Research has concentrated on the development of pain assessment instruments and clinical trials investigating the safety and efficacy of several analgesics in neonates [2,3]. Knowledge about analgesic effect has increased, using these newly developed pain assessment tools in randomised trials comparing different dose regimens and different agents [4].

The pharmacokinetic and pharmacodynamic behaviour of analgesics in neonates is dependent on the maturation of enzyme and receptor systems and physiological processes responsible for absorption, distribution, metabolism and elimination [5].

Analgesics in neonatal patients can be largely divided into opioids and non-opioids.

5.5.2 OPIOIDS

Mechanisms of Action

Opioid analgesics include naturally occurring agents (opium-alkaloids) and synthetic opioid-agonists that elicit morphine-like activity. The analgesic effects of opioids occur by activation of µ (mu), κ (kappa), and/or δ (delta) receptors in the CNS [6]. Each class of receptors is divided into subtypes that have different clinical effects. Analgesia is obtained by spinal or supraspinal activation of opioid-receptors, leading to decreased neurotransmitter release from nociceptive neurons inhibiting the ascending neuronal pain pathways and altering the perception and response to pain [7]. Opioid receptors also exist outside the central nervous system in the dorsal root ganglia and on peripheral terminals of primary afferent neurons [8].

Opioids in neonates are primarily used for moderate to severe pain, such as postoperative pain. Furthermore, opioids may be used in the Neonatal Intensive Care Unit (NICU) for pain or stress related to artificial ventilation, surgical procedures [chest tube placement, vessel cannulation for extracorporeal membrane oxygenation (ECMO)] or painful conditions, such as necrotising enterocolitis. The most frequently used opioids are fentanyl and morphine as well as fentanyl derivatives such as alfentanil, sufentanil and remifentanil.

Opioids produce adverse effects that may be minimised by appropriate drug selection and dosing. Respiratory depression, hypotension, glottic and chest wall rigidity, constipation, urinary retention, seizures, sedation, and bradycardia are well described. Continuous monitoring and frequent assessment of vital signs should be performed during opioid administration. Naloxone is a competitive opioid receptor agonist that reverses many of these side effects, if used in an appropriate dosage.

Morphine

Pharmacokinetics

Morphine is metabolised by the enzyme UDP-glucuronosyl transferase 2B7 (UGT2B7) into morphine-3-glucuronide (M3G) and morphine-6-glucuronide (M6G) [9]. M6G has been shown to have higher analgesic potency than morphine [10,11] and also respiratory depressive effects [12,13]. M3G has been suggested to antagonise the anti-nociceptive and respiratory depressive effects of morphine and M6G [14,15], and contributes to the development of tolerance. The enzyme responsible for morphine glucuronidation, UGT2B7 [16], is mainly found in the liver, but is also present in the intestines and kidneys [17]. Sulphation is a minor pathway [18,19]. The metabolites are cleared by the kidneys and partly by biliary excretion (19). Some enterohepatic circulation of morphine occurs due to β-glucuronidase activity in the gut [20]. Impaired renal function has been shown to lead to accumulation of M3G and M6G [21]. Morphine clearance increases with postconceptual age [16,22] and reaches adult values at 6 to 12 months of age [3,23-25].

Pharmacodynamics

The intravenous administration of 10 to 30 mcg/kg morphine has been shown to decrease pain in premature neonates requiring artificial ventilation [4,26,27]. Chay *et al.* have reported that mean morphine concentrations of 125 ng/ml are needed to produce adequate sedation in 50% of neonates [28], but target plasma concentrations for analgesic purposes are thought to be around 15 to 20 ng/ml [25,29]. Respiratory depression may occur at concentrations of 20 ng/ml [25]. Hypotension, bradycardia and flushing are associated with rapid intravenous bolus administration.

Fentanyl

Fentanyl is a synthetic opioid with a wide margin of safety, beneficial effects on haemodynamic stability, a rapid onset and a short duration of action [30,31]. Because of its rapid onset of action and short duration of effect, fentanyl efficiently alleviates procedural pain [32]. It has been used in neonates on artificial ventilation [33] with broncho-pulmonary dysplasia, pulmonary hypertension and/or diaphragmatic hernia.

Fentanyl has been shown to effectively prevent surgical stress responses in preterm neonates and to improve postoperative outcome [34]. Fentanyl significantly reduced physiological and behavioural measures of pain and stress during mechanical ventilation in preterm neonates [35,36], comparable with the effects seen after the use of morphine [33]. The use of transdermal and transmucosal fentanyl has not been studied in newborns. The clearance of fentanyl appears to be somewhat immature at birth but increases rapidly after birth. The volume of distribution of fentanyl at steady state is greatest in the neonatal period (5.9 L/kg) [37].

Fentanyl, alfentanil and sufentanil are all metabolised by CYP3A4, and other drugs that also use this enzyme (e.g., ciclosporin, erythromycin) may decrease clearance leading to increased plasma concentrations [38,39].

Research investigating DNA polymorphisms has shown genetic variability in the different enzymes involved in the metabolism of morphine and other opioids, frequently used in the neonatal population. Therefore, the use of pharmacogenetic tools might lead in the near future to more individualised use of opioids in the neonatal population [40].

Alfentanil

Alfentanil is a synthetic opioid that is chemically a derivate of fentanyl. Sufficient analgesia during endotracheal intubation and suctioning has been found using 10-20 mcg/kg alfentanil in preterm neonates [41-43]. Alfentanil plasma protein binding varies from 65% in preterm neonates to 79% in term infants (44,45). Clearance of alfentanil is reduced in the neonate (20-60 ml·min^{-1}·70kg^{-1}). Consequently, the elimination half-life is higher in the neonatal period, especially in premature neonates (6-9 hours) [46,47]. Alfentanil cannot be used without neuromuscular blocking drugs in newborns, because of a very high incidence of rigidity [41,48].

5.5.3 PARACETAMOL

The relative bioavailability of rectal paracetamol is higher in neonates where suppository insertion height may result in a different rectal venous drainage pattern. Hopkins *et al.* [49] noted that the AUC in neonates after suppository administration was significantly greater than in infants and in older children. Paracetamol absorption depends on gastric emptying which is slow and erratic in the neonate [50]. Clearance is reduced and the volume of distribution is greatest in the neonatal period.

Neonates can produce hepatotoxic metabolites (e.g. NAPQI), but there are suggestions of a lower activity of cytochrome P-450 in neonates. This may explain the low occurrence of hepatotoxicity seen in neonates [51,52], despite reports of high serum concentrations [51,53].

REFERENCES

1. Anand, K. J., Hickey, P. R., *N. Engl. J. Med., 317*, 1321 (1987).
2. Van Dijk, M., Tibboel, D., *Clin. Perinatol., 29*, 469 (2002).
3. Van Lingen, R., Simons, S., Anderson, B., et al., *Clin. Perinatol., 29*, 511 (2002).
4. Anand, K. J., Barton, B. A., McIntosh, N., et al., *Arch. Pediatr. Adolesc. Med., 153*, 331 (1999).
5. Kearns, G. L., Abdel-Rahman, S. M., Alander, S. W., et al., *N. Engl. J. Med., 349*, 1157 (2003).
6. Inturrisi, C. E., *Clin. J. Pain, 18*, S3 (2002).
7. Suresh, S., Anand, K. J., *Semin. Perinatol., 22*, 425 (1998).
8. Stein, C., Machelska, H., Binder, W., et al., *Curr. Opin. Pharmacol., 1*, 62 (2001).
9. Coffman, B. L., Rios, G. R., King, C. D., et al., *Drug Metab. Dispos., 25*, 1 (1997).
10. Osborne, P. B., Chieng, B., Christie, M. J., *Br. J. Pharmacol., 131*, 1422 (2000).
11. Murthy, B. R., Pollack, G. M., Brouwer, K. L., *J. Clin. Pharmacol., 42*, 569 (2002).
12. Osborne, R., Thompson, P., Joel, S., et al., *Br. J. Clin. Pharmacol., 34*, 130 (1992).
13. Thompson, P. I., Joel, S. P., John, L., et al., *Br. J. Clin. Pharmacol., 40*, 145 (1995).
14. Gong, Q. L., Hedner, J., Bjorkman, R., et al., *Pain, 48*, 249 (1992).
15. Smith, M. T., Watt, J. A., Cramond, T., *Life Sci., 47*, 579 (1990).
16. Faura, C. C., Collins, S. L., Moore, R. A., et al., *Pain, 74*, 43 (1998).
17. Fisher, M. B., Vandenbranden, M., Findlay, K., et al., *Pharmacogenetics, 10*, 727 (2000).
18. Choonara, I., Ekbom, Y., Lindstrom, B., et al., *Br. J. Clin. Pharmacol., 30*, 897 (1990).
19. McRorie, T. I., Lynn, A. M., Nespeca, M. K., et al., *Am. J. Dis. Child., 146*, 972 (1992).
20. Koren, G., Maurice, L., *Pediatr. Clin. North Am., 36*, 1141 (1989).

21. Choonara, I. A., McKay, P., Hain, R., et al., *Br. J. Clin. Pharmacol., 28*, 599 (1989).
22. Kart, T., Christrup, L. L., Rasmussen, M., *Paediatr. Anaesth., 7*, 5 (1997).
23. Anderson, B. J., McKee, A. D., Holford, N. H., *Clin Pharmacokinet., 33*, 313 (1997).
24. Van Lingen, R. A., Simons, S. H., Anderson, B. J., et al., *Clin. Perinatol., 29*, 511 (2002).
25. Lynn, A., Nespeca, M. K., Bratton, S. L., et al., *Anesth. Analg., 86*, 958 (1998).
26. Quinn, M. W., Wild, J., Dean, H. G., et al., *Lancet. 342*, 324 (1993).
27. Scott, C. S., Riggs, K. W., Ling, E. W., et al., *J. Pediatr., 135*, 423 (1999).
28. Chay, P. C., Duffy, B. J., Walker, J. S., *Clin. Pharmacol. Ther., 51*, 334 (1992).
29. Kart, T., Christrup, L. L., Rasmussen, M., *Paediatr. Anaesth., 7*, 93 (1997).
30. Yaster, M., Koehler, R. C., Traystman, R. J., *Anesthesiology, 66*, 524 (1987).
31. Hickey, P. R., Hansen, D. D., Wessel, D. L., et al., *Anesth. Analg., 64*, 1137 (1985).
32. Barrington, K. J., Byrne, P. J., *Am. J. Perinatol., 15*, 213 (1998).
33. Saarenmaa, E., Huttunen, P., Leppaluoto, J., et al., *J. Pediatr., 134*, 144 (1999).
34. Anand, K. J., Sippell, W. G., Aynsley-Green, A., *Lancet, 1*, 62 (1987).
35. Guinsburg, R., Kopelman, B. I., Anand, K. J., et al., *J. Pediatr., 132*, 954 (1998).
36. Lago, P., Benini, F., Agosto, C., et al., *Arch. Dis. Child., Fetal Neonatal Ed., 79*, F194 (1998).
37. Johnson, K. L., Erickson, J. P., Holley, F. O., et al., *Anesthesiology, 61*, A441 (1984).
38. Touw, D. J., *Drug Metabol. Drug Interact., 14*, 55 (1997).
39. Tanaka, E., *J. Clin. Pharm. Ther., 23*, 403 (1998).
40. Roses, A. D., *Lancet, 355*, 1358 (2000).
41. Saarenmaa, E., Huttunen, P., Leppaluoto, J., et al., *Arch. Dis. Child., Fetal Neonatal Ed., 75*, F103 (1996).
42. Pokela, M. L., *Biol. Neonate, 64*, 360 (1993).
43. Pokela, M. L., Koivisto, M., *Acta Paediatr., 83*, 151 (1994).
44. Meuldermans, W., Woestenborghs, R., Noorduin, H., et al., *Eur. J. Clin. Pharmacol., 30*, 217 (1986).
45. Wilson, A. S., Stiller, R. L., Davis, P. J., et al., *Anesth. Analg., 84*, 315 (1997).
46. Marlow, N., Weindling, A. M., van Peer, A., et al., *Arch. Dis. Child., 65*, 349 (1990).
47. Killian, A., Davis, P. J., Stiller, R. L., et al., *Dev. Pharmacol. Ther., 15*, 82 (1990).
48. Pokela, M. L., Ryhanen, P. T., Koivisto, M. E., et al., *Anesth. Analg., 75*, 252 (1992).
49. Hopkins, C. S., Underhill, S., Booker, P. D., *Arch. Dis. Child., 65*, 971 (1990).
50. Gupta, M., Brans, Y., *Pediatrics, 62*, 26 (1978).
51. Roberts, I., Robinson, M. J., Mughal, M. Z., et al., *Br. J. Clin. Pharmacol., 18*, 201 (1984).
52. Levy, G., Garrettson, L. K., Soda, D. M., *Pediatrics, 55*, 895 (1975).
53. Lederman, S., Fysh, W. J., Tredger, M., et al., *Arch. Dis. Child., 58*, 631 (1983).

5.6 Prevention and management of pain in neonates

E. Jacqz-Aigrain[1], K. J. S. Anand[2]

[1]Department of Paediatric Pharmacology and Pharmacogenetics, Hospital Robert Debré and Faculty of Medicine Bichat / Robert Debré, University Paris VII, France
[2]Department of Paediatrics, University of Arkansas for Medical Sciences, Arkansas Children's Hospital, Little Rock, Arkansas, USA

5.6.1 INTRODUCTION

All neonates, whether they be sick, premature or healthy term babies, routinely experience pain and may have an increased sensitivity to pain [1-3]. They exhibit greater cardiovascular, hormonal and metabolic responses to pain than older children and more clinical complications [4]. Pain often remains under-recognised, under-evaluated and undertreated. In addition, there are very few drugs evaluated and approved for the treatment of pain in neonates, and management of pain remains very heterogeneous among clinicians, health care providers and between countries [5-9].

The Consensus statement for the prevention and management of pain in the newborn (10) was developed by the International Evidence-Based Group for Neonatal Pain in order to improve pain management in neonates. Disciplines represented were paediatrics, neonatalogy and intensive care, anaesthesiology, pharmacology (and others). Experts were from fourteen different countries and all primarily involved in the evaluation of pain and in behavioural and pharmacological methods for the management of pain.

The group performed systematic reviews of all the studies published either in neonates and/or in older children on their specific topic and evaluated the quality of published data. The group met twice in April 1998 and August 1999 and the guidelines are based on a combination of critically evaluated published evidence and discussions.

5.6.2   CONSENSUS STATEMENT

1. *Painful procedures commonly performed in the neonatal intensive care unit:*
   All the painful procedures commonly performed in neonatal intensive care units were listed. They were diagnostic procedures (vein and arterial punctures, heel lancing, etc.), therapeutic procedures (tracheal intubation, mechanical ventilation, central line insertion/removal, chest tube insertion/removal, etc.) and surgical procedures (circumcision, and others) and may also include areas of inflammation and hyperalgia around tissue injury, postoperative pain, local infection or inflammation, skin burns.

2. *Assessment of pain in newborns:*
   The assessment of pain requires validated methods for infants of different gestational ages and/or with acute or continuous pain (Table 1). It must be undertaken routinely, every 4 to 6 hours or as recommended by the pain score or the clinical condition of the neonate [11-14]. Pain assessment should be comprehensive and should include contextual, behavioural and physiological indicators [3,11,13,15-20].

3. *Suggested management approaches for neonatal pain:*
   Management of pain should include strategies for prevention of pain, such as environmental approaches to reducing acoustic and thermal stresses [21,22], behavioural [23-25] and pharmacological approaches (Table 2) in the different clinical situations. Recommended analgesic doses were made and potential side-effects were discussed.

By this initiative, the group wanted to underline that neonates feel pain and that the management of pain must be considered an important part of neonatal healthcare. Other aims include convincing clinicians to train in order to evaluate pain and treat it appropriately, stimulating further research in the areas where current evidence is not available for defining the efficacy/safety of specific therapeutic approaches and increasing drug studies in neonates.

Adapted from the "Consensus statement for the prevention and management of pain in the newborn" [10].

Table 1

Commonly used methods for assessment of pain in newborns

	Premature Infant Pain Profile (PIPP) [22,21]	Neonatal Facial Coding Scale (NFCS) [22,23]	Neonatal Infant Pain Scale (NIPS) [24]	CRIES Score [25]
variables assessed	gestational age behavioral state heart rate oxygen saturation brow bulge eye squeeze naso-labial furrow	brow bulge eye squeeze naso-labial furrow open lips stretch mouth lip purse taut tongue chin quiver tongue protrusion	facial expression cry breathing patterns arms legs state of arousal	Crying Requires increased oxygen administration Increased vital signs Expression Sleeplessness
reliability data	inter-rater and intra-rater reliability >0.93	inter-rater and intra-rater reliability >0.85	inter-rater reliability >0.92	inter-rater reliability > 0.72
forms of validity established	face, content, construct (in preterm and term neonates)	face, content, construct, and convergent (r = 0.89)	face, construct, and concurrent (r = 0.53-0.84)	face, content, discriminant and concurrent (r=0.49-0.73)
clinical utility	feasibility and utility established at bedside	feasibility established at bedside	not established	nurses preferred CRIES over another scale

Table 2

Suggested use of behavioural and pharmacological approaches for the management approaches for neonatal pain

	Evidence based	Possibly effective
pacifier with sucrose concentration 12–24% given 2 minutes before the procedure	heel lance [26,28-31] percutaneous venous and arterial catheter insertion (similar approach for venepuncture) [26,27,33-35] central line placement [26,27] lumbar puncture [26,27] circumcision [26,27,32,36]	percutaneous arterial catheter insertion [26,27] umbilical catheter insertion [26,27] peripherally inserted central catheter placement [26,27] subcutaneous or intramuscular injection [26,27] endotracheal suction [26,27] gastric tube insertion [26,27] chest tube insertion [26,27]
EMLA cream eutectic mixture of local anesthetics lidocaine and prilocaine hydrochloride in an emulsion base	percutaneous venous catheter insertion (38) central line placement [38,39] circumcision [36-38]	percutaneous arterial catheter insertion [38] central venous line placement [38,39] peripherally inserted central catheter placement lumbar puncture [38] subcutaneous and intramuscular injection [40]
local lidocaine (subcutaneous, topical spray)		percutaneous arterial puncture and catheter insertion [38] lumbar punction [41,42] endotracheal intubation [43,44] chest tube insertion [42]

Table 2 *(continued)*

Evidence based	Possibly effective
opioid administration (dose or continuous infusion) data are available with morphine and/or fentanyl)	
percutaneous venous catheter insertion (if intravenous access available) [46]	chest tube insertion [5,41]
central venous line placement [46,48]	
endotracheal succion[45,49,50]	
ongoing analgesia for routine procedures in ventilated patients [45,47]	
anesthetic agents	
central venous line placement	
chest tube insertion	

Remarks:

1) Heel lance is a very painful procedure [51] and venepuncture should be considered instead in full-term and premature neonates when possible [33,34,52]. EMLA. paracetamol are ineffective.

2) Efficacy and safety of repeated paracetamol doses are unknown. Consider paracetamol for post-operative pain [53].

3) Avoid subcutaneous and intramuscular injections and give drugs intravenously when possible.

4) The use of midazolam, a widely used sedative agent, is not recommended as there is no evidence to show that neonates can be safely sedated for several weeks [54,55].

REFERENCES

1. Anand, K. J. S., *Biol. Neonate, 73*, 1 (1998).
2. Johnston, C. C., Stevens, B. J., Yang, F., Horton, L., *Pain, 61*, 471 (1995).
3. Craig, K. D., Whitfield, M. F., Grunau, R. V. E., Linton, J., Hadjistavropoulos, H. D., *Pain, 52*, 201 (1993).
4. Anand, K. J. S., Brown, M. J., Causon, R. C., *et al., J. Pediatric. Surg., 20*, 41 (1985).
5. Yaster, M., *Anesthesiology, 56*, 433 (1987).
6. Greeley, W. J., de Broijn, N. P., *Anesth. Analg., 57*, 86 (1988).
7. Chay, P. C. W., Duffy, B. J., Walker, J. S., *Clin. Pharmacol. Ther., 51*, 334 (1992).
8. Olkkola, K. T., Hamunen, K., "Pharmacokinetic-pharmacodynamic of analgesic drugs," in *Pain in Neonates* (2nd ed.), Anand, K. J. S., Stevens, B. J., McGrath, P. J. (eds.), Elsevier Science, pp.135-158 (2000).
9. Scott, C. S., Riggs, K. W., Ling, E. W., et al., *J. Pediatr., 135*, 423 (1999).
10. Anand, K. J. S., *Arch. Pediatr. Adolesc. Med., 155*, 173 (2001).
11. Chiswick, M. L., *Lancet, 355*, 6 (2000).
12. Abu-Saad, H. H., Bours, G. J., Stevens, B., Hamers, J. P., *Semin. Perinatol., 22*, 402 (1998).
13. Stevens, B., Johnston, C., Gibbins, S., "Assessment of pain in the neonate," in *Pain in Neonates* (2nd ed.), Anand, K. J. S., Stevens, B., McGrath, P. J. (eds.), pp.101-134 (2000).
14. Franck, L. S., Greenberg, C. S., Stevens, B., *Pediatr. Clin. North Am., 47*, 487 (2000).
15. Stevens, B., Johnston, C. C., Petryshen, P., Taddio, A., *Clin. J. Pain, 12*, 13 (1996).
16. Ballantyne, M., Stevens, B., McAllister, M., et al., *Clin. J. Pain, 15*, 297 (1999).
17. Grunau, R. V. E., Oberlander, T. F., Holsti, L., et al., *Pain, 76*, 277 (1998).
18. Guinsburg, R., Berenguel, R. C., Xavier, R. C., Almeida, M. F. B., Kopelman, B. "Are behavioral scales suitable for preterm and term pain assessments?" in the *Proceedings of the 8th World Congress on Pain*, Jensen, T. S., Turner, J. A., Wiesenfeld-Hallin, Z. (eds.), Seattle, WA, International Association for the Study of Pain, pp. 893-902 (1997).
19. Lawrence, J., Alcock, D., McGrath, P., Kay, J., McMurray, S. B., Dulberg, C., *Neonatal Network, 12*, 59 (1993).
20. Krechel, S. W., Bildner, J., *Paediatr. Anesth., 5*, 53 (1995).
21. Sauve, R., Saigal, S. (eds.) *Optimizing the neonatal intensive care environment*. Report of the 10th Canadian Ross Conference in Pediatrics, Montreal, Quebec, GCI Communications, (1995).
22. American Academy of Pediatrics, Committee on Environmental Health, "Noise: a hazard for the fetus and the newborn," *Pediatrics, 100*, 724 (1997).
23. Gray, L., Watt, L., Blas, E. M., *Pediatrics, (serial online), 105*, e14 (2000).
24. Corff, K. E., *Néonatal Network, 12*, 74 (1993).
25. Corff, K. E., Seideman, R., Venkataraman, P., Lutes, L., Yates, B., *J. Obstet. Gynecol. Neonatal Nurs., 24*, 143 (1995).
26. Stevens, B.., Ohlsson, A. *Sucrose for analgesia in newborn infants undergoing painful procedures* (Cochrane Review on CD-Rom). Oxford, England, Cochrane Library, 2000, 2CD001069.
27. Stevens, B., Taddio, A., Ohlsson, A., Einarson, T., *Acta Paediatr., 86*, 837 (1997).
28. Abad, F., Diaz, N., Domenech, E., Robayna, M., Rico, J., *Acta Paediatr., 85*, 854 (1996).
29. Bucher, H. U., Moser, T., von Siebenthal, K., Keel, M., Wolf, M., Duc, G., *Pediatr. Res., 38*, 332 (1995).
30. Haouri, N., Wood, C., Griffiths, G., Levene, M., *Br. Med. J., 310*, 1498 (1995).
31. Carbajal, R., Chauvet, X., Couder, S., Olivier-Martin, M., *Br. Med. J., 319*, 1393 (1999).
32. Herschel, M., Knoshnood, B., Ellman, C., et al., *Arch. Pediatr. Adolesc. Med., 152*, 279 (1998).
33. Larsson, B. A., Tannfeldt, G, Lagercrantz, H., Olsson, G. L., *Pediatrics, 101* 882 (1998).

34. Shah, V., Ohlsson, A. *Venipuncture versus heel lance for blood sampling in term neonates* (Cochrane Review on CD-ROM), Oxford, England, Cochrane Library, 2000, 2:CD001452.

35. Ohlsson, A., Taddio, A., Jadad, A. R., Stevens, B. J. "Evidence-based decision making, systematic reviews and the Cochrane collaboration: implications for neonatal analgesia," in *Pain in Neonates* (2nd ed.), *Pain Research & Clinical Management* (vol. 10), Anand, K. J. S., Stevens, B. J., McGrath, P. J. (eds.), Elsevier Science, pp. 251-268, (2000).

36. Taddio, A., Pollock, N., Gilbert-MacLeod, C., Ohlsson, K., Koren, G., *Arch. Pediatr. Adolesc. Med., 154,* 620 (2000).

37. Taddio, A., Stevens, B., Craig, K., *N. Engl. J. Med., 336,* 1197 (1997).

38. Taddio, A., Ohlsson, A., Stevens, B., Einarson, T. R., Koren, G., *Pediatrics (serial online), 101,* e1 (1998).

39. Garcia, O. C., Reichberg, S., Brion, L. P., Schulman, M., *J. Perinatol., 17,* 477 (1997).

40. Uhzri, M., *Pediatrics, 92,* 719 (1993).

41. Menon, G., Anand, K. J. S., McIntosch, N., *Semin. Perinatol., 22,* 417 (1998).

42. Larsson, B. A., *Res. Clin. Forums, 20,* 63 (1998).

43. Mostafa, S. M., Murthy, B. V., Barrett, P. J., McHugh, P., *Eur. J. Anaesthesiol, 16,* 7 (1999).

44. Lehtinen, A. M., Hovorka, J., Widholm, O., *Br. J. Anaesth., 56,* 239 (1984).

45. Anand, K. J. S., McIntosh, N., Lagercrantz, H., *et al., Arch. Pediatr. Adolesc. Med., 153,* 331 (1999).

46. Moustogiannis, A. N., Raju, T. N. K., Roohey, T., McCulloch, K. M., *Neurol. Res., 18,* 440 (1996).

47. Orsini, A. J., Leef, K. H., Costarino, A., Dettore, M. D., Stefano, J. L., *J. Pediatr., 129,* 140 (1996).

48. Cordero, L., Gardner, D. K., O'Shaughnessy, R., *Am. J. Perinatol., 8,* 284 (1991).

49. Pokela, M. L., *Pediatrics, 93,* 379 (1994).

50. Saarenmaa, E., Huttunen, P.,, Leppaluoto J., Fellman, V., *Arch. Dis. Child Fetal Neonatal. Ed., 75,* F103-(1996).

51. Taddio, A., Shah, V., Gilbert-MacLeod, C., Katz, J., *JAMA, 288,* 857 (2002).

52. McIntosh, N., Van Veen, L., Brameyer, H., *Arch. Dis. Child., 70,* F177 (1994).

53. Howard, C. R., Howard, F. M., Weitzman, M. L., *Pediatrics, 93,* 641 (1994).

54. Ng, E., Taddio, A., Ohlsson, A. *Intravenous midazolam infusion for sedation of infants in the neonatal intensive care unit.* (Cochrane Review on CD-Rom), Oxford, England, Cochrane Library, 2002, CD002052.

55. Jacqz-Aigrain, E., Daoud, P., Burtin, P., et al., *Lancet, 344,* 646 (1994).

5.7 Pharmacology and use of methylxanthines and doxapram for the treatment of neonatal apnoea

Marie-Jeanne Boutroy

Neonatal Intensive Care and Perinatal Pharmacology, INSERM, The Henri Poincaré University, Maternité Régionale Universitaire, 10, rue Heydenreich 54042 Nancy, France

Apnoea is defined as a transient cessation of breathing. In neonates, it may be attributed to immaturity of the respiratory command systems or to specific disorders. The incidence and severity of apnoea is inversely related to maturation and one of the most frequent problems in a neonatal unit. The first step in its management is to look for identifiable causes and to treat them. Thereafter, drugs such as methylxanthines or doxapram are used. Mechanical ventilation is the last step. Treatment discontinuation starts after complete cessation of significant apnoeic episodes, apnoeas having to be monitored for 4 to 5 days after treatment weaning, in order to detect a possible recurrence.

5.7.1 METHYLXANTHINES

The methylxanthines (theophylline and caffeine) are the first-line therapeutic agents for treating apnoea in neonates. They are central nervous system stimulants and they act by increasing chemoreceptor sensitivity to CO_2, improving respiratory muscle contraction and metabolic homeostasis. They increase the ventilation by increasing the respiratory centre output. Moreover, they are antagonists of endogenous adenosine, which is a respiratory depressant [1].

Efficacy

Kuzemko and Paala introduced the methylxanthines into the management of apnoeic premature infants in the early 1970s [2]. Others have shown theophylline and caffeine to both be effective [3-6]. Severe apnoeic spells were significantly decreased [7] or completely controlled [8]. There was also a decrease in the total duration of hypoxaemia and of hyperoxaemia [5], as fewer interventions with oxygen for apnoeic spells were required. Controlled studies have demonstrated a reduction in the need for assisted ventilation and in the cardio-respiratory altera-tions [10,11]. No extensive placebo controlled trials on methylxanthine efficacy have been performed since these initial studies.

Therapeutic drug monitoring (TDM)

Effective plasma concentrations range from 10 to 20 mg/l, for both methy-lxanthines [9]. Blood samples should be taken for TDM: 4 to 5 days after com-mencing oral treatment, when steady-state is reached, then once a week with caffeine, and 2 or 3 times a week with theophylline, or at any time if the drug is given intravenously. In case of therapeutic failure, a trough sample should be taken. If toxicity is suspected, a sample 2-4 hours post dose should be collec-ted.

Pharmacokinetics

Some pharmacokinetic characteristics are the same for theophylline and caf-feine in the neonatal period: excellent digestive absorption, maximal plasma concentration peak within 1 hour, protein binding lower than in adults and wide inter-individual variability in plasma concentrations. Plasma half-lives are much longer in neonates (theophylline, 30 h; caffeine, 60-100 h) than in adults (5 h and

Table 1

Pharmacokinetic parameters of caffeine and theophylline in neonates

Pharmacokinetic parameters (mean)	Neonate	Adult
Caffeine		
volume of distribution (l/kg)	0.9	0.6
clearance (1/kg/h)	8.9	94
elimination half-life (h)	60 -100	6
Theophylline		
volume of distribution (l/kg)	0.7	0.5
clearance (1/kg/h)	22	66
elimination half-life (h)	30	6.7

Table 2

Dosage schedules of respiratory stimulants administered for neonatal apnoeas

	Caffeine	Theophylline	Doxapram
loading dose (mg/kg)	10	6	2
maintenance dose (mg/kg/day)	2.5	4	24-30
therapeutic range (mg/l)	6-20	5-15	0.5-3
therapeutic drug monitoring (maintenance)	once weekly	2-3 times weekly	once weekly

6 h respectively) (Table 1). Caffeine clearance is influenced by postnatal age, bodyweight and gestational age [12], but the plasma concentrations are more stable and predictable for caffeine than for theophylline [13].

Dose guidelines

The current recommended dosages are:

- theophylline: 6 mg/kg loading dose, then 2 mg/kg twice daily. Plasma concentrations range from 6 to 9 mg/l with this regimen [11].
- caffeine base: 10 to 12.5 mg/kg loading dose, then 2.5-3 mg/kg once daily (Table 2).

The methylxanthine of choice is caffeine, as it is easier to use and has fewer side effects.

Toxicity

The most frequent adverse effects observed with methylxanthines are: irritability, tachycardia, high blood pressure, increased gastric aspiration and gastro-oesophageal reflux, polyuria and excessive natriuria [8,14-17]. They usually only require a dose reduction. The role of methylxanthines in the occurrence of necrotising enterocolitis remains controversial [18]. Dramatic adverse effects have been reported with overdoses (plasma drug concentrations ranging from 26 to 346 mg/l): tachypnoea, opisthotonos and seizures [19-21]. The question of potential long-term adverse effects on cerebral and neuronal development remains open, since these drugs are known to reduce cerebral blood flow in animals and adult humans [22,23] and to antagonise adenosine, which is cerebro-protective against hypoxic insult [1]. Studies on cerebral blood flow in infants given theophylline or caffeine have resulted in conflicting conclusions [24,25].

A prospective controlled study of 21 children treated for apnoeas in the neonatal period has shown that caffeine given at conventional doses had no apparent harmful effects 18 months to 3 years later, in term of growth and psychomotor development.

5.7.2 DOXAPRAM

Doxapram is a respiratory stimulant with similar properties to methylxanthines. It is indicated as a second-line agent, when methylxanthines have proved ineffective. The mechanism of its action seems related to the dose: working peripherally (carotid bodies) at low doses and centrally (brain and medulla system), at high doses [26,27]. Its major metabolite, ketodoxapram, is active and in animal studies appears to be better tolerated than doxapram [28].

Efficacy

Efficacy of intravenous doxapram in treating apnoeas has been shown in a few open studies. In 12 infants given doxapram intravenously (2-2.5 mg/kg/h), total disappearance of apnoea episodes was obtained in 5 infants within 6 hours, and the mean number of episodes decreased significantly in all 12 patients [29]. In another study, cessation of apnoeas was obtained in 9 of 12 patients given 1 to 1.5 mg/kg/h [30]. A dose/effect relationship has been shown with complete cessation of apnoeas in 47% of infants given 0.5mg/kg/h and in 89% of infants given 2.5 mg/kg/h [31]. The mean frequency of apnoeas decreased in 48% infants given 0.25 mg/kg/h and in 75 % when given 1 mg/kg/h [32]. Placebo controlled, blinded, randomised trials have shown doxapram to be more effective than placebo for the treatment of apnoea [33] and to stimulate spontaneous breathing in infants weaned from assisted ventilation [34]. However, prophylactic infusion of doxapram, given prior to extubation, did not increase the chance of successful extubation [35].

Therapeutic drug monitoring

Plasma concentrations effective in controlling apnoea in premature infants vary greatly, ranging from 0.47 to 5 mg/l [31,36-38]. In view of the wide inter-individual variability of the pharmacokinetic characteristics [39], plasma concentrations should be measured 48 hours after the beginning of the infusion and then weekly. As the therapeutic index is very narrow, measurement of the plasma concentration of the active metabolite, ketodoxapram, may be beneficial [39].

Pharmacokinetics

Several studies have calculated the elimination half-life of doxapram in infants. It is longer in premature neonates: (6.6 to 9.9 h) [37,38,40] than in adults (3.4 h) [41]. Oxidation to ketodoxapram occurs rapidly after oral or intravenous administration of doxapram. Absorption is fast and the maximal plasma concentration is attained within 1 hour. Bioavailability is 60% in adults.

Dosage schedules

The recommended dosage is 0.5 mg/kg/h to 1.5 mg/kg/h for the maintenance intravenous dose [31,33,37]. A loading dose of 3 mg/kg may be given [33]. The mean plasma concentrations obtained with these dosages range from 1.3 to 3.1 mg/l. Doxapram may be given orally by nasogastric tube over 1 hour, at a dose calculated as the daily intravenous dose of more than 50% [40], which leads to plasma concentrations within the therapeutic range [42]. To our knowledge, the only doxapram formulation marketed currently is the intravenous formulation (also used for the oral route), containing chlorbutanol as preservative.

Toxicity

Side effects of doxapram in neonates are similar to those reported for methylxanthines. Digestive troubles were the most frequently described disorders: vomiting, abdominal distension and necrotising enterocolitis [42]. Adverse CNS effects have been observed: jitteriness, seizures, irritability and an increase in the time spent awake. Cardiovascular side effects include an increase in mean arterial blood pressure [34] and the QTc interval lengthening [43].

Drug toxicity is dose-related. Severe side effects were noted in infants with doxapram plus ketodoxapram concentrations over 9 mg/l [39]. Mild adverse effects have been observed with plasma concentrations as low as 0.39, 0.47 and 1.0 mg/l [36,38]. As for the methylxanthines, an extensive study of the follow-up of neonates treated with doxapram is needed to assess the safety on neurological development.

5.7.3 MISCELLANEOUS

A number of other drugs have been proposed as "antiapnoeic" drugs (antenatal betamethasone, acetazolamide, diphemanil methylsulfate, primidone), but they have not been properly evaluated. Adverse effects may be significant, for example, "torsades de pointes", QT interval lengthening and atrioventricular blocks [44].

5.7.4 CONCLUSIONS

In summary, the use of respiratory stimulants in neonates should be kept for treating primary apnoeas linked to immaturity of the ventilatory command, or apnoeas still present after management of the causal disease, such as metabolic disorder, infection, or gastro-oesophageal reflux. Long-term outcome of children treated for apnoea in the neonatal period has not been reported.

REFERENCES

1.   Rubio, R.,Berne, R. M., Bockman, E. L., Curnish, R. R., *Am. J. Physiol., 228,* 1896 (1975).
2.   Kuzemko, J. A., Paala, J., *Arch. Dis. Child., 48,* 404 (1973).
3.   Shannon, D. C., Gotay, F., Stein, I. M., Rogers, M. C., Todres, I. D., Moylan F. M., *Pediatrics, 55,* 589 (1975).
4.   Bednarek, E. J., Roloff, D. W., *Pediatrics, 58,* 335 (1976).
5.   Peabody, J. L. , Neese, A. L., Philip, A. G., Lucey J. F., Soyka, L. F., *Pediatrics, 62,* 698 (1978).
6.   Aranda, J. V., Gorman, W., Bergsteinsson, H., Gunn, T., *J Pediatr., 90,* 467 (1977).
7.   Uauy, R., Shapiro, D. L., Smith, B., Warshaw, J. B., *Pediatrics, 55,* 595 (1975).
8.   Aranda, J. V., Turmen, T., *Clin. Perinatol., 6,* 87 (1979).
9.   Aranda, J. V., Cook, C. E., Gorman, W., et al., *J. Pediatr., 94,* 663 (1979).
10. Gunn, T. R., *J. Pediatr., 94,* 106 (1979).
11. Bairam, A., Boutroy, M. J., Badonnel, Y., Vert, P., *J. Pediatr., 110,* 636 (1987).
12. Lee, T. C., Charles, B., Steer, P., Flenady, V., Shearman, A., *Clin. Pharmacol. Ther., 61,* 628 (1997).
13. Scanlon, J. E. M., Chin, K. C., Morgan, M. E. I., Durbin, G. M., Hale, K. A., Brown, S. S., *Am. J. Dis. Child., 67,* 425 (1992).
14. Boutroy, M. J., Vert, P., Royer, R. J. , et al., *J. Pediatr., 94,* 996 (1979).
15. Gorodischer, R., Karplus, M., *Eur. J. Clin. Pharmacol., 22,* 47 (1982).
16. Walther, F. J., Erickson, R., Sims, M. E., *Am. J. Dis. Child. 144,* 1164 (1990).
17. Vandenplas, Y., Dewolf, D., Sacre, L., *Pediatrics, 77,* 807 (1986).
18. Novicki, P. T., *J. Pediatr. Gastro. Enterol. Nutr., 9,* 137 (1989).
19. Kulkarni, P. B., Dorand, R. D., *Pediatrics, 64,* 254 (1979).
20. Banner, W., Czajka, P. A., *Am. J. Dis. Child. 134,* 495 (1980).
21. Van Den Anker, J. N., Jengejan, H. T. M., Sauer, P. J. J., *Eur. J. Pediatr., 151,* 466 (1992).
22. Wechsler, R. L., Kleiss, L.M., Kety, S.S., *J. Clin. Invest., 29,* 28 (1950).
23. Cameron, O. G., Modell, J. G., Hariharan, M., *Life Sci., 47,* 1141 (1990).
24. Lundstrom, K. E., Pryds, O., Greisen, G., *Acta Paediatr., 84,* 6 (1995).
25. Saliba, E., Relier, J. P., *Progrès en néonatalogie, 16,* 43 (1996).
26. Kato, H., *J. Pharmacol. Exp. Ther., 144,* 260 (1964).
27. Funderburk, W. H., Alphin, R. S., *Fed. Proc., 21,* 324, (1962).
28. Bairam, A., Blanchard, P. W., Mullahoo, K., Beharry, K., Laudignon, N., Aranda, J. V., *Pediatr. Res., 28,* 142 (1990).
29. Barrington, K. J., Finer, N. N., Peters, K. L., Barton, J., *J. Pediatr., 108,* 125 (1986).
30. Hayakawa, F., *J. Pediatr., 109,* 138 (1986).
31. Barrington, K. J., Finer, N. N., Torok-Both, G., Jamali, F., Coutts, R. T., *Pediatrics, 80,* 22 (1987).
32. Bairam, A., Faulon, M., Monin, P., Vert, P., *Biol. Neonate, 61,* 1209 (1992).
33. Peliowski, A., Finer, N. N., *J. Pediatr., 116,* 648 (1990).

34. Huon, C., Rey, E., Mussat, P., Parat, S., Moriette, G., *Acta Paediatr.*, *87*, 1180 (1998).
35. Barrington, K. J., Muttitt, S. C., *Acta Paediatr.*, *87*, 191 (1998).
36. Tay-Uyboco, J., Kwiatkowski, K., Cates, D. B., et al., *Biol. Neonate, 59*, 190 (1991).
37. Jamali, F., Coutts RT, Malek F, Finer NN, Peliowski A. *Dev. Pharmacol. Ther., 16*, 78 (1991).
38. Kumita, H., Mizuno, S., Shinohara, M., et al., *Acta Paediatr. Scand., 80*, 786 (1991).
39. Barbé, F., *Ther. Drug Monitor., 21*, 547 (1999).
40. Bairam, A., Akramoff-Gershan, L., Beharry, K., Laudignon, N., Papageorgiou, A., Aranda, J. V., *Am. J. Perinatol., 8*, 110 (1991).
41. Robson, R. H., Prescott, L. F., *J. Chromatogr., 143*, 527 (1977).
42. Hascoet, J. M., Hamon, I., Boutroy, M. J., *Drug Safety, 23*, 363 (2002).
43. Maillard, C., *Clin. Pharmacol. Ther., 70*, 540 (2002).
44. Bennasr, S., Baumann, C., Casadevall, I., Bompard, Y., Jacqz-Aigrain, E., *Arch. Fr. Pediatr., 50*, 413 (1993).

5.8 Therapy for persistent ductus arteriosus

Bart Van Overmeire

Department of Paediatrics, Division of Neonatology, Antwerp University Hospital, Belgium

5.8.1 INTRODUCTION

During fetal life, the ductus arteriosus allows most of the combined ventricular output to be diverted from the pulmonary circulation to the fetal systemic and placental circulation. In healthy term babies, the ductus is functionally closed in 20% of babies on day 1, 82% on day 2, 96% on day 3 and 100% on day 4 [1]. Closure of the ductus involves the interaction of oxygen, prostaglandins, nitric oxide and the unique musculature of the vessel wall [2-4]. In prematurely born infants, the ductus frequently remains open. This is observed in about 40% of infants < 1000 g birth weight on day 3 [5]. In infants < 28 weeks gestation, spontaneous closure occurs in only 27% of infants by day 5 [6]. A patent ductus arteriosus (PDA) results in a significant left to right shunt and an increase in left ventricular output, an increased pulmonary perfusion and a reduced renal and gastrointestinal blood flow. A PDA has been thought to increase the severity of respiratory distress [7,8] and the risk of bronchopulmonary dysplasia [9].

Surgical ligation and the use of non-steroidal antiinflammatory drugs (NSAIDs), such as indomethacin and more recently also ibuprofen [10-13], are used to treat infants with a haemodynamically significant PDA. The pharmacological closure of PDA in premature infants was first reported in 1976 [14,15]. Subsequent studies showed NSAIDs to be effective in about 80% of treated infants [12,13]. Indomethacin has also been used to prevent the development of a symptomatic PDA [16]. Unfortunately, the optimum regimen to manage PDA in premature infants has still not been unequivocally established, despite more than 30 years of clinical trials using different approaches.

5.8.2 INDOMETHACIN

Pharmacokinetics

Indomethacin is a methylated indole derivative and a non-selective potent inhibitor of the prostaglandin-forming cyclo-oxygenase. It is highly bound to serum proteins. Metabolism via O-methylation, deacylation and conjugation to inactive metabolites allows for elimination via the biliary route; less than 10% is excreted unchanged in the urine [17].

Pharmacokinetic parameters in the neonate vary widely, due to the many physiological changes occurring after birth. With oral administration, absorption of indomethacin is incomplete and variable [18]. For intravenous indomethacin, various studies have used different doses (0.2 and 0.3 mg/kg), but all show large interindividual variations and important differences compared with the adult data: plasma half-life (11-36 h) is much longer and clearance rate (7-16 mL/kg/h) much less [19-22]. There are also large inter individual variations in the apparent volume of distribution and the duct, being open or closed, may in part contribute to this [22]. Increasing postnatal age, more than gestational age, is an important determinant of increasing indomethacin clearance and volume of distribution [23].

The relationship between the plasma concentration of indomethacin and the desired effect of ductal closure is controversial. In some studies plasma concentration did not correlate with efficacy [10,24]; however, others have found a relationship [21,23,25-28]. Various effective concentrations were suggested: 1 mg/L 10 h post dose [28]; 0.5 mg/L 8 h post dose [29]; 0.25 mg/L over a 3 day course [31]. From population pharmacokinetic data, the mean serum concentration at PDA closure was 1.43 mg/L [23]. Furthermore, low plasma levels of indomethacin, short serum half-life, and more efficient clearance are associated with unsuccessful indomethacin infusions [21].

Therapeutic versus prophylactic approach

Indomethacin may be administered prophylactically (usually within 24 h of birth), early symptomatic (2-3 days) or late symptomatic (8-10 days). Clyman evaluated these different strategies and found that the group receiving early symptomatic treatment with indomethacin had a reduction in bronchopulmonary dysplasia (OR=0.39; 95% CI 0.21-0.76; p<0.005), need for mechanical ventilation (p<0.005) and necrotising enterocolitis (NEC) (OR=0.24; CI 0.06-0.96; p<0.05) compared with late symptomatic treatment [29]. Prophylactic treatment did not have any advantage in long-term pulmonary outcomes or NEC compared with early symptomatic treatment; however, the incidence of grade 3 and 4 intraventricular haemorrhage (IVH) was reduced (OR=0.51; CI 0.28-0.95; p<0.05). A recent meta-analysis that combined 19 eligible trials randomising 2872 infants [16], concluded that prophylactic indomethacin reduced the incidence of symptomatic PDA (pooled RR=0.44; CI 0.38-0.50), but showed no evidence that treatment influenced respiratory outcomes. No difference in mortality at latest follow-up between infants receiving prophylactic indomethacin and controls was found, (RR=0.96; CI 0.81-1.12). Prophylaxis reduced the need for surgical PDA ligation (RR=0.51; CI 0.37-0.71), the incidence of grades 3 and 4 IVH, (RR=0.66; CI 0.53-0.82), but increased the incidence of oliguria (RR=1.90; CI 1.45-2.47).

Length of treatment

The most commonly used course of indomethacin is three doses at 12 hourly intervals. In the National Collaborative Study, an initial duct closure of about 80% with a subsequent reopening in responders <30% was obtained with 3 doses of 0.2 mg/kg for infants at 2-7 days of life and 200 mcg/kg followed by 2 doses of 0.25 mg/kg for infants >8 days, which was broadly comparable with previous studies [10].

To improve sustained closure rates, prolonged courses of indomethacin have been suggested as an alternative [25,30,31]. The additional administration of 0.2 mg/kg/day for 5 days after the conventional 3 doses of indomethacin, reduced PDA recurrence rate and the need for surgical ligation [30]. In a trial of 121 babies, with clinical detection of PDA, 0.1 mg/kg/day for 6 days was compared with 3 doses of 0.2 mg/kg 12 hourly [31]. For the prolonged course, the initial response rate appeared to be better (90% *vs.* 77%), and the relapse rate (21% *vs.* 40%) was lower. Other studies have not confirmed the advantages of a prolonged course. In 70 low birth weight infants randomised to either 2 doses of 0.15 mg/kg at 12 h interval or a maintenance regimen of 0.1 mg/kg/day for 5 days after the initial doses, the initial closure rate appeared better in the maintenance group, but the incidence of reopening was not different [32]. More recently, short *versus* prolonged indomethacin therapy was compared in 61 pre-term infants with echocardiographically confirmed PDA [33]. Those receiving a short course had shorter duration of oxygen supplementation and less frequent symptoms of NEC and sustained closure rates (74% *vs.* 60%); mortality and other neonatal morbidity rates were similar. In a study that randomised 140 infants to either 3 doses of 0.2 mg/kg every 12 h *versus* prolonged low dose indomethacin (6 daily doses of 0.1 mg/kg), no difference in efficacy and a trend toward more NEC was observed in the prolonged low-dose group [34]. Exis-ting evidence seems inconclusive as to which regimen offers the most optimal benefit/side effect ratio.

Side effects

Indomethacin reduces bloodflow in different organ systems. A decreased renal perfusion leads to the clinically frequently observed renal impairment: a transient decrease in glomerular filtration rate, urine production, fractional excretion of sodium and chloride [35,36]. A careful fluid balance can partly prevent this side effect. Many centres consider a serum creatinine above 1.2 to 1.6 mg/dL, a rising serum creatinine concentration or a urine production of less than 1 mL/kg per hour as a contraindication for starting indomethacin treatment. Furosemide has been tried to prevent the renal side effects but may increase circulating prostaglandin levels and promote ductal patency [37]. Sufficient data are lacking to recommend its use during treatment of PDA [38,39]. Irrespective of this, diuretics seem to have little value for the treatment of PDA in preterms, unless there is congestive heart failure.

Indomethacin reduces cerebral blood flow, which may lead to impaired cerebral oxygenation [40,41]. This effect may not entirely be due to its inhibitory effects on prostaglandin synthesis [42,43]. The reduction of mesenteric blood flow may explain the observed association between indomethacin use and localised bowel perforations [44-48]. Although an increased occurrence of occult blood loss from the gastrointestinal tract was noted [10], the available overall evidence does not confirm an increased incidence of NEC after indomethacin use. Platelet aggregation is impaired by indomethacin and bleeding time may be prolonged [49]. It is widely accepted not to administer the drug when there is increased bleeding tendency or frank bleeding. Because of the remaining concern about the side effects of indomethacin, alternative NSAIDs, e.g. aspirin and ibuprofen, and other related compounds, have been studied for the prevention and treatment of PDA.

5.8.3   IBUPROFEN

Since 1995, ibuprofen has been studied extensively as an alternative treatment for PDA [11,13,53,54,55].

Pharmacokinetics

Ibuprofen is a non-steroidal antiinflammatory agent with a 2-arylpropionic acid structure, that exerts its pharmacological effect in the S(+)isomer configuration by inhibiting cyclo-oxygenase [50]. An unidirectional conversion from almost 60% of the R-form to S-form has been reported in adults. Indomethacin and ibuprofen are both non-selective cyclo-oxygenase (COX) inhibitors reducing the activity of both COX 1 and COX 2. Ibuprofen is insoluble in water. In most studies, the lysine salt of d,l-2-(4-isobutylphenyl)-propionic acid (ibuprofen-lysine) was used in order to obtain a formulation suited for intravenous infusion. In one study, ibuprofen-tri(hydroxymethyl)aminomethane (THAM) formulation was used. This trial was prematurely stopped because of the occurrence of severe side effects [51]. Very recently, it has been shown that oral administration might also be effective for inducing ductal closure in preterm infants [52].

There are few studies on the pharmacokinetics of ibuprofen in neonates, some papers providing only plasma levels (Table 1). The most frequently applied dosing schedule consists of a first dose of 10 mg/kg followed by a second and third dose of 5 mg/kg after 24 and 48 h [10-13,53-56]. As compared to indomethacin, ibuprofen has a prolonged serum half-life. Marked age-related differences and a wide interpatient variability in plasma concentrations and in pharmacokinetics are reported. In very low birth weight infants, the clearance of ibuprofen is slower and serum half-life is significantly prolonged as compared with older children [57,58]. In a study of 27 preterm infants with PDA, serum half-life decreased from 43 h (SD 7.8) on the third day to 26.8 h (SD 6.5) on the fifth day of life [56].

Loading doses varied from 5 to 20 mg/kg, successive doses from 5 to 8 mg/kg and these were given after intervals from 12 to 24 h.

Ibuprofen is more than 95% bound to serum albumin [54]. Because of its competitive binding with bilirubin, there is a risk of displacing bilirubin and increasing the unbound fraction of bilirubin in jaundiced infants [58]. A 4-fold increase of the fraction of free bilirubin was reported in an *in vitro* experiment at ibuprofen levels of 750 µmol/L [59]. *In vivo* ibuprofen serum levels in preterm infants ranged from 25 to 55 mg/L one hour after the infusion of 10 mg/kg of ibuprofen (1 mg/L ibuprofen = 4,8 µmol/L). With these levels, an insignificant increase in the unbound bilirubin fraction was observed [60,61].

Therapy and prophylaxis

Actual evidence from 8 studies including 509 patients, demonstrates that ibuprofen therapy is equally effective as indomethacin for inducing closure of PDA in preterm infants (RR=0.92; 95%CI 0.69-1.22) [12]. The occurrence of oliguria (urine production <1mL/kg/h) was significantly lower in the ibuprofen group than the indomethacin group. Mortality, surgical ligation of the ductus, duration of ventilatory support, intraventricular haemorrhage and NEC were not significantly different statistically between the two drugs.

When administered prophylactically, ibuprofen reduces the development of a PDA [65-67]. Acute hypoxaemia and pulmonary hypertension were observed immediately after the infusion of ibuprofen-tham solution in 3 infants [51].

Table 1

Overview of ibuprofen pharmacokinetic reports in neonates

Infants (n)	Method	Birth weight (g)	Gestation (weeks)	Age (days)	Doses (mg/kg)	Ibuprofen level (mg/L)	Ref.
15	plasma-levels	1210 (±550)	28.4 (± 3.1)	4-5	10-5-5 (q 24 h)	24.3 (±11.5)	61
43	sequential bayesian approach	<1500	<30	3-5	5 10 15 20	22 (n=7) (peak) 40.5 (n=6) 57 (n=6) 100 (n=2)	62
21	serum-levels 1 h after 1st and 2nd dose	929 (±213)	27.6 (±2.3)		8-8-8 (q 24 h)	20.2 (±9.5) 1st dose 25.6 (±13.5) 2nd dose	63
27	conventional pharmacokinetics	1250 (±460)	28.6 (±1.9)	3-5	10-5-5 (q 24 h)	43 (±11.2) 1st dose 42.4 (±22.3) 3rd dose	56
18	plasma-levels		23-28	>7	10-5-5 (q 12 h) po or iv	10.3 – 36 (3 h after 1st dose)	64
21	conventional pharmacokinetics	575-1450	22-31		10 (n=11) 10-5-5 (n=10) (q 24 h)	180.6 (±11.1) 1st 116.6 (54.5) 2nd 113.6 (58.2) 3rd	54

Although there is no clear explanation for this phenomenon, it has not been reported in trials using the ibuprofen-lysine formulation. Moreover, such "effect" is possibly not unique for ibuprofen, as an increase in oxygen requirement was also observed after the prophylactic administration of indomethacin [68]. Prophylaxis with ibuprofen seems to not reduce the occurrence of intraventricular haemorrhage (IVH) [65], in contrast with a pronounced and statistically significant reduction of severe IVH after early indomethacin administration [16,69].

Renal and mesenteric perfusion is less influenced by ibuprofen than by indomethacin [70] although, animal experiments could not confirm these observations [71]. Ibuprofen causes no reduction of cerebral blood flow and cerebral oxygen delivery was not significantly altered, an effect which contrasts with the effects of indomethacin [72-74]. Minimal changes that are observed in cerebral blood volume or blood flow after ibuprofen prophylaxis were comparable to those provoked with saline infusion [74].

5.8.4 ALTERNATIVE DRUGS

Early experience suggested that other cyclo-oxygenase inhibitors could also be used as therapy for PDA. Aspirin (acetylsalicylic acid) appeared to induce fewer side effects, however its efficacy was less than that of indomethacin [75]. In a prospective, randomised, controlled, multicentre study in 75 preterm infants, PDA closure was obtained in 92% of patients in the indomethacin group and in only 43% in the aspirin group (p<0.001) [76].

Mefenamic acid was studied in 16 preterm infants and compared with 30 historical indomethacin-treated control infants. Its efficacy was comparable to that of indomethacin, as was its induction of feeding intolerance [77-79]. No randomised studies are available.

Ethamsylate seemed to be able to prevent symptomatic PDA: only 1 in 96 neonates as compared with 8 in 38 historical control patients [80]. In a placebo-controlled trial of 20 premature infants, it appeared effective to obtain PDA closure (80% *versus* 20%; p<0.02) [81,82].

Sulindac had a comparable efficacy, as compared to indomethacin, in a case control study in 16 closely matched infants. Severe unexpected gastrointestinal complications were encountered in the sulindac group [83]. In a case report, fatal haemorrhagic gastritis was described after oral administration [84].

5.8.5 SURGICAL TREATMENT

Since the increasing use of indomethacin in the 1980s, surgical closure of the ductus is reserved for infants with contraindications to, or failure of, medical treatment. Perioperative complications associated with surgery include hypertension, pneumothorax, chylothorax, minor wound infections and vocal cord paralysis. Morbidity rates reported are 1-16% and mortality 0-10%. In experienced centres with low morbidity and mortality, surgery may be an appropriate alternative in extremely low birth weight infants where indomethacin has a high failure rate and significant adverse effects [85-88].

5.8.6 CONCLUSION

Indomethacin is still the most frequently used drug for therapy for PDA. The most optimal dosing scheme, with favourable benefit/risk ratio, still needs to be confirmed. Available data do indicate that ibuprofen(-lysine) is an effective alternative with less renal toxicity and less disturbance on cerebral and gastrointestinal circulations.

REFERENCES

1. Lim, M. K., Hanretty, K., Houston, A. B., Lilley, S., Murtagh, E. P., *Arch. Dis. Child., 67,* 1217 (1992).
2. Clyman, R. I., Chan, C. Y., Mauray, F., et al., *Pediatr. Res., 45,* 19 (1999).
3. Clyman, R. I., Waleh, N., Black, S. M., Riemer, R. K., Mauray, F., Chen, Y. Q., *Pediatr. Res., 43,* 633 (1998).
4. Hammerman, C., *Clin. Perinatol., 22,* 457 (1995).
5. Ellison, R. C., Peckham, G. J., Lang, P., et al., *Pediatrics, 71,* 364 (1983).
6. Narayanan, M., Cooper, B., Weiss, H., Clyman, R.I., *J. Pediatr., 136,* 330 (2000).
7. Thibeault, D. W., Emmanouilides, G. C., Nelson, R. J., Lachman, R. S., Rosengart, R. M., Oh, W., *J. Pediatr., 86,* 120 (1975).
8. Siassi, B., Blanco, C., Cabal, L. A., Coran, A. G., *Pediatrics, 57,* 347 (1976).
9. Brown, E. R., *J. Pediatr., 95,* 865 (1979).
10. Gersony, W. M., Peckham, G. J., Ellison, R. C., Miettinen, O. S., Nadas, A. S., *J. Pediatr., 102,* 895 (1983).
11. Van Overmeire, B., Smets, K., Lecoutere, D., et al., *N. Engl. J. Med., 343,* 674 (2000).
12. Ohlsson, A., Walia, R., Shah, S., *The Cochrane Library Issue 2, Oxford: Update Software* (2003).
13. Patel, J., Marks, K. A., Roberts, I., Azzopardi, D., Edwards, A.D., *Lancet,* 346 (1995).
14. Heymann, M. A., Rudolph, A. M., Silverman, N. H., *N. Engl. J. Med., 295,* 530 (1976).
15. Friedman, W. F., Hirschklau, M. J., Printz, M. P., Pitlick, P. T., Kirkpatrick, S. E., *N. Engl. J. Med., 295,* 526 (1976).
16. Fowlie, P. W., Davis, P. G., *The Cochrane Library Issue 1 2003, Oxford: Update Software* (2003).
17. Hammerman, C., *Clin. Perinatol., 22,* 457 (1995).
18. Evans, M. A., Bhat, R., Vidyasagar, D., Vadapalli, M., Fisher, E., Hastreiter, A., *Clinical Pharmacology & Therapeutics,26,* 746 (1979).
19. Thalji, A. A., Carr, I., Yeh, T. F., Raval, D., Luken, J. A., Pildes, R. S., *J. Pediatr., 97,* 995 (1980).
20. Yaffe, S. J., Friedman, W. F., Rogers, D., Lang, P., Ragni, M., Saccar, C., *J. Pediatr., 97,* 1001 (1980).

21. Brash, A. R., Hickey, D. E., Graham, T. P., Stahlman, M. T., Oates, J. A., Cotton, R.B., *N. Engl. J. Med., 305,* 67 (1981).
22. Gal, P., Ransom, J. L., Weaver, R. L., et al., *Thera. Drug Monitor., 13,* 42 (1991).
23. Wiest, D. B., Pinson, J. B., Gal, P. S., et al., *Clin. Pharmacol. Ther., 49,* 550 (1991).
24. Ramsay, J. M., Murphy, D. J. Jr., Vick, G. W, Courtney, J. T., Garcia-Prats, J. A., Huhta, J. C., *Am. J. Dis. Child., 141,* 294 (1987).
25. Seyberth, H. W., Knapp, G., Wolf, D., Ulmer, H. E., *Eur. J. Pediatr., 141,* 71 (1983).
26. Lewis, I. G., Harvey, D. P., Maxwell, G. M., *Austr. Paediatr. J., 21,* 181 (1985).
27. Yeh, T. F., Achanti, B., Patel, H., Pildes, R. S., *Dev. Pharmacol. Ther., 12,* 169 (1989).
28. Gal, P., Ransom, J. L., Schall, S., Weaver, R. L., Bird, A., Brown, Y., *J. Perinatol., 10,* 20 (1990).
29. Clyman, R. I., *J. Pediatr., 128,* 601 (1996).
30. Hammerman, C., Aramburo, M. J., *J. Pediatr., 117,* 771 (1990).
31. Rennie, J. M., Cooke, R. W., *Arch. Dis. Child., 66,* 55 (1991).
32. Rhodes, P. G., Ferguson, M. G., Reddy, N. S., Joransen, J. A., Gibson, J., *Eur. J. Pediatr., 147,* 481 (1988).
33. Tammela, O., Ojala, R., Iivainen, T., et al., *J. Pediatr., 134,* 552 (1999).
34. Lee, J., Rajadurai, V. S., Tan, K. W., Wong, K. Y., Wong, E. H., Leong, J. Y., *Pediatrics, 112,* 345 (2003).
35. Betkerur, M. V., Yeh, T. F., Miller, K., Glasser, R. J., Pildes, R.S., *Pediatrics, 68,* 99 (1981).
36. Rennie, J. M., Doyle, J., Cooke, R. W., *Arch. Dis. Child., 61,* 233 (1986).
37. Green, T. P., Thompson, T. R., Johnson, D. E., Lock, J. E., *N. Engl. J. Med., 308,* 743 (1983).
38. Romagnoli, C., Zecca, E., Papacci, P., et al., *Clin. Pharmacol. Ther., 62,* 181 (1997).
39. Brion, L. P., Campbell, D. E., *Cochrane Library of Reviews, 3* (1999).
40. Cowan, F., *J. Pediatr., 109,* 341 (1986).
41. Edwards, A. D., Wyatt, J.S., Richardson, C., et al., *Lancet, 335,* 1491 (1990).
42. Malcolm, D. D., Segar, J. L., Robillard, J. E., Chemtob, S., *J. Appl. Physiol., 74,* 1672 (1993).
43. Chemtob, S., Beharry, K., Barna, T., Varma, D. R., Aranda, J. V., *Pediatr. Res., 30,* 106 (1991).
44. Vanhaesebrouck, P., Thiery, M., Leroy, J. G., et al., *J. Pediatr., 113,* 738 (1988).
45. Wolf, M. M., Snover, D. C., Leonard, A. S., *J. Pediatr. Surg., 24,* 409 (1989).
46. Meyers, R. L., Alpan, G., Lin, E., Clyman, R. I., *Pediatr. Res., 29,* 569 (1991).
47. Scholz, T. D., McGuinness, G. A., *J. Pediatr. Gastroenterol. Nutr., 7,* 773 (1988).
48. Grosfeld, J. L., Chaet, M., Molinari, F., et al., *Ann. Surg., 224,* 350 (1996).
49. Corazza, M. S., Davis, R. F., Merrit, T. A., Bejar, R., Cvetnic, W. *J. Pediatr., 105,* 292 (1984).
50. Adams, S. S., Bresloff, P., Mason, C. G., *J. Pharm. Pharmacol., 28,* 256 (1976).
51. Gournay, V., Savagner, C., Thiriez, G., Kuster, A., Rozé, J.-C., *Lancet, 359,* 1486 (2002).
52. Heyman, E., Morag, I., Batash, D., Keidar, R., Baram, S., Berkovitch, M. *Pediatrics, 112,* 354 (2003).
53. Varvarigou, A., Bardin, C. L., Beharry, K., Chemtob, S., Papageorgiou, A., Aranda, J.V., *JAMA, 275,* 539 (1996).
54. Aranda, J. V., Varvarigou, A., Beharry, K., et al., *Acta Paediatr, 86,* 289 (1997).
55. Van Overmeire, B., Follens, I., Hartmann, S., Creten, W. L., Van Acker, K. J., *Arch. Dis. Child., 76,* F179 (1997).
56. Van Overmeire, B., Touw, D., Schepens, P. J. C., Kearns, G. L., van den Anker, J. N., *Clin. Pharmacol. Ther., 70,* 336 (2001).
57. Kauffman, R. E., Nelson, M.V., *J. Pediatr., 121,* 969 (1992).
58. Hansen, T. W., *Eur. J. Pediatr., 162,* 356 (2003).
59. Cooper-Peel, C., Brodersen, R., Robertson, A., *Pharmacol. Toxicol., 79,* 297 (1996).

60. Ahlfors, C. E., *J. Pediatr., 144*, 386 (2004).
61. Van Overmeire, B., Vanhagendoren, S., Schepens, P. J., Ahlfors, C. E., *Pediatr. Res., 55,* 474A (2004).
62. Desfrere, L., Zohar, S., Morville, P., et al., *Pediatr. Res., 53,* 420A (2003).
63. Plavka, R., Svihovec, P., Borek, I., et al., *Pediatr. Res., 49,* 375A [2153] (2001).
64. Raju, N. V., Bharadwaj, R. A., Thomas, R., Konduri, G. G., *J. Perinatology, 20,* 13 (2000).
65. Van Overmeire, B., Casaer, A., Allegaert, K., et al., *Pediatr. Res., 51,* 379A (2002).
66. De Carolis, M. P., Romagnoli, C., Polimeni, V., et al., *Eur. J. Pediatr., 159,* 364 (2000).
67. Shah, S. S., Ohlsson, A., *The Cochrane Library, Issue 2,* Oxford (2003).
68. Schmidt, B., Wright, L. L., Davis, P., Solimano, A., Roberts, R. S., *Lancet, 360,* 492 (2002).
69. Schmidt, B., Davis, P., Moddemann, D., et al., *N. Engl. J. Med., 344,* 1966 (2001).
70. Pezzati, M., Vangi, V., Biagiotti, R., Bertini, G., Cianciulli, D., Rubaltelli, F. F., *J. Pediatr., 135,* 733 (1999).
71. Guignard, J. P., *Semin. Perinatol., 26,* 398 (2002).
72. Mosca, F., Bray, M., Lattanzio, M., Fumagalli, M., Tosetto, C., *J. Pediatr., 131,* 549 (1997).
73. Patel, J., Roberts, I., Azzopardi, D., Hamilton, P., Edwards, A. D., *Pediatr. Res., 47,* 36 (2000).
74. Naulaers, G., Delanghe, G., Allegaert, K., Casaer, P., Devlieger, H., Van Overmeire, B., *Pediatr. Res., 51,* 338A (2002).
75. Heymann, M. A., Rudolph, A. M., Silverman, N. H., *N. Engl. J. Med., 295,* 530 (1976).
76. Van Overmeire, B., Brus, F., van Acker, K. J., et al., *Pediatr Res 38,* 886 (1995).
77. Sakhalkar, V. S., Merchant, R. H., *Indian Pediatr., 29,* 313 (1992).
78. Fujiwara, T., in *Pulmonary Surfactant,* Robertson, B., Van Golde, L. M. G. (eds.), Elsevier Science, p. 495, (1984).
79. Ito, K., Niida, Y., Sato, J., Owada, E., Ito, K., Umetsu, M., *Acta Paediatr. Jpn., 36,* 387 (1994).
80. Rosti, L., Piva, D., Rosti, D., *Arch. Pediatr. Adolesc. Med., 148,* 1103 (1994).
81. Amato, M., Hüppi, P. S., Markus, D., *Acta Paediatr., 81,* 351 (1992).
82. Amato, M., Hüppi, P. S., Markus, D., *J. Perinatol., 13,* 2 (1993).
83. Ng, P. C., So, K. W., Fok, T. F., Yam, M. C., Wong, M. Y., Wong, W., *J. Paediatr. Child. Health, 33,* 324 (1997).
84. Ng, P. C., So, K. W., Fok, T. F., To, K. F., Wong, W., Liu, K., *Acta Paediatr., 85,* 884 (1996).
85. Trus, T., Winthrop, A. L., Pipe, S., Shah, J., Langer, J. C., Lau, G. Y., *J. Pediatr. Surg., 28,* 1137 (1993).
86. Palder, S. B., Schwartz, M. Z., Tyson, K. R., Marr, C. C., *J. Pediatr. Surg., 22,* 1171 (1987).
87. Zerella, J. T., Spies, R. J., Deaver, D. C., Dailry, W. J., Haple, D. C., Trump, D. S., *J. Pediatr. Surg.,18,* 835 (1983).
88. Robie, D. K., Waltrip, T., Garcia-Prats, J. A., Pokorny, W. J., Jaksic, T., *J. Pediatr. Surg., 31,* 1134 (1996).

5.9 Antibiotics in neonates: the need for a more rational approach

Jean-Paul Langhendries[1], John N. van den Anker[2]

[1]NICU, CHC-Site St Vincent, B-4000 Rocourt-Liège, Belgium

[2]Department of Paediatrics, Erasmus MC-Sophia, Rotterdam, The Netherlands

5.9.1 INTRODUCTION

The incidence of documented severe bacterial infection in the neonate within the first 48 hours of life (early-onset infection) varies from 0.5 to 1% of total deliveries [1]. Whether the membranes are ruptured or not, the very preterm delivery (< 30 weeks gestation) is the final result of an infectious process in more than 50% of preterm births, even if the baby is not infected at all [2]. On the other hand, an invasive infection occurs later (late-onset infection) in more than 25% of hospitalised very low birth weight (VLBW) infants.

These epidemiological data, the immaturity of the immune system of the newborn and the occult clinical symptoms of infection explain the frequent overuse of antibiotics in neonatal intensive care units (NICUs) [3]. A more rational approach in the perinatal use of antibiotics is needed because of two observations: 1) the emergence of multi-resistant bacteria in the NICU resulting in severe morbidity and the increase in neonatal fungal infections [4,5]; 2) the perinatal anti-bacterial treatment that might interfere with the delicate process of the bacterial colonisation of the bowel, initiating the innate immunologic response in the early stage [6,7]. It has been hypothesised that this interference early in life may give rise to further disturbances of the adaptive immune responses in adults [8]. The proposed rational approach of the perinatal anti-bacterial treatment not only aims at optimal use of antibiotics to reduce the incidence of opportunistic infections, but also focuses on the pharmacokinetic/pharmacodynamic (PK/PD) relationship of the particular antibiotics used in the NICU [9]. As a consequence, new dosing regimens of antibiotics in neonates have been proposed in order to ameliorate their PK/PD relationship [10-17]. Despite the fact that several studies investigating some of these "old" drugs are being presently conducted, most antibiotics in the newborn are still used in an off-label, undesirable fashion [18]. Finally, the optimal use of anti-microbial drugs also includes the prompt discontinuation of the treatment in case infection cannot be demonstrated. A recent survey showed a large variation between different neonatal units in the use of antibacterial agents for the treatment of late-onset septicaemia in the neonate [19].

The aim of this review is therefore to give an overview of antibiotic use in neonates and to give guidelines for clinical use.

5.9.2 BACTERIAL EPIDEMIOLOGY

During the last few decades, the organisms responsible for neonatal early-onset sepsis have changed very little since *Streptococcus agalactiae* and *Escherichia coli* represent the major causative pathogens [20]. On the other hand, coagulase negative staphylococci, most often methicillin resistant, have steadily increased as the primary cause of late-onset neonatal septicaemia and invasive infections. However, a recent survey has indicated that there may be an increase in multi-resistant *E. coli* as the leading causative agent of early onset septicaemia in VLBW infants [21]. This increase in multi-resistant *E.coli* may be related to the increase in use of broad-spectrum antibiotics during labour and delivery. Regarding late-onset sepsis, some recent reports have shown an emerging number of cases where gram negative bacilli have been found as the causative agent [22]. These findings regarding changes in the bacteria responsible for infectious diseases in the neonate are of concern. These preliminary findings need confirmation, but it shows that this problem cannot be solved by a more aggressive antibiotic regimen with even broader spectrum antibiotics. Moreover, these reports should motivate perinatologists to use a more rational approach with narrow-spectrum antibiotics and an optimal antibiotic combination [23]. In addition, implementing good clinical practice procedures and guidelines in each intensive care unit is needed. Among those measures, consequent hand washing by care providers and isolating and cohorting the patients colonised with multi-resistant bacteria should be reinforced [24].

5.9.3 WHICH ANTIBIOTICS SHOULD BE PRESCRIBED?

An appropriate combination of antibiotics should be initiated as soon as an infectious disease is suspected, and accordingly adjusted depending on the bacterial sensitivity, or discontinued if the presumed infection is not confirmed.

Early-onset septicaemia

An antibiotic combination is always needed and should consist of antibiotics with a time-dependent (beta-lactam) and a concentration-dependent bactericidal activity (aminoglycoside). In the absence of antibiotics given to the mother and/or the lack of serious clinical conditions, such as meningitis or septic shock, penicillin-G or ampicillin remain the β-lactams of choice. The former is preferred if infection caused by *Streptococcus agalactiae* is suspected. Systematic use of third-generation cephalosporins with or without another β-lactam, in combination or not with an aminoglycoside, has to be avoided, as these agents are strong inducers of β-lactamases, notably AmpC produced by most species of *Enterobacteriaceae* which can move from chromosome to plasmid. In addition, third-generation cephalosporins favour the emergence of gram negative bacteria producing plasmid mediated extended spectrum β-lactamases (ESBL) [25]. These antibiotics should be therefore reserved, either in a bi-or tri-therapeutic approach, in clinical situations highly suggestive of meningitis or septic shock. When an EBSL strain is suspected antenatally, the new carbapenems should be the treatment of choice. Although these agents are equally strong inducers of β-lactamases, they are less sensitive to enzymatic hydrolysis and remain able to kill the bacteria.

Late-onset septicaemia

After the first days of life, the antibiotic prescription should take into account different important items [26]: 1) the clinical presentation based on specific signs (shock, suspected meningitis, digestive symptoms, etc.); 2) the existing bacteriological data into the unit and in the patient; 3) and the available epidemiological data in the literature regarding neonatal nosocomial infection. Intravenous cloxacillin and an aminoglycoside as first line treatment is to be recommended in a baby facing suspected sepsis without other specific organ injury [27]. This combination is effective in most cases of coagulase negative *staphylococcus* infection which rarely causes fulminant sepsis. Subsequent switch to vancomycin is advised if it is demonstrated that the strain is resistant to cloxacillin and aminoglycoside. Proceeding in such a way prevents the overuse of vancomycin, which is the main risk factor for the emergence of vancomycin-resistant enterococci. Indeed, the latter strain has been directly implicated in the widespread dissemination of resistance between bacterial species. If gastrointestinal pathology is suspected, cephalosporins should be used in association both with an aminoglycoside and an antibiotic against anaerobic bacteria (for instance metronidazole). The choice of cephalosporins will depend on the clinical presentation. Second-generation cephalosporins are less able to induce β-lactamases release but are more sensitive to their hydrolysis [28]. Third-generation cephalosporins should be preferred in all life-threatening

Table 1

Doses for beta-lactams

Antibiotic	Postconceptional age (PCA) (weeks)	Dose /kg	Interval (hours)	Comments*
beta-lactams	<32	50.000 IU	12	300'000 IU /kg/
penicillin G	32-45	50.000 IU	8	day in life threate-
	>45	50.000 IU	6	ning infections
ampicillin	<32	50 mg	12	200 mg /kg/day in
	32-45	50 mg	8	case of meningitis
	>45	50 mg	6	
oxacillin	<32	50 mg	12	
	32 – 45	50 mg	8	
	>45	50 mg	6	
methicillin	<32	50 mg	12	
	32 – 45	50 mg	8	
	>45	50 mg	6	
piperacillin	<32	75 mg	12	
	32 -45	75 mg	8	
	>45	75 mg	6	
cefoxitin	<32	30 mg	12	
	32-45	30 mg	8	
	>45	30 mg	6	
cefuroxime	<32	50 mg	12	
	32-45	50 mg	8	
	>45	50 mg	6	
cefotaxime	<32	50 mg	12	slow infusion
	32-45	50 mg	8	over 30 minutes.
	>45	50 mg	6	200 mg /kg/day is recommended in case of meningitis
ceftazidime	<32	50 mg	12	
	32-45	50 mg	8	
	>45	50 mg	6	
meropenem	<32	20 mg	12	40 mg/kg in case
	32-45	20 mg	8	of infection by
	>45	30 mg	8	Pseudomonas

* Continuous infusion of all penicillin-containing compounds not validated in neonates. The same total daily dose per kg may also be divided into smaller doses with shorter intervals between doses. Avoid all C3 cephalosporins if EBLS bacteria.

situations or suspected meningitis due to its better penetration into the cerebrospinal fluid (CSF). When an EBSL strain is colonising the baby suspected to be infected, cephalosporins should be replaced by imipenem or meropenem. Their systematic use is, however, not recommended, as it would take the risk of selecting bacterial species (*pseudomonas*) and/or favour bacterial resistance.

Other specific infections in the neonate

Chlamydia trachomatis and Ureaplasma urealyticum infection

A neonatal infection caused by these specific pathogens need to be suspected in case of late neonatal dacro-cystitis infection for the former pathogen and in case of late-onset pneumonia for both. The antibiotic of choice in the neonatal period is erythromycin.

Fungal infections

If specific clinical symptoms (i.e., suggestive rash, irritability, temperature instability, etc.) are present in a patient with *Candida* strain carriage, a neonatal fungal infection needs to be considered. This suspicion of infection also has to be addressed in every neonatal progressive clinical deterioration not ameliorated by usual antibiotics. Amphotericin B, with or without flucytosine, still remains the drug(s) of choice despite the fact that the antifungal azoles (ketoconazole, fluconazole, itraconazole) are increasingly used as an alternative. For all these drugs, including amphotericin B and flucytosine, the optimal dose as well as the duration of treatment are still a matter of discussion due to the lack of sufficient pharmacokinetic data [29].

Parasitic infections

The risk of a congenital *Toxoplasma* infection is increased when the maternal infection occurs late in gestation. In the case of a congenital infection, sulfadiazine and pyrimethamine are usually given in combination, as these drugs act synergistically. It still remains a matter of debate how long these drugs have to be given and if they have to be administered without any interruption, or on a 21-day-course base alternating with 4 to 6 week courses of spiramycin. However, folinic acid must be given along with the combination of sulfadiazine and pyrimethamine.

5.9.4 PHARMACOKINETICS (PK)

While the pharmacodynamic action on the bacterial target is obviously the same in neonates, dramatic differences exist in terms of neonatal antibiotic pharmacokinetics, as compared to children and adults [30-32]. These changes are even more pronounced in sick neonates and have been detailed in other chapters. It renders the administration of antibiotics in this population difficult, especially for drugs with a narrow therapeutic index. In addition, larger inter- and intra-patient variations exist.

Oral absorption of most of the antibiotics is erratic in neonates, decreasing the bioavailability of the drug. This route of administration is, therefore, not recommended in neonates with haemodynamic instability. Data that suggest that the antifungal azoles can be given orally in neonates need to be confirmed [29]. IM administration is painful and the bioavailability of the drug given in such a way also depends on muscle perfusion. Intravenous infusion is therefore the preferred route of administration in neonates. In extremely premature babies, total body water (TBW) may account for 85% of body weight with a higher proportion of extracellular fluid (45%), which is quite different when compared to term babies (TBW 75%) [33]. The high proportion of extracellular fluid is the most important parameter which explains the high volume of distribution of hydrophilic drugs in neonates. The high volume of distribution and longer half-life of aminoglycosides affects the loading dose, which is proportionally higher in neonates. On the other hand, the levels of plasma proteins are lower in the newborn, which result in lower binding capacity and higher levels of free drug. Hepatic metabolism matures with advancing gestational age and wide inter-patient variation in drug hepatic metabolism has been reported.

Renal function, however, is the most important determinant in respect to the elimination of antibiotics. The process of birth is the signal of a dramatic increase in glomerular filtration rate (GFR), in both term and preterm infants, while this increase is slower in the latter group until nephrogenesis is achieved by 34 weeks of postconceptional age (PCA) [34-35]. The clearance of aminoglycosides is proportional to the PCA [36]. The immature kidney exhibits a reduced capacity (20 to 30% of adult values) to excrete organic anions, such as the β-lactams. This excretion is done through specific transporters implementing the secretion pathway [37] and it matures rapidly in preterm as well as term babies [38]. Clinical situations such as sepsis, hypoxia and PDA may interact with the renal elimination of potentially toxic antibiotics.

Co-administration of drugs which act on renal perfusion (e.g. NSAIDs, furosemide) or which modify the anionic charge of proximal phospholipids (vancomycin) may increase aminoglycoside nephrotoxicity [39]. Monitoring of aminoglycoside serum concentrations is indicated in order to avoid drug accumulation.

5.9.5   PHARMACOKINETIC/PHARMACODYNAMIC (PK/PD) RELATIONSHIP

General principles

Bactericidal efficacy of antibiotics has been assessed from *in vitro* studies, using both the minimal inhibitory concentration (MIC) and the minimal bactericidal concentration (MBC). These *in vitro* parameters have been found useful in determining the best choice of antibiotics, but it is only in the last two decades that experimental studies have highlighted the relationship between their efficacy and their *in vivo* concentration at the site of infection [40-43]. In particular, the post-antibiotic effect (PAE) has been determined, i.e. the time it takes for the bacteria to start re-growing when the antibiotic concentration falls below the MIC [44]. This has resulted in a new antibiotic classification [45-47]: 1) time-dependent bactericidal activity: β-lactams (penicillins, cephalosporins, carbapenems), glycopeptides, erythromycin; 2) concentration-dependent bactericidal activity: aminoglycosides, metronidazole; 3) mixed bactericidal activity: quinolones.

β-lactams and glycopeptides

Both antibiotics target bacterial membrane synthesis and are inhibitors of the peptidoglycan synthesis, although their modes of action are distinctive. Both the killing effect of β-lactams and minimising bacterial resistance are maximal, while their concentration at the site of infection is continuously above the MIC. A concentration four times greater than the MIC is recommended (Table 1). β-lactams, except carbapenems, show no PAE for gram negative bacilli. Glycopeptides have a longer bacteriostasis for gram positive bacteria than β-lactams.

Aminoglycosides

Aminoglycosides need to enter the bacterial cell to inhibit protein synthesis. Uptake of aminoglycosides across the cytoplasmic membrane entails two energy-dependent phases, the former being driven by the transmembrane electrical potential. The second phase is stimulated by the binding of aminoglycoside molecules to the ribosome process, which gives rise to the incorporation of defective proteins in the bacterial wall [48]. Their rapid killing effect is directly related to the maximum serum concentration (C_{max}). A C_{max}/MIC ratio of at least 8 is now recommended. The toxicity (nephro- and ototoxicity) of aminoglycosides is directly related to the total dose and the length of treatment [39]. It has been postulated that maximising C_{max} should render these drugs more efficacious and could allow decreasing the length of treatment with the antibiotic combination (Table 2).

Table 2

Doses for aminoglycosides

Antibiotic	Postconceptional age (PCA) (weeks)	Dose (mg/kg)	Interval (hours)	Recommended trough levels (μg/ml)
amikacin	<28	20	48	<5
	28 - 30	18	42	
	31 - 34	16	36	
	34 – 37	16	30	
	> 37	15	24	
gentamicin	<28	5	48	<2
	28 - 30	4.5	42	
	31 - 34	4	36	
	34 – 37	4	30	
	> 37	4	24	
netilmicin	<28	5	48	<2
	28 - 30	4.5	42	
	31 - 34	4	36	
	34 – 37	4	30	
	> 37	4	24	
tobramycin	<28	5	48	<2
	28 - 30	4.5	42	
	31 - 34	4	36	
	34 – 37	4	30	
	> 37	4	24	

* Administer aminoglycosides over 20 minutes at a strictly separate time from all penicillin-containing compounds. Check the terminal infusion device to be sure that the entire dose is given at the proposed time. For all aminoglycosides, increase the interval by 6 hours or according to the blood trough level, whatever the PCA in the presence of asphyxia, haemodynamic instability or concomitant nephrotoxic drugs.

Antifungal drugs

Amphotericin B acts by binding with sterols of the fungal cell membrane, resulting in a loss of potassium channel integrity. There is considerable inter- and intra-patient variability in pharmacokinetics. Dosing and dosing intervals are summarised in Table 3. The major side effects are transient nephrotoxicity, hepatotoxicity and bone marrow suppression, in particular in very small infants where the drug appears to accumulate.

Flucytosine is de-aminated into the fungal cell and acts as an inactive pyrimidine substitute interfering with DNA synthesis. High concentrations are thought

Table 3

Doses for antifungals

Antifungal	Postconceptional age (PCA) (weeks)	Dose mg/kg	Interval (hours)	Administration	Comments
amphotericin B	<29	0.75	48	stepwise progression related to the renal tolerance: daily increase of 0.25 mg/kg to achieve a daily dose of 1 mg/kg whatever the PCA. minimal total cumulative dose of 25 mg/kg in disseminated candidal disease.	infusion over 8 hours; check renal function, blood count and electrolytes.
	30 - 36	0.75	24		
	≥ 37		24		
flucytosine	<29	25	12	oral administration in combination with amphotericin B or fluconazole.	not to be used alone; check the hepatic tests.
	30 - 36	25	8		
	≥ 37	25	6		
fluconazole	<29	6	48	oral administration is possible if the baby is haemodynamically stable.	one hour IV infusion; check the hepatic tests.
	30 - 36	6	24		
	≥ 37	6	12		

to be hepatotoxic. The mechanism of action is the inhibition of the synthesis of ergosterol, through the inhibition of the cytochrome P-450 system.

Studies on the pharmacokinetics of fluconazole have been conducted in neonates, but the full PK/PD relationship is still unknown. Flucytosine is available both as oral and intravenous formulations. Some 80% of the administered dose of fluconazole is excreted unchanged in the urine and it accounts for the rapidly increasing clearance with the postnatal age. Fluconazole has good penetration into the CSF. Itraconazole and ketoconazole are only available as oral formulations.

Drugs that act against anaerobes and parasites

Metronidazole belongs to the nitro-5 imidazoles family and acts on anaerobes and parasitic cells through the inhibition of DNA synthesis (Table 4). Its action is concentration-dependent, as this drug has to be metabolised by the micro-organisms.

5.9.6 ADMINISTRATION OF ANTIBIOTICS

The mode of infusion will depend both on the PCA and the type of antibiotics. The PCA is a reflection of the postnatal maturation of elimination pathways and will determine the dosing interval; the bacteriocidal effect of the antibiotic will determine the dose. Taking this classification into account, new modes of antibiotic administration have been proposed in order to improve efficacy, reduce toxicity and minimise the emergence of bacterial resistance. This approach, however, has not been validated in neonates [49]. Most studies have been performed in adults and children.

β-lactams and glycopeptides

In the premature neonate, the majority of PK studies have demonstrated adequate trough concentrations with the recommended doses and intervals [49]. The same total daily dose per kg may also be divided into smaller doses with shorter intervals between doses. Activity against multi-resistant bacteria is the only reason why intermittent high plasma concentrations are needed. β-lactams are eliminated through the renal route using tubular transporter processes which undergo rapid postnatal maturational processes, even in preterm neonates. The therapeutic index is large, eliminating the need for therapeutic drug monitoring.

Monitoring of plasma levels is recommended for vancomycin [50]. Vancomycin has a time-dependent killing effect. Sufficient concentrations at the site of infection, therefore, have to be reached in order to optimise bactericidal activity. This can be achieved with plasma concentrations continuously between 15 and 20 mcg/ml. Continuous infusion of vancomycin, after a loading dose, has been proposed in neonates as a new approach, in order to maintain permanently sufficient plasma levels [17] and decrease the incidence of red man syndrome. Prophylactic use of vancomycin is not recommended due to the increased risk of inducing vancomycin bacterial resistance.

Aminoglycosides and metronidazole

The concentration-dependent bactericidal activity of aminoglycosides, their high volume of distribution and the finding that their binding to the brush membrane of the proximal tubule, as well as the cochlear cell, is a saturable process, has given rise to the concept that relatively larger doses with much longer intervals between doses should be used in neonates, as has been demonstrated in adults [10-16]. Initial results are encouraging. Therapeutic drug monitoring is recommended. The optimal trough blood levels are as follows: amikacin: <5

Table 4

Doses for antibiotics

Antibiotic	Postconceptional age (PCA) (weeks)	Dose mg/kg	Interval (hours)	Administration	Comments
glycopeptides					
vancomycin	<29	10	12	IV infusion over 60 minutes	or continuous infusion after a loading dose
	30 - 36	10	8		
	≥37	7.5	6		
macrolides					
erythromycin	whatever the PCA	5 to 10	6	IV when severe infections. Infuse slowly a diluted concentration of 1 to 5 mg/ml over at least 60 minutes	do not use with cisapride; caution with methyl- xanthines
		12.5	*6*	*oral dose*	
anti-anaerobes					
metronidazole	<29	15	48		concentration dependent bactericidal effect
	30 - 36	15	24		
	≥ 37	10	12		

mcg/dl; netilmicin: <2 mcg/dl; gentamicin: <2 mcg/dl: tobramycin: <2 mcg/dl. Lower trough concentrations are probably indicated if the dosing intervals are prolonged. The concentration dependent killing effect of metronidazole allows once daily administration (Table 4).

Stability of antibiotics given in combination

It is important to realise that an inactivation of both antibiotics may result from the combination of aminoglycosides and β-lactams into the tubing devices [51]. Fortunately, this problem appears limited in the neonate [52]. It is recommended, however, to separate the time of administration of both antibiotics or to flush the tubing device. The degree of inactivation is not only dependent on the type of aminoglycoside but also on the temperature, the concentration and the type of solution being used [51,53].

REFERENCES

1. Klein, J. O., Remington, J. S., "Current concepts of infections of the fetus and newborn infant," in *Infectious diseases of the fetus and newborn infant,* Remington, J. S., Klein, J. O. (eds.), W. B. Saunders, pp. 1-19 (1995).
2. Gomez, R., Ghezzi, F., Romero, R., Munoz, H., Tolosa, J. E., Rojas, I., *Clin. Perinatol., 22,* 281 (1995).
3. Fonseca, S. N., Ehrenkranz, R. A., Baltimore, R. S., *Infect. Control Hosp. Epidemiol., 15,* 156 (1994).
4. Royle, J., Halasz, S., Eagles, G., et al., *Arch. Dis. Child. Neonatal Ed., 80,* F64 (1999).
5. Finnstrom, O., Isaksson, B., Haeggman, S., Burman, L. G., *Acta Paediatr., 87,* 1070 (1998).
6. Braback, L., Hedberg, A., *Clin. Exp. Allergy, 28,* 936 (1998).
7. Droste, J. H. J., Wieringa, M. H., Weyler, J. J., Nelen, V. J., Vermeire, P. A., Van Bever, H. P., *Clin. Exp. Allergy, 30,* 1547 (2000).
8. Bach, J. F., *N. Engl. J. Med., Sep 19, 347,* 930 (2002).
9. Vogelman, B., Craig, W. A., *J. Pediatr., 108(2),* 835 (1986).
10. Langhendries, J. P., Battisti, O., Bertrand, J. M., et al., *Dev. Pharmacol. Ther., 20,* 220 (1993).
11. Skopnik, H., Heimann, G., *Pediatr. Infect. Dis. J., 14,* 71 (1995).
12. Langhendries, J. P., Battisti, O., Bertrand, J. M., et al., *Biol. Neonate, 74,* 351 (1998).
13. Hayani, K. C., Hatzopoulos, F. K., Frank, A. L., et al., *J. Pediatr., 131,* 76 (1997).
14. De Hoog, M., Schoenmaker, R. C., Mouton, J. W., van den Anker, J. N., *Clin. Pharmacol. Ther., 62,* 393 (1997).
15. Thureen, P. J., Reiter, P. D., Gresores, A., Stolpman, N. M., Kawato, K., Hall, D. M., *Pediatrics, 103,* 594 (1999).
16. De Alba Romero, C., Castillo, E. G., Secades, C. M., Lopez, J. R., Lopez, L. A., Valiente, P. S., *Pediatr. Infect. Dis. J., 17,* 1169 (1998).
17. Pawlotsky, F., Thomas, A., Kergueris, M. F., Debillon, T., Roze, J. C., *Br. J. Clin. Pharmacol., 46,* 163 (1998).
18. Abramson, J. S., Holland, M. E., *Pediatr. Infect. Dis J., 17,* 739 (1998).
19. Rubin, L. G., Sanchez, P. J., Siegel, J., Levine, G., Saiman, L., Jarvis, W. R., *Pediatrics, 110,* e42 (2002).
20. Aujard, Y., *Arch. Pediatr., 5(2),* 200s (1998).
21. Stoll, B. J., Hansen, N., Fanaroff, A. A., et al., *N. Engl. J. Med., 347(4),* 240 (2002).
22. Nambiar, S., Singh, N., *Pediatr. Infect. Dis. J., 21,* 839 (2002).
23. de Man, P., Verhoeven, B. A., Verbrugh, H. A., Vos, M. C., van den Anker, J. N., *Lancet, 355,* 973 (2000).
24. Langer, M., Caretto, E., Haeusler, E. A., *Intensive Care Med., 27,* 1561 (2001).
25. Sahm, D. F., Storch, G., *Pediatr. Infect. Dis. J., 17,* 421 (1998).
26. Langhendries, J.P., Kalenga, M., Rousseaux, D., "Du bon usage des antibiotiques en Néonatologie," In : *Les médicaments en réanimation néonatale,* Pons, G, Huon, C, Moriette, G. (eds.), Collection: Recherche Clinique et Décision Thérapeutique, Springer-Verlag, France, , 119-139, (1999).
27. Isaacs, D., *Arch. Dis. Child. Fetal Neonatal Ed., 82,* F1 (2000).
28. Tullus, K., Burman, L. G., *Lancet, 24, 1,* 1405 (1989).
29. Van den Anker, J. N., van Popele, N. M. L., Sauer, P. J. J., *Antimicrob. Agents Chemother., 39,* 1391 (1995).
30. Mulhall, A., *Clin. Exp. Obstet. Gynecol., 13,* 129 (1986).
31. Paap, C. M., Nahata, M. C., *Clin Pharmacokinet., 19,* 280 (1990).
32. de Louvois, J. "Pharmacology of Antibiotics in the newborn", in *Neonatal clinical pharma-*

cology and therapeutics, Rylance, G., Harvey, D., Aranda, J. (eds.), Butterworth-Heinemann Ltd., pp. 153-165 (1991).

33. Friis-Hanssen, B., *Acta Paediatr. Scand., 296 (Suppl.),* 44 (1982).
34. Arant, B. S., *J. Pediatr., 92,* 705 (1978).
35. Guignard, J. P., *Pediatr. Clin. North Am., 29,* 777 (1982).
36. Pons, G., d'Athis, P., Rey, E., et al., *Therapeut. Drug Monitor.,10,* 421 (1988).
37. Dresser, M. J., Leabman, M. K., Giacomini, K. M., *J. Pharm. Sci., 90,* 397 (2001).
38. Van den Anker, J. N., Schoemaker, R. C., Hop, W. C., et al., *Clin. Pharmacol. Ther., 58,* 650 (1995).
39. Rybak, M. J., Abate, B. J., Kang, S. L., Ruffing, M. J., Lerner, S. A., Drusano, G. L., *Antimicrob. Agents Chemother., 43,* 1549 (1999).
40. Moore, R. D., Lietman, P. S., Smith, C. R., *J. Infect. Dis., 155,* 93 (1987).
41. Craig, W. A., Leggett, J., Totsuka, K., Vogelman, B., *J. Drug Dev., 1(3),* 7 (1988).
42. Vogelman, B., Gudmundsson, S., Leggett, J., Turnidge, J., Ebert, S., Craig, W. A., *J. Infect. Dis., 158,* 831 (1988).
43. Craig, W. A., Ebert, S. C., *J. Infect. Dis., 74(Suppl).,* 63 (1991).
44. Craig, W. A., Gudmundsson, S. "Postantibiotic Effect," in *Antibiotics in Laboratory Medicine,* Lorian, V. (ed.), The Williams and Wilkins Co., pp. 403-431.
45. Drusano, G. L., *Scand. J. Infect. Dis., 74 (Suppl.),* 235 (1991).
46. Craig, W. A., *Eur. J. Clin. Microbiol. Infect. Dis., 12(1),* S6 (1993).
47. Craig, W. A., *Clin. Infect. Dis., 26,* 1 (1998).
48. Livermore, D. M., *Scand. J. Infect. Dis., 74 (Suppl.),* 15 (1991).
49. McCracken, G. H., Nelson, J. D. *Antimicrobial therapy for newborn. Practical application of pharmacology to clinical usage,* Grune and Stratton, (1977).
50. James, A., Koren, G., Milliten, J., Soldin, S., Prober, C., *Antimicrob. Agents Chemother., 31,* 52 (1987).
51. Walterspiel, J. N., Feldman, S., Van, R., Ravis, W. R., *Antimicrob. Agents Chemother., 35,* 1875 (1991).
52. Daly, J. S., Dodge, R. A., Glew, R. H., Keroack, M. A., Bednarek, F. J., Whalen, M., *J. Perinatol., 17,* 42 (1997).
53. Schuetz, D. H., King, J. C., *Am. J. Hosp. Pharm., 35,* 33 (1978).

5.10 The use of antiepileptic drugs in neonates

Alexis Arzimanoglou

Head of the Epilepsy Program, Child Neurology and Metabolic Diseases Department, University Hospital Robert Debré, Paris, France

5.10.1 INTRODUCTION

When dealing with neonatal seizures, a number of considerations may make early management decisions difficult. Although the initial management of neonates with seizures is based upon the usual principles of general medical management and cardiovascular-respiratory stabilisation, transposition of acquired knowledge from epilepsy treatment of older children and adults is not always possible. This is due to several specific factors that influence treatment choices for paroxysmal events observed in neonates:

a. The difficulty to recognise and interpret motor phenomena and autonomic signs in the newborn.

b. Classification of seizure type and epilepsy syndrome in newborns is less straightforward than in older children and adults.

c. The role and impact of electrical discharges without clinical manifestations.

d. Difficulties related to the determination of aetiology and pathophysiology of paroxysmal events.

e. Limited knowledge on the effect that true epileptic seizures and/or antiepileptic drugs (AEDs) may have on the developing brain.

f. Lack of controlled studies in newborns on the use of new AEDs.

This chapter discusses the characteristics of neonatal seizures and of recognised epilepsy syndromes in newborns. The relationship of the various clinical phenomena with EEG events is briefly considered, followed by a discussion of diagnosis, aetiology and treatment issues.

5.10.2 CLINICAL AND EEG FEATURES OF NEONATAL SEIZURES

The clinical and EEG phenomena that characterise neonatal seizures are different from the usual epileptic patterns of older age. Typical absences, jacksonian attacks, and generalised tonic-clonic convulsions are not observed in neonates. However, some of the rapidly migrating movements that occur in multifocal clonic seizures or in bouts of shudders and tremors may mimic motor seizures. The frequently atypical manifestations and most types of bizarre or unusual transient event in the neonatal period may be epileptic seizures, especially if stereotyped, not sensitive to stimuli or restraint and periodically recurring [1-3].

The differences between epileptic attacks in newborn babies, and those in older children, probably reflect the incomplete neuroanatomical and neurophysiological development of the neonatal brain. Detailed study of the mechanisms responsible for the generation and propagation of epileptic discharges in neonates is beyond the scope of this chapter. The absence of generalised tonic-clonic seizures probably reflects both the lack of a sufficient degree of cortical organisation necessary to propagate and sustain the electrical discharge and the failure of interhemispheric transmission resulting from commissural immaturity [4-7]

The seizures of newborns require a different classification from those applicable at other ages. Most authors recognise four main types of seizure: subtle ([8] or minimal, clonic (focal or multifocal), tonic, and myoclonic. Several seizure types are frequently associated in the same infant; subtle seizures are frequently associated with other types in severely ill neonates. The possibility that some seizures may occur in the absence of simultaneous EEG seizure activity is widely accepted. Increasing use of EEG-polygraphic-video monitoring techniques during the last decade allows for a more precise description and classification of neonatal seizures and for a better understanding of their pathophysiology.

Unifocal clonic seizures usually show good correlation with EEG but do not necessarily imply focal pathology. Multifocal discharges usually accompany multifocal seizures. Careful direct observation and manipulation manoeuvres (restraint, repositioning, stimulation) allow distinction from non-seizure states [1,2] such as jitteriness, tremor, shudders, hypnic jerks and benign myoclonus of sleep.

The term *subtle seizure* is used to describe various behavioural phenomena that may involve the limbs, the axial muscles, or the face and eyes. Lip smacking, sucking, or swallowing movements, mouth puckering, grimacing, eye deviations (lateral or vertical), repetitive blinking, and staring can all be observed. Some authors [1] prefer the term *motor automatisms* to describe these phenomena that they consider as usually nonepileptic in nature. Although this may be true for a number of cases, the use of video-EEG suggests caution when interpreting such phenomena [3]. Such motor behaviours or automatisms may occur in premature or encephalopathic babies. When non-ictal in nature they are usually not accompanied by autonomic changes, can be stopped by restraint or repositioning or can be triggered by stimulation [2]. *Apnoeic seizures* are common, usually in association with ocular or autonomic signs, although they can occur alone [8-10].

Tonic seizures are most often *generalised*, featuring tonic extension of all limbs or, occasionally flexion of the upper limbs with extension of the legs. These symmetric tonic postures are rarely true seizures. They, more commonly, represent "release" phenomena and can be triggered by stimulation. Background activity is usually abnormal, but ictal paroxysmal activity is not observed. However, abrupt tonic limb extension/flexion with abduction may represent true epileptic spasms. *Focal tonic epileptic seizures* consist of sustained asymmetric posturing of one limb with flexion of the trunk toward the involved side. Tonic eye deviation may be associated [11].

Myoclonic seizures are uncommon. They can be erratic or fragmentary. More often, myoclonic jerks are generalised and may be associated with tonic spasms or with multifocal clonic patterns. They may be provoked by stimulation and, from a pathophysiological point of view, they may be epileptic or not. Overall neurological context, usually in favour of a severe neurologic insult, allows easy distinction from shudders or benign neonatal sleep myoclonus [3].

Isolated seizures are relatively uncommon in the neonatal period. The occurrence of at least a few attacks is the rule. In a significant proportion of the patients, seizures go on for long periods [12]. However, neonatal seizures tend to be self-limiting and last 24 to 96 hours [13,14], which also complicates the assessment of therapy.

Neonatal status epilepticus has been defined as the repetition of clinical and/or purely electrical seizures with the interictal persistence of an abnormal neurological status [15]. The term *serial seizures* is perhaps preferable to that of status epilepticus in the neonatal period because it does not refer to an abnormal interictal neurological state, which may be impossible to assess accurately due to drug therapy [16].

5.10.3 THE ROLE AND IMPACT OF ELECTRICAL DISCHARGES

Not all clinical phenomena in neonates share the same mechanisms [2,14,16]. Some are unassociated, or only inconsistently associated, with paroxysmal EEG changes, whereas others are regularly associated with rhythmic EEG discharges of clear epileptic nature.

In the first group (mainly tonic and subtle paroxysmal phenomena), the mechanism is uncertain. Cortical epileptic discharges may not be picked up on the scalp. Some of them are probably not epileptic but instead represent "release" phenomena due to liberation of subcortical structures from cortical control due to extensive cortical destruction or dysfunction [14,17,18].

In the second group of paroxysmal phenomena, (mainly clonic seizures), the origin of the ictal discharges appears to be focal cortical and may often be correlated with discrete cortical insults.

Electrical discharges can occur without clinical manifestations [12,19-22] and these may be much more common than electroclinical seizures.

Epileptic phenomena can also be generated at subcortical levels [23-26]. Almost all paroxysmal electrical activity in the neonate begins focally, except for the more generalised activity associated with myoclonic jerks or infantile spasms. Ictal discharges in the full-term neonate are exceedingly variable in appearance, voltage, frequency, and polarity [27-30]. Changes occur between different discharges in the same infant and even within the same discharge. Two main elements constitute the EEG discharges in newborns: *abnormal paroxysmal rhythms* and *repetitive spikes* or *sharp waves*. Both are commonly associated. In premature infants, the EEG findings tend to be more stereotyped [31].

The distinction between epileptic and non-epileptic seizures is of practical significance for therapy and prognosis. However, clinical differentiation of these two types of seizure may be extremely difficult. EEG seizure discharges without detectable clinical manifestations are common in neonates. This is especially common in infants receiving antiepileptic drugs, particularly phenobarbital, which often produces "uncoupling" of clinical and EEG seizure manifestations [12,32]. An undefined proportion of newborns may have such discharges without having seizures. Conversely, clinical seizures deemed to be epileptic but without typical EEG abnormality are also encountered [15,26].

5.10.4 AETIOLOGY

As previously discussed, the most difficult issues in the diagnosis of neonatal seizures are (a) to decide which atypical ictal behavioural events can be regarded as epileptic seizures and (b) whether subclinical EEG seizures are occurring. For both purposes, EEG monitoring is essential and should be prolonged to answer the question regarding subclinical seizures. Aetiological context and global neurological evaluation provide a solid, and indispensable, basis for decision-making [11].

During the last decade, development of neonatology, emergency facilities and a more rapid access to sophisticated investigations and techniques have changed the relative importance of various aetiologic factors [2]. The survival rate of very sick neonates has significantly improved, creating a larger population of infants at risk to develop seizures. The main aetiological factors of neonatal seizures are listed in Table 1 [3].

The probable diversity of mechanisms responsible for neonatal paroxysmal events makes it difficult to determine whether some types of seizures are specifically related to certain aetiologic factors. Moreover, several factors are often operative in the same patient, e.g., hypocalcaemia, hypoxia, and infection [2]. Each factor may require prompt recognition to adopt specific therapies when possible. Although hypoxic-ischaemic encephalopathy and metabolic disorders are more often a cause of subtle seizures of "nonepileptic" nature, they can also produce typical epileptic events.

Table 1

Causes of neonatal seizures
(From ref. [3] modified)

Abnormalities of cortical development
(clinical presentation of seizures depending upon the type and extension of the malformation)

Hypoxic-ischaemic encephalopathy
(may produce both clearly epileptic attacks and seizures probably nonepileptic in nature)

Intracranial haemorrhage
  - subarachnoid haemorrhage (clonic seizures in term infants 1-5 days of age)
  - intraventricular haemorrhage (mainly tonic seizures and episodes of apnoea without EEG correlates, occasionally typical EEG discharges)
  - intracerebral haematoma (fixed localised clonic seizures)

Intracranial infections
  - bacterial meningitis and/or abcess
  - viral meningo-encephalitis

Metabolic causes
  - hypocalcaemia (clonic, multifocal seizures)
  - hypoglycaemia
  - hyponatraemia
  - inborn errors of amino acids or organic acids and ammonia metabolism (often atypical, usually not associated with EEG discharges)
  - molybdenum cofactor deficiency
  - bilirubin encephalopathy (atypical, no EEG discharges)
  - pyridoxine dependency
  - biotinidase deficiency
  - carbohydrate-deficient glycoprotein syndrome

Toxic or withdrawal seizures *(probably nonepileptic phenomena in most cases)*

Familial neonatal convulsions

"Benign" neonatal seizures of unknown origin

Table 2

Drugs for the treatment of neonatal seizures

Drug	Loading dose (IV)	Maintenance (mg/kg/day)	Therapeutic blood conc (µg/ml)	Apparent half-life (h)
phenobarbital	20 mg/kg	3-4	20-40	100*
phenytoin	20 mg/kg	2-5	15-35	100 (40-200)*
diazepam	250 mcg/kg – 1 mg/kg	may be repeated 1-3 times	200-800	31-84
lorazepam	50 mcg/kg	may be repeated	-	17
clonazepam	100 mcg/kg (infusion over 5 minutes)	–	–	–

*declines after day 5-10

5.10.5 RECOGNISED EPILEPSY SYNDROMES IN NEONATES

In most cases, the overall neurological context provides valuable clues about the nature of paroxysmal phenomena and orientates treatment attitudes. To facilitate diagnosis and treatment choices, Aicardi [33] suggested a clinical approach, by separating neonates with neurological damage from well babies.

In newborns with neurological damage, seizures that occur during the first 3 days of life are often fragmentary or subtle in type and tend to occur in long series or to constitute episodes of status epilepticus. It is in this group that the more abnormal EEG patterns, i.e., flat records, *tracé paroxystique,* and inactive tracing, are often obtained [34,35]. Many cases of early-onset neonatal convulsions are the result of hypoxic-ischaemic encephalopathy or intraventricular haemorrhage. Metabolic disturbances (especially hypoglycaemia, hypocalcaemia, or hypomagnesaemia) are frequent and should be corrected, but they are less common.

Within this group, two syndromes have been more clearly delineated and included in the ILAE classification: *early myoclonic encephalopathy* or *neonatal myoclonic encephalopathy* (NME), first described by Aicardi and Goutières and *early-infantile epileptic encephalopathy* (EIEE) with suppression burst described by Ohtahara [36].

EIEE is characterised by a very early onset, frequent tonic spasms and suppression-burst pattern in the EEG. Most cases are associated with structural brain damage. Treatment is disappointing, although ACTH and/or corticosteroids are occasionally helpful [36]. Other agents (vitamin B6 and sodium valproate) have also been used. The possible role of new AEDs has not been assessed. One case treated with zonisamide has been reported [37], and some cases were treated surgically.

NME is also characterised by early onset of seizures. The main ictal phenomena are partial or fragmentary erratic myoclonus, partial motor seizures, sometimes massive myoclonus and often the late occurrence of repetitive tonic spasms. The neurological status is very poor at birth or as soon as seizures appear and the course is always severe. Treatment with AEDs, corticosteroids or ACTH has not been effective [57].

In neurologically well newborns, *early onset seizures*, beginning most commonly on the second day of life, are relatively infrequent [7]. They may remain focalised to the same site, shift from one area to another, or be multifocal. The EEG is usually normal between seizures. Usual causes are neonatal strokes, due to localised vascular obstruction, primary subarachnoid hemorrhage and localised intracerebral bleeding. *Late onset seizures*, with onset after the third day of life, are infrequently associated with neurological signs, except in the case of bacterial meningitis. Late-onset seizures may be due to late hypocalcaemia, where the infants are jittery and have increased tendon reflexes. More commonly, no cause is found. The prognosis is better than in early onset seizures, even in cases without known cause.

Two age-related syndromes, both belonging to the category of generalised idiopathic epilepsies, are recognised by the ILAE: *benign familial neonatal convulsions* and *benign neonatal convulsions* (also called the syndrome of "fifth-day fits"). The main difference between the two is the presence or absence of a family history.

Benign neonatal familial convulsions (BFNC) have their onset, in most cases, within 2 to 15 days of birth, most commonly on day 2 or 3. The prevalence of the syndrome is unknown. Plouin recently reviewed the data from 334 cases of BFNC belonging to 38 families, published since the initial description [38]. The seizures, usually clonic, are frequently repeated (up to 30-40 per day) but spontaneously stop after a variable duration. The EEG confirms the epileptic nature of the fits [40,41]. The interictal EEG is either normal or shows minimal focal or multifocal abnormalities or a pattern of *théta pointu alternant*. The overall rates of secondary epilepsy and febrile convulsions were 11% and 5% respectively. The rate of febrile convulsions is comparable to that of the general population [38].

Benign idiopathic neonatal convulsions (BINC) were first described by Dehan et al. under the name of *fifth-day fits* [42]. The attacks are of two main types: clonic focal or multifocal convulsions and apnoeic spells. They last on average 20 hours. The interictal EEG shows preserved rhythms and bursts of alternating θ-rhythms, or *théta pointu alternant*. Many AEDs have been used, but the seizures usually stop without treatment. Sodium valproate leads to a rapid cessation of seizures. If initiated, treatment can be discontinued within a few months.

5.10.6 MANAGEMENT OF NEONATAL SEIZURES

The most important predictor of prognosis for infants with neonatal seizures is the cause of the convulsions. It is, therefore, of primary importance to establish a firm aetiological diagnosis and treat accordingly. A reasonable degree of certainty on the epileptic nature of the paroxysmal manifestations and on their recurring character is a prerequisite to a decision to treat with antiepileptic drugs. Neither the effect of seizures nor of AEDs, on the developing brain, is well known.

Monitoring of respiration and heart rate is essential. Continuous EEG recording is extremely useful, as many neonatal seizures occur without clinical manifestations. Ultrasound scanning will indicate if there is bleeding, infarction, or gross malformation in the CNS. An intravenous line should be established for administration of glucose following blood glucose determination and for rapid correction of any metabolic derangement.

Levels of electrolytes, pH, calcium, and magnesium should be obtained and fluid administration restricted to 75% of the normally required amount to avoid dilutional hyponatraemia. Proper ventilation and maintenance of body temperature are essential, and a lumbar puncture is imperative. Treatment of "cerebral oedema" with dexamethasone was considered essential by some [43], but the evidence in favour of such a treatment is limited, as the oedema is usually of cytotoxic mechanism and responds poorly to steroids [2,44]. Such treatment, however, may be indicated for the treatment of established oedema, due to the cause of seizures.

All infants with early onset intractable seizures or status should receive a trial of pyridoxine (usual dose: IV administration of 100 mg within 30 minutes) *before* the use of another anticonvulsant so that the rare case of pyridoxine dependency is not missed [45,46]. Biotinidase deficiency is only exceptionally a cause of early seizures but responds dramatically to the intravenous administration of 10 mg of biotin.

5.10.7 TREATMENT FOR ELECTROLYTE ABNORMALITIES

If hypoglycaemia is present, 2 ml/kg of a 10 % solution of glucose is given intravenously during the acute phase, followed by up to 8 mg/kg/min as maintenance therapy. If hypocalcaemia is found, 2 ml/kg of a slow (10 minutes) intravenous injection of 2.5 to 5% calcium gluconate is administered with electrocardiographic monitoring. After restoration of normocalcaemia, tapering dosage may help in preventing rebound hypocalcemia [47]. Magnesium sulphate, 2 to 8 ml of a 2 to 3% solution, intravenously, or 0.2 ml/kg of a 50% solution, intramuscularly, is added when there is associated hypomagnesaemia. Serum levels of magnesium should be monitored to avoid its potential curare-like effect. Other metabolic derangements are uncommon and are more often responsible for non-epileptic abnormal movements than for true seizures [3].

5.10.8 ANTIEPILEPTIC DRUG THERAPY

There is no agreement, at the moment, as to which seizures require AED treatment, the timing for starting them, or which drug to use preferentially.

Those authors, who think that neonatal seizures of whatever cause can harm the brain [48], recommend immediate therapy with large doses of long-acting anticonvulsants (phenobarbital, phenytoin). Those investigators who believe that only prolonged seizures are harmful [17,2] tend to postpone initiation of long-acting antiepileptic agents until diagnosis is clarified, unless there is evidence of systemic effects. Lombroso [2] favours the initial use of short-acting drugs, especially in cases in which seizure seem likely to be transient, such as mild hypoxic-ischaemic encephalopathy, sepsis, or cryptogenic seizures.

Lorazepam is rapidly effective and has the advantage of a prolonged action and absence of secondary release from brain and fat tissue [49]. The usual dose is 100 mcg/kg once a day.

Diazepam can be used intravenously, single dose of 300 mcg/kg or continuous infusion at a rate of 300 mcg/kg/h for up to 24 h.

Clonazepam in dosages of 100 mcg/kg has been used successfully in neonates. Recently, Sheth and co-workers [50] reported administration of *midazolam* (loading dose of 150 mcg/kg followed by maintenance doses of 100-400 mcg/kg/h IV) to six neonates.

When a long-acting anticonvulsant is to be administered, the drug of first choice is usually *phenobarbital*. Therapeutic blood levels are at least 20 mcg/ml, and to achieve this level promptly, the drug is administered intravenously as a loading dose of 20 mg/kg. Some advocate doses up to 35 to 40 mg/kg for reaching levels of about 40 mg/ml. Maintenance doses are 4 mg/kg/day given intravenously once daily. Because the apparent half-life is very long (average 100 h after day 5-7) in the first 1 or 2 weeks of life and shortens thereafter, monitoring of blood levels to permit dosage adjustment is mandatory.

Phenytoin is usually the second drug used. It is often combined with phenobarbital. A recent study [51] compared the effectiveness of acute administration of phenobarbital versus phenytoin in seizure control and found no significant difference between the two drugs. Phenytoin should be used parenterally to obtain plasma levels of 15-20 mcg/ml. A loading dose of 15-20 mg/kg is usually effective, followed by a maintenance dose of 3-5 mg/kg/day. The loading dose should be administered slowly, at a rate not exceeding 50 mg/min, to avoid disturbances of cardiac rhythm.

Fosphenytoin, a prodrug of phenytoin, was administered to 11 neonates, in a study [52], which included 75 children and adolescents. The range of values for conversion half-life of *fosphenytoin* and resultant plasma total and free phenytoin concentration-time profiles following IV administration were greater in neonates, although globally similar to values in older children and adults. Kriel and Cifuentes [53] reported on the use of fosphenytoin in two low birth weight infants.

Valproate was used at doses of 50 mg/kg, rectally or orally, to achieve blood levels of 60 to 80 mcg/ml [54,2], but the safety and efficacy of this drug remain to be fully assessed. Alfonso et al. [55] measured serum valproate concentrations in two neonates, 45 minutes and 3 hours after initiation of the infusion. They found that each 1 mg/kg of intravenous valproate increases the 45 minute and 3 hour postinfusion serum valproic acid concentrations by approximately 4 mcg/ml, and 3 mcg/ml, respectively. Valproate should not be used when an inborn error of metabolism is suspected.

Other agents used have been *primidone* (loading dosages between 15-20 mg/kg followed by maintenance dosages of 12-20 mg/kg/day) and *carbamazepine* (loading dosage of 5 mg/kg every 12 hours), but few data are available for these agents.

Data is scarce concerning newer antiepileptic drugs. *Lamotrigine* was given to a 17 day old neonate, after failure of several agents, at a single daily dose of 4.4 mg/kg/day for 3 days, followed by divided daily doses every 12 hours. Control of seizures was obtained. An open-label study in 13 infants of lamotrigine showed age-dependent kinetics [56]. Apparent clearance increased during the first year of life, with a break point at 2 months of age. Neonates (aged 3-4 weeks) showed a half-life of 23 ± 3 hours.

A number of neonates continue to have seizures despite drug therapy. Many of them have severe underlying brain damage and interictal EEG patterns of unfavourable significance. In such cases, increasing antiepileptic dosage may do more harm than good [2], especially as at least some of these events are not epileptic seizures of cortical origin [14,17]. Whether electrical discharges without clinical manifestations justify antiepileptic treatment is not established [2,7].

No specific guidelines have been established regarding the *optimal duration of maintenance treatment*, following neonatal convulsions [47]. Discontinuation of AEDs, after a period of clinical seizure control, should be individualised. Our opinion is that continuation of antiepileptic treatment for more than a few weeks is justified only when there is a high likelihood of recurrent seizures, specifically in cases featuring abnormalities of cortical development [3]. For acute conditions such as haemorrhages, mild or moderate hypoxic-ischemic encephalopathy, and cryptogenic neonatal seizures, there is no need to continue therapy. In cases of severe hypoxic-ischaemic encephalopathy or other forms of acquired brain damage, most authors advise maintenance therapy [2,7,16], although the frequency of later epilepsy is poorly known [49] and the feasibility of preventing later epilepsy is at best uncertain.

5.10.9 CONCLUSIONS

The value of anticonvulsant treatment remains controversial because there is no absolute clinical evidence that neonatal seizures are a hazard to the brain. It should be borne in mind, however, that there is no evidence that neonatal seizures are innocuous, and common sense would suggest that prompt and active treatment should be instituted.

The various modalities of treatment are still being debated, but awareness of pharmacokinetic properties unique to the neonatal period permits a more rational use of anticonvulsant drugs.

REFERENCES

1. Mizrahi E. M., Kellaway P., *Diagnosis and Management of Neonatal Seizures*. New York, Lippincott-Raven (1998).
2. Lombroso, C. T., "Neonatal seizures," in *The Medical Treatment of Epilepsy*, Resor, S. R., Kutt, H. (eds), pp. 115-125, Marcel Dekker (1992).
3. Arzimanoglou A., Guerrini R., Aicardi J.; *Aicardi's Epilepsy in children* (3rd ed.), Lippincott Williams & Wilkins, (2004).
4. Moshé, S. L., *Epilepsia, 28*, S3-S15 (1987).
5. Holmes, G.L., Sarkisian, M., Ben-Ari, Y., Chevassus-au-Luis, N., "Effects of recurrent seizures in the developing brain," in *Childhood epilepsies and brain development*, edited by Nehlig, A., Motte, J., Moshé, S. L., Plouin, P. (eds.), John Libbey, pp. 263-276, (1999).
6. Ben-Ari, Y., Cherubini, E., Krnjevic, K., *Neurosci. Lett. 94*, 88 (1988).
7. Volpe, J. J., *Neurology of the Newborn* (3rd ed.), WB. Saunders (2001) .
8. Fenichel, G. M., "Seizure in newborns," in *Neonatal Neurology*, Fenichel, G. M., (ed.) p. 25, Churchill Livingstone, (1985).
9. Navelet, Y., Wood, R. C., Robieux, C., Tardieu, M., *Arch. Dis. Childh., 64*, 357 (1989).
10. Watanabe, K., Hara, K., Miyazaki, S., Hakamada, S., and Kuroyanagi, M., *Am. J. Dis. Child., 15*, 584 (1982).
11. Arzimanoglou, A., Aicardi, J., "Seizure disorders of the neonate and infant," in *Fetal and Neonatal Neurology and Neurosurgery*. Levene, M. I., Chervenak, F. A., Whittle, M. (eds.) Churchill Livingstone, 647-656 (2001).
12. Clancy, R. R., Legido, A., Lewis, D., *Epilepsia, 29*, 256-261 (1988).
13. Bout, F., Plouin, P., Jalin, C., Frenkel, A., Dulac, O., Bonifas, P., *Rev. EEG Neurophysiol. Clin., 13*, 162 (1983).
14. Camfield, P. R., and Camfield, C. S., *J. Child Neurol., 2*, 244 (1987).

15. Dreyfus-Brisac, C., and Monod, N., "Neonatal status epilepticus," in *Handbook of Electroencephalography and Clinical Neurophysiology*, Vol. 15, Part B, Remond, A., (ed.) pp. 39-52. Elsevier (1977).

16. Aicardi J., "Epilepsy and other seizure disorders," in *Aicardi's Diseases of the Nervous System in Childhood* (2nd ed.), MacKeith Press, 575-637 (1998).

17. Kellaway, P. M., Mizrahi, E. M., "Clinical, electroencephalographic, therapeutic, and pathophysiologic studies of neonatal seizures," in *Neonatal Seizures*, Wasterlain, C. G., Vert, P., (eds.) pp. 1-13. Raven Press (1990).

18. Volpe, J. J., "Neonatal seizures: Clinical overview" in *Neonatal Seizures*, Wasterlain, C. G., Vert, P., (eds.), pp. 27-39. Raven Press, (1990).

19. Bridgers, S.L., Ebersole, J.S., Ment, L.R., Ehrenkranz, R.A., and Silva, C.G., *Arch. Neurol.*, *43*, 49 (1986).

20. Connell, J. A., De Vries, L. S., Dubowitz, L. M. S., and Dubowitz, V., *Arch. Dis. Child.*, *64*, 452 (1989).

21. Glauser, T. A., and Clancy, R. R., *J. Child Neurology*, *7*, 215 (1992).

22. Hellstrom-Westas, L., Rosen, I., and Svenningsen, S. W., *Acta Paediatr. Scand.*, *74*, 741, (1985).

23. Harvey S., Jayakar P., Duchowny M., et al., *Ann. Neurol.*, *40*, 91 (1996).

24. Arzimanoglou, A., Salefranque, F., Goutières, F., Aicardi, J., *Epileptic Disord.* *1*, 121 (1999).

25. Scher, M. S., Aso, K., Beggarly, M., Hamid, M. Y., Steppe, D. A., Painter, M. J., *Pediatrics*, *91*,128 (1993).

26. Weiner, S. P., Painter, M. J., Geva, D., Guthrie, R. D., and Scher, M. S., *Pediatr. Neurol.*, *7*, 363 (1991).

27. Dreyfus-Brisac, C., Peschanski, N., Radvanyi, M. F., Cukier-Hémeury, F., and Monod, N., *Rev. EEG Neurophysiol.*, *11*, 367 (1981).

28. Estivill, E., Sanmarti, F., and Fernandez-Alvarez, E., *Rev. EEG Neurorphysiol. Clin.*, *13*, 145 (1983).

29. Rowe, J. C., Holmes, G. L., Hafford, J., et al., *Electroencephalogr. Clin. Neurophysiol.*, *60*, 183 (1985).

30. Hrachovy, R. A., Mizrahi, E. M., Kellaway, P., "Electroencephalography of the newborn," in *Current Practice of Clinical Electroencephalography* (2nd ed.) Daly, D., Pedley, T. A., (eds), Raven, 210-242 (1990).

31. Scher, M. S., Aso, K., Painter, M. J., *Ann. Neurol.*, *24*, 344A (1988).

32. Shewmon, D. A., *Ann. Neurol.*, *14*, 368 (1983).

33. Aicardi, J., *Epilepsy in Children*, Raven Press (1986).

34. Dreyfus-Brisac, C., "Neonatal electroencephalography" in *Reviews of Perinatal Medicine* (vol. 3), Scarpelli, E. M., Cosmi, E. V. (eds.), Raven Press, 627 (1979).

35. Marret, S., Parain, D., Ménard, et al., *Electroencephalography and Clinical Neurophysiology 102*, 178 (1997).

36. Ohtahara, S., Ohtsuka, Y., Yamatogi, Y., Oka, E., and Inoue, H., "Early infantile epileptic encephalopathy with suppression-bursts," in *Epileptic Syndromes in Infancy, Childhood and Adolescence* (2nd ed.), Roger, J., Bureau, M., et al., (eds.) pp. 25-34, John Libbey (1992).

37. Ohno, M., Shimotsuji, Y., Abe, J., Shimada, M., Tamiya, H., *Pediatr Neurol. 23 (4)*, 341-4 (2000).

38. Plouin P., "Benign idiopathic neonatal convulsions (familial and non-familial). Open questions about these syndromes" in *Epileptic Seizures and Syndromes*, Wolf, P. (ed.), John Libbey, pp. 193-202 (1994).

39. Rett, A., Teubel, R., *Wien. Klin. Wschr. 76*, 609 (1964).

40. Camfield, P. R., Dooley, J., Gordon, K., and Orlik, P., *J. Child Neurol.*, *6*, 340 (1991).

41. Hirsch, E., Velez, A., Sellal, F., et al., *Annals Neurol. 34*, 835 (1993).

42. Dehan, M., Quilleron, D., Navelet, Y., et al., *Arch. Franç. Pédiatr., 34*, 730 (1977).
43. Brown, J. K., and Minns, R. A., "Epilepsy in neonates," in *The Treatment of Epilepsy*, Tyrer, J. H. (ed.), pp. 161-202, MTP Press, (1980).
44. Hill, A., and Volpe, J. J., *Ann. Neurol., 10*, 109 (1981).
45. Goutières, F., and Aicardi, J., *Ann. Neurol, 17*, 117 (1985)
46. Baxter, P., *Arch Dis Child, 81*, 431 (1999).
47. Mizrahi, E. M., Kellaway, P., "Neonatal seizures" in *Pediatric Epilepsy; Diagnosis and therapy*. Pellock, J. M., Dodson, W. E., Bourgeois, B., (eds.), Demos Medical Publishing, 145 (2001).
48. Volpe, J. J., *Pediatrics, 84*, 422 (1989).
49. Maytal, J., Novak, G. P, King, K. C., *J. Child. Neurol., 6*, 319 (1991).
50. Clancy, R. R., and Legido, A., *Epilepsia, 32*, 69 (1991).
51. Painter M. J., Scher, M. S., Stein, A. D. et al., *N. Engl. J. Med., 341*, 485 (1999).
52. Morton, L. D., *J. Child. Neurol. 13* (Suppl.1), S19 (1998).
53. Kriel, R. L., Cifuentes, R. F., *Pediatr Neurol., 24*, 219 (2001).
54. Gal, P., Oles, K. S., Gilman, J. T., Weaver, J., *Neurology, 38*, 467 (1988).
55. Alfonso, I., Alvarez, L. A., Gilman, J., et al., *J. Child. Neurol., 15*, 827 (2000).
56. Mikati, M. A., Fayad, M., Koleilat, M., et al., *J Pediatr., 141*, 31 (2002).
57. Aicardi J., Otahara S., "Severe neonatal epilepsies with suppression-burst pattern," in *Epileptic Syndromes*, Bureau, M. (ed.), John Libbey, pp.33-44 (2002).

Infections

6.1 Antibacterial agents

Patricia Mariani-Kurkdjian, Edouard Bingen

Department of Microbiology, Hospital Robert Debré, 48 Boulevard Serurier, 75019 Paris, France

6.1.1 INTRODUCTION

Paediatric patients with serious infections are a challenge to the clinicians treating them. The pharmacokinetics and disposition of antibacterials are influenced by the changes taking place in the course of normal growth and development [1,2]. Clinicians may also face failures secondary to bacterial resistance [3,4]. New solutions are required to successfully treat patients with high risk infections, including the development of new antibiotics [5]. In this review, we summarise information for the most commonly used antibacterials in paediatric patients [6].

6.1.2 INHIBITORS OF CELL WALL SYNTHESIS

Beta-lactams

Compounds of this family all contain the beta-lactam ring [7]. The different groups within this family are distinguished by the structure of the ring attached to the beta-lactam ring and by the side chains attached to these rings. Penicillins have a five-membered ring (Figure 1), and cephalosporins a six-membered ring (Figure 2).

Mechanism of action

Beta-lactams inhibit cell wall synthesis by binding to enzymes known as 'penicillin binding proteins' (PBPs). These proteins are carboxypeptidases and

Penicillin nucleus

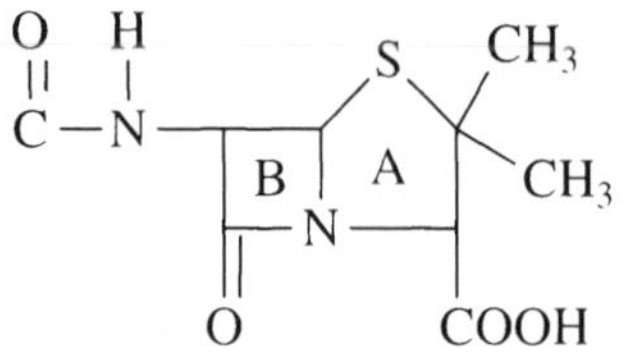

A : thiazolidine ring
B : ß lactam ring

Figure 1

Cephalosporin nucleus

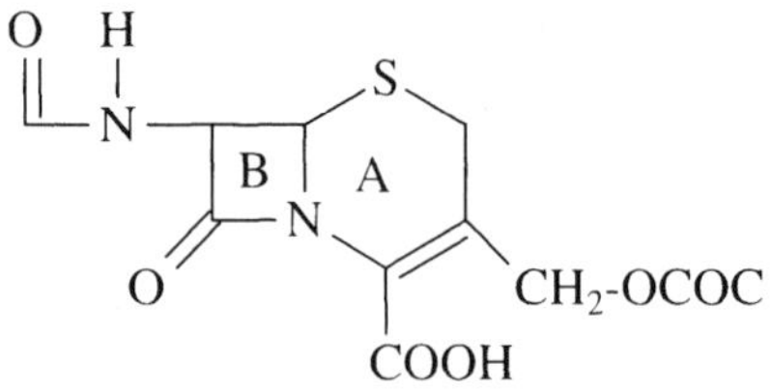

A : dihydrothiazine ring
B : ß lactam ring

Figure 2

transpeptidases responsible for the final stages of cross-linking the bacterial cell wall structure [8,9].

Indications

There are more than 40 different beta-lactam antibiotics currently registered for clinical use. Penicillin is active mainly against Gram-positive organisms, whereas other beta-lactams (ampicillin, ticarcillin, ureidopenicillins) have been developed for their effect against Gram-negative rods (Figure 3, 4). Cephalosporins are classified as first and second generation (Figure 5), and third generation (Figure 6, 7, 8). Only ceftazidime (Figure 9) and imipenem (Figure 10) are active against organisms such as *Pseudomonas aeruginosa* [10].

Toxicity

Serious allergy to beta-lactam drugs in the form of an immediate hypersensitivity reaction occurs in approximately 0.004-0.015% of treatment courses. Mild

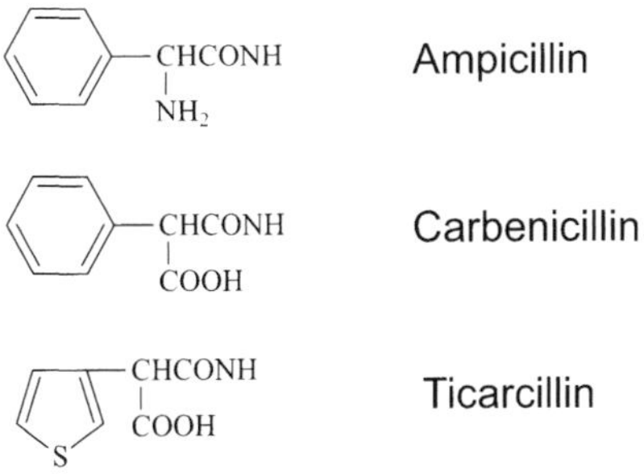

Figure 3

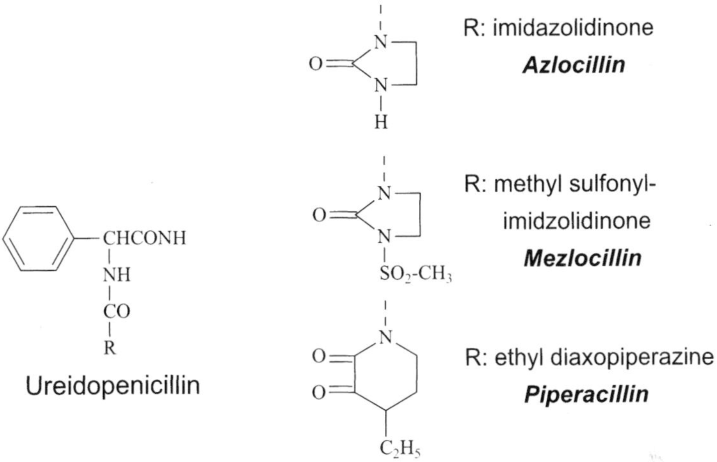

Figure 4

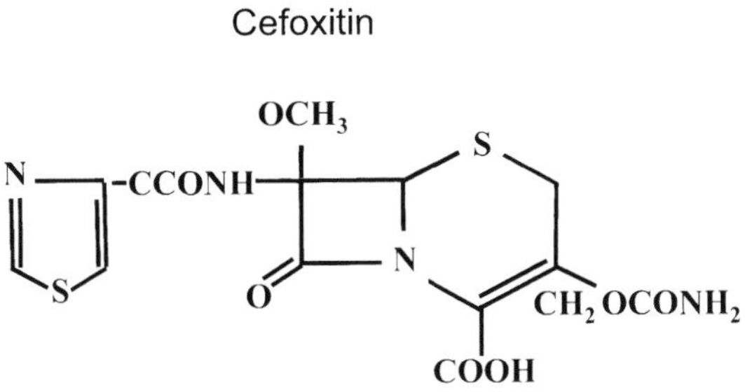

Figure 5

Cefotaxime

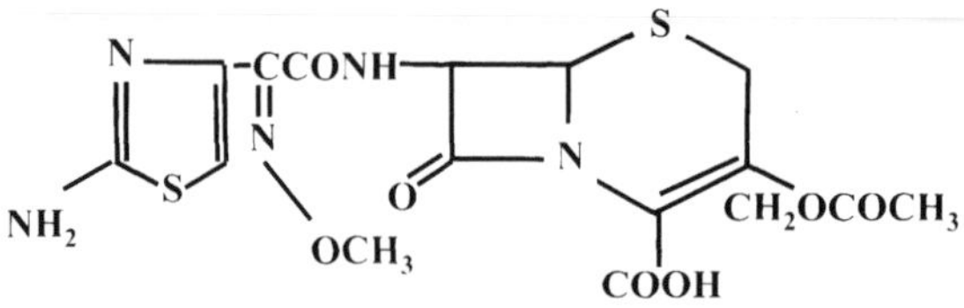

Figure 6

Ceftriaxone

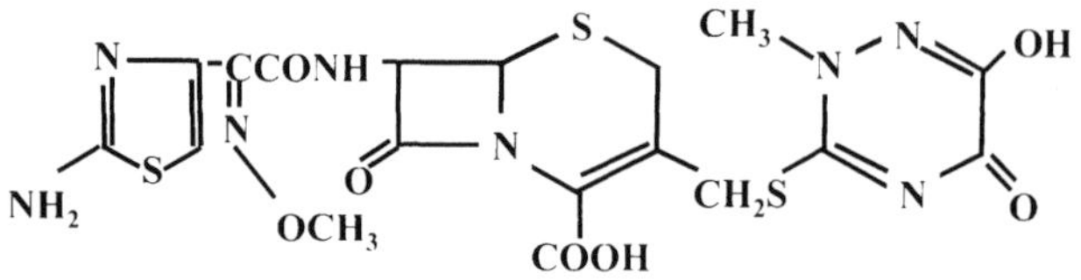

Figure 7

Cefpirome

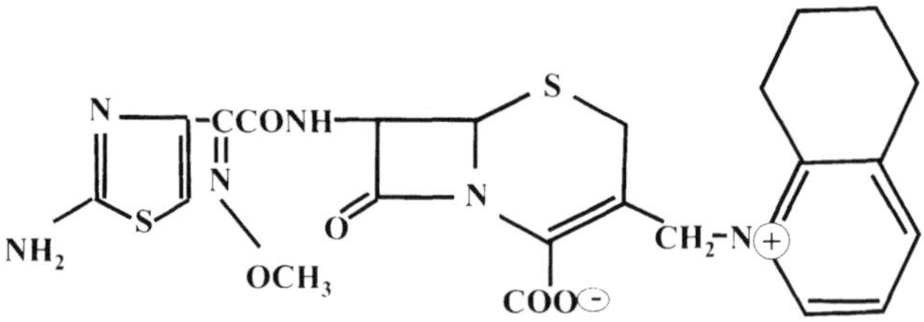

Figure 8

Ceftazidime

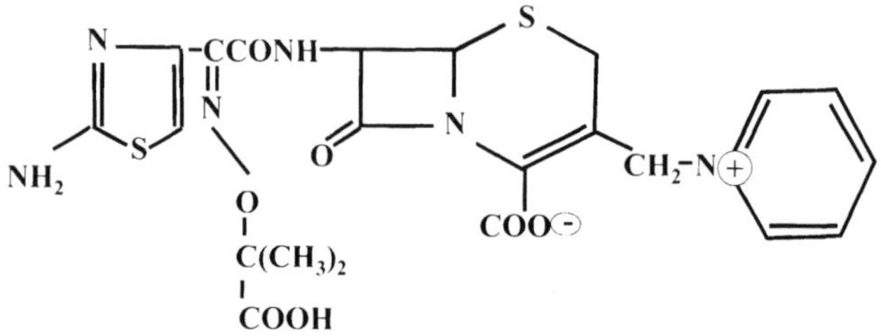

Figure 9

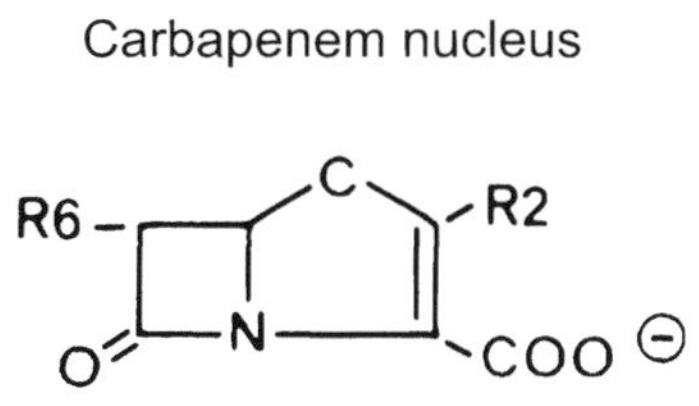

Carbapenem nucleus

Figure 10

idiopathic reactions, usually in the form of a rash, occur more frequently (23% of treatment courses), and especially with ampicillin [11,12]. About 10% of the patients who are allergic to penicillin are also allergic to cephalosporins [10], but aztreonam, a monobactam, shows negligible cross-reactivity.

Resistance

Clinical isolates resistant to beta-lactams may exhibit any one, or more of three mechanisms of resistance.

Alteration in target site. Methicillin-resistant staphylococci synthesise an additional PBP which has a much lower affinity for beta-lactams than the normal PBPs, and is thus able to continue cell wall synthesis, even when the other PBPs are inhibited. Methicillin-resistant staphylococci are resistant to all beta-lactams [13].

Alteration in access to the target site. This mechanism is found in Gram-negative cells where beta-lactams gain access to their target PBPs by diffusion through protein channels (porins) in the outer membrane. Mutations in porin genes result in a decrease in permeability of the outer membrane, and hence, resistance. Strains resistant to this mechanism may exhibit cross-resistance to unrelated antibiotics which use the same porins.

Production of beta-lactamases. Beta-lactamases are enzymes which cata-lyse the hydrolysis of the beta-lactam ring to yield microbiologically inactive products. Genes encoding these enzymes are found on the chromosome or on plasmids. In Gram-positive bacteria, beta-lactamases are released into the extra-cellular environment. In Gram-negative cells, the beta-lactamases remain within the periplasm [14,15]. There are many different beta-lactamase enzymes which have the same function, but differ in their affinity for different beta-lactam subs-trates. Beta-lactamase inhibitors, such as clavulanic acid, are molecules which contain a beta-lactam ring and which act as 'suicide inhibitors', binding to beta-lactamases and preventing them from destroying beta-lactams (Figure 11) [15].

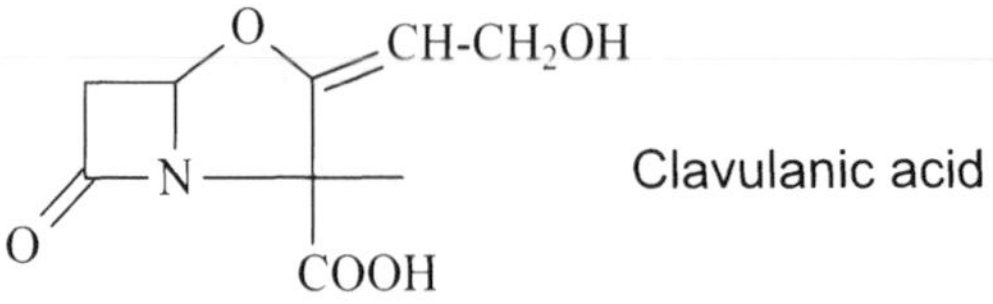

Clavulanic acid

Figure 11

Glycopeptides

Glycopeptides include vancomycin and teicoplanin (Figure 12). Teicoplanin is a complex of five different, but closely related, molecules.

Mechanism of action

Glycopeptides interfere with cell wall synthesis by binding to terminal D-ala-D-ala at the end of pentapeptide chains that are part of the growing bacterial cell wall structure. This binding inhibits the transglycosylation reaction and prevents incorporation of new subunits into the growing cell wall. Glycopeptides act at an earlier stage than beta-lactams.

Indications

Both vancomycin and teicoplanin are only active against Gram-positive organisms. They are mainly used for the treatment of infections caused by Gram-positive cocci and Gram-positive rods that are resistant to beta-lactam drugs, or

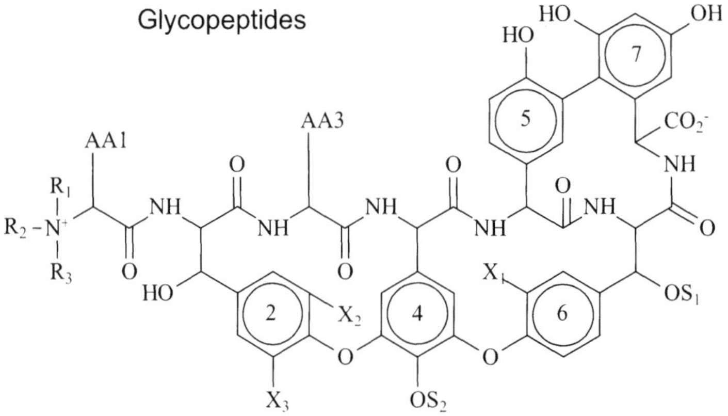

Figure 12

in patients who are allergic to beta-lactams. Oral administration is used for the treatment of *Clostridium difficile* in antibiotic-associated colitis.

Toxicity

Vancomycin must be given by slow intravenous infusion to avoid 'red man' syndrome (due to histamine release) [6]. The glycopeptides are potentially oto-toxic and nephrotoxic [17,18]. Teicoplanin is less toxic than vancomycin.

Resistance

Acquired resistance has now been reported among enterococci, where resistance is plasmid-mediated and transmissible [19,20].

6.1.3 INHIBITORS OF PROTEIN SYNTHESIS

Aminoglycosides

This is a family of related molecules containing either streptomycin or 2-deoxystreptamine, e.g gentamicin (Fig 13, 14).

Mechanism of action

Aminoglycosides inhibit and kill organisms by interfering with the binding of formylmethionyl-tRNA to the ribosome, and thereby, preventing the formation of initiation complexes from which protein synthesis proceeds.

Indications

Gentamicin, amikacin and netilmicin are used in combination with other antibiotics for the treatment of serious Gram-negative infections, including those caused by *P. aeruginosa*. They are not active against streptococci, but

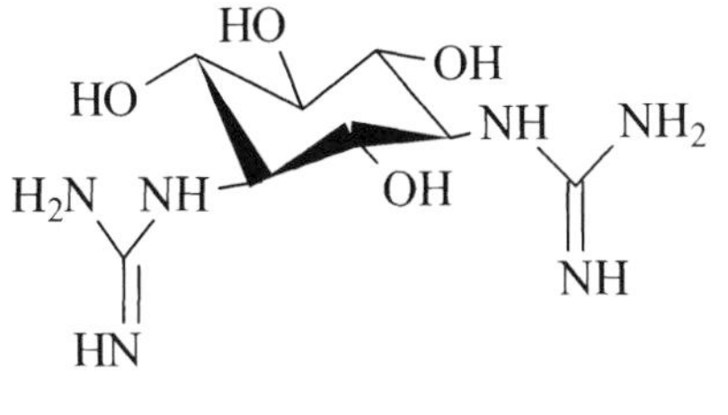

Streptomycin

Figure 13

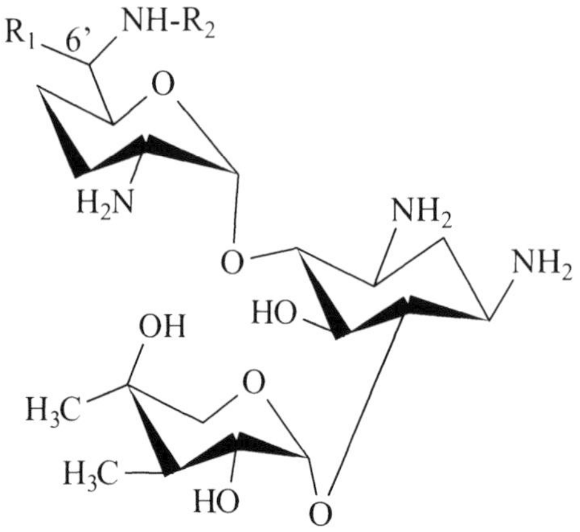

Gentamicin

Figure 14

are synergistic in combination with β-lactams. They are active against staphylococci. Tobramycin is slightly more active than gentamicin against *P. aeruginosa*. Streptomycin is now reserved almost entirely for the treatment of mycobacterial infections. Spectinomycin is used to treat beta-lactam resistant *Neisseria gonorrhoeae* infections.

Toxicity

The aminoglycosides are potentially nephrotoxic and ototoxic, and the therapeutic 'window' between serum concentrations required for successful treatment and those that are toxic, is small. Blood concentrations should be monitored regularly, particularly in patients with renal impairment. Aminoglycosides given once a day are more active and less nephro- and ototoxic than aminoglycosides given 8 or 12 hourly [22,23].

Resistance

Production of aminoglycoside-modifying enzymes is the most important mechanism of acquired resistance. The genes for these enzymes are often plasmid-mediated and transferable from one bacterial species to another. The effect of the enzymes is to alter the structure of aminoglycoside molecule which consequently changes the uptake of drugs by the bacteria. Resistance may also arise in Gram-negative rods through alterations in cell wall permeability through the porins.

Tetracyclines

Tetracyclines are a family of large cyclic structures which have several sites for chemical substitutions. The members of the family differ mainly in their pharmacological properties, rather than in their antibacterial spectra (Figure 15).

Mechanism of action

Tetracyclines inhibit protein synthesis by preventing aminoacyl transfer RNA from entering the acceptor sites on the ribosome.

Indications

Tetracyclines are usd in the treatment of infections caused by *Mycoplasma, Chlamydiae and Rickettsiae*. Tetracyclines are active against a wide variety of different bacterial species, but their use is now restricted by widespread resistance.

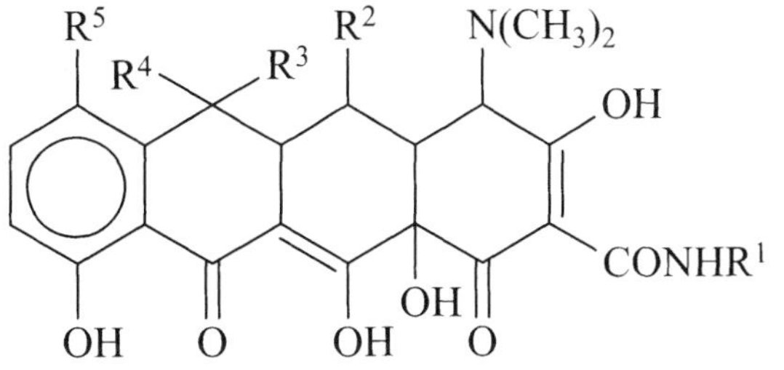

Tetracyclines

	R1	R2	R3	R4	R5
Tetracycline	H	H	CH_3	OH	H
Chlortetracycline	H	H	CH_3	OH	Cl
Oxytetracycline	H	OH	CH_3	OH	H
Doxycycline	H	OH	CH_3	H	H
Minocycline	H	H	H	H	$N(CH_3)_2$

Figure 15

Toxicity

Suppression of normal gut flora causes gastrointestinal upset and diarrhoea and leads to overgrowth by resistant bacteria, (e.g *Staphylococcus aureus*) and fungi, e.g Candida. Brown staining of teeth occurs in the fetus and in children, and thus these drugs should be avoided in pregnancy and in children under 8 years of age.

Resistance

Resistance is due partly to the widespread use of these drugs in humans as growth promotors in animal feedstuffs. The resistance genes are carried on a transposon. The mechanism of resistance is associated with the efflux mechanism which positively pumps out the antibiotic of the cells [24].

Chloramphenicol

Chloramphenicol contains a nitrobenzene nucleus which is responsible for some of the toxic problems associated with the drug (Figure 16).

Mechanism of action

Chloramphenicol blocks the action of peptidyl transferase thereby preventing peptide bond synthesis and thus inhibits bacterial protein synthesis.

Indications

Chloramphenicol is active against a wide variety of bacterial species: Gram-positive and Gram negative, aerobes and anaerobes, including intra cellular organisms such as *Salmonella typhi, Chlamydiae* and *Rickettsiae* [25].

Toxicity

Dose-dependent bone marrow suppression, occurs when the drug is given for long periods and is reversible when treatment is stopped.

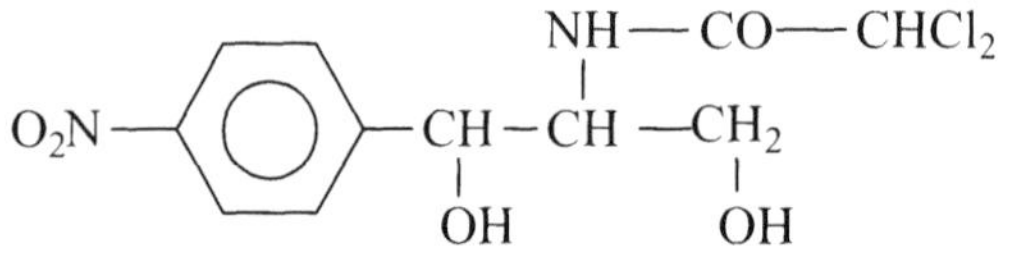

$$O_2N-\underset{}{\bigcirc}-\underset{\underset{OH}{|}}{CH}-\underset{\underset{OH}{|}}{\overset{\overset{NH-CO-CHCl_2}{|}}{CH}}-CH_2$$

Chloramphenicol

Figure 16

An idiosyncratic, not dose-dependent reaction, causes aplastic anaemia. Chloramphenicol is also toxic to neonates particularly premature babies whose liver enzyme systems are incompletely developed. This can result in the "grey baby syndrome". Chloramphenicol serum concentrations should be monitored in infants and children [26,27].

Resistance

The most common mechanism involves inactivation of the drug by plasmid-mediated chloramphenicol acetyl transferases produced by resistant bacteria. Acetylated chloramphenicol fails to bind to the ribosomal target. Resistance is becoming increasingly common and this, together with the potential toxicity, has significantly reduced the use of the drug.

Macrolides

Macrolides, lincosamides, ketolides, and streptogramins share overlapping binding sites on ribosomes, and have similar antimicrobial activities, as well as similar mechanisms of resistance [4]. The macrolides are a family of large cyclic molecules, all containing a macrocyclic lactone ring. The macrolide class of antibiotic agents includes compounds with 14-membered (erythromycin, clarithromycin) (Figure 17), 15-membered (azithromycin), and 16-membered (rokitamycin, spiramycin and josamycin) ring structures. Streptogramin antibiotics, such as pristinamycim or dalfopristin-quinupristin, contain two active components, type A and type B, which synergistically inhibit peptide elongation. Streptogramin B agents include quinupristin (Figure 18) and pristinamycin IA. Streptogramin A agents include dalfopristin and pristinamycim IIA.

The ketolide telithromycin is generated from erythromycin by removing a cladinose sugar and replacing it with a keto group. In addition, the C-11/C-12 region of the erythromycin molecule is bridged by a carbamate. The side group is an aryl-alkyl extension of the carbamate.

Mechanism of action

Erythromycin binds to the 23S rRNA in the 50S subunit of the ribosome and blocks the trans-location step in protein synthesis, thereby preventing the release of tRNA after peptide bond formation.

Indications

Erythromycin is active against Gram-positive cocci and is an important alternative treatment for infections caused by streptococci in patients allergic to peni-

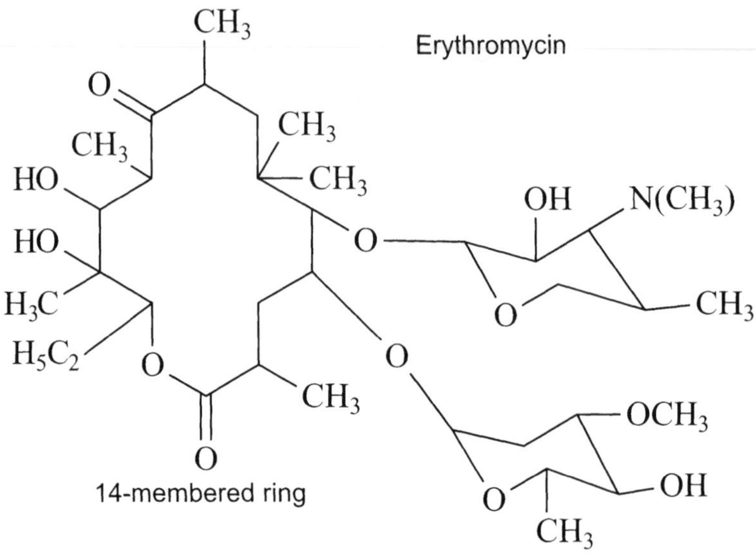

Figure 17

cillin. It is active against *Legionella pneumophila* and *Campylobacter jejuni*. It is also active against *Mycoplasma*, *Chlamydiae* and *Rickettsiae* and is therefore an important drug for treatment of atypical pneumonia and chlamydial infections of the urogenital tract. Spiramycin is used, almost exclusively, for the treatment of cryptosporidiosis and in the prevention of congenital toxoplasmosis.

Toxicity

Erythromycin is a relatively non-toxic drug, although it causes nausea and vomiting after oral administration in a significant number of patients.

Resistance

Two main mechanisms of resistance are described. The first mechanism is due to alteration in the 23S rRNA target by methylation of two adenine nucleotides in the RNA encoded by *erm* genes. At least 12 classes of *erm* genes have been identified by nucleic acid hybridisation analysis and nucleotide sequence comparison. The *erm*B gene confers co-resistance to 14-, 15-, and 16-membered-ring macrolides and lincosamides (MICs at which 90% of the isolates tested

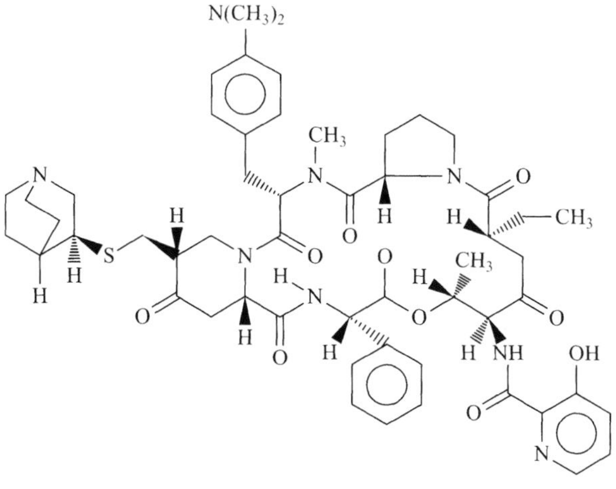

Quinupristin

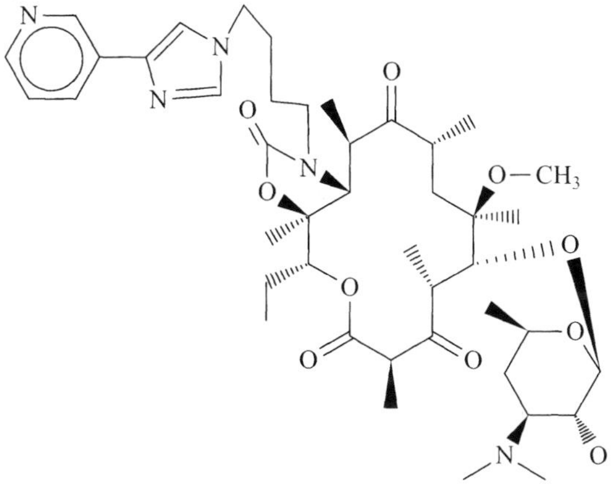

Telithromycin

Figure 18

are inhibited exceeding 128 μg/ml for both three drugs), and to streptogramin B (MIC90 = 64 μg/ml). The second mechanism involves the presence of an efflux pump encoded by the *mef* gene, which removes drugs from the bacterial cell. The *mef*A gene encodes a hydrophobic 44-2kDa protein sharing homology with membrane-associated pump proteins. The *mef*A gene confers resistance only to 14- and 15-membered-ring macrolides (MIC$_{90}$ = 8 μg/ml). Mutation of the ribosomal target of macrolides remains a rare resistance mechanism.

Lincosamides

This group contains lincomycin and clindamycin (Figure 19).

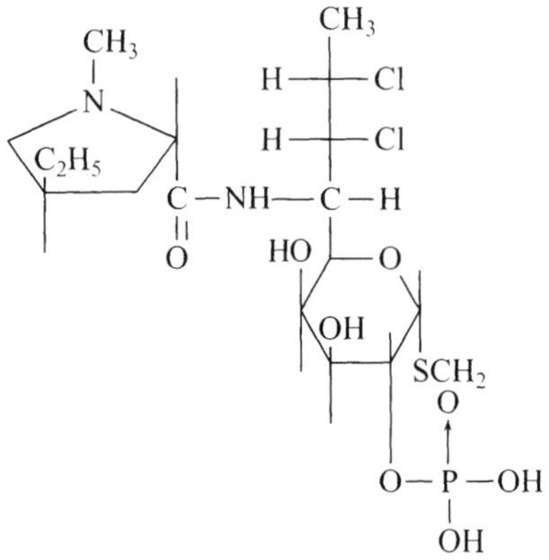

Clindamycin

Figure 19

Mechanism of action

Lincosamides bind to the 50S ribosomal subunit and inhibit protein synthesis by inhibiting peptide bond formation.

Indications

Clindamycin has a spectrum of activity similar to erythromycin, but it is much more active against anaerobes, both Gram-positive, e.g. *Clostridium* spp, and Gram-negative, e.g. *Bacteroides*. However, *Cl. difficile* is resistant and may overgrow in the gut, causing pseudomembranous colitis. The activity of clindamycin against *Staphylococcus aureus* and its penetration into bone makes it a valuable drug in the treatment of osteomyelitis [25].

Toxicity

The association between antibiotic administration and pseudomembranous colitis caused by *Cl. difficile* was first noted following clindamycin treatment.

Fusidic acid

Fusidic acid is a compound which inhibits protein synthesis by forming a stable complex with elongation factor EF-G, guanosine diphosphate and the ribosome.

Indications

Fusidic acid is active against Gram-positive cocci and is used is in the treatment of staphylococcal infections, which are resistant to beta-lactams, or in patients who are allergic to alternative staphylococcal agents. Fusidic acid should be given in combination with another antistaphylococcal agent, e.g rifampicin or β-lactams, to prevent the emergence of resistant mutants.

Toxicity

Fusidic acid may occasionally cause jaundice and gastrointestinal upset [28].

Resistance

Resistant mutants with altered EF-G emerge rapidly in staphylococcal populations exposed to the drug in monotherapy.

6.1.4   INHIBITORS OF NUCLEIC ACID SYNTHESIS

This category comprises antibacterial agents which act as inhibitors of nucleic acid synthesis.

Sulphonamides

This group of molecules are all structural analogues of para-amino benzoic acid (PABA).

Mechanism of action

Sulphonamides act in competition with PABA for the active site of dihydropteroate synthetase, an enzyme which catalyses a step in the synthetic pathway of tetrahydrofolic acid (THFA), required for the synthesis of purines and pyrimidines and thus for nucleic acid synthesis.

Indications

The sulphonamides have a spectrum of activity primarily against Gram-negative organisms (except *Pseudomonas*). Thus, they are used in the treatment of urinary tract infection. However, resistance is widespread.

Toxicity

Sulphonamides are relatively free of toxic side effects, but rashes and bone marrow suppression may occur.

Resistance

Plasmid-mediated genes code for an altered dihydropteroate synthetase, which has a greatly decreased affinity for the sulphonamide, but is essentially unchanged in its affinity for PABA.

Trimethoprim (and cotrimoxazole)

Trimethoprim is one of a group of pyrimidine-like structures analogous to the aminohydroxypyrimidine moiety of the folic acid molecule (Figure 20).

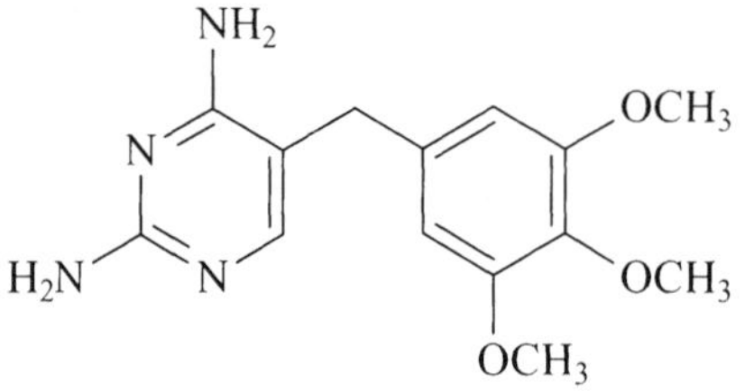

Trimethoprim

Figure 20

Mechanism of action

Trimethoprim, like sulphonamides, also prevents the synthesis of THFA, but at a later stage by inhibiting dihydrofolate reductase. Trimethoprim is often given in combination with sulphamethoxazole (cotrimoxazole). The advantages of this combination over either drug alone are :

• mutant bacteria which are resistant to one agent are unlikely to be resistant to the other, i.e double mutation;
• the two agents act synergistically against some bacteria.

Indications

Trimethoprim alone is active against Gram-negative rods with the exception of *Pseudomonas* species and its main use is in the treatment (and long-term prophylaxis) of urinary tract infection. Cotrimoxazole is active against a wide range of urinary tract pathogens and *S. typhi*. When given intravenously in high doses, this combination is valuable for the treatment of *Pneumocystis carinii* pneumonia.

Toxicity

Trimethoprim alone and in combination with sulphamethoxazole can cause neutropenia. Nausea and vomiting may occur.

Resistance

Plasmid-encoded dihydrofolate reductases with altered affinity for trimetho-prim, allow the synthesis of THFA to proceed unhindered by the presence of trimethoprim.

Quinolones

This is a large family of synthetic agents. Nalidixic acid is one of the earlier molecules, but the synthesis of fluoroquinolones has led to a number of chemical derivatives with improved antibacterial activity (Figure 21).

Mechanism of action

Quinolones act by inhititing the activity of DNA gyrase and thereby preven-ting supercoiling of the bacterial chromosome. As a result, the bacterial cell can no longer 'pack' its DNA into the cell.

Indications

Nalidixic acid is only active against enterobacteria and its use is limited to the treatment of urinary tract infection. Ciprofloxacin has a greater degree of activity than nalidixic acid against Gram-negative rods and is particularly active against

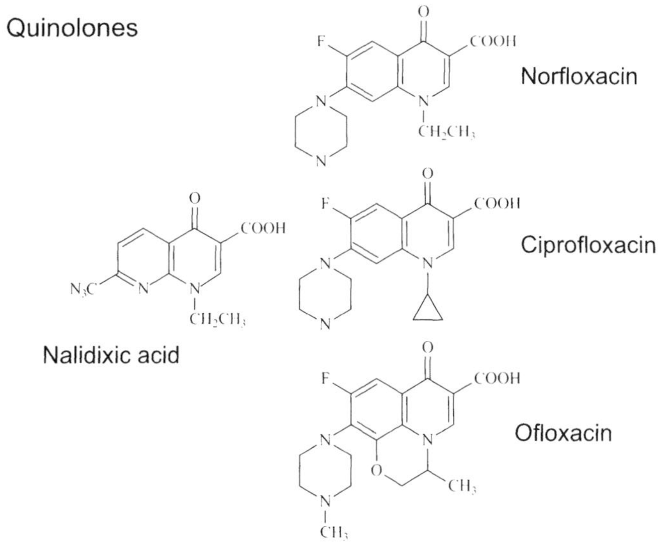

Figure 21

P. aeruginosa. Dosage may be adapted according to pharmacokinetics [29]. The newer fluoroquinolone moxifloxacin is also active against *Streptococcus pneumoniae.*

In addition to the treatment of urinary tract infection, the newer quinolones are useful for systemic Gram-negative infections and may find a role in the treatment of chlamydial and rickettsial infections. They may also be useful in infections caused by other intracellular organisms such as *L. pneumophila* and *S. typhi,* and in combination with other agents for 'atypical' mycobacteria. They are active against staphylococci, but less so against streptococci. Enterococci are resistant. [30]. Ciprofloxacin, ofloxacin and moxifloxacin can all be administered orally.

Toxicity

Gastro-intestinal disturbances are the most common side effects. Quinolones are usually avoided in children because of the risk of arthropathy. Seizure threshold may be reduced by ciprofloxacin [31,32].

Rifamycins

Rifampicin is a large molecule with a complex structure (Figure 22).

Mechanism of action

Rifampicin binds to the β-subunit of the RNA polymerase and blocks the synthesis of proteins.

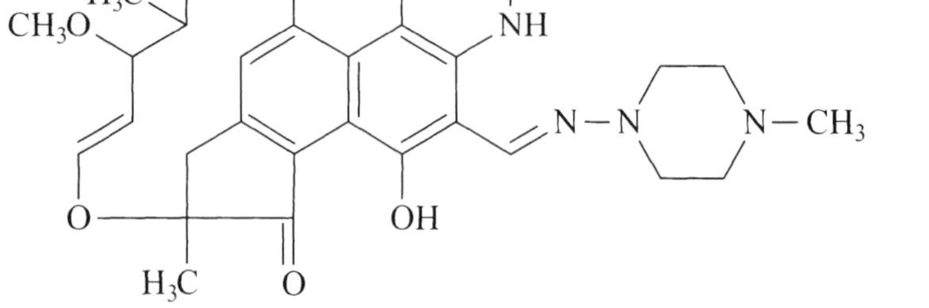

Figure 22

Indications

Rifampicin is used in treatment of mycobacterial infections. Rifampicin is the drug of choice for the prophylaxis of close contacts of meningococcal and *Haemophilus meningitis*. The drug is used in combination in the treatment of staphylococci infections.

Toxicity

Rashes, hypersensitivity reactions and jaundice occur with rifampicin treatment.

Resistance

Chromosomal mutations produce alterations in the RNA polymerase target which induces lowered affinity for rifampicin.

TEXTBOOKS

Courvalin, P.; Goldstein, F.; Philippon, A.; Sirot J. *L'antibiogramme*, Editions MPC-Videom, Paris, Bruxelles, 1987.

Duval, J.; Soussy, C. J. *Abrégé d'antibiothérapie*, Masson, 1985.

Gale, E. F.; Cundliffe, E.; Reynolds, P. E. *et al. The molecular basis of antibiotic action*, John Wiley and sons, London, 1981.

Lorian, V. *Antibiotics in Laboratory Medicine*, Fourth edition., Williams & Wilkins, Baltimore, 1996.

Kucers, A; Bennet, N. *The use of antibiotics*, Fourth edition, Heinemann Medical books, Oxford, 1987.

Michel-Briand, Y. *Mécanismes moléculaires de l'action des antibiotiques*, Masson, Paris, 1986.

REFERENCES

1. Butler, D. R., Kuhn, R. J., Chandler, M. H., *Clin Pharmacokinet.*, *26*, 374 (1994).
2. Paap, C. M., Nahata, M. C., *Clin Pharmacokinet.*, *19*, 280 (1990).
3. Lieberman, J. M., *Pediatr. Infect Dis J.*, *22*, 11433 (2003).
4. Sefton, A. M., *Drugs*, *62*, 557 (2002).
5. Darst, S. A., *Trends Biochem Sci.*, *29*, 159 (2004).
6. Singh, J., Burr, B., Stringham, D., Arrieta, A., *Paediatr. Drugs*, *3*, 733 (2001).
7. Rolinson, G., *J. Antimicrob. Chemother.*, *17*, 5 (1986).
8. Nikaido, H., Vaara, M., *Microbiol Rev.*, *49*, 1 (1985).
9. Nikaido, H., *Microbiol. Mol. Biol. Rev.*, *67*, 593 (2003).
10. Shing, J., Arrieta, A. C., *Sem. Paeditr. Infect. Dis.*, *10*, 14 (1999).
11. Torres, M. J., Blanca, M., Fernandez, J., et al., *Allergy*, *58*, 961 (2003).
12. Cerny, A., Pichler, W., *Pharmacoepidemiol. Drug Saf.*, *7*, S23 (1998).
13. Berger-Bachi, B., Rohrer, S., *Arch. Microbiol.*, *178*, 165 (2002).
14. Medeiros, A. A., *Br. Med. Bull.*, *40*, 18 (1984).
15. Neu, H. C., *Am. Med.*, *79*, 2 (1985).
16. Sivagnanam, S., Deleu, D., *Crit. Care*, *7*, 119 (2003).

17. Torel Ergur, A., Onarlioglu, B., Gunay, Y., Cetinkaya, O., Eray Bulut, H., *Acta. Paediatr. Jpn., 39*, 422 (1997).

18. Rocha, J. L., Kondo, W., Baptista, M. I., Da Cunha, C. A., Martins, L. T., Bratz, J., *Infect. Dis., 6*, 196 (2002).

19. Appelbaum, P. C., Bozdogan, B., *Clin. Lab. Med., 24*, 381 (2004).

20. Weigel, L. M., Clewell, D. B., Gill, S. R., et al., *Science, 28*, 302 (2003)

21. Gilbert, D., *Antimicrobial Agents Chemother., 35*, 3 (1991).

22. Tulkens, P. M., *J. Antimicrobial. Chemother., 27*, 49 (1991).

23. Galimand, M., Courvalin, P., Lambert, T., *Antimicrob Agents Chemother., 47*, 2565 (2003).

24. Huys, G., D'haene, K., Collard, J. M., Swings, J., *Appl. Environ. Microbiol., 70*, 1555 (2004).

25. Kasten, M. J., *Mayo Clin. Proc., 74*, 825 (1999).

26. Holt, D., Harvey, D., Hurley, R., *Adverse Drug React. Toxicol. Rev., 12*, 83 (1993).

27. Sack, C. M., Koup, J. R., Opheim, K. E., Neeley, N., Smith, A. L., *Pediatr. Pharmacol., (New York), 2*, 93 (1982).

28. Christiansen, K., *Int. J. Antimicrob. Agents, 12*, S3 (1999).

29. Payen, S., Serreau, R., Munck, A., et al., *Antimicrob. Agents Chemother., 47*, 3170 (2003).

30. Koyle, M. A., Barqawi, A., Wild, J., Passamaneck, M., Furness, P. D., *3rd Pediatr. Infect. Dis. J., 22*, 1133 (2003).

31. Grady, R., *Pediatr. Infect. Dis. J., 22*, 1128 (2003).

32. Cuzzolin, L., Fanos, V., *Expert Opin. Drug Saf., 1*, 319 (2002).

6.2 Urinary tract infections

Ulf Jodal

The Paediatric Uro-Nephrologic Centre, The Queen Silvia Children's Hospital, Göteborg University, SE-416 85 Göteborg, Sweden

6.2.1 CLINICAL PRESENTATION

Urinary tract infection (UTI) is common in children and during the first 6 years of life, 8% of girls and 2% of boys have a symptomatic UTI [1,2]. The symptoms depend upon the age of the patient, the level of the infection (bladder or kidney involvement), as well as the intensity of the inflammatory reaction that the bacteria induce in the patient. Bacteria can be present in the urinary tract without symptoms (asymptomatic or covert bacteriuria), a state that has been shown to be innocent and best being left untreated [3,4]. In the bladder, bacteria may cause acute cystitis, characterised by symptoms such as dysuria and frequency, together with an inflammatory reaction manifested by pyuria and sometimes haematuria. Infection in the kidneys results in acute pyelonephritis with classical features of high fever, loin pain and acute phase responses. In some patients bacteraemia may occur [5].

6.2.2 BACTERIA AND THE HOST REACTION

Escherichia coli cause 80 to 90% of UTIs in children. They are so called primary pathogens for the urinary tract. Most of the *E. coli* that cause acute pyelonephritis carry virulence factors that facilitate invasion of the urinary tract and initiate an inflammatory host response [6]. Capacity to adhere to uroepithelial cells is the virulence factor that is most characteristic of *E. coli* strains capable of inducing pyelonephritis. These strains often carry several adhesins, but the most important type is so called P-fimbriae. These favour colonisation of the large intestine that is the major reservoir for bacteria that enter the urinary tract by ascending up through the urethra, and bind to specific receptors on uroepithelial cells. This binding activates a cytokine cascade, leading to a local and systemic inflammatory response [7]. It is the extent and severity of the inflammatory reaction in the kidney that determines if the patient will develop permanent renal damage (renal scarring) or if the inflammation will heal without sequelae [6].

Individuals with congenital defects of the urinary tract or impaired emptying of the bladder have an increased susceptibility to UTI. Bacteria lacking specific virulence factors, e.g. secondary pathogens that include *Klebsiella*, *Pseudomonas*, *Staphylococci* and *Enterococci*, can invade their urinary tracts. Thus, the finding of such non-*E. coli* bacteria in urine cultures from a patient with high fever suggests abnormalities of the urinary tract [8,9]. Abnormalities of clinical significance are vesico-ureteric reflux with dilatation of the upper urinary tract (grade III-V) that is found in 10 to 15% of children with acute pyelonephritis, and malformations causing obstruction of the urinary flow found in about 10% of boys and 2% of girls with UTI [10].

6.2.3   RENAL DAMAGE

Permanent renal damage can be congenital (primary) or acquired, i.e. secondary to acute pyelonephritis. Congenital damage (renal dysplasia) is most frequently found in boys and associated with severe forms of vesico-ureteric reflux or obstruction of the urinary flow [11]. Acquired damage is most common in girls and is related to the number of attacks of acute pyelonephritis [12]. A long delay before start of treatment will also increase the risk that acute pyelonephritis will lead to permanent renal damage. Since the long-term consequence of damage can be development of hypertension and even chronic renal failure, adequate treatment is essential to reduce the risk of progressive damage.

6.2.4   ANTIBACTERIAL TREATMENT

There are several issues that need to be taken into consideration in the treatment of children with symptomatic UTI, e.g. choice of antibiotic, route of administration and duration of treatment. It is obvious that the severity of the infection also is of importance. In children with high fever and suspected to have renal involvement, treatment should be started without delay; results of culture and sensitivity cannot be waited for. This means that knowledge of the resistance pattern of the uropathogens in the region is essential to allow an appropriate choice of antibiotic; international reports are of little value in this respect. Recent antibacterial treatment should also be taken into consideration, as this may have changed the gut flora towards more resistant strains.

Bacterial resistance

The uropathogenic bacteria have become increasingly resistant to antibiotic drugs. Although the local variations are considerable, sulphonamides and ampicillin derivatives are no longer useful for empirical treatment of children with UTI because of resistance rates that are at the best 30 to 50%. Resistance to trimethoprim and co-trimoxazole is in some regions at a low level of 10%. If higher, then the drug is of more doubtful value as a first choice antibiotic. The best sensitivity, with resistance <10%, is found for modern cephalosporins, e.g. the oral drugs cefadroxil, cefixime and ceftibutene, and for the parenteral preparations cefepime, cefotaxime, ceftazidime and ceftriaxone. It should be emphasised that the report of the urine culture is of importance, not only to confirm the diagnosis of UTI, but also to monitor the use of an appropriate antibiotic, depending on the sensitivity of the isolated bacterial strain.

Route of administration in pyelonephritis

There are only a few studies on treatment of children with UTI that fulfil modern standards for clinical trials. Mostly local traditions and beliefs guide treatment recommendations. There are many difficulties in performing randomised clinical trials in small children and a major obstacle is to include adequate numbers of patients. The most relevant end-point to study is probably development of permanent renal damage, both incidence and extent, which with the techniques available today will require repeated scintigraphies. Other possibilities are to look at sterilisation of urine, time for fever resolution, and incidence of early recurrence of UTI. Hoberman et al. studied all these factors in children 1 to 24 months of age with febrile UTI [13]. A total of 306 children were allocated to either oral cefixime for 14 days or initial intravenous cefotaxime for 3 days, followed by oral cefixime for 11 days. The overall results were excellent with sterilisation of urine after 24 hours in all patients and a very low incidence of permanent renal damage at 6 months, 9% as determined by scintigraphy. More importantly, there was no difference in outcome between the treatment groups for any of the end-points. The same results were found in two smaller studies of children with febrile UTI [14,15]. However, children who were seriously ill with suspected or evident hypotension, or who vomited, were excluded in these studies. Certainly, children with such symptoms should be given parenteral treatment. Otherwise, drugs such as trimethoprim and the modern oral cephalosporins, rapidly attain high concentrations in urine and renal tissue and should be as effective in eradicating bacteria as antibiotics given parenterally to children with acute pyelonephritis.

Duration of treatment in pyelonephritis

A 10 day course is adequate to treat a child with pyelonephritis; there are no reports on shorter courses in this type of patients. Instead, there has been speculation that uptake defects on an acute scintigram should lead to more aggressive antibacterial treatment. There is no evidence to support such a hypothesis. There is a correlation between extent of inflammation and uptake defects on scintigraphy. However, killing of the bacteria is usually a rapid process and prolonged antibacterial treatment is not likely to influence the inflammatory process, the extent of which is decisive for the development of permanent damage [6]. Rather, we need new methodology to moderate inflammation to be able to decrease or prevent damage. It is, however, important to start antibacterial treatment without delay, because there is a correlation between the duration of infection and extent of inflammation [16].

Bladder infection (acute cystitis)

There are many reports on small, randomised, controlled trials of short course therapy in children with lower UTI. Results have been variable and 3 recent meta-analyses, comparing 4 day or shorter treatment with longer courses, have reached different conclusions [17-19]. When short courses are used, it is important to take into account that the antibiotics have different pharmacokinetic properties. Antibiotics that are rapidly excreted (serum half-life of about 1 hour) have no antibacterial activity in urine after 8 to 12 hours; examples are some oral cephalosporins and nitrofurantoin. Trimethoprim, on the other hand, has a serum half-life of about 10 hours that results in effective antibacterial concentrations in urine for several days. This means that a 3 day course with trimethoprim, with or without sulphonamide, is equivalent to at least 5 days treatment with cefadroxil or nitrofurantoin. Single dose or 1 day courses are not recommended, whereas a 3-day course of co-trimoxazole was as effective as longer courses. In a single centre study of 300 consecutive children with acute cystitis, treated with nitrofurantoin or trimethoprim for 5 days, the bacteriological cure rate was 99%; recurrences within 1 month occurred in 2% [20]. With such clear-cut results, it can be firmly concluded that children with acute cystitis need treatment for a maximum of 5 days; there is no need for a controlled trial. But the importance of checking for the resistance pattern of the bacteria is emphasised again.

REFERENCES

1. Hellström, A.-L., Hanson, E., Hansson, S., Hjälmås, K., Jodal, U., *Arch. Dis. Child.*, 66, 232 (1991).
2. Mårild, S., Jodal, U., *Acta Paediatr.*, 87, 549 (1998).

3. Kemper, K. J., Avner, E. D., *Am. J. Dis. Child., 146,* 343 (1992).
4. Linshaw, M., *Kidney. Int., 50,* 312 (1996).
5. Hansson, S., Martinell, J., Stokland, E., Jodal, U., *Infect. Dis. Clin. N. Am., 11,* 499 (1997).
6. Wullt, B., Bergsten, G., Fischer, H., et al., *Infect. Dis. Clin. N. Am., 17,* 279 (2003).
7. de Man, P., Jodal, U., Lincoln, K., Svanborg Edén, C., *J. Infect. Dis., 158,* 29 (1988).
8. de Man, P., Claeson, I., Johanson, I., Jodal, U., Svanborg Edén, C., *J. Pediatr., 115,* 915 (1989).
9. Honkinen, O., Lehtonen, O. P., Ruuskanen, O., Huovinen, P., Mertsola, J., *Br. Med. J., 318,* 770 (1999).
10. Winberg, J., Andersen, H. J., Bergström, T., Jacobsson, B., Larson, H., Lincoln, K., *Acta Paediatr. Scand., Suppl 252,* 3 (1974).
11. Risdon, R. A., *Pediatr. Nephrol., 7,* 361 (1993).
12. Wennerström, M., Hansson, S., Jodal, U., Stokland, E., *J. Pediatr., 136,* 30 (2000).
13. Hoberman, A., Wald, E. R., Hickey, R. W., et al., *Pediatrics, 104,* 79 (1999).
14. Levtchenko, E., Lahy, C., Levy, J., Ham, H., Piepsz, A., *Pediatr. Nephrol., 16,* 878 (2001).
15. Baker, P. C., Nelson, D. S., Schunk, J. E., *Arch. Pediatr. Adolesc. Med., 155,* 135 (2001).
16. Miller, T., Phillips, S., *Kidney Int., 19,* 654 (1981).
17. Tran, D., Muchant, D. G., Aronoff, S. C., *J. Pediatr., 139,* 93 (2001).
18. Keren, R., Chan, E., *Pediatrics, 109,* E70 (2002).
19. Michael, M., Hodson, E. M., Craigh, J. C., Martin, S., Moyer, V. A., *Arch. Dis. Child., 87,* 118 (2002).
20. Abrahamsson, K., Hansson, S., Larsson, P., Jodal, U., *Acta Paediatr., 91,* 55 (2002).

6.3 Bacterial meningitis

Antoine Bourillon[1], Edouard Bingen[2]

[1]Department of Paediatrics, Hospital Robert Debré, Paris, France
[2]Department of Microbiology, Hospital Robert Debré, Paris, France

6.3.1 PATHOPHYSIOLOGY OF MENINGEAL INFECTION

The blood-brain barrier consists of two distinct structures. The first one is the endothelium of the meningeal capillaries, characterised by tight junctions between the endothelial cells. The relative lack of pinocytosis vesicles reflects the low transcytosis activity of these cells. The second structure is the choroid plexus that produce the cerebrospinal fluid (CSF) and are located in the ventricles. They are made of epithelial cells with tight junctions that lie on the basal membrane and are associated with a fenestrated endothelium [1, 2].

The invasion of the meningeal spaces probably requires a number of preliminary conditions:

- Colonisation of the oropharyngeal or intestinal mucosa (depending on the type of bacteria).
- Translocation into the bloodstream (bacteraemia).
- Resistance of the bacteria against defences of the host and multiplication in the blood (septicaemia).
- Crossing of the blood-brain barrier and multiplication in the cerebrospinal fluid.

A history of head trauma or surgery at the base of the skull, in the presence of rhinorrhoea; ear infection, sinusitis, asplenia or HIV infection all suggest that pneumococcal infection should be considered. Numerous clinical and experimental studies have suggested that a relationship exists between the magnitude of the bacteraemia and the occurrence of meningitis. The contamination of the CSF requires a prolonged, elevated bacteraemia [3]. Thus, meningitis due to *S. pneumoniae* is observed more frequently in children with a bacterial load that exceeds 10^2 colony forming units (CFU)/ml in blood (86%) than in those with a load below this level (4%) [4, 5]. Similarly, a threshold close to 10^2 CFU/ml has been reported with meningitis due to *Haemophilus influenzae b (Hib)* [4, 5]. The existence of a critical bacterial load threshold appears therefore to be required (Table 1). The capacity of *Hib* and *S. pneumoniae* to cause meningitis is associated with their specific capacity to resist the defences of the host during the systemic phase [3]. In the case of *Neisseria meningitidis*, the role of a load threshold is less well established [5, 6].

Blood cultures therefore represent an important diagnostic tool for bacterial meningitis. Treatment must have systemic bactericidal activity (in addition to bactericidal activity within the CSF) in order to ensure efficacy.

Table 1

Correlation between the magnitude of the bacteraemia and meningitis

Bacteria	CFU/ml of blood	Number of patients	Number of patients with meningitis		Reference
			n	**%**	
S. pneumoniae	≤ 100	20	1	(5)	4
	> 100	4	3	(75)	
	< 500	30	1	(3)	5
	≥ 500	3	3	(100)	
H. influenzae type b	< 100	12	2	(17)	5
	≥ 100	22	11	(50)	
	≤ 100	11	7	(63)	4
	> 100	15	11	(73)	
N. meningitidis	< 500	22	7	(32)	6
	≥ 500	13	8	(62)	
	< 1,000	8	5	(63)	5
	≥ 1,000	4	1	(25)	

6.3.2 TREATMENT

The antibiotic treatment of bacterial meningitis must satisfy the following criteria:

- It must be initiated as rapidly as possible because a delay in the administration of treatment may affect the prognosis. A delay in the diagnosis and treatment has been associated with higher concentrations of organisms in the CSF, and with a poor prognosis [7]. However, a direct correlation between the duration of the symptoms preceding the diagnosis of meningitis and the prognosis has not been shown [8, 9]. Some patients treated early after the onset of symptoms develop sequelae, while others treated less rapidly do not suffer from complications [10]. The relationship between the early initiation of treatment and prognosis is complex and apparently related to the variability and duration of the symptoms during the three phases of evolution: (i) non-specific incubation; (ii) systemic dissemination; (iii) meningeal invasion. A relationship has been established with the bacteriological aetiology of paediatric meningitis. For *S. pneumoniae*, early treatment was significantly associated with a reduced rate of complications. For *H. influenzae* and *N. meningitidis*, however, there is no direct relation.

• It must be bactericidal as the CSF has no natural bactericidal potency, due to the absence of macrophages, antibodies and complement [12, 13]. Opsonisation is therefore ineffective when the organisms responsible for paediatric meningitis express their virulence through the capsule [14]. This is the case for the capsular antigen of *N. meningitidis* rich in sialic acid, for the polyribose-phosphate antigen of *Hib* and for the polysaccharide capsular antigen of *S. pneumoniae*.

• Bactericidal activity should be rapid. A delay in obtaining the sterilisation of the CSF correlates with sequelae [16, 17]. Lebel et al [16] have shown a significantly higher rate of neurological sequelae among children who had posi-

Table 2

Relationship between delayed sterilisation of the cerebrospinal fluid and outcome [16]

Outcome	Repeat CSF culture (18 to 36 hours after the beginning of treatment)			
	Positive (n = 20)		Negative (n= 281)	
	n	%	n	%
neurological abnormalities at discharge;	9	45	52	19*
follow-up after 6 weeks				
ataxia	3	18	7	3*
hemiparesis	3	18	6	2*
developmental delay	5	20	18	7*
hearing impairment	6[†]	35	34[‡]	15*

* p < 0.05 comparison with positive culture; [†](n = 17); [‡](n = 234)

tive cultures of the CSF 18 to 36 hours after the start of treatment (Table 2). Auditory complications were more common (17%) in children receiving cefuroxime, for whom sterilisation had been slower than in children treated with ceftriaxone (4%) [18].

These criteria, however, are difficult to establish, especially for the β-lactams as their *in vivo* bactericidal activity is reduced because of the slow multiplication rate in the CSF (the generation time for *S. pneumoniae* is 60 minutes in the CSF while it is 20 minutes in culture) and because of the usual presence of a strong bacterial inoculum in the CSF.

6.3.3   PREDICTIVE EFFICACY CRITERIA OF ANTIBIOTICS DURING MENINGITIS

During experimental studies of meningitis, maximal bactericidal activity with β-lactams or vancomycin can only be obtained with concentrations 10 to 30 times higher than the minimal bactericidal concentration (MBC) [21, 22]. Therefore, an inhibitor quotient (IQ = *in situ* concentration at peak / MBC for the bacterium) in the CSF ≥10 appears to be required to obtain maximal bactericidal activity. Thus, the activity of β-lactams in the CSF appears to be concentration dependent. However, recent studies have concluded that their activity is also time dependent [23]. In a *S. pneumoniae* experimental model in rabbits, the best predictive factor of the bactericidal activity of ceftriaxone in the CSF was the time during which the antibiotic concentration was above the MBC of the bacterium (T > CMB) [24]. An interval between two administrations that allowed one to achieve T > 100% was the only parameter to correlate with maximal bactericidal activity in the CSF and rapid sterilisation. A higher bactericidal activity with rapid sterilisation of the CSF was obtained by giving ceftriaxone twice daily on the first day and then once daily [24]. However, in humans the elimination half-life of ceftriaxone is much longer in humans than in rabbits (8 h *vs.* 3 h) and the direct extrapolation of these results to humans requires additional studies [23].

The peak concentration in the CSF is probably an approximate indirect measurement of the T > CMB, in particular for antibiotics with a long half life. However, for those antibiotics with a short half life, the dosage schedule should be strictly followed, and an increase of the number of daily injections may be beneficial. Therefore, an IQ > 10 or a T > MBC of 100% are the objectives to be reached in order to achieve maximal effectiveness in the CSF.

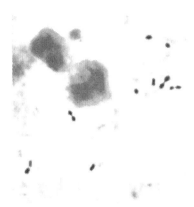

Figure 1. Gram staining of the CSF showing *Streptococcus pneumoniae*.

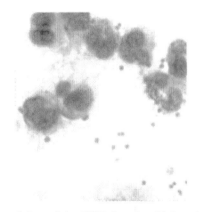

Figure 2. Gram staining of the CSF showing *Neisseria meningitidis*.

6.3.4   EPIDEMIOLOGY OF THE ORGANISMS INVOLVED IN BACTERIAL MENINGITIS

The optimal treatment for meningitis depends upon knowledge of the epidemiology of the infecting organisms and of their sensitivity to antibiotics. Meningitis affects the young child, with two thirds of all cases occurring in children under the age of five years. At all ages, the disease is associated with a significant risk of mortality or neurological lesions with permanent sequelae [12].

In France there are about 150 to 200 cases of meningitis per year due to *S. pneumoniae* in children, 67% occurring in children under 2 years [25], and about 300 cases per year due to *N. meningitidis* among all ages. Among children aged 3 months to 2 years, the predominant strains are: *Hib*, *S. pneumoniae* and *N. meningitidis*. Vaccination against *Hib* has been associated with a very important reduction of *Hib* meningitis. The number of cases was reduced from 500 in 1991 to 50 in 1999. The near disappearance of this type of meningitis over an 8 year period in France has considerably reduced the incidence of meningitis in paediatrics and thus modified its epidemiology. *S. pneumoniae* is now the organism responsible for 60% of cases of meningitis in children between 2 months and one year of age [26] (Figure 1). *N. meningitidis* is the organism responsible for 60% of cases of meningitis between one and two years (Figure 2), while it is 33% for *S. pneumoniae* [26]. Table 3 shows the data for all cases of childhood bacterial meningitis in France.

Table 3

Cases of bacterial meningitis in children in France over 3 years (2001-3)

Bacteria	Number of cases	(%)	
N. meningitidis	599	55	
group B	310		*52*
group C	203		*34*
others	86		*14*
S. pneumoniae	362	33	
S. agalactiae	55	5	
H. influenzae b	27	3	
E. coli	13	1	
others	28	2	
Total	1683	100	

6.3.4.1 N. meningitidis

N. meningitidis is sensitive to a large number of antibiotics but about 50% of strains are resistant to sulfanomides. Penicillin has previously been considered as the treatment of choice for meningitis due to *N. meningitidis*, as all strains had a minimal inhibitory concentration (MIC) of penicillin of 0.05 mg/l. The first strains that had an intermediate sensitivity to penicillin (MIC of penicillin G between 0.1 and 1 mg/l) were isolated in Spain in 1985 and then in South Africa and Great Britain. In 1998, they accounted for 3 to 4% of all strains in Holland and 6% in Belgium. In England, prevalence increased from 8 to 10% before 1995 to 18% in 1996 [27]. In Spain, they accounted for 5% of strains in 1986 and 67% in 1993 [28]. In France, the same phenomenon has been observed, with the progression of strains of limited sensitivity from 4% in 1994 to 32% in 1999 [29].

6.3.4.2 Mechanisms of penicillin resistance

Two mechanisms have been described. The first is production of β-lactama-ses, a characteristic of several strains that have a MIC for penicillin > 1 mg/l. Fortunately these strains have not spread much. The third-generation cephalos-porins remain active against these strains.

The second mechanism of resistance is related to a reduced affinity of the PLP (penicillin linking proteins) for penicillin. Analysis of the genes that encode for the PLP show the presence of mosaic genes. This mechanism is similar to the one already described for *S. pneumoniae*. These mosaic genes contain fragments of the PLP-2 gene of *N. meningitidis* sensitive to penicillin as well of the homo-logous PLP genes of *Neisseria* that co-exists in the rhinopharyngeal flora and is naturally resistant to penicillin, such as *N. flava*, *N. subflava*, or *N. lactamica* [27].

6.3.4.3 S. pneumoniae

Meningitis due to *S. pneumoniae* has significant morbidity with 30% of patients having permanent neurological or auditory sequelae [30]. The first penicillin resistant strains isolated from patients with pneumococcal meningitis were isolated in 1977 in South Africa, where 3 fatal cases were reported with a high level of penicillin resistance (MIC of 4 to 8 mg/l). A strain is considered sensitive to penicillin G if the MIC is below 0.125 mg/l. Strains of *S. pneumo-*

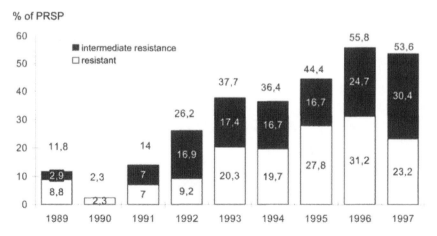

Figure 3. Evolution of the prevalence of penicillin-resistant *S. pneumoniae* (PRSP) as isolated from children with meningitis from 1989 through 1997 (n = 539) [31].

niae with reduced penicillin sensitivity (PRPS) are divided into two groups of low or intermediate level of resistance (MIC between the values of 0.125 and 1.0 mg/l) and high-level of resistance (MIC > 1.0 mg/l). From 1989 to 1997, the annual rate of PRPS isolated from children with meningitis in France increased from 11.8% to 53.6% [31] (Figure 3), while it was only 31% in adults in 1997 [31]. In 1997, 47.7% of the PRPS strains isolated in the CSF were sensitive to cefotaxime, 51.2 % showed an intermediate sensitivity and only 1.2% showed high-level resistance, with an MIC of 4 mg/l [31, 32]. The high prevalence of *S. pneumoniae* strains resistant to β-lactams in paediatrics requires one to determine the MICs for penicillin, amoxicillin, cefotaxime and ceftriaxone for each of the isolated strains.

6.3.4.4 Mechanism for penicillin resistance in S. pneumoniae

The resistance of *S. pneumoniae* to penicillin G is related to a reduction of the affinity of PLP. This resistance is of chromosomal origin and results from structural changes of one or several PLP. Penicillin resistant strains contain genes that encode for PLP 1a, 2b, and 2x having a mosaic type structure, with regions identical to sensitive strains and highly variable regions; resulting from transformation and inter- or intra-species recombinations. Original gene segments of PLP in penicillin sensitive strains of *S. pneumoniae* are thereby replaced with corresponding segments originating from PLP genes of related species (*S. mitis, sanguis* and *oralis*). In addition to these transformations and recombinations, additional point mutations can also contribute to increase the level of resistance [33]. The emergence of this resistance in France is related to the spread of resistant clones and to the horizontal transfer of resistant genes.

6.3.5 THERAPEUTIC STRATEGY

6.3.5.1 Antibiotic strategy for meningitis by N. meningitidis

The MIC for penicillin for strains of *N. meningitidis* isolated in France is between 0.1 and 0.7 mg/l [29]. No β-lactamase producing strain has been isolated in France. The reduction of penicillin sensitivity has been associated with a reduction in activity of most of the other β-lactams. The MIC 90 of penicillin, ampicillin, and first-generation cephalosporins towards strains of reduced penicillin sensitivity is 0.3, 0.7, and 2.4 mg/l, respectively [35]. The injectable third-generation cephalosporins still remain very sensitive, with a MIC 90 for cefotaxime or ceftriaxone < 0.01 mg/l [35]. The emergence of intermediate penicillin-resistant strains has not been associated with treatment failures if high doses of penicillin are used [36]. Some authors, however, have reported a more severe illness [37]. In France, a third generation cephalosporin (C3G) is recommended as the drug of choice.

6.3.5.2 Antibiotic strategy for meningitis due to S. pneumoniae

C3Gs, such as cefotaxime or ceftriaxone are recommended in the treatment of bacterial meningitis. Clinical failures have only been reported for strains with MICs for the C3Gs ≥ 0.5 mg/l [20, 25, 30, 38, 39]. Using cefotaxime at the dose of 200 mg/kg/24 h, or ceftriaxone at the dose of 80 to 100 mg/kg/24 h, average concentrations in the CSF are close to 5 mg/l with the corresponding IQ for these strains close to 10.

Two strategies have been used to enhance bacterial eradication of penicillin resistant strains of S. pneumoniae.

(1) Increase the local concentration of antibiotics, by increasing the dose (assumes that penetration into the CSF is linear) [38].

(2) Reduce the MIC of the bacterium by the use of synergetic associations [38, 39].

The combination of these two strategies has been put forward in the following recommendations. In 1997, the Committee of Infectious Diseases of the American Academy of Pediatrics [40] advised for children over one month of age with suspected *pneumococcal* meningitis, either cefotaxime (225 mg/kg/day in three or four injections) or ceftriaxone (100 mg/kg/day in one or two injections). The treatment is then adapted according to the MIC of the β-lactams of the isolated strain (Table 4).

In France similar recommendations were published in 1996 [41]. For *pneumococcal* meningitis, a higher dose of cefotaxime (300 mg/kg/day) in combination with vancomycin is recommended (40 to 60 mg/kg/day in 4 doses). A repeat

Table 4

Treatment of pneumococcal meningitis in children: modification based on antibiotic susceptibilities of the *S. pneumoniae* isolates

Penicillin MIC (mg/l)	Cefotaxime or ceftriaxone MIC (mg/l)	Therapy	Recommended dosage (mg/kg/day)
≥ 0.1	≤ 0.5	cefotaxime	200-225
		or ceftriaxone	100
	1	add vancomycin	60
	≥ 2	add vancomycin	60
		and rifampicin	20

lumber puncture at 36-48 hours is advocated because delays in sterilisation have been associated with poor prognosis in *pneumococcal* meningitis [41]. If the organism is highly sensitive, MIC of C3G < 0.5mg/l, the vancomycin can be discontinued, the dose of C3G reduced or changed to amoxicillin (150 to 200 mg/kg/day) if the MIC of amoxicillin is < 0.5 mg/l.

The possibility of complications such as subdural effusions should be considered in patients who fail to improve. Cranial imaging may be required [43].

The optimal duration of treatment for bacterial meningitis has not been clearly established [44]. The duration of treatment is dependent on the organism; 7 days for *Hib*-based meningitis, 14 days for *S. pneumoniae* and 5 days for *N. meningitidis* [45, 50].

Faced with the emergence of resistance, only the development of prevention strategies, notably vaccination, are likely to reduce the mortality and the morbidity of bacterial meningitis. In the near future in France, the use of heptavalent conjugated vaccines against *pneumococcus* for children under two years of age should reduce the number of cases of meningitis in the age range where the incidence is the highest [51].

REFERENCES

1. Durand, M. L., Calderwood, S. B., Weber, D. J., et al., *N. Engl. J. Med.*, *328*, 21 (1993).
2. Tuomanen, E., *FEMS Microbiol. Rev.*, *18*, 289 (1996).
3. Kim, K. S., *Infect. Immun.*, *69*, 5217 (2001).
4. Bell, L., Alpert, G., Campos, J. M., Plotkin, A., *Pediatrics*, *76*, 901 (1985).
5. Sullivan, T. D., Lascolea, L. J., Neter, E., *Pediatrics*, *69*, 699 (1982).
6. Sullivan, T. D., Lascolea, L. J., *Pediatrics*, *80*, 63 (1987).
7. Feldmann, W. E., Richmond, V. A., *J. Pediatr.* *88*, 549 (1976).
8. Radestski, M., *Pediatr. Infect. Dis. J.*, *11*, 694 (1992).
9. Short, W. R., Tunkel, A., *Curr. Infect. Dis. Rep.*, *3*, 360 (2001).
10. Bonadio, W. A., *Am. J. Emerg. Med.*, *15*, 420 (1997).
11. Bonsu, B. K., Harper, M. B., *Clin. Infect. Dis.*, *32*, 566 (2001).
12. Schuchat, A., Robinson, K., Wenger, J. D., et al., *N. Engl. J. Med.*, *337*, 970 (1997).
13. Taüber, M. G., Martin, G., Merle, A., Sande, M. A., *Infect. Dis. Clin. North Am.*, *4*, 661 (1990).
14. Zwahlen, A., Nydegger, U. E., Vaudaux, P., et al., *J. Infect. Dis.*, *14*, 635 (1982).
15. Scheld, W. M., Sande, M. A., *J. Clin. Invest.*, *71*, 411 (1983).
16. Lebel, M. H., McCracken, G. H., Jr., *Pediatrics*, *83*, 161 (1989).
17. Aronin, S. I., Peduzzi, P., Quagliarello, V. J., *Ann. Intern. Med.*, *129*, 862 (1998).
18. Schaad, U. B., Sutter, S., Gianella-Borradori, A., *N. Engl. J. Med.*, *322*, 141 (1990).
19. Bingen, E., Lambert-Zechovsky, N., Mariani-Kurkdjian P., et al., *Eur. J. Clin. Microbiol. Infect. Dis.*, *9*, 278 (1990).
20. Bourrillon, A., Doit, C., Bingen, E., *Press Med.*, *27*, 1183 (1998).
21. Aronin, S. I., *Curr. Infect. Dis. Rep.*, *2*, 337 (2000).
22. Tauber, M. G., Hackbarth, C. J., Scott, K. G., et al., *Antimicrob. Agents Chemother.*, *27*, 340 (1985).
23. Lutsar, I., Friedland, I. R., *Clin. Pharmacokinet.*, *39*, 335 (2000).

24. Lutsar, I., Ahmed, A., Friedland, I. R., et al., *Antimicrob. Agents Chemother., 41,* 2414 (1997).
25. Olivier, C., Begue, P., Cohen, R., Floret, D. (for the GPIP), *BEH, 16* (2000).
26. de Benoist, A. C., Laurent, E., Goulet, V., *BEH, 15* (1999).
27. Bingen E., *MT Pédiatrie, 3,* 19 (2000).
28. Pascual, A., Joyanes, P., Martinez-Martinez, L., Suarez, A. I., Perea, E. J., *J. Clin. Microbiol., 34,* 588 (1996).
29. Alonso, J. M., Taha, M. K., *Annual Report, National Reference Centre* (1999).
30. Olivier, C., Cohen, R., Begue, P., Floret, D., *Pediatr. Infect. Dis. J., 19,* 1015 (2000).
31. Geslin, P., *Annual Report, National Reference Centre* (1997).
32. Gaudelus, J., Cohen, R., Reinert, P., *Acta Paediatr., 89 (Suppl.),* 27 (2000).
33. Sibold, C., Wang, J., Henrichsen, J., Hakenbeck, R., *Infect. Immun., 60,* 4119 (1992).
34. Doit, C., Picard, B., Loukil, C., Geslin, P., Bingen, E., *J Infect Dis* 181 : 1971-1978 (2000).
35. Saez Nieto, J. A., Vasquez, J. A., *Microbiologia, 13,* 337 (1997).
36. Brunen, A., Peetermans, W., Verhaegen, J., et al., *Eur. J. Clin. Microbiol. Infect. Dis., 12,* 969 (1993).
37. Luaces Cubells, C., Garcia, J. J., Roca Martinez, J., et al., *Acta Paediatr., 86,* 26 (1997).
38. Bingen, E., Bourrillon, A., *Presse Med., 23,* 763 (1994).
39. Bingen, E., Doit, C., Bouillie, C., Blanchard, B., Geslin, P., Bourrillon, A., *Lancet, 346,* 311 (1995).
40. Committee of infectious diseases of the American academy of pediatrics, *Pediatrics, 99,* 289 (1997).
41. Conférence de consensus, *Med. Mal. Infect., 26,* 952 (1996).
42. Rocha, P., Baleeiro, C., Tunkel, A. R., *Curr. Infect. Dis. Rep., 2,* 399 (2000).
43. Aujard, Y., *Encyclopédie médico-chirurgicale, Pédiatrie,* (Elsevier) 4-120-B10 (1996).
44. Radetsky, M., *Pediatr. Infect. Dis. J., 9,* 2 (1990).
45. Lin, T. Y., Chrane, D. F., Nelson, J. D., McCraken, G. H., *JAMA, 253,* 3559 (1985).
46. Friedland, I. R., Paris, M., Shelton, S., McCracken, G. H., *J. Antimicrob. Chemother., 34,* 231 (1994).
47. Quagliarello, V. J., Scheld, V. M., *N. Engl. J. Med., 336,* 708 (1997).
48. Roine, I., Ledermann, W., Foncea, L. M., Banfi, A., Cohen, J., Peltoloa, H., *Pediatr. Infect. Dis. J., 19,* 219 (2000).
49. Viladrich, P. F., Pallares, R., Ariza, J., *Arch. Intern. Med.,* 1986; 146: 2380-2382.
50. Martin, E., Hohl, P, Guggi, F., et al., *Infection, 18,* 70 (1990).
51. Black, S., Shinefield, H., Fireman, B., et al., *Pediatr. Infect. Dis. J., 19,* 187 (2000).

6.4 Antivirals

Pierre Lebon

*Virology Department, Hôpital Cochin Saint Vincent de Paul,
University of Paris V, Paris, France*

6.4.1 INTRODUCTION

Even though vaccination has led to the disappearance of several viral infections such as poliomyelitis, smallpox, measles, mumps, and rubella, chemotherapy is still required for infections by many viruses. Progress over the past twenty years in the discovery of compounds with antiviral properties has been considerable. However, few of these are currently licensed and until recently few of these were indicated for children. Progress in this area is the consequence of progress in macromolecular biology, specifically in the precise identification of viral protein structures that are different from cellular proteins that sometimes have the same function. The objective of this chapter is to describe biological and chemical antiviral substances that have been approved and used in children.

6.4.2 MECHANISM OF ACTION

Viruses need to invade cells in order to replicate. The infection of a cell by a virus results in a narrow molecular adaptation of one relative to the other, and the mechanisms by which a virus assures its multiplication are therefore dependent on the different functions and structures of the cell. In order to be useful clinically, inhibitors of viral multiplication must not interfere with the metabolic pathways of the cell, i.e. they must be non-toxic.

The virus penetrates into the cell after a phase of absorption onto the membrane receptors, which are relatively specific to the different families of virus. These receptors are not only structural cellular proteins but they also frequently have a biological function: enzyme, chemokine receptor, growth factor or alternate complement pathway factor. While the virus attaches itself to the receptor, it interacts with other cellular co-receptors, including those biosensors that are toll-like receptors leading to the stimulation of an ensemble of cellular genes that favour cellular and viral multiplication and also leading to the synthesis of secreted proteins that act on the neighbouring cells (such as interferon).

One of the active pathways of antiviral agents during this phase is based on the saturation of external proteins of the virus, either by antibodies directed against surface proteins of the virus or by using soluble proteins that mimic the cellular receptor, or via receptors and truncated co-receptors that inhibit the fixation of the virus onto its cellular receptor. Applications have been carried out for the HIV-1 virus, for which a number of compounds are in the testing phase.

Other approaches have been developed based on molecules that attach onto a specific hydrophobic site of the viral envelope that is normally used to attach to a cellular receptor and thus blocking the attachment of the virus onto the cell membrane. It is in this fashion that pleconaril blocks the cellular penetration of more than 90% of enterovirus types and 80% of the 101 rhinoviruses.

Once attached to the membrane of the cell, the virus is internalised after the fusion of the viral surface proteins and the cellular membrane. Certain substances are capable of blocking this phase, such as the derivatives of chloroquine and other substances which can alkalinise the vesicles during endocytosis for those viruses that require an acidic environment to unmask the site of fusion for a viral protein.

One part of the virus is thus at the interior of the cytoplasm more-or-less decapsidated. If the multiplication site is strictly cytoplasmic, the viral replication will begin; if a nuclear step is required, the viral components will follow the path of the cytoskeleton by using the molecular dynein-motor into the nucleus and pass through the nuclear pores. Certain compounds are capable of interrupting this transfer, but molecules that destabilise the microtubules are toxic to the cell.

The following targets are the viral proteins that prepare and carry out the replication of the viral nucleic acid. The RNA polymerases, transcriptases or replicases, and the DNA- or RNA-dependent DNA polymerases are molecules whose functions are blocked by nucleosidal analogues, or by non-nucleosidal inhibiting molecules. New antiviral compounds have been obtained with a powerful blocking action of the viral helicases, but these are not yet licensed [1].

After the replication of the viral nucleic acid and the synthesis of viral proteins, the phase of viral envelope maturation, i.e. the assembly of the various components to build the virus, offers additional targets. This deals mostly with viral enzymes of the protease type which are required to break down viral polypeptides, and for which we now know of inhibitors used against the HIV and HCV viruses; viruses formed in the presence of these anti-proteases are non-infectious.

Finally, there is a phase which consists of the freeing of the viral particles to the exterior of the cell. Certain viral enzymes, such as the neuraminidase of the influenza virus are required for this stage of the cycle. Inhibitors of neuraminidase of the flu virus have been shown to be effective. The discovery that the RNA of 21 complementary nucleotides (siRNA) of the viral messenger RNA is capable of blocking transcription and translation of the viral genes during different phases of the viral cycle represents a new therapeutic pathway. The development of applications is limited by the stability and the administration of this family of RNA.

Other antiviral substances of biological origin have yet another mode of action. For example, the type-1 interferons (IFN alpha or beta) do not have an effect directly on the virus but rather act via the infected cell. These substances activate numerous cellular genes that render the cell impermissible to viral infection. Certain substances such as the derivative of imiquimod can induce alpha- and beta- interferons.

6.4.3 INHIBITORS OF CELLULAR ENTRY

6.4.3.1 Polyclonal or monoclonal antibodies

Standard immunoglobulins are prepared from pooled donor blood, and they have shown their effectiveness for different indications for the prevention of viral disease. Their effectiveness depends on their specific activity, i.e., on the level of specific antibodies of a virus relative to the weight of immunoglobulin of the preparation. It is important to note that the levels have dropped since the introduction of vaccination for certain viruses. This is the case for antibodies to hepatitis A, which is the consequence of a reduction of infection in young adults following a reduction of the virus in the population.

Indications for the use of immunoglobulins have decreased following improved immunisation programmes. There are, however, occasions when immunoglobulins are recommended and these include the following:

- The prevention of measles in immunocompromised children following exposure to measles.
- Contact with hepatitis A in immunocompromised children.
- Replacement therapy for patients with agammaglobulinaemia or hypogammaglobulinaemia.

Specific immunoglobulins are indicated in the following situations:

- For infants born to mothers who have become infected with hepatitis B during the pregnancy or who are high risk carriers.
- Varicella–zoster immunoglobulin following contact with chickenpox in an immunocompromised child or a neonate.

Biotechnology now allows the *in vitro* manufacture of anti-RSV monoclonal antibodies, of which the variable part is directed against a neutralisation epitope of one of the viral surface proteins and for which the fixed part has been "humanised".

The antibody directed against the G protein of the RS virus shows a strong capacity for the neutralisation of the A and B variants of the RS virus *in vitro*. A preparation of these antibodies is commercially available under the name of palivizumab. *Palivizumab* is prescribed for premature infants under six months of age and for children under two years of age with bronchopulmonary dysplasia during the six months that precede the epidemic season. The recommended dose is 15 mg/ kg /month IM. The half life is 30 days and the duration of protection is between one and two months. Its effectiveness has been proven in a number of studies [2].

Pleconaril

This substance, 3-[3,5dimethyl-4[[3-(3-methyl-5-isoxazolyl)propyl]oxy]phe nyl]-5-trifuoromethyl)-1,2,4-oxadiazole, has not yet received official approval. Randomised studies with adults and children have shown that it reduces the severity and duration of symptoms for meningitis due to enterovirus. A study with children under the age of 12 months suffering from enteroviral meningitis has shown good tolerance for the drug at concentrations greater than that required to cause a 90% inhibition effect for the enterovirus [3]. However, an accumulation of the drug between day 2 and day 7 and some secondary effects that were more frequent than for placebo, underscore the need to reinforce surveillance of the toxicity. Indications for this product should be limited to the rare serious cases of enterovirus infection and not for those types that are spontaneously resolving.

6.4.4   INHIBITORS OF VIRAL EXCRETION: VIRAL INHIBITORS OF INFLUENZA A AND B

The first molecules used in the treatment for flu, amantadine and rimantadine, were confirmed to have had a limiting effect on the A virus, but their effectiveness was limited because of the emergence of resistant strains. The design of a new class of inhibitors for viral neuraminidase has resulted in the creation of effective antiviral compounds free of toxicity [4]. Neuraminidase is an enzyme present in the envelope of the virus that facilitates its liberation after its budding. It does this by cleaving the residues of sialic acid (Neu5 Ac) that are present in the glucoconjugates of the viral receptor at the surface of the respiratory epithelium. Neuraminidase also facilitates the diffusion of the virus through the bronchial tree by modifying the structure of the mucus, rich in muccoploysac-charides. The blocking of this enzymatique activity leads to the inhibition of the formation of lysis zones and reduces the titre of viral particles in the animal and in humans. Two of these compounds have received market authorisation in several countries: zanamivir (Relenza®) and oseltamivir (Tamiflu®) [5].

Zanamivir is administered by aerosol. The bioavailability following inhalation is 4-17%. It is eliminated by the kidneys. *Oseltamivir* exists in the form of a prodrug and is administered orally; 75% of the drug is absorbed and distributed systemically after esterification in the liver [6]. The half-life of zanamivir is 2.5 to 5.1 hours, and that of oseltamivir is between 6 and 10 hours [6]. Experimentally, zanamivir has been shown to prevent infection (in 82% of cases), fever (95%) and to reduce the carrying of the virus (96%) if it is administered 4 hours before the inoculation of the A virus (H1N1) and then for 5 days. If the product is given for a period of 4 days starting at 26 to 32 hours after the nasal inoculation of the virus, the local viral titre is significantly reduced and the carrying of the virus is reduced by three days. It is also active against the B virus [7, 8].

Oseltamivir, under similar experimental conditions, has an effectiveness of 61% and 100% for the prevention of infection and fever, respectively. It has been shown to be effective against the A and B viruses [9, 10]. Administered 28 hours after infection, oseltamivir reduced the average duration of symptoms from 95 to 53 hours; the viral titre in nasal aspirations is reduced by a $\log_{10}$ factor of 2.1 to 3.5. These two compounds thus have a comparable effectiveness to what has been confirmed for adults in several studies of natural infection conditions [11].

In terms of understanding the curative effects in children, studies are fewer than for adults; a reduction of the average length of symptoms of 1.25 days has been observed with zanamivir [12]. Oseltamivir also reduces symptoms (by 36 hours) compared to placebo: fever, coughing, secondary complications requiring the prescription of antibiotics (reduction of 40%), and formation of secondary ear infections (reduction by 44%) [13]. Studies in adults suggest that both drugs are effective in preventing infection [14]. Zanamivir is recommended for curative treatment of influenza A and B for adults and adolescents older than 12 years of age, administered by inhalation twice daily for 5 days, i.e. 20 mg/day. The drug is effective during the first 48 hours upon first appearance of symptoms [12].

Oseltamivir is indicated for children older than one year of age that show symptoms of flu in times of viral epidemics. It is administered in doses of 75 mg for 5 days for children aged 13 years and older. The same dose is given for 7 days as prophylaxis during an epidemic.

Resistance to neuraminidase inhibitors has been demonstrated *in vitro* but appears to be rare in humans. The pathogenic strength of mutant strains is reduced in animals and remains unknown in humans. Gastrointestinal side effects have been reported in 10% of adults. Vomiting may be reduced in children if the oseltamivir is taken during meals. Zanamivir by aerosol must be used with care for asthmatic patients, for whom it would be better to prescribe oseltamivir.

6.4.5 INHIBITORS OF VIRAL MULTIPLICATION: INHIBITORS OF THE REPLICATION OF VIRAL NUCLEIC ACIDS

Selective nucleoside inhibitors against herpes have been used clinically for 24 years.

6.4.5.1 Acyclovir

Acyclovir or 9-(2-hydroxyethoxy)-methyl) guanine is an analogue of desoxyguanosine. Valacyclovir is the L-valine ester of acyclovir, rapidly transformed in the liver into acyclovir. The compound stops the elongation of the nucleoside chain of viral DNA, due to the absence of a hydroxyl group on the acyclic chain, thus preventing attachment of the next nucleotide. The DNA polymerase of the herpes simplex virus (HSV) is blocked, and the multiplication stops. The compound is only active in infected cells. Indeed, acyclovir cannot undergo its first phosphorylation except by a viral protein, thymidine kinase coded by the virus; the two other phosphorylations are done using cellular enzymes. This first phosphorylation is also carried out by the thymidine kinase of the chickenpox virus. The acyclovir concentration that leads to a 50% viral inhibition varies from 0.1 to 1 micromolar and from 3 to 20 micromolar, depending on the techniques used. Acyclovir only has an effect on the virus in the process of multiplication and not on the latent virus.

Acyclovir can be given to children either orally or intravenously. It also exists in ophthalmologic gel (3 %) and as a skin cream (5 %) for external use. The half-life of the compound by IV administration is 3 hours and depends on renal function [15]. The levels in the cerebrospinal fluid (CSF) are 50% lower than in the plasma. For children older than three months, the dose is 500 mg/m^2 every eight hours for VZV infections and for herpes encephalitis, and 250 mg/m^2 for cutaneous HSV infections. For the newborn, the dose is 10 mg/kg every 8 hours in case of infections to HSV and VZV. Acyclovir during pregnancy has not caused teratogenic effects. Oral acyclovir is well tolerated, but the bioavailability is only between 10 and 15 %. The most frequent side effects are gastrointestinal disturbance [16]. Transient neutropenia has been reported in neonates [16].

Treatment of chickenpox

In four different studies [17-20] oral acyclovir was given to immunocompetent children aged between 2 and 18 years with chickenpox, in doses between 10 and 20 mg/kg four times each day (maximum single dose of 800 mg) for 5 to 7 days. Treatment accelerated by 1 to 2 days the healing of the skin rash compared to placebo, regardless of whether the treatment was for 5 or 7 days. No reduction in the rate of complications or in the transmission of the virus has been observed [20-23]. The American Academy of Pediatrics (AAP) recommends a dose of 20 mg/kg, four times a day for five days [24] in children older than 12 years with chickenpox, with pulmonary or chronic skin disease, or if receiving corticosteroids or salicylates, or suffering from a metabolic disease or complications from chickenpox.

Neonatal chickenpox

The risk of neonatal chickenpox is greatest if the mother first shows symptoms in the period between 5 days before the birth to 2 days following. Several case reports have described the benefit of acyclovir [25, 26]. There have not, however, been any controlled studies. An IV dose of 20 mg/kg every eight hours for those infants who contract chickenpox in the first month of life is recommended [27, 28].

Chickenpox in the immunodeficient child

Acyclovir is given intravenously in a dose of 30 mg/kg every 8 hours for 7 to 14 days or in the dose of 1500 mg/m^2 every 8 hours. Treatment leads to a significant reduction in VZV-induced pneumonia [29]. The maximum benefit of the treatment is obtained if it is begun within 72 hours from the start of the rash [28]. Treatment that is begun after viral dissemination is often unsuccessful. Valacyclovir has not yet been licensed for this indication in children.

Treatment of herpes zoster

Oral acyclovir accelerates the healing of the eruption, reduces the acute pain and the dissemination if it is given during the first days of the eruption [30]. Valacyclovir works 30 times faster than oral acyclovir on the lesions in immunocompetent adults.

However, zoster is more benign in the immunocompetent child than in the adult [31]. Treatment can be reserved for the first and second branch of the trigeminal nerve, in view of the risk of ophthalmic zoster and its complications.

HSV infections of the newborn

Despite doses of 30 mg/kg/day for 10 to 14 days, herpes simplex infection, in its generalised form or with central nervous system involvement, has a mortality near 50%. Higher doses, 20 mg/kg every eight hours for 21 days increases the survival rate of newborns. Treatment is recommended for 14 days for mucocutaneous infection, and 21 days for disseminated and cerebral involvement.

Mucocutaneous forms in immunocompetent children

10 to 30 % of HSV primary infections in children present as gingivostomatitis. Doses of 15 mg/kg, orally five times per day for 7 days reduce the duration of the fever, the carrying of the virus, and the duration of refusal to eat [32, 33]. In immunodepressed patients, the benefits are more pronounced if the administration is systemic instead of local or given orally.

Herpetic encephalitis

Acyclovir has been shown superior to vidarabine with greater survival [34]. Treatment must be given intravenously at doses of 10 mg/kg every 8 hours and begun as soon as possible after the appearance of neurological symptoms [35]. A minimum of 14 days treatment is recommended.

6.4.5.2 Valacyclovir

Valacyclovir has been prescribed in three paediatric studies. It has a bioavailability of 45-48 %, which is comparable with that measured in adults and is given orally twice daily [16]. There are no safety data available for children under 12 years of age. For children 13 years or older, valacyclovir is well tolerated at doses of 8 g/ day for a period of 90 days. This drug is not yet available in a stable form that would permit its oral administration for newborns.

6.4.5.3 Pencyclovir

Pencyclovir is not recommended for children under the age of 16 years, and its effectiveness *in vitro* is less than that of acyclovir.

6.4.5.4 Gancyclovir

Gancyclovir is indicated for systemic infections and ocular infections of the central nervous system due to cytomegalovirus (CMV). 5-10 mg/kg/day for immunocompetent children can eliminate the CMV antigen load in 80% of cases after 14 days of treatment [36]. It is also given for prophylactic purposes after tissue graft because of treatment with immunosuppressants. Treatment is given for a period of 14 days after the tissue graft, and the dose must be adjusted in the case of renal insufficiency. The results of controlled trials for the treatment of the congenital form of the CMV infection are still rare. One study attempted to evaluate the effectiveness of gancyclovir on the hearing of newborns affected by congenital CMV infection [37]. 42 newborns were randomised of whom 25 received gancyclovir intravenously for a period of six weeks. 21 (84%) experienced an improvement or no changes in their hearing against only 10 out of 17 of the newborns in the control group during the 6 months of follow-up (p=0.06). Upon stopping treatment, a deterioration in hearing was only observed in the untreated infants (7 / 17, 41%). 5 out of 24 (21%) of the treated newborns suffered from a deterioration in their hearing one year later, compared to 13 out of 19 (68 %) for the non-treated newborns. Finally, two-thirds of the treated newborns developed neutropenia compared to 21 % for the untreated control group, during the period of treatment [37].

6.4.5.5 *Foscarnet*

Foscarnet, or phosphonoformic acid, directly inhibits viral DNA polymerases of the herpes virus. It is given in cases of disseminated CMV infections, retinitis, or for HSV and VZV infections that are resistant to acyclovir in immunodepressed patients. The doses used are 80 mg/kg IV, because the drug is only poorly absorbed orally. It is not recommended for children under the age of 16 years except in extremely serious situations.

6.4.5.6 *Cidofovir*

Cidofovir, (S)-1-[3-hydroxy-2(phosphonyle-methoxy)-propyl]cytosine), an analogue of cytidine carrying a phosphonate group, is a nucleotide analogue phosphorylated by cellular kinases. It is prescribed as a treatment in CMV retinitis, administered intravenously in doses of 5 mg/kg (because of its renal toxicity); or in infections of HSV and VZV having a deficient thymidine kinase. This drug is however not recommended for children. Tests are underway for its use in the treatment of infections due to adenovirus in the immunodepressed patient.

6.4.6 RIBAVIRIN

Ribavarin is a guanine analogue and has a large spectrum of activity *in vitro* on both DNA viruses (adenovirus, herpes) and RNA viruses (VRS, influenza, paramyxovirus, arenavirus, bunyavirus, and reovirus). The molecule is triphosphorylated by cellular kinases; it inhibits viral and cellular transcription, which explains its toxicity towards bone marrow and its teratogenic effect [38]. It is prescribed in cases of hepatitis C in association with alpha interferon in doses of 800 mg to 1200 mg per day for adults; it is prescribed for children over the age of 16 years orally. A trial using this drug to treat severe forms of adenovirus disease was carried out in 5 immunodepressed children; two of the children were cured of the infection after intravenous treatment [39]. It has been used on a trial basis for chronic infections of the central nervous system measles virus, with only modest results. Some positive results have been obtained in the treatment of Lassa fever after intravenous injections of 1 g/day. It has been prescribed in aerosol form for serious infections of RSV and measles interstitial pneumonia.

6.4.6.1 Interferons

Interferons are proteins of human origin. They display an anti-viral, immuno-modulating and cytotoxic activity. They act after interaction with specific receptors and cause a cascade of biochemical events, of which the result is the synthesis and activation of various proteins with antiviral activity (RNA-dependent protein kinase, 2,5-oligoadenylate synthetase, Mx protein).

Interferons of type 1 alpha / beta are being tested in trials for multiple sclerosis in children (beta interferon). They have also been used for varicella infections in immunocompromised children. Alpha-2-interferon is administered in the case of infections by the hepatitis B and C viruses at doses of 3 million units /m^2 in adults every two days or every week using the slow-release form (pegylated interferon) for a duration of 6 to 12 months. For children, there is no official indication for the prescription of alpha and beta interferons for the treatment of hepatitis. However in one study, 100 children were treated with interferon alpha for a period of twelve months, and this was well tolerated with a response rate of 46%, i.e. elimination of the Hbe antigen and normalisation of liver enzymes. Complete cure with disappearance of the ag HBs was seen in 14% of the cases only [40]. Trials of alpha interferon and an inhibitor of reverse transcriptase (lamivudine) are now underway in children [41].

Alpha- and beta-interferons may be proposed in association with immunotherapy for the prevention of rabies in conjunction with facial wounds that lead to short incubation times for encephalitis.

Gamma-interferon is prescribed for infectious septic granulomatosis and for chronic mycobacterial infections, even though its effectiveness has not been shown. There are no indications for its use in viral infections.

REFERENCES

1. Kleymann, G., *Herpes, 10*, 2 (2003).
2. Razafimahefa, H., Lacaze-Masmonteil, T., *Virologie, 7*, 162 (2003).
3. Abzug, M. J., Cloud, G., Bradley, J., Sanchez, P. J., *Pediatr. Infect. Dis. J., 22*, 335 (2003).
4. McClellan, K., Perry, C. M., *Drugs, 61*, 263 (2001).

5. Dreitlein, W. B., Maratos, J., Brocavich, J., *Clin. Ther., 23*, 327 (2001).

6. Chidiac, C., *Virologie, 6*, 139 (2002).

7. Hayden, F. G., Treanor, J. J., Betts, R. F., Lobo, M., Esinhart, J. D., Hussey, E. K., *JAMA, 275*, 295 (1996).

8. Hayden, F. G., Lobo, M., Hussey, E. K., "Efficacy of intranasal GG167 in experimental human influenza A and B virus infection," in *Options for the control of influenza III*, Brown, L. E., Hampson, A. W., Webster, R. G, eds., Elsevier Science, 718-25 (1996).

9. Hayden, F. G., Treanor, J. J., Fritz, R. S., et al., *JAMA, 282*, 1240 (1999).

10. Hayden, F. G., Jennings, L., Robson, R., et al., *Antivir. Ther., 5*, 205 (2000).

11. Hayden, F. G., Atmar, R. L., Schilling, M. et al., *N. Engl. J. Med., 34*, 1336 (1999).

12. Hedrick, J. A., Barzilai, A., Behre, U., et al., *Pediatr. Infect. Dis. J., 19*, 410 (2000).

13. Whitley, R. J., Hayden, F. G., Reisinger, K. S., et al., *Pediatr. Infect. Dis. J., 20*, 127 (2001).

14. Monto, A. S., Robinson, D. P., Herlocher, M. L., Hinson, J. M., Jr., Elliott, M. J., Crosp, A., *JAMA, 282*, 31 (1999).

15. Eksborg S., *Herpes, 10*, 3 (2003).

16. Tyring, S. K., Baker, D., Snowden, W., *J. Infect. Dis., 186 (Suppl)*, S40 (2002).

17. Arvin, A. M., *Semin. Pediatr. Infect Dis., 13(1)*, 12 (2002).

18. Balfour, H. H., Jr., Edelman, C. K., Anderson, R. S., et al., *Pediatr. Infect. Dis. J., 20*, 919 (2001).

19. Dunkle, L. M., Arvin, A. M., Whitley, et al., *N. Engl. J. Med., 325*, 1539 (1991).

20. Klassen, T. P., Belseck, E. M., Wiebe, N., Hartling, L., *BMC Pediatr., 2*, 9 (2002).

21. Balfour, H. H., Jr., Rotbart, H. A., Feldman, S., et al., *J. Pediatr., 120*, 627 (1992).

22. Balfour, H. H., Jr., Kelly, J. M., Suarez, C. S., et al., *J. Pediatr., 116*, 633 (1990).

23. Klassen, T. P., Belseck, E. M., Wiebe, N., Hartling, L., *Cochrane Database Syst. Rev.,* CD002980 (2002).

24. American Academy of Pediatrics Committee on Infectious Diseases, *Pediatrics, 91*, 674 (1993).

25. King, S. M., Gorensek, M., Ford-Jones, E. L., Read, S. E., *Pediatr. Infect. Dis., 5*, 588 (1986).

26. Williams, H., Latif, A., Morgan, J., Ansari, B. M., *J. Infect., 15*, 65 (1987).

27. Englund, J. A., Fletcher, C. V., Balfour, H. H., Jr., *J. Pediatr., 119*, 129 (1991).

28. Prober, C. G., Gershon, A. A., Grose, C., McCracken, G. H., Jr., Nelson, J. D., *Pediatr. Infect. Dis. J., 9*, 865 (1990).

29. Enright, A. M., Prober, C., *Herpes, 10*, 2 (2003).

30. Wood, M. J., Kay, R., Dworkin, R. H., Soong, S. J., Whitley, R. J., *Clin. Infect. Dis., 22*, 341 (1996).

31. Smith, C. G., Glaser, D. A., *Pediatr. Dermatol., 13*, 226 (1996).

32. Amir, J., Harel, L., Smetana, Z., Varsano, I., *BMJ, 314*, 1800 (1997).

33. Aoki, F. Y., *33rd Interscience Conference on Antimicrobial Agents and Chelotherapy, New Orleans*, Abstract 399 (1993).

34. Whitley, R. J., Alford, C. A., Hirsch, M. S., et al., *N. Engl. J. Med., 314*, 144 (1986).

35. Rachilas, F., Wolff, M., Delatour, F., et al., *Clin. Infect. Dis., 35*, 254 (2002).

36. Avila-Aguero, M. L., Paris, M. M., Alfaro, W., Avila-Aguero, C. R., Faingezio, B., *Int. J. Infect. Dis., 7*, 278 (2003).

37. Kimberlin, D. W., Lin, C. Y., Sanchez, P. J., et al., *J. Pediatr., 143*, 13 (2003).

38. Tam, R. C., Lau, J. Y., Hong, Z., *Antivir. Chem. Chemother.,12*, 261 (2001).

39. Gavin, P. J., Katz, B. Z., *Pediatrics, 110(1 Pt 1)*, e9 (2002).

40. Liberek, A., Luczak, G., Korzon, M., et al., *J. Paediatr. Child Health., 40*, 265 (2004).

41. Saltik-Temizel, I. N., Kocak, N., Demir, H., *Pediatr. Infect. Dis. J., 23*, 466 (2004).

6.5 Antiretroviral drugs in paediatric patients

Jean Marc Tréluyer

Perinatal and Paediatric Pharmacology, Cochin University,
Hospital Saint Vincent of Paul, Paris, France

6.5.1 INTRODUCTION

Antiretroviral drugs are used in children for two very different purposes. During the perinatal period, they are used to prevent mother-to-child transmission of human immunodeficiency virus (HIV). They are also used to treat children already infected with the virus. Although the pathogenesis of human immunodeficiency virus (HIV) infection and the general virological and immunological principles underlying the use of antiretroviral therapy are similar for all patients infected with HIV, there are unique considerations for HIV-infected infants, children, and adolescents. Children, like adults, have benefited from recent progress in therapy due to the development of increasingly effective antiretroviral molecules. However, as in the case of many other drugs used to treat children, most antiretroviral drugs have still not been adequately assessed in this population [1]. It is, therefore, important to determine the pharmacokinetic and pharmocodynamic characteristics of each molecule in children to optimise the risk/benefit ratio and to improve the treatment of these patients.

This review article addresses paediatric pharmacological characteristics and their effects on the principal antiretroviral drugs used in neonates, infants, and children. A search of the literature from January 1983 through December 2001 was performed on MEDLINE using the search terms children, neonates, HIV, pharmacokinetics. Relevant book chapters and conference presentations were also included. Clinical trials are still needed to determine the optimal dosing schedules of antiretroviral drugs in all age groups. Prenatal transmission from mother to fetus and the treatment of infected children are addressed as separate issues.

6.5.2   THE HUMAN IMMUNODEFICIENCY VIRUS–HIV

The Human Inmmunodeficiency Virus (HIV) is a retrovirus. The virus contains 2 identical copies of a positive sense (i.e. mRNA) single-stranded RNA strand about 9,500 nucleotides long. These may be linked to each other to form a genomic RNA dimer. The RNA dimer is in turn associated with a basic nucleocapsid (NC) protein (p9/6). The ribonucleoprotein particle is encapsidated by a capsid made up of a capsid protein (CA), p24. The capsid environment also contains other viral proteins such as integrase and reverse transcriptase. It also contains a wide variety of other macromolecules derived from the cell including tRNAlys3, which serves as a primer for reverse transcription. The capsid has an icosahedral structure. The capsid is in turn encapsidated by a layer of matrix protein (MA), p17. This matrix protein is associated with a lipid bilayer or envelope.

The HIV envelope is derived from the host cell plasma membrane and is acquired when the virus buds through the cell membrane. The major HIV protein associated with the envelope is gp120/41. This functions as the viral antireceptor or attachment protein. gp41 traverses the envelope, gp120 is present on the outer surface and is noncovalently attached to gp41. The precursor of gp120/41 (gp160) is synthesised in the endoplasmic reticulum and is transported via the golgi body to the cell surface.

The genetic information in a retrovirus particle is encoded by RNA. Upon entry into the host cell this RNA is copied into DNA by the virus enzyme reverse transcriptase. This cDNA copy of the virus' genetic information can integrate into the host cell chromosomes in the nucleus. This provirus can lay dormant for many cell divisions before being reactivated and producing more infectious retrovirus particles.

6.5.3   CLASSIFICATION OF ANTIRETROVIRAL DRUGS

There are 4 major classes of antiretroviral drugs: *nucleoside* and *nucleotide analogue reverse transcriptase inhibitors* (NRTIs), *nonnucleoside reverse transcriptase inhibitors* (NNRTIs), *protease inhibitors* (PIs), and *fusion inhibitors* (Figure 1).

1. NRTIs function by inhibiting the synthesis of DNA by reverse transcriptase, the viral enzyme that copies viral RNA into DNA in the newly infected cell. Nucleoside analogues bear a structural resemblance to the natural building blocks of DNA: the purine nucleosides adenosine (A) and guanosine (G), and the pyrimidine nucleosides thymidine (T) and cytidine (C). Nucleoside analogues are triphosphorylated within the cell, and some undergo further

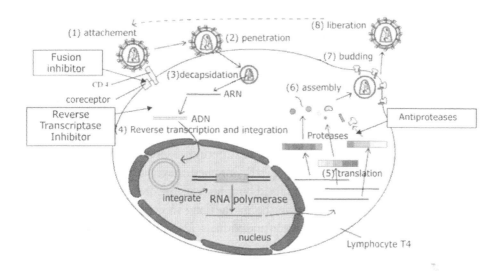

Figure 1. Mode of action of antiretroviral drugs

modifications. Nucleotide analogues resemble monophosphorylated nucleo-sides, and therefore require only two additional phosphorylations to become active inhibitors of DNA synthesis. Reverse transcriptase fails to distinguish the phosphorylated NRTIs from their natural counterparts, and attempts to use the drugs in the synthesis of viral DNA. When an NRTI is incorporated into a strand of DNA being synthesised, the addition of further nucleotides is prevented, and a full length copy of the viral DNA is not produced.

2. NNRTIs also inhibit the synthesis of viral DNA, but rather than acting as false nucleotides, the NNRTIs bind to reverse transcriptase in a way that inhibits the enzyme's activity.

3. PIs bind to the active site of the viral protease enzyme, preventing the proces-sing of viral proteins into functional forms. Viral particles are still produced when the protease is inhibited, but these particles are ineffective at infecting new cells.

4. Fusion inhibitors prevent HIV from entering target cells, by binding to the HIV envelope protein gp41, which is involved in viral entry. By blocking the interactions between regions of the gp41 molecule, fusion inhibitors interfere with the conformational change (folding) of the envelope molecule required for fusion with the target cell membrane.

6.5.4 ANTIRETROVIRAL DRUGS IN NEONATES

The first nucleoside inhibitor of reverse transcriptase introduced commercially was zidovudine, for which an oral solution is available for Paediatric patients. Like other nucleoside inhibitors of reverse transcriptase, zidovudine is a prodrug that must undergo 3 intracellular phosphorylation steps to become active. The elimination of zidovudine depends on hepatic glucuronidation, which is not fully developed in neonates.

Zidovudine is used during the perinatal period to prevent HIV transmission from mother to fetus. A study of 28 HIV-infected pregnant women treated with zidovudine showed that the plasma and lymphocyte concentrations of triphosphorylated zidovudine were similar in the mother and in cord blood, but that there was considerable variation between individuals [2]. These results confirm that zidovudine crosses the placental barrier and suggest that there is no immaturity in the processes required for the phosphorylation of zidovudine in neonates.

The first Phase III study (PACTG 076) demonstrating the efficacy of zidovudine in the prevention of HIV transmission from mother to fetus was published in 1994 [3]. The trial enrolled 477 women infected with HIV. The transmission rate was 8% in the group treated with zidovudine and 25% in the group given placebo. The dosing schedule used in this study was 100 mg zidovudine orally, 5 times daily, administered until delivery, followed by a 2 mg/kg per-partum intravenous infusion and the administration of 2 mg/kg 4 times daily orally to the newborn from birth until age 6 weeks. This dosing schedule was based on the fact that the plasma half-life of zidovudine is very short (~1 hour) [4]. When the protocol was designed, it was thought that having a short interval between doses would ensure that the molecule was continuously present in the plasma. Later, after publication of the results the recommended dosing schedule for adults was changed to 300 mg BD, based on intracellular pharmacokinetic studies [5] that showed that the intracellular half-life of the phosphorylated metabolites was much longer than the half-life of zidovudine in plasma. Thus, in subsequent studies[6] on the transmission of HIV from mother to fetus, zidovudine 300 mg BD was given to the mother as recommended for the curative treatment of adults. However, the dosing schedule recommended for neonates was not modified, and it remains 2 mg/kg QDS.

This dosing schedule is not justifiable in pharmacokinetic terms. It was recently shown that exposure to zidovudine (area under the plasma concentration time curve and maximum plasma concentration) was of the same order of magnitude with dosing schedules of 4 mg/kg BD and 2 mg/kg QDS [7]. However, data on the intracellular pharmacokinetics of zidovudine in neonates using either of the 2 dosing schedules are not available. It may be difficult for parents to administer zidovudine to their infants every 6 hours; administration of zidovudine to neonates in 2 daily doses, as is done in adults, might be preferable. This dosing schedule (4 mg/kg BD) was used in a randomised, placebo-controlled comparative study (the PETRA study)[8]. Similar results were obtained in reducing perinatal HIV-1 transmission in the PETRA treated groups and in the PACTG 076 treated group. However, aspects of treatment, other than the dosing schedule for zidovudine, differed between the PETRA and PACTG 076 trials (eg, duration of treatment, combined treatment with lamivudine), making it impossible to specifically identify the consequences of the change in dosing schedule. From the results currently available, it is not possible to determine whether zidovudine administered BD is as effective as zidovudine administered QDS for preventing HIV transmission from mother to fetus.

The recommendations concerning dosing schedules for zidovudine in neonates should also take postnatal age into account. Zidovudine is eliminated by glucuronidation, for which the involved enzymes are only weakly active at birth[9]. Mirochnick *et al.* [10] showed that zidovudine clearance rates are very low during the first few days of life and then increase gradually until the age of 4 to 8 weeks. Therefore, administering zidovudine 2 mg/kg QDS (the currently recommended dosing schedule for newborns) produces higher plasma zidovudine concentrations in newborns than in adults; this is especially true of premature infants [11,12]. Because zidovudine has been shown to display concentration-dependent haematologic toxicity [13], it would appear logical to adapt doses for neonates, according to the development of the systems involved in the elimination of the drug. However, it is currently difficult to recommend a different dosing schedule given the absence of results from prospective trials of the efficacy and toxicity of other dosing schedules (Table 1).

Few data are available concerning the pharmacokinetics of other nucleoside inhibitors of reverse transcriptase in neonates [14,15]. The 2 studies of the pharmacokinetics of didanosine in newborns enrolled a total of only 12 patients [14,15], and interindividual variability makes it difficult to interpret the results. Two studies [7,16] have been published on the pharmacokinetics of lamivudine in neonates. In these studies, a lower clearance rate was observed in newborns than in children. The clearance of abacavir also appears to be less efficient during the neonatal period compared to children and adults [17].

Similarly, little research is available concerning the use of protease inhibitors in pregnant women and neonates. However, these drugs do not cross the placental barrier efficiently [18] and thus are rarely administered during the perinatal period.

Studies have been carried out on non-nucleoside inhibitors of reverse transcriptase, particularly nevirapine. This molecule crosses the placental barrier efficiently and has a long half-life, making it possible to prevent mother-to-fetus transmission of HIV, using a dosing schedule that is simpler than that used for

Table 1

Presentation and recommended dose of antiretroviral drugs in neonates, children and adults (adapted from http://www.aidsinfo.nih.gov/guidelines/Paediatric). No recommended doses are available in children for emcitrabine, tenofovir, atazanavir, fosamprenavir.

	Preparations	Neonatal dose	Paediatric dose	Adult dose
Nucleoside Reverse Trascriptase Inhibitors (NRTIs)				
abacavir	solution : 20 mg/mL tablets: 300 mg	not approved for neonates	not approved for infants less than 3 months of age after 3 months: 8 mg/kg BD, maximum dose 300 mg/BD	300 mg BD
didanosine	paediatric powder for oral solution: 10 mg/mL; chewable tablets with buffers: 25, 50, 100, 150 and 200 mg; buffered powder for oral solution: 100, 167, and 250 mg; delayed-release capsules (enteric-coated beadlets): Videx EC 125, 200, 250 and 400 mg	neonates and infants aged<90 days: 50 mg/m² BD	90 mg/m² BD	body weight>60 kg: Videx EC : 400 mg OD body weight <60 kg : Videx EC : 250 mg OD
lamivudine	solution: 10 mg/mL; tablets: 150 mg	2 mg/kg BD	4 mg/kg BD	150 mg BD
stavudine	solution: 1 mg/mL capsules: 15, 20, 30 and 40 mg	not approved for neonates	1 mg/kg BD (up to weight of 30 kg)	body weight> 60 kg: 40 mg BD body weight<60 kg: 30 mg BD
zidovudine	syrup: 10 mg/mL; capsules: 100 mg; tablets: 300 mg	premature: 1.5 mg/kg BD from birth to 2 weeks of age: 2 mg/kg every 8 hours after 2 weeks of age neonates: 2 mg/kg every 6 hours	90 mg/m² every 6 to 8 hours	300 mg BD

Table 1 (*continued*)

NonNucleoside Reverse transcriptase Inhibitors (NNRTIs)

efavirenz	capsules: 50, 100, and 200 mg	not approved	10-15 kg: 200 mg 15-20 kg: 250 mg 20-25 kg 300 mg 25-32.5 kg: 350 mg 32.5-40 kg: 400 mg >40 kg: 600 mg OD	600 mg OD
nevirapine	suspension: 10 mg/mL; tablets: 200 mg	120 mg/m² OD for 14 days followed by 120 mg/m² BD for 14 days followed by 200 mg/m² BD	120 mg/m² OD for 14 days followed by 120 mg/m² BD	200 mg BD

Protease Inhibitors (Pis)

amprenavir	solution: 15 mg/mL capsules: 50 and 150 mg	not recommended<3 years of age	>3 years: 22.5 mg/kg BD or capsules 20 mg/kg BD	1200 mg BD
indinavir	capsules 200 and 400 mg	not approved	500 mg/m² every 8 hours	800 mg every 8 hours
lopinavir/ritonavir	solution: 80 mg lopinavir and 20 mg ritonavir per mL; capsules: 133 mg lopinavir/33.3 mg ritonavir	not approved	7-15 kg: 12 mg/kg lopinavir BD 15-40 kg: 10 mg/kg BD >40 kg: 400 mg lopinavir BD	400 mg BD
nelfinavir	powder for oral suspension: 50 mg per one level gram scoop full; tablets: 250 mg	40 mg/kg every 12 hours	30 to 45 mg/kg TDS or 50 mg/kg BD	1250 mg BD
ritonavir	solution: 80 mg/mL; capsules: 100 mg	not approved	400 mg/m² BD	600 mg BD
saquinavir	capsules: 200 mg	not approved	50 mg/kg TDS	1600 mg BD

zidovudine [19]. This may make it useful in the treatment of patients in countries with poorly developed medical and paramedical infrastructures [20]. With a maternal dose of 200 mg during labour, followed by a dose of 2 mg/kg for the newborn at 72 hours, plasma concentrations of nevirapine remain >100 mg/L in neonates (10 times the concentration that will inhibit 50% of HIV replication *in vitro*) for >7 days. A prospective Phase III trial [20], comparing nevirapine with zidovudine administered orally during labour and after delivery, showed that nevirapine treatment was significantly more effective at preventing mother-to-fetus HIV transmission.

In the past few years, progress in terms of efficacy has been made in the prevention of HIV transmission from mother to fetus. However, undesirable effects of the therapies used are now beginning to surface (i.e. mitochondrial and neurological toxixcity [21-23]). Future research should focus on decreasing the risk of these effects, increasing our knowledge of the pharmacokinetics of antiretroviral drugs in neonates, and using the molecules currently available more effectively, with dosing schedules adapted to the developing systems of newborns.

6.5.5 ANTIRETROVIRAL DRUGS IN INFANTS AND CHILDREN

In HIV-infected infants and children the goal of treatment is to limit viral replication in the same way as in adults. Thus, it is important to obtain plasma drug concentrations high enough for treatment to be effective. Suboptimal plasma concentrations of antiretroviral drugs favour viral replication and may, therefore, result in the emergence of viral resistance to treatment. A significant relationship has been demonstrated between residual concentrations of ritonavir, measured 15 days after the initiation of treatment, and the frequency of HIV mutations *in vivo* [24]. Studies have demonstrated a relationship between treatment efficacy and plasma concentrations of protease inhibitors, such as saquinavir [25] indinavir [26], ritonavir [27], and nelfinavir [28]. In a study by Durant *et al.* [29], it was shown that, for predicting changes in viral load, plasma concentrations of protease inhibitors were as important as adaptation to viral genotype. Results are also available for nevirapine [30,31], efavirenz [32] and some nucleoside inhibitors of reverse transcriptase, such as zidovudine [33]. For some molecules, a relationship has been demonstrated between plasma concentration and undesirable effects. This is the case for zidovudine, indinavir, ritonavir, and efavirenz [13,32,34,35].

Thus, it is important to determine the pharmacokinetic characteristics of each antiretroviral molecule used in children, with the aim of identifying the plasma concentration at which the efficacy/tolerance ratio is most favourable.

Published studies concerning the pharmacological properties of antiretroviral drugs in children have found major differences with respect to what is known about these drugs in adults. The maximal plasma concentrations of abacavir, zidovudine, and zalcitabine are lower in children than in adults for identical doses expressed in milligrams per kilogram [2,4,36,37]. Lewis *et al.* [38] reported lower plasma lamivudine concentrations in children than in adults who took equivalent doses. The clearance rate of stavudine is higher in children than in adults, with lower plasma concentrations and a smaller area under the plasma concentration time curve for identical doses with respect to body mass [39]. To obtain the same plasma nelfinavir concentrations in children as in adults, it is necessary to triple the dose with respect to body mass in children [40-44]. The clearance rate of nevirapine is higher in children than in adolescents [45]. Consistent results were also obtained with emcitrabine and tenofovir [46,47].

However, the available data are patchy, often obtained from a limited number of patients, and include only a few molecules and age groups. Antiretroviral drugs are often prescribed for children without adequate pharmacologic data, such as are required for the use of such drugs in adults [48].

Thus, monitoring plasma antiretroviral drug concentrations in practice is even more important in children than in adults, given the considerable variability between individuals in pharmacokinetic parameters [49,50] and the demonstration of a concentration/effect relationship for most of the antiretroviral drugs currently available.

6.5.6   CONCLUSIONS

Most of the antiretroviral drugs available for use in adults are also available in paediatric formulations, with studies specifically dedicated to their use in children. However, much progress remains to be made in determining the dosing schedules that will provide optimal plasma concentrations with the best possible risk/benefit ratio. Clinical trials should take into account all age groups. In the absence of adequate pharmacokinetic data, the monitoring of plasma concentrations during follow-up should make it possible to detect patients with plasma concentrations of the drug that are too low to be effective or too high to prevent toxicity.

REFERENCES

1. Impicciatore, P., Choonara, I., *Br. J. Clin. Pharmacol.*, *48*, 15 (1999).
2. Rodman, J. H., Flynn P. M., Robbins, B., et al., *J. Infect. Dis.*, *180*, 1844 (1999).
3. Connor, E. M., Sperling, R. S., Gelber, R., et al., *N. Engl. J. Med.*, *331*, 1173 (1994).
4. Wintergerst, U., Rolinski, B., Vocks-Hauck, M., et al., *Infection*, *23*, 344 (1995).
5. Stretcher, B. N., Pesce, A. J., Wermeling, J. R., Hurtubise, P. E., *Ther. Drug Monit.*, *14*, 281 (1992).

6.   Lallemant, M., Jourdain, G., LeCoeur, S., et al., *N. Engl. J. Med., 343*, 982 (2000).
7.   Moodley, D., Pillay, K., Naidoo, K., et al., *J. Clin. Pharmacol., 41*, 732 (2001).
8.   Petra study, *Lancet, 359 (9313),* 1178 (2002).
9.   de Wildt, S. N., Kearns, G. L., Leeder, J. S., et al., *Clin. Pharmacokinet., 36*, 439 (1999).
10. Mirochnick, M., Capparelli, E., Connor, J., *Clin. Pharmacol. Ther., 66,* 16 (1999).
11. Mirochnick, M., Capparelli, E., Dankner, W., et al., *Antimicrob. Agents Chemother., 42*, 808 (1998).
12. Capparelli, E. V., Mirochnick M., Dankner, W. M., et al., *J. Pediatr., 142*, 47 (2003).
13. Mentre, F., Escolano, S., Diquet, B., Golmard, J. L., Mallet, A., *Eur. J. Clin. Pharmacol., 45,* 397 (1993).
14. Wang, Y., Livingston, E., Patil, S., et al., *J. Infect. Dis., 180,* 1536 (1999).
15. Rongkavilit, C., Thaithumyanon, P., Chuenyam, T., et al., *Antimicrob. Agents Chemother., 45,* 3585 (2001).
16. Moodley, J., Moodley, D., Pillay, K., et al., *J. Infect. Dis., 178,* 1327 (1998).
17. Johnson, G., *et al. Preliminary analysis of abacavir succinate pharamacokinetics in neonates differs from adults and young children,* in 7th Conference in Retroviruses and Opportunistic Infections, San Francisco (2000).
18. Marzolini, C., Rudin, C., Decosterd, L. A., et al., *Aids, 16*, 889 (2002).
19. Musoke, P., Guay, L. A., Bagenda, D., et al., *Aids, 13*, 479 (1999).
20. Guay, L.A., Musoke, P., Fleming, T., et al., *Lancet, 354*, 795 (1999).
21. Blanche, S., Tardieu, M., Rustin, P., et al., *Lancet, 354*, 1084 (1999).
22. Barret, B., Tardieu, M., Rustin, P., et al., *Aids, 17*, 1769 (2003).
23. Landreau-Mascaro, A., Barret, B., Mayaux, M. J., Tardieu, M., Sebire, G., *Lancet, 359*, 583 (2002).
24. Molla, A., Korneyeva, M., Gao, Q., et al., *Nat. Med., 2*, 760 (1996).
25. Gieschke, R., Fotteler, B., Buss, N., Steiner, J.-L., *Clin. Pharmacokinet., 37*, 75 (1999).
26. Murphy, R. L., Sommadossi, J., P., Lamson, M., et al., *J. Infec. Dis., 179*, 1116 (1999).
27. Mueller, B. U., Zeichner, S. L., Kuznetsov, V. A., Heath-Chiozzi, M., Pizzo, P. A., *Aids, 12,* F191 (1998).
28. Hoetelmans, R. M., Reijers, M. H., Weverling, G. J., et al., *Aids, 12*, F111 (1998).
29. Durant, J., Clevenbergh, P., Garraffo, R., et al., *Aids, 14*, 1333 (2000).
30. Havlir, D., Cheeseman, S. H., McLaughlin, M., et al., *J. Infect. Dis., 171*, 537 (1995).
31. Veldkamp, A. I., Weverling, G. J., Lange, J. M., et al., *Aids, 15*, 1089 (2001).
32. Marzolini, C., Telenti, A., Decosterd, L. A., Greub, G., Biollaz, J., Buclin, T., *Aids, 15*, 71 (2001).
33. Fletcher, C. V., Acosta, E. P., Henry, K., et al., *Clin. Pharmacol. Ther., 64*, 331 (1998).
34. Dieleman, J. P., Gyssens, I. C., van der Ende, M., et al., *Aids, 13*, 473 (1999).
35. Gatti, G., Di Biagio, R., Casazza, R. M., et al., *Aids, 13*, 2083 (1999).
36. Chadwick, E. G., Nazareno, L. A., Nieuwenhuis, T. J., et al., *J. Infect. Dis., 172*, 1475 (1995).
37. Hughes, W., McDowell, J. A., Shenep, J., et al., *Antimicrob. Agents Chemother., 43*, 609 (1999).
38. Lewis, L. L., Venzon, D., Church, J., et al., *J. Infect. Dis., 174*, 16 (1996).
39. Kline, M. W., Dunkle, L. M., Church, J. A., et al., *Paediatrics, 96 (2 Pt. 1)*, 247 (1995).
40. Krogstad, P., Wiznia, A. A., Luzuriaga, K., et al., *Clin. Infect. Dis., 28*, 1109 (1999).
41. Litalien, C., Faye, A., Compagnucci, A., et al., *Pediatr. Infect. Dis. J., 22*, 48 (2003).
42. Floren, L. C., Wiznia, A., Hayashi, S., et al., *Paediatrics, 112 (3 Pt 1)*, e220 (2003).
43. Bergshoeff, A. S., Fraaij, P. L., van Rossum, A. M., et al., *Antivir. Ther., 8*, 215 (2003).
44. Gatti, G., Castelli-Gattinara, G., Cruciani, M., et al., *Clin. Infect. Dis., 36*, 1476 (2003).
45. Luzuriaga, K., Bryson, Y., McSherry, G., et al., *J. Infect. Dis., 174*, 713 (1996).

46. Wang, L. H., Wiznia, A. A., Rathore, M. H., et al., *Antimicrob. Agents Chemother., 48,* 183 (2004).
47. Hazra, R., Balis, F. M., Tullio, A. N., et al., *Antimicrob. Agents Chemother., 48,* 124 (2004).
48. Bonati, M., Choonara, I., Hoppu, K., Pons, G., Seyberth, H., *Lancet, 353,* 1625 (1999).
49. Merry, C., Barry, M. G., Mulcahy, F., et al., *Aids, 11,* F29 (1997).
50. Vanhove, G. F., Kastrissios, H., Gries, J. M., et al., *Antimicrob. Agents Chemother., 41,* 2428 (1997).

6.6 Problems of drug therapy in developing countries

J. Brian S. Coulter

Liverpool School of Tropical Medicine, Pembroke Place, Liverpool L3 5QA and Royal Liverpool Children's NHS Trust, Alder Hey, Eaton Road, Liverpool L12 5AA, UK

6.6.1 INTRODUCTION

There are a number of differences between industrialised and developing countries in the organisation, availability and requirement for drug therapy, some of which are briefly outlined below. In developing countries, bacterial infections and malnutrition are common, and in tropical regions, parasitic diseases, especially malaria, are major problems requiring drug therapy. Availability is limited by the ability of parents to afford optimal treatment. The introduction of fees-for-service and cost recovery may exclude the poorest from treatment [1]. Across-the-counter prescribing is the norm, and amongst the less educated, knowledge and understanding of therapy is limited.

6.6.2 ADMINISTRATION

Intramuscular (IM) injections are the usual route for sick children and a common sight is mothers queuing up for their IM injections in the ward. Intravenous (IV) therapy is often limited by lack of (or inability of mothers to afford) cannulae and drip chambers, and nurses to monitor therapy safely. IM oil-based chloramphenicol (1 dose) is commonly used for meningococcal meningitis during epidemics and monthly IM benzathine penicillin for prevention of rheumatic fever. Chloramphenicol (IM followed by oral) is one of the commonest drugs used for severe pneumonia, meningitis, typhoid fever and other forms of septicaemia. Serious complications of IM injections, administered by untrained health workers, include sciatic nerve palsy, polio induced paralysis (now fortunately uncommon), abscesses, hepatitis B and C and less likely HIV infections.

6.6.3   ESSENTIAL DRUG SCHEME

WHO produces an essential drug list (EDL) which is updated every 2 years. There is also a WHO Model Formulary [2]. The EDL contains over 300 drugs. Problems include outdated drugs, and where resistance to first line drugs is known to be common, e.g., malaria, shigellosis and tuberculosis (TB), failure to provide suitable alternatives. Drugs for HIV/AIDS and some common malignancies are limited [3,4]. However, in developing countries, 60-90% of drug costs are met by patients and many cannot afford the more expensive second-line drugs. In some countries, governments can only provide 2 Euros per capita per annum for drug treatment. Some countries, e.g. India and South Africa, are now manufacturing generic drugs for HIV/AIDS at a cost affordable by, at least, the moderately well off in developing countries. Major problems in drug supply include expired, obsolete (often donated), poor quality and counterfeit drugs. There is a lack of trained pharmacists in small hospitals resulting in inefficiency, inadequate procurement, monitoring of stocks and wastage. Although international pharmaceutical companies are accused of not investing in research and developing of drugs for parasitic diseases in developing countries, e.g. trypanosomiasis and leishmaniasis, recent changes in attitude are encouraging [5]. Antiretroviral drugs, e.g. nevirapine, are being provided free, or at low cost, for control of mother-to-child HIV infection. Initiatives by WHO and other international agencies, combined with free drug donations, have had a dramatic impact on control of onchocerciasis and are now being applied to lymphatic filariasis [6]. The directly observed therapy short course (DOTS) programme is important in the maintenance of first line TB drugs and hopefully should prevent development of resistance. The establishment in 2002 of the Global Fund which supports programmes for control of HIV/AIDS, malaria and TB is a further advance in improving availability of drugs in low resource countries.

6.6.4   DRUG RESISTANCE

Antibiotic resistance is a global problem, but certain conditions and practices in low-resource countries facilitate its development and spread. Important causes include availability of non-prescription antimicrobials, sub-optimal therapeutic regimes, substandard counterfeit drugs, poor hygiene and cross infection in hospitals [7].

In addition to resistance patterns common in industrialised countries, e.g. MRSA and penicillin-resistant *Streptococcus pneumoniae*, in developing countries major problems are reduced sensitivity or multidrug resistance to *Salmonellae typhi* (including nalidixic acid and ciprofloxacin), *Shigella dysenteriae* (including naladixic acid), *Mycobacterium tuberculosis* (especially in HIV-infected subjects), and *Neisseria gonorrhoea* (including penicillin, co-trimoxazole and ciprofloxacin). Chloramphenicol resistance in *Haemophilus influenzae* meningitis may be as high as 20%, and many parents cannot afford alternatives such as ceftriaxone. Resistance to third generation cephalosporins is emerging [8]. Poor organisation, supervision and service in medical microbiology laboratories and lack of facilities to test bacteria for sensitivity exacerbate the problem.

Of parasitic infections, resistance of *Plasmodium falciparum* on a global scale has a major impact on management of malaria, especially in sub-Saharan Africa. For severe malaria chloroquine is obsolete, and there is resistance to alternatives such as sulfadoxine-pyrimethamine and mefloquine. Combination therapy is now being advised, e.g. sulfadoxine-pyrimethamine combined with amodiaquine or artesunate, and dapsone-chlorproguanil (Lapdap) [9]. Promotion of insecticide-impregnated bednets is an essential programme to reduce morbidity and mortality (up to 25%) of young children from malaria.

6.6.5 DRUG METABOLISM IN MALNUTRITION

A limited amount of research on metabolism of drugs has been undertaken in severely malnourished children, most of whom also had a systemic infection [10-12]. However, few studies have compared kwashiorkor with marasmus. Effects of malnutrition on drug metabolism differ between adults and children. Low albumin levels in kwashiorkor result in increased plasma levels for drugs that are highly protein-bound. Constraints in absorption, metabolism, distribution and excretion are to be expected in severe malnutrition and are outlined in Table 1 [12]. A number of experiments have used antipyrine to investigate drug metabolism in malnourished children. A limited number of studies have

Table 1

Factors that may affect absorption, metabolism and excretion of drugs in malnutrition

Gastrointestinal tract	Liver metabolism	Distribution	Renal excretion
vomiting diarrhoea hypochlorhydria villous atrophy	reduction in endoplasmic reticulum and activity of mixed function oxidases; decreased albumin and lipoprotein production	decrease in muscle and fat mass; increased total body water; hypoalbuminaemia, oedema	hypotonic acid urine; impaired concentration

used chloramphenicol to examine conjugation to glucosamic acid, paracetamol for conjugation by sulphation and glucuroxidation, sulfadiazine and isoniazid for acetylation and penicillin and gentamicin for renal excretion. Studies have generally only used a single dose, and thus have not looked at accumulation following multiple doses or variations in drug dosage. Patients were used as their own controls with follow up studies after nutritional rehabilitation.

Although most of the drugs investigated in malnourished children show abnormal accumulation due to impaired liver and, to a lesser extent, renal function, no clear evidence of toxicity has been demonstrated. Even though there is evidence that lower dosage of some drugs would be appropriate, at least in the first few days of treatment, malnourished children are usually given standard doses.

REFERENCES

1. Riddle, V., *Bull. WHO, 81*, 532 (2003).
2. WHO Model Formulary. Couper, M.R., Mehta, D. K., (eds.) Geneva: World Health Organisation (2002).
3. Pécoul, B., Chirac, P., Trouilla, P., Pinel, J., *JAMA 281*, 361 (1999).
4. Debate. Royal Society of Tropical Medicine, *Trans. R. Soc. Trop. Med. Hyg., 97*, 1-15 (2003).
5. Anabwani, G. M., *Paediatr. Pernat. Drug Therapy, 5*, 4 (2002).
6. Molyneux, D. H., Taylor, M. J., *Curr. Opin. Infect. Dis.,14*, 155 (2001).
7. Shears, P., *Trans. R. Soc. Trop. Med. & Hyg., 95*, 127, (2001).
8. Peltola, H., *Clin. Infect. Dis., 32*, 64 (2001).
9. Annotation, *Lancet, 360*, 1998 (2002).
10. Mehta, S., *J. Pediatr. Gastroenterol, Nutr., 2*, 407 (1983).
11. Buchanan, N., *Wld. Review Nutr. Dietetics, 43*, 129 (1984).
12. Krishnaswamy, K., *Clin. Pharmacokin., 17 (Suppl 1)*, 68 (1989).

6.7 Immunisations and drug therapy

Robert Cohen

Department of Microbiology, Centre Hospitalier Intercommunal de Créteil, 40 Avenue de Verdun, 94000, Créteil, France

6.7.1 INTRODUCTION

The goals of immunisation are the prevention of disease in individuals and, herd immunity for a group, or eradication of the disease. High immunisation rates have reduced dramatically or almost eliminated, diphtheria, measles, mumps, polio, rubella, tetanus, and *Haemophilus influenzae type b* diseases in developed countries. Because organisms that are responsible for these diseases persist around the world, continued immunisation efforts must be maintained and reinforced. In some circumstances, the immunisation programme has to be modified due to an underlying disease or concurrent medical treatment. The aim of this chapter is to focus on interactions between vaccines and medicines. Contraindications to immunisation are often misunderstood and many common conditions, circumstances and treatment are not contraindications for vaccinations.

6.7.2 ANTIBIOTIC TREATMENT

Antimicrobial therapy and mild acute illness (with low-grade fever or mild diarrhoea) do not interfere with any vaccines and are not contraindications for any immunisation [1].

6.7.3 SPACING OF DIFFERENT VACCINES

Generally, vaccines should be administered following the recommended schedule programme of the country [2]. However, in some circumstances (international travels, exposure to vaccine preventable diseases, lapsed immunisations with doubt about the return of the recipient for further vaccination), multiple vaccines can be given concurrently. When simultaneous vaccines are administered, separate syringes and sites should be used, and injections into the same extremity should be separated by at least 2.5 cm, so that any local reactions can be differentiated [1].

Simultaneous administration of vaccines neither results in an increase in adverse drug reactions nor immunological problems. Inactivated vaccines do not interfere significantly with the immune response to other inactivated vaccines or live attenuated vaccines. An inactivated vaccine can be administered either simultaneously to, or at any time before or after, a different inactivated vaccine or live vaccine. Two doses of different live vaccine can be administered simultaneously. To minimise the risk of interference, parenteral live attenuated virus vaccines should be administered on the same day or at least one month apart [1].

A tuberculin skin test can be performed at the same time as vaccines are administered. Measles vaccine, however, can temporarily suppress tuberculin reactivity, and therefore tuberculin testing should not be done at the same time as measles immunisation, but should be postponed for 4 to 6 weeks. The effect of live-virus varicella and yellow fever vaccines on tuberculin skin test reactivity is not known [1].

6.7.4 CORTICOSTEROIDS

Corticosteroid treatment may reduce the immunogenicity (and consequently the efficacy) of vaccines and raise concerns about vaccine safety for live virus vaccines (Table 1). However, the minimal dose and duration of systemic corticosteroids sufficient to cause immunosuppression is not well defined. The frequency and route of administration of corticosteroids, the underlying disease, and concurrent other therapy are additional factors to be taken into account. Despite these unresolved questions, sufficient experience exists to recommend empirical rules for administration of vaccines to children suffering from non-immunocompromising conditions [1,3]. In addition, when deciding whether to administer live virus vaccines, the potential benefits and risks of immunisation for an individual patient and specific circumstances should be considered. Nevertheless, unless immunisation can be deferred temporarily until corticosteroids are discontinued, children should be immunised in the case of exposure to disease.

Table 2 summarises the main rules of use of live vaccines in children receiving corticosteroid treatment, according to the recommendations of the American Academy of Pediatrics [1].

Table 1

Live attenuated and inactivated vaccines

Inactivated vaccines and toxoids	Live vaccines
pertussis	BCG
diphtheria	measles
injectable polio	mumps
tetanus	rubella
haemophilus b	varicella
hepatitis B	yellow fever
pneumococcus	nasal influenza
meningococcus	rotavirus
typhoid	oral polio
influenza	
hepatitis A	
rabies	
Japanese encephalitis	
tick-born encephalitis	

Table 2

Use of live attenuated vaccines for patients receiving corticosteroids

No contra-indication	Contra-indication
• physiological maintenance doses of corticosteroids • topical therapy or local injections of corticosteroids* • children receiving <2 mg/kg per day of prednisone or its equivalent, or <20 mg/day if they weigh more than 10 kg)	• children receiving 2 mg/kg per day of prednisone or its equivalent, or 20 mg/day if they weigh more than 10 kg, *for more than 2 weeks* should not receive live-virus vaccines until corticosteroid therapy has been discontinued for at least 1 month • children receiving 2 mg/kg per day of prednisone or its equivalent, or 20 mg/day if they weigh more than 10 kg, given daily or on alternate days for less than 2 weeks can receive live-virus vaccines immediately after discontinuation of treatment**

* Live-virus vaccines should not be administered, if clinical or laboratory evidence of systemic immunosuppression results from prolonged application, until corticosteroid therapy has been discontinued for at least 1 month.

** Some experts would delay immunisation until 2 weeks after corticosteroid therapy has been discontinued.

6.7.5 IMMUNOSUPPRESSIVE THERAPY

Inactivated vaccines should be used when appropriate to reduce the risk of complications. However, immune responses may vary and may be inadequate. In children in whom immunosuppressive therapy is discontinued, an adequate response occurs usually between 3 months and 1 year after discontinuation.

Live vaccines generally are contraindicated, for an interval of at least 3 months after immunosuppressive therapy has been discontinued, because of an increased risk of serious adverse effects. However, the interval may vary with the underlying disease, the type and the dosage of immunosuppressive therapy and other factors. Therefore, it is often not possible to make a definitive recommendation for an interval after cessation of immunosuppressive therapy when live virus vaccines can be administered safely and effectively. *In vitro* testing of immune function, notably the measurement of serum antibody titres after immunisation with inactivated vaccine, may guide the safe timing of immunisations in individual patients.

Immunocompetent siblings and other household contacts of people with an immune deficiency should not receive oral poliovirus vaccine, because vaccine virus may be transmitted to immunocompromised people. However, siblings and household contacts should receive measles, mumps, rubella (MMR), varicella (for susceptible contacts) and influenza vaccine to reduce the risk of infection of immunocompromised children.

6.7.6 IMMUNOGLOBULIN AND BLOOD PRODUCTS

Live virus vaccines may have reduced immunogenicity when given shortly before (two weeks) or several months after the administration of immunoglobulin (Ig). This is the case for measles, mumps, rubella and varicella. Following the recommendations of the American Academy of Pediatrics, the interval between Ig administration and these types of immunisation will vary with the specific product (Table 3) [1].

Administration of Ig preparations does not interfere with antibody responses to other live virus immunisation: yellow fever vaccine and oral polio vaccine. Administration of Ig does not cause significant inhibition of the responses to inactivated vaccines and toxoids (Table 2). The respiratory syncytial virus monoclonal antibody (palivizumab) does not interfere with the response to inactivated or live vaccines.

Table 3
Intervals recommended between Ig or blood product administrations

Product	Intervals with MMR and varicella vaccine (months)
palivizumab	0
red blood cells washed	
specific Ig (tetanus, hepatitis A, HBS, rabies, varicella)	3 to 5
red blood cells	
whole blood	6
packed red blood cells	
plasma	
platelets	
intravenous Ig	8 to 11

6.7.7 ANTICOAGULANT THERAPY

Children with bleeding disorders, e.g. haemophilia, or on anticoagulants are at increased risk of bleeding after intramuscular injection (IM). The smallest needle (23-gauge or less) should be used and firm pressure, without rubbing, should be applied to the injection site for several minutes. All live viral attenuated vaccines should be administered subcutaneously. Furthermore, vaccines recommended by IM route (inactivated vaccine with adjuvant) could be administered subcutaneously if the immune response and clinical reaction are expected to be comparable e.g. polysaccharide vaccines or Hib conjugate vaccines.

REFERENCES

1. American Academy of Pediatrics. *Active immunisation*, in Pickering, L. (ed.), Red Book: Report of the Committee on Infectious Diseases, 26th Edition, Elk Grove Village, IL, American Academy of Pediatrics, pp. 7-53 (2003).
2. Atkinson, W.; Pickering, L.; Watson, J.; Peter, G., *General Immunisation Practices*, in Plotkin, S.; Oreinstein, W. (eds.), Vaccines (4th), Saunders, pp. 91-122 (2004).
3. Marshall, G., *The vaccine handbook*, Lippincott Williams & Wilkins, pp. 309-26 (2004).

Critical care, neurology and analgesia

7.1 Nitric oxide

Jean-Christophe Mercier[1], Anh Tuan Dinh-Xuan[2]

[1] *Department of Paediatric and Neonatal Intensive Care, Hospital Robert Debré, 75019 Paris, France*

[2] *Laboratory of Respiratory Physiology, Faculty of Medicine, University of Paris V, 750014 Paris, France*

7.1.1 CLINICAL SETTING

Normally, within minutes after birth, pulmonary vascular resistance (PVR) rapidly falls from high fetal levels. This allows pulmonary blood flow to increase nearly tenfold and enables the lung to assume its postnatal role in gas exchange. Severe hypoxaemia characterises the course of severe neonatal respiratory failure. Hypoxaemia may result from intrapulmonary or extrapulmonary shunting. Intrapulmonary shunting is associated with severe parenchymal lung disease (e.g., surfactant deficiency), bacterial pneumonia, and meconium aspiration pneumonitis; generally amenable to exogenous surfactant, and/or ventilator strategies aimed at adequate lung volume recruitment. Severe hypoxaemia commonly relates to persistent pulmonary hypertension of the neonate (PPHN) [1]. Persistently raised PVR results in extrapulmonary shunting of desaturated blood across the patent ductus arteriosus, foramen ovale, or both, and sustained pulmonary hypertension in right ventricular dysfunction and critical hypoxia. To cut the vicious cycle of hypoxia and increased barotraumatism due to hyperventilation commonly used in this setting, there was no other alternative than to put the lung "at rest" using extracorporeal membrane oxygenation and waiting for the PVR to spontaneously decrease. Therefore, interventions aimed at selective pulmonary vasodilation could be beneficial.

7.1.2   INHALED NITRIC OXIDE THERAPY

Nitric oxide (NO) is a highly reactive molecule with many biological effects. In 1980, the discovery of the obligatory role of endothelial cells in the relaxation of arterial smooth muscle by acetylcholine led one to hypothesise the existence of an "endothelium-derived relaxing factor (EDRF)" [2], that was later identified as a gaseous compound, i.e., nitric oxide [3,4]. A few years before this discovery, it had been observed that nitrovasodilators were metabolised into NO, and that these compounds relaxed vascular smooth muscle by elevating cyclic guanosine monophospate levels [5]. As a result, NO was named "Molecule of the Year 1992 " [6], and Robert Furchgott, Louis Ignarro and Freid Murrad were given the Nobel Prize for Medicine in 1998 [7].

NO is synthesised from the amino acid L-arginine and molecular oxygen by a family of enzymes, the nitric oxide synthases (NOS). This process occurs through the L-arginine-NO pathway that requires a number of cofactors, namely, NADPH, flavin mononucleotide (FMN), flavin adenin dinucleotide (FAD) and tetrahydrobiopterin, as well as calmomodulin [8]. Several isoforms of NOS exist. Constitutive NOS were first located in the vascular endothelium (endothelial or type-3), and subsequently in platelets, neurons of the central nervous system (neuronal or type-1), and in nonadrenergic, noncholinergic (NANC) nerves. Calmomodulin-dependent NOS generate small quantities of NO that carry out a variety of physiological functions, via activation of the soluble guanylate cyclase. Inducible NOS (iNOS or type-2) are induced by immunological or inflammatory stimuli, including LPS or various cytokines, and give rise to the sustained release of substantial amounts of NO. NOS have been characterised, their cDNA cloned, and their genes identified on various chromosomes [9].

Inhaled NO was envisioned as having potential therapeutic application as a potent pulmonary vasodilator in the early 1990s. This was almost simultaneous with its identification as the EDRF [10,11]. Soon after, inhaled NO was shown to reverse hypoxaemia in a few of infants with persistent pulmonary hypertension of the newborn (PPHN) [12,13]. Clinical benefit has been demonstrated in infants with PPHN in two large studies [14,15]. Both the EAMA and the FDA have approved inhaled NO, as a treatment to reverse hypoxaemia and decrease the use of extracorporeal membrane oxygenation (ECMO), in hypoxic newborns with PPHN.

Mechanisms leading to severe PPHN are poorly understood, but they include altered pulmonary vascular reactivity and structural remodeling. A PPHN animal model has been developed in fetal lambs by ductal constriction or ligation a few days before delivery. In this model, eNOS mRNA protein and its acti-

vity have been found to be significantly decreased [16,17], suggesting a rationale for inhaled NO replacement therapy. However, several other critical steps
of the NO downsignaling pathway may be altered in PPHN, including reduced
soluble guanylate cyclase activity and high type-5 phosphodiesterase activity,
which suggest the potential for using specific inhibitors, such as sildenafil [18].
Impaired prostacyclin (PgI_2), or conversely increased thromboxane A_2 release,
elevated endothelin-1 (ET-1) levels and/or modified ET-1 receptor subtypes,
may also play a significant role in PPHN. From a clinical perspective, failure of
inhaled NO therapy may occur when the lung is not adequately inflated, thereby
explaining the synergistic effects of inhaled NO with exogenous surfactant therapy and high-frequency oscillatory ventilation (HFOV). It is often associated
with either pulmonary hypoplasia, such as in congenital diaphragmatic hernia
[19], or misalignment of the pulmonary vessels, such as in alveolar capillary
dysplasia [20].

   Inhaled NO therapy has been based on impressive observations of reduced pulmonary artery pressures, improved ventilation/perfusion matching,
and increased oxygenation. On the basis of the dose-curve effect achieved in
hypoxic newborn lambs, the initial recommendation was to use 80 parts per
million (ppm) [21]. However early clinical use in newborns showed that both
20 or 80 ppm were equally effective. Furthermore, studies showed that 5 and 80
ppm were equally effective in terms of increase in oxygenation in infants referred for extracorporeal membrane oxygenation [22], or in infants less severely
hypoxaemic [23]. Indeed, in the NINOS trial, no infant with PPHN who did
not respond to 20 ppm actually benefited from higher doses, i.e., 40 or 80 ppm,
whereas in the second pivotal trial, CINRGI, the hypoxic newborns were treated
with 20 ppm for a maximum of 24 h followed by 5 ppm for no more than 96 h.
Lower doses of inhaled NO may be effective [24,25]. In contrast, in adults with
evolving ARDS, the dose-response curve clearly differs in terms of oxygenation
in that it is significantly improved at about 0.1 ppm and decreased in PVR that
occurs at about 1 ppm [26].

7.1.3 PHARMACOLOGY OF NO

   The great advantage of inhaled NO therapy over infusion of vasodilatators lies
in the apparently selective pulmonary vasodilator effect, with a lack of systemic
hypotensive effects. This was attributed to the destruction of residual bioactive
NO that enters the pulmonary circulation after transit of the alveoli and their
associated blood vessels [27]. However, considerable recent interest and controversy have focused on the role of intravascular NO-derived molecules that con-
serve and stabilise NO bioactivity and may contribute to blood flow and oxygen

delivery [28]. Such molecules include plasma low- and high-molecular-weight S-nitrosothiols and nitrite. In addition, NO reacts reversibly with haemoglobin to form an NO-heme adduct, nitrosyl(heme)haemoglobin, and can also nitrosate a surface thiol on cystein-93 of the β-globin chain to form nitrosohaemoglobin (SNO-Hb). The role of haemoglobin as a NO transporter is particularly appealing in that delivery of NO may be linked energetically to oxygen binding, promoting the allosteric delivery of oxygen and NO to peripheral tissues with low oxygen tension [29].

Inhaled NO is thought to immediately react with oxyhaemoglobin to form bioinactive nitrates, thereby limiting its diffusion to the nearby pulmonary vascular endothelium. Animal studies, however, have shown that inhaled NO reduces systemic vascular resistance [30], and restores blood flow to the intestine after an infusion of a NOS inhibitor [31]. Similarly, inhaled NO has been shown to increase coronary blood flow in patients with coronary artery disease [32], and peripheral blood flow after blockade of NO followed by forearm exercise test in normal humans [33]. While these observations are consistent with intravascular biostabilisation, transport and delivery of NO, the therapeutic relevance remains uncertain. Recently, cell-free haemoglobin induced by intravascular haemolysis has been shown to limit NO bioavailability in sickle-cell disease, thereby favouring the frequent vascular occlusive crisis complications observed in this disease [34]. As a consequence, low blood levels of arginine have been found in this disease, and oral L-arginine supplementation has been shown to reduce estimated pulmonary artery systolic pressure [35]. Likewise, infants with PPHN have been shown to have low plasma concentrations of arginine and NO metabolites, and a genetically predetermined capacity of the urea cycle, in particular the efficiency of carbamoyl-phosphate synthetase, which may contribute to the availability of precursors for nitric oxide synthesis [36].

7.1.4 INHALED NO THERAPY SIDE EFFECTS

Use of doses higher than 20 ppm in the rich oxygen environment, usually required to alleviate hypoxaemia, was shown to be associated with increased levels of potentially toxic nitrogen dioxide (NO_2) > 0.5–2 ppm [22,37]. Furthermore, it favours the formation of methaemoglobinemia, which is harmful when >5%. Finally, it may favour life-threatening rebound of pulmonary hypertension and hypoxaemia, whenever there is a sudden NO withdrawal due, for example, to ventilator disconnection for tracheal aspiration [38]. Thus, it is recommended to progressively wean inhaled NO by using a highly reliable delivery system with back-up manual ventilation [39]. Pulsed delivery of inhaled NO during spontaneous ventilation, using nasal cannula, has been reported both in adults

[40] and children [41], using delivery systems that have yet to be approved. New delivery systems are currently being tested for use with continuous positive airway pressure systems.

7.1.5 CONCLUSION

Since the identification of NO, there has been a great increase in understanding of its ubiquitous role. In contrast, its clinical application so far remains limited to the restrictive licensing of inhaled NO to treat hypoxic infants with PPHN, thereby alleviating the need for ECMO. However, new fields of application are currently being explored, such as the prevention of the alveolar arrest that characterises the "new" chronic lung disease observed in the premature infants, or the therapy of acute chest syndrome, which constitutes a severe complication of sickle cell disease. Moreover, new NO donors or adducts are being developed for aerosol therapy. Furthermore, other pharmacological compounds, such as PDE-5 inhibitors, may enhance or prolong the therapeutic effects of NO. Despite this progress, further research is needed to determine the benefits of this highly reactive drug which can have both beneficial or harmful effects, depending upon its environment, i.e., oxygen radical species.

REFERENCES

1. Gersony, W. M., *Clin Perinatol, 11*, 517 (1984).
2. Furchgott, R. F., Zawadski, J.V., *Nature, 288*, 373 (1980).
3. Palmer, R. M., Ferrige, A. G., Moncada, S., *Nature, 327*, 524 (1987).
4. Ignarro, L. J., Byrns, R. E., Buga, G. M., Wood, K. S., *Circ. Res., 61*, 866 (1987).
5. Katsuki, S., Arnold, W., Mittal, C., Murad, F., *J. Cyclic Nucleotide Res., 3*, 23 (1977).
6. Molecule of the year of *Science*, (1992).
7. Nobel Foundation, Stokholm, Sweden, 210 (1999).
8. Moncada, S., Higgs, A., *N. Engl. J. Med., 329*, 2002 (1993).
9. Moncada, S., Higgs, E. A., *F.A.S.E.B.J., 9*, 1319 (1995).
10. Pepke-Zaba, J., Higgenbottam, T., Dinh-Xuan, A. T., Stone, D., Wallwork, J., *Lancet, 338*, 1173 (1991).
11. Frostell, C., Fratacci, M. D., Wain, J. C., Jones, J. C., Zapol, W. M., *Circulation, 83*, 2038 (1991).
12. Roberts, J. D., Polaner, D. M., Lang, P., Zapol, W. M., *Lancet, 340*, 818 (1992).
13. Kinsella, J. P., Neish, S. R., Shaffer, E., Abman, S. H., *Lancet, 340*, 819 (1992).
14. The Neonatal Inhaled Nitric Oxide Study Group, *N. Engl. J. Med., 336*, 597 (1997).
15. Clark, R. H., Kueser, T. J., Walker, et al., and the Clinical Inhaled Nitric Oxide Research Group, *N. Engl. J. Med., 342*, 469 (2000).
16. Shaul, P. W., Yuhanna, I. S., German, Z., Chen, Z., Steinhorn, R. H., Morin, F. C., *Am. J. Physiol., 272*, L1005 (1997).
17. Villamor, E., Le Cras, T. D., Horan, L. P., Halbower, A. C., Tuder, R. M., Abman, S. H., *Am. J. Physiol., 272*, L1013 (1997).
18. Abman, S. H., Kinsella, J. P., Mercier, J. C., in *Lung development*, Gaultier, C., Bourbon, J., Post, M. (eds.) Oxford University Press, p. 196 (1999).

19. The Neonatal Inhaled Nitric Oxide Study Group, *Pediatrics, 99*, 838 (1997).

20. Steinhorn, R. H., Cox, P. N., Fineman, J. R., et al., *J. Pediatr., 130*, 417 (1997).

21. Roberts, J. D., Chen, T. Y., Kawai, N., et al., *Circulation, 72*, 246 (1993).

22. Finer, N. N., Etches, P. C., Kamstra, B., Tierney, A. J., Peliowski, A., Ryan, A., *J. Pediatr., 124*, 302 (1994).

23. Davidson, D., Barefield, E. S., Kattwinkel, J., et al., and the INO/PPHN Study Group. *Pediatrics, 101*, 325 (1998).

24. Cornfield, D. N., Maynard, R. C., de Regnier, R. O., Guiang, S. F., Barbato, J. E., Milla, C. E. *Pediatrics, 104*, 1089 (1999).

25. Finer, N. N., Sun, J. W., Rich, W., Knodel, E., Barrington, K., *J. Pediatrics, 108*, 949 (2001).

26. Gerlach, H., Rossaint, R., Papert, D., Falke, K. J., *Eur. J. Clin. Invest., 23*, 499 (1993).

27. Rimar, S., Gillis, C. N., *Circulation, 88*, 2884 (1993).

28. Jia, L., Bonaventura, C., Bonaventura, J., Stamler, J. S., *Nature, 380*, 221 (1996).

29. McMahon T. J., Moon, R. E., Lushinger, B. P., et al., *Nat. Med., 8*, 711 (2002).

30. Takahashi, Y., et al., *Am. J. Physiol., 274*, H349 (1998).

31. Fox-Robichaud, A., *J. Clin. Invest., 101*, 2497 (1998).

32. Adrie, C., Bloch., K. D., Moreno, et al., *Circulation, 94*, 1919 (1996).

33. Cannon, R. O., Schechter, A. N., Panza, J. A., et al., *J. Clin. Invest., 108*, 279 (2001).

34. Reiter, C. D., Wang, X., Tanus-Santos, J. E., et al., *Nat. Med. 12*, 1383 (2002).

35. Morris, C. R., Morris, S. M., Hagar, W., et al., *Am. J. Respir. Crit. Care Med., 168*, 63 (2003).

36. Pearson, D. L., Dawling, S., Walsh, W. F., et al., *N. Engl. J. Med., 344*, 1832 (2001).

37. Bouchet, M., Renaudin, M. H., Raveau, C., Mercier, J. C., Dehan, M., Zupan, V. *Lancet, 241*, 968 (1993).

38. Davidson, D., Barefield, E. S., Kattwinkel, J., et al., and the INO/PPHN Study Group. *Pediatrics, 104*, 231 (1999).

39. Kirmse, M., Hess, D., Fujino, Y., Kacmarek, R. M., Hurford , W. E., *Chest, 113*, 1650 (1998).

40. Channick, R. N., Newhart, J. W., Johnson, F. W., et al., *Chest, 109*, 1545 (1996).

41. Ivy, D. D., Griebel, J. L., Kinsella, J. P., Abman, S. H., *J. Pediatr., 133*, 453 (1998).

7.2 Drug disposition during extra-corporeal membrane oxygenation (ECMO)

Hussain Mulla

AstraZeneca R&D Charnwood, Loughborough, Leics LE11 5RH, UK

7.2.1 INTRODUCTION

Extracorporeal membrane oxygenation (ECMO) is a complex life support technique for severe pulmonary or cardiopulmonary failure, developed through modification of the heart lung bypass machine [1]. The technique of ECMO involves oxygenating blood outside the body and, thus, obviates the need for gas exchange in the lungs. The technique is categorised as either veno-venous (VV) or veno-arterial (VA), depending on the type of cannulation. In VV ECMO, deoxygenated blood is drained and oxygenated blood re-infused via venous sites. In neonates, this is achieved by placing a double lumen cannula in the right internal jugular vein (Figure 1). In VA ECMO, deoxygenated blood drawn from the right internal jugular vein is returned oxygenated via the right common carotid artery. While VV ECMO provides support purely with gas exchange, VA ECMO also supports the heart.

7.2.2 DRUG DISPOSITION DURING ECMO

Pharmacological therapy in the ECMO patient presents a challenge since the continuous extracorporeal circulation of blood may conceivably impact on pharmacokinetics and pharmacodynamics. Table 1 shows some of the potential ECMO circuit related effects on drug disposition.

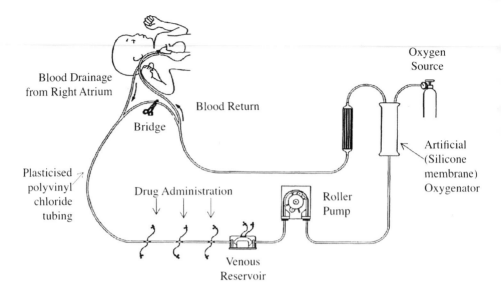

Figure 1. Schematic diagram of a paediatric ECMO circuit

Table 1

ECMO specific effects that may impact on drug disposition

Specific effect	Potential impact on drug disposition
expansion of circulating volume haemodilution protein binding changes	increased volume of distribution changes in free fraction of drug
sorption to polymeric components of circuit (plasticised polyvinyl chloride tubing and silicone membrane oxygenator)	reversible increased volume of distribution irreversible increased clearance
drug delivery via the circuit	stagnation of drug in venous reservoir at low flow rates (<250ml/min)
non-pulsatile blood flow	reduced clearance?
recirculation of blood during VV cannulation	increased volume of distribution reduced clearance
photodegradation	increased clearance (decreased stability of drugs)

Haemodilution

The most obvious alteration to pharmacokinetics occurs on initiation of ECMO, when the patients own blood volume mixes with the priming volume in the extracorporeal circuit. For example, on initiation of ECMO the effective circulating volume will be almost doubled in neonates. One possible effect of this acute haemodilution is a decrease in the total blood concentration of any drug present. The pharmacological impact will depend on the apparent volume of distribution of the drug, the degree of protein binding and the extent of equilibration between tissue concentrations and plasma concentrations on initiation of ECMO.

Table 2

A comparison of mean pharmacokinetic values determined
in non-ECMO and ECMO neonates[a]

Drug	Study Group	V (L/kg)	CL (L/kg/h)	$t_{1/2}$(h)
gentamicin[b]	non-ECMO[d] [6]	0.47	0.05	
	ECMO [2]	0.75	during ECMO 0.24 litre/h	9.24
		0.47	post- ECMO 0.35 litre/h	3.87
vancomcyin[b]	non-ECMO[e] [7]	0.57	0.08	4.90
	ECMO [4]	0.67	0.04	8.44
midazolam[c]	non-ECMO [8]	1.01	0.07[f] 0.11[g]	10.0 6.36
	ECMO [9]	4.10	0.08	33.30

[a] All ECMO neonates were term
[b] Routinely monitored antibiotics with small volume of distribution
[c] Commonly used sedative with large volume of distribution
[d] Post conceptional age > 34 weeks, Apgar ≥ 7.
[e] Term neonates
[f] Gestational age <39 weeks.
[g] Gestational age >39 weeks
V = Apparent volume of distribution
CL= Plasma clearance

The impact of an enlarged circulating volume on the pharmacokinetics of three routinely monitored drugs with small volumes of distribution (aminophylline, gentamicin and vancomycin) has been investigated in term ECMO neonates [2-4]. Results from these studies reveal that volume of distribution tends to increase, clearance is reduced and half-life is inevitably prolonged (Table 2). This has led investigators to suggest the need for an initial loading dose, followed by an extended interval maintenance regimen [2]. In contrast, drugs such as fentanyl with a large volume of distribution would be expected to show only a slight change following the expansion of plasma volume, the initial lowering of plasma concentration from haemodilution being counteracted by the back diffusion of the drug into plasma from the large tissue reservoirs.

The acute haemodilution on initiation of ECMO also produces a large reduction in circulating plasma protein concentration such as albumin. For drugs that are highly protein bound, decreased concentration of binding proteins will lead to an increase in the fraction of unbound drug. This favours transfer of drug from the plasma to the tissues and contributes to the lowering of plasma concentration. The pharmacodynamic result of this may be an increased effect because of an increased free fraction at the receptor sight. Another effect is acute anaemia, which may affect the degree of drug binding to red blood cells [10]. These effects would be transient however, since following cannulation it is standard practice to normalise the effects of haemodilution by transfusing blood and related products, including albumin.

Drug sorption

Reduction of the amount of drug available to a patient, resulting from interactions with drug administration devices constructed from plastics, is a well recognised and documented phenomenon [11]. The potency loss appears to be dependent on the physico-chemical characteristics of the drug and solution, the nature of the plastic and the dynamics of the system [12]. However, ECMO circuits are unique in sorption studies since, unlike plastic containers and administration sets used for drug delivery, they are an extension of the patient's circulatory system. Therefore, in comparison to sorption in intravenous delivery devices, drug-circuit interactions *in vivo* will have a significant influence on pharmacokinetics and hence therapeutic efficacy. For drug sorption to occur, not only will the drug have to distribute from the blood phase, but the process will compete with binding to circulating plasma proteins and red blood cells.

The component materials of an ECMO circuit are mainly organic in nature. The circuit tubing is made from plasticised polyvinyl chloride, whilst currently the most popular gas exchange membrane in the oxygenator is a silicone mem-

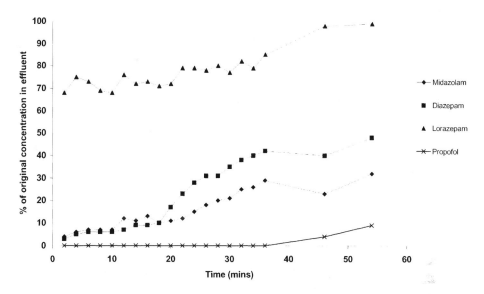

Figure 2. Sedative drugs were continuously infused at a constant flow rate of 360 ml/min through an isolated linear neonatal ECMO circuit. Concentration of drugs in the effluent was analysed at various time points and plotted as a percentage of the original.

brane construct [13]. Significant drug sorption with both of these materials is possible since the patient's blood is exposed to a large surface area of each plastic. However, to date there are few reports and limited to *in vitro* investigations. Rosen demonstrated fentanyl uptake by the membrane oxygenator, whilst others in their preliminary *in vitro* evaluation revealed significant decreases in circuit concentration of various drugs with time [14,15]. We have reported a significant capacity for sorption of commonly used sedatives, diazepam, midazolam, lorazepam and propofol during an *in vitro* simulation of drug infusion through an ECMO circuit (Figure 2) [16].

The impact of midazolam sorption on pharmacokinetics in neonates supported on ECMO has also been investigated [9]. A population pharmacokinetic model suggested a significantly enlarged volume of distribution, three to four times larger than previously reported in neonates, attributed to reversible sorption of midazolam to the circuit. Consequently, half-life was substantially prolonged, and hence midazolam must not be considered a short acting sedative agent in neonatal ECMO patients (Table 2).

Flow rates and injection sites

A fundamental difference between ECMO and non-ECMO patients is the site of drug delivery. The ECMO circuit has multiple ports available for infusion of drugs, blood products, parenteral nutrition, for blood sampling and for attachment of haemofiltration circuits, although the exact location of these ports varies amongst centres. Though it is possible to administer drugs via direct access into the patient, systemic heparinisation necessitates minimal direct interventions, such as intravenous line insertion, in order to avoid excessive bleeding. Most drugs are therefore administered directly into the ECMO circuit. The effects of circuit injection sites and pump rates on the flow of drugs through an ECMO circuit have been investigated in a single *in vitro* study [17]. This study suggested that drug injected proximal and directly into the venous reservoir may pool at the top of the reservoir at flow rates of 75 ml/minute (such as during the weaning phase of ECMO) and some pooling of drug was possible at flow rates less than 250 ml/min. In contrast, drug injected distal to the reservoir will not pool at any flow rate.

Pulsatile versus non-pulsatile blood flow

Whereas VV ECMO results in pulsatile blood flow, VA ECMO at high flow rates (> 100 ml/kg/min) may produce non-pulsatile flow. Non-pulsatile blood flow can alter perfusion of tissues, reducing capillary circulation and aerobic metabolism [18]. Under experimental conditions, pulseless perfusion of the kidneys of dogs resulted in reduced urine production and impaired sodium excretion, although glomerular filtration was not affected [19]. The kidneys interpret pulseless blood flow as hypotension and activate the renin-angiotensin system [20]. Regional blood flow changes in the liver can also affect drug clearance, in particular those drugs with a high extraction ratio, e.g. propranolol, lignocaine [21].

Recirculation of blood

An additional but thus far little investigated phenomenon during VV ECMO is recirculation of blood, where a fraction of oxygenated blood from the circuit flows directly from the re-infusion site to the drainage catheter and back into the circuit, instead of the patient's circulation [13]. At high flow rates (~120 ml/kg/min) as much as 60% of blood returning from the circuit may recirculate. During the early period of ECMO with high flow rates, recirculation of blood will significantly affect distribution of drug into systemic circulation and hence elimination of drug. However, as ECMO support is weaned with lower flow rates, recirculation is significantly reduced (<10%) and no significant influence is expected. This phenomenon also suggests that pharmacokinetic parameters will necessarily change with time.

Photodegradation

The catalysis by light of drug degradation reactions such as oxidation or hydrolysis, has been reported for a number of drugs; amphotericin B, furosemide, dacarbazine, doxorubicin, vitamin A and sodium nitroprusside [11]. During ECMO, drug in circulating blood passes through transparent polyvinyl chloride tubing, exposed to light for a length of time dependent on the pump flow rates and length of circuit. The nature of the light will on the whole be fluorescent, though ECMO circuits may be exposed to some sunlight depending on their position in the intensive care unit. Such exposure to light of drugs in blood has not been investigated, but it is plausible that the stability of susceptible drugs may be affected.

REFERENCES

1. Peek, G., Killer, H., Sosnowski, A., Firmin, R., *Hospital Medicine (London) 59*, 304 (1998).
2. Dodge, W., Jelliffe, R., Zwischenberger, J., Bellanger, R., Hokanson, J., Snodgrass, W., *Ther. Drug Monit., 16*, 522 (1994).
3. Mulla, H., Nabi, F., Nichani, S., Lawson, G., Firmin, R., Upton, D., *Br. J. Clin. Pharmacol., 55*, 23 (2003).
4. Mulla, H., Pooboni, S., *Br. J. Clin. Pharmacol.* (in press).
5. Gilman, J., Gal, P., Levine, R., Hersh, C., Vildan Erkan, N., *Ther. Drug Monit., 8*, 4 (1986).
6. Thomson, A., Way, S., Bryson, S., McGovern, E., Kelman, A., Whiting, B., *Develop. Pharmacol. Ther., 11*, 173 (1988).
7. McDougal, A., Ling, E., Levine, M. *Ther. Drug Monit., 17*, 319 (1995).
8. Burtin, P., Jacqz-Aigrain, E., Girard, P., et al., *Clin. Pharmacol. Ther., 56*, 615 (1994).
9. Mulla, H., McCormack, P., Lawson, G., Firmin, R., Upton, D., *Anesthesiology, 99*, 275 (2003).
10. Hynynen, M., *Acta Anaesth. Scand., 31*, 706 (1987).
11. Trissel, L., *Handbook on Injectable Drugs*, American Society of Health System Pharmacists, 2001.
12. Kowaluk, E., Roberts, M., Polack, A., *Am. J. Hosp. Pharm., 39*, 460 (1982).
13. Zweischenberger, J., Steinhorn, R., Bartlett, R., *ECMO: Extracorporeal Cardiopulmonary Support in Critical Care*, Extracorporeal Life Support Organization, Ann Arbor, Michigan, 2000.
14. Rosen, D., Rosen, K., Davidson, B., Broadman, L., *J. Cardiothor. Anaesth., 2*, 619 (1988).
15. Dagan, O., Klein, J., Gruenwald, C., Bohn, D., Barker, G., Koren, G., *Ther. Drug Monit., 15*, 263 (1993).
16. Mulla, H., Lawson, G., von Anrep, C., et al., *Perfusion, 15*, 21 (2000).
17. Hoie, E., *Am. J. Hosp. Pharm., 50*, 1902 (1993).
18. Shevde, K., Dubois, W., *J. Cardiothor. Anaesth., 1*, 165 (1987).
19. Many, M., Soroff, S., Birtwell, W., Wise, H., Deterling, R. *Arch. Surg., 97*, 917 (1986).

20.   Bartlett, R., *Curr. Probl. Surg.*, *27*, 621 (1990).
21.   Mckindley, D., Hanes, S., Boucher, B., *Pharmacotherapy*, *18*, 759 (1998).

7.3 Muscle relaxants

Stephen D. Playfor

Paediatric Intensive Care Unit, Royal Manchester Children's Hospital, Hospital Road, Pendlebury, Manchester M27 4HA, UK

7.3.1 INDICATIONS FOR NEUROMUSCULAR BLOCKADE

There are many indications for the administration of neuromuscular blocking agents in the paediatric intensive care unit (PICU) and these are summarised in Table 1. The most common indications are to facilitate endotracheal intubation and to prevent patient-ventilator dyssynchrony, particularly during "unphysiological" techniques such as high frequency oscillatory ventilation, inverse ratio ventilation or controlled hypoventilation. Other indications centre around the management of specific clinical conditions.

Table 1

Indications for neuromuscular blockade

facilitation of endotracheal intubation
prevention of patient-ventilator dyssynchrony
management of raised intracranial pressure
management of pulmonary hypertension
prevention of shivering during induced hypothermia
reduction of metabolic demand and oxygen consumption
management of tetanus
management of malignant hyperthermia
management of neuroleptic malignant syndrome
facilitation of patient transfer
protection of specific postoperative surgical wounds

7.3.2 PHYSIOLOGY OF NEUROMUSCULAR BLOCKADE

In most mammalian muscles, each muscle fibre has a single area of contact with the axon of the motor neurone that supplies it. This specialised structure is the neuromuscular junction, which facilitates transmission of the electrical impulse from the nerve terminal to the motor end plate of the muscle. This is achieved by the transmission of acetylcholine molecules across the 60-100 nanometre synaptic cleft. Acetylcholine binds to receptors on the motor end plate, causing depolarisation and contraction of the muscle fibre. After acetylcholine dissociates from the receptor, it is degraded by acetylcholinesterase into acetate and choline, which is reabsorbed into the nerve terminal for recycling into further acetylcholine [1].

7.3.3 COMMONLY USED NEUROMUSCULAR BLOCKING AGENTS

Features of commonly used neuromuscular blocking agents, when used in children, are shown in Table 2. It can be seen that suxamethonium is the only depolarising neuromuscular blocking agent in clinical use. A useful review of the pharmacodynamics of neuromuscular blocking agents in children can be found in the paper by Martin and colleagues [2].

With regard to neuromuscular blocking agents, there are significant pharmacokinetic differences between infants, older children and adults. There are several reasons for this variation, including differences in pseudocholinesterase activity, variation in the proportion of muscle mass, changes in the extracellular fluid volume, the relative number of acetylcholine receptors and the disposition of muscle fibre types. Plasma pseudocholinesterase levels in young infants are about 50% that of older children and adults until the age of 12 months. The muscle mass of an infant is 20-25% of body weight, compared to 40-50% in an adult. Extracellular fluid volume is 45% of body weight in the newborn, reducing to 30% at two months of age, 20% at five years of age and 16% in adults. Children have a larger muscle mass:fat ratio than infants and adults and, as a result, have relatively more acetylcholine receptors. With greater numbers of receptors, the plasma concentration of a neuromuscular blocking agent required to maintain paralysis also increases. There are also differences in the proportion and distribution of types of muscle fibres. The neonatal diaphragm has fewer type I (slow-twitch, high-oxidative) fibres which are more sensitive to neuromuscular blocking agents than in the older child or adult. The overall result of these differences is that the weight-indexed dosage requirements of neuromuscular blocking agents are higher in children than in infants and adults.

Table 2

Characteristics of commonly used neuromuscular blocking agents

Drug	Initial IV dose (mg/kg)			Onset time (seconds)	IV infusion dose (mcg/kg/min)			Total clearance (ml/kg/min)		
	Infant	Child	Adult		Infant	Child	Adult	Infant	Child	Adult
suxamethonium	3	2	1	35	-	-	-	-	-	-
atracurium	0.3	0.5	0.4	60-180	10-20	10-20	5-9	7.9	6.8	5.3
vecuronium	0.1	0.15	0.1	60-180	1.0-1.5	1.5-2.5	1.5-2.0	5.6	5.9	5.2
rocuronium	0.5	0.8	0.6	50-90	-	-	10	-	5.8	3.7
pancuronium	0.1	0.15	0.15	120-240	0.4-0.6	0.5-1.0	0.4-0.6	-	-	1.9
mivacurium	0.2	0.2	0.2	60-120	10	16	7	-	-	63-100

Suxamethonium

This agent produces rapid, profound and short-lived neuromuscular blockade. Its use is restricted to emergency endotracheal intubation because of the many complications associated with its use. The structural similarity between suxamethonium and acetylcholine is responsible for many of the more serious of these adverse effects, including bradycardias, tachycardias and arrhythmias. The depolarisation of muscles following administration can result in painful muscle fasciculation, myalgia and myoglobinuria. Hyperkalaemia can be a clinically important problem associated with the administration of suxamethonium. In normal circumstances, an increase in serum potassium of 0.5 mmol/l can be expected following the administration of suxamethonium. In certain circumstances, however, this increase can be much greater. A large population of extra-synaptic, immature-type acetylcholine receptors develops after denervation of mammalian muscle, after prolonged immobility and following the administration of neuromuscular blocking agents. In these circumstances, a standard dose of suxamethonium can result in large increases in the serum potassium concentration, which may result in cardiac arrhythmias or cardiac arrest.

Administration of suxamethonium can result in the generation of malignant hyperthermia in susceptible individuals, and will produce prolonged neuromuscular blockade in those with plasma cholinesterase deficiency.

Atracurium

Atracurium is a bisquaternary tetrahydropapaverum derivative and three isomers predominate in the commercial preparation; trans-trans, cis-trans and cis-cis. Atracurium has a unique metabolism, which makes it particularly useful in PICU. It is broken down by two purely chemical processes; Hofman degradation and non-specific ester hydrolysis via plasma cholinesterase. Atracurium does not accumulate, and its actions are not prolonged, in renal or hepatic failure. The development of tolerance is a clinically important feature associated with prolonged administration of atracurium, and there may be cross-resistance to other non-depolarising neuromuscular blocking agents.

Vecuronium

Vecuronium is a monotertiary, monoquaternary derivative of pancuronium, which is notable for its lack of adverse effects even, at high doses. Some accumulation of active metabolites may occur in renal failure.

Rocuronium

Rocuronium has a relatively rapid onset of action and, as such, represents the best non-depolarising neuromuscular blocking agent available for the facilitation of endotracheal intubation. It should be remembered, however, that its long duration of action might cause significant problems in the setting of a difficult airway where rapid endotracheal intubation is impossible.

Pancuronium

Although pancuronium is an older drug, it continues to be widely used in clinical practice because it is an effective and inexpensive product with a known side-effect profile. Pancuronium causes tachycardia, which can lead to an increase in myocardial oxygen demand.

Mivacurium

Mivacurium is a benzyllisoquinolinium derivative of atracurium, which is hydrolysed by plasma cholinesterase at 88% of the rate of suxamethonium. Consequently, it has the shortest duration of action of any non-depolarising neuromuscular blocking agent.

7.3.4 TOXICITY AND COMPLICATIONS OF NEUROMUSCULAR BLOCKADE

Many neuromuscular blocking agents can cause ganglion and vagal blockade and cause the liberation of considerable amounts of histamine. These actions can produce adverse cardiovascular effects, including hypotension and bradycardia or tachycardia. A further common problem with the administration of neuromuscular blocking agents is prolonged muscle weakness after their discontinuation. In addition the prolonged physical immobility caused by these agents may result in pressure sores, muscle atrophy, pulmonary atelectasis and associated pneumonia, joint contractures and corneal damage.

The incidence of these complications can be reduced by restricting the use of neuromuscular blocking agents to circumstances where their benefits clearly outweigh the risks associated with their administration. If these agents are to be used, then their use should be kept to a minimum; they should either be given by intermittent dosing schedules or, when given by continuous infusion, these should be routinely discontinued to allow for recovery of neuromuscular function and assessment of the underlying degree of sedation. Monitoring of the depth of neuromuscular blockade using train-of-four transcutaneous stimulation of the ulnar nerve, can allow for the titration of doses of neuromuscular blocking agents when given by continuous infusion and allow for the minimisation of the total doses received.

REFERENCES

1. Playfor, S. D., *Paed. Perinat. Drug Ther.*, 5, 35 (2002).
2. Martin, L. D., Bratton, S. L., O'Rourke, P. P., *Crit. Care Med.*, 27, 1358 (1999).

7.4 Sedation

Stephen D. Playfor

Paediatric Intensive Care Unit, Royal Manchester Children's Hospital, Hospital Road, Pendlebury, Manchester M27 4HA, UK

7.4.1 THE AIMS OF SEDATION

The aim of sedation on the paediatric intensive care unit (PICU) is to care for both the physical and the psychological comfort of mechanically ventilated, critically ill children. The specific goals of sedation are shown in Table 1.

It must be emphasised that the administration of analgesic and sedative drugs is no substitute for the sympathetic handling of a patient and it is important to identify correctable causes of distress. Attention to simple factors may help provide comfort and non-pharmacological measures including massage, music, noise reduction, temporal orientation with natural lighting and visible clocks, may decrease the requirement for sedative agents. Promotion of a normal sleeping pattern is also important, with increasing evidence available demonstrating that even modest periods of sleep deprivation can have significant negative physiological effects.

Table 1

The goals of sedation

reduced anxiety and distress in the child
facilitation of mechanical ventilation
tolerance of therapeutic and diagnostic procedures
reduced metabolic rate and oxygen demand
enhanced analgesia
enhanced sleep pattern
reduced patient recall

Admission to PICU is a traumatic event for individual children and their families, and there is increasing evidence of a high incidence of psychological morbidity amongst PICU survivors. It is unknown to what degree effective sedation and reduced patient recall may reduce the incidence of these problems.

7.4.2   COMMONLY USED SEDATIVE AND ANALGESIC AGENTS

Surveys of PICU practice have shown that the most commonly used sedation regimes involve the administration of a sedative agent; usually a benzodiazepine such as midazolam, and an analgesic agent; usually an opiate such as morphine or fentanyl [3]. Enteral sedative agents, such as chloral hydrate, triclofos sodium and antihistamines are frequently introduced at an early stage and have been demonstrated to provide satisfactory levels of sedation. Dosing ranges and descriptions of commonly used agents are shown in Table 2.

Midazolam

Midazolam is a water-soluble benzodiazepine that produces effective sedation and reduces the recall of critically ill patients. Tolerance and withdrawal phenomena are the most frequently encountered problems with midazolam. The incidence of adverse events following midazolam discontinuation in critically ill children is quoted as between 17% and 30% and usually occurs within a few hours of stopping the drug. There may be central nervous system symptoms such as seizures, paradoxical agitation, hallucinations and psychotic reactions or somatic manifestations such as tachycardia, vomiting and fever.

In critically ill children, the total clearance of midazolam shows enormous inter-individual variation, 1.7-52 ml/kg/min, compared to 4.0-8.8 ml/kg/min in critically ill adults. The volume of distribution of midazolam in critically ill children is 0.2-3.5 L/kg compared to 1.6-4.8 L/kg in critically ill adults. The elimination half-life of midazolam in critically ill children is 0.23-10.9 h compared to 3.8-7.7 h in critically ill adults. The elimination half-life of midazolam has been shown to be prolonged in the critically ill compared to healthy volunteers, whilst clearance is unaffected. This effect is thought to be due to a marked increase in the volume of distribution, which has been shown to be three times higher in the critically ill than that in healthy volunteers of similar body weights. The cause of this increase is thought to be due to a combination of factors, including an expanded extracellular fluid volume and changes in the characteristics of binding proteins. The marked interpatient variability in pharmacokinetics accounts for the wide range of dosing requirements for midazolam when used for sedation in the PICU.

Table 2

Characteristics of commonly used sedative and analgesic agents

Drug	Dose	Advantages	Disadvantages
midazolam	0.1 mg/kg load IV 2-10 mcg/kg/min IV	effective and familiar; reduced patient recall	tolerance; withdrawal; hypotension
morphine	0.1 mg/kg load IV 10-60 mcg/kg/h IV	powerful analgesia; antitussive action	caution in renal and hepatic dysfunction; tolerance; withdrawal
fentanyl	5-10 mcg/kg load IV 5-10 mcg/kg/h IV	less histamine release; rapid onset	reduce dose in hepatic impairment
clonidine	0.1-2 mcg/kg/h IV 3-5 mcg/kg/day NG (4 divided doses)	anxiolysis; analgesia	hypotension
ketamine	1-2 mg/kg load IV 5-20 mcg/kg/min IV	analgesia; cardiovascular stability; bronchodilatation	possible catecholamine depletion
chloral hydrate	25-100 mg/kg/dose 6 hrly NG max. 2 g per dose	enteral route; rapid onset	gastric irritation; paradoxical excitement
trimeprazine	2-4 mg/kg/dose 6 hrly NG max. 50 per dose	enteral route	avoid in renal and hepatic failure

Clonidine

Clonidine is an α_2-adrenoreceptor agonist, increasingly used in the PICU, which produces sedation, anxiolysis and analgesia. Dexmedetomidine is an α_2-adrenoreceptor agonist with seven times greater specificity to the α_2-adrenoreceptor than clonidine, and a significantly shorter half-life. Its use has been described in small series of children in North America. There also remains the prospect of developing agonists specific for subtypes of the α_2-adrenoreceptor, offering reversible sedation without adverse cardiovascular effects.

Remifentanil

Remifentanil is a synthetic opioid with a similar potency to fentanyl and with cardiorespiratory effects similar to other opioids. It is broken down by plasma esterases and has a very short half-life, which is unaffected by the duration of its administration, offering the prospect of very rapid wakening in all age groups. Remifentanil is an expensive agent and prolonged use is characterised by the rapid development of tolerance, which further adds to the cost. There has been only limited success in using remifentanil as a single agent for the prolonged sedation of mechanically ventilated children in PICU, but it may prove effective as an agent for procedural sedation in this environment.

7.4.3   TOXICITY AND COMPLICATIONS OF SEDATIVE AGENTS

There are many potential complications associated with the administration of sedative agents and an 'ideal' sedative agent does not exist. The most common complications involve the development of tolerance and withdrawal phenomena, and of cardiovascular depression. Renal dysfunction, immunosuppression and hepatotoxicity may all occur and it is important to remember that the handling of sedative agents is complex in the critically ill, with marked variability between individuals. There is also recent data suggesting that the administration of sedative agents is associated with prolonged periods of mechanical ventilation and with prolonged admission in critical care units.

There are specific problems associated with some sedative agents and these often become apparent when the use of agents becomes commonplace in critical care without a full assessment of their safety profile in this patient population. Etomidate was widely used for the induction of anaesthesia during the 1970s and became a commonly used agent for sedation on the intensive care unit. In 1984, however, Watt and Ledingham demonstrated that the introduction of etomidate in this setting was associated with an increase in mortality. It was later demonstrated that etomidate causes profound suppression of the adrenocortical axis and that this effect was probably responsible for the observed increase in mortality.

Propofol (2,6-diisopropylphenol) is registered in many countries for the induction and maintenance of anaesthesia in adults and children over the age of three years and was for many years used for prolonged sedation in PICU. There were early reports of adverse neurological events following the use of propofol in children and in 1992 the deaths of five children were reported as a result of increasing metabolic acidosis, bradycardia and progressive myocardial failure. In March 2001, the results of an unpublished clinical trial were circulated, where

327 PICU patients had been randomised to receive 2% propofol, 1% propofol, or standard sedative agents. During the trial and up to 28 days of follow-up, the mortality rates were 11% in the group receiving 2% propofol, 8% in the group receiving 1% propofol and 4% in the group receiving standard sedative agents. Propofol is therefore contraindicated for prolonged sedation in the PICU [1]. Recent data has suggested that the so-called propofol infusion syndrome may result from specific disruption of fatty acid oxidation caused by impaired entry of long-chain acylcarnitine esters into the mitrochondria and failure of the mitochondrial respiratory chain at complex II [2].

7.4.4   ASSESSMENT OF SEDATION

The level of sedation may be assessed at the bedside using scoring systems or by "measuring" sedation using neurophysiological techniques. Scoring systems for sedation assess either the arousability of the patient by documenting their responsiveness to noxious stimulation, or score physiological variables such as heart rate and respiratory pattern.

First published almost 30 years ago, but still widely used, is the arousability scale derived by Ramsay and colleagues [4]. This scale involves stimulating the patient with glabellar taps and auditory stimulation. Following on from the work of Ramsay, many other simple scales of arousability have been constructed and these are widely used in clinical practice.

The COMFORT scale is an observational scoring system of eight behavioural and physiological variables, which can be carried out without disturbing the patient [5]. Mean arterial blood pressure, heart rate, muscle tone, facial tension, alertness, calmness/agitation, respiratory behaviour and physical movement are scored after a two minute period of observation. A major drawback of the COMFORT scale is its complexity, and many groups have worked to produce simpler scoring systems, usually by excluding the cardiovascular variables.

Assessment of the electroencephalogram (EEG) offers the potential for continuous, non-invasive measurement of brain function. The bispectral index is derived from the EEG, using a sophisticated algorithm, that uses the advanced signal processing of bispectral and power spectral variables in a multivariate analysis to produce a bispectral index number. The bispectral index monitor uses bispectral indexing to calculate a processed multivariate parameter on a scale from 0 (no brain activity) to 100 (fully awake) and has proved useful in monitoring the depths of anesthesia and sedation in the intensive care unit. Poor skin contact, muscle activity or rigidity, head and body motion, sustained eye movements, improper sensor placement or skin preparation, and unusual or excessive interference may cause potential artefacts. Some drugs also affect the bispectral index number; for any given level of sedation, the administration of ketamine causes an increase in the bispectral index number, whilst the administration of nitrous oxide causes a decrease in the bispectral index number.

Whilst there is insufficient evidence to support its routine use, the bispectral index monitor may be useful in monitoring the level of sedation in patients receiving neuromuscular blocking agents.

REFERENCES

1. MCA/CSM, *Current problems in pharmacovigilance*, August, 27, 10 (2001).
2. Wolf, A., Weir, P., Segar, P., et al., *Lancet*, *357*, 606 (2001).
3. Playfor, S. D., Thomas, D. A., Choonara, I., *Paediatr. Anaesth.*, *13*, 147 (2003).
4. Ramsay, M. A. E., Savage, T. M., Simpson, B. R. J., et al., *Br. Med. J.*, *2*, 656 (1974).
5. Ambuel, B., Hamlett, K. W., Marx, C. M., et al., *J. Pediatr. Psychol.*, *17*, 95 (1992).

7.5 Epilepsy and antiepileptic drugs

Richard Appleton

The Roald Dahl EEG Unit, Department of Neurology, Alder Mey Children's Hospital, Liverpool, L12 2AP, UK

7.5.1 INTRODUCTION

Epilepsy is the most common treatable neurological condition of childhood. The term 'epilepsies', rather than 'epilepsy'. better defines and characterises the heterogeneous nature of the condition, particularly in childhood. Although there is considerable knowledge and understanding of the epilepsies–which is continuing to expand–there remains much that is both speculative and empirical. Often, this knowledge is based on experience rather than science, including in the areas of the diagnosis, classification and drug management of the different epilepsy syndromes [1].

There are no precise data on the prevalence of epilepsy in children. The estimate of 0.7 to 0.8% of all school children (aged 5-17 years) is often quoted, and is similar to adult data [2]. The lifetime cumulative incidence of epilepsy, derived from a large population based study, is approximately 3%. The discrepancy between lifetime cumulative incidence and prevalence reflects the transient nature of the condition in many patients, particularly in children. It is generally considered that approximately one third of patients who develop epilepsy in childhood will remit spontaneously by puberty or early adult life; however, this generalisation is of little practical use for any individual because the chance of remission depends primarily on the type of epilepsy (epilepsy syndrome) and underlying cause.

The diagnosis of epilepsy must always be considered at four levels:

- recognition of epileptic seizures;
- classification of the seizure type or types;
- identification of the epilepsy syndrome;
- determination of aetiology.

Recognising epileptic seizures is obviously essential in making a diagnosis of epilepsy and differentiating it from other paroxysmal, but non-epileptic, conditions. In most cases it is possible to classify the seizure type. For children and adolescents, it is then important to try and identify a specific syndrome. Finally, a cause of the epilepsy should always be considered and, when appropriate, actively looked for.

7.5.2 RECOGNITION OF EPILEPTIC SEIZURES

The recognition and diagnosis of epileptic seizures is almost entirely dependent on the history. The examination will often be normal, and the results of any investigations can only be interpreted with reference to the history. A detailed description is required of events occurring before, during and after a suspected epileptic seizure. The accurate account of any eyewitness is essential and will be the only history available when young children present with seizures. Video-recordings of suspected seizures may be very helpful and should be encouraged whenever there is any continuing doubt or confusion from the history. If the diagnosis remains uncertain, then it is appropriate to await further episodes, because a delay in making a diagnosis of epilepsy is, in most children, unlikely to be harmful.

7.5.3 EPILEPTIC SEIZURES

Seizures are classified into either generalised or focal (also called partial) seizures. The whole brain (or at least the whole of the cerebral cortex) is involved in generalised seizures, whereas focal seizures only affect part of the brain, and often, only one part of one lobe of the brain. Generalised seizures are broadly classified into 'absence', 'myoclonic', 'atonic', 'tonic', 'clonic' and 'tonic-clonic'. Focal or partial seizures are classified as 'simple', in which consciousness is retained, or 'complex', in which consciousness is impaired or lost; it is possible if not likely, that this division of 'simple' and 'complex' may be dropped from future classifications of the epilepsies. Simple partial seizures with sensory, autonomic or psychic symptoms may be easily overlooked in younger children unable to describe such symptoms. Focal or partial seizures may become secondarily generalised, resulting in a tonic-clonic convulsion. The symptoms of a simple partial seizure, prior to secondary generalisation, constitute the epileptic aura. A recent diagnostic scheme proposal has been submitted to the International League Against Epilepsy (ILAE). The scheme encompasses five levels or axes: seizure description, seizure type, epilepsy syndrome, aetiology and associated physical and/or learning impairments [3].

7.5.4   EPILEPSIES AND EPILEPSY SYNDROMES

The epileptic seizure type is one criterion used to define epileptic syndromes. Epileptic syndromes are determined by:

- seizure type(s);
- age of onset;
- EEG findings (interictal and ictal);
- associated features, such as neurological findings, family history.

Epileptic syndromes are important in the management of epilepsy in terms of:

- predicting prognosis (seizure-control and seizure-remission);
- selecting anti-epileptic treatment;
- defining the likelihood of identifying an underlying aetiology.

The classification of epilepsies and epileptic syndromes is divided, according to the anatomical origin of seizures, into those where the origin of the seizures is focal or partial and those that are generalised. A further subdivision is made aetiologically into symptomatic, in which the cause of the epilepsy is known; cryptogenic, in which there is a likely, but unidentified, cause; and idiopathic, in which there is no underlying cause apart from perhaps a genetic predisposition. International classifications of epilepsy are unlikely to be entirely satisfactory (due to our limited understanding of the basic pathophysiology and neuropharmacology of epilepsy), and reservations have been expressed, particularly with respect to seizure classification in infants. However, these classifications are of practical value in permitting a common dialogue on the diagnosis and treatment of epilepsy and in facilitating collaborative research.

7.5.5   AETIOLOGY OF EPILEPSY

Although a cause of epilepsy must always be considered in every child with epilepsy, in only approximately 30 to 35% of school age children will a cause be found. In children with epilepsy starting within the first 12 months of life, a cause will be found in approximately 70% – and commonly due to cerebral dysgenesis or sequelae of intracranial haemorrhage, hypoxic-ischaemic encephalopathy and congenital/neonatal meningo-encephalitis. Inevitably, a number of the idiopathic epilepsies and epilepsy syndromes will ultimately be found to have a genetic basis.

7.5.6   ASSOCIATED IMPAIRMENTS

The fifth level or axis of the diagnostic classification proposed by the ILAE is to include any associated physical or cognitive (educational) impairment [3]. This is justified because many epilepsies and epilepsy syndromes are frequently accompanied by such impairments (e.g., Lennox-Gastaut syndrome, severe myoclonic epilepsy in infancy and mesial temporal lobe epilepsy). The association of epilepsy with physical or cognitive impairments is often termed, 'epilepsy plus'.

The often (and usually severe) physical, learning and behavioural difficulties that accompany these epilepsy syndromes typically cause more problems than the epilepsy, not just for the individual, but for their family and community, as well as placing excessive demands on social, educational and public health resources.

7.5.7   PROGNOSIS OF EPILEPSY

Whether seizures will respond to treatment and whether the epilepsy will spontaneously remit is determined by the specific epilepsy syndrome, underlying aetiology and, almost certainly (and as yet unidentified), genetic factors [4]. There is no evidence that the use of antiepileptic drugs influences the natural history of the vast majority of the epilepsies and epilepsy syndrome; the drugs simply suppress the seizures, whilst the epilepsy is active within an individual's brain. It is possible that, in some of the rare, early onset, infantile epileptic encephalopathies (specifically West syndrome, characterised by infantile spasms and hypsarrhythmia on the EEG [5] and the Landau-Kleffner syndrome), early and aggressive treatment may improve developmental and cognitive outcome. Further data, however, are required to confirm or refute these limited findings.

7.5.8   DRUG TREATMENT OF EPILEPSY

There are a number of decisions that must be taken regarding the use of antiepileptic drugs:

- when to start a drug?
- which drug and in what dose?
- when to change the drug?
- when to add a second drug (and which one)?

- when to seek a specialist opinion?
- when to stop the drug?
- when to measure blood levels of the drug?
- when to consider epilepsy surgery?

It is not within the scope of this Chapter, however, to discuss all of these decisions.

7.5.9   WHEN TO START A DRUG?

Most clinicians would not recommend starting treatment after a single gene-ralised tonic-clonic seizure, but would after a cluster of seizures or two seizures within a period of one or two months. Similarly, a child with severe physical and learning difficulties, with infrequent myoclonic or brief focal seizures may not require an anticonvulsant, in contrast to a child attending a normal school who experiences frequent generalised tonic-clonic seizures on waking. There are two main reasons why clinicians – and sometimes parents – are keen to start medica-tion after one or only two seizures. Firstly, there has been the theoretical concern that one seizure may lead to a second, a second to a third and, eventually, to a state of chronic epilepsy that may then be more difficult to treat. This process is termed 'kindling', where one seizure – which may be clinically or only elec-troencephalographically evident – 'begets another seizure' and so on; the evi-dence for this is primarily derived from animal (predominantly rodent) data and has not been convincingly demonstrated in humans. Nevertheless, some clini-cians still believe that early treatment, after just one or two seizures, may prevent the risk of the development of chronic and drug-resistant epilepsy. Recent data [6] would suggest that this is most unlikely, providing the number of tonic-clonic seizures is 10 or less [7]. Secondly, is the concern that there may be an increased incidence and risk of injuries with further seizures, and therefore early treatment may reduce this risk. Again, recent evidence has suggested that this is unlikely [8], and in fact, physical injuries are probably more likely to occur in children already diagnosed with epilepsy and receiving antiepileptic medication [9].

Once drug therapy is started, the objective is to use this as monotherapy and to achieve (complete) seizure control without unacceptable side effects and using the most appropriate formulation that can be taken by the child [10].

7.5.10 WHICH DRUG AND IN WHAT DOSE?

The specific epilepsy syndrome or (if no specific syndrome can be determined) seizure type and safety profile of the drug determine the choice of antiepileptic drug. The preparation of drug available is an additional factor when deciding the choice of drug – particularly in infants and young children. In the UK, cost is not currently regarded as an important factor in the choice of the antiepileptic drug, despite the fact that there may be at least a six or even 10-fold difference in cost between an 'older' (e.g., carbamazepine or sodium valproate) and 'newer' drug (e.g., lamotrigine or topiramate). Whichever drug is chosen is introduced gradually to avoid any dose-related side effects, and increased slowly to its target maintenance dose based on the child's body weight and recommended guidelines. The dose of this drug should be increased to the maximally tolerated level before either adding a second drug (if the first drug has had a partial effect) or substituting another drug (if the first drug was completely ineffective).

The currently recommended first-line drugs in treating the majority of childhood epilepsies are sodium valproate (VPA) for generalised epilepsies and syndromes, and carbamazepine (CBZ) for generalised and partial (focal) seizures/epilepsy syndromes. A randomised clinical trial (RCT) showed that VPA and CBZ were equally effective in both primary generalised tonic-clonic seizures and focal seizures, with or without secondary generalisation [11]. However, despite this single study, it is the experience of many (including the author) that VPA is not particularly effective in treating focal (partial) seizures.

Although ethosuximide may be effective for typical absences, it does not suppress tonic-clonic seizures, which may develop in 10 to 20% of children with childhood or juvenile-onset typical absence epilepsy. Ethosuximide may occasionally be helpful in treating myoclonic seizures. Carbamazepine exacerbates myoclonic and typical absence seizures and should therefore be avoided in juvenile myoclonic epilepsy and childhood-onset absence epilepsy.

The only other syndrome/seizure type for which VPA and CBZ are not drugs of first choice is West's syndrome, which is characterised by infantile spasms. In Europe, vigabatrin is usually the preferred drug, whereas in the USA, adrenocorticotrophic hormone (ACTH) is generally regarded as the drug of choice [12]. In Japan, the drug of choice is pyridoxine (vitamin B6), for which it is reported that up to 10 to 15% of infants will respond. This does not imply that these infants have pyridoxine dependency, but simply that pyridoxine may occasionally be effective in treating some cases of infantile spasms. The mechanism of action of ACTH (in reducing spasms and normalising the EEG in West syndrome) is unclear. It has to be given by intramuscular injection (which is painful), and is frequently associated with severe, and potentially fatal, side effects. Prednisolone or hydrocortisone, commonly used alternative, oral steroids to ACTH/tetracosactrin, generally have fewer and less serious side effects. Vigabatrin is likely to suppress spasms in approximately 60% of patients, and prednisolone or ACTH in 65 to 70% of patients, depending on the cause. Vigabatrin is particularly effective in treating infantile spasms caused by tuberous sclerosis. If spasms show no reduction after 10 days of the maximum dose of vigabatrin (120-150 mg/kg/day), then it is very unlikely that this drug will be effective and it should be withdrawn and replaced with another drug (e.g., prednisolone, pyridoxine, nitrazepam, levetiracetam or topiramate), depending on the specific clinical situation; (these specific drugs reflect the author's preference based on clinical experience.). A systematic review of the treatment of infantile spasms gave inconclusive results [13], and the Practice Parameter, published jointly by the American Academie of Neurology and Child Neurology Society, has implied no clear preference for ACTH or vigabatrin in treating infantile spasms [12].

Phenytoin and phenobarbitone must no longer be used as first-line maintenance drugs because of side effects, particularly on behaviour, cognitive performance and bone mineralisation and, with phenytoin, the cosmetic side effects of gingival hyperplasia and the developmental of facial and limb hair. These drugs should be considered for oral therapy only when other drugs have failed, and where seizure control is the over-riding–if not the only–priority.

7.5.11 WHEN TO CHANGE A DRUG OR ADD A SECOND DRUG?

If unacceptable side effects develop, or if control has been sub-optimal with the first drug, then the child will require either a different drug or an additional drug ('polytherapy'). Where the first drug has been ineffective, it would be appropriate to replace this drug with an alternative; where the first drug has been partly effective, it would seem appropriate to add a second drug and to consider withdrawing the first drug if seizure-freedom is subsequently achieved, in order to maintain monotherapy. The choice of the second drug is based on the same criteria as for the first drug, namely, seizure type or syndrome and safety profile. A single drug (monotherapy) will achieve total seizure control in only 65 to 70% of children. Two drugs in combination will result in further significant (even complete) control in an additional 5 to10% of children. Three drugs rarely (if ever) result in any additional control, and frequently cause more side effects – and should therefore be avoided wherever possible.

7.5.12 MECHANISM OF ACTION OF ANTIEPILEPTIC DRUGS

Most of the proposed mechanisms of action of the antiepileptic drugs are either speculative or unknown. Many of the older drugs were, almost accidentally, found to have antiepileptic or anticonvulsant actions, including phenobarbitone, phenytoin, carbamazepine and sodium valproate. Table 1 summarises the currently available oral antiepileptic drugs and their proposed or believed

Table 1

Mechanisms of action and indications for antiepileptic drugs

Drug	Mechanism	Indications
acetazolamide	inhibits brain carbonic anhydrase activity	focal (often in conjunction with carbamazepine)
benzodiazepines (clobazam, clonazepam, diazepam, nitrazepam)	allosteric enhancement of $GABA_A$-receptor mediated chloride channels	generalised (myoclonic and absence); focal (clobazam); infantile spasms (nitrazepam)
carbamazepine (and oxcarbazepine)	limits or inhibits repetitive firing of voltage-gated sodium channels	focal and primary or secondary generalised tonic-clonic (but exacerbates myoclonic and absence seizures)
ethosuximide	inhibits low-threshold T-type voltage-gated calcium channels in the thalamus; possible enhancement of non-GABA mediation neuronal inhibition	generalised (absence and occasionally, myoclonic); no effect on tonic-clonic and focal seizures
felbamate (no longer available in the UK and most of Europe)	uncertain; multiple mechanisms possibly including inhibition of voltage-gated sodium channels; indirect inhibition of calcium channels through NMDA-receptor antagonism and possible GABA potentiation	generalised (all types) and focal

Table 1 (*continued*)

Mechanisms of action and indications for antiepileptic drugs

Drug	Mechanism of action	Indications
gabapentin	binds to a novel calcium channel receptor	focal, primary and secondarily generalised tonic-clonic; no effect on absence and myoclonic seizures
lamotrigine	inhibits voltage-gated sodium channels; reduces release of excitatory amino-acids (specifically glutamate)	generalised (all types; not very effective for myoclonic seizures) and focal
levetiracetam	unknown; antiepileptogenic rather than anticonvulsant effect; no obvious effect on the GABA (inhibitory system), benzodiazepine or glutamate/aspartate (excitatory system) amino acid receptors	generalised (all types) and focal (limited paediatric data)
phenobarbitone (and primidone)	enhances GABA-mediated inhibition	generalised (tonic-clonic, tonic, myoclonic, absence); focal rarely
phenytoin	inhibits sustained repetitive firing of voltage-gated sodium channels	generalised (tonic-clonic; clonic); focal; (exacerbates myoclonic seizures)
sodium valproate	uncertain; possible inhibition of voltage-gated sodium channels; possible enhancement of GABA-mediated inhibition by inhibiting GABA transaminase; as yet unidentified novel action (? on T-type calcium channels)	generalised (absence, myoclonic, tonic-clonic, photosensitive); focal (rarely)
striripentol	unknown	myoclonic; absence (limited data)

Table 1 (*continued*)

Mechanisms of action and indications for antiepileptic drugs

sulthiame	uncertain; possible carbonic anhydrase inhibitor	focal; myoclonic seizures (for both, often as adjunctive therapy with other drugs)
tiagabine	inhibits the neuronal and pre-synaptic glial uptake of GABA after its release from post-synaptic GABA receptors (the drug therefore enhances GABA-mediated inhibition)	focal and secondary generalised tonic-clonic
topiramate	multiple mechanisms; blocks voltage-activated sodium channels; stimulates $GABA_A$-receptor mediated chloride currents; blocks glutamate receptors; weak carbonic anhydrase inhibitor (this latter mechanism is unlikely to exert any significant anticonvulsant/antiepileptic effect)	generalised (tonic-clonic; tonic, atonic); focal
vigabatrin	enzyme-activated suicidal inhibitor of GABA-aminotransaminase therefore increasing GABA-mediated inhibition	infantile spasms (one of the drugs of first choice); focal with or without secondary generalised tonic-clonic seizures
zonisamide (not yet licensed for use in the UK)	multiple; blocks voltage-dependent sodium channels; inhibits voltage-dependent T-type calcium channels; facilitates dopaminergic and serotoninergic transmission	generalised (tonic-clonic, myoclonic, absence); focal; *possibly* infantile spasms (limited paediatric data)

mechanisms of action and the main seizure type(s) for which they are used; some or all of these mechanisms may need to be revised, pending the results of further neuropharmacological and neurophysiological research [14, 15].

7.5.13 WHEN TO STOP THE DRUG

Most clinicians would consider withdrawing drug therapy once the patient has been seizure-free for two, or at most, three years, which is an arbitrary period. The risks and benefits of drug withdrawal and implications of seizure recurrence are different for children and teenagers/young adults and this must be discussed with both the child and parents [16]. There is no evidence that repeating an EEG prior to drug withdrawal is informative and therefore of any clinical benefit in deciding whether anticonvulsant therapy should be stopped. Anticonvulsant withdrawal should be undertaken at a convenient time in term of the child's education and social circumstances and each drug should be withdrawn gradually, over two or three months, to prevent withdrawal seizures. This may be a problem with the benzodiazepines, but possibly not phenobarbitone [17]; limited data suggests that the newer antiepileptic drugs can also be safely withdrawn over two to three months.

7.5.14 TOXICITY

Most antiepileptic drugs may be associated with adverse side effects that may be acute and idiosyncratic (allergic), dose-related or chronic, developing after many years of use. Most are mild and may be acceptable to the child and the family; rarely, the side effects may be more serious and even life-threatening [18]. Cognitive and behavioural side effects are frequently of most concern to parents of children with epilepsy and may be difficult to recognise in children who have 'epilepsy plus', and particularly those with learning difficulties. These children also usually have the more severe or malignant epilepsy syndromes that often require the use of two antiepileptic drugs to achieve 'acceptable' seizure control. In these children, a therapeutic compromise has to be reached (and accepted by all concerned), where the priority is to control the major and head-injuring types of seizures, whilst accepting more minor seizures and without producing excessive sedation or loss of function. Clearly, in many situations, this therapeutic compromise may be difficult to achieve and may be repeatedly influenced by the fluctuating and non-static natural history of epilepsy, particularly in the first decade of life.

The use of multiple antiepileptic drugs (polytherapy) increases both the risk and incidence of side effects and this again emphasises the point that no more than two antiepileptic drugs should be used simultaneously. Side effects are more commonly seen with the older drugs (phenobarbitone, phenytoin, carbamazepine, sodium valproate and the benzodiazepines), but may also be seen with the newer drugs. Adverse side effects have been linked with the premature deaths of both paediatric and adult patients with epilepsy [18]. Although side effects tend to be identified in clinical trials of new drugs (phase I-IV), some may not be recognised for some time and after years of routine clinical use. Felbamate and vigabatrin, two of the 'newer' generation of drugs illustrate this problem. Within months of the pivotal clinical trials of felbamate being completed (and published), it became clear that the drug caused severe, including fatal aplastic anaemia and hepatitis, resulting in the drug being withdrawn from the UK and many European countries. Approximately 10 years after vigabatrin was first prescribed, a characteristic visual field defect (symmetrical, bilateral constriction with relative temporal sparing) was reported, which appears to be specific to the use of this drug and may occur in up to 40% of adults, although in most cases the defect is asymptomatic and detected only on detailed visual field perimetry. The precise incidence of this defect in children is not known, but is thought to be lower, possibly 25%. Current evidence suggests that early visual field constriction may be seen after a minimum of six months' exposure to the drug, but far more typically, after at least two years of continued use. Long-term follow-up data will clarify whether the visual field deficit is likely to be permanent and irreversible.

Some children and paediatric populations are recognised to be at an increased risk of developing individual and usually serious side effects. Specific examples include:

- sodium valproate and hepatotoxicity: *children under three years of age with a severe epilepsy with multiple seizure types (including myoclonic seizures) with an onset under 12 months of age, global developmental delay and receiving at least one other antiepileptic drug (it is possible that the children in this high-risk group have a pre-existing metabolic defect of fatty acid oxidation);*
- lamotrigine and rash: *an idiosyncratic, but also dose-related, rash (including Stevens-Johnson syndrome) developed in 25 to 30% of patients in early clinical trials with the drug; a lower starting dose and more gradual dose increase has reduced the incidence to a consistent level of 3 to 5%;*
- phenytoin and pseudo-ataxia and pseudo-dementia: *the use of this drug in children with myoclonic seizures, and particularly with an evolving and progressive myoclonic epilepsy typically causes a pseudo-ataxia and pseudo-dementia that may confuse the clinical picture and may not be completely irreversible.*

Advances in molecular genetics and the application of this to clinical pharmacology (pharmacogenomics) may hopefully identify those patients who are at greatest risk of developing the more serious and potentially life-threatening side effects of antiepileptic drugs. However, it remains uncertain whether this could lead to the identification of a simple screening test to identify those patients who are at a high risk of developing any serious side effects before prescribing a specific drug. Table 2 outlines the more common side effects of the antiepileptic drugs. Many are well-recognised, whilst others are only now emerging, due to the relatively short periods they have been available.

Almost certainly, additional antiepileptic drugs will be developed in the future, although these are unlikely to match the dramatic appearance of the seven 'new' antiepileptic drugs that occurred between 1989 and 2001 [19,20]. It is also unlikely that any future drug will prove effective and free of any side effects for

Table 2

Toxicity of antiepileptic drugs

Drug	Side effects
acetazolamide	malaise and fatigue; paraesthesiae of limbs and face; metabolic acidosis; leucopenia
benzodiazepines	drowsiness and sedation; irritability; tolerance; excessive salivation (clonazepam and nitrazepam)
carbamazepine (and oxcarbazepine)	nausea, dizziness and diplopia; allergic rash; agranulocytosis; teratogenic
ethosuximide	nausea and diarrhoea; drowsiness; chronic headache; leucopenia
felbamate	nausea and anorexia; insomnia; severe and potentially fatal aplastic anaemia and hepatitis
gabapentin	behavioural changes (aggression); sedation (high dose)
lamotrigine	allergic and dose-related rash (particularly when used simultaneously with sodium valproate); tremor and headache (when used in conjunction with sodium valproate)
levetiracetam	drowsiness and dizziness; behavioural changes (aggression)
phenobarbitone (and primidone)	drowsiness and irritability; cognitive slowing; allergic rash; osteomalacia/osteopenia; tolerance; teratogenic
phenytoin	allergic rash; cosmetic effects (gingival hypertrophy and facial hair often developing after 3-4 months of use); mild cognitive slowing and involuntary movements (chorea, athetosis); osteomalcia/osteopenia; leucopenia and pancytopenia; teratogenic

Table 2 *(continued)*
Toxicity of antiepileptic drugs

Drug	Side effects
sodium valproate	drowsiness; increased appetite and weight gain; tremor; alopecia; menstrual irregularities; mild cognitive slowing and behavioural changes; hepatitis and pancreatitis; teratogenic (neural tube defects and 'fetal valproate syndrome')
stiripentol	drowsiness; behavioural changes (hyperactivity and aggression)
sulthiame	somnolence; respiratory changes (hyperpnoea and dyspnoea); paraesthesiae of limbs; psychosis
tiagabine	dizziness, somnolence and fatigue; complex partial status epilepticus in some patients
topiramate	sedation; anorexia (may be severe leading to weight loss); impaired concentration; impaired short-term memory and word-finding difficulties; behavioural changes (either aggression or withdrawal and depression); renal calculi; teratogenic
vigabatrin	sedation and hypotonia (high dose); behavioural changes (irritability); bilateral visual field constriction
zonisamide	somnolence, confusion and ataxia; anorexia; renal calculi

treating all seizure types in all patients, due to the marked heterogeneity of the epilepsies and the effect of pharmacogenetics in determining both drug-responsiveness and drug-resistance.

7.5.15 HOLISTIC TREATMENT

Clearly, the management of epilepsy extends far beyond the prescription of antiepileptic medication. It is obviously important to correctly identify the seizure type and epilepsy syndrome and to prescribe the most appropriate antiepileptic drug to obtain optimal control of seizures without unacceptable side effects. However, for many patients and their families, social, educational and psychological features far outweigh the problem of controlling seizures. This always requires an inter-disciplinary team approach within a specialist epilepsy clinic that can bring together experience and advice from many sources, including nursing, psychology and social work. Finally, each family should be informed about the existence of all relevant and national voluntary organisations that are frequently able to provide invaluable support and an educative role, as well as contributing to epilepsy research.

7.5.16 STATUS EPILEPTICUS (CONVULSIVE AND NON-CONVULSIVE)

This is a major and important topic that cannot be dealt with thoroughly within this chapter [21]. The following medicines have all been shown to be effective in treating children with acute seizures: rectal diazepam, intravenous lorazepam and buccal midazolam. For children with intravenous access, lorazepam 100 microg/kg should be given intravenously; for those with no immediate intravenous access, the drug of first choice could be either rectal diazepam (500 microg/kg) or buccal midazolam (500 microg/kg). Lorazepam is considered to be more effective than diazepam and is associated with less respiratory depression [21-23]. Midazolam, administered by either the buccal or nasal route, has been shown to be effective in small prospective studies. Paraldehyde is a cheap and effective anticonvulsant for the child with an acute seizure and is best administered rectally. It can be administered intramuscularly, but there is a risk of sterile abscess formation.

7.5.17 CONCLUSION

The mini-explosion of numerous antiepileptic drugs within the past 15 years has unequivocally improved the management of many children with epilepsy. However, not surprisingly, this has resulted in a degree of therapeutic confusion and specifically in knowing which drug to use first, which drug to use next and what are the most (and least) effective – and potentially harmful – drug combinations. This greater choice and use of antiepileptic drugs may also result in an inappropriate delay in both the consideration and performance of potentially curative surgery.

The mechanisms underlying the actions of antiepileptic drugs are heterogeneous and complex. This is also true of the responsiveness (and resistance) that children show to the antiepileptic drugs and the development of serious side effects, issues that almost certainly reflect the influence of pharmacogenetic factors.

Despite the advent of the new antiepileptic drugs, there remain a significant number (at least 25%) of children with persistent and refractory seizures. The majority, but not all, of these children will have 'epilepsy plus'. Polytherapy is commonly undertaken in these children but rarely results in seizure-freedom and is frequently accompanied by adverse effects on learning, behaviour and day-to-day functioning. Consequently, the option of no medication is reasonable and appropriate in this population, although there is the theoretical risk of a marked deterioration in seizure control and convulsive (tonic-clonic) status epilepticus.

As with all management decisions and irrespective of the drug on intervention, this must be discussed openly and honestly with the child's family – and, where appropriate, the children themselves.

REFERENCES

1. Camfield, P., Camfield, C., *J. Child. Neurol., 18*, 272 (2003).
2. Appleton, R. E., Gibbs, J. *Epilepsy in Childhood and Adolescence*, 3rd Edition, Martin Dunitz, pp. 2-4, (2003).
3. Engel, J., *Epilepsia, 42*, 796 (2001).
4. Ramachandran, V., *Epilepsia, 44 (Suppl. 1)*, 33 (2003).
5. Eisermann, N. M., DeLaRaillere, A., Dellatolas, G., et al., *Epilepsy Res., 55*, 21 (2003).
6. Van Donselaar, C. A., Brouwer, O. F., Geerts, A. T., Arts, W. F., Stroink, A. H., Peters, A. C., *Br. Med. J., 314*, 410 (1997).
7. Camfield, C., Camfield, P., Gordon, K., Dooley, J., *Neurology, 46*, 41 (1996).
8. Appleton, R. E., *Epilepsia, 43*, 764 (2002).
9. Wirrell, E. C., Camfield, P. R., Camfield, C. S., Dooley, J. M., Gordon, K. E., *Arch. Neurol., 53*, 929 (1996).
10. Newton, R., *J. R. Soc. Med., 97*, 15 (2004).
11. Verity, C. M., Hosking, G., Easter, D. J., *Dev. Med. Child. Neurol., 37*, 97 (1995).
12. Mackay, M. T., Weiss, S. K., Adams-Weber, T. et al., *Neurology, 62*, 1668 (2004).
13. Hancock, E., Osborne, J., Milner, P., *Cochrane Database Syst. Rev., 3*, CD001770 (2003).
14. Rho, J. M., Sankar, R., *Epilepsia, 40*, 1471 (1999).
15. McAuley, J. W., Biederman, T. S., Smith, J. C., Moore, J. L., *Ann. Pharmacother., 36*, 119 (2002).
16. Appleton, R. E., *Seizure, 8*, 381 (1999).
17. Chadwick, D., *Brain, 122*, 441 (1999).
18. Clarkson, A., Choonara, I., *Arch. Dis. Childhood, 87*, 462 (2002).
19. Marson, A. G., Chadwick, D. W., *J. Neurol. Neurosurg. Psychiatry, 70*, 143 (2001).
20. Trevathan, E., *Epilepsia, 44 (Suppl. 7)*, 19 (2003).
21. Appleton, R., Choonara, I., Martland, T., Phillips, B., Scott, R., Whitehouse, W., *Arch. Dis. Child., 83*, 415 (2000).
22. Appleton, R., Sweeney, A., Choonara, I., Robson, J., Molyneux, E., *Dev. Med. Child. Neurol., 37*, 682 (1995).
23. Norris, E., Marzouk, O., Nunn, A. J., McIntyre, J., Choonara, I., *Dev. Med. Child. Neurol., 41*, 340 (1999).

7.6 Sedatives

William P. Whitehouse

School of Human Development, University of Nottingham, and Department of Paediatric Neurology, Queen's Medical Centre, Nottingham NG7 2UH, UK

7.6.1 INTRODUCTION

Sedation is a neurological process, a state of altered brain function. Patients benefit from sedation for procedures such as magnetic resonance imaging (MRI) of the brain or spinal cord, auditory evoked potentials (AEPs), elective lumbar puncture, needle muscle biopsy, botulinum toxin injections, and occasionally for electroencephalography (EEG).

What are sedatives?

Sedatives are not a *class* of drug, but rather various types of drugs that are *used* to induce sedation. This concept is fundamental to the safe, effective and tolerable prescribing, dispensing, administration, care and monitoring of patients.

For example, midazolam, a short-acting benzodiazepine, acting at the GABA-A receptor, can be given buccally (the buccal fossa is between the inside of the cheek and the teeth and gums when the jaw is closed), intranasally or rectally for trans-mucous membrane absorption; orally or by gastric tube for enteral absorption; intramuscularly (IM); intravenously (IV) by bolus or infusion; or subcutaneously (SC) by infusion [1]. The clinical setting and particularly the dose administered will determine if it is being used as a sedative.

In the appropriate dose and route, midazolam is a safe, effective and well-tolerated sedative (e.g. 100 microgram/kg bolus IV or IM, 200 microgram/kg buccal or intranasal, 500 microgram/kg rectal or oral). It is also used as an anxiolytic, in low dose, e.g. as pre-medication prior to general anaesthesia (e.g. 70 micrograms/kg IM). Large doses result in anaesthesia.

The state achieved cannot be wholly predicted by the dose administered, even when adjusted per kg body weight. Some patients are more sensitive and become sedated on midazolam 500 nanogram/kg/min IV or anaesthetised at 5 microgram/kg/min IV Other patients will appear fully conscious and might even be anxious after a typical "sedation" dose.

7.6.2   WHAT IS SEDATION?

There are a range of drug-induced and environmentally-cued states that need to be considered.

Awake

An awake patient is responsive to everyday stimuli in an age or developmentally appropriate way, encompassing a range of moods, states of arousal, calmness, anxiety or alertness. The borderline state between awake and asleep, *drowsiness*, is familiar to us from everyday experience and corresponds to sleep stage 1 [2]. Anxiolytic treatment can be given and the patient remains awake; however, all anxiolytic drugs can induce sedation in a sensitive patient, or at a higher dose. Other pathological borderline states between awake and *coma* have also been defined [3].

Sleep

Sleep is a natural, spontaneous state of reduced responsiveness re-occurring in regular cycles over time, in infants, children and adults. Specialised polysomnographic recordings can be used to operationally define the transition from wakefulness to sleep and the various stages of sleep, which usually occur in a particular sequence, cycle and pattern, i.e. *normal sleep architecture* [2].

Casual observation of a patient allows natural sleep to be identified: typically the eyes are closed, the limbs are still, breathing is deep and regular and there are intermittent brief changes in state, such as sighing, sniffing, and limb movements. These patterns are recognisable from everyday experience. If in doubt, then attempts at arousal, by verbal or non-painful tactile stimuli, would almost always be successful, even if only transiently.

It is recognised that, initially, it may be difficult to distinguish natural sleep from sedation and pathological impairment, such as coma (for the Glasgow Coma Scale (GCS) [4] and for the child's Glasgow Coma Scale (CGCS) [5]). However, even in deep natural sleep, unrestrained patients with normal airways and brain stem respiratory drive are in no danger and require no more supervision or monitoring than would be reasonable for someone of a similar age at home.

Natural sleep may be induced by suggestion, environmental cues (e.g. a warm, cosy, quiet room) and after a satisfying meal, sleep deprivation [6], or melatonin [7].

Patients who are asleep will be as safe during the procedure (e.g. EEG, MRI) as they would be at home, asleep in their own bed, i.e. no particular monitoring will be needed, unless they have a history of obstructive sleep apnoea. Also, they can be easily woken up afterwards. *Sleep is not sedation* but can be used as an alternative to sedation in settings without pain or other alerting stimuli, e.g. if patients are required to lie fairly still (e.g. MRI [8]) or refrain from removing equipment (e.g. EEG [9]). Melatonin is less effective for AEPs because of the arousing effect of the auditory stimulation [10]. Doses of melatonin from 2.5 to 10 mg have been extensively used in children and are effective and well tolerated [9,11,12].

Sleep can be encouraged by environmental cues, e.g. an MRI can be undertaken in an infant who falls asleep after a feed in a quiet, warm and dimly lit room. The noise of the MRI might or might not wake the child up. Sleep deprivation is a standard method for achieving sleep for EEG, but can be difficult in older children with impulsiveness, hyperactivity, learning disabilities or autism spectrum disorders. Sleep deprivation is not very well tolerated by families [13].

Sedation

Sedation comprises sleep-like states following the administration of medication. These have been defined as *conscious sedation*, *deep sedation* and *general anaesthesia* [14,15], and are adapted here to distinguish deep sedation from general anaesthesia on the basis of a self-maintained airway and airway clearing reflexes (cough) in the former and not in the latter.

Conscious sedation

Conscious sedation is a medically controlled state of depressed consciousness that

- permits appropriate response by the patient to physical stimulation or verbal command, corresponding to a GCS or CGCS of 10-14 [14,15]

and

- allows airway protective reflexes to be maintained, including retention of the patient's ability to maintain a patent airway independently and continuously, including an adequate cough reflex, an adequate gag reflex, and a sufficient degree of upper airway muscle tone to allow un-assisted adequacy of the upper airway.

Deep sedation

Deep sedation is a medically controlled state of depressed consciousness or unconsciousness from which the patient is not easily aroused [14]. It is accompanied by:

- the inability of the patient to respond purposefully to physical stimulation or verbal command, corresponding to a GCS or CGCS of 3-9

and

- intact airway protective reflexes, including retention of the patient's ability to maintain a patent airway independently and continuously.

There may be rapid loss of these, however, as sedation deepens, the patient drifting imperceptibly into anaesthesia.

General anaesthesia

General anaesthesia is a medically controlled state of unconsciousness accompanied by:

- the inability of the patient to respond purposefully to physical stimulation or verbal command, corresponding to a GCS or CGCS of 3-9

and

- loss of airway protective reflexes, including loss of the patient's ability to maintain a patent airway independently.

In general anaesthesia, one or more of these reflex mechanisms becomes inadequate. Loss of the cough reflex leads to increased danger of aspiration, as does loss of the gag reflex, producing an unsafe swallow. Pharyngeal hypotonia compromises upper airway patency leading to obstruction or hypoventilation that can be overcome by holding the jaw forward or by an airway.

7.6.3 WHAT IS SEDATION FOR?

Sedation may be used for one or more specific purposes. It may be essential for the patient to lie still, e.g. for neuroimaging, or for delicate invasive procedures e.g., lumbar puncture, or to keep a sterile surgical field, e.g. for a needle muscle biopsy in a young child. It may be needed to ensure the child is not frightened and as an adjunct to analgesia for invasive or painful procedures, e.g. venous cannulation, lumbar puncture, needle muscle biopsy. It should not be used as a substitute for analgesia. It may be used to assist the recording of biological signals, e.g.

AEPs, EEGs, although sleep is better for EEG in most cases. It may be used to suppress oro-pharyngeal reflexes, e.g. for upper gastrointestinal endoscopy, i.e. to induce general anaesthesia according to the proposed definition.

7.6.4 CHOICE OF DRUG(S)

It is important to be clear which specific aims apply in each case, so that an appropriate option can be selected. Sedation regimes have developed and spread over the years with little systematic evaluation. Individual physicians, departments and specialties tend to have their own traditional regimes, even for similar aims on the same ward. There seems little evidence base from which to chose some drugs over others.

Propofol, an alkyl phenol "intravenous anaesthetic agent", acting at the GABA-A receptor, is used safely in many Emergency Departments during short painful procedures, by specifically trained non-anaesthetists [16]. Typical doses are 1-2.5 mg/kg IV with 500 microgram/kg IV top ups. However, its use in children has declined following reports of serious adverse events, including delayed convulsions, acidosis, myocardial depression.

Ketamine, an NMDA receptor antagonist, is another "intravenous anaesthetic agent" which combines sedation/anaesthesia with analgesia and amnesia, for short painful procedures. Typical doses range from 500 microgram to 4.5 mg/kg IV. It can raise intracranial pressure and nowadays is usually only used by anaesthetists and intensivists.

Chloral hydrate, metabolised in the liver to the active trichloroethanol, at 50-100 mg/kg (maximum 2 g) has been compared with midazolam at 500 microgram/kg (maximum 10 mg) for oral sedation for neuroimaging in a randomised controlled trial [17]. At these doses chloral was significantly more effective with no adverse effects. Chloral may be less anxiolytic than other "sedatives" and is not analgesic.

Barbiturates, which act at the GABA-A receptor, remain popular in sedation regimes. "Short acting" barbiturates, quinalbarbitone (Secobarbital) 7.5-10 mg/kg oral (maximum 200 mg) in the UK, or pentobarbital 1-2 mg/kg then 1-2 mg/kg/h IV in the USA, are used for sedation, particularly for non-painful procedures as they are not analgesic.

Benzodiazepines, particularly the shorter acting midazolam (at various doses by different routes, see above) or, less often now, oral temazepam 1 mg/kg are used. One British group has used chloral hydrate or temazepam and droperidol for neuroimaging sedation [18]. As many as 1094 to 1155 (95%) procedures were judged successful and there were no adverse events relating to airway or breathing (95%, CI 0.35%). Droperidol is a butyrophenone major tranquilliser / neuroleptic, which has also been traditionally used in sedation for AEPs. However, there is a significant risk of it inducing movement disorders, e.g. an oculogyric crisis, typically a few hours after the child has woken up and gone home. It no longer appears in the British National Formulary [19] or Medicines for Children [1].

7.6.5 ADVERSE EFFECTS AND SAFETY

A small risk of adverse effects from any drug is inevitable. However, the production of an unintended anaesthetised state in a child intended for sedation can constitute an unacceptable risk. The child needs to be carefully monitored by trained nurses or doctors able to recognise the problem and secure the child's airway and breathing effectively and promptly [20]. The boundary between deep sedation and general anaesthesia is especially treacherous. In a systematic critical incident investigation [21], no particular drug, class of drug, or route was more dangerous than another, however overdosage, the use of polypharmacy (3 or more sedating drugs at once) and the administration by non-trained personnel, e.g. the child's family, outside of a medical facility, were dangerous.

Although there has been a move towards formally employing anaesthetists and their assistants to provide general anaesthesia instead of sedation [22], this is very costly in terms of often hard pressed medical and nursing resources and has not been demonstrated to be more effective, better tolerated and safer than a well organised and trained sedation team [23].

7.6.6 CONCLUSIONS

There are many historical sedation regimes. Sleep is a viable alternative in some situations. For painful procedures analgesia should be given. There have been very few RCTs to compare drug efficacy and tolerability and few large systematic observational studies of safety. What evidence there is suggests that the team caring for the sedated child is critical for safety and must be well trained and operate within a safe clinical environment.

REFERENCES

1. Royal College of Paediatrics and Child Health, Neonatal and Paediatric Pharmacists Group. *Medicines for Children, 2nd edition*, R.C.P.C.H. Publications, London, (2003).
2. Rechtschaffen, A., Kales A., *A manual of standardized terminology, techniques and scoring systems for sleep stages of human subjects*. U.C.L.A. Brain Information Service/Brain Research Institute, Los Angeles, (1968).

3. Plum, F., Posner, J. B., *The diagnosis of stupor and coma*. F.A. Davis Co., Philadelphia, (1982).

4. Jennet, B., Teasdale, G., *Lancet, i,* 878 (1977).

5. Tatman, A., Warren, A., Williams, A., Powell, J. E., Whitehouse, W., *Arch. Dis. Chld., 77,* 519 (1997).

6. Wassmer, E., Whitehouse, W., Quinn, E., Seri, S., *Seizure, 8,* 371 (1999).

7. Wassmer, E., Quinn, E., Whitehouse, W., Seri, S., *Clin. Neurophys., 112,* 683 (2001).

8. Johnson, K., Page, A., Williams, H., Wassmer, E., Whitehouse, W., *Clin. Radiol., 57,* 502 (2002).

9. Wassmer, E., Carter, P. F. B., Quinn, E., et al., *Dev. Med. Child. Neurol., 43,* 735 (2001).

10. Espezel, H., Graves, C., Jan, J. E., et al., *Dev. Med. Child. Neurol., 42,* 646 (2000).

11. Lin-Dyken, D. C., Dyken, M. E., *Infs. Yng. Chld., 15,* 20 (2002).

12. Wassmer, E., Ross, C., Whitehouse, W., *Paed. Perinat. Drug Ther., 4,* 45 (2000).

13. Wassmer, E., Whitehouse, W., Quinn, E., Seri, S., *Seizure, 8,* 371 (1999).

14. American Academy of Pediatrics Committee on Drugs, *Pediat., 89,* 1110 (1992).

15. Royal College of Anaesthetists and Royal College of Radiologists, *Sedation and anaesthesia in radiology: report of a joint working party Royal College of Anaesthesia and Royal College of Radiologists,* The Royal College of Anaesthetists, London, 1992.

16. Barbi, E., Gerarduzzi, T., Marchetti, F., et al., *Arch. Pediatr. Adoles. Med., 157,* 1097 (2003).

17. D'Agostino, J., Terndrup, T. E., *Pediatr. Emerg. Care, 16,* 1 (2000).

18. Sury, M. R., Hatch, D. J., Deeley, T., Dicks-Mireaux, C., Chong, W. K., *Lancet, 353,* 1667 (1999).

19. British Medical Association, Royal Pharmaceutical Society of Great Britain, *British National Formulary,* B.M.A. & R.P.S.G.B., London, (2003).

20. Blike, G., Cravero, J., Nelson, E., *Qual. Manage. Hlth. Care, 10,* 17 (2001).

21. Cote, C. J., Karl, H. W., Notterman, D. A., Weinberg, J. A., McCloskey, C., *Pediatr., 106,* 633 (2000).

22. Morton, N. S., Oomen, G. I., *Paediat.Anaesth., 8,* 65 (1998).

23. Murphy, M. S., *Arch. Dis. Chld., 77,* 281 (1997).

7.7 Opioid analgesic drugs

Greta M. Palmer[1], Brian J. Anderson[2]

[1]Department of Anaesthesia and Pain Medicine, Royal Children's Hospital, Melbourne, Victoria 3052, Australia

[2]Departments of Anesthesia and Intensive Care, Auckland Children's Hospital, Park Road, Grafton, Auckland, New Zealand

7.7.1 INTRODUCTION

Opioids have analgesic effects by activation of μ (mu), κ (kappa), and/or δ (delta) receptors in the central nervous system (CNS) [1]. Each receptor class has subtypes that have different clinical effects. Analgesia occurs by activation of spinal or supraspinal opioid receptors, leading to decreased neurotransmitter release from nociceptive neurons, inhibiting the ascending neuronal pain pathways and altering the perception and response to pain [2]. Opioid receptors also exist outside the CNS, in the dorsal root ganglia and on primary afferent neurons' peripheral terminals [3].

The effects and side effects of opioids depend upon their affinities for different receptor types. Alteration of the stereochemical structure of an opioid changes this binding affinity. Morphine has a rigid pentacyclic structure that conforms to a T-shape (Figure 1) [4]. Important features include a tertiary positively charged basic nitrogen, a quaternary carbon (C-13) separated from the basic nitrogen by an ethylene (-CH_2-CH_2-) chain and attached to a phenyl group, and the presence of phenolic hydroxyl (at C-3) or a ketone group (as in pethidine and methadone). The basic amino group is essential for opioid activity, where short-chain alkyl group substitution results in mixed opioid agonist-antagonists [4].

Correlations between the analgesic drug plasma concentrations and validated pain scores are weak [5,6]. Pharmacodynamic inter-individual variability, assessed by electroencephalography (EEG), varies 3- to 5-fold. Opioids produce adverse effects that may be minimised by appropriate drug selection and dosing. Respiratory depression, nausea and vomiting, histamine release, hypotension,

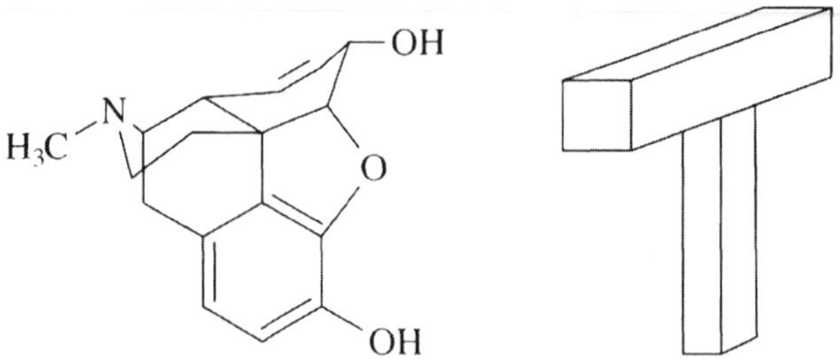

Figure 1. The morphine molecule. From Thorpe, D. H. [4], with permission. The pentacyclic structure conforms to a "T" shape with a piperidine ring forming one crossbar and a hydroxylated aromatic ring lying in the vertical axis.

glottic and chest wall rigidity, constipation, urinary retention, seizures, sedation, pruritis, tachy- and bradycardia are well described. Naloxone is a competitive opioid receptor antagonist that reverses many of these side effects.

7.7.2   NATURALLY OCCURRING OPIOIDS

Morphine

Morphine {7,8-didehydro-4, 5-epoxy-17-methylmorphinan-3, 6-diol} is obtained from the poppy, *Papaver somniferum*, and is the most commonly used opioid in children. Morphine's main analgesic effect is by μ-receptor activation [7-9]. Morphine (named after the Greek god of dreams, Morpheus) is soluble in water, but lipid solubility is poor compared with other opioids (Table 1). Morphine's low oil/water partition coefficient of 1.4 and its pKa of 8 (10-20% unionised drug at physiologic pH) contribute to delayed onset of peak action with slow CNS penetration (Table 2). Morphine is available as elixir, immediate release tablets, slow release tablets or granules and parenteral formulations, as the sulphate or hydrochloride salt.

Pharmacodynamics

Target analgesic plasma concentrations are believed to be 10-20 ng/ml after major surgery [10,11]. The concentration required for sedation during mechanical ventilation may be higher. Mean morphine concentrations of 125 ng/ml were required to produce adequate sedation in 50% of neonates [12]. The

Table 1

Onset, peak and duration of effects after intravenous administration and lipid solubility of opioids

	Pethidine	Morphine	Methadone	Hydro-morphone	Fentanyl	Sufentanil	Alfentanil	Remifentanil
relative potency	0.1	1	1:1 acute* 4-12:1 chronic	5	50-100	500-1000	10	~40
IV loading	0.5-1 mg/kg	0.05-0.1 mg/kg	0.05-0.1 mg/kg	0.01-0.02 mg/kg	0.5-1.0 mcg/kg	0.025-0.05 mcg/kg	0.005-0.01 mg/kg	0.1-1 mcg/kg/min
peak effect (min)	20	45-90	30-60	15-30	3-4	5-6	1-2	1
duration	2-4 h	4-5 h	4-8	2-4 h	30 min	30 min	15 min	5-10 min context sensitive
oil / H$_2$O partition coefficient	39	1.4	>morphine	1.28	860	1.8	13.4	17.9

Adapted from: www.anaesthetist.com and Volles, D. F. [96].

*acute pain management: 4hly dosing schedule; dose should be adjusted down after 48-72h due to long half life (15-40h) chronic pain management: 6-12hly dosing schedule; dose depends on previous opioid tolerance/dose.

Table 2

Onset, peak and duration of effect after epidural administration

	Morphine	Hydromorphone	Fentanyl
onset of effect (min)	20-60	15	5-10
peak effect (min)	40-120	30	10-30
duration (h)	12-24	5-12	3-6

Adapted from Sinatra, R. S. [214] and Volles, D. F. [96].

large pharmacokinetic and pharmacodynamic variability means that morphine is often titrated to effect, using small incremental doses (20 mcg/kg), in children suffering postoperative pain [13]. Neonates are thought to have increased sensitivity to morphine, attributed to an immature blood-brain barrier (BBB) [14]. This effect may, however, be due to decreased clearance and consequent higher effect site concentrations in this age group when compared to older children and adults given the same weighted dose [15]. The effect compartment equilibration half-time (T_{eq}) for morphine is ~17 min in adults [16] and has been predicted for younger age groups, based on allometric modelling (Table 3) [17].

The principle metabolites of morphine, morphine-3-glucuronide (M3G) and morphine-6-glucuronide (M6G), have pharmacological activity. M6G has greater analgesic potency than morphine [18,19] and also respiratory depressant effects [20,21]. It has been suggested that M3G antagonises morphine and M6G has anti-nociceptive and respiratory depressant effects [22,23] and contributes to the development of tolerance. M6G/morphine ratios increase with age, from 0.8 in neonates to 4.2 in children [23,24]. These ratio changes are attributed to the maturation of hepatic and renal clearance with age, but the clinical effect of these ratio changes is probably minimal. Reduced morphine clearance and receptor numbers (opioid, GABA, acetylcholine) in the neonate will have greater impact on pain perception [25].

Pharmacokinetics

Morphine is mainly metabolised by the hepatic enzyme uridine 5′-diphosphate glucuronosyl transferase-2B7 (UGT2B7) into M3G and M6G [26,27], although the enzyme also exists in the intestines and kidneys [28]. Sulphation is a minor pathway [29,30]. Clearance is perfusion limited, with a high hepatic extraction

Table 3

Morphine clearance changes with age

Age	Vd L/70 kg	CL L/h/kg	CL std (%CV) L/h/70 kg	Teq (min)
24-27 weeks		0.14	3.4	
28-31 weeks		0.19	5.1	
term neonate	84	0.44	14.5	8
3 months	131	1.14	43.1	
6 months	136	1.43	57.3	
1 year	136	1.57	67.8	10
3 years	136	1.51	71.1	
5 years	136	1.42	71.1	12
10 years	136	1.25	71.1	14
14 years	136	1.11	71.1	
15 years				16
adult	136	1.01	71.1	17

The increased clearance (l/h/kg) observed during infancy is a size artifact. Clearance reaches adult levels (equivalent to hepatic blood flow), using an allometric size model (l/h/70kg), by the end of infancy. Data from Bouwmeester *et al.* [37]. Premature neonatal data from Scott *et al.* [215]. Predictions using the allometric model (Anderson *et al.* [17]) are similar to those observed by Lynn [11] and McRorie [30]. The effect compartment equilibration half time (Teq) adult data are from Inturrisi *et al.* [16].

ratio. Oral bioavailability is ~35% due to this first pass effect. The metabolites are cleared renally and partly by biliary excretion [30]. Some recirculation of morphine occurs due to gastrointestinal β-glucuronidase activity [31]. Impaired renal function leads to M3G and M6G accumulation [32].

Fetuses are capable of metabolising morphine from 15 weeks gestation [33,34]. Morphine clearance matures with post conceptual age [27,35], reaching adult values at 6-12 months [15,36]. The increased clearance observed in children, when expressed per kilogram, is a size artifact and not attributable to this age group's increased liver size (Table 3 and Figure 2). Pharmacokinetics in children have been described: volume of distribution 136 l/70kg (percentage clearance volume, 59.3%), formation clearance to M3G 64.3 l/h/70kg (58.8%), formation clearance to M6G 3.63 l/h/70kg (82.2%), morphine clearance by other routes 3.12 l/h/70kg, elimination clearance of M3G 17.4 l/h/70kg (43.0%), elimination clearance of M6G 5.8 l/h/70kg (73.8%). The volume of distribution of morphine increased exponentially with a maturation half-life of 26 days from 83 L/70kg at birth; formation clearance to M3G and M6G at birth also increased with a maturation half-life of 88.3 days from 10.8 and 0.61 l/h/70kg respectively. Metabolite clearance increased with age (maturation half-life 129 days), similar to that described for glomerular filtration rate (GFR) maturation in infants. M3G elimination clearance is greater than GFR, suggesting tubular secretion and non-renal elimination [37].

Morphine pharmacokinetic parameters show large inter-individual variability, contributing to the range of morphine serum concentrations observed during constant infusion. Clinical circumstances, such as type of surgery and concurrent illness [11,30,38], also influence morphine pharmacokinetics. Protein binding of morphine is low (from 20% in premature neonates [30, 39] to 35% in adults [40]), but this has minimal impact on disposition changes with age [41].

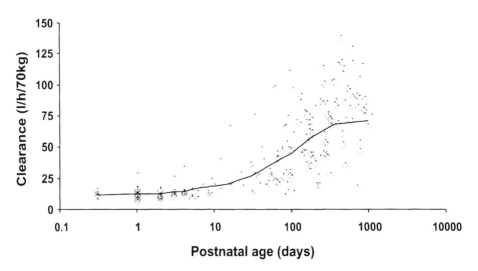

Figure 2. Maturation of morphine clearance with age. Total body morphine clearance is 80% that of 'adult' values by 6 months. Data from Bouwmeester *et al.* [37].

Side effects and tolerance

Respiratory depression may occur at concentrations of 20 ng/ml and is similar in children aged from 2 to 570 days at the same morphine concentration [42]. Intrathecal dosing causes similar respiratory depression at similar cerebro-spinal fluid (CSF) concentrations in children 4 months to 15 years [43]. Hypotension, bradycardia and flushing reflect morphine's histamine releasing property and are associated with rapid intravenous bolus administration [44]. The incidence of vomiting in postoperative children is related to morphine dose. Doses above 100 mcg/kg were associated with a greater than 50% incidence of vomiting [45,46].

Withdrawal symptoms are observed in children after cessation of continuous morphine infusion for greater than 2 weeks, and possibly shorter periods, if doses > 40 mcg/kg/h are administered. Prevention strategies include the use of neuraxial analgesia, nurse-controlled sedation management protocols, ketamine or naloxone mixed with morphine infusion and the use of alternate agents (e.g. methadone) with lower potential for tolerance [2,47].

Dosing

There is variation in opioid requirements depending on the nature of the surgical insult. Morphine requirements after tonsillectomy (50 mcg/kg) are less than after lateral thoracotomy (mean initial dose 230 mcg/kg) [48-50] or lower extremity orthopaedic surgery (170 mcg/kg) [51]. A mean steady-state serum concentration of 10 ng/ml can be achieved in children with a morphine hydrochloride infusion of 5 mcg/kg/h at birth (term neonates), 8.5 mcg/kg/h at 1 month, 13.5 mcg/kg/h at 3 months, 18 mcg/kg/h at 1 year and 16 mcg/kg/h for 1-3 year-old children [37]. After major surgery, continuous morphine doses of 10 to 40 mcg/kg/h have been shown to be effective in alleviating pain in children aged 0-14 years [52-56].

Subcutaneous intermittent morphine boluses, through an indwelling needle or catheter, offer an alternative administrative route [57], but 'stinging' may reduce tolerability. Intramuscular injections are avoided in children as the fear and pain associated with them may be greater than that produced by the underlying pathology. Rectal administration has the disadvantage of large variability in plasma concentrations [58] and has been associated with fatal outcome following a medication error (incorrect dose) [59]. Oral morphine offers a good alternative despite a high first pass effect.

Epidural (25-50 mcg/kg) or intrathecal (2-5 mcg/kg) morphine [60] provides good analgesia for up to 24 h, but may cause delayed respiratory depression for up to 18 h [43,61,62]. This is attributed to morphine's reduced lipophilicity and consequent greater rostral CSF migration. Epidural morphine infusions have also been used in children [63,64]. Patient-controlled analgesia (PCA, 20 mcg/kg bolus) is possible in some children aged as young as six years and nurse-controlled analgesia (NCA) [65] can be used effectively in younger children.

Codeine

Codeine (methylmorphine, {7,8-didehydro-4,5-epoxy-3-methoxy-*17*-methyl morphinan-6-ol monohydrate}) differs in structure from morphine, with a methyl group replacing the hydroxyl group at C-3. Codeine is less potent than morphine (potency ratio 1:10), but has a high oral-parenteral potency ratio (2:3) because hepatic conversion to morphine (O-demethylation, CYP2D6) is necessary for codeine's activity [66]. Around 10% of codeine is metabolised to morphine. Side effects are similar to those described for morphine whether used as an analgesic, anti-tussive or anti-diarrhoea agent. A parenteral formulation (for intramuscular use), elixir (used orally or rectally) and oral tablets are available. Intravenous codeine is not recommended because of hypotensive effects [66,67].

Pharmacodynamics

Codeine has lower affinity for μ-receptors than morphine and reduced effectiveness. An analgesic ceiling effect is apparent with increasing dose, only causing increased side effects. Analgesia may be mediated through morphine, but codeine receptor occupancy contributes to side effects [68]. There is little evidence for the broad belief that codeine causes less side effects, such as sedation and respiratory depression, compared with other opioids [66]. Combination paracetamol 1000 mg/codeine 60 mg in adults provides better analgesia than codeine 60 mg, illustrating the synergistic effect of analgesic combinations [69].

Pharmacokinetics

Oral codeine is rapidly absorbed with approximately 50% of the dose undergoing first pass metabolism. Peak plasma concentrations occur at ~1 h and the plasma half-life is 3-3.5 h in adults. A neonatal half-life will be longer due to immature clearance (e.g. 4.5 h [70]), while that of an infant shorter (e.g. 2.6 h [70]), attributable to size factors [17]. Intramuscular absorption is faster and rectal administration is associated with slower and variable absorption. A volume of distribution (Vd) of 3.6 l/kg and a clearance of 0.85 l/h have been described in adults, but there are few data detailing paediatric pharmacokinetic developmental changes.

Glucuronidation at the 6-OH position is the major metabolic clearance pathway, but minor pathways are N-demethylation to norcodeine (10-20%) and

O-demethylation (CYP2D6) to morphine (5-15%). Codeine is also excreted unchanged in the urine (5-15%) [66]. Genetic polymorphism of the CYP2D6 enzyme results in three broad phenotypes - ultra-rapid extensive, extensive and slow codeine metabolisers in the population [71,72]; 7-10% of Caucasians are slow metabolisers; this percentage varies with ethnicity [72-74]. Codeine has no analgesic effect in the poor metabolisers, but side effects persist [74].

Dosing

In children, codeine is generally given in doses of 0.5-1 mg/kg 4 hourly for analgesia. Antitussive doses are lower. Medications competing for the CYP2D6 enzyme (e.g. quinidine, tramadol) may decrease codeine's analgesic effect.

7.7.3 SEMI- SYNTHETIC OPIOIDS

Diamorphine (Heroin)

Diamorphine (3,6-diacetylmorphine) is essentially a prodrug of morphine. The increased lipid solubility of diamorphine and its immediate metabolite, 6-acetylmorphine, means these agents can cross the brain-blood barrier (BBB) more rapidly than morphine and have a more rapid onset of analgesia [75]. Respiratory depression may be reduced after extradural administration compared to morphine, because less diamorphine is available for rostral CSF spread; a greater proportion of diamorphine localises to the spinal cord [76].

Intranasal diamorphine in adults is rapidly absorbed as a dry powder, with peak plasma concentrations occurring within 5 min, and has a similar pharmacokinetic profile to that of intramuscular diamorphine. It is rapidly converted to 6-acetylmorphine (peak concentrations within 5-10 min) and then hydrolysed to morphine (peak concentrations within 1 h) [77]. Nasal diamorphine 100 mcg/kg has a more rapid onset and similar analgesia to intramuscular morphine 200 mcg/kg [78,79]. Intravenous and subcutaneous infusions of diamorphine were equally effective after abdominal surgery in children at doses similar to those used for morphine.

Oxycodone

Oxycodone {dihydrohydroxycodeinone} is derived from the naturally occurring opioid, thebaïne, and has a similar potency (variably quoted as 1:1 or 2:3) to morphine. There are few paediatric pharmacodynamic data. Adult data suggests that an oxycodone 10 mg/paracetamol 325 mg combination results in enhanced analgesia and an improved side effect and safety profile compared to oxycodone 20 mg alone [80].

Mean values of drug clearance and volume of distribution (Vss) were 15.2 ml/min/kg (SD 4.2) and 2.1 l/kg (SD 0.8) in children after ophthalmic surgery [81]. The relative bioavailability of intranasal, oral and rectal formulations was approximately 50% that of intravenous in adults [82-84].

Oxycodone (100 mcg/kg IV), given to children after ophthalmic surgery, caused greater ventilatory depression compared to other opioids [81]. Maximum mean end-tidal carbon dioxide concentration and minimum mean ventilatory rate occurred 8 min after administration of oxycodone, but the minimum mean peripheral arteriolar oxygen saturation occurred at 4 min.

Hydromorphone

Hydromorphone, a ketone congener of morphine (dihydromorphinone), has ~5-7.5 times morphine's potency [85]. Hydromorphone is used for chronic cancer pain and for post-operative analgesia, but it does not convincingly demonstrate clinical superiority in adults over other strong opioid analgesics [86]. Its intermediate hydrophilic and lipophilic activity, lying between that of morphine and fentanyl (Tables 1 and 2), may offer advantages for epidural use [87-92]. Patient-controlled analgesia with hydromorphone results in similar analgesia and side effects compared to morphine in children for the management of mucositis pain after bone marrow transplantation. Effective plasma concentrations of around 4.7 ng/ml (range 1.9-8.9 ng/ml) were in these children [85].

Hydromorphone is metabolised to hydromorphone-3-glucuronide (neuroexcitation and psychomimetic effects) and, to a lesser extent, to dihydroisomorphine and dihydromorphine [93-95]. The IV:oral dose ratio is ~1:5, as there is high first-pass metabolism [96]. A clearance of 51.7 ml/min/kg (range 28.6 - 98.2) is reported in children [85,97].

Commonly used doses are 15 mcg/kg 4-6 h intravenously and 30-80 mcg/kg 4-6 h orally. For epidural administration, 10 mcg/ml in combination with local anaesthetic and 20 mcg/ml alone is administered at 0.1-0.2 ml/kg/h in both paediatric and adult populations, or as single shot 20-100 mcg/kg. Patient-controlled analgesia bolus dose is 5 mcg/kg and continuous IV infusion rates are 2-8 mcg/kg/h.

7.7.4 SYNTHETIC OPIOIDS

Diphenylpropylamine Series

Methadone

Methadone is a synthetic opioid (μ-agonist with NMDA receptor antagonist properties [98]) with analgesic potency similar to morphine but with more rapid distribution, slower elimination [99] and greater lipid solubility [100].

Methadone has high oral bioavailability (80-90%). Adult data describe a clearance of 178 ml/min, Vss of 410 l and an elimination half-life of 25 h (SD 22) [99], but there are few paediatric pharmacokinetic data. Neonatal data report a slow elimination half-life with large inter-individual variability (3.8-62 h) [101].

Methadone is widely used for the treatment of opioid withdrawal in neonates and children [47,102]. Intravenous methadone (200 mcg/kg) has been shown to be an effective analgesic for postoperative pain relief in children aged 3-7 years [103]. Oral administration (200-600 mcg/kg/day) has been recommended as the first-line opioid for severe and persistent pain in children [104].

Phenylpiperidine Series
Pethidine (Meperidine)

Pethidine is 7-10 times less potent than morphine. Pethidine has anticholinergic effects, histamine liberation and spasmogenic effects on smooth muscle, but anticholinergic effects on the biliary and renal tracts have not been demonstrated *in vivo*. The increase in biliary pressure, secondary to biliary spasm is, in contrast to other opioids, not reversible with naloxone [105,106].

Pharmacodynamics

The analgesic effects are detectable within 5 min of intravenous administration and peak effect is reached within 10 min [31,107]. An adult blood pethidine concentration of 0.7 mcg/ml provides relief from severe pain in approximately 95% of cases. The slope of the concentration-response curve is steep, with a difference of as little as 0.05 mcg/ml between the mean concentrations associated with severe pain and those associated with effective analgesia [108].

Pharmacokinetics

Pethidine undergoes hepatic metabolism by N-demethylation and diesterification to norpethidine, meperidinic acid and normeperidinic acid. Less than 5% is renally excreted. Pethidine clearance in infants and children is approximately 10 ml/min/kg [109,110], but Vss is reported as 8 l/kg in infants [110] and 2.8 l/kg in children [109]. Elimination in neonates is greatly reduced and elimination half-time in neonates, who have received pethidine by placental transfer, may be 2-7 times longer than that in adults [111]. Rectal bioavailability is half that of intravenous [109].

Side effects

Accumulation of renally excreted norpethidine results in seizures and dysphoria [112]. The ventilatory effects of single intravenous dose of morphine 100 mcg/kg and pethidine 670 mcg/kg were compared in children aged 3-8 years. The decrease in ventilatory frequency and acute decrease in oxygen saturation was greater and more transient with pethidine versus morphine [113].

Dosing

Intravenous pethidine 0.5-1 mg/kg provides effective analgesia. Pethidine's local anaesthetic properties and moderate lipid solubility have been found useful for epidural techniques [114] and peritonsillar infiltration for tonsillectomy.

Fentanyl

Fentanyl {phenylethyl-propionyl-anilinopiperidine} is a potent μ-receptor agonist, with a potency 70-125 times that of morphine. Fentanyl offers greater haemodynamic stability than morphine [115,116], rapid onset (T_{eq} = 6.6 min) and short duration of effect. Its relative increased lipid solubility and small molecular conformation enables efficient penetration of the BBB and redistribution.

Pharmacodynamics

A plasma concentration of 15-30 ng/ml is required to provide total intravenous anaesthesia in adults, while the IC_{50}, based on EEG evidence, is 10 ng/ml [117,118]. The intra-operative use of fentanyl 3 mcg/kg in infants did not result in respiratory depression or hypoxaemia in a placebo controlled trial [119]. Only 3 out of 2000 non-intubated infants and children experienced short apnoeic episodes after fentanyl 2-3 mcg/kg for the repair of facial lacerations [120]. Fentanyl has similar respiratory depression in infants and adults when plasma concentrations are similar [121].

Pharmacokinetics

Fentanyl is metabolised by oxidative N-dealkylation (CYP3A4) into nor-fentanyl and hydroxylised [122,123]. All metabolites are inactive and a small amount of fentanyl is renally eliminated unmetabolised [124]. Fentanyl clearance is 70-80% of adult values in term neonates and, standardised to a 70-kg person, reaches adult values (approx. 50 l/h/70kg) within the first 2 weeks of life [15,125,126]. The increased clearance observed in infancy, when expressed per kilogram, is a size artifact and not attributable to this age group's increased hepatic blood flow. Fentanyl's Vss is ~5.9 l/kg in term neonates and decreases with age to 4.5 l/kg during infancy, 3.1 l/kg during childhood, and 1.6 l/kg in adults [127]. Fentanyl clearance may be impaired with decreased hepatic blood flow, e.g. from increased intra-abdominal pressure in neonatal omphalocoele repair [126].

Fentanyl is widely distributed with short duration of effect due to redistribution to deep, lipid rich compartments. Fentanyl redistributes slowly from lipid-rich tissues after discontinuation of therapy, resulting in prolonged periods of sedation and respiratory depression [125]. The context sensitive half-time after 1 h infusion is ~20 min, but after 8 h is 270 min [128].

Transdermal fentanyl is used for severe cancer-related pain [129] or in paediatric palliative care [130]. Fentanyl plasma concentrations are not measurable until 2 h after application and there is 8-16 h latency until maximal effects are observed. Following 'patch' removal, serum fentanyl concentrations decline gradually and fall to 50% in ~16 h. This prolonged apparent elimination half-life occurs because fentanyl continues to be absorbed from the skin depot [131]. The systemic availability of fentanyl by this route is ~30% of intravenous [132].

Side effects

Tolerance to synthetic opioids develops more rapidly (3-5 days) compared to morphine (2 weeks) and heroin (>2 weeks) [101,133,134]. Fentanyl also has a propensity for muscular rigidity [135]. Other drugs metabolised by CYP3A4 (e.g. ciclosporin, erythromycin) may compete for clearance and increase fentanyl plasma concentrations [136,137].

Fentanyl's respiratory depressant effects may outlast the analgesic effect (35-45 min) due to the prolonged context sensitive half-life and/or recirculation of fentanyl bound to the stomach's acid medium (up to 20% of an IV dose) or delayed release from reperfused skeletal muscle beds.

Dosing

Single fentanyl doses (1-3 mcg/kg) attenuate haemodynamic reactions during intubation [138], and infusion (0.5-2.5 mcg/kg/h) reduces physiological and behavioural measures of pain and stress during mechanical ventilation in neonates [139-141] and infants [138]. Fentanyl (1.5-3 mcg/kg) is suitable for procedural pain because of its rapid onset of action and short duration of effect [119,142], but respiratory depression may occur [143,144]. Bolus administration of 10-75 mcg/kg for cardiac surgical procedures results in minimal adverse cardiovascular effects [125,145,146]. Oral transmucosal fentanyl (5-15 mcg/kg) provides consistent analgesia for brief painful procedures after 15-20 min or as anaesthetic pre-medication, but can be associated with vomiting and desaturation [147,148]. Fentanyl (0.25-2 mcg/ml) is commonly added to local anaesthetic epidural infusions to improve analgesic quality.

Alfentanil

Alfentanil, N-{1-[2-(4-ethyl-5-oxo-2-tetrazolinyl)ethyl]-4-methoxymethyl-4-piperidyl} propionamide, has a potency one-tenth that of fentanyl [149]. Alfentanil has lower lipid solubility balanced by a pKa 6.5 with 89% unionised at plasma pH. It consequently has rapid onset (T_{eq} = 0.9 min) and brief duration of action. It causes less histamine release than fentanyl. It is used as a procedural analgesic for paediatric patients [150].

A target plasma concentration of 400 ng/ml is used in anaesthesia. A lower mean plasma concentration (79 ng/ml, SD 23) is reported for maintenance of sedation during paediatric cardiac catheterisation [151].

Metabolism is by oxidative N-dealkylation (CYP3A4) [122] and O-dealkylation (phase 1), and then conjugation to renally excreted metabolites [152]. Vss is larger in adults (457 ml/kg, SD 160) than in children (163 ml/kg, SD 110) [153]. Clearance of alfentanil, standardised to a 70-kg person, is similar at different ages (250 - 500 ml/min/70kg) except in neonates, where clearance is decreased (20-60 ml/min/70kg) [17]. Consequently, elimination half-life in children (40-68 min) is lower than in the neonatal period. In premature neonates, the half-life is 6-9 h [154,155]. Clearance may be reduced in cyanotic patients [151], but not in children with chronic renal failure, chronic hepatic disease [156] or raised intra-abdominal pressure. Alfentanil plasma protein binding (principally alpha-1 acid glycoprotein {AAG}) increases from 65 and 79% in premature and term infants to ~ 90% in adults) [157,158].

There is a high incidence of thoracic rigidity associated with alfentanil administration [150,159] and alfentanil can induce electroencephalographic seizures in epileptic patients [160].

A dose of 10-20 mcg/kg provides good conditions for tracheal intubation, while doses up to 100 mcg/kg are considered safe (LD_{50}:ED_{50} 1080:1) with controlled ventilation [150,161-165]. Infusion doses of 1-3 mcg/kg/min are commonly used in anaesthesia [149,165,166].

Sufentanil

Sufentanil has 5-10 times fentanyl's potency with a T_{eq} of 6.2 min. It is a highly lipophilic compound (octanol-water partition coefficient 1,757) imparting rapid and extensive distribution, with a high margin of safety (LD_{50}:ED_{50} 10:1). The high lipid solubility imparts rapid and extensive distribution.

A plasma concentration of 5-10 ng/ml is required for total intravenous anaesthesia; 0.2-0.4 ng/ml for analgesia.

Elimination of sufentanil is by O-demethylation and N-dealkylation {CYP3A4} and aromatic hydroxylation [167,122]. The Vss in children 2-8 years is 2.9 l/kg (SD 0.6) – 1½ times greater than adults [168]. An increased clearance in childhood, when expressed per kilogram (30.5, SD 8.8 ml/min/kg), is a function of size [17,168].

The amount of free sufentanil decreases with age (neonates: 19%, infants: 11%, children/adults: 8%) and is strongly correlating with AAG plasma concentration [157]. Although the synthetic opioids have high protein binding (>70%) and high hepatic (or non-hepatic for remifentanil) extraction ratios, protein binding changes are probably clinically unimportant [41], because dose is titrated to effect and clearance variability has greater impact.

Sufentanil has a profound parasympathotonic effect (bradycardia, hypotension) that may be reversed by atropine. Intranasal sufentanil (1-2 mcg/kg) has been used for sedation during minor procedures under local anaesthesia but is associated with prolonged recovery room stay [169,170]. Doses of 0.5-1 mcg/kg are used to supplement volatile agents during general anaesthesia, with up to 20 mcg/kg used during cardiac surgery with minimal cardiac depressant effect [171,172]. Epidural sufentanil 0.75 mcg/kg produces an analgesic effect in 3 min lasting up to 200 min in children 4-12 years [173].

Remifentanil

Remifentanil, 3-{4-methoxycarbonyl-4-[(1-oxopropyl)-phenylaminol]-1-piperidine}-propanoic acid methylester, is a selective μ-receptor agonist with rapid onset (T_{eq}=1.16 min), short duration of action and short elimination half-life of 3 to 6 min [174] .

A plasma concentration of 2-4 ng/ml supplements anaesthesia for major surgery. Analgesic plasma concentrations are 0.2-0.4 ng/ml.

Remifentanil's ester linkages are metabolised by nonspecific tissue and red blood cell esterases [175,176] to carbonic acid, which is renally excreted. Clearance data, expressed per kilogram, show decreasing clearance with age; 90 ml/kg/min in infants less than 2-years-old, 60 ml/kg/min between 2-12 years and 40 ml/kg/min in adults [61, 174]. Vss in children is smaller (213 ml/kg, SD 85) compared with adults (390 ml/kg, SD 250), and possibly increased in young infants (450 ml/kg) [61,174]. Remifentanil's context-sensitive half-life is independent of infusion duration (3.2 min, SD 0.9).

There is a high incidence of life-threatening respiratory depression when used for conscious sedation [177]. Rapid development of μ-receptor tolerance occurs after infusion during anaesthesia, resulting in increased postoperative pain and morphine consumption [178]. The inhibitory neurotransmitter glycine is the carrier for remifentanil and the formulation is not used spinally or epidurally [179]. Thoracic rigidity is reported after rapid intravenous injection [179].

Remifentanil is generally infused because of its short duration of action [180,181]. Administration of 1 mcg/kg intravenously, followed by 0.1-1.0 mcg/kg/min, results in sufficient analgesia during surgery in children and neonates [165,182-185].

7.7.5　MIXED OPIOID AGONIST/ANTAGONISTS

Nalbuphine

Nalbuphine retains oxymorphone and naloxone in its structure and has a low side effect profile and low abuse potential. Effective through kappa receptor agonism and partial μ-receptor antagonism [186], nalbuphine has ~0.5-0.7 morphine's potency and antagonist effect 25 times weaker than naloxone [187].

Nalbuphine 100 mcg/kg is equivalent to pethidine 1 mg/kg [188-190]. A ceiling effect for respiratory depression is seen after 30 mg in adults [191]. Nalbuphine does not reverse respiratory depression after morphine [192], but may be effective after fentanyl [193].

Nalbuphine has 50% oral bioavailability. Elimination half-life was significantly shorter in children aged 1.5-5 years (0.9 h) than it was in children 5-8.5 years (1.9 h) and in adults (2.3 h). Systemic clearance of nalbuphine decreases significantly with age [194]. A half-life of 4 h has been reported in neonates, reflecting immaturity of hepatic glucuronide metabolism [195].

The drug is usually given intravenously 150-300 mcg/kg or 300 mcg/kg rectally [196-198].

Pentazocine

Pentazocine was the first opioid agonist/antagonist to be used in humans and is a weak kappa and delta agonist and mu antagonist [199]. Pentazocine has 25-50% morphine's potency with a ceiling effect for both analgesia and respiratory depression at 30-70 mg/kg in adults [200]. The duration of analgesia after paediatric ophthalmic surgery was 164 min (SD 59) after 500 mcg/kg [201].

Elimination half-life was 3.0 h (SD 1.5) and clearance 21.8 ml/min/kg (SD 5.9) in children 4-8 years. The values for Vc, Vss and V beta were 0.73 (SD 0.21), 4.0 (SD 1.2) and 5.3 (SD 2.1) l/kg, respectively [201].

Prolonged use can lead to physical dependence [202]. Emetic side effects limit use. Respiratory depression after pentazocine 900 mcg/kg is similar to pethidine 1 mg/kg [203]. A single dose of pentazocine 500 mcg/kg in children 4-8 years after ophthalmic surgery caused 4 of 10 children's oxygen saturation to decrease below 90% for up to 2 min [201].

Buprenorphine

Buprenorphine is a partial μ-agonist with 30-50 times the potency of morphine [204]. Buprenorphine shows slow receptor association (30 min), but with high affinity to multiple sites from which dissociation is very slow (T_{eq} = 166 min) and incomplete (50% binding after 1 h) [204]. Peak effect may not occur until 3 h and duration of effect in adults is prolonged (<10 h). High doses of naloxone are required to antagonise buprenorphine because of its relative inability to displace buprenorphine from opioid receptors [200,205]. Buprenorphine's metabolite, norbuprenorphine, has dose-dependent analgesic activity and efficacy similar to its parent [206].

Buprenorphine's sublingual and buccal biovailability is estimated as 51.4% and 27.8%, respectively [207]. Mean intranasal bioavailability is 48.2% [208].

Glucuronide metabolism occurs in the liver. An adult clearance of 20 ml/min/kg and Vss of 2.8 l/kg is reported [200]. Clearance in children (4-7 years) is higher than those in adults with similar Vss [209]. Buprenorphine clearance was 0.0138 ml/min/kg (SD 0.0042), the elimination half-life was 20 h (SD 8) and the volume of distribution was 6.2 l/kg (SD 2.11) in premature neonates [210].

Although buprenorphine reverses fentanyl induced postoperative respiratory depression, it may also worsen respiratory depression after fentanyl anaesthesia [211].

The intravenous dose achieving initial analgesia was 5.2 mcg/kg (SD 2.8) in children 4-14 years. The duration of effect was significantly longer with buprenorphine (248 min, SD 314) than with morphine 166 mcg/kg (114 min, SD 109). Buprenorphine (4 mcg/kg) has been used by caudal [212] and epidural routes [213]. Epidural analgesia is prolonged (25.6 h) but respiratory depression can not be reversed by naloxone [213].

REFERENCES

1. Inturrisi, C. E., *Clin. J. Pain*, *18*, S3 (2002).
2. Suresh, S., Anand, K. J., *Semin. Perinatol.*, *22*, 425 (1998).
3. Stein, C., Machelska, H., Binder, W., Schafer, M., *Curr. Opin. Pharmacol.*, *1*, 62 (2001).
4. Thorpe, D. H., *Anesth. Analg.*, *63*, 143 (1984).
5. Suri, A., Estes, K. S., Geisslinger, G., Derendorf, H., *Int. J. Clin. Pharmaco.l Ther.*, *35*, 307 (1997).
6. Olkkola, K. T., in *Pain in Neonates 2nd Revised and Enlarged Edition*. Anand, K. J., Stevens, B., McGrath, P. (eds.), Elsevier, 135 (2000).
7. Loh, H. H., Liu, H. C., Cavalli, A., Yang, W., Chen, Y. F., Wei, L. N., *Brain Res. Mol. Brain Res.*, *54*, 321 (1998).
8. Matthes, H. W., Maldonado, R., Simonin, F., et al., *Nature*, *383*, 819 (1996).
9. Sora, I., Takahashi, N., Funada, M., et al., *Proc. Natl. Acad. Sci. USA*, *94*, 1544 (1997).
10. Kart,T., Christrup, L. L., Rasmussen, M., *Paediatr. Anaesth.*, *7*, 93 (1997).
11. Lynn, A., Nespeca, M. K., Bratton, S. L., Strauss, S. G., Shen, D. D., *Anesth. Analg.*, *86*, 958 (1998).

12. Chay, P. C., Duffy, B. J., Walker, J. S., *Clin. Pharmacol. Ther.*, *51*, 334 (1992).
13. Anderson, B. J., Persson, M., Anderson, M., *Acute Pain*, *2*, 59 (1999).
14. Way, W. L., Costley, E. C., Way, E. L., *Clin. Pharmacol. Ther.*, *6*, 454-461 (1965).
15. Anderson, B. J., McKee, A. D., Holford, N. H., *Clin. Pharmacokinet.*, *33*, 313 (1997).
16. Inturrisi, C. E.,Colburn, W. A., in *Advances in Pain Research and Therapy. Opioid Analgesics in the Management of Clinical Pain*, Foley, K. M., Inturrisi, C. E. (eds.), Raven Press, 441 (1986).
17. Anderson, B. J., Meakin, G. H., *Paediatric Anaesthesia.*, *12*, 205 (2002).
18. Osborne, P. B., Chieng, B., Christie, M. J., *Br. J. Pharmacol.*, *131*, 1422 (2000).
19. Murthy, B. R., Pollack, G. M., Brouwer, K. L., *J. Clin. Pharmacol.*, *42*, 569 (2002).
20. Osborne, R., Thompson, P., Joel, S., Trew, D., Patel, N., Slevin, M., *Br. J. Clin. Pharmacol.*, *34*, 130 (1992).
21. Thompson, P. I., Joel, S. P., John, L., Wedzicha, J. A., Maclean, M., Slevin, M. L., *Br. J. Clin. Pharmacol.*, *40*, 145 (1995).
22. Gong, Q. L., Hedner, J., Bjorkman, R., Hedner, T., *Pain*, *48*, 249 (1992).
23. Smith, M. T., Watt, J. A., Cramond, T., *Life Sci.*, *47*, 579 (1990).
24. Barrett, D. A., Barker, D. P., Rutter, N., Pawula, M., Shaw, P. N., *Br. J. Clin. Pharmacol.*, *41*, 531 (1996).
25. Coyle, J. T., Campochiaro, P., *J. Neurochem.*, *27*, 673 (1976).
26. Coffman, B. L., Rios, G. R., King, C. D., Tephly, T. R., *Drug Metab. Dispos.*, *25*, 1 (1997).
27. Faura, C. C., Collins, S. L., Moore, R. A., McQuay, H. J., *Pain*, *74*, 43 (1998).
28. Fisher, M. B., Vandenbranden, M., Findlay, K., et al., *Pharmacogenetics*, *10*, 727 (2000).
29. Choonara, I., Ekbom, Y., Lindstrom, B., Rane, A., *Br. J. Clin. Pharmacol.*, *30*, 897 (1990).
30. McRorie, T. I., Lyn,n A. M., Nespeca, M. K., Opheim, K. E., Slattery, J. T., *Am. J. Dis. Child.*, *146*, 972 (1992).
31. Koren, G., Maurice, L., *Pediatr. Clin. North Am.*, *36*, 1141 (1989).
32. Choonara, I. A., McKay, P., Hain, R., Rane, A., *Br. J. Clin. Pharmacol.*, *28*, 599 (1989).
33. Pacifici, G. M., Sawe, J., Kager. L., Rane, A., *Eur. J. Clin. Pharmacol.*, *22*, 553 (1982).
34. Pacifici, G. M., Franchi, M., Giuliani, L., Rane, A., *Dev. Pharmacol. Ther.*, *14*, 108 (1989).
35. Kart, T., Christrup, L. L., Rasmussen, M., *Paediatr. Anaesth.*, *7*, 5 (1997).
36. van Lingen, R., Simons, S. H., Anderson, B. J., Tibboel, D., *Clin. Perinatol.*, *29*, 511 (2002).
37. Bouwmeester, N. J., Anderson, B. J., Tibboel, D., Holford, N. H., *Br. J. Anaesth.*, *92*, 208 (2004).
38. Pokela, M. L., Olkkola, K. T., Seppala, T., Koivisto, M., *Dev. Pharmacol. Ther.*, *20*, 26 (1993).
39. Bhat, R., Abu-Harb, M., Chari, G., Gulati, A., *J. Pediatr.*, *120*, 795 (1992).
40. Olsen, G. D., *Clin. Pharmacol. Ther.*, *17*, 31 (1975).
41. Benet, L. Z., Hoener, B. A., *Clin. Pharmacol. Ther.*, *71*, 115 (2002).
42. Lynn, A. M., Nespeca, M. K., Opheim, K. E., Slattery, J. T., *Anesth. Analg.*, *77*, 695 (1993).
43. Nichols, D. J., Yaster, M., Lynn, A. M., et al., *Anesthesiology*, *79*, 733 (1993).
44. Anand, K. J., Stevens, B. J., McGrath, P. J., *Pain in Neonates, Vol 10*, Elsevier, Amsterdam, pp.159 (2000).
45. Anderson, B. J., Ralph, C. J., Stewart, A. W., Barber, C., Holford, N. H., *Anaesth. Intensive Care*, *28*, 155 (2000).
46. Weinstein, M. S., Nicolson, S. C., Schreiner, M. S., *Anesthesiology*, *81*, 572 (1994).
47. Suresh, S., Anand, K. J. S., *Paediatric Anaesthesia*, *11*, 511 (2001).
48. Krane, E. J., Jacobson, L. E., Lynn, A. M., Parrot, C. , Tyler, D. C., *Anesth. Analg.*, *66*, 647 (1987).
49. Hoffman, G. M., Remynse, L. C., Casale, A. J., et al. *Anesthesiology*, *17*, 3A (1989).
50. Maunuksela, E. L., Kokki, H,, Bullingham, R. E., *Clin. Pharmacol. Ther.*, *52*, 436 (1992).
51. Maunuksela, E. L., Korpela, R., Olkkola, K. T., *Anesth. Analg.*, *67*, 233 (1988).

52. van Dijk, M., Bouwmeester, N. J., Duivenvoorden, H. J., et al., *Pain*, *98*, 305 (2002).
53. Beasley, S. W., Tibballs, J., *Aust. N. Z. J. Surg.*, *57*, 233 (1987).
54. Millar, A. J., Rode, H., Cywes, S., *S. Afr. Med. J.*, *72*, 396 (1987).
55. Farrington, E. A., McGuinness, G. A., Johnson, G. F., Erenberg, A., Leff, R. D., *Am. J. Perinatol.*, *10*, 84 (1993).
56. Esmail, Z., Montgomery, C., Courtrn, C., Hamilton, D., Kestle, J., *Paediatr. Anaesth.*, *9*, 321 (1999).
57. Lamacraft, G., Cooper, M. G., Cavalletto, B. P., *J. Pain Symptom Manage.*, *13*, 43 (1997).
58. Lundeberg, S., Beck, O., Olsson, G. L., Boreus, L. O., *Acta Anaesthesiol. Scand.*, *40*, 445 (1996).
59. Gourlay, G. K., Boas, R. A., *Br. Med. J.*, *304*, 766 (1992).
60. Gall, O., Aubineau, J. V., Berniere, J., Desjeux, L., Murat, I., *Anesthesiology*, *94*, 447 (2001).
61. Reich, A., Beland, B., van Aken, H., in *Pediatric Anesthesia*, Bissonnette, B.; Dalens, B. J. (eds.), McGraw-Hill, pp. 259, (2002).
62. de Beer, D. A., Thomas, M. L., *Br. J. Anaesth.*, *90*, 487 (2003).
63. Shayevitz, J. R., Merkel, S., O'Kelly, S. W., Reynolds, P. I., Gutstein, H. B., *Journal of Cardiothoracic & Vascular Anesthesia*, *10*, 217 (1996).
64. Malviya, S., Pandit, U. A., Merkel, S., et al., *Regional Anesthesia & Pain Medicine*, *24*, 438 (1999).
65. Berde, C. B., Lehn, B. M., Yee, J. D., Sethna, N. F., Russo, D., *J. Pediatr.*, *118*, 460 (1991).
66. Williams, D. G., Hatch, D. J., Howard, R. F., *Br. J. Anaesth.*, *86*, 413 (2001).
67. Parke, T. J., Nandi, P. R., Bird, K. J., Jewkes, D. A., *Anaesthesia*, *47*, 852 (1992).
68. Poulsen, L., Brosen, K., Arendt-Nielsen, L., Gram, L. F., Elbaek, K., Sindrup, S. H., *Eur. J. Clin. Pharmacol.*, *51*, 289 (1996).
69. Cunliffe, M., *Br. J. Anaesth.*, *86*, 329 (2001).
70. Quiding, H., Olsson, G. L., Boreus, L. O., Bondesson, U., *Br. J. Clin. Pharmacol.*, *33*, 45 (1992).
71. Chen, Z. R., Somogyi, A. A., Bochner, F., *Lancet*, *2*, 914 (1988).
72. Sindrup, S. H., Brosen, K., *Pharmacogenetics*, *5*, 335 (1995).
73. Williams, D. G., Patel, A., Howard, R. F., *Br. J. Anaesth.*, *89*, 839 (2002).
74. Eckhardt, K., Li, S., Ammon, S., Schanzle, G., Mikus, G., Eichelbaum, M., *Pain*, *76*, 27 (1998).
75. Boerner, U., *Drug Metab. Rev.*, *4*, 39 (1975).
76. Wilson, P. T., Lloyd-Thomas, A. R., *Anaesthesia*, *48*, 718 (1993).
77. Kendall, J. M., Latter, V. S., *Clin. Pharmacokinet.*, *42*, 501 (2003).
78. Wilson, J. A., Kendall, J. M., Cornelius, P., *J. Accid. Emerg. Med.*, *14*, 70 (1997).
79. Kendall, J. M., Reeves, B. C., Latter, V. S., *Br. Med. J.*, *322*, 261 (2001).
80. Gammaitoni, A. R., Galer, B. S., Bulloch, S., et al., *J. Clin. Pharmacol.*, *43*, 296 (2003).
81. Olkkola, K. T., Hamunen, K., Seppala, T., Maunuksela, E. L., *Br. J. Clin. Pharmacol.*, *38*, 71 (1994).
82. Leow, K. P., Smith, M. T., Watt, J. A., Williams, B. E., Cramond, T., *Ther. Drug Monit.*, *14*, 479 (1992).
83. Leow, K. P., Cramond, T., Smith, M. T., *Anesth. Analg.*, *80*, 296 (1995).
84. Takala, A., Kaasalainen, V., Seppala, T., Kalso, E., Olkkola, K. T., *Acta Anaesthesiol. Scand.*, *41*, 309 (1997).
85. Collins, J. J., Geake, J., Grier, H. E., et al., *J. Pediatr.*, *129*, 722 (1996).
86. Quigley, C., *J. Pain Symptom Manage.*, *25*, 169 (2003).
87. Goodarzi, M., *Paediatr. Anaesth.*, *9*, 419 (1999).
88. Chaplan, S. R., Duncan, S. R., Brodsky, J. B., Brose, W. G., *Anesthesiology*, *77*, 1090 (1992).
89. Brodsky, J. B., Chaplan, S. R., Brose, W. G., Mark, J. B., *Annals of Thoracic Surgery*, *50*,

888 (1990).

90. Hammer, G. B., Ngo, K., Macario, A., *Anesthesia & Analgesia*, *90*, 1020 (2000).
91. Hammer, G. B., *Anesthesia & Analgesia*, *92*, 1449 (2001).
92. Moon, R. E., Clements, F. M., *Anesthesiology*, *63*, 238 (1985).
93. Cone, E. J., Phelps, B. A., Gorodetzky, C. W., *J. Pharm. Sci.*, *66*, 1709 (1977).
94. Babul, N., Darke, A. C., *Pain*, *51*, 260 (1992).
95. Hagen, N., Thirlwell, M. P., Dhaliwal, H. S., Babul, N., Harsanyi, Z., Darke, A. C., *J. Clin. Pharmacol.*, *35*, 37 (1995).
96. Volles, D. F., McGory, R., *Critical Care Clinics*, *15*, 55 (1999).
97. Babul, N., Darke, A. C., Hain, R., *J. Pain Symptom Manage.*, *10*, 335 (1995).
98. Davis, A. M., Inturrisi, C. E., *J. Pharmacol. Exp. Ther.*, *289*, 1048 (1999).
99. Gourlay, G. K., Wilson, P. R., Glynn, C. J., *Anesthesiology*, *57*, 458 (1982).
100. Berkowitz, B. A., *Clin. Pharmacokinet.*, *1*, 219 (1976).
101. Chana, S. K., Anand, K. J., *Arch. Dis. Child. Fetal Neonatal Ed.*, *85*, F79 (2001).
102. Tobias, J. D., *Crit. Care Med.*, *28*, 2122 (2000).
103. Berde, C. B., Beyer, J. E., Bournaki, M. C., Levin, C. R., Sethna, N. F., *J. Pediatr.*, *119*, 136 (1991).
104. Shir, Y., Shenkman, Z., Shavelson,, V., Davidson, E. M., Rosen, G., *Clin. J. Pain*, *14*, 350. (1998).
105. Goldberg, M., Vatashsky, E., Haskel, Y., Seror, D., Nissan, S., Hanani, M., *Anesth. Analg.*, *66*, 1282 (1987).
106. McCammon, R. L., Stoelting, R. K., Madura, J. A., *Anesth. Analg.*, *63*, 139 (1984).
107. Jaffe, J. H. in *The pharmacological basis of therapeutics*, Goodman-Gilman, A., Rall, T. W., Nies, A. S., Taylor, P. (eds.) Pergamon Press, pp. 485 (1990).
108. Austin, K. L., Stapleton, J. V., Mather, L. E., *Anesthesiology*, *53*, 460 (1980).
109. Hamunen, K., Maunuksela, E. L., Seppala, T., Olkkola, K. T., *Br. J. Anaesth.*, *71*, 823 (1993).
110. Pokela, M. L., Olkkola, K. T., Koivisto, M., Ryhanen, P., *Clin. Pharmacol. Ther.*, *52*, 342 (1992).,
111. Caldwell, J., Wakile, L. A., Notarianni, L. J., et al., *Life Sci.*, *22*, 589 (1978).
112. Berde, C. B., Sethna, N. F., *N. Engl. J. Med.*, *347*, 1094 (2002).
113. Hamunen, K., *Br. J. Anaesth.*, *70*, 414 (1993).
114. Ngan Kee, W. D., *Anaesth. Intensive Care*, *26*, 137 (1998).
115. Yaster, M., Koehler, R. C., Traystman, R. J., *Anesthesiology*, *66*, 524 (1987).
116. Hickey, P. R., Hansen, D. D., Wessel, D. L., Lang, P., Jonas, R. A., Elixson, E. M., *Anesth. Analg.*, *64*, 1137 (1985).
117. Scott, J. C., Stanski, D. R., *J. Pharmacol. Exp. Ther.*, *240*, 159 (1987).
118. Wynands, J. E., Townsend, G. E., Wong, P., Whalley, D. G., Srikant, C. B., Patel, Y. C., *Anesth. Analg.*, *62*, 661 (1983).
119. Barrier, G., Attia, J., Mayer, M. N., Amiel-Tison, C., Shnider, S. M., *Intensive Care Med.*, *15*, S37 (1989).
120. Billmire, D. A., Neale, H. W., Gregory, R. O., *J. Trauma.*, *25*, 1079 (1985).
121. Hertzka, R. E., Gauntlett, I. S., Fisher, D. M., Spellman, M. J., *Anesthesiology*, *70*, 213 (1989).
122. Tateishi, T., Krivoruk, Y., Ueng, Y. F., Wood, A. J., Guengerich, F. P., Wood, M., *Anesth. Analg.*, *82*, 167 (1996).
123. Labroo, R. B., Paine, M. F., Thummel, K. E.,, Kharasch, E. D., *Drug Metab. Dispos.*, *25*, 1072. (1997).
124. Jacqz-Aigrain, E., Burtin, P., *Clin. Pharmacokinet.*, *31*, 423 (1996).
125. Koehntop, D. E., Rodman, J. H., Brundage, D. M., Hegland, M. G., Buckley, J. J., *Anesth. Analg.*, *65*, 227 (1986).
126. Gauntlett, I. S., Fisher, D. M., Hertzka, R. E., Kuhls, E., Spellman, M. J., Rudolph, C., *Anes-

thesiology, *69*, 683 (1988).

127. Johnson, K. L., Erickson, J. P., Holley, F. O., et al., *Anesthesiology*, *61*, A441 (1984).
128. Hughes, M. A., Glass, P. S., Jacobs, J. R., *Anesthesiology*, *76*, 334 (1992).
129. Collins, J. J., Dunkel, I. J., Gupta, S. K., et al., *J. Pediatr.*, *134*, 319 (1999).
130. Hunt, A., Goldman, A., Devine, T., Phillips, M., *Palliat. Med.*, *15*, 405 (2001).
131. Lehmann, K. A., Zech, D., *J. Pain Symptom Manage.*, *7*, S8 (1992).
132. Sebel, P. S., Barrett, C. W., Kirk, C. J., Heykants, J., *Eur. J. Clin. Pharmacol.*, *32*, 529 (1987).
133. Franck, L. S., Vilardi, J., Durand, D., Powers, R., *Am. J. Crit. Care*, *7*, 364 (1998).
134. Arnold, J. H., Truog, R. D., Scavone, J. M., Fenton. T., *J. Pediatr.*, *119*, 639 (1991).
135. Taddio, A., *Clin. Perinatol.*, *29*, 493 (2002).
136. Touw, D. J., *Drug Metab. Drug Interact.*, *14*, 55 (1997).
137. Tanaka, E., *J. Clin. Pharm. Ther.*, *23*, 403 (1998).
138. Anand, K. J., Sippell, W. G., Aynsley-Green, A., *Lancet.*, *1*, 62 (1987).
139. Guinsburg, R., Kopelman, B. I., Anand, K. J., de Almeida, M. F., Peres, C. de A., Miyoshi, M. H., *J. Pediatr.*, *132*, 954 (1998).
140. Lago, P., Benini, F., Agosto, C., Zacchello, F., *Arch. Dis. Child. Fetal Neonatal Ed.*, *79*, F194 (1998).
141. Saarenmaa, E., Huttunen, P., Leppaluoto, J., Meretoja, O., Fellman, V., *J. Pediatr.*, *134*, 144 (1999).
142. Barrington, K. J., Byrne, P. J., *Am. J. Perinatol.*, *15*, 213 (1998).
143. Friesen, R. H., Alswang, M., *Paediatr. Anaesth.*, *6*, 15 (1996).
144. Hart, L. S., Berns, S. D., Houck, C. S., Boenning, D. A., *Pediatr. Emerg. Care.*, *13*, 189 (1997).
145. Koren, G., Goresky, G., Crean, P., Klein, J., MacLeod, S. M., *Anesth. Analg.*, *63*, 577 (1984).
146. Yaster, M., *Anesthesiology*, *66*, 433 (1987).
147. Schechter, N. L., Weisman, SJ, Rosenblum M, Bernstein B, Conard PL *Pediatrics*, *95*, 335 (1995).
148. Dsida, R. M., Wheeler M, Birmingham PK, et al *Anesth. Analg.*, *86*, 66 (1998).
149. Sfez, M., Le Mapihan Y, Gaillard JL, Rosemblatt JM. *Acta Anaesthesiol Scand.*, *34*, 30 (1990).
150. Saarenmaa, E., Huttunen, P., Leppaluoto, J., Fellman, V., *Arch. Dis. Child. Fetal Neonatal Ed.*, *75*, F103 (1996).
151. Rautiainen, P., *Can. J. Anaesth.*, *38*, 980 (1991).
152. Davis, P. J., Cook, D. R., *Clin. Pharmacokinet.*, *11*, 18 (1986).
153. Meistelman, C., Saint-Maurice, C., Lepaul, M., Levron, J. C., Loose, J. P., Mac Gee, K., *Anesthesiology*, *66*, 13 (1987).
154. Marlow, N., Weindling, A. M., Van Peer, A., Heykants, J., *Arch. Dis. Child.*, *65*, 349 (1990).
155. Killian, A., Davis, P. J., Stiller, R. L., Cicco, R., Cook, D. R., Guthrie, R. D., *Dev. Pharmacol. Ther.*, *15*, 82 (1990).
156. Davis, P. J., Stiller, R. L., Cook, D. R., Brandom, B. W., Davis, J. E., Scierka, A. M., *Anesth. Analg.*, *68*, 579 (1989).
157. Meuldermans, W., Woestenborghs, R., Noorduin, H., Camu, F., van Steenberge, A., Heykants, J., *Eur. J. Clin. Pharmacol.*, *30*, 217 (1986).
158. Wilson, A. S., Stiller, R. L., Davis, P. J., et al., *Anesth. Analg.*, *84*, 315 (1997).
159. Pokela, M. L., Ryhanen, P. T., Koivisto, M. E., Olkkola, K. T., Saukkonen, A. L., *Anesth. Analg.*, *75*, 252 (1992).
160. Keene, D. L., Roberts, D., Splinter, W. M., Higgins, M., Ventureyra, E., *Can. J. Neurol. Sci.*, *24*, 37 (1997).
161. Pokela, M. L., *Biol. Neonate.*, *64*, 360 (1993).

162. Pokela, M. L., Koivisto, M., *Acta Paediatr.*, *83*, 151 (1994).
163. McConaghy, P., Bunting, H. E., *Br. J. Anaesth.*, *73*, 596 (1994).
164. Senel, A. C., Akturk, G., Yurtseven, M., *Middle East J. Anesthesiol.*, *13*, 605 (1996).
165. Davis, P. J., Lerman, J., Suresh, S., et al., *Anesth. Analg.*, *84*, 982 (1997).
166. Davis, P. J., Chopyk, J. B., Nazif, M., Cook, D. R., *J. Clin. Anesth.*, *3*, 125 (1991).
167. Lavrijsen, K., Van Houdt, J., Van Dyck, D., et al., *Drug Metab. Dispos.*, *18*, 704 (1990).
168. Guay, J., Gaudreault, P., Tang, A., Goulet, B., Varin, F., *Can. J. Anaesth.*, *39*, 14 (1992).
169. Zedie, N., Amory, D. W., Wagner, B. K., O'Hara, D. A., *Clin. Pharmacol. Ther.*, *59*, 341 (1996).
170. Abrams, R., Morrison, J. E., Villasenor, A., Hencmann, D., Da Fonseca, M., Mueller, W., *Anesth. Prog.*, *40*, 63 (1993).
171. Barankay, A., Richter, J. A., Henze, R., Mitto, P., Spath, P., *J. Cardiothorac. Vasc. Anesth.*, *6*, 185 (1992).
172. Anand, K. J., Hickey PR *N. Engl. J. Med.*, *326*, 1 (1992).
173. Benlabed, M., Ecoffey, C., Levron, J. C., Flaisler, B., Gross, J. B., *Anesthesiology*, *67*, 948 (1987).
174. Ross, A. K., Davis, P. J., Dear, G. L., et al., *Anesth. Analg.*, *93*, 1393 (2001).
175. Egan, T. D., *Clin. Pharmacokinet.*, *29*, 80 (1995).
176. Dershwitz, M., Hoke, J. F., Rosow,. C. E., et al., *Anesthesiology*, *84*, 812 (1996).
177. Litman, R. S., *J. Pain Symptom Manage.*, *19*, 468 (2000).
178. Guignard, B., Bossard, A. E., Coste, C., et al., *Anesthesiology*, *93*, 409 (2000).
179. Thompson, J. P., Rowbotham, D. J., *Br. J. Anaesth.*, *76*, 341 (1996).
180. Patel, S. S., Spencer, C. M., *Drugs*, *52*, 417 (1996).
181. Duthie, D. J., *Br. J. Anaesth.*, *81*, 51 (1998).
182. Prys-Roberts, C., Lerman, J., Murat, I., et al., *Anaesthesia*, *55*, 870 (2000).
183. Donmez, A., Kizilkan, A., Berksun, H., Varan, B., Tokel, K., *J. Cardiothorac. Vasc. Anesth.*, *15*, 736 (2001).
184. Chiaretti, A., Pietrini, D., Piastra, M., et al., *Pediatr. Neurosurg.*, *33*, 83 (2000).
185. Davis, P. J., Galinkin, J., McGowan, F. X., et al., *Anesth. Analg.*, *93*, 1380-6 (2001).
186. De Souza, E. B., Schmidt, W. K., Kuhar, M. J., *J. Pharmacol. Exp. Ther.*, *244*, 391 (1988).
187. Schaffer, J., Piepenbrock, S., Kretz, F. J., Schonfeld, C., *Anaesthesist.*, *35*, 408 (1986).
188. Habre, W., McLeod, B., *Anaesthesia*, *52*, 1101 (1997).
189. Viitanen, H., Annila, P., *Br. J. Anaesth.*, *86*, 572 (2001).
190. van den Berg, A. A., Halliday, E., Lule, E. K., Baloch, M. S., *Acta Anaesthesiol. Scand.*, *43*, 28 (1999).
191. Romagnoli, A., Keats, A. S., *Clin. Pharmacol. Ther.*, *27*, 478 (1980).
192. Bailey, P. L., Clark, N. J., Pace. N. L., Isern, M., Stanley, T. H., *Anesth. Analg.*, *65*, 605 (1986).
193. Latasch, L., Probst, S., Dudziak, R., *Anesth. Analg.*, *63*, 814 (1984).
194. Jaillon, P., Gardin, M. E., Lecocq, B., et al., *Clin. Pharmacol. Ther.*, *46*, 226 (1989).
195. Nicolle, E., Devillier, P., Delanoy, B., Durand, C., Bessard, G., *Eur. J. Clin. Pharmacol.*, *49*, 485 (1996).
196. Littlejohn, I. H., Tarling, M. M., Flynn, P. J., Ordman, A. J., Aiken, A., *Eur. J. Anaesthesiol.*, *13*, 359 (1996).
197. Bessard, G., Alibeu, J. P., Cartal, M., Nicolle, E., Serre Debeauvais, F., Devillier, P., *Fundam. Clin. Pharmacol.*, *11*, 133 (1997).
198. van den Berg, A. A., Montoya-Pelaez, L. F., Halliday, E. M., Hassan, I., Baloch, M. S., *Eur. J. Anaesthesiol.*, *16*, 186 (1999).
199. Matsumoto, R. R., Bowen, W. D., Walker, J. M., et al. *Eur. J. Pharmacol.*, *301*, 31 (1996).
200. Zola, E. M., McLeod, D. C., *Drug Intell. Clin. Pharm.*, *17*, 411 (1983).
201. Hamunen, K., Olkkola, K. T., Seppala, T., Maunuksela E. L., *Pharmacol. Toxicol.*, *73*, 120

(1993).

202. Jasinski, D. R., Martin, W. R., Hoeldtke, R. D., *Clin. Pharmacol. Ther.*, *11*, 385 (1970).

203. Iisalo, E. U., Iisalo, E., *Ann. Chir. Gynaecol.*, *67*, 123 (1978).

204. Boas, R. A., Villiger, J. W., *Br. J. Anaesth.*, *57*, 192 (1985).

205. Gal, T. J., *Clin. Pharmacol. Ther.*, *45*, 66 (1989).

206. Cowan, A., *Int. J. Clin. Pract. Suppl.*, 3 (2003).

207. Kuhlman, J. J. Jr., Lalani, S., Magluilo, J., Jr., Levine, B., Darwin, W. D., *J. Anal. Toxicol.*, *20*, 369 (1996).

208. Eriksen, J., Jensen, N. H., Kamp-Jensen, M., Bjarno, H., Friis, P., Brewster, D., *J. Pharm. Pharmacol.*, *41*, 803 (1989).

209. Olkkola, K. T., Maunuksela, E. L., Korpela, R., *Br. J. Clin. Pharmacol.*, *28*, 202 (1989).

210. Barrett, D. A., Simpson, J., Rutter, N., Kurihara-Bergstrom, T., Shaw, P. N., Davis, S. S., *Br. J. Clin Pharmacol.*, *36*, 215 (1993).

211. Zanette, G., Manani, G., Giusti, F., Pittoni, G., Ori, C., *Paediatr. Anaesth.*, *6*, 419 (1996).

212. Kamal, R. S., Khan, F. A., *Paediatr. Anaesth.*, *5*, 101 (1995).

213. Miwa, Y., Yonemura, E., Fukushima, K., *Can. J. Anaesth.*, *43*, 907 (1996).

214. Sinatra, R. S. in *Neural Blockade in Clinical Anesthesia and Management of Pain*, Cousins, M. J., Bridenbaugh, P. O. (eds.), Lippencott-Raven, pp. 793 (1998).

215. Scott, C. S., Riggs, K. W., Ling, E. W., et al., *J. Pediatr.*, *135*, 423 (1999).

7.8 Non-opioid analgesics

Greta M. Palmer[1], Brian J. Anderson[2]

[1]*Royal Children's Hospital, Flemington Rd., Parkville (Melbourne) 3052, Victoria, Australia*

[2]*Department of Anaesthesia, University of Auckland, School of Medicine, Auckland, New Zealand*

7.8.1 TRAMADOL

Tramadol is a moderately potent analgesic [1, 2]. It is an aminocyclohexanol derivative or phenylpiperidine analogue of codeine and its analgesic effect is mediated through noradrenaline re-uptake inhibition, both increased release and decreased re-uptake of serotonin in the spinal cord and very weak μ-opioid receptor effect [1, 3-5]. Tramadol is available in parenteral, suppository, oral drops, and tablet formulations.

Pharmacodynamics

Tramadol's affinity for opioid receptors is ~6000 times weaker than morphine, but the active *O*-desmethyltramadol (+)-M1 metabolite has an affinity ~200 times greater than tramadol [5] - mediating tramadol's attributed opioid effect. Both the μ-receptor effect and reduced serotonin uptake in descending spinal cord pathways may contribute to tramadol's emetic effect. A serum tramadol concentration above 100 ng/ml is associated with satisfactory analgesia [6].

Intradermal tramadol 5% can provide local anaesthesia similar to prilocaine 2%, but with higher incidence of local adverse effects [7].

Pharmacokinetics

The (+)-M1 is formed via the genetically polymorphic CYP2D6 iso-enzyme system responsible for codeine metabolism [5] and individuals may be classified as extensive or poor metabolisers of tramadol. Higher concentrations of the (+)-M1 metabolite and greater analgesic efficacy of tramadol is reported in extensive metabolisers with reduced nausea, vomiting and tiredness amongst poor metabolisers [5].

An elimination half-life of 3.6 h (SD 1.1), a serum clearance 5.6 ml/kg/min (SD 2.7) and a volume of distribution 4.1 l/kg (SD 1.2) is reported in 5 year-old children given oral tramadol drops [6]. The (+) enantiomer concentration was 14.2 % (SD 4.9) greater than that of the (-) enantiomer. The M1 metabolites had a (-) enantiomer concentration 92.3 % (SD 75.1) greater than the (+) enantiomer. The M1 elimination half-life was 5.8 h (SD 1.7) [6]. In children 1-12 y, similar data are reported after intravenous administration. The elimination half-life of tramadol was 6.4 h (SD 2.7), with a Vd of 3.1 l/kg (SD 1.1) and total plasma clearance of 6.1 ml/kg/min (SD 2.5) [8].

Tramadol's oral bioavailability is ~75%, rising with continued administration, resulting from hepatic enzyme saturation. Plasma concentrations are detectable ~30 minutes and peak at 2 hours after oral administration. The half-life of tramadol is 5-7 hours in young adults and 15-30% of tramadol is excreted unchanged renally.

Side effects

Tramadol has a greatly reduced potential for sedation, respiratory depression (<0.5% in children and neonates) [9] and dependence compared to conventional opioids [10]. Tramadol's adverse effect profile includes nausea, vomiting, constipation, dizziness, somnolence, fatigue, sweating and pruritus. Tramadol is associated with a high incidence of postoperative nausea and vomiting (PONV) in adults and children (up to 50%) limiting its usefulness [11-14]. PONV is generally managed with the anti-emetic ondansetron. Ondansetron is a serotonin antagonist (anti 5-HT$_3$) and the CYP2D6 iso-enzyme is also involved in the metabolism of ondansetron [4]. Concurrent use is associated with reduced analgesic effect of tramadol, as ondansetron decreases conversion to M1 and is itself antalgesic. Simultaneously, as tramadol causes nausea and vomiting via 5-HT, ondansetron's antiemetic efficacy is reduced. [3,4]. Tramadol is not associated with histamine release, but has been associated with seizures [15] [16].

Dosing

An oral elixir 1.5 mg/kg, given to children (5.3 y, SD 1.1) undergoing dental surgery, provided effective analgesia for 7 h [6]. Oral tramadol 1-2 mg/kg is well tolerated and effective in teenage orthopaedic surgery [17] and postoperative children (7-16 y) ready to transition from morphine PCA [18]. Oral tramadol 1-2 mg/kg 4-6 h (maximal dose: 8 mg/kg/day, not to exceed 400 mg/day) is a safe and effective analgesic when used for up to 30 days [19].

Intravenous/intramuscular tramadol 1-3 mg/kg is an effective analgesic after tonsillectomy [13,20] and general surgery [9] and can be infused postoperatively 0.21 mg/kg/h. Caudal tramadol (2 mg/kg) provided reliable postoperative analgesia in children for hypospadias and inguinal hernia repair [21,22], but this effect may be attributable to systemic absorption [8,23].

7.8.2 CLONIDINE

Clonidine is an imadazoline α_2-adrenergic receptor agonist (potency α_2:α_1 of 200:1) that inhibits adenylate cyclase and the consequent formation of cyclic AMP. Clonidine exerts its therapeutic effects with central reduction of noradrenergic outflow at the level of the locus coeruleus, by stimulation of α_2-adrenoreceptors within several brain stem nuclei implicated in analgesia [24] and via primary afferent terminals of neurones in the superficial laminae of the spinal cord. The analgesic effect is more pronounced when adminstered neuraxially than systemically [25]. Clondine's clinical uses encompass anti-hypertensive and anti-migraine therapy, sedation, anxiolysis, attention deficit hyperactivity disorder, analgesia for somatic and neuropathic pain states and drug withdrawal regimes [24-29]. It is available as tablet, transdermal patch and parenteral form, the latter used intravenously, intrathecally and epidurally.

Pharmacodynamics

Plasma concentrations within the range 0.2-2.0 ng/ml are believed to have clinical effect [30]. Sedation is dose-dependent and does not usually occur below 0.3 ng/ml [31].

Pharmacokinetics

Children (14-48 months) given rectal clonidine 2.5 mcg/kg achieved a median maximum plasma concentration of 0.77 (95%CI, 0.62-0.88) ng/ml, time to maximum plasma concentration 51 min (29-70), terminal half-life 12.5 (8.7-19.5) h, and bioavailability 95 (73-119) % [30]. Oral bioavailability is also high ~90%, but there are few clearance data available for children.

Side effects

Dose-dependent sedation and dose-independent decrease in heart rate (5-20%) and reduction in blood pressure (10-15%) are common [29,32]. Bradycardia, hypotension, respiratory depression, miosis and coma are described [33, 34]. Respiratory depression is uncommon, unless the dose is excessive [29,35].

Doses

Clonidine 1-2 mcg/kg added to epidural bupivacaine 0.25% or ropivacaine 0.2% prolongs the duration of postoperative analgesia [36-39]. Higher doses are associated with increased sedation (3-5 mcg/kg orally or regionally causes mild sedation lasting 3-9 h) [25]. Clonidine premedication, 2-4 mcg/kg orally, decreased analgesic requirements after minor surgery in children 5-7 y [40]. A small intravenous dose 0.2-0.3 mcg/kg added to morphine decreases opioid consumption [41].

7.8.3 KETAMINE

The analgesic properties of ketamine are mediated by multiple mechanisms at central and peripheral sites. The contribution from N-methyl-D-aspartate (NMDA) receptor antagonism and interaction with cholinergic, adrenergic, serotonergic, opioid pathways and local anaesthetic effects remains to be fully elucidated.

Pharmacodynamics

Ketamine is available as a mixture of two enantiomers; the S(+)-enantiomer has four times the potency of the R(-)-entantiomer [42]. S(+)-ketamine has approximately twice the potency of the racemate. The metabolite norketamine has a potency one third that its parent [43]. Plasma concentrations associated with hypnosis and amnesia during surgery are 0.8-4 mcg/ml; awakening usually occurs at concentrations lower than 0.5 mcg/ml. Pain thresholds are elevated at 0.1 mcg/ml [44].

Pharmacokinetics

Ketamine has high lipid solubility with rapid distribution. Ketamine undergoes N-demethylation to form norketamine. Racemic ketamine elimination is complicated by R(-)-ketamine inhibiting the elimination of S(+)-ketamine [45]. Clearance in children is similar to adult rates (80 l/h/70kg, i.e. liver blood flow) within the first six months of life, when corrected for size using allometric models [46,47]. Clearance in the neonate is reduced (26 l/h/70 kg) [46,48]. Volume of distribution at steady state (Vss) decreases with age (in l/kg): 3.46 at birth; 3.03 in infancy; 1.18 at 4 years; and 0.75 in adulthood. [46]. There is a high hepatic extraction ratio and the relative bioavailability of nasal and rectal formulations is 50% and 30% respectively [49,50].

Side effects

Psycholergic emergence reactions can cause distress. These can be ameliorated by the benzodiazepines. An antisialagogue may be required to diminish copious secretions [51]. Tolerance in children is described with repeated use [52]. Ketamine generally maintains good airway tone, but this is not guaranteed.

Dosing

Ketamine 3-10 mg/kg by the oral, nasal and rectal routes has been used as a premedicant and analgesic for operative procedures [49,53-55]. Intravenous ketamine (200-800 mcg/kg) provides short-term analgesia. It is also used as a co-analgesic infusion 50 mcg-1 mg/kg/h. Adult data suggests a 1:1 morphine/ketamine PCA mixture achieves best analgesia [56]. Preservative free ketamine 500 mcg/kg, added to *caudal* bupivacaine 0.25% (1 ml/kg), produces analgesia for 11 h with minimal behavioural side effects [57,58]. S(+)-ketamine may provide a better alternative [36,59].

REFERENCES

1. Bamigbade, T. A., Langford, R. M., *Hosp. Med.*, *59*, 373 (1998).
2. Turturro, M. A., Paris, P. M , Larkin, G. L., *Annals of Emergency Medicine.*, *32*, 139 (1998).
3. De Witte, J. L., Schoenmaekers, B., Sessler. D. I., Deloof, T., *Anesth. Analg.*, *92*, 1319 (2001).
4. Stamer, U. M., Stuber. F., *Anesth. Analg.*, *93*, 1626. (2001).
5. Poulsen, L., Arendt-Nielsen, L., Brosen, K., Sindrup, S. H., *Clin. Pharmacol. Ther.*, *60*, 636 (1996).
6. Payne, K. A., Roelofse, J. A., Shipton, E. A., *Anesth. Prog.*, *49*, 109 (2002).
7. Altunkaya, H., Ozer, Y., Kargi, E., Babuccu, O., *Br. J. Anaesth.*, *90*, 320 (2003).
8. Murthy, B. V., Pandya, K. S., Booker, P. D., Murray, A., Lintz, W., Terlinden, R., *Br. J. Anaesth.*, *84*, 346 (2000).
9. Bosenberg, A. T., Ratcliffe, S., *Anaesthesia.*, *53*, 960 (1998).
10. Broome, I. J., Robb, H. M., Raj, N., Girgis, Y., Wardall, G. J., *Anaesthesia.*, *54*, 289 (1999).
11. Pang, W. W., Mok, M. S., Lin, C. H., Yang, T. F., Huang, M. H., *Can. J. Anaesth.*, *46*, 1030 (1999).
12. Pang, W. W., Mok, M. S., Huang, S., Hung, C. P., Huang, M. H., *Can. J. Anaesth.*, *47*, 968 (2000).
13. van den Berg, A. A., Halliday, E., Lule, E. K., Baloch, M. S., *Acta Anaesthesiol. Scand.*, *43*, 28 (1999).
14. Torres, L. M., , Rodriguez, M. J., Montero, A., et al., *Reg. Anesth. Pain Med.*, *26*, 118 (2001).
15. Tobias, J. D., *South Med. J.*, *90*, 826 (1997).
16. Gibson, T. P., *Am. J. Med.*, *101*, 47S (1996).
17. Tobias, J. D., *Am. J. Pain Manag.*, *6*, 51 (1996).
18. Finkel, J. C., Rose, J. B., Schmitz, M. L., et al., *Anesth. Analg.*, *94*, 1469 (2002).
19. Rose, J. B., Finkel, J. C., Arquedas-Mohs, A., Himelstein, B. P., Schreiner, M, Medve, R. A., *Anesth. Analg.*, *96*, 78 (2003).

20. Viitanen, H., Annila, P., *Br. J. Anaesth.*, *86*, 572 (2001).
21. Batra, Y. K., Prasad, M. K., Arya, V. K., Chari, P., Yaddanapudi, L. N., *Int. J. Clin. Pharmacol. Ther.*, *37*, 238 (1999).
22. Ozcengiz, D., Gunduz, M., Ozbek, H., Isik, G., *Paediatr. Anaesth.*, *11*, 459 (2001).
23. de Beer, D. A., Thomas, M. L., *Br. J. Anaesth.*, *90*, 487 (2003).
24. Yaksh, T. L., Pogrel, J. W., Lee, Y. W., Chaplan, S. R., *J. Pharmacol. Exp. Ther.*, *272*, 207 (1995).
25. Eisenach, J. C., De Kock, M., Klimscha, W., *Anesthesiology*, *85*, 655 (1996).
26. Eisenach, J. C., , DuPen, S., Dubois, M., Miguel, R., Allin, D., *Pain*, *61*, 391 (1995).
27. Gowing, L., Ali, R., White, J., *Cochrane Database of Systematic Reviews*, CD002021 (2000).
28. Hoder, E. L., Leckman, J. F., Ehrenkranz, R., Kleber, H., Cohen, D. J., Poulsen, J. A., *N. Engl. J. Med.*, *305*, 1284 (1981).
29. Nishina, K., Mikawa, K., Shiga, M., Obara, H., *Paediatr. Anaesth.*, *9*, 187 (1999).
30. Lonnqvist, P. A., Bergendahl, H. T., Eksborg, S., *Anesthesiology.*, *81*, 1097 (1994).
31. Ivani, G., Bergendahl, H. T., Lampugnani, E., et al., *Acta Anaesthesiol. Scand.*, *42*, 306 (1998).
32. Bergendahl, H. T., Eksborg, S., Lonnqvist, P. A., *Acta Anaesthesiol. Scand.*, *41*, 381 (1997).
33. Wiley, J. F., *J. Pediatr.*, *116*, 654 (1990).
34. Kappagoda, C., Schell, D. N., Hanson, R. M., Hutchins, P., *J. Paediatr. Child Health*, *34*, 508 (1998).
35. Penon, C., Ecoffey, C., Cohen, S. E., *Anesth. Analg.*, *72*, 761 (1991).
36. De Negri, P., Ivani, G., Visconti, C., De Vivo, P., *Paediatr. Anaesth.*, *11*, 679 (2001).
37. Jamali, S., Monin, S., Begon, C., Dubousset, A. M., Ecoffey, C., *Anesth. Analg.*, *78*, 663 (1994).
38. Klimscha, W., Chiari, A., Michalek-Sauberer, A., et al., *Anesth. Analg.*, *86*, 54 (1998).
39. Lee, J. J., Rubin, A. P., *Br. J. Anaesth.*, *72*, 258 (1994).
40. Mikawa, K., Nishina, K., Maekawa, N., Obara, H., *Anesth. Analg.*, *82*, 225 (1996).
41. Lyons, B., Casey, W., Doherty, P., McHugh, M., Moore, K. P., *Intensive Care Med.*, *22*, 249 (1996).
42. Geisslinger, G., Hering, W., Thomann, P., Knoll, R., Kamp, H. D., Brune, K., *Br. J. Anaesth.*, *70*, 666 (1993).
43. Leung, L. Y., Baillie, T. A., *J. Med. Chem.*, *29*, 2396 (1986).
44. Grant, I. S., Nimmo, W. S., McNicol, L. R., Clements, J. A., *Br. J. Anaesth.*, *55*, 1107 (1983).
45. Ihmsen, H., Geisslinger, G., Schuttler, J., *Clin. Pharmacol. Ther.*, *70*, 431 (2001).
46. Cook, R. D., in *Pediatric Cardiac Anesthesia*, Lake, C. L., (ed.) Appleton & Lange, 134 (1993).
47. Anderson, B. J., Meakin, G. H., *Paediatr. Anaesth.*, *12*, 205 (2002).
48. Hartvig, P., Larsson, E., Joachimsson, P. O., *J. Cardiothorac. Vasc. Anesth.*, *7*, 148 (1993).
49. Malinovsky, J. M., Servin, F., Cozian, A., Lepage, J. Y., Pinaud, M., *Br. J. Anaesth.*, *77*, 203 (1996).
50. Pedraz, J. L. A., Calvo, M. B., Lanao, J. M., Muriel, C., Santos Lamas, J., Dominguez-Gil, A., *Br. J. Anaesth.*, *63*, 671 (1989).
51. Hollister, G. R., Burn, J. M., *Anesth. Analg.*, *53*, 264 (1974).
52. Byer, D. E., Gould, A. B., Jr., *Anesthesiology*, *54*, 255 (1981).
53. Tobias, J. D., Phipps, S., Smith, B., Mulhern, R. K., *Pediatrics*, *90*, 537 (1992).
54. Tobias, J. D., *Pediatr. Emerg. Care.*, *15*, 173 (1999).
55. Gutstein, H. B., Johnson, K. L., Heard, M. B., Gregory, G. A., *Anesthesiology.*, *76*, 28 (1992).

56. Sveticic, G., Gentilini, A., Eichenberger, U., Luginbuhl, M., Curatolo, M., *Anesthesiology.*, *98*, 1195 (2003).
57. Semple, D., Findlow, D., Aldridge, L. M., Doyle, E., *Anaesthesia.*, *51*, 1170 (1996).
58. Naguib, M., Sharif, A. M., Seraj, M., el Gammal, M., Dawlatly, A. A., *Br. J. Anaesth.*, *67*, 559 (1991).
59. Koinig, H., Marhofer, P., Krenn, C. G., et al., *Anesthesiology.*, *93*, 976 (2000).

7.9 Paracetamol

Brian J. Anderson

Department of Anaesthesiology, University of Auckland, School of Medicine, Auckland, New Zealand

7.9.1 MECHANISM OF ACTION

Paracetamol (acetaminophen) has been used in clinical practice for over one hundred years and is the most commonly prescribed paediatric medicine. The popularity of paracetamol over the non-steroidal antiinflammatory agents increased after the reported association between Reye's syndrome and aspirin in the 1980s. Paracetamol is widely used in the management of pain and fever, but is lacking antiinflammatory effects. The mechanism of action of paracetamol analgesia is multifactorial. It is a potent inhibitor of prostaglandin synthesis within the central nervous system but also acts peripherally by blocking impulse generation within the bradykinin-sensitive chemoreceptors responsible for the generation of nociceptive impulses. Paracetamol is also thought to have an analgesic effect by antagonising N-methyl-D-aspartate (NMDA) and substance P in the spinal cord, with some inhibitory action on spinal nitric oxide mechanisms [1-3] and serotonergic pathways [4].

7.9.2 PHARMACODYNAMICS

Neither antipyretic nor analgesic effects are directly related to serum concentrations. There is a delay of approximately one hour, attributable to effect compartment kinetics. This delay can be quantified by an effect compartment equilibration half-time (T_{eq}). Pharmacodynamic effects have been estimated using Sigmoid E_{max} models.

Antipyresis

A paracetamol serum target concentration range of 10-20 mg/L has been postulated for fever control [5]. Brown *et al.* [6] have reported a T_{eq} of 71 SE 7 min; the Sigmoid E_{max} component yielded an EC_{50} of 4.63 g/L and a Hill coefficient (N) of 3.98 SE 0.42. The predicted maximum effect response was $1.38°C$, but this was possibly constrained by the temperature enrolment criteria and a failure to explore doses higher than 12.5 mg/kg. However, temperature reduction was also dependent on the initial temperature and an underlying cyclic component, attributable to the underlying disease and/or physiologic temperature regulation.

Analgesia

Korpela *et al.* [7] studied children having day-stay surgery, randomised to receive a single dose of 0, 20, 40, or 60 mg/kg of rectal paracetamol. Pain scores after surgery were lower in the 40 and 60 mg/kg groups, compared with placebo and 20 mg/kg groups. The calculated dose of paracetamol, at which 50% of the children did not require a rescue opioid, was 35 mg/kg.

Anderson *et al.* [8] investigated children after tonsillectomy. Pharmacodynamic population parameter estimates (population variability CV) for an E_{max} model (N=1), in which the greatest possible pain relief (VAS 0-10) equates to an E_{max} of 10, were E_{max} 5.17 (64%) and EC_{50} 9.98 (107%) mg/L. The T_{eq} of the analgesic effect compartment was 53 (217%) min. A target effect compartment concentration of 10 mg/L was associated with a pain reduction of 2.6 from a possible score of 10.

7.9.3   PHARMACOKINETICS

Bioavailability

Paracetamol has low first pass metabolism and the hepatic extraction ratio is 0.11-0.37 in adults [9]. Essentially all of an oral dose is absorbed across the gut wall [10]. The relative bioavailability of rectal compared with oral paracetamol formulations ($F_{rectal/oral}$) has been reported as 0.52 (range 0.24-0.98) [11] and even as low as 0.3 [12]. The relative bioavailability of rectal formulations appears to be age-related. The bioavailability of the capsule suppository, relative to elixir, decreased with age from 0.92 (22%) at 28 weeks post-conception, to 0.86 at 2 years age; while the triglyceride base decreased from 0.86 (35%) at 28 weeks post-conception to 0.5 at 2 years age. The relative bioavailability of a rectal solution was 0.66 [13].

Absorption

Oral

Paracetamol has a pKa of 9.5 and it is unionised in the alkaline medium of the duodenum. Consequently, absorption of the unionised form from the duodenum to the systemic circulation is rapid ($t_{1/2}$ abs 2.7, SE 1.2 min; t_{lag} 4.2, SE 0.4 min in febrile children given elixir orally [14]). Similar absorption half-lives have been estimated in children given paracetamol as an elixir before tonsillectomy ($t_{1/2}$ abs 4.5 min, CV 63%, t_{lag} 0) [15]. Absorption in children under the age of 3 months is delayed 3.68 times [16]. Paracetamol absorption depends on gastric emptying and gastric emptying is slow and erratic in the neonate. Normal adult rates may not be reached until 6-8 months. Oral absorption was considerably delayed in premature neonates in the first few days of life [17].

Rectal

Absorption through the rectal route is slow, erratic and has large variability. Absorption parameters for the triglyceride base and capsule suppository were $t_{1/2}$abs 1.34 h (CV 90%), t_{lag} 0.14 h (31%) and $t_{1/2}$abs 0.65 h (63%), t_{lag} 0.54 h (31%) respectively. The absorption half-life for rectal formulations was prolonged in infants under three months (1.51 times greater), compared with those seen in older children [16].

Clearance

Several studies have shown that the rate constant for paracetamol glucuronide formation in neonates is considerably smaller than in adults but the rate constant for sulphate formation is similar to adults when size is standardised to a 70-kg person using an allometric model [18-21]. Glucuronide/sulphate ratios range from 0.12 in premature neonates of 28-32 weeks post conception, 0.28 in those at 32-36 weeks post conception [20] and 0.34 in term neonates 0-2 days old [19]. Ratios of 0.75 in children 1-9 years [21], 1.61 in those aged 12 years and 1.8 in adults are reported [19]. Approximately 4% of paracetamol is excreted in urine unmetabolised and the amount is dependent on urine flow [21,22].

Clearance (CL/F_{oral}) standardised to a 70 kg human, using an allometric 3/4 power model [13], increased from 28 weeks post-conception (0.74 L/h/70kg) with a maturation half-life of 11.3 weeks to reach 10.9 L/h/70kg by 60 weeks. Clearance in children out of infancy is 13.5 L/h/70kg, CV 46% [14-16,23]. These estimates are at the lower range of those reported for adults (CL 12-21 L/h/70kg) [24].

Volume of distribution

The volume of distribution for paracetamol in mammals, including humans, is 49-70 L/70kg. The volume of distribution decreases exponentially with a maturation half-life of 11.5 weeks from 109.7 L/70kg at 28 weeks post-conception to 72.9 L/70kg by 60 weeks. A V/F$_{oral}$ of 66.6 L/70kg is reached out of infancy [17].

Disease states

The effects of altered physiology such as fever [25], anaesthesia [26], or mild hepatic dysfunction [27] on pharmacokinetic parameters have received little attention, but appear to have minimal impact in children.

Parenteral formulation

Propacetamol (N-acetylpara-aminophenoldiethyl aminoacetic ester) is a water-soluble prodrug of paracetamol that can be used intravenously over 15-30 min. It is rapidly hydroxylated into paracetamol (1 g propacetamol = 0.5 g paracetamol). Propacetamol may rarely be painful on infusion but is generally well tolerated.

7.9.4   TOXICITY

Chronic use of paracetamol

Paracetamol is the most common analgesic used in childhood. It is a safe medication in children, even with chronic use, when dose recommendations are not exceeded. This is not true of adults where the effects of alcohol, concomitant drug use and disease processes increase the potential for hepatotoxicity. Paracetamol overdose results in increased production of highly reactive electrophilic arylating metabolites by the hepatic cytochrome P450-dependent mixed function oxidase enzyme system (CYP2E1). The toxic metabolite of paracetamol, N-acetyl-p-benzoquinone imine (NAPQI), binds to intracellular hepatic macromolecules to produce cell necrosis and damage.

Significant hepatic and renal disease, malnutrition and dehydration increase the propensity for toxicity. Medications that induce the hepatic P450 CYP2E1, 1A2 and 3A4 systems (e.g. phenobarbitone, phenytoin, and rifampicin) may also increase the risk of hepatotoxicity. It is currently impossible to predict which individuals have an enhanced susceptibility to cellular injury from paracetamol. The co-ingestion of therapeutic drugs, foodstuffs or other xenobiotics has potential to induce these enzymes [28]. The influence of disease on paracetamol toxicity is unknown. It has been speculated that ingestion of paracetamol increases the potential for liver injury by another cause, such as a viral agent [29]. Hepatotoxicity causing death or requiring liver transplantation has been reported with doses above 75 mg/kg/day in children and 90 mg/kg/day in infants [28,30-32]. It is possible that even these regimens may cause hepatotoxicity if used for longer than 2-3 days [28]. Kozer *et al.* [33] have demonstrated that ill children receiving repeated large doses of paracetamol (>90 mg/kg/day) may show abnormalities in liver function associated with red cell glutathione depletion. It is unknown if there is a difference in the propensity to toxicity between children given paracetamol for fever and those given paracetamol for postoperative analgesia.

Single dose of paracetamol

The plasma concentration associated with toxicity after a single dose of paracetamol in children is extrapolated from adult data. The Rumack-Matthew paracetamol toxicity nomogram is widely used to guide management of paracetamol overdose in adults and children [34]. Paracetamol concentrations of more than 300 mg/L at 4 hours were always associated with severe hepatic lesions in adults, but none were observed with concentrations less than 150 mg/L. The half-life was less than 4 hours in all patients without liver damage.

Clearance is a non-linear function of weight, while volume of distribution is a linear function of weight. Dose is usually expressed as a linear function of weight. The 4-hour concentration is determined by clearance, not volume of distribution in children, because absorption is rapid after oral elixir. As a consequence younger children (1-5 years) require larger doses than older children and adults to achieve similar concentrations at 4 hours [35,36]. Children (1-5 years) with reported accidental ingestion of greater than 250 mg/kg (c.f. 150 mg/kg in adults) can have serum concentration measured at 2 h after ingestion for Rumack-Matthew nomogram plots rather than the 4 hour time point recommended in adults.

Neonates can produce hepatotoxic metabolites (e.g. NAPQI), but there are suggestions of a lower activity of cytochrome P450 in neonates [37,38].

7.9.5 PARACETAMOL DOSING

The maintenance dose of paracetamol for a patient can be calculated as follows:

Paracetamol maintenance dose =
paracetamol clearance × target concentration

Table 1

Suggested paracetamol dosing*

Age	Typical bodyweight (kg)	Clearance (L/h/kg)	Dose (oral)
neonate	3.3	0.21	50 mg 8 h
3 months	6	0.25	90 mg 6 h
6 months	7.5	0.27	120 mg 6 h
1 year	10	0.29	120 mg 4 h
5 year	18	0.25	250 mg 4 h
8 years	24	0.24	300 mg 4 h
12 years	38	0.21	375 mg 4 h
16 years	50	0.3	500 mg 4 h
adult	70	0.3	1000 mg 6 h

*Modified from Holford, N. H. G., *New Ethicals*, C9 (2002).

Individual clearance values can be predicted from weight and age [16,17]. Most of the age-related changes in clearance are complete by 1 year of age. There are also size-related changes in clearance so that the value per kg continues to decrease as weight increases in children. Clearance in adults is reported as higher than in children.

The doses in Table 1 have been calculated to maintain a target trough serum paracetamol concentration of 10 mg/L [9]. Doses in adults are limited by concerns about hepatotoxicity. For patients more than 20% overweight, ideal bodyweight should be used. Children prefer elixir to capsules or tablets. It is important to note that elixir formulation is available in different formulation strengths when prescribing the dose by volume. Proprietary cough and cold medications containing paracetamol are also available. Concurrent use of such medications can result in inadvertent paracetamol toxicity. The relative bioavailability of rectal formulations is reduced out of the neonatal period and dose can be modestly increased (20-30%). Dose is reduced in premature neonates (e.g. 24 mg/kg/day and 45 mg/kg/day at 30 weeks and 34 weeks post conception age, respectively) [17].

REFERENCES

1. Piletta, P., Porchet H. C., Dayer, P., *Clin. Pharmacol. Ther., 49.* 350 (1991).
2. Bjorkman, R., Hallman, K. M., Hedner, J., Hedner, T., Henning, M., *Pain, 57,* 259 (1994).
3. Bjorkman, R., *Acta Anaesthesiol. Scand. Suppl., 103,* 1 (1995).
4. Courade, J. P., Chassaing, C., Bardin, L., Alloui, A., Eschalier, A., *Eur. J. Pharmacol., 432,* 1

(2001).
5. Rumack, B. H., *Pediatrics, 943,* 943 (1978).
6. Brown, R. D., Kearns, G. L., Wilson, J. T., *J. Pharmacokinet. Biopharm., 26,* 559 (1998).
7. Korpela, R., Korvenoja, P., Meretoja, O. A., *Anesthesiology, 91,* 442 (1999).
8. Anderson, B. J., Woollard, G. A., Holford, N. H. G., *Eur. J. Clin. Pharmacol., 57,* 559 (2001).
9. Rawlins, M. D., Henderson, B. D., Hijab, A. R., *Eur. J. Clin. Pharmacol., 11,* 283 (1977).
10. Manyike, P. T., Kharasch, E. D., Kalhorn, T. F., Slattery, J. T., *Clin. Pharmacol. Ther., 67,* 275 (2000).
11. Montgomery, C. J., McCormack, J. P., Reichert, C. C., Marsland, C. P., *Can. J. Anaesth., 42,* 982 (1995).
12. Dange, S. V., Shah, K. U., Deshpande, A. S., Shrotri, D. S., *Indian Pediatr., 24,* 331 (1987).
13. Anderson, B. J., Meakin, G. H., *Paediatr. Anaesth., 12,* 205 (2002).
14. Brown, R. D., Wilson, J. T., Kearns, G. L., Eichler, V. F., Johnson, V. A., Bertrand, K. M., *J. Clin. Pharmacol., 32,* 231 (1992).
15. Anderson, B. J., Holford, N. H., Woollard, G. A., Kanagasundaram, S., Mahadevan, M., *Anesthesiology, 90,* 411 (1999).
16. Anderson, B. J., Woollard, G. A., Holford, N. H. G., *Brit. J. Clin. Pharmacol., 50,* 125 (2000).
17. Anderson, B. J., van Lingen, R. A., Hansen, T. G., Lin, Y. C., Holford, N. H. G., *Anesthesiology, 96,* 1336 (2002).
18. Levy, G. O., Khanna, N. N., Soda, D. M., Tsuzuki, O., Stern, L., *Pediatrics, 55,* 818 (1975).
19. Miller, R. P., Roberts, R. J., Fischer, L.T., *Clin. Pharmacol. Ther., 19,* 284 (1976).
20. van Lingen, R. A., Deinum, J. T., Quak, J. M., et al., *Arch. Dis. Child. Fetal Neonatal Ed., 80,* F59 (1999).
21. van der Marel, C., Anderson, B. J., Van Lingen, R. A., et al., *Eur. J. Clin. Pharmacol., 22,* Epub May 23 (2003).
22. Miners, J. O., Osborne, N. J., Tonkin, A. L., Birkett, D. J., *Br. J. Clin. Pharmacol., 34,* 359 (1992).
23. Anderson. B. J., Holford, N. H. G., Woollard, G. A., Chan, P. L,. *Br. J. Clin. Pharmacol., 46,* 237 (1998).
24. Prescott, L. F., *Paracetamol (acetaminophen), A Critical Bibliographic Review.* Taylor and Francis Publishers, (1996).
25. Wilson, J. T., Brown, R. D., Bocchini, J. A., Jr., Kearns, G. L., *Ther. Drug. Monit., 4,* 147 (1982).
26. Reilly, C. S., Nimmo, W. S., *Anaesthesia, 39,* 859 (1984).
27. al Obaidy, S. S., McKiernan, P. J., Li Wan Po, A., Glasgow, J. F., Collier, P. S., *Eur. J. Clin. Pharmacol., 50,* 69 (1996).
28. Kearns, G. L., Leeder, J. S., Wasserman, G. S., *J. Pediatr., 132,* 5 (1998).
29. Alonso, E. M., Sokol, R. J., Hart, J., et al., *J. Pediatr., 127,* 888 (1995).
30. Heubi, J. E., Bien, J. P., *J. Pediatr., 130,* 175 (1997).
31. Heubi, J. E., Barbacci, M. B., Zimmerman, H. J., *J. Pediatr., 132,* 22 (1998).
32. Rivera Penera, T., Gugig, R., Davis, J., et al., *J. Pediatr., 130,* 300 (1997).
33. Kozer, E., Barr, J., Bulkowstein, M., et al., *Vet. Hum. Toxicol., 44,* 106 (2002).
34. Rumack, B. H., Matthew, H., *Pediatrics, 55,* 871 (1975).
35. Anderson, B. J., Holford, N. H. G., Armishaw, J. C., Aicken, R., *J. Pediatrics, 135,* 290 (1999).
36. Mohler, C. R., Nordt, S. P., Williams, S. R., Manoguerra, A. S., Clark, R. F., *Ann. Emerg. Med., 35,* 239 (2000).
37. Roberts, I., Robinson, M. J., Mughal, M. Z., Ratcliffe, J. G., Prescott, L. F., *Br. J. Clin. Pharmacol., 18.* 201 (1984).
38. Levy, G., Garrettson, L. K., Soda, D. M., *Pediatrics., 55,* 895 (1975).

Immunosuppressants, rheumatic and gastrointestinal topics

Clinical section (A)

8.1 Pharmacology and use of immunosuppressants

Evelyne Jacqz-Aigrain[1], Pierre Wallemacq[2]

[1]*Department of Paediatric Pharmacology and Pharmacogenetics, Hospital Robert Debré, 48 Bvd. Sérurier, 75019 Paris, France*

[2]*Department of Clinical Chemistry, University Hospital St. Luc, Université Catholique de Louvain, 10 Ave. Hippocrate, B-1200 Brussels, Belgium*

8.1.1 INTRODUCTION

It was not long ago that great concern was expressed over the relative deficiency of information regarding immunosuppressive drug therapy in children. Most clinical trials after kidney, liver, heart or bone marrow transplantation are conducted in adult populations. However, several groups have gained experience in paediatric transplantation and generated data to change the status of "the therapeutic orphan" [1]. Organ transplantation is currently considered as a reasonable and viable therapeutic option in paediatrics, with an impressive improvement in the success rates reached. According to data for 1999 from the United Network for Organ Sharing Scientific Registry, 1 year patient survival for paediatric transplantation ranges from 97 to 100%, 80 to 91% and 80 to 91% for cadaveric kidney, liver and heart transplants, respectively [2]. This achievement has been reached in spite of both the significantly smaller number of paediatric transplantations in comparison to the adult population, and the specificity of this population. In 2001, within the Eurotransplant area (Belgium, Netherlands, Luxembourg, Germany, Austria and Slovenia), 207 children underwent an organ transplantation (kidney, liver, lung or heart), corresponding to about 5% of the transplantations in the adult population.

Specific paediatric concerns

The mean duration of the immunosuppressive therapy is longer than in adults, and the optimisation of therapy should consequently be even more rigorous.

Paediatric patients differ from adults in relation to body composition (water/fat), membrane permeability, plasma proteins and metabolic activity [3]. The administration of medicines, both oral and IV, is a potential problem [4, 5]. Corticosteroids may affect growth in children [6]. Post-transplant lymphoproliferative disorders (PTLD) are more common in children [7]. Compliance is a significant problem in adolescents [8].

Ontogeny and drug disposition

The maturation of organ systems during childhood exerts a profound effect on drug disposition. Most immunosuppressive agents are metabolised by the CYP3A subfamily, and the maturation of drug metabolising enzymes is a major factor accounting for age-associated changes in non-renal drug clearance [9]. Clinicians should consider the impact of ontogeny on CYP3A when prescribing immunosuppressants to children, in order to avoid both sub-therapeutic and toxic doses [10].

Therapeutic drug monitoring

Individualising a patient's drug therapy to obtain the ideal balance between therapeutic efficacy and occurrence of adverse effects is the physician's goal. Pharmacokinetic and pharmacodynamic approaches contribute in helping physicians to individualise long term and life threatening therapies. Pharmacogenetics may appear in the close future as an additional approach [11]. Therapeutic drug monitoring (TDM) helps clinicians to maintain blood or plasma concentrations within the target ranges, avoiding reduced efficacy or toxicity. Pharmacokinetic parameters such as trough concentrations (C_0), selected post-dose concentrations (e.g., two hours postdose C_2), area under the concentration time profile (AUC), sometimes monitored by using a limited sampling strategy [12], are the most widely used tools for investigation of the relation between drug concentration and effect and to validate target therapeutic ranges [13]. It is generally accepted that after the initial post-transplant period, reduced immunosuppression is required to prevent rejection. In addition, the impact of co-medications may have an effect on the pharmacokinetics of reference compounds [14].

8.1.2 PHARMACOLOGY AND USE OF IMMUNOSUPPRESSANTS

Initially, ciclosporin (CsA), the cornerstone of immunosuppressive treatment, was given in association with azathioprine and corticosteroids to prevent acute and chronic rejection. Tacrolimus (Tac) is more potent and somewhat less toxic than CsA, and may replace CsA as the first-line drug in paediatric transplantation. Mycophenolate mofetil (MMF) is increasingly used in paediatric organ (both liver and kidney) transplantation, in association with CsA or Tac in patients with deteriorating renal function, or to provide additional immunosuppression in patients resistant to conventional medication. Monoclonal antibody therapy in combination with standard immunosuppressive therapy has given promising preliminary data in children.

Anticalcineurines

Ciclosporin

Mechanism of action. CsA selectively inhibits T lymphocyte activation in an early stage (transition from the G_0 to the G_1 phase of the cell cycle). It complexes and inhibits the rotamase activity of its major cytoplasmic binding protein, namely cyclophilin. The immunophilin-drug complex binds specifically to the calcineurin/calmodulin/calcium complex, thereby inhibiting the calcineurin serine/threonine phosphatase activity, resulting in the immunosuppressive effect (transcription of T-cells genes coding for interleukin-2 and additional cytokines and cytokine receptors). CsA also increases expression of TGF-β, a potent inhibitor of IL2 stimulated T-cell proliferation and generation of antigen-specific cytotoxic T lymphocytes.

Pharmacokinetics. CsA can be administered intravenously or orally (Neoral® microemulsion). It is a highly lipophilic molecule with wide inter- and intra-individual variability in pharmacokinetics and a low therapeutic index. CsA has a low and variable oral bioavailability, mainly related to a hepatic first pass-effect and to the differential expression of P-glycoprotein (P-gp) and CYP3A family in the intestinal epithelial cell. CYP3A4 is involved in the intestinal first-pass effect of the drug and the P-gp protein acts as an efflux pump limiting the amount of CsA absorbed [15]. The drug is highly bound to plasma proteins and to erythrocytes. The volume of distribution is large. CsA is extensively metabolised by the liver (CYP3A4 and CYP3A5). Clearance is higher in infants and children than in adults and the elimination half-life may appear shorter. Major drug interactions observed with CsA involve CYP3A inhibitors (ketoconazole, erythromycin, nicardipine) and inducers (rifampicin, phenobarbital, carbamazepine).

Adverse drug reactions. Side effects are, in most cases, reversible by dose reduction. Nephrotoxicity is of particular significance to renal transplantation, and critical to distinguish from rejection episode. Additional complications include neurological side effects (tremor, seizures, cortical blindness), hirsutism, gingival hyperplasia, hypertension, hypomagnesaemia and hypercholesterolaemia.

Dosage recommendations and monitoring. Inter-individual variation make oral dosing management of CsA difficult. As CsA displays a low therapeutic index, monitoring is mandatory to avoid reduced efficacy or toxicity. Similarly to adult data, monitoring should not be based on the predose concentration but on concentrations two (C_2) or three hours (C_3) after dosing [16-17]. Cole *et al.* proposed some recommendations focused on a step-wise approach to implementation of Neoral® C_2 monitoring [18]. The target CsA C_2 levels for liver transplant patients are 1000 µg/l (months 0-3 post-transplantation) and show a progressive reduction down to 600 µg/l (>6 months). Higher levels are proposed for renal transplant patients, starting from 1500-1700 µg/l (months 0-2) and reduced to about 800 µg/l (>12 months). Most of these data were obtained from adult studies [16-18]. The AUC, that represents the patient's exposure to the drug, is however a better tool for CsA monitoring than C_0 or C_2, and limited sampling strategies were developed to estimate AUC based on a small number of samples collected at specific times [12]. Inter-individual variation in pharmacokinetics is reflected in the wide range of Neoral® doses, ranging from 2-10 mg/kg/day [10], to be adjusted according to blood concentrations.

Tacrolimus (FK506)

Mechanism of action. Tacrolimus (Tac) is a lipophilic macrolide displaying similar but more potent immunosuppressive properties. Similarly to CsA, Tac inhibits calcineurin and calcineurin dependent events, although via a distinct cytosolic protein receptor (FK506 binding protein-12 or FKBP12), resulting in inhibition of T-cell activation [19].

Pharmacokinetics. Tac is also a substrate of P-gp and CYP3A4 [20]. Due to first-pass effect, drug absorption is poor and variable, the peak concentration occurring between 0.5 and 4 hours. The drug is highly bound to plasma proteins (75 – 90%) and red blood cells, with a high blood to plasma ratio (12:1). Metabolism by CYP3A, both in the liver and the gut, is followed by biliary excretion. At least 10 metabolites are formed, some of which retain significant activity [21]. Paediatric patients have a higher volume of distribution and a higher clearance than adults. The mean total body clearance normalised to bodyweight in paediatric patients is 0.14 L/h/kg after intravenous administration following liver transplantation (Table 1). This was approximately double the value observed in adult liver transplant patients of 0.06 L/h/kg [21].

Adverse drug reactions. The side effects are similar to those of CsA although hirsutism and gingivitis are less frequent. Cardiotoxicity has been reported in a few patients [22]. Many paediatric centres favour this drug, due to its potent activity together with the absence of hirsutism.

Table 1

Pharmacokinetic parameters of
ciclosporin, tacrolimus and mycophenolate mofetil

	Ciclosporin	Tacrolimus	Mycophenolate mofetil
substrate of P-gp and CYP3A	yes	yes	no
bioavailability (%)	21-24	25 (3-77)	95
volume of distribution (l/kg)	4.3	2.6 (0.5-4.7)	3 - 6
systemic clearance (l/h/kg)	0.37	0.14 (0.07-0.21)	0.16
elimination half-life (hours)	8.1	12.4 (8-16.8)	12-18
references	46	20, 21, 47	48

Dosage recommendations and drug monitoring. As for CsA, due to high inter-individual variation, TDM of Tac is mandatory. Published pharmacokinetic data have shown a weak correlation between dose and AUC and a satisfactory correlation between trough level and AUC [23]. At Tac $C_0 < 5$ μg/l, the incidence of acute rejection was around 30% with hardly any toxicity. At levels > 15 μg/l the incidence of rejection was about 3% but there was a 45% incidence of toxicity, resulting in the proposed target concentration range of 5-15 μg/l for optimal efficacy and minimal toxicity [24]. However, the same pharmacokinetic/ pharmacodynamic rationale proposed for CsA C_2 could probably be applied to Tac. Indeed, recent data question the predictive value of single C_0 and suggest that a measurement made at an additional time point during the absorption phase would be useful [25].

Oral Tac doses range from 40-300 mcg/kg/day, depending upon the blood levels, the concomitant therapies, the age and the period post-transplantation [10]. The consequence of the higher clearance observed in paediatric patients will have an impact on the dose required to reach target blood concentrations. The comparison with the adult population is illustrated in Figure 1. A recent paper described the use of a steroid-free combined tacrolimus-basiliximab treatment [26].

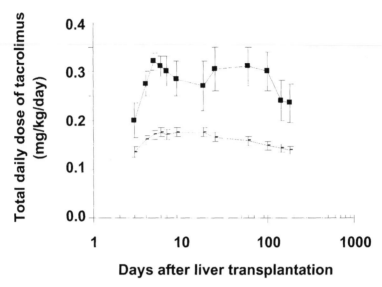

Days after liver transplantation

Figure 1. Comparison of mean dosages (± SEM) of oral tacrolimus required to maintain similar trough blood concentrations in adult and paediatric (mean age 4.6, range 0.7-13.2 years) liver transplant patients. Paediatric dosages were approximately 2-fold higher than the corresponding adult dosages because of the faster clearance of tacrolimus in children. (With permission of Clinical Pharmacokinetics [21]. (**I**, adult population; ■, paediatric population).

Sirolimus

Sirolimus (Siro) was introduced recently in clinical practice for the immunotherapy of organ transplantation in adults, and drug evaluation in paediatric population is very limited.

Mechanism of action. Siro binds, within the target cells, to the same member of the immunophilin family of cytosolic binding proteins as Tac, FKBP-12. The sirolimus-FKBP-12 complex binds to a specific cell-cycle regulatory protein – mTor (mammalian target of rapamycin) – and inhibits its activation. Even though Siro binds to FKBP-12, the complex has no effect on calcineurin activity [27]. The inhibition of mTor results in suppression of cytokine-driven (ie, cytokines such as interleukins IL-2, -4 and -15) T-lymphocyte proliferation, inhibiting the progression from G_1 to the S phase of the cell cycle.

Pharmacokinetics. Studies in adults have reported a large interindividual variation in the pharmacokinetic parameters of the drug. It is a substrate of the P-gp transporter and metabolised by CYP3A subfamily. Gastro-intestinal absorption is rapid but variable, and the volume of distribution is large. Elimination half-life is long, between 35 and 95 hours. This should therefore reduce the frequency of Siro dosage, as compared to CsA or Tac, but also increases the importance of rapidly reaching the optimal blood concentration by using a loading dose, or perhaps, pharmacogenetics investigations prior to transplantation [28]. Siro binds and is sequestered to an even larger extent to red blood cells, since the whole blood/plasma ratio is approximately 30:1.

Adverse drug reactions. In contrast to CsA and Tac, Siro is not nephrotoxic. Hyperlipidaemia is a very common side effect that although manageable, remains a limiting factor for use in paediatric patients. Additional side effects observed under Siro therapy include thrombocytopenia and leukopenia.

Dosage recommendations and drug monitoring. TDM of Siro is essential. Dose is a poor predictor of drug exposure and clearance is dependant on liver function and blood flow. Biodisposition depends on the activity of CYP3A4, CYP3A5 and P-gp subject of potential drug interactions and genetic polymorphism. In paediatric patients, higher dosage may be required, as clearance by bodyweight appears higher in patients aged from 5-11 years than in older patients (12-18 years) [29]. Target concentrations for Siro appear to be between 4-12 μg/l, when in combination with either CsA or Tac [30]. Due to the long half-life, blood samples for TDM should not be collected for at least one week after changes in dosage.

Cytotoxic agents

Azathioprine

Azathioprine (Aza) is a prodrug of 6-mercaptopurine (6MP), widely used until recently in combination with CsA and corticosteroids for the prevention of allograft rejection in organ transplantation. Immunosuppression is dependent upon the active 6-thioguianine nucleotides.

Pharmacokinetics. Aza is well absorbed after oral administration. It is distributed through the body with 30% of the drug being protein bound. The metabolism of 6MP by the cytosolic enzyme thiopurine methyltransferase (TPMT) is under pharmacogenetic control. Elimination half-life of both Aza and 6MP is less than 2 hours.

Adverse drug reactions. The pharmacogenetics of 6MP account for the genotypic differences in the risk of side effects. Bone marrow suppression was initially reported in poor metabolisers, homozygous for a deficient TPMT allele. Other adverse effects include rash, alopecia, hepatotoxicity, pancreatitis and intestinal disorders (vomiting and diarrhoea).

The monitoring of Aza plasma concentrations is of minimal value, as the efficacy of the drug is dependent on the concentrations of the active intracellular metabolite.

Mycophenolate mofetil

Mycophenolate mofetil (MMF) is a potent immunosuppressive drug, replacing Aza in most renal transplant protocols as it reduces the incidence of acute rejection episodes and improves long term graft survival in adults and children [31].

Mechanism of action. MMF is an ester prodrug of mycophenolic acid (MPA), a potent and reversible inhibitor of the *de novo* synthesis pathway of guanosine monophosphate, resulting in DNA synthesis inhibition [32].

Pharmacokinetics. MMF can be administered orally or intravenously and the prodrug is rapidly converted to MPA. Peak concentrations occur 0.5-1.3 hours after oral administration. MPA is highly protein bound. It is eliminated by hepatic and extrahepatic glucuronidation. Enterohepatic recirculation might occur, resulting in a second MPA peak during the elimination phase [33].

Drug interactions affecting MPA have been reported, via different mechanisms. Tac increases MPA concentrations by inhibition of glucuronidation, while CsA decreases MPA concentrations by MPA biliary excretion [34]. MPA binding is not affected by drugs such as CsA, Tac, paracetamol or ibuprofen, but is decreased by sodium salicylate and furosemide.

The pharmacodynamic response to MPA may be assessed by monitoring inosine monophosphate dehydrogenase [35].

Dosage recommendations and drug monitoring. TDM is recommended due to the high inter-individual variation in pharmacokinetics [36]. High performance liquid chromatography (HPLC) remains the reference method that allows one to quantify plasma concentrations of MPA and glucuronidated metabolites. The EMIT method, however, is faster and suitable for TDM. The recommended trough concentration is 1-3.5 mg/l. The total MPA AUC values of 30-60 mg.h/l is a reasonable target in the early post-transplant period, when associated with CsA. Advised oral doses of MMF range from 490-1150 mg/m^2 12 hourly [10]. The possible steroid-sparing potential of MMF is an important issue in paediatric renal transplantation. Preliminary data demonstrate improved longitudinal growth, less Cushingoid habitus and lower blood pressure after steroid-withdrawal in paediatric renal transplant recipients on MMF and CsA therapy [37].

Corticosteroids

Classification. Glucocorticoids belong to an old class of drugs, widely used in children with allergic and inflammatory diseases. The different compounds are presented in Table 2.

Mechanism of action. The effects are mediated through corticosteroid receptors, members of the large steroid-nuclear receptor family that also include receptors for mineralocorticoids, sexual hormones, vitamin D, thyroid hormone and retinoic acid. The corticosteroid receptor is located in the cytoplasm and enters the nucleus of the cell after activation by different hormones. The corticosteroid receptor regulates gene transcription in target genes by binding, as a homodimer, to corticosteroid responsive elements. Corticosteroids enhance the transcription of several genes, limiting expression of inflammatory cytokines through transcription factors such as NK-kB and AP-1, associated with the induction of numerous genes, encoding for cytokines, chemokines and adhesion molecules (TNF-α, IL-1β, IL-6, ICAM-1), and resulting in the anti-inflammatory effects of corticosteroids. For some genes (IL-11, GM-CSF and cyclooxygenase), regulation of expression by corticosteroids is both transcriptional and posttranscriptional [38].

Table 2

Glucocorticoids

Glucocorticoids	Equivalent dose (mg)	Gluco-corticoid potency	Mineralo-corticoid potency	Plasma half-life (min)	Biologic half-life (h)
short-acting					
cortisol	20	1	2	2	90
cortisone	25	0.8	2	2	80-118
intermedate-acting					
prednisone	5	4	1	1	60
prednisolone	5	4	1	1	115-200
triamcinolone	4	5	0	0	30
methylprednisolone	4	5	0	0	180
long-acting					
dexamethasone	0.5	25-50	0	0	15-20
betamethasone	0.6	25-50	0	0	200

Adverse drug reactions. Chronic treatment has been associated with multiple side effects, including Cushing's syndrome during therapy and adrenal insufficiency after discontinuation of treatment. In children, the major side effect is growth retardation. This effect is reduced by alternate day administration.

Corticosteroid administration in organ transplantation. Corticosteroids are prescribed as first-line drugs for the prevention and treatment of acute rejection. Prednisone, a prodrug of the active prednisolone, is the most frequently used molecule in reference immunosuppressive regimen, in association with CsA and MMF. Due to side effects in paediatric patients, the trend is now to discontinue their use early after transplantation or even to avoid corticosteroids in the more recent immunosuppresive protocols [26]. Methylprednisone pulses are administered to control acute graft rejection [39].

Antibodies

Polyclonal and monoclonal antibodies belong to a class of immunosuppressants for which conventional pharmacological evaluation is inappropriate, as the mechanism of action is blockage of a cell target. In addition, evaluation has been predominantly conducted in adult patients, and data regarding efficacy and tolerance in paediatric organ transplant patients are limited.

Polyclonal antithymocyte (ATG) and antilymphocyte globulin (ALG) were the first antibodies included in the initial post-transplantation phase. Monoclonal antibodies, which are more homogenous and more specific, are better tools for immunosuppression than polyclonal antibodies.

Muromonad - OKT3 was the first monoclonal to be used in transplanted patients, demonstrating efficacy in the treatment of acute rejection in adult organ transplanted patients [40].

Daclizumab is an anti-IL2 receptor antibody, primarily used to prevent acute rejection in organ transplanted patients with renal dysfunction [41].

Basiliximab is a chimeric human/mouse monoclonal antibody against the α chain of the IL-2 receptor CD25. It was demonstrated to reduce acute rejection in adult renal transplanted patients under dual or triple immunotherapy, without increasing the incidence of adverse events (including infection and malignancy) [42]. Although involving a limited number of children, immunoprophylaxis with basiliximab given intravenously at day 0 (pre-operatively) and 4, associated with dual therapy, reduced acute rejection in the first six months and maintained graft survival and function [43-44]. Adverse drug reactions were similar in baxiliximab and placebo treated adult renal transplant recipients receiving immunosuppressants. Tolerability data are however limited in paediatric patients. Anaphylactic shock is a very rare event, reported after a second course of basiliximaab [45].

8.1.3 CONCLUSION

Immunosuppressive therapy in paediatric organ transplant recipients is changing, as a consequence of the increasing number of available immunosuppressive agents. The rationale for monitoring the four immunosuppressants CsA, Tac, MMF and Siro is due to the significant inter-individual variation in pharmacokinetics, together with a narrow therapeutic index. Another issue may be the emergence of generic formulations. Bioequivalence guidelines recommend studying bioequivalence in subpopulations of patients who are known to be poor absorbers or who exhibit marked pharmacokinetic variability, such as children. Once again, most studies are, or will be, conducted in adult populations. The potentially different pharmacokinetic profiles of generic drugs in the paediatric population and the appearance of new formulations has raised several safety concerns.

Population methods should be particularly appropriate in children because they allow the modelling of pharmacokinetic responses in a relatively large group of patients with very sparse data obtained during routine TDM. Population models would offer the possibility of better dosage individualisation, both *a priori* and *a posteriori*, using Bayesian forecasting.

REFERENCES

1. Shirkey, H., *J. Pediatr., 72*, 119 (1968).
2. *Transplant Patient Datasource*, Richmond (VA) : United Network for Organ sharing. http://www.patients.unos.org/data.htm, (2000).
3. Milsap, R. L., Hill, M. R., Szefler, S. J. *Chap 10, Applied Pharmacokinetics. Principles of Theraputic Drug Monitoring* (3rd ed.), Evans, W. E., Schentag, J. J., Jusko, W. J., Applied Therapeutic Inc, Vancouver, WA, (1992).
4. Reding, R., Sokal, E., Paul, K., et al., *Pediatr. Transplant., 6,* 124 (2002).
5. Firdaous, I., Hassoun, A., Clement de Cletty, S., Reding, R., Otte, J. B., Wallemacq, P. E., *Transplantation, 57*, 1821 (1994).
6. Benfield, M. R., Stablein, D., Tajani, A., *Pediatr. Tranplant., 1*, 27 (1999).
7. Sokal, E. M., Antunes, H., Beguin, C., et al., *Transplantation, 64*, 1438 (1997).
8. Swanson, M. A., Palmeri, D., Vossler, E. D., Bartus, S. A., Hull, D., Schweizer, R. T., *Pharmacotherapy, 11*, 1735 (1991).
9. de Wildt, S. N., Kearns, G. L., Leeder, J. S., van den Anker, J. N., *Clin. Pharmacokinet., 37*, 485 (1999).
10. Fernandez de Gatta, M. , Santos-Buelga, D. , Dominguez-Gil, A. , Garcia, M. J., *Clin. Pharmacokinet., 41*, 115 (2003).
11. Haufroid, V., Mourad, M., Van Kerckhove, V., et al., *Pharmacogenetics, 14*, 147 (2004).
12. David, J. O., Johnston, A., *Ther. Drug Monit., 23*, 103 (2001).
13. Shaw, L. M., Holt, D. W., Keown, P., Venkataramanan, R., Yatscoff, R. W., *Clin. Ther., 21*, 1632 (1999).
14. Brown, N. W., Aw, M. M., Mieli-Vergani, G., Dhawan, A., Tredger, J. M., *Ther. Drug Monitor., 24*, 598 (2002).
15. Saeki, T., Ueda, K., Tanigawara, Y., Hori, R., Komano, T., *J. Biol. Chem., 268*, 6077 (1993).

16. Cole, E., Midtvedt, K., Johnston, A., Pattison, J., O'Grady, C., *Transplantation, 73*, S19-S22 (2000).
17. Levy, G., Thervet, E., Lake, J., Uchida K., *Transplantation, 73*, S12 (2002).
18. Cole, E., Midtevedt, K., Johnston, A., Pattison, J., O'Grady, C., *Transplantation, 73,* S19 (2002).
19. Peters, D. H., Fitton, A., Plosker, G. L., Faulds, D., *Drugs, 46,* 746 (1993).
20. Lampen, A., Christians, U., Guenguerich, F. P., et al., *Drug Metab. Dispos., 23,* 1315 (1995).
21. Wallemacq, P. E., Verbeeck, R. K., *Clin. Pharmacokinet., 40,* 283 (2001).
22. Atkinsoon, P., Joubert, G., Barron, A., et al., *Lancet, 345,* 894 (1995).
23. Filler, G., Grygas, R., Mai, I., et al., *Nephrol. Dial. Transplant., 12,* 1668 (1997).
24. Oellerich, M., Armstrong, V. W., Schütz, E., Shaw, L. M., *Clin. Biochem., 31,* 309 (1998).
25. Macchi-Andanson, M., Charpiat, B. , Jelliffe, R.W. , Ducerf, C., Fourcade, N., Baulieux, J., *Ther. Drug Monit., 23,* 129 (2001).
26. Reding, R., Gras, J., Sokal, E., Otte, J. B., Davies, H. F., *Lancet, 362,* 2068 (2003).
27. Dumont, F. J., Staruch, M. J., Koprak, S. L., Melino, M. R., Sigal, N. H., *J. Immunol., 144,* 251 (1990).
28. Thummel, K. E., *Pharmacogenetics, 14,* 145 (2004).
29. Tejani, A., Alexander, S., Ettenger, R., et al., *Pediatr. Transplant., 8,* 151 (2004).
30. Holt, D. W., Denny, K., Lee, T. D., Johnston, A., *Transplant. Proc., 35,* 157S (2003).
31. European Mycophenolate Study Group. *Transplantation, 68,* 391 (1999).
32. Allison, A. C., Eugui, A. M., *Immunopharmacology, 47,* 85 (2000).
33. Jacqz-Aigrain, E., Khan Shaghaghi, E., Baudouin, V., et al. *Pediatr. Nephrol.,14,* 95 (2000).
34. Shaw, L. M., Kaplan, B., DeNofrio, D., Korecka, M., Brayman, K. L., *Ther. Drug Monit., 22,* 14 (2000).
35. Langman, L. J., Legatt, D. F., Halloran, P. F., Yatscoff, R. W.,*Transplantation, 62,* 666 (1996).
36. Shaw, L. M., Holt, D. W., Oellerich, M., Meiser, B., van Gelder, T., *Ther. Drug Monitor., 23,* 305 (2001).
37. Leung, D. Y., Bloom, J. W., *J. Allergy Clin. Immunol., 111,* 3 (2003).
38. Weber, L. T., Hocker, B., Mehls, O., Tonshoff, B., *Minerva Urol. Nefrol., 55,* 91 (2003).
39. Debray, D., Furlan, V., Baudouin, V., et al., *Pediatr. Drugs, 5,* 81 (2003).
40. Hooks, M. A., Wade, C. S., Millikan, W. J., *Pharmacotherapy, 11,* 26 (1991).
41. Nashan, B., Light, S., Hardie, I. R., Lin, A., Johnson, J. R., *Transplantation, 15,* 110 (1999).
42. Chapman, T. M., Keating, G. M. *Drugs, 63,* 2803 (2003).
43. Kovarik, J. M., Offner, G., Broyer, M., et al., *Transplantation, 74,* 966 (2002).
44. Clark, G., Walsh, G., Deshpande, P., Koffman, G., *Nephrol. Dial. Transplant., 17,* 1304 (2002).
45. Baudouin, V., Crusiaux, A., Haddad, E., et al., *Transplantation, 76,* 459 (2003).
46. Wallemacq, P. E., Reding, R., Sokal, E., et al., *Transplant. Int., 10,* 466 (1997).
47. Wallemacq, P. E., Furlan, V., Moller, A., et al., *Eur. J. Drug Metab. Pharmacokinet., 23,* 367 (1998).
48. Fulton, B., Markham, A. *Drugs, 51,* 278 (1996).

8.2 Rheumatic disorders

A. G. Cleary[1], H. Venning[2]

[1]Department of Paediatric Rheumatology, Royal Liverpool Children's Hospital, Eaton Road, Liverpool LI2 2AP, UK

[2]Department of Paediatric Rheumatology, Queen's Medical Center, University Hospital, Nottingham NG7 2NH, UK

8.2.1 INTRODUCTION

The spectrum of rheumatic disorders in childhood is varied and includes several disorders sharing the same broad aetiopathogenesis (i.e. that of a complex genetic trait interacting with an as yet undefined environmental trigger). The phenotypical features of each disease are determined by an organ-specific autoimmune process. Until the recent advent of biological therapies, the pharmacological management of rheumatic disorder in childhood involved the use of agents with a broad antiinflammatory or immunosuppressant mode of action. The improved outcome in many rheumatic diseases has been associated with earlier and more aggressive use of such immunosuppressants which may be considered as disease modifying antirheumatic drugs (DMARDs). Targeted biological therapy is now available to the paediatric rheumatologist in the form of anti-tumour necrosis factor alpha (TNF-α) treatment, with the realistic expectation that other biological agents will soon complete the transition from experimental to licensed therapy.

The incidence of all rheumatic disorders in the UK is 32-42/100,000 children under age 16 [1]. There is, however, considerable worldwide variation in this figure [2]. Juvenile idiopathic arthritis (JIA) has been proposed as an umbrella term to unify the two previous classifications for childhood inflammatory arthritis, juvenile rheumatoid arthritis (JRA) and juvenile chronic arthritis (JCA) [3]. Other conditions that will be considered here include connective tissue diseases with the potential for multi-organ involvement, vasculitic disorders, and inflammatory uveitis. This chapter will highlight the pharmacological aspects of management of rheumatic disease, but it must be emphasised that management of these diseases in children requires a multidisciplinary approach with focus on medical, physical and psychosocial aspects. There is unfortunately a limited evidence base in the form of randomised controlled trials to guide pharmacological management.

8.2.2 TREATMENT OPTIONS

Intra-articular corticosteroids

In JIA intra-articular corticosteroid injections are now considered in many centres as first-line therapy in the oligoarticular sub-type, or adjunctive therapy before DMARDs have become effective in polyarthritis. It is often possible to induce rapid, early control of synovitis, relief of symptoms and rapid rehabilitation. Complications, such as leg length discrepancy, may be avoided [4]. The mechanisms of action of corticosteroids are discussed below.

The procedure can often be performed in older children (usually over 7 years) with local anaesthesia and/or conscious sedation with, more and more commonly, nitrous oxide [5]. Alternatively, some centres prefer to sedate with intravenous benzodiazepines such as midazolam. A short general anaesthetic may be necessary in younger or more anxious patients [6]. Triamcinolone hexacetonide is the most effective agent [7], but, at the time of writing, there are worldwide manufacturing difficulties. Triamcinolone acetonide is the most appropriate alternative. The most common adverse effect is subcutaneous atrophy at the site of injection [8]. This can be minimised by good injection technique. Methylprednisolone and hydrocortisone are short-acting, and not generally recommended for intra-articular use, but frequently used for injection into tendon sheaths, where the risk of subcutaneous atrophy is high. Systemic absorption sufficient to

Table 1

Dose regime for intra-articular corticosteroid injection in JIA

	Triamcinolone hexacetonide	Triamcinolone acetonide
large joint (knee, ankle, hip, shoulder)	1 mg/kg (max. 40 mg)	2 mg/kg (max. 80 mg)
smaller joint (wrist, elbow)	0.5 mg/kg (max. 20 mg)	1 mg/kg (max. 40 mg)
MCPs / MTPs	1-2 mg/joint	2-4 mg/joint
PIPs	0.6-1 mg/joint	1.2 – 2 mg/joint

induce a Cushingoid appearance has been reported, and is more common after multiple injections and with the use of less soluble agents, such as triamcinolone acetonide [9]. A typical dose regime for intra-articular steroid injection is shown in Table 1.

Nonsteroidal anti-inflammatory drugs

NSAIDs are used for their analgesic and antiinflammatory effects. A summary of the NSAIDs most frequently used is shown in Table 2. The dosage and frequency of administration required to achieve antiinflammatory and antipyretic effect are generally higher than those required to achieve analgesic effect. NSAIDs may be used in all rheumatic disease, but most frequently in JIA where they are occasionally used as monotherapy in mild disease, but more frequently are used as an adjunct to DMARDs, both in the lag phase before the latter has become effective or where DMARDs are not fully effective at achieving control.

Table 2

Non-steroidal antiinflammatory drugs

Drug	Total daily dose	Frequency (times daily)	Notes
diclofenac	1-3 mg/kg	2-3	max. 150 mg/day; SR, dispersible available
naproxen	10-20 mg/kg	2-3	max. 1g/day; suspension available
ibuprofen	20-60 mg/kg	3-4	suspension available; melt preparation available
indomethacin	1-2 mg/kg	2	max. 200 mg/day
piroxicam	body weight dose < 15kg 5 mg 16-25 kg 10 mg 26-45 kg 15 mg > 46 kg 20 mg	Once daily	sublingual preparation available
celecoxib	200 mg	1-2	not licensed for children. (<14 years)

The pro-inflammatory mediators, prostaglandins and leukotrienes, are produced from arachidonic acid by the action of the cyclo-oxygenases 1 and 2 (COX-1 and COX-2). NSAIDs inhibit the COX enzyme. The so-called "constitutive" COX-1 is expressed in many cells and, in particular, regulates the production of cytoprotective gastric mucus and also contributes to control of renal blood flow [10]. COX-2 is expressed selectively at sites of inflammation, and as such, is considered "inducible".

Adverse effects associated with NSAIDs accord to the relative potency of each agent. NSAIDs are generally well tolerated in children, although potentially serious adverse gastrointestinal, hepatic, cutaneous, CNS and renal adverse effects are recognised [11]. Ibuprofen has the lowest incidence of gastro-intestinal side effects (discomfort, nausea, diarrhoea and ulceration), although the risk increases with more potent agents such as diclofenac, naproxen, piroxicam and indomethacin. The COX-2 selective drugs rofecoxib and celecoxib have theoretically less gastro-intestinal toxicity, but their role remains unclear in paediatric practice. Rofecoxib is being withdrawn by the manufacturer due to its longterm cardiovascular toxicity. Other side effects include bronchospasm, headache, dizziness, blood disorders, nervousness, depression and sleep disturbance. Naproxen has been associated with scarring pseudoporphyria skin reactions on the face [12]. Care needs to be taken in children with pre-existing renal impairment, as NSAIDs may provoke renal failure, although this is rare and tends to be reversible on discontinuation of the drug [13].

Systemic corticosteroids

Corticosteroids have an established role in the management of inflammatory disease with major organ involvement, but the significant short- and long-term risk of side effects make judicious and carefully considered use mandatory. Corticosteroids bind to cytoplasmic receptors and are translocated into the nucleus where they inhibit the expression genes coding for inflammatory proteins (cytokines, enzymes, receptors and adhesion molecules) and up-regulate the expression of genes coding for anti-inflammatory proteins. These complex mechanisms are reviewed in detail elsewhere [14]. Corticosteroids are metabolised in the liver, and their effects may be reduced by drugs that induce liver enzymes. The relative potency of corticosteroid preparations is shown in Table 3.

Systemic corticosteroids are rarely used in JIA. The principle indication is to treat the inflammatory features seen in the systemic arthritis sub-type. High dose intravenous "pulses" of methylprednisolone (30 mg/kg/day) may be used, but regimes vary and include a daily regime for 3 consecutive days, possible repeated the following week, or weekly for several weeks, according to res-

Table 3

Equivalent anti-inflammatory doses of corticosteroids

Drug	Equivalent anti-inflammatory dose (mg)
betamethasone	0.75
dexamethasone	0.75
hydrocortisone	20
methylprednisolone	4
prednisolone	5
triamcinolone	4

Adapted from British National Formulary March 2003.

ponse [15]. Oral prednisolone, in doses ranging from 500 mcg to 2 mg/kg/day, may be necessary in systemic arthritis, but the dose should be tapered rapidly and, if not clinically possible to do so, then adjunctive immunosuppressant therapy should be considered. Some paediatric rheumatologists will use pulses of methylprednisolone, or strictly defined courses of oral prednisolone, for rapid symptom relief in conjunction with the initiation of second line treatment of inflammatory polyarthritis. Systemic corticosteroids are indicated in connective tissue disease with major organ involvement, frequently in conjunction with disease modifying agents, such as methotrexate, azathioprine, ciclosporin or cyclophosphamide.

Topical corticosteroids and mydriatics are employed in the management of idiopathic uveitis or uveitis associated with JIA. However, if the inflammation cannot readily be controlled, many centres will consider the early use of DMARDs, such as methotrexate [16], ciclosporin [17] or etanercept [18], in order to minimise the risk of steroid-induced cataract and glaucoma.

The risk of adverse effects of corticosteroids increases with increasing daily dose (Table 4). All children treated for periods greater than 3 weeks should have the dose tapered, rather than stopped abruptly, because of the risk of adrenal suppression.

Methotrexate

Methotrexate is currently the DMARD of first choice in JIA, with its benefit proven in a randomised placebo-controlled double-blind trial [19]. It should be considered and commenced early in polyarthritis, and appears to be particularly

Table 4

Adverse effects of corticosteroids
growth failure
osteoporosis
Cushing's syndrome
weight gain
oedema
hypertension
gastro-intestinal disturbance
impaired glucose tolerance
mood change
myopathy
glaucoma
cataract
increased susceptibility to infection
adrenal suppression

effective in the extended oligoarthritis sub-group [20]. Methotrexate should be considered in oligoarticular disease, when there is inadequate response to intra-articular steroid injections. In addition, methotrexate may have a role in the treatment of all the paediatric rheumatic diseases.

The mechanism by which methotrexate mediates its antiinflammatory effect is unknown. Methotrexate is a folic acid antagonist that inhibits DNA synthesis by inhibition of thymidylate synthetase [21]. Methotrexate may interfere with *de novo* purine biosynthesis by inhibition of aminoimidazocarboxamide transfor-mylase [22]. More recent work suggests that the primary mechanism of action of methotrexate in JIA is mediated by adenosine acting at one or more of its receptors [23].

Methotrexate is used in doses of up to 30 mg/m^2 per week in children with JIA, and it is generally accepted that children tolerate doses that are higher than those usually given to adult patients (7.5-15 mg) [24]. Children have been shown to clear methotrexate via the renal tract more rapidly than adults [25]. It is routine to commence therapy with doses in the region of 10-15 mg/m^2 per week. The convention for weekly administration in rheumatic disease is extrapolated from oncological practice. Methotrexate is frequently administered parenterally (subcutaneous, intramuscular or intravenous) if the oral route is not tolerated or with sub-optimal response. Indeed, at doses beyond 15 mg/m^2 per week the parenteral route is preferable because of the decreased oral bioavailability of the drug [26]. A recent multi-centre international trial showed no additional benefit of a larger dose of 30 mg/m^2 per week over a dose of 15 mg/m^2 per week, in those patients who showed inadequate response to a conventional dose of 8-12.5 mg/m^2 per week [27]. Patients or carers can be trained to administer subcutaneous methotrexate at home.

Nausea, vomiting and dyspepsia are the most common side effects. This may occasionally be extreme and necessitate discontinuation of the drug. Strategies that may help to overcome these effects include the use of folic acid, omitting concomitant NSAID on the day of methotrexate administration, splitting the methotrexate dose over 12 hours and cognitive behavioural management [28]. Full blood count and liver monitoring is required fortnightly for one month and then monthly thereafter. It is extremely rare for abnormalities to develop within the first month and this regime is considered by some to be over cautious. The monitoring interval may be increased to 3 monthly for low dose methotrexate, when a stable dose is achieved. If the liver enzymes rise to twice normal levels, the drug should be discontinued or the dose reduced temporarily until restoration of normal levels. Lahdenne *et al.* reported only minor and reversible histopathological changes in a series of liver biopsies in children with JIA treated with methotrexate at 20-30 mg/m^2 per week [29]. Impairment of lung function is extremely rare [30]. Rash, severe oral ulcers, new or increasing dyspnoea and cough or haematological abnormalities, such as a platelet count of <150 × 10^9/l or neutrophil count of < 1.5 × 10^9/l require discontinuation of the drug and discussion with a paediatric rheumatologist [22].

The need to administer concomitant folic or folinic acid is controversial and there is no consensus view. When used, folic acid is given in doses ranging from 1 mg/day to 2.5-5 mg weekly, as a single dose. Methotrexate is teratogenic, and appropriate contraceptive counselling of patients of child-bearing potential is mandatory, as is advice to limit the use of alcohol whilst taking methotrexate, to minimise liver enzyme disturbance [22].

Anti-tumour necrosis factor alpha therapy

Tumour necrosis factor alpha (TNF-α) is considered to be a pivotal cytokine involved in the pathogenesis of the inflammatory arthritides. AntiTNF-α therapy is now recommended for children over the age of 4 years with inflammatory arthritis, who fail to achieve adequate disease control with methotrexate [31]. Etanercept is a chimeric fusion molecule of two human soluble P75 TNF receptor molecules, fused to a "Fab" fragment of human immunoglobulin G, and efficacy in MTX-resistant JIA has been shown in a placebo-controlled randomised trial [32]. Etanercept neutralises TNF by binding with an affinity 50 to 1000 times that of naturally occurring TNF receptors and may also exert its effect by binding other cytokines, including IL-1 α and TNF-β [33]. Etanercept is administered twice weekly, at a dose of 400 mcg/kg by subcutaneous injection. Simultaneous treatment with methotrexate is recommended [34]. An international randomised placebo-controlled trial of an alternative anti-TNF agent, infliximab (monoclonal antibody to TNF-α), is currently being undertaken.

The side effect profile of etanercept appears acceptable at the time of writing. Major side effects have been reported in the form of isolated case reports: musculoskeletal infection, including tuberculosis [35,36]; systemic opportunistic infections in adult patients [37]; diabetes mellitus [38] and cutaneous vasculitis [39]. Urticarial reactions have been reported [40]. The largest follow-up study to date concludes that, in general, adverse events are of the type seen in a general paediatric population, but that children taking etanercept should be monitored closely for infections [41]. As a European licence requirement, all children treated with biological therapies should be on a register, although national arrangements may vary.

Sulphasalazine

Sulphasalazine has been used to treat patients with HLA B27-associated joint disease and arthritis associated with inflammatory bowel disease. Use in oligo- and polyarticular onset disease is supported by evidence from a placebo-controlled trial [42]. Sulphasalazine is metabolised by intestinal bacteria, resulting in the release of sulfapyridine and 5-aminosalicylate. The mechanism of action may relate to its effects on prostaglandin synthesis or interference with arachidonic acid metabolism by the lipoxygenase pathway. Monitoring of blood count and liver function is required, as sulphasalazine is rarely associated with blood dyscrasias and hepatitis [28]. The drug is generally well tolerated. Dose-related side effects include nausea, malaise, and headache. Hypersensitivity reactions such as rash, fever, urticaria, arthralgia, hepatitis may improve with a desensitisation program [43].

Ciclosporin

Ciclosporin, a fungal macrolide, may be considered as an adjunct to methotrexate in resistant JIA. It is used specifically in the treatment of macrophage activation syndrome in systemic JIA [44]. Dosages of 3-5 mg/kg/day are frequently employed. Ciclosporin may be effective in juvenile dermatomyositis resistant to first-line therapy [45]. Blood pressure is monitored weekly for 4 weeks and monthly thereafter. Full blood count and renal profile is monitored monthly. Twelve hour trough blood levels are monitored at 1 week then monthly (range 95-205 mcg/ml, although in rheumatology practice, levels significantly below this range are frequently sufficient to achieve an adequate effect). Some centres monitor trough levels 4-6 monthly when a patient is established on a stable low dose of ciclosporin. Adverse effects include impaired renal function (increased risk with concomitant administration of NSAIDs), hypertension, hepatic toxicity, tremor, gingival hyperplasia and hypertrichosis.

Antimalarials

Antimalarials (chloroquine and hydroxychloroquine) are widely used in SLE, in particular for treatment of cutaneous manifestations and arthritis. They are occasionally used as adjunctive therapy in JIA. These drugs inhibit cell division and RNA transcription and translation [46]. Efficacy was confirmed in a prospective study of patients with SLE who showed disease exacerbation after discontinuing hydroxychloroquine [47]. Side effects are rare and include gastro-intestinal discomfort, malaise, rash, dizziness and insomnia [46]. Screening for retinopathy is recommended only in patients who have taken the drug for 6 years or more [48].

Cyclophosphamide

Cyclophosphamide is an alkylating agent that inhibits DNA replication and is cytotoxic to dividing cells. It is primarily indicated for the treatment of SLE with glomerulonephritis or cerebral disease. It may be administered either as a daily oral dose, or as intermittent intravenous pulses. A more detailed review of this and other immunosuppressant drugs used in paediatric rheumatic disease is available elsewhere [49].

8.2.3 IMMUNISATION

There appears to be wide variation in practice regarding immunisation in children with rheumatic disease [48]. Children with inflammatory diseases treated with immunosuppressant agents should not receive live vaccines but should be given non-live vaccines, according to relevant national schedules. It is advisable to check varicella antibody status prior to treatment with immunosuppressive agents and, where appropriate, varicella zoster vaccine should be given at this time [50].

REFERENCES

1. Symmons, D. P., Jones, M., Osborne, J., et al., *J Rheumatol.*, *23*, 1975 (1996).
2. Gare, B. A., *Curr. Opin. Rheumatol.*, *8*, 449 (1996).
3. Petty, R. E., Southwood, T. R., Baum, J., et al., *J. Rheumatol.*, *25* , 1991 (1998).
4. Sherry, D. D., Stein, L. D., Reid, A. M., et al., *Arthritis Rheum.*, *42* , 2330 (1999).

5.   Cleary, A. G., Ramanan, A.V., Baildam, E., et al., *Arch. Dis. Child.*, *86*, 416 (2002).

6.   Cleary, A. G., Murphy, H. D., Davidson, J. E., *Arch. Dis. Child.*, *88*, 192 (2003).

7.   Martini, G., Gobber, D., Agosto, C., et al., *Ann. Rheum. Dis.*, *60 (Suppl. 11):ii*, 12 (2001).

8.   Job-Deslandre, C., Menkes, C. J., *Clin. Exp. Rheumatol.*, *8*, 413 (1990).

9.   Jansen, T. L., Van Roon, E. N., *Neth. J. Med.*, *60*, 151 (2002).

10. Giovanni, G., Giovanni, P., *J. Nephrol.*, *15*, 480 (2002).

11. Lindsley, C. B., *Am. J. Dis. Child.*, *147*, 229 (1993).

12. Lang, B. A., Finlayson, L. A., *J. Pediatr.*, *124*, 639 (1994)

13. Cuzzolin, L., Dal Cere, M. , Fanos, V., *Drug Saf.*, *24*, 9 (2001).

14. Barnes, P. J., *Clin. Sci. (Lond.)*, *94*, 557 (1998).

15. Adebajo, A. O., Hall, M. A., *Br. J. Rheumatol.*, *37*, 1240 (1998).

16. Smith, J. R., *Pediatr. Drugs.*, *4*, 183 (2002).

17. Kilmartin, D. J., Forrester, J. V., Dick, A. D., *Br. J. Ophthalmol.*, *82*, 737 (1998).

18. Reiff, A., Takei, S. , Sadeghi, S. , et al., *Arthritis Rheum.*, *44*, 1411 (2001).

19. Giannini, E. H., Brewer, E. J. , Kuzmina, N., et al., *N. Eng. J. Med.*, *326*, 1043 (1992).

20. Woo, P., Southwood, T. R., Prieur, A. M., et al., *Arthritis Rheum.*, *43*, 1849, (2000).

21. Kremer, J. M., *J. Rheumatol.*, *21*, 1 (1994).

22. Ramanan, A. V., Whitworth, P., Baildam, E. M., *Arch. Dis. Child.*, *88*, 197 (2003).

23. Chan, E. S., Cronstein, B. N., *Arthritis Res.*, *4*, 266 (2002).

24. Wallace, C. A., *Arthritis Rheum.*, *41*, 381 (1998).

25. Albertioni, F., Flato, B., Seideman, P., et al., *Eur. J. Clin. Pharmacol.*, *47*, 507 (1995).

26. Jundt, J. W., Browne, B. A., Fiocco, G. P., et al., *J. Rheumatol.*, *20*, 1845 (1993).

27. Ruperto, N. , Murray, K. J., Gerloni, V., et al., *Ann. Rheum. Dis.*, *61*, 60 (2002).

28. Murray, K. J., Lovell, D. J., *Best Pract. Res. Clin. Rheumatol.*, *16*, 361 (2002).

29. Lahdenne, P., Rapola, J., Ylijoki, H., et al., *J. Rheumatol.*, *29*, 2442 (2002).

30. Schmeling, H., Stephan, V., Burdach, S., et al., *Z. Rheumatol. 61*, 168 (2002).

31. National Institute for Clinical Excellence. Technology Appraisal Guidance – No 35. March 2002.

32. Lovell, D. J., Giannini, E. H., Reiff, A., et al., *N. Engl. J. Med.*, *342*, 763 (2000).

33. Wilkinson, N., Jackson, G., Gardner-Medwin, J., *Arch. Dis. Child.*, *88*, 186 (2003).

34. Schmeling, H., Mathony, K., John, V., et al., *Ann Rheum Dis.*, *60*, 410 (2001).

35. Elwood, R. L., Pelszynski, M. M., Corman, L. I., *Pediatr. Infect. Dis. J.*, *22*, 286 (2003).

36. Myers, A., Clark, J., Foster, H., *N. Engl. J. Med.*, *345*, 623 (2002).

37. Silfman, N. R., Gershon, S. K., Lee, J. H., et al., *Arthritis Rheum. 48*, 319 (2003).

38. Bloom, B. J., *Arthritis Rheum.*, *43*, 2606 (2000).

39. Livermore, P. A., Murray, K. J., *Rheumatology 41*, 1450 (2002).

40. Skytta, E., Pohjankoski, H., Savolainen, A., *Clin. Exp. Rheumatol.*, *18*, 533 (2000).

41. Lovell, D. J., Giannini, E. H., Reiff, A., et al., *Arthritis Rheum.*, *48*, 218 (2003).

42. Van Rossum, Fiselier T. J., Franssen, M. J., et al., *Arthritis Rheum.*, *41*, 808 (1998).

43. Tolia, V., *Am. J. Gastroenterol.*, *87*, 1029 (1992).

44. Sawnhey, S., Woo, P., Murray, K. J., *Arch. Dis. Child.*, *85*, 421 (2001).

45. Reiff, A., Rawlings, D. J., Shaham, B., et al., *J. Rheumatol.*, *24*, 2436 (1997).

46. Carreno, L., Lopez-Longo, F. J., Gonzalez., et al., *Pediatr. Drugs.*, *4*, 241 (2002).

47. The Canadian Hydroxychloroquine Study Group, *N. Engl. J. Med.*, *324*, 150 (1991).

48. Mavrikakis, I. , Sfikakis, P. P., Mavrikakis, E., et al., *Ophthalmology, 110*, 1321 (2003).

49. Davies, K. , Woo, P., et al., *Rheumatology (Oxford), 41*, 937 (2002).

50. Royal College of Paediatrics and Child Health. *Immunisation of the immunocompromised child.* Best Practice Statement. February 2002.

8.3 Treatment of inflammatory bowel diseases in children

Jean Pierre Cezard, Jean Pierre Hugot

*Department of Paediatric Gastroenterology and Nutrition,
Hospital Robert Debré, 75019 Paris, France*

8.3.1 INTRODUCTION

Inflammatory bowel diseases (IBDs) include Crohn's disease (CD) and ulcerative colitis (UC). When the diagnosis is uncertain, the disease is classified as undetermined colitis (10%). Twenty percent of all patients with IBDs are children. IBDs are uncommon in children less than 10 years old and are most prevalent in North America, Europe and in Caucasians from South Africa and Australia. A north to south gradient has been described in America and Europe. The incidence of IBDs in children is variable and estimated at 2.5 to 6/100 000 children/year [1,2]. CD is twice as common as UC in children.

The pathogenesis is multifactorial, including genetic and environmental factors. One gene has been identified for CD, which codes for a protein implied in NFKβ activation and is stimulated by bacterial products. Other genes have been localised on chromosomes 7 and 12 for CD and chromosome 2 for UC [3]. Many environmental factors have been suggested, including gut microbial flora and tobacco [1,2].

Clinical symptoms, gut localisation, type of lesions, microscopic finding and specific biological data are summarised in Table 1. Biopsy is usually required to confirm the diagnosis. IBDs share many extra intestinal manifestations: skin (erythema nodosum), joints (arthralgia or arthritis), liver (sclerosing cholangitis), eye (episcleritis, uveitis) and bone (osteoporosis). They represent 20 to 25% symptoms of IBDs in adults and children. IBDs are chronic diseases with variable periods of relapses and remission. Nutritional complications include

653

Table 1

Main characteristics of IBDs

	Crohn's disease	Ulcerative colitis	Undetermined colitis
clinical symptoms	abdominal pain diarrhoea perianal disease growth failure	bloody diarrhoea abdominal pain	UC type
location	whole gut (70% ileum ± right colon)	colon rectum + left colon 58% pancolitis 43%	colon
macroscopic lesions	segmental discontinuous	continuous from rectum (100%) to caecum (40%)	discontinuous
microscopic lesions	transmural aphthous ulcers linear ulcers, stenosis granuloma 50%	mucosa mucus depletion crypt abscesses	As UC
specific biology	ASCA 40-50 % CARD 15 (40%)	pANCA 50-60 % CARD15 (5%)	?

weight loss and malnutrition. Massive haemorrhagia, toxic megacolon in UC or stenosis, abscess and fistula in CD often require surgery.

One complication is specific in paediatric CD: growth and pubertal delay, which may precede the intestinal symptoms. It is present in 40 to 80% of children at diagnosis or during the follow up of the disease. It is rare in UC (<10%).

8.3.2 TREATMENT

Treatment of a child or adolescent with IBD may include dietary advice, drug therapy and surgery. The goals are to stop a flare up, prevent relapses but also to ensure optimal growth and puberty, as well as to prevent disease-related complications [1,2].

Surgery

Surgery is reserved for severe complicated IBD. Colectomy or proctocolectomy with a reservoir (ileal pouch) cure UC. Surgery should be reserved for intractable complications despite medical therapy and as limited as possible because of the high rate of recurrence in CD [1,2].

Drug Therapy

Drugs employed in the treatment of IBDs are similar and based on the nature and localisation of the disease.

Aminosalicylates

The active ingredient in all aminosalicylates is 5 aminosalicylic acid (5-ASA). Sulfazalazine is a 5-ASA molecule and sulfapyridine acts as a carrier for 5-ASA. 5-ASA is released in the colon by bacterial cleavage. 5-ASA preparations are also available in three different prodrugs, in order to deliver 5-ASA to the distal gut. Olsalasine is two 5-ASA molecules with an azo bond, released in the colon by bacteria. Mesalazine is 5-ASA included in acrylic based resin, which breaks in the terminal ileum and colon releasing 5-ASA. Mesalamine is 5-ASA coated to a semi-permeable ethyl-cellulose membrane, with a continuous release along the small intestine and colon [4]. The localisation of the disease will help determine the choice of treatment. 5-ASA is the active molecule and mainly inhibits leukotriene biosynthesis.

The toxicity of sulfazalazine is much higher than with other 5-ASA preparations and adverse reactions are seen in 20 to 25% of patients. Most of the adverse reactions are due to the sulfapyridine component. They are dose related (digestive symptoms, headache, haemolysis), reflecting the acetylator status or due to an hypersensitivity (fever, exanthems, pulmonary fibrosis, hepatotoxicity and rarely agranulocytosis). Ninety percent of patients intolerant or allergic to sulfazalazine will usually tolerate other 5-ASA preparations [5].

Sulfazalazine (60 mg/kg/day) and oral 5-ASA formulations (50 mg/kg/day) are effective in mild to moderate UC, with a remission rate of 60 to 70% in adult patients. A meta-analysis demonstrated that both molecules reduce the 60 to 70% natural annual relapse rate to 30% [6].

Two multicentre studies confirmed the efficacy of sulfazalazine in CD notably when the colon is involved [7,8]. Time-release and pH-dependant 5-ASA have demonstrated the advantages in the treatment of CD involving the terminal ileum. However, their effect is modest in comparison with a placebo.

A meta-analysis by Camma showed a modest therapeutic effect of 5-ASA as maintenance therapy after surgery and ileal localisation [9]. A recent multicentre study in children did not find a difference with a placebo, although a positive effect of 5-ASA was found in one subgroup during the first period of the study [10]. There is insufficient data to recommend the use of amniosalicylates to maintain remission in either paediatric or adult CD.

Corticosteroids

Systemic corticosteroids. Corticosteroids are the reference treatment for a flare up in IBD. Prednisone, prednisolone and methylprednisolone have all been used. Prednisone is rapidly converted to prednisolone by the liver. Both are promptly absorbed. The mechanism of action is mainly inhibition of cell-mediated immunity ($NFK\beta$ inhibition) and antiinflammatory effects on neutrophil chemotaxis and eicosanoid production [11].

Adverse effects of corticosteroids are common and described in detail in the previous chapter. Complications such as cataracts necessitate regular ophthalmologic assessment [12,13].

Corticosteroids are the standard treatment for moderate to severe IBD, proving effective in 70 to 90% of patients. Dosage in children is 1 mg/kg/d with a maximum of 60 mg for 3 to 6 weeks, depending on the severity of the flare up and the rapidity of response. The dose is gradually decreased at weekly intervals. In fulminant colitis or oral steroid resistance, IV methylprednisolone is indicated in conjunction with bowel rest. There is no evidence that higher dosage (2 mg/kg/day) is more efficient.

IBD with low-dose alternate-day prednisolone therapy has not proved effective in adults and should be avoided in children because of its effects on growth and bone mineralisation [13,14].

Topical therapy. Budesonide is a glucocorticoid with a high topical antiinflammatory potency, rapid metabolism in the liver and a low systemic bioavailability (10% compared to 80% for prednisolone). In order to be delivered to the distal ileum and right colon, budesonide is bound to ethylcellulose and encapsulated by endragil I resin. The incidence of adverse systemic effects is significantly decreased in comparison with classic corticosteroids in adults and children [15,16]. Indications for budesonide are CD of the ileum and right colon. The dosage is 9 mg/d in children and adults [15,16]. Efficacy is slightly less than with systemic steroids and there is also a significant delay in action. The use of budesonide for maintenance of remission in children is not indicated, as it has not proved effective in adults [15-18].

Rectal formulations of topical 5 ASA, corticosteroid or budesonide have less systemic effects than oral formulations and are indicated in distal colitis. It has been shown that an enema can reach the left colon and suppositories the rectosigmoid colon.

Antibiotics

Metronidazole and other antibiotics (ciprofloxacin) are used in CD with perianal disease (abscess, fistula). Adverse effects include peripheral neuropathy with metronidazole and arthropathies with ciprofloxacin. These antibiotics are not indicated in UC [19].

Immunosuppressors

Azathioprine and 6 Mercaptopurine (6MP). Azathioprine and 6 MP are being used more extensively in children and adults with IBD. Their delayed onset of action (2 to 4 months) explain their use to prevent relapses and reduce corticosteroid requirements [20].

Adverse effects of azathioprine and 6 MP are allergic or dose/metabolism dependant. Hypersensitivity effects are pancreatitis (4%), fever, rash, arthralgia or digestive symptoms (7%). They occur during the first weeks of therapy and usually require the cessation of treatment. Non-allergic adverse events are haematological (4%) (leukopenia, thrombocytopenia and rarely bone marrow aplasia) due to a genetic polymorphism of thiopurine methyltransferase. Other dose-dependant adverse effects are hepatitis, infection (2%) and a slightly increased malignancy risk (EBV-induced lymphoma) [21,22].

Placebo-controlled trials using azathioprine (2.0-2.5 mg/kg/d) or 6MP (1.5 mg/kg/d) have been shown to be effective in preventing relapse in 60 to 70% of patients [23]. This has been confirmed in two paediatric studies with a better linear growth [24,25]. Determination of the active metabolite may be useful to adjust drug treatment and compliance [24].

In UC data are more limited. In one study, azathioprine was found to be more effective than placebo in maintaining remission [26].

Methotrexate. Methotrexate has been used recently, when azathioprine or 6MP are not effective or tolerated. Folic acid should be given once a week (5-10 mg) to prevent haematoxicity. Its antiinflammatory effect is related to reduction of IL_2 and apoptosis of T cell population. Adverse effects are minor digestive effects or a transient increase in liver enzymes [27]. Hypersensitivity pneumonitis and liver fibrosis are exceptional. In CD, methotrexate prevents relapse in 65% of patients with chronic active disease, *versus* 39% with a placebo. The onset of action is 6 weeks to 2 months. After 4 months of full dosage, methotrexate can be reduced by 40% [28]. The IM route should be maintained, because of great variation in its oral absorption. Methotrexate has failed to show efficacy in UC [29].

Ciclosporin and tacrolimus. Ciclosporin has been proposed in acute severe UC resistant to corticosteroids. Its efficacy has been demonstrated in 82% of patients, with an onset of action of 7 days. 2 mg/kg/d appears to be as effective as 4 mg/kg/d IV with less side effects (nephrotoxicity and hypertension). A switch to oral ciclosporin after remission has been proposed, but evidence of long-term efficacy is lacking, with the rate of relapse and colectomy remaining high at one year (50%) [30,31]. Data on tacrolimus are rare but it seems to have the same efficacy as ciclosporin [32]. The use of these molecules has been disappointing in CD, except for anal and perianal lesions [33].

Biologic Agents

Antitumour necrosis factor α is the first biologic agent which has proven efficacy in CD, with infliximab licensed in 1993. Randomised placebo-controlled trials in adults with moderate to severe CD resistant to conventional treatment, have shown that a single infusion of infliximab 5 mg/kg produced a clinical response in 81% of patients and remission in 50% [34]. One study showed that endoscopic and histologic appearance of the intestine improved in patients given infliximab, but not in those given a placebo [35]. Patients with perianal fistula who were given three injections showed positive effects [36]. Relapses were frequent in all these studies, occuring in approximatiely 50% of patients 3 to 6 months after the first infusion.

Retreatment every 2 months maintained remission at a higher rate (72%) than with placebo, with no increase in side effects [37,38]. The main short-term side effects described are pulmonary infections, infusion reaction and reduced duration of response to treatment, due to antibodies against infliximab [37-39]. Concomitant immunosuppressive therapy and hydrocortisone premedication reduced these two last side effects [37-40]. Tuberculosis, ongoing infection and cardiac insufficiency are contra-indications. Long-term side effects are still not known. Possible toxicity includes auto-immune disease, EBV-induced lymphoma and cancer [38-41].

Reports of paediatric use are increasing [42-44]. As in adults, relapses are frequent (50% at 6 months and 90% at 1 year) [43,44], implying the need for repeated courses [42,44]. Data on the efficacy of infliximab on severe UC are rare but promising. Other biologic agents are under investigation.

Enteral Nutrition (EN)

Although not considered as a drug treatment, EN with a liquid diet is effective in CD but not in UC. In children, it has been shown that EN is nearly as effective as steroids in active CD [45,46]. Its mechanism remain conjectural. Unlike cortiosteroids, it does not have a negative effect on growth [47,48]. EN is considered as first line therapy in some paediatric centers, especially when steroid dependance occurs or when growth failure is present. EN has no direct adverse effects but is unpleasant.

8.3.3 CONCLUSION

Treatment of paediatric IBD has improved the last two decades. New biological agents are under evaluation. The same pharmacological agents are used in paediatric IBD as in adult IBD, but EN should be considered as a primary therapy in paediatric CD.

REFERENCES

1. Griffiths, A.M., Buller A.B., in *Ped. Gastro, Int. Dis., Eds. Hamilton*, 41, 613 (2000).
2. Cezard, J. P., Hugot, J. P., in *Gastro Ped. Eds. Flamm. Med. et Sciences*, 25, 354 (2000).
3. Hugot, J. P., Chamaillard, M., *Nature, 411*, 599 (2001).
4. Laursen, S. L., *Gut, 31*, 1271 (1990).
5. Rao, S.S., Cann, P.A., *Scand. J. Gastroenterol.* 22, 332 (1987).
6. Sutherland, L. R., May, G. R., *Ann. Intern. Med. 118*, 540 (1993).
7. Summer, R.W., Switz, D. M., Sessions, J. T., et al., *Gastroenterology, 77*, 847 (1979).
8. Malchow, H., Ewe, K., Brandes, J. W., et al., *Gastroenterology, 86*, 249 (1984).
9. Camma, C., Giunta, M., Rosselli, M., et al., *Gastroenterology, 113*, 1465 (1997).
10. Cezard, J.P., Munck, A., *Gastroenterology, 124*, A379 (2003).
11. Zimmerman, M. J., Jewell, D. P., *Aliment. Pharmacol. Ther., 10*, 93 (1996).
12. Kusunoki, M., Moeslein, G., Shoji, Y., et al., *Dis. Colon. Rectum, 35*, 1003 (1992).
13. Hyams, J. S., Carey, D. E., *J. Pediatr., 113*, 249 (1988).
14. Lennard-Jones, J. E., *Gut, 24*, 177 (1983).
15. Rutgeerts, P., Lofberg, R., Michon, H., et al., *N. Engl. J. Med., 331*, 842 (1994).
16. Escher, J. C., Lindquist, B., *Gastroenterology, 122*, A12 (2002).
17. Kundhal, P., *Can. J. Gastroenterol., 14*, 46A (2000).
18. Greenberg, G. R., Feagan, B. G., Martin, F., et al., *Gastroenterology, 110*, 45 (1996).
19. Prantera, C., Zannoni, F., Scribano, M. L., et al., *Am. J. Gastroenterol., 91*, 328 (1996).
20. Lennard, L., *Eur. J. Clin. Pharmacol. 43*, 329 (1992).
21. Sandborn, W. J., *Am. J. Gastroenterol., 91*, 423 (1996).
22. Dayhsarh, G. A., Loftus, E. V., *Gastroenterology, 122*, 72 (2002).
23. Pearson, D. C., May, G. R., Fick, G. H., et al., *Ann. Intern. Med., 122*, 132 (1995).
24. Cuffari, C., Theoret, Y., Latour, S., Seidman, G., *Gut, 39*, 401 (1996).
25. Markowitz, J., Grancher, K., Kohn, N., et al., *Gastroenterology, 119*, 895 (2000).
26. Hawthorne, A. B., Logan R. F., Hawkey C. J., et al., *Br. Med. J., 305*, 20 (1992).
27. Egan, L. J., Sandborn, W. J., *Mayo. Clin. Proc., 71*, 69 (1996).
28. Feagan, B., Rochon, J., Fedorak, R. N. , et al., *N. Engl. J. Med., 342*, 1627 (2000).
29. Oren, R., Oren, R., Arber, N., Odes, S., et al., *Gastroenterology, 110*, 1416 (1996).
30. Van Assche, G., D'Haens, G., Noman, M., et al., *Gastroenterology, 125*, 1025 (2003).
31. Rayner, C. K., McCormack, G., Emmanuel, A. V., Kamm, M. A., *Aliment. Pharmacol. Ther., 18*, 303 (2003).
32. Fellermann, K., Tanko, Z., Herrlinger, K. R., et al., *Inflamm. Bowel Dis., 8*, 317 (2002).
33. Present, D. H., Lichtiger, S., *Dig. Dis. Sci., 39*, 374 (1994).
34. Targan, S. R., Hanauer, S. B., van Deventer, S. J. H., et al. *N. Engl. J. Med., 337*, 1029 (1997).
35. D'Haens, G., Van Deventer, S., Van Hogezand, R., et al. *Gastroenterology, 116*, 1029 (1999).
36. Present, D. H., Rutgeerts, P., Targan, S., et al., *N. Engl. J. Med., 340*, 1398 (1999).
37. Rugeerts, P., D'Haens, G., Targan, S., et al., *Gastroenterology, 117*, 761 (1999).

38. Hanauer, S. B., Feagan, B. G., *The Lancet, 359*, 1541 (2002).
39. Baert, F., Noman, M., *N. Engl. J. Med., 348*, 601 (2003).
40. Farrell, J., Alsahli, M., Jeen, Y. T., Falchuk, K. R., Peppercorn, M. A., Michetti, P., *Gastroenterology, 124*, 917 (2003).
41. Bickston, S. J., Lichtenstein, G. R., Arseneau, K. O., Cohen, R. B., Cominelli, F., *Gastroenterology, 117*, 1433 (1999).
42. Hyams, J. S., Markowitz, J., Wyllie, R., *J. Peditric., 137*, 192 (2000).
43. Stephens, M. C., Shepanski, M. A., Mamula, P., Markowitz, J. E., Brown, K. A., Baldassano, R. N., *Am. J. Gastroenterol., 98*, 104 (2003).
44. Cezard, J. P., Nouaili, N., *J. Ped. Gastroenterol. Nutr., 36*, 632 (2003)
45. Heuschkel, R. B., Menache, C. C., Megerian, J. T., Baird, A. E., *J. Ped. Gastroenterol. Nutr., 31*, 8 (2000).
46. Cezard, J. P., Messing, B., *Clin. Nutr. 1*, 875 (1993).
47. Belli, D. C., Seidman, E., Bouthillier, L., et al., *Gastroenterology, 94*, 603 (1988).
48. Wilschanski, M., Sherman, P., Pencharz, P., Corey, M., Griffiths, A. M., *Gut, 38*, 543 (1996).

8.4 Gastro-oesophageal reflux

Marc Bellaiche, Prévost Jantchou, Jean Pierre Cezard

Gastroenterology and Paediatric Nutrition Service, Hospital Robert Debré, 75019 Paris, France

8.4.1 INTRODUCTION

Gastro-oesophageal reflux (GOR) symptoms are present in up to 30% of infants [1]. The pathophysiology of GOR is mainly the result of transient relaxation of the lower oesophageal sphincter. In infants, there are several reasons which also explain the importance of GOR: short abdominal oesophagus, increased oesophageal clearance, delayed gastric emptying associated with a large liquid intake (120 ml/kg/d) and activities of the infant. The frequency of GOR is thought to have risen by 50% in the last ten years [2].

The most common symptoms of GOR are regurgitation and/or vomiting with no other clinical manifestations. Sometimes, GOR is complicated by persistent crying, dysphagia (suggesting oesophagitis), upper airway problems (chronic laryngitis) [3], chronic respiratory disease [4] or failure to thrive.

The prognosis is good in most cases. GOR disappears within a year in 80% of cases, without any treatment during childhood [5]. Treatment helps eliminate symptoms in infants and prevents complications.

8.4.2 NON-PHARMACOLOGICAL THERAPIES

Dietary changes, such as frequent feeding of small amounts or using casein-predominant infant formulae or those with low osmolality, may help. A reduction of the food volume results in a decrease in the number of episodes of vomiting but no change in acid reflux [6]. Milk thickening agents have been included in anti-regurgitation infant formulae. A systematic review of ten controlled trials demonstrates a reduction in vomiting in infants, but there is no effect on pH [7]. Side effects have been described with guar gum, carob bean gum and soybean polysaccharides, such as abdominal pain, diarrhoea or a decrease in intestinal absorption [8]. In atopic families, hydrolysate formulae can be used if cow's milk allergy is suspected.

8.4.3 ANTACIDS

Alginate-antiacids form a viscous fluid with surface-active properties, acting as a barrier against reflux from the stomach to the oesophagus. Dimethicone, gaviscon and algicon are all used. Gaviscon has a high sodium carbonate content and algicon a significant amount of aluminium. Dimethicone is more a feed thickener than an antacid. The use of antacids is limited by their toxicity, and efficacy is not fully proven [9].

8.4.4 PROKINETICS

Prokinetics affect the lower oesophageal sphincter, oesophageal clearance and peristalsis and gastric emptying. Metoclopramide and domperidone have a dopamine-receptor blocking action (anti-emetic properties) and cisapride is a mixed serotonergic agent.

Metoclopramide

No clinical trial has demonstrated significant efficacy. Many side effects have been reported: drowsiness, asthenia, sleepiness, methaemoglobinaemia, galactorrhea and extrapyramidal effects [10].

Domperidone

There are no trials of efficacy in infants [9]. Adverse effects are less common than metoclopramide but still include dystonic reactions, somnolence, anxiety and prolongation of QT interval [11].

Erythromycin

This macrolide antibiotic has some prokinetic effects at the doses of 3 mg/kg/d. Erythromycin possesses adverse cardiac effects such as prolongation of QT interval [12].

Cisapride

A Cochrane review of 7 clinical trials suggested that cisapride reduces symptoms and improves pH in children [13]. Side efffects are rare, but the potential risk of serious cardiac arrhythmia led to its contraindication and withdrawal in many countries.

8.4.5 ACID SUPPRESSANTS

These include histamine-2-receptor antagonists (H_2RA) and proton pump inhibitors (PPIs). They are recommended for oesophagitis diagnosed by endoscopy. The efficacy of H_2RA appears better in *mild* oesophagitis and side effects are uncommon (1 to 6%): dizziness, headache, dyspepsia, abdominal pain and diarrhoea. PPIs have been shown to be more effective than H_2RA [14]. The effects of lansoprazole and omeprazole have been studied in children. Lansoprazole has shown its efficacy in reducing symptoms and healing oesophagitis in children. The availability of a liquid formulation or fast disintegrating tablets facilitates its use in most countries [15]. There is no placebo-controlled trial with omeprazole, but acute and chronic acid-related disorders are well treated in children aged 2 months to 18 years [16].

The pharmacokinetic profiles of PPIs are similar to those in adults but there is significant interindividual variability. Dosage should be reduced in infants less than 3 months and neonates. Doses used are 2 to 3.5 mg/kg/d with omeprazole and 1.5 mg/kg/d with lansoprazole [16,17]. Most PPIs are metabolised by a genetically polymorphic enzyme, CYP2C19, that is absent in approximately 20% of Asians and 3% of Caucasians [18]. Side effects are reported in 10% of cases: headache, dizziness, constipation, diarrhoea, cutaneous reactions, gynaecomastia and proliferation of gastric flora. Hypergastrinemia occurs in all patients on long term PPI. Theoretically, this could lead to gastric malignancy, but fortunately this never has been observed!

8.4.6 SURGICAL THERAPY

Surgical management by Nissen's fundoplication is reserved for children with severe reflux resistant to medical treatment. About 10% had complications such as dysphagia or dumping syndrome. Moreover, the operation failed in 5 to 20% of cases[19].

8.4.7 THERAPEUTIC ENDOSCOPIC PROCEDURES

Endoscopic gastroplasty, radiofrequency delivery at the cardia and injection therapy have been reported in adults, but not yet in children.

8.4.8 CONCLUSION

There are commonly 4 different levels of treatment [8]:

- Level 1 : reassurance, diet and lifestyle changes
- Level 2 : prokinetics
- Level 3 : acid suppressants
- Level 4 : surgery

In practice, GOR is treated initially with a non pharmacological approach. Drugs and/or surgery are usually reserved for infants with complications of GOR, such as chronic respiratory disease or failure to thrive.

REFERENCES

1. Chouhou, D., et al., *Arch. Fr. Pediatr.*, *49*, 843 (1992).
2. Callahan, C. W., *Acta Paediatr.*, *87*, 1219 (1998).
3. Contencin, P., et al., *Ann. Otolaryngol. Chir. Cervicofac.*, *116*, 2 (1999).
4. Blecker, U., et al., *Acta Gastroenterol. Belg.*, *58*, 348 (1995).
5. Carre, I. J., *Arch. Dis. Child.*, *34*, 344 (1989).
6. Khoshoo, V., Ross, G., Brown, S., et al., *J. Pediatr. Gastroenterol. Nutr.*, *31*, 554 (2000).
7. Caroll, A.; Garrison, M.; Christakis, D. et al., *Arch. Pediatr. Adolesc. Med.*, *156*, 109 (2002).
8. Vandenplas, Y., Belli, D.. Cadranel, S., et al., *Acta Paediatr.*, *87*, 462 (1998).
9. Rudolph, C., Mazur, I., Lipatak, G., et al., *J. Pediatr. Gastroenterol. Nutr.*, *32*, S1 (2001).
10. Cinquetti, M., Bonetti, P., Bertramini, P., *Pediatr. Med. Chir.*, *22*, 1 (2000).
11. Drolet, B., Rousseau, G., Daleau, P., et al., *Circulation*, *102*, 1883 (2000).
12. Guerin, J. M., Leibinger, F., *Chest.*, *121*, 301 (2002).
13. Augood, C., McLennan, S., Gilbert, R., et al., *Cochrane Database Syst. Rev.*, *3*, CD002300 (2000).
14. Vandenplas, Y., Belli, D., Benhamou, P., et al., *Eur. J. Pediatr.*, *156*, 343 (1997).
15. Scott, L., *Paediatr. Drugs*, *5*, 57 (2003).
16. Zimermann, A., Walters, J., Katona. B., et al., *Clin. Ther.*, *23(5)*, 660, (2001).
17. Faure, C., Michaud, L., Shagagi, E. et al. , *Ailment. Pharmacol. Ther.*, *15*, 1397 (2001).
18. Flockhart, D., Desta, Z., Mahal, S., et al., *Clin. Pharmacokinet.*, *39*, 295 (2000).
19. Fonkalsrud, E., Burstorff-Silva, J., Perez, C., et al., *J. Pediatr. Surg.*, *34*, 527 (1999).

8.5 Diarrhoea

Ulrich Meinzer[1], Marc Bellaiche[2]

[1] 68, rue de Lhomond, 75005 Paris, France
[2] Gastroenterology and Paediatric Nutrition Service, Hospital Robert Debré, 75019 Paris, France

8.5.1 INTRODUCTION

Diarrhoea can be defined as an altered balance between absorption and secretion of water and electrolytes in the intestine. A more clinical definition, often used in epidemiological studies, is passage of loose or watery stools and increase of stool frequency to more than 3 per 24 h.

Worldwide, diarrhoea kills approximately three million children in developing countries each year. More than thirty years ago, oral rehydration therapy was proven to be effective in the management of patients with severe dehydrating diarrhoea caused by cholera. Since then, therapy with oral dehydration solution (ORS) has become the mainstay of the World Health Organsiation (WHO) to decrease the morbidity and mortality of diarrhoea. ORS is considered to be the most famous discovery of the twentieth century [1]. In developed countries, despite important progress in hygiene, education and medical care, acute diarrhoea remains one of the most frequent diseases in children. In the USA, there are still 55,000 hospitalisations per year and 50 deaths due to rotavirus, with a global cost of 500 million dollars [2]. The composition of ORS is shown in Table 1 [3].

8.5.2 DRUGS

Pathological mechanisms lead to hypersecretion of water and electrolytes by intestinal crypts and altered absorption of water and electrolytes by intestinal villi. Therefore the appropriate treatment of gastroenteritis in children is

Table 1

Components of ORS components

Constituent	Concentration (mmol/L)
Glucose	74-111
Sodium	60
Potassium	20
Chloride	>25
Citrates	10
Osmolarity	200-250 mosmol/kg

to re-establish and maintain normal water and electrolyte balance with ORS. Early feeding (between 4 to 6 hours) of age appropriate foods is recommended [4]. Complementary therapies are used to relieve discomfort of the disease by decreasing water and electrolyte losses or shortening the course of the disease [5,6]. Possible adverse effects should always be considered before use as most episodes of gastroenteritis are self-limited.

8.5.3 CLASSIFICATION OF DRUGS

Alteration of intestinal motility

Drugs that alter the intestinal motility include *opiates* and *anticholinergic* agents. They reduce intestinal motility and may act on secretion. Due to side effects on the central nervous system [7], **most of these drugs are contraindicated in children**. A well known example is *loperamide*, a piperadine derivate that binds preferentially to intestinal μ- and δ-opiate receptors and has fewer effects on the central nervous system in adults [8]. Loperamide may also influence calmodulin a protein involved in intestinal transport. It reduces the intestinal motility and therefore increases the possibility of the gut retaining fluids and may also have antisecretory properties [9]. Controlled studies in children showed that loperamide at a dose of 800 mcg/kg/d had a beneficial effect on the duration of acute diarrhoea [10]. Due to important side effects including ileus, somnolence and consecutive death after using higher doses than recommended, the drug is contraindicated in children under 2 years of age [11]. It is also contraindicated in bacterial diarrhoea where it can lead to aggravation of diarrhoea and to bacterial translocation.

Alteration of secretion

Racecadotril is a specific inhibitor of enkephalinase, a cell membrane peptidase located in various tissues, notably the epithelium of the small bowel. Studies in animals and humans show that oral racecadotril has an effect on secretory diarrhoea caused by cholera without influencing the intestinal transit time. These effects were anatagonised by the opioid-receptor antagonist naloxone suggesting an involvement of endogenous opioid peptides. In randomised controlled trials in children with acute diarrhoea racecadotril (1.5 mg/kg administered orally 3 times daily) used as adjuvant therapy to oral rehydration was well tolerated and reduced the output of stools by up to 50% [12,13].

In another study, racecadotril and rehydration were compared with rehydration alone in children aged 3 months to 3 years who had acute diarrhoea and were evaluated in the emergency department [14]. The treated group had a significantly lower number of stools, and a faster recovery. Moreover, the children receiving racecadotril needed less additional visits for the same episode.

The mechanism of action of *bismuth*-containing agents (e.g. bismuth subsalicylate, bismuth subnitrate or bismuth subgallate) remains largely unknown. Studies have shown that bismuth subsalicylate inhibits intestinal secretion caused by cholera toxins and enterotoxic *E. coli*. In children with acute diarrhoea, studies have shown that treatment with bismuth subsalicylate reduces the frequency of unformed stools and the duration of diarrhoea [15]. Reports of encephalopathy occurring during ingestion of bismuth-containing compounds [16] led to prohibition of bismuth containing agents in France.

Absorption of fluids and toxins

Silicates such as actapulgite or diosmectite have hydrophilic action, can absorb or fix a large number of molecules such as bacterial toxins and may influence production of glycoproteins in the mucus. Controlled studies with diosmectite [17] and actapulgite [18] confirm these effects. Nevertheless, diosmectide did not change global stool output. Therefore, current data suggest that absorbants may change the presentation of stools without reducing the loss of liquid and electrolytes.

Alteration of the intestinal flora

Treatment with prebiotics and probiotics alters the bacterial microflora of the colon. *Prebiotics* are defined as nondigestible food ingredients that beneficially affect the host by selectively stimulating the growth and/or activity of one or a little number of bacteria in the colon and thus improve host health [19]. *Pro-*

biotics have been defined as a live microbial food ingredient that is beneficial to health [20]. A range of probiotics including *lactobacillus acidophilus, lactobacillus rhamnosus, bifidobacterium bifidum, streptococcus thermophilus* and others have been evaluated for their antidiarrhoeal properties. The mechanism of action of probiotics has not been fully elucidated, but may involve decrease of intestinal pH by digestion of residual lactose, inhibition of bacterial adhesion, influence of mucosal growth by production of polyamines, synthesis of bacteria-growth- inhibiting substances, stimulation of immune response or inhibition of bacterial proliferation by competitive consummation of nutriments [21]. Probiotics are generally reported as safe and well tolerated. However, further evaluation for possible side effects, such as infection in high-risk groups is needed.

Antibiotics

In developed countries, only 10 to 15% of cases of diarrhoea are due to bacteria. The use of antibiotics increases the cost of treatment and may prolong the disease. In some situations it may increase the duration of carrier state (e.g. *salmonella* infection) and increases the risk of secondary disease such as haemolytic uraemic syndrome in enterohaemorrhagic *E. Coli*. Recommendations for the use of antibiotics are based on the virulence of the microorganism and the clinical context. Antibiotic treatment is always indicated in *salmonella typhi* or *paratyphi, shigella*, and invasive *amoebiasis*. Other agents have a relative indication depending on the clinical context such as prolonged course of disease caused by *E. Coli, salmonella* infections in very young children, *yersinia* infection in patients with sickle cell disease, or severe bacterial diarrhoea in immunocompromised children. Ceftriaxone is recommended (50 to 80 mg/kg/d) but ciprofloxacin may also be used [6].

Others

Enteral *immunoglobulins* are extremely expensive and should probably be reserved for immunodeficient children with protracted diarrhoea.

8.5.4   CONCLUSION

Treatment of acute gastroenteritis [22] includes the following:
1: use of ORS for rehydration
2: hypotonic solution
3: fast oral rehydration over 3-4 hours
4: rapid realimentation with normal feeding thereafter

5: use of special formula is unjustified
6: use of diluted formula is unjustified
7: continuation of breast feeding at all times
8: supplementation with ORS for ongoing losses
9: no unnecessary medications

REFERENCES

1. Avery, M. H., Snyder, J. D., *N. Engl. J. Med., 323,* 891 (1990).
2. Glass, R. J., Kilgore, P. E., Holman, R. C., et al., *J. Infect. Dis., 174 (Suppl.),* 5 (1996).
3. ESPGAN, *J. Pediatr. Gastroenterol. Nutr., 14,* 113 (1992).
4. Sandhu, B. K., et al., *J. Pediatr. Gastroenterol. Nutr., 24,* 522 (1997).
5. World Health Organization, *The rational use of drugs in the management of acute diarrhea in children,* Geneva, Word Health Organization, 1999, WHO/CDD/SER/80.2.
6. Cézard, J. P., et al., *Arch. Pédiatr., 9,* 620 (2002).
7. US Pharmacopeia. *Anticholinergic/Antispasmodics System.* Rockville, MD, *US Pharmacopeia Dispensing Information, 1,* 312 (1992).
8. Schiller, L. R., Santa Ana, C. A., Morawski, S. G., Fordtran, J. S., *Gastroenterology, 85,* 1475 (1984).
9. Diarrhea disease study group (UK), *Br. Med. J., 289,* 1263 (1984).
10. Motala, C., Hill, I. D., *J. Pediatr., 117,* 467, 50 (1990).
11. Bhutta, T. I., Tahir, K. I., *Lancet, 335,* 363 (1990).
12. Cézard, J. P., et al., *Gastroenterology, 120,* 799 (2001).
13. Salazar-Lindo, E., et al., *N. Engl. J. Med., 343,* 363 (2000).
14. Cojocaru, B., Bocquet, N., Timsit, S., et al., *Arch. Pediatr., 9(8),* 774 (2002).
15. Figueroa-Quintanilla, D., Salazar-Lindo, E., Sack, R. B., et al., *N. Engl. J. Med., 328,* 1653 (1993).
16. Mendelowitz, P. C., Hoffman, R. S., Weber, S., *Ann. Intern. Med., 112,* 140 (1990).
17. Madkour, A. A., Madina, M., El-Assouni, O. E., et al., *J. Pediatr. Gastroenterol. Nutr., 32,* 71 (2001).
18. Charritat, J. L., Corbineau, S., Guth, S., Meunier, M., *Ann. Pediatr. (Paris), 5,* 326 (1992).
19. Gibson, G. R., Roberfroid, M. B., *J. Nutr., 125,* 1401 (1995).
20. Aggett, P. J., et al., *Br. J. Nutr., 81,* S1 (1999).
21. Vanderhoof, J. A., Young, R. J., *J. Pediatr. Gastroenterol., 27,* 323 (1998).
22. Guandalini, S., *J. Pediatr. Gastroenterol. Nutr., 30,* 486 (2000).

8.6 Insulin

Rachel M. Williams, David B. Dunger

Department of Paediatrics, University of Cambridge, Addenbrooke's Hospital, Cambridge, UK.

8.6.1 INTRODUCTION

All children with type one diabetes mellitus (T1DM) require insulin replacement therapy in order to normalise blood glucose concentrations and minimise the risk of the long-term complications of diabetes. Insulin is almost invariably administered via subcutaneous injection, either as a combination of long- and short-acting insulin preparations in a 2-, 3- or 4-injection per day regime, or by the continuous infusion of short-acting insulin via a pump [1].

Insulin was first isolated from the dog's pancreata by Banting and Macleod. The first patient to receive the extract, in January 1922 at the Toronto General Hospital, with immediate success, was Leonard Thomson; a 14-year-old boy with diabetes. Early insulins were derived from pig and cow pancreata; modified by the addition of protamine to delay absorption and prolong the action of insulin preparations. Subsequently, recombinant DNA technology has allowed the production of insulin identical to human insulin from recombinant strains of *Escherichia coli* and yeast, and animal insulins are now rarely used.

More recently, the native human insulin molecule has been genetically modified to alter the absorption kinetics of insulin in an attempt to achieve more physiological insulin concentrations, both basally and post prandially [2]. Insulin preparations are divided into 2 forms, based on their pharmacokinetic profiles: long-acting preparations that are used for basal insulin replacement, and faster-acting preparations that are used to **control** post-prandial rises in blood glucose. An ideal long-acting insulin preparation would produce relatively constant plasma concentrations of insulin over a 24-hour period, while an ideal

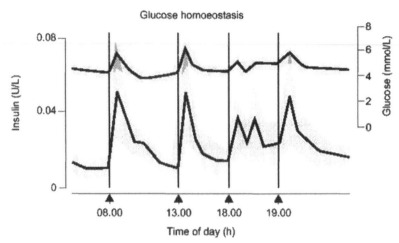

Figure 1. 24-hour plasma glucose and insulin profiles in a group of 12 healthy individuals (© Elsevier Science. Reproduced with permission from Elsevier Science. Owens D. R., et al., *Lancet, 358,* 739 (2001)).

short-acting preparation would have a rapid onset and peak of action, but a short overall duration, allowing quick and effective glucose disposal following a meal, whilst reducing the risk of hypoglycaemia due to residual absorption several hours following a meal; the combination mimicking normal physiological insulin secretion (Figure 1).

8.6.2   HUMAN INSULIN

Human insulin preparations consist of the native and unmodified human insulin molecule produced in recombinant strains of bacteria or yeast. Manipulation of the amount of zinc or protamine added to the crystalline insulin solution permits modification of the absorption profile leading to different pharmacokinetic profiles.

Soluble insulin

Soluble insulin was, until recently, the most widely used fast-acting insulin. It comprises zinc-insulin crystals in a clear fluid of concentration 100 units per ml of solution. Following subcutaneous injection, it has an onset of action of between 30 and 60 minutes, peaks at between 2 and 4 hours and has a duration of between 6 and 8 hours [2]. Soluble insulin should be injected approximately 30 minutes before eating, in order to provide adequate post-prandial amelioration of blood glucose.

Isophane insulin

Isophane (or NPH) insulin is an intermediate acting insulin comprising a crystalline suspension of insulin in combination with protamine and zinc, which leads to a delay in absorption following subcutaneous injection, resulting in a later onset of action (between 1 and 3 hours), and a peak at between 6 and 8 hours. Total duration is between 12 and 16 hours.

Lente insulin

The addition of zinc to a crystalline suspension of human insulin molecules results in an alternative intermediate-acting insulin; insulin lente which has an onset of action between 1 and 4 hours after subcutaneous injection, peak of action between 6 and 10 hours and total duration of action between 14 and 18 hours.

Ultralente insulin

Ultralente is the longest-acting of the human insulin preparations, again consisting of a crystalline solution of insulin and zinc. It has an onset of action of between 2 and 4 hours, peak action between 8 and 10 hours and a total duration of action that varies between 16 and 24 hours, depending on the individual.

Insulin mixtures

A vast array of mixed insulin preparations are available which are designed to allow the single injection of a mixture of short- and intermediate-acting insulin preparations. The proportion of short-acting insulin varies from 10 to 50% and should be selected according to individual requirements. Fixed mixtures are simple to use and are commonly administered via a pen device, where the appropriate dose is selected and then injected subcutaneously, using a short needle. The disadvantage of fixed mixtures is that they are less flexible. The proportion of short- and long-acting insulin cannot be varied according to day-to-day variations in routine; most commonly diet and exercise. An alternative is to draw up the intermediate- and short-acting components separately into a syringe, a technique known as free-mixing, which offers more day-to-day flexibility but is more time consuming, requires more thought by the family and cannot be administered via a pen.

8.6.3 INSULIN ANALOGUES

The ideal short-acting insulin would have an onset of action almost immediately following injection and a total duration of only around 90 minutes, in order to provide good control of blood glucose following a meal, but avoiding the late hypoglycaemia which may occur secondary to ongoing absorption of insu-

1A

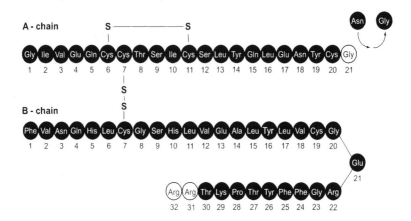

1B

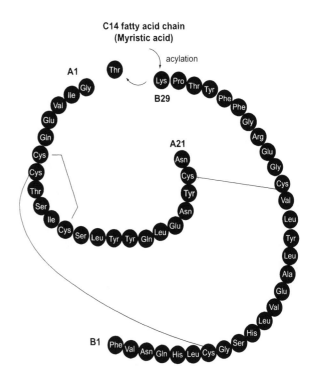

Figure 2. Molecular structures of the insulin analogues insulin glargine (A), insulin detemir (B), insulin lispro (C), and insulin aspart (D).

1C

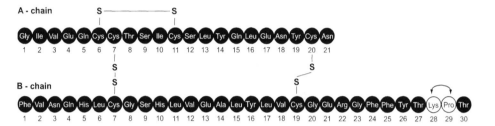

1D

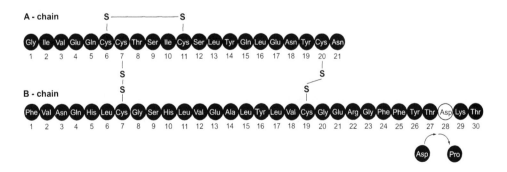

Figure 2. (*cont.*) Molecular structures of the insulin analogues insulin glargine (A), insulin detemir (B), insulin lispro (C), and insulin aspart (D).

lin from the injection site. In contrast, an ideal long-acting insulin preparation would lead to an almost constant concentration of background insulin concentrations over a 24-hour period. In an attempt to achieve this, insulin preparations are now available in which the native human insulin molecule has been modified in some way in order to alter the pharmacokinetic profile following subcutaneous injection. Both short- and long-acting insulin analogues, are now available and licensed for use in Europe (Fig. 2).

In the case of rapid-acting analogues, such as lispro and aspart, substitution of an alternative amino acid for a proline at position B28, prevents subcutaneous hexamer formation resulting in more rapid absorption [3,4].

Insulin lispro

Insulin lispro is a rapid-acting insulin analogue of recombinant origin. Substitution of a lysine residue for proline at position B28 on the native human insulin molecule results in a rapid onset of action (between 5 and 20 minutes), a peak of action at around 60 minutes and a shorter total duration of action (between 2 and 3 hours) than is seen with soluble human insulin.

Insulin aspart

Substitution of an aspartate residue for the native proline, at position B28 again, results in a rapid-acting insulin preparation, which has onset of action 10 minutes following injection, peaks at 40 minutes in the paediatric population and has a total duration of action of between 2 and 3 hours [5].

Insulin glargine

Insulin glargine has two additional positive charges in the form of 2 arginine molecules, which alter the isoelectric point of the insulin molecule leading to precipitation in the subcutaneous tissues, thus prolonging absorption [6]. Insulin glargine has onset of action at around 6 hours and has a broad peak of activity lasting approximately 24 hours. Substitution of insulin glargine for isophane insulin has been shown to reduce nocturnal hypoglycaemia in children and adolescents, when used in conjunction with a rapid-acting analogue [7,8].

Insulin detemir

Insulin detemir is the most recently licensed insulin analogue and has only just become available in Europe. The addition of a fatty acid side-chain promotes hexamer formation at the injection site and delays absorption [9]. Albumin binding

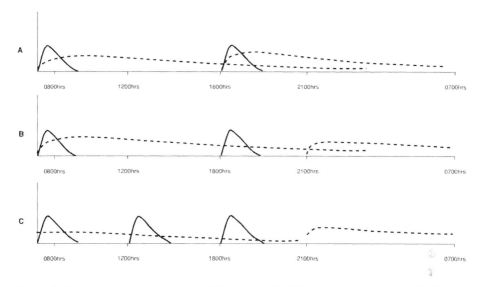

Figure 3. Schematic representation of alternative insulin regimen; A twice daily, B three times daily and C four times daily injections. Soluble or rapid acting analogues are represented in solid lines and basal insulin as a dotted line.

is also enhanced which further prolongs insulin action, but results in lower molar availability, meaning doses need to be around 2.4 times the isophane dose for equivalent glycaemic control [10,11]. Recent data, comparing the pharmacokinetic profile of insulin detemir in children, adolescents and adults, reported consistent pharmacokinetic profiles across age groups, with an onset of action between 1 and 2 hours, peak of action at 6 hours and total duration of action of 18 hours [12].

Insulin regimens

Traditionally, long- and short-acting insulins were given in combination twice daily (Fig. 3). More recently, 3- or 4-injection regimens have been introduced, which can provide more intensive insulin therapy. These regimens comprise combinations of soluble plus isophane / lente / ultralente insulin. Short-acting analogues, aspart and lipro, can be substituted for soluble insulin in these regimens. The new long-acting insulins, detemir / glargine, are most often used in 4 times a day or multiple-injection regimens, as they cannot be freely mixed with short-acting insulins.

8.6.4   SIDE EFFECTS OF INSULIN THERAPY

Hypoglycaemia

Hypoglycaemia is the most commonly encountered, unwanted effect of exogenous insulin administration in young people with T1DM and is secondary to a relative excess of insulin at a given time. Asymptomatic nocturnal hypoglycaemia in children may be profound and prolonged, and has been reported in up to 45% of prepubertal children [13].

Hypoglycaemia is most commonly mild, resulting in symptoms that allow intervention in the form of administration of oral carbohydrate. Rarely, hypoglycaemia may be severe, and the individual may require intramuscular administration of glucagon (dose 1 mg) in order to treat hypoglycaemia. As glycaemic control improves, so the incidence and severity of hypoglycaemia increases; and it is often an obstacle to the achievement of tight control in young people, as parental anxiety regarding increased risk of hypoglycaemia, particularly at night, may result in reluctance to intensify insulin regimen in order to improve glycaemic control [14].

Allergic reactions

With the widespread use of recombinant human insulins, the development of clinically significant allergy to the insulin molecule itself has become rare, although nearly all patients who have received human insulin will produce antiinsulin autoantibodies that can be detected by sensitive assays. However, individuals may develop allergies to the excipients in the insulin preparation, such as protamine, or to preservatives used in preparations. In most cases, switching to an equivalent insulin, from an alternative manufacturer, will solve the problem.

Lipoatrophy and lipohypertrophy

The habitual and repetitive use of the same injection site by young people with T1DM can lead to alterations in the proliferation of adipose tissue at the site of injection. Lipohypertrophy refers to the proliferation of adipose tissue, which results in the appearance of lumps, which, while unsightly, are not in themselves dangerous. However, the absorption of insulin from these sights is impaired and can lead to increased unpredictability in glycaemic control. Lipoatrophy is less commonly seen and refers to the excessive loss of adipose tissue at a repetitively used injection site.

Weight gain

The lipogenic actions of insulin, mediated via its inhibitory effect on hormone sensitive lipase, means that treatment with excessive amounts of insulin may result in inappropriate weight gain, in the form of fat mass. Particularly in girls and young women, this can be very undesirable, and manipulation of insulin dose to prevent weight gain can lead to problems with compliance and increased risk of complications [15].

8.6.5 STRATEGIES FOR INSULIN THERAPY

Ultimately, insulin therapy must be tailored to suit the individual child and their family; what is acceptable and efficacious to one may be of no benefit to another, and so the physician should be prepared to try different preparations and regimen in different individuals. It is also important to remember that, as the child enters puberty and then young adulthood, the total daily insulin requirements will increase from around 0.7 u/kg in the pre-pubertal child to up to 1.5 u/kg during mid-puberty, before falling back to pre-pubertal values in adult life.

Daily insulin requirements

In the individual without diabetes, while insulin is released in bursts following meals in order to maintain normoglycaemia, it is also constantly released from the pancreas to suppress hepatic glucose production, even during a fast. In T1DM insulin is administered exogenously to try to mimic normal physiology as closely as possible, with short-acting preparations used at mealtimes and intermediate and long-acting preparations as basal insulins. Basal insulin requirements vary depending on the time of day and are at their lowest at around 2 a.m. and then increase, being maximal at dawn. This is known as the 'dawn' phenomenon, which is particularly pronounced during puberty [16].

In children, intermediate- or long-acting insulin preparations are injected once or twice daily to mimic basal insulin production by the healthy pancreas. When given at bedtime as part of a multiple-injection regime, the peak of action of intermediate preparations, such as isophane insulin, occurs in the early part of the night, when insulin requirements are low. Conversely, the actions are wearing off by dawn when insulin requirements rise and may result in morning hyperglycaemia. Increasing the dose of bedtime isophane to control morning blood glucose may exacerbate lengthy periods of nocturnal hypoglycaemia, and may be compounded by ongoing absorption of soluble insulin injected with the evening meal. The long-acting insulin analogues, which have more of a peakless profile, may reduce this problem. Substitution of insulin glargine and a rapid-acting insulin analogue for a regime of isophane and soluble insulins leads to reductions in nocturnal hypoglycaemia in adolescents with T1DM [8]. Preliminary data in adults indicate that substitution of insulin detemir for isophane may reduce the risk of hypoglycaemia [11], however there is as yet no published data pertaining to rates of hypoglycaemia in children treated with insulin detemir.

Mealtime insulin requirement depends on the amount of carbohydrate consumed and on its glycaemic index. Historically, carbohydrate counting, carbohydrate exchanges and food weighing were encouraged, but are now less popular in an attempt to normalise the lifestyles of children with diabetes. Instead, children are now encouraged to eat a balanced and healthy diet. However, carbohydrate, and often fat exchange diets do remain the norm in many European and US centres. Either short-acting insulin, or a rapid-acting analogue is injected at mealtimes. The option to inject rapid-acting insulin analogues immediately before, or even shortly after meals, allows greater flexibility and freedom and can be useful in very young children in whom food refusal following insulin injection can be a source of anxiety and conflict for many parents.

Choice of regime

The results of the diabetes control and complications trial (DCCT), which compared intensive insulin therapy (in the form of a multiple-injection regime or pump therapy) with conventional twice daily therapy, demonstrated intensive insulin treatment led to improvement in glycaemic control and reduced risk of the complications of diabetes in adolescents [17]. This has led to a move away from conventional twice-daily regimen towards multiple-injection therapy and pump therapy [18]. In the intensively treated arm of DCCT, there was an increased risk of severe hypoglycaemia and excessive weight gain, particularly in the girls. However, as experience with multiple-injection regimen and pump therapy grows, improvements in control are now being achieved without unwanted side effects [19].

Insulin pump therapy

The continuous administration of insulin into the subcutaneous tissues using a pump device is becoming increasingly more popular both in young children and teenagers in North America and may improve glycaemic control. DCCT reported a higher incidence of hypoglycaemia with the use of insulin pumps. However, as experience grows, there are now reports that hypoglycaemia may be less common in children using insulin pumps, with more sustained improvements in glycaemic control than when using a multiple-injection regime [19].

A basal infusion of insulin is administered over a 24-hour period and can be varied to accommodate differing insulin requirements at different times (for example, reduced during the early hours of the night and then increased again towards dawn, thus reducing risk of nocturnal hypoglycaemia while providing adequate insulin concentrations at dawn). In addition to the basal infusion rate, patients can administer boluses of appropriate dose before meals. The choice of insulin for pump therapy is almost always a rapid-acting insulin analogue, such as aspart or lispro, as their pharmacokinetic profiles mean that a change in the infusion rate of the pump has a rapid effect on blood glucose concentrations and leads to greater flexibility for the patient. However, patients using pumps need to recognise a failure in insulin delivery rapidly and be trained to act accordingly, as with only rapid-acting insulin, a mechanical failure which goes unnoticed will quickly result in hyperglycaemia and diabetic ketoacidosis unless appropriate steps are taken to restore insulin provision.

8.6.6 SUMMARY

In summary, there are many insulin preparations available with a range of pharmacokinetic profiles that may be used in the management of young people with type one diabetes. It is vital to treat each child and family individually when suggesting an insulin regimen. One must also bear in mind that the transition from childhood to adulthood will result both in changes in insulin requirement and in a change in lifestyle requiring adaptations to therapy in order to optimise glycaemic control in the young person.

REFERENCES

1. Tamborlane, W. V., Press, C. M., *Pediatr. Clin. North. Am., 31*, 721 (1984).
2. Bolli, G. B., Di Marchi, R. D., Park, G. D., Pramming, S., Koivisto, V. A., *Diabetologia, 42*, 1151 (1999).
3. Howey, D. C., Bowsher, R. R., Brunelle, R. L., et al., *Clin. Pharmacol. Ther., 58*, 459 (1995).
4. Heinemann, L., Weyer, C., Rave, K., Stiefelhagen, O., Rauhaus, M,, Heise, T., *Exp. Clin. Endocrinol. Diabetes, 105*, 140 (1997).
5. Mortensen, H. B., Lindholm, A., Olsen, B. S., Hylleberg, B., *Eur. J. Pediatr., 159*, 483 (2000).
6. Heinemann, L., Linkeschova, R., Rave, K., Hompesch, B., Sedlak, M., Heise, T. , *Diabetes Care, 23*, 644 (2000).
7. Mohn, A., Strang, S., Wernicke-Panten, K., Lang, A. M., Edge, J. A., Dunger, D. B. , *Diabetes Care, 23*, 557 (2000).
8. Murphy, N. P., Keane, S. M., Ong, et al., *Diabetes Care, 26*, 799 (2003).
9. Kurtzhals, P., Havelund, S., Jonassen, I., et al., *Biochem. J., 312(Pt 3)*, 725 (1995).
10. Kurtzhals, P., Havelund, S., Jonassen, I., Markussen, J., *J. Pharm. Sci., 86*, 1365 (1997).

11. Hermansen, K., Madsbad, S., Perrild, H., Kristensen, A., Axelsen, M., *Diabetes Care, 24,* 296 (2001).
12. Danne, T., Lupke, K., Walte, K., Von Schuetz, W., Gall, M. A., *Diabetes Care, 26,* 3087 (2003).
13. Amin, R., Ross, K., Acerini, C. L., Edge, J. A., Warner, J., Dunger, D. B., *Diabetes Care, 26,* 662 (2003).
14. Marrero, D. G., Guare, J. C., Vandagriff, J. L., Fineberg, N. S., *Diabetes Educ., 23,* 281 (1997).
15. Bryden, K. S., Neil, A., Mayou, R. A., Peveler, R. C., Fairburn, C. G., Dunger, D. B., *Diabetes Care, 22,* 1956 (1999).
16. Edge, J.A, Matthews, D.R, and Dunger, D.B., *Clin Endocrinol (Oxf) 1990, 33,* 729 (1990).
17. Diabetes Control and Complications Trial Research Group, *J. Pediatr., 125,* 177 (1994).
18. Holl, R. W., Swift, P. G., Mortensen, H. B., et al., *Eur. J. Pediatr., 162,* 22 (2003).
19. Boland, E. A., Grey, M., Oesterle, A., Fredrickson, L., Tamborlane, W. V., *Diabetes Care, 22,* 1779 (1999).

8.7 Human growth hormone

Paul Czernichow

Endocrinology Service, Hospital Robert Debré, 75019 Paris, France

8.7.1 GH PHARMACOLOGY

Human GH (GH) is produced by the pituitary as a single chain, 191-amino-acid protein. Approximately 75% of GH secreted in the circulation is a 22 kDA protein. A 20 kDA form normally accounts for 5% to 10% of pituitary GH. The remainder includes deamidated and N-acetylated protein, as well as various oligomers. GH can be identified in fetal serum by the end of the first trimester. Plasma concentrations tend to decline during infancy. Secretion increases during adolescence and declines thereafter throughout adult life. GH secretion is pulsatile and under the control of two major hypothalamic regulatory peptides: GH releasing hormone and somatostatin.

The pulsatile nature of GH secretion is readily demonstrable by frequent serum sampling. Maximal GH secretion occurs during the night, at the onset of slow wave sleep. GH exerts its action through its binding to a membrane associated GH receptor. The extracellular human binding domain of this receptor is cleaved and is present in plasma. This GH binding protein binds GH with high specificity and affinity, but relatively low capacity.

In the liver and other target tissues, GH induces the production of insulin-like growth factor (IGF-I and IGF-II). These factors are found in plasma bound to a family of proteins called IGF binding proteins. The IGFs are important metabolic and mitogenic factors involved in cell growth and metabolism. They are produced in the liver, in the bone cells and in other tissues, at least partially under the control of GH. Most of the anabolic actions of GH are mediated through the IGF peptides production and action. However, GH also stimulates a variety of metabolic effects, some of which appear to occur independent of IGF such as lipolysis, amino acid transport and production of liver specific proteins. The biological action of GH depends on its secretion rate, as well as sensitivity at the receptor level [1].

Pharmacokinetic studies in healthy adults and in patients with various disorders have shown an inverse correlation between metabolic clearance rate and serum GH levels. Since the renal clearance of GH is independent of GH levels, the observed saturation most likely occurs in the extra renal clearance pathway. Receptor mediated uptake of GH in the liver constitutes the major extra renal mechanism. Extra renal elimination of GH reaches half its maximal saturation at a plasma concentration close to values obtained during peak secretion [2].

The GHBP complex (GH bound to proteins) is cleared from the circulation at a tenfold slower rate than free GH, both because the complex is too large for glomerular filtration and because the receptor mediated clearance is impaired [3]. At a physiological GHBP concentration, the half life of "free" GH has been estimated at 7 minutes and that of bound GH at 30 minutes [3]. The measured half life of 15-20 minutes [2] represents a composite of these two clearance components. The relationship between GHBP concentration or pharmacokinetic parameters and the pharmacodynamic effect of GH is uncertain [4,5].

8.7.2   GH AS A THERAPEUTIC AGENT

The only GH that is active in humans is primate GH. For many years, the only source of GH for treatment of GH deficiency was human cadaver pituitary glands which were first used in the late 1950s. However, the occurrence of Creutzfeld-Jakob disease, a rare and fatal spongiform encephalopathy reported from pituitary glands, halted the use of this product [6].

Nowadays, recombinant human GH (hGH) has universally replaced pituitary derived GH as the treatment of choice for children with GH deficiency. This preparation has not only been proven to be safe but has increased the amount of GH available and, therefore, has opened new possibilities in the treatment of short stature of non-GH deficient conditions. Most hGH preparations come in lyophilised powder reconstituted with a solute before injection. Subcutaneous daily injection is the recommended regimen. Continuous subcutaneous 24 h-injection, using mini pump, has proven to be effective; however for practical reasons, this regimen is not in use in clinical practice. Devices allowing transdermal injections are also available.

In documented GH deficiency, the recommended starting dosage is 25 to 40 mcg/kg/day. On this regimen, a child with typical GH deficiency will grow from a pretreatment rate of 3 to 4 cm/year to 10 to 12 cm in the first year of treatment, and 7 to 9 cm/year in the second and third years. This waning effect of GH is not understood. Treatment at an early age and GH dosage are good predictors of treatment efficacy [7].

Since the early experience with GH treatment in GH deficiency, other conditions with short stature have been investigated for GH treatment efficacy. In many countries, GH is a registered indication for Turner syndrome, renal insufficiency and short children born small for gestational age (SGA). More recently, obesity linked to Prader Willi syndrome has been recognised as an indication for hGH treatment. In these situations GH secretion is normal or subnormal and dosage of GH is higher than that used in GH deficient children. In these situations IGF-I is increased above normal values. The value of GH therapy has been demonstrated in Turner syndrome, renal insufficiency and short children born SGA [7].

8.7.3 SIDE EFFECTS

More than 15 years of experience with rDNA-derived hGH have been extremely encouraging since no major side effects have been observed. This evaluation has been greatly facilitated by extensive databases established by hGH manufacturers.

Development of leukaemia

More than 30 cases of leukaemia in hGH treated children have been reported with a disproportionate number coming from Japan. One of the difficulties in assessing the role of hGH treatment in these cases is that many children with GH deficiency have clinical conditions that predispose them to the development of leukaemia, such as histories of prior malignancies, histories of irradiation or underlying syndromes known to predispose to leukaemia (Bloom syndrome, Fanconi anaemia).

At present this issue is not entirely resolved but the risk, as it is, appears minimal. Particular care should be used in prescribing hGH in patients at risk for leukaemia. This is particularly the case for patients with Fanconi syndrome [8].

Recurrence of central nervous system tumours

Since GH is a potent mitogen and because several patients treated with hGH have acquired GH deficiency from central nervous tumours or from this treatment (irradiation), the possibility of tumour recurrence is of obvious concern. However, extensive analyses of the data in large groups of children in Europe and in the US [9,10], treated for a central nervous tumour and who received hGH, has not indicated an increased risk.

Pseudotumour cerebri

Idiopathic intra-cranial hypertension (pseudotumour cerebri) has been reported in rare occasions (11) at the onset of hGH treatment. Its mechanism is unclear but may reflect changes in fluid dynamics and sodium retention. Decreasing hGH dosage results in resolution of the intracranial hypertension.

Metabolic effects

In non-GH deficient patients, GH induces a state of insulin resistance with increased insulin secretion, as evaluated by oral glucose tolerance test. Glucose tolerance, however, is normal and insulin status is restored to pretreatment value after the end of GH treatment.

In conclusion, extensive clinical experience has shown that GH is a safe therapeutic agent. However, one should remember that GH has various metabolic and mitogenic actions. All patients receiving hGH must be carefully monitored for side effects.

REFERENCES

1. Hansen, T. K., *Growth Hormone & IGF Research*, *12*, 342 (2002).
2. Haffner, D., Schaefer, F., Girard, J., Ritz, E., Mehls, O., *J. Clin. Invest.*, *93*, 1163 (1994).
3. Hansen, T. K., Fisker, S., Hansen, B., et al., *Clinical Endocrinology*, *57*, 779 (2002).
4. Sun, Y-N., Lee, H. J., Almon, R. R., Jusko, W. J., *J. Pharmacol. Exp. Ther.*, *289*, 1523 (1999).
5. Bright, G. M., Veldhuis, J. D., Iranmanesh, A., Baumann, G., Maheshwari, H., Lima, J. *J. Clin. Endocrinol. Metab.*, *84*, 3301 (1999).
6. Preece, M., *Horm. Res.*, *39*, 95 (1993).
7. Rosenfeld, R. G., in *Pediatric Endocrinology*, Sperling, M. A., (ed.), (1996).
8. Cowell, C. T., Dietsch, S., *J. Pediatr. Endocrinol. Metab.*, *8*, 243 (1995).
9. Ogilvy-Stuart, A. L., Ryder, W. D. J., Galtamaneni, H. R., Clayton, P. E., Shalet, S. M. *BMJ*, *304*, 601 (1992).
10. Rappaport, R., Oberfield, S. E., Robison, L., et al., *J. Pediatr.*, *126*, 759 (1995).
11. Malozowski, S., Tanner, L., Wysowski, D., Fleming, G., *N. Engl. J. Med.*, *329*, 665 (1993).

8.8 Antipsychotropic drugs

Marie France LeHeuzey

Department of Child and Adolescent Psychiatry, Hospital Robert Debré, Paris, France

8.8.1 INTRODUCTION

In the field of child and adolescent psychiatry, drug therapy is not usually first-line treatment for historical and theoretical reasons; priority is given to psychosocial treatments. Our knowledge regarding the use of psychotropic drugs in children is inadequate due to the scarcity of clinical trials. Many psychotropic drugs are not approved for paediatric use.

8.8.2 STIMULANT MEDICATIONS

Attention-Deficit/Hyperactivity Disorder (ADHD)

Attention-Deficit/Hyperactivity Disorder (ADHD) is a major problem in children, affecting up to 5% of the school-age population. It is characterised by three groups of symptoms: inattention, motor hyperactivity, and impulsivity. Three clinical forms have been identified: the predominantly inattentive type, the predominantly hyperactive-impulsive type, and the combined type [1-4].

Mechanism of action

The central mechanism of action of stimulant medications in the treatment of ADHD is not completely understood. Stimulants have putative effects on central dopamine and norepinephrine pathways, and they increase the intrasynaptic concentration of dopamine. The pharmacokinetics of stimulant medications are characterised by rapid absorption (30 minutes), weak protein binding, and rapid metabolism.

Efficacy

Since Bradley's publication in 1937 [5], reporting behavioural improvement in children treated with benzedrine, studies have increased in number and, over the last 60 years, confirmed the benefit of stimulant medication in the treatment of this disorder. Drugs most often studied in the international community are: methylphenidate (MPH), dextro-amphetamine (DEX), mixed-salts amphetamine (AMP), and pemoline (PEM). These drugs are the most widely prescribed psychotropic medications for children.

Numerous short-term clinical trials have demonstrated the efficacy of these drugs on symptoms of inattention, hyperactivity and impulsivity. Improvement is noticeable at school, with a decrease in disruptive behavior and an improvement of the child's attention span in the classroom; at home, parent-child relationships and compliance are improved; and in the social sphere, peer relationships, as well as attention in sports and leisure activities are enhanced. The immediate release preparations of stimulants have a rapid (30 minutes), but brief (3 to 5 hours) action. This requires the use of multiple doses during the day to maintain improvement.

The efficacy of stimulant pharmacotherapy was initially demonstrated in school-age hyperactive children of both sexes (in over 75% of the cases), and has subsequently been confirmed in pre-school children and adolescents [6].

Symptoms of ADHD are also improved in the presence of co-morbid disorders, such as conduct disorders, oppositional disorders and anxiety disorders. More recently, some long-term studies (12 to 24 months) [7,8] have confirmed the maintenance of improvement over time. For instance, the MTA study (Multimodal Treatment Study of Children with ADHD) showed that stimulants (either by themselves or in combination with behavioural treatments) lead to stable improvements in ADHD symptoms, as long as the drug continues to be taken.

Dosage

The starting dose is 5 mg for MPH. A stepwise increase of 5 to 10 mg per dose is recommended; the maximum daily dosage is 60 mg. The starting dose for DEX/AMP is 2.5 mg, increasing by 2.5 - 5.0 mg, with a maximum dose of 40 mg.

Adverse effects

Mild adverse effects are frequent and include: appetite suppression, sleep disturbances, headaches, tics, abdominal pain. These adverse effects can generally be decreased by modifying both the time of administration during the day and the daily dosage. Staring, irritability, anxiety, daydreaming, representing pre-existing symptoms rather than side effects, may all decrease when increasing the dose. Serious adverse reactions (acute psychotic episode, hallucinations) are rare, generally brief, and subside when treatment is discontinued.

Contraindications

Contraindications include hypersensitivity to stimulant medication, severe cardiovascular conditions, hyperthyroidism, glaucoma, hypertension, treatment with non-selective monoamine oxidase inhibitors (MAOIs). Some contraindications are relative: individual or family history of substance abuse, individual and/or family history of tics and Tourette's disorder. Epilepsy is not a contraindication, as long as the epilepsy is under control.

New formulations

Over the last decade, new formulations of methylphenidate with prolonged action have become available in some countries. Unfortunately their efficacy has been questionned. OROS®MPH is one of them [9,10]. One morning dose is as effective as standard methylphenidate administered three times a day. A single daily administration increases compliance and suppresses the constraints of in-school dosing.

Other indications

Narcolepsy

When the disorder is severe enough to disturb the child's daily functioning, particularly when the disorder is associated with catalepsia, modafinil or methyphenidate may be prescribed. Modafinil may be administered at a dose of 100 mg/day and, progressively increased every other week, to reach a daily dose of 200 mg.

Brain damage

Methylphenidate has been shown to be more effective than placebo in improving the attention span and behaviour of children who display sequelae of traumatic brain injuries.

8.8.3   ANTIDEPRESSANTS

Depression

The prevalence rate of depression is 2% in children (5 to 8% in adolescents). Two different types of depression are generally distinguished: major depressive episode and dysthymic disorder. A major depressive episode is defined by the presence of specific signs of depression (depressed mood, loss of pleasure, sleep disorders, change in appetite, etc.) over a period of at least two weeks. A dysthymic disorder is a chronic mood disorder, characterised by mild to moderate symptoms of depression. A double depression is characterised by the joint presence of both disorders. Findings from the pharmacotherapy of depression in children are far from complete and concern almost exclusively major depressive episodes [12,13].

Several controlled studies have been conducted with imipramine, nortryptiline, amitryptiline and desipramine, usually yielding non-significant results. This overall trend has recently been confirmed by a meta-analysis indicating that tricyclic agents are no more effective than placebo [14].

Two controlled studies, carried out by Emslie [15,16] showed that fluoxetine was more effective than placebo and was well tolerated by the patients. Clinical evaluation scores improved significantly during the 9 weeks therapy. A study by Simeon [17], however, yielded negative results.

Two multicentre studies [18] compared the effects of sertraline to placebo on 376 patients between 6 and 17 years of age (189 on sertraline/187 on placebo) over a period of 10 weeks. Sertraline was superior to placebo and was well tolerated by the subjects (5% adverse effects: primarily digestive problems, agitation, and weight gain).

A controlled study of paroxetine (20 to 40 mg) vs. placebo vs. imipramine (200-300 mg) [19] in the treatment of 275 adolescents showed paroxetine to be superior to both placebo and imipramine on clinical evaluation scores, but not on self-evaluation and parent-evaluation scores. Paroxetine was better tolerated than imipramine. In Braconnier's study [20], paroxetine and clomipramine yielded similar results in the treatment of adolescent depression, but the lack of a placebo group limit the scope of the conclusions that can be drawn from this trial. Finally, fluoxetine [21] appears to be a promising drug that is well tolerated in both dysthymic disorders and double depression, but controlled studies are too scarce. Other existing studies are open-label or retrospective studies.

A controlled study of venlafaxine [22] was conducted on 33 children and adolescents: they all received cognitive-behavioural treatment and were either placed in a placebo or an active drug group. There were no significant differences between the two groups, who both displayed significant improvement.

Tricyclics

Tricyclic agents are the older forms of antidepressants. They include imipramine, clomipramine, amitryptiline and desipramine.

Pharmacology

Their action involves the inhibition of synaptic reuptake of noradrenaline and serotonin. In children, tricyclic agents are rapidly absorbed and metabolised, with a shorter half-life than in adults, thus necessitating fragmented administration of the dose. Plasma levels can vary in children and correlate well with cardiotoxicity.

Dosage

The initial dose is 25 mg, which is slowly increased, until a dose of 1 to 3 mg/kg/day is reached, taken three times a day. Monitoring is accomplished through clinical surveillance, checking of plasma levels, and electrocardiograms (ECGs). Drug discontinuation must be accomplished progressively over several weeks, in order to avoid withdrawal symptoms of anxiety, agitation, and sleep disturbances.

Adverse effects

Adverse reactions are frequent: asthenia, drowsiness, anticholinergic effects (dryness of the mouth, disorders of accommodation, constipation). Seizures may occur, as well as abnormal movements and mood switching. Cardiovascular effects include orthostatic hypotension, and diastolic hypertension, lengthening of the QRS complex on the electrocardiogram. Severe effects are generally dose dependent and include states of mental confusion and a prolongation of QTc interval on the electrocardiogram. Overdose may be lethal due to arrhythmias, seizures or collapse. Prior to initiating treatment, a medical examination must be performed, including a cardiovascular examination and an ECG, as well as an evaluation of liver function.

Selective serotonin reuptake inhibitors (SSRIs)

These agents include fluoxetine, fluvoxamine, paroxetine, sertraline, and citalopram.Their action involves the inhibition of the synaptic reuptake of serotonin. Fluoxetine has a half-life of two days, and that of its active metabolite is 7 to 19 days. Other SSRIs have a half-life of 12 to 36 hours and do not produce active metabolites. Plasma levels vary and have little therapeutic usefulness.

Recent studies in pharmacotherapy suggest an increased risk of suicide in young depressed subjects. Regulatory authorities have advised against the use of these drugs in depressed patients below the age of 18 years.

Adverse effects

Adverse reactions occur primarily at the beginning of treatment: nausea, diarrhoea, dyspepsia, weight loss, irritability, insomnia, sedation, fidgetiness. A syndrome of loss of motivation may occur: lack of initiative, amnesic disorders, and the sensation of being distant from one's environment [11]. The occurrence of a serotoninergic syndrome is often associated with the combination of several serotoninergic agents: shivering, agitation, tremors, gastro-intestinal problems; occasionally the syndrome may be more severe, with fever, coma and epilepsy.

Other antidepressant agents

First generation monoamine oxidase inhibitors (MAOI) are not often used in children. Moclamine, a reversible and selective MAOI, has rarely been studied in children. Serotonin and noradrenaline reuptake inhibitors, such as venlafaxine, have properties that are similar to tricyclic antidepressants without having their side effects.

Mianserine, a tetracyclic antidepressant, and mirtazapine are sedative molecules that frequently cause significant weight gain; they have not been studied in children.

Recommendations

At the present time, the first-line treatment of depression in children remains psychotherapy. Antidepressant drugs should only be considered when depressive signs are severe and resist psychotherapy. In this case, preference should be given to SSRIs, and more specifically to fluoxetine. Two controlled studies on this agent have yielded positive results, and TADS controlled trials confirmed these results [23].

Obsessive-Compulsive Disorder (OCD)

The obsessive-compulsive disorder is characterised by the presence of obsessions or compulsions that create significant distress. Most common obsessions involve fear of contamination and fear of disasters. Most frequent compulsions include washing, repetition and checking. The most common comorbid disorders are depression, other anxiety disorders, and Tourette's syndrome. Treatment involves a combination of cognitive-behavioural therapy and pharmacological treatment with antidepressants.

Clomipramine has been the first antidepressant drug to demonstrate its efficacy in comparison to placebo, and several placebo-controlled studies have confirmed this finding. But adverse effects are relatively frequent and potentially serious. For this reason, the prescription of SSRIs is currently preferred.

Fluvoxamine, fluoxetine and sertraline have all been submitted to controlled clinical trials, showing better efficacy than placebo [24-27]. Citalopram and paroxetine have been studied in open studies [28].

In children with comorbid tics, the efficacy of the SSRIs is not as marked. In addition, SSRIs, particularly when they are administered at high doses, can exacerbate or even induce tics in some patients.

Recommendations

The initial treatment should be the prescription of a SSRI, but in the current state of affairs, we cannot recommend one drug over another [29]. The target dosage is 5 to 40 mg for fluoxetine, 50 to 200 mg for fluvoxamine and sertraline. Maintenance is reached by a process of stepwise increases. Therapeutic effects of the drug are observable within 3 to 8 weeks. Because long-term studies are rare in children, the optimal duration of treatment is not clearly established [30]. One study [31] showed that sertraline remained useful and well tolerated after 52 weeks of treatment. It is generally advised that treatment be maintained during a period of 12 to 18 months after a therapeutic response has been obtained; discontinuation of treatment is best accomplished during a period of vacation, when the child is minimally stressed.

Other indications

Anxiety disorders

Tricyclic and SSRI studies have yielded variable results for a range of anxiety disorders, as well as in manifestations of anxiety in autistic children and in selective mutism. Complementary studies are needed.

ADHD

Antidepressant drugs are less effective than stimulant medication. Atomoxetine, a selective norepinephrine reuptake inhibitor, is, on the other hand, a promising drug that is currently under study in Europe and is already licensed in the United States [32].

8.8.4   NEUROLEPTICS AND ATYPICAL ANTIPSYCHOTICS

Typical neuroleptics

Neuroleptics that are used in children are: chlorpromazine, levomepromazine, thioridazine, cyamemazine, propericiazine, haloperidol, pimozide, and pipamperone. They block dopamine receptors. There is no specific correlation between their clinical efficacy, plasma levels and dosage; however, there is a correlation between plasma levels and side effects [33,34].

Adverse effects

These agents are not well tolerated. Most frequent undesirable effects include sedation, drowsiness, tiredness with loss of energy and abulia, cognitive impairment related to those effects, vegetative signs (low blood pressure, dryness of the mouth, constipation, disorders of accommodation), weight gain, and neurological signs (acute dyskinesia, akathisia, parkinsonism).

Potentially dangerous adverse reactions are rare: they include the neuroleptic malignant syndrome, disturbed cardiac conduction, blood and liver function abnormalities, and tardive dyskinesia associated with long-term use (to be distinguished from more frequent, but transient, forms of withdrawal dyskinesia).

Pre-treatment monitoring includes recording the patient's weight and height, listing all abnormal movements, liver function tests, and an ECG. During treatment, monitoring should be conducted at least once a week.

Rules of prescription are: use of a single neuroleptic agent, progressive increase of dosage, find the minimal dosing that is therapeutically optimal, administer the drug in divided doses (two or three times daily).

Atypical antipsychotics

These agents are serotonin-dopamine antagonists that block both 5HT 2A serotonin receptors and D2 dopamine receptors. Antipsychotic drugs that are most often used in children are risperidone, olanzapine, quietapine, clozapine, amisulpride.

Adverse effects

Atypical antipsychotic drugs are better tolerated than neuroleptics: they have no cardiovascular effects and fewer neurological effects. Most common effects are sedation, a transient phenomenon that abates when dosing is adjusted; and weight gain which is observed with all agents.

Childhood onset schizophrenia

This disorder is a progressive-deteriorating developmental disorder whose prevalence rate is very low (1/10 000) before the age of 12, and increases with puberty and the beginning of adolescence. The clinical picture combines in variable degrees "positive" symptoms such as delusional ideas and hallucinations, and "negative" symptoms such as affective flattening, alogia and avolition.

Drugs most often prescribed are classical neuroleptics (haloperidol, chlorpromazine); but, in spite of the fact these drugs have been used for a long time, controlled studies are very rare and the few that have been conducted on young subjects did not specifically target early onset forms of schizophrenia.

A study of 12 children (aged 5-12 years) over a 10-week period showed that haloperidol is superior to other agents at a dose of 20 to 120 mcg/kg/day (0.5 to 3.5 mg/day) [35]. Therapeutic response to haloperidol is negatively correlated to duration of illness and positively correlated to the age of the patient and his/her level of intellectual functioning. Pool's study [36] of 75 adolescents (aged 13-18 years) compared haloperidol (2 to 6 mg/day), loxapine (510 to 200 mg/day), and placebo. Both active agents were superior to placebo, particularly for those patients who were most severely disturbed; haloperidol had milder sedative effects.

The initiation of pharmacological treatment from the very first psychotic episode in young patients is recommended, since it has been demonstrated that long-term prognosis of schizophrenia is correlated with the duration of the untreated initial episode. Treatment must be instigated in a specialised setting and re-evaluated on a regular basis.

Although studies remain scarce and there are no licensed paediatric products, preference should presently be given to new antipsychotic drugs, as is the case for adults. In open studies involving children and adolescents, risperidone is effective for both positive and negative symptoms of schizophrenia, at doses of 2 to 10 mg. A study of 16 patients (aged 9-20 years) suggested that risperidone was effective for schizophrenia [37].

Olanzapine has also been assessed through open studies. Symptomatic improvement in 15 children between the ages of 6 and 13, at doses of 2.5 to 10 mg/day, was reported [38]. Similarly, negative and positive symptoms of 16 adolescents (12 to 17 years of age) in Findling's study [39] were improved by doses of 2.5 to 20 mg/day, with weight gain and sedation as the two main side effects.

Open and controlled studies [40,41] have shown that clozapine is an effective antipsychotic drug in early onset schizophrenia, but it must only be prescribed in cases of resistant forms of schizophrenia, due to its potentially serious adverse effects: haematological (leukopenia, agranulocytosis), cardiovascular (tachycardia, arrhythmia), cerebral (epilepsy), and hepatic effects. Clozapine has never been used as a first-line treatment.

Pervasive developmental disorders

Pervasive developmental disorders (PDDs), including autism and Asperger's syndrome are neurodevelopmental disorders characterised by social impairments, communication abnormalities, restricted interests and repetitive behaviours, Its most severe form, autism, has a prevalence rate of 20 per 10 000.

Typical neuroleptics are the most often prescribed class of drugs; more specifically haloperidol, which has been the subject of numerous studies. Haloperidol reduces agitation, aggressiveness, stereotyped behaviors and self-mutilation, by improving social interaction and without excessive sedation. Dosage varies between 20 and 200 mcg/day. Pimozide is also effective in the treatment of PDDs, but problems with drug tolerance tend to encourage the use of new antipsychotic medications [42].

Risperidone [43] is the drug that has been most studied, in several open studies. One recent controlled study, involving 101 children (82 boys and 19 girls) with a mean age of 8.8 years (± 2.7) over a period of 8 weeks, showed that 34 out the 49 children treated by risperidone (69%) were improved (irritability and temper tantrums), as opposed to 6 out of 52 (12%) in the placebo group. In two-thirds of the children with a positive response to risperidone at eight weeks, the benefit was maintained at six months. Increased appetite, fatigue, drowsiness, dizziness and drooling were more common in the risperidone group.

Olanzapine [44], quietapine [45] and ziprasidone [46] also seem to be promising drugs, but placebo-controlled studies are needed.

Tics and Tourette's syndrome

Traditionally, haloperidol and pimozide have been considered to be the most beneficial drugs; but over the past few years, risperidone has stood out as the most effective treatment [47]. The efficacy of this drug, suggested by several open studies, has been confirmed by a controlled study carried out over 8 weeks on 34 subjects, including 26 children. Long-term studies are needed to evaluate the durability and safety of the drug over time.

Behaviour and conduct disorder

In addition to psychosocial treatment, new antipsychotic agents are suggested as adjunct therapy in the treatment of aggressive and antisocial behavior in adolescents. Some recommendations have been published, awaiting controlled studies [48].

Anorexia nervosa

Some authors suggest that olanzapine be used as auxiliary treatment of young hospitalised anorexic patients: in addition to weight gain, olanzapine reduces agitation and pre-meal anxiety [49].

8.8.5 CONCLUSIONS

Our knowledge regarding the use of effective and safe psychotropic drugs in the field of child and adolescent psychopathology remains limited, and controlled clinical trials are needed in many areas. Indeed, approximately half of psychotropic medicines are prescribed off-label or unlicensed in a specialised psychiatric ward treating children [50] and an early evaluation of these drugs should be undertaken. Regulatory agencies (UK and USA) have recently issued warnings about a possible link between suicidal thoughts and attempts and SSRI use in paediatric populations. Additional clinical trials should be performed to evaluate the safety and efficacy of antidepressants in children and to establish guidelines for the treatment of depressive children and adolescents.

Table 1

Classification of the psychotropic medications prescribed in children and status in France

Therapeutic class categories	ICU	License status in children in France	Approved age (years) in France
Antipsychotics (neuroleptics)			
phenotiazines	alimemazine	yes	1 and older*
	chlorpromazine	yes	3 and older*
	cyamemazine	yes	6 and older*
	levomepromazine	yes	3 and older*
	propericiazine	yes	3 and older*
butyrophenomes	haloperidol	yes	3 and older*
benzamides	amisulpride	no	CI under 15*
	tiapride	yes	6 and older*
dibenzodiazepines	clozapine	no	NP
	olanzapine	no	NR
others	loxapine	no	CI under 16*
	risperidone	no	NR*

*: liquid formulation available, CI: contra-indicated, NR: not recommended and/or nor evaluated, NP: No conditions of prescriptions in paediatrics (Adapted from [50]).

Table 1 (continued)

Classification of the psychotropic medications prescribed in children and status in France

Therapeutic class categories	ICU	License status in children in France	Approved age (years) in France
Anxiolytics			
benzodiazepins	alprazolam	yes	6 and older
	clonazepam	yes	all ages*
	chlorazepate diazepam	yes	6 and older
	lorazepam	yes	all ages *
		yes	6 and older
hydroxyzine	hydroxyzine	yes	30 months and older
Antidepressants			
SSRIs	venlafaxine	no	CI under 18
	citalopram	no	CI under 15*
	fluoxetine	no	CI under 15*
	paroxetine	no	CI under 15*
	sertraline	yes	6 and older
	mianserine	no	CI under 15
	tianeptine	nn	CI under 15
Others antidepressors	mirtazapine	no	NR
Stimulants			
modafinil	modafinil	no	NP
methylphenidate	methylphenidate	yes	6 and over
Normothymics			
valproate derivates	valproïc acid	yes	all ages
	valpromide	no	NP
	sodium di valproate	no	NR under 18
Others			
opioids antagonists	naltrexone	no	NP
	tropatepine	no	NP

*: liquid formulation available, CI: contra-indicated, NR: not recommended and/or nor evaluated, NP: No conditions of prescriptions in paediatrics (Adapted from [50]).

REFERENCES

1. Greenhill, L. L., Halperin, J. M., Abikoff, H., *J. Am. Acad. Child. Adolesc. Psychiatry, 38,* 503 (1999).
2. Greenhill, L. L., Pliszka, S., Dulcan, M. K., et al., *J. Am. Acad. Child. Adolesc. Psychiatry, 41,* 26S (2002).
3. Le Heuzey, M. F., Ferrari, P., et al. (eds.), *Actualités en psychiatrie de l'enfant et de l'adolescent,* Paris. Flammarion, 2002, pp. 361-7.
4. Le Heuzey, M. F., Mouren-Simeoni, M. C., *Encycl. Med. Chir.,* 37-218-A-50 (2003).
5. Bradley, C., *Am. J. Psychiatry, 94,* 577 (1937).
6. Findling, R. L., Short, E. J., Manos, M. J., *J. Am. Acad. Child. Adolesc. Psychiatry, 40,* 1441 (2001).
7. Arnold, L. E., Abikoff, H. B., Cantwell, D. P., et al., *Arch. Gen. Psychiatry, 56,* 1073 (1999).
8. Gillberg, C., Melander, H., von Knorring, A. L., et al., *Arch. Gen. Psychiatry, 54,* 857 (1997).
9. Swanson, J., Gupta, S., Lam A., et al., *Arch. Gen. Psychiatry, 60,* 204 (2003).
10. Wolraich, M. L., Greenhill, L. L., Pelham, W., et al., *Pediatrics, 108,* 883 (2001).
11. Garland, E. J., Baerg, E. A., *J. Child. Adolesc Psychopharmacol., 11,* 181 (2001).
12. Ambrosini, P. J., *Psychiatr. Serv., 51(5),* 627 (2000).
13. Martin, A., Kaufman, J., Charney, D., *Child. Adolesc. Psychiatr. N. Am. 9,* 135 (2000).
14. Hazell, P., O'Connell, D., Heathcote, D., Henry, D., *Cochrane Database Syst. Rev. 2,* CD 002317, (2002).
15. Emslie, G. J., Heilligenstein, J. H., Wagner, K. D., et al., *J. Am. Acad. Child. Adolesc. Psychiatry, 41,* 1205 (2002).
16. Emslie, G. J., Rush, A. J., Weinberg, W. A., et al., *Arch. Gen. Psychiatry, 54,* 1031 (1997).
17. Simeon, J. G., Dinicola, V. F., Ferguson, H. B., Copping, W., *Prog. Neuropsychopharmacol. Biol. Psychiatry, 14,* 791 (1990).
18. Wagner, K. D., Ambrosini, P., Rynn, M., et al., *JAMA, 290,* 1033 (2003).
19. Keller, M. B., Ryan, N. D., Strober, M., et al., *J. Am. Acad. Child. Adolesc. Psychiatry, 40,* 762 (2001).
20. Braconnier, A., Le Coent, R., Cohen, D., *J. Am. Acad. Child. Adolesc. Psychiatry, 42,* 22 (2003).
21. Walslick, B. D., Walsh, T., Greenhill, L., Eilenberg, M., Capasso, L., Lieber, D., *J. Affective Disorders, 56,* 227 (1999).
22. Mandoki, M. W., Tapia, M. R., Tapia, M. A., Summer, G. S., Parker, J. L., *Psychopharmacol. Bull., 33,* 149 (1997).
23. Glass, R. M., *JAMA, 292,* 861 (2004).
24. Geller, D. A., Hoog, S. L., Heiligenstein, J. H.,et al., *J. Am. Acad. Child. Adolesc. Psychiatry, 40,* 773 (2001).
25. Liebowitz, M. R., Turner, S. M., Piacentini, J., et al., *J. Am. Acad. Child. Adolesc. Psychiatry, 41,* 1431 (2002).
26. Riddle, M. A., Scahill, L., King, R., et al., *J. Am. Acad. Child. Adolesc Psychiatry, 31,* 1062 (1992).
27. Riddle, M. A., Reeve, E. A., Yaryura-Tobias, J. A., et al., *J. Am. Acad. Child. Adolesc. Psychiatry, 40,* 222 (2001).
28. Thomsen, P. H., Ebbesen, C., Persson, C., *J. Am. Acad. Child. Adolesc. Psychiatry, 40,* 895 (2001).
29. Grados, M. A., Riddle, M. A., *J. Clin. Child. Psychol., 30,* 67 (2001).
30. Pine, D. S., *J. Child. Adolesc. Psychopharmacol., 12,* 189 (2002).

31. Cook, E. H., Wagner, K. D., March, J. S., et al., *J. Am. Acad. Child. Adolesc. Psychiatry, 40*, 1175 (2001).
32. Wernicke, J. F., Kratochvil, C. J., *J. Clin. Psychiatry, 63(12)*, 50 (2002).
33. Campbell, M., Rapoport, J. L., Simpson, G. M., *J. Am. Acad. Child. Adolesc. Psychiatry, 38*, 537 (1999)
34. Le Heuzey, M. F., Ferrari, P., et al. (eds.), *Actualités en psychiatrie de l'enfant et de l'adolescent*, Paris. Flammarion pp. 396-404 (2002).
35. Spencer, E. K., Campbell, M., *Schizophr. Bull., 20*, 713 (1994).
36. Pool, D., Bloom, W., Mielke, D. H., Roniger, J. J., Gallant, D M., *Curr. Ther. Res., 19*, 99 (1976).
37. Grcevich, S. J., Findling, R. L., Roxane, W. A., Friedman, L., Schulz, S. C., *J. Psychopharmacology Child. Adoles., 6(4)*, 251 (1996).
38. Sholevar, E. H., Baron, D. A., Hardie, T. L., *J. Child. Adolesc. Psychopharmacol., 10*, 69 (2000).
39. Findling, R. L., McNamara, N. K., Youngstrom, E. A., Branicky, L. A., Demeter, C. A., Schulz, S. C., *J. Am. Acad. Child. Adolesc. Psychiatry, 42*, 170 (2003).
40. Chalasani, L., Kant, R., Chengappa, K. N. R., *Can. J. Psychiatry, 46*, 965 (2001).
41. Remschmidt, H., Fleischhaker, C., Hennighausen, K., Schulz, E., *Pediatr. Drugs, 2*, 253 (2000).
42. Barnard, L., Young, A. H., Pearson, J., Geddes, J., O'Brien, G., *J. Psychopharmacol., 16*, 93 (2002).
43. McCracken, J. T., McGough, J., Shah, B., et al., *N. Engl. J. Med., 347*, 314 (2002).
44. Malone, R. P., Cater, J., Sheikh, R. M., Choudhury, M. S., Delaney, M. A., *J. Am. Acad. Child. Adolesc. Psychiatry, 40*, 887 (2001).
45. Findling, R. L., *J. Clin. Psychiatry, 63*, 27 (2002).
46. McDougle, C. J., Kem, D. L., Posey, D. J., *J. Am. Acad. Child. Adolesc. Psychiatry, 41*, 921 (2002).
47. Scahill, L., Leckman, J. F., Schultz, R. T., Katsovich, L., Peterson, B. S., *Neurology, 60*, 1130 (2003).
48. Pappadopulos, E., McIntyre, J. C., Crismon, M. L., et al., *J. Am. Acad. Child. Adolesc. Psychiatry, 42*, 145 (2003),
49. Boachie, A., Goldfield, G. S., Spettigue, W., *Int. J. Eat. Disord., 33*, 98 (2003).
50. Serreau, R., LeHeuzey, M. F., Gilbert, A., Mouren, M. C., Jacqz-Aigrain, E. , *Paed. Perinat. Drug Ther., 6*, 14 (2004).

Respiratory, endocrine, cardiac, and renal topics

Clinical section (B)

9.1 Asthma

Kristine Desager

Department of Paediatrics, University Hospital Antwerp, Wilrijkstraat 10, 2650 Antwerp, Belgium

9.1.1 CLINICAL SETTING

Asthma is a chronic inflammatory disorder, which makes the airways sensitive to stimuli such as allergens, tobacco smoke, cold air or exercise. When exposed to these stimuli, the airways become swollen, constricted and filled with mucus. The resulting airflow limitation is reversible, either spontaneously, or with treatment. Asthmatic children suffer from recurrent episodes of wheezing, difficulty of breathing and/or (nocturnal) cough.

9.1.2 THERAPY

Treatment aims at preventing exacerbations and at reducing symptoms. Asthma is well controlled if urgent visits to the hospital are rare, the need for bronchodilators is minimal and if no limitations occur with physical activities. The patient should have a nearly normal lung function and should have no, or minimal, side effects from the medication.

The GINA guidelines propose a 6-point programme to reach the goal of optimal asthma control (www.ginasthma.com).

1. Patient's education

The patient will learn how to avoid risk factors, how to inhale medication correctly and how to seek medical help in case of worsening asthma. They will learn the difference between anti-asthma drugs.

2. Evaluation

Treatment is monitored by evaluating symptoms and use of rescue medication and, if possible, by measuring lung function. Adherence to treatment and technical skills of inhaling medication is checked.

3. Avoidance of exposure to risk factors

To improve the control of asthma and reduce the need for medication, the patient should avoid exposure to allergens (house dust mites, pets) and tobacco smoke. Physical activity is recommended.

4. Individual treatment plan

Anti-asthma medication is divided into two groups. The bronchodilators act within 15 minutes in case of symptoms and are called *reliever* or *rescue medication*, whereas the antiinflammatory agents or *controller medication* will prevent symptoms from occurring and are therefore administered continuously (Table 1). Antiinflammatory agents, particularly inhaled corticosteroids, are currently the most effective long-term preventive medications.

Inhaled drugs are preferred because of their high therapeutic-toxic ratio, i.e. high concentrations of drug are delivered directly to the airways, with potent

Table 1

Anti-asthma drugs

Controllers		Relievers	
Class	Name	Class	Name
systemic corticosteroids	methylprednisolone	short acting beta-agonists	fenoterol salbutamol
inhaled corticosteroids	beclomethasone budesonide fluticasone	anticholinergics	ipratropium
long acting beta agonists	formoterol salmeterol		
leukotriene antagonists	montelukast zafirlukast		

therapeutic effects and few systemic side effects. Several systems for inhalation exist: metered dose inhaler, dry powder inhaler or nebuliser. The use of a spacer, which facilitates the use of a metered dose inhaler, increases the deposition and decreases the side effects. For these reasons, a spacer is the treatment of choice in children below the age of 4 years. In infants and young children a spacer with a mask is indicated. Several factors will guide the choice of a spacer. Spacer and metered dose inhaler should fit properly, but also the mask or mouth piece should fit to the child. The volume of the spacer should be adapted to the lung volume of the patient and the child should be capable of moving the valves of the spacer. For children aged from 4 to 6 years, dry powder inhalers can be used, depending on the cooperation and inspiratory effort of the child. During attacks, since inspiratory effort is decreased, bronchodilators can be administered using metered dose inhaler and spacer, or using a nebuliser.

The first goal in the treatment is to bring the asthma symptoms under control. Two approaches are possible. In the first approach, either a high dose of inhaled or systemic corticosteroids, or a combination of inhaled corticosteroids and long acting beta agonists, is prescribed to control the symptoms. In the second approach, treatment is initiated according to one of the 4 classes of severity of the disease, ranging from mild to severe persistent (see GINA guidelines in Table 2). If the asthma is not brought under control, therapy is increased or "stepped up".

The second goal is to bring the symptoms under control with minimal drug treatment. Treatment is then tapered down or "stepped down", but not discontinued, if the asthma is under control for at least 3 months.

Table 2

Anti-asthma drugs

Grade	Daily controller drug	Alternative
intermittent	none	
mild persistent	low dose ICS	LTA
moderate persistent	medium dose ICS	medium dose ICS + LABA medium dose ICS + LTA
severe persistent	high dose ICS + LABA or LTA	

ICS = Inhaled corticosteroids
LABA = Long acting beta agonists
LTA = Leukotriene antagonists

5. Treatment plan in case of exacerbation

Patients should seek medical advice in case of progressive increase in shortness of breath, coughing, wheezing and if short-acting beta-agonists do not work.

6. Regular follow-up care should be organised.

9.1.3 PHARMACOLOGY AND SIDE EFFECTS OF INHALED CORTICOSTEROIDS

The pharmacokinetics of the various inhaled corticosteroids differ markedly, and, together with differences in the inhalation systems used with the various products, may lead to important differences in both clinical and systemic effects. The topical effects of an inhaled corticosteroid depend on the glucocorticosteroid activity of the molecule and, probably also, on the local pharmacokinetics in the target tissue. The systemic effects are related to the glucocorticosteroid activity of the molecule, the total amount that is absorbed (becomes systemically available) and the rate of clearance from the body.

After inhalation, an amount of the drug will reach the systemic circulation and can result in systemic effects. The amount of drug reaching the systemic circulation is determined by the fraction deposited in the airways (pulmonary fraction) and in the mouth (oral fraction). The oral fraction will be swallowed and absorbed. How much of this fraction becomes available for the systemic circulation depends on the first pass metabolism in the liver. The more recently developed inhaled corticosteroids (e.g. fluticasone proprionate, mometasone furoate and budesonide) have a low oral bioavailability. 50 to 70% of the swallowed fraction of beclomethasone dipropionate, triamcinolone and flunisolide is metabolised. The pulmonary fraction is likely to be more or less completely absorbed in active form to the systemic circulation. The systemic concentration will be reduced by continuous recirculation and inactivation of the drug by the liver.

When selecting an inhaled corticosteroid, the following factors have to be taken into account:

- Select an inhaled corticosteroid with a high first pass metabolism.
- Be aware that, if the deposition increases because of a more effective inhaler or an aerosol with smaller particle size, the risk of systemic side effects increases.
- Children have a higher oropharyngeal deposition than adults, resulting in an increased oral fraction. A spacer can resolve this problem.

Effect on growth

Asthma may influence the growth of a child in two ways. The disease and its level of control may directly affect the growth rate, usually towards the end of the first decade [1-6]. This reduced growth rate continues into the mid-teens and is associated with a delay in the onset of puberty and delay in skeletal maturation, so that the bone age of the child corresponds to the height. There is no decrease in final adult height, though it is reached at a later age than normal. This growth pattern seems to be unrelated to the use of inhaled corticosteroids and is more pronounced in children with severe asthma. Recent studies have confirmed that poorly controlled asthma may itself adversely affect growth [7]. Severe asthma may adversely affect final height [8].

The level of lung function in asthma may affect the systemic availability of inhaled corticosteroids. Systemic bioavailability and effects of an inhaled drug are more pronounced in patients with mild asthma than in patients with more severe disease [9]. This is probably due to differences in the deposition pattern caused by a smaller airway diameter in patients with more severe disease.

Studies evaluating the effect of exogenous corticosteroids on growth are traditionally divided into:

- Growth marker studies, which measure corticosteroid-induced changes in various serum markers that are thought to reflect bone and collagen formation/degradation or growth.
- Short-term studies that assess growth over periods of 3 months or less.
- Intermediate-term studies that evaluate growth over periods longer than 3 months but do not assess final height.
- Long-term studies that assess growth for many years and also include final adult height in relation to predicted adult height.

It is important to remember these characteristics when assessing the clinical relevance of the results of clinical trials. Changes in growth markers or change in growth velocity on short term does not imply an effect on long term or on final height.

The following conclusions can be drawn from short and intermediate term studies [10]:

- Controlled studies have not reported any statistically or clinically significant adverse effect on growth with inhaled corticosteroids at doses of 100-200 μg/day budesonide or equivalent.
- Growth retardation can be seen with all inhaled corticosteroids when a high dose is administered, without any adjustment for disease severity and control.

- Growth retardation is dose-dependent.
- Important differences seem to exist between growth-retarding effects of various inhaled corticosteroids and inhalers.
- Different age groups seem to differ in their susceptibility of inhaled corticosteroids: children 4 to 10 years of age are more susceptible than pubertal children,
- The growth retarding effect of inhaled corticosteroid treatment seems to be more marked at the beginning of treatment and becomes attenuated with continued treatment.

Most physicians consider final adult height to be the most important growth outcome and the key question with regard to the effects on growth is whether the slowing in short-term growth rates leads ultimately to a diminished final adult height. At present, few prospective long-term studies of attained adult height have been conducted [1,2,11]. No effect on final height of inhaled corticosteroids was found. Some possible explanations for the apparent discrepancy between the findings of some intermediate-term studies and the conclusions on final height are as follows [10]:

- Correlation between two consecutive annual height velocity values for normal prepubertal children is poor.
- A low gain in one year is not necessarily followed by a low gain the next year and vice versa [12].
- The growth retarding effect of inhaled corticosteroids is most pronounced during the first year of treatment.
- Conclusions from short-term studies cannot be extrapolated to the long term situation.
- Corticosteroids seem to retard bone maturation. If this occurs to the same extent as the retardation of growth, then final height should not be adversely affected. These children will grow for a longer period and eventually attain normal final height.
- Results obtained in prepubertal children may not be valid in other age groups, because prepubertal children may be more sensitive to the growth-retarding effect of inhaled corticosteroids.
- In real life situations, children with mild asthma are not treated continuously with a rather high, fixed dose.

Dose-response curves

A number of dose-response studies in children with moderate to severe asthma demonstrated a significant improvement in symptoms and lung function with a dose of 100 μg/day budesonide [13-16]. When the dose was increased, little additional improvement was found.

In recent years, with the advent of more potent steroids and more efficient delivery systems, the relative doses commonly used have increased. There have been several reports of serious adverse events resulting from doses of inhaled corticosteroids in excess of those recommended. These include growth failure, suppression of the hypothalamo-pituitary-adrenal axis [17], resulting in acute hypoglycaemia, altered consciousness, coma, convulsions and death [18-21]. While the majority of these effects have been reported at higher doses, some have occurred at a dose within the recommended range. This suggests that individual susceptibility may also be important. These effects are more commonly associated with more potent inhaled corticosteroids, but this could be due to the more frequent prescription of that drug in more severe asthma.

Is there any evidence that inhaled corticosteroids are overused? A recent meta-analysis, examining the dose response to inhaled corticosteroids in adolescents, reported that 90% of the maximum benefit was achieved at a daily dose of 250 μg fluticasone [22]. Minimal further improvement resulted from increases to 600 μg/day. The dose response curve for systemic side-effects reaches a plateau at 2000 μg/day. The introduction of long acting beta-agonists or leukotriene antagonists at low doses of inhaled corticosteroids can achieve improved asthma control, avoiding the need for higher doses of inhaled corticosteroids.

When asthma is not brought under control by a dose of inhaled corticosteroids equivalent to 1000 μg budesonide or its equivalent, consideration should be given to issues of adherence to the treatment regimen, inhaler technique or an alternative diagnosis. In the British survey of adrenal crisis due to inhaled corticosteroids, three of the 28 children did not have asthma, and in five children, asthma did not account for all the respiratory symptoms [20].

9.1.4   PRACTICAL PROPOSAL: TREATMENT OF ASTHMA

If there are signs of an acute attack, treatment consists of administration of oxygen, inhaled beta-agonists and systemic corticosteroids. When the condition of the patient is stable, asthma severity is classified. If the complaints occur less than once a week and the night time asthma symptoms occur less than twice a month, symptomatic treatment with beta-agonists or continuous treatment with leukotriene antagonists can be initiated. In all other cases, inhaled corticosteroids are prescribed in a dose of maximum 500 μg budesonide or equivalent (e.g., 250 μg/day fluticasone or beclomethasone CFC 500 μg/day or beclomethasone HFA 250 μg/day). The patient is instructed according to the 6-point programme and is asked to record symptoms during one month.

After 1 month, the doctor can evaluate if asthma is under control. There are two possibilities:

- Asthma is under control. Treatment should continue, but the dose will be decreased in steps of 100 μg/day budesonide or equivalent.
- Asthma is not under control, therefore two possibilities exist. Patients on intermittent short-acting beta-agonists will be switched to daily inhaled corticosteroids (max. budesonide 500 μg/day or equivalent). If the patient is already taking budesonide 500 μg/day or equivalent, long acting beta agonists or leukotriene antagonists will be associated.

REFERENCES

1. Balfour-Lynn, L., *Arch. Dis. Child, 61*, 1049 (1986).
2. Balfour-Lynn, L., *Pediatrician, 14*, 237 (1987).
3. Fergusson, D. M., Hons, B. A., Horwood, L. J., *Pediatr. Pulmonol., 1*, 99 (1985).
4. Hauspie, R., Susanne, C., Alexander, F., *J. Allergy Clin. Immunol., 59*, 200 (1977).
5. Hauspie, R., Susanne, C., Alexander F., *Hum. Biol. 48*, 271 (1976).
6. Martin, A. J., Landau, L. I., Phelan, P. D., *Acta Paediatr. Scand. 70*, 683 (1981).
7. Ninan T. K., Russell, G., *Arch. Dis. Child 67*, 703 (1992).
8. Van Bever, H. P., Desager, K. N., Lijssens, N., Weyler, J. J., Du Caju M.V., *Pediatr. Pulmonol. 27*, 369 (1999).
9. Lipworth B. J., Clark D. J., *Lancet, 348*, 820 (1996).
10. Pedersen, S., *Am. J. Respir. Crit Care Med., 164*, 521 (2001).
11. Agertoft L., Pedersen, S., *N. Engl. J. Med., 343*, 1064 (2000).
12. Karlberg, J., Gelander, L., Albertsson-Wikland, K., *Acta Paediatr., 82*, 631 (1993).
13. Agertoft L., Pedersen, S., *J. Allergy Clin. Immunol., 99*, 773 (1997).
14. Pedersen S., Hansen, O. R. *J. Allergy Clin. Immunol., 95*, 29 (1995).
15. Katz, Y., Lebas, F. X., Medley, H. V., Robson, R., *Clin. Ther., 20*, 424 (1998).
16. Shapiro, G., Bronsky, E. A., LaForce, C. F., *J. Pediatr., 132*, 976 (1998).
17. Fitzgerald, D., Van Asperen, P., Mellis, C., Honner, M., Smith, L., Ambler, G. *Thorax, 53*, 656 (1998).
18. Patel, L., Wales, J. K., Kibirige, M. S., Massarano, A. A., Couriel, J. M., Clayton, P. E., *Arch. Dis. Child, 85*, 330 (2001).
19. Todd, G. R., Acerini, C. L., Buck, et al., *Eur. Respir. J., 19*, 1207 (2002).
20. Todd, G. R., Acerini, C. L., Ross-Russell, R., Zahra, S., Warner, J. T., McCance, D., *Arch. Dis. Child, 87*, 457 (2002).
21. Drake, A. J., Howells, R. J., Shield, J. P., Prendiville, A., Ward, P. S., Crowne, E. C. *BMJ, 324*, 1081 (2002).
22. Holt, S., Suder, A., Weatherall, M., Cheng, S., Shirtcliffe, P., Beasley, R. *BMJ, 323*, 253 (2001).

9.2 Cystic fibrosis

Alan Smyth

City Hospital, Hucknall Road, Nottingham NG5 1PB, UK

9.2.1 CLINICAL SETTING

Cystic fibrosis (CF) is a genetic disorder in which sufferers have a failure of pancreatic enzyme secretion and malabsorption. They are also susceptible to pulmonary infection. The majority of patients in the United Kingdom have chronic pulmonary infection with the organism *Pseudomonas aeruginosa* by their late teens. Much CF therapy is directed towards improving nutrition and preventing, or managing, chronic infection with *P. aeruginosa*.

CF is inherited in an autosomal recessive fashion. The carrier incidence is around 1/25 with 1/2500 live births affected in Western Europe. The cystic fibrosis gene is located on chromosome 7. The commonest gene mutation in CF patients in the UK is ΔF508, with 52% of CF patients homozygous for this mutation. A defective form of the gene product, the cystic fibrosis transmembrane conductance regulator (CFTR), leads to abnormal transport of chloride ions at mucosal surfaces. This leads to the production of thick, sticky secretions in the lung, gastrointestinal tract, hepatobiliary system, pancreas and reproductive system. This presents some particular problems for drug delivery. Drugs (including bronchodilators, inhaled steroids, antibiotics and mucolytics) are frequently given by the inhaled route. Drug delivery to the lung may be inefficient in the normal airway, but in patients with CF, there are added obstacles of mucus plugging and ventilation-perfusion mismatch. There may be problems with the absorption of oral medication, such as fat soluble vitamins, due to fat malabsorption, liver disease and interruption of enterohepatic circulation by antibiotics. Frequent intravenous antibiotic treatment may be necessary and this may require the placement of an indwelling intravenous access device (e.g., Portacath®).

9.2.2 PHARMACOLOGY AND DRUG THERAPY

Nutritional therapy

The presence of pancreatic malabsorption can be confirmed by measuring stool chymotrypsin or pancreatic elastase. Over 90% of CF infants have detectable fat malabsorption by the age of 1 year. The management of pancreatic malabsorption has steadily improved over the last 30-40 years, mainly due to a better understanding of the nutritional needs of CF patients and improved formulations of pancreatic enzyme supplements. The resting energy expenditure of CF patients is increased and may be up to 150% of the normal for an unaffected individual of similar size [1]. An increased calorie intake is therefore required. The increased energy expenditure may be due to the increased work of breathing and the catabolic effects of chronic inflammation. A high calorie, high fat diet is needed for normal growth in children. Proprietary calorie supplements or enteral feeding are indicated for some patients [2].

Early pancreatic enzyme supplements were crude extracts of animal origin. These were partially denatured by gastric acid and caused oral and perianal irritation. Newer formulations use enteric coated microspheres or microtablets (porcine pancreatin), which pass unchanged through the stomach and are activated in the duodenum [3]. However, even with these improved formulations, effective digestion of food may be impaired by lack of physiological "mixing" of enzymes with food [4]. The presence of viscid secretions in the pancreas not only impairs the secretion of digestive enzymes into the duodenum, but also the secretion of bicarbonate which is necessary to neutralise gastric acid and create the alkaline environment in the duodenum which is necessary for optimum enzyme action.

CF patients with symptoms of malabsorption or poor nutrition, in spite of pancreatic enzyme therapy, may benefit from a proton pump inhibitor, such as omeprazole (Table 1) [5]. Biliary sludging occurs in CF patients and absorption of fat may be further impaired by the absence of the emulsifying effect of bile. Chronic biliary obstruction may ultimately lead to cirrhosis and portal hypertension (which affects 2% of adults with CF in the UK). Many CF units treat patients with ursodeoxycholic acid (Table 1), a hydrophilic bile acid which improves the flow of bile and shows promise in preventing liver disease in CF [6]. Treatment should be started early in the course of liver disease and should be for life.

The dose of pancreatic enzyme supplements is titrated to achieve symptomatic control of malabsorption and normal growth. Typical starting doses are given here for newly diagnosed infants, using as an example one pancreatic enzyme brand (Creon, Solvay Healthcare Ltd, Southampton, UK). Infants should start with one half scoop of Creon Micro (or ¼ Creon 10 000 capsule) with every 60-120 ml of milk feed. The dose can gradually be increased, until bowel symptoms are well-controlled and weight gain satisfactory - to a maximum of 10 000 lipase units/kg/day. The microspheres can be mixed with a little fruit puree and given to the infant from a teaspoon [2].

At birth, the presence of viscid meconium may lead to small bowel obstruction (meconium ileus), requiring reduction with radiological contrast medium or surgery [7]. Later in life, the accumulation of bulky, partially digested material at the ileo-caecal valve can cause a form of small bowel obstruction known as distal intestinal obstruction syndrome (DIOS). Most patients will respond to a stepwise approach, using oral lactulose and a radiological contrast medium, such as gastrografin [8]. However, some patients with DIOS will require surgery [7].

Adverse effects of treatment

Pancreatic enzyme supplements were believed to have no serious adverse effects until reports of colonic strictures first appeared in the early nineties [9]. This was a new condition in CF patients, with characteristic histological findings and was termed "fibrosing colonopathy." A subsequent case-control study showed a dose dependent association between the risk of fibrosing colonopathy and the use of high strength formulations of pancreatic enzyme supplements [10]. The authors speculated that undissolved microspheres or mintablets might pass through the ileocaecal valve into the colon where lipase, protease and amylase might be released in high concentrations. There might also be a toxic effect on the colon of the inert components of the microspheres or minotablets. The UK Committee on Safety of Medicines subsequently issued a warning that enzyme doses should not exceed 10 000 units of lipase/kg/day.

9.2.3   TREATMENT OF PULMONARY INFECTION

Certain organisms are associated with pulmonary infection in CF patients, in different age groups. Infection with *Staphylococcus aureus* is present in infants as young as 3 months, is accompanied by evidence of inflammation and is implicated in the development of lung damage in CF [11]. The use of a prophylactic anti-staphylococcal antibiotic, such as flucloxacillin (Table 1), continuously up to 3 years of age, results in fewer children having this organism, though not in improved clinical outcomes [12].

Later, children may acquire chronic pulmonary infection with *P. aeruginosa*. Initially, the organism is present in a free-living, planktonic form. Later, infection becomes chronic and the organism forms an adherent biofilm, within which the organism is resistant to antibiotic treatment. Very different antibiotic strategies are used at these two stages. At the planktonic stage, eradication treatment

Table 1

A formulary of drugs used in cystic fibrosis [2,25,26]

Drug	Route	Age/Wt.	Dose	Doses/day	Duration	Comments
ceftazidime	IV	all	50 mg/kg (max 3 g tds)	3	2 weeks	
ciprofloxacin	PO	<5 years	15 mg/kg	2	3 weeks (step 1 & 2)	eradication treatment for *P. aeruginosa*
		>5 years	20 mg/kg		3 months (step 3)	
colistin	NEB	all	1 MU	2	3 weeks (step 1)	eradication
			2 MU	3	3 weeks (step 2)	eradication
			2 MU	3	3 months (step 3)	eradication
		< 1 year	0.5 MU	2	maintenance	
		1–10 years	1 MU	2	maintenance	
		>10 years	2 MU	2	maintenance	
dornase alfa	NEB	>5 years	2.5 mg	1	maintenance	after physiotherapy
flucloxacillin	PO	all	25 mg/kg (max 1 g qds)	4	2 week short course	treatment
flucloxacillin	PO	<3 years	125 mg	2	continuous until 3 years	prophylaxis
gastrografin	PO	<15 kg	15–30 ml	1		distal intestinal obstruction
		15–25 kg	50 ml			
		>25 kg	100ml			
lactulose	PO	<7 years	10 ml	2	may need maintenance	
		>7 years	20 ml	2		

Table 1 *(continued)*

Drug	Route	Age/Wt.	Dose	Doses/ day	Duration	Comments
omeprazole	PO	<12 years	700 mcg/kg increasing to 3 mg/kg	1	maintenance	as an adjunct to pancreatic enzyme supplements
		12-18 years	20-40 mg			
pancreatic enzymes	PO	all	maximum 10 000 U/kg/day	with meals	life long	
prednisolone	PO	all	1-2 mg/kg/dose	1	2-4 weeks	monitor for adverse effects of steroids
tobramycin	IV	all	10 mg/kg (max starting dose 660 mg od)	1	2 weeks	monitor renal function & trough tobramycin levels during course (target trough <1 mg/L)
ursodeoxycholic acid	PO	all	10mg/kg/dose	2	life long	
vitamin A	PO	<1 year	4000 IU (as Abidec®)	1	maintenance	measure vitamin levels annually & adjust
		>1 year	4000–10 000 IU (as Abidec®)	1		
vitamin E	PO	<1 year	10-50 mg	1	maintenance	measure vitamin levels annually & adjust
		1-10 year	50-100 mg	1		
		>10 year	100-200 mg	1		
vitamin K	PO / IV	>1-11 years	5 mg	1		
		>12 years	10 mg	1		

is proposed, usually with a combination of oral ciprofloxicin and the macrolide antibiotic colistin, given in nebulised form. Treatment may be given in a stepwise fashion, with a progressively increasing dose of colistin and longer duration of treatment [13]. However, by the late teens, over half of CF patients will have acquired chronic infection with *P. aeruginosa*. At this point, the emphasis shifts to long term maintenance treatment with a nebulised anti-pseudomonal antibiotic (colistin or the aminoglycoside antibiotic tobramycin). Here, the aim is not to eradicate the organism but rather to improve pulmonary function and to reduce the frequency of acute pulmonary infection with *P. aeruginosa* [14]. Patients with chronic infection may also need intermittent courses of intravenous antibiotics. There is controversy over whether such intravenous treatment should be given regularly (conventionally at 3 monthly intervals), or only when the patient has worsening symptoms [15]. Antibiotic resistance amongst *P. aeruginosa* has become an increasing problem and resistant strains may be transmitted between patients [16]. Recombinant DNase (Dornase alfa) is used to reduce sputum viscidity and improve pulmonary function [17].

Many clinicians recommend that combined treatment is given with two anti-pseudomonal antibiotics to discourage the emergence of resistance [16]. A commonly used combination is the aminoglycoside antibiotic tobramycin, given intravenously, combined with ceftazidime. For each drug, a pharmacological strategy has been proposed to allow more effective bacterial killing. Tobramycin shows concentration dependent killing and a post antibiotic effect, and therefore, a once daily regimen with high peaks and low troughs should give better bacterial killing. In practice, a once daily regimen appears to be equally effective and less nephrotoxic than the traditional three times daily approach [18]. The reduced nephrotoxicity is likely to be due to the pharmacokinetics of aminoglycoside uptake into the proximal tubule, which are saturable at high serum concentration. In the case of ceftazidime, bacterial killing is enhanced if the concentration of the antibiotic exceeds the minimum inhibitory concentration of ceftazidime for *P. aeruginosa* for most of the time. A continuous infusion should therefore be more effective [19]. However, a controlled trial of continuous ceftazidime, with sufficient statistical power to detect a clinically important difference in outcome, has not yet been undertaken.

The pharmacokinetics of many drugs in CF patients differs from other individuals, with a larger volume of distribution and more renal rapid elimination [20]. Some drugs, metabolised in the liver, are metabolised more rapidly due to an increase in the activity of the cytochrome P-450 enzyme CYP1A2. This effect can be demonstrated in young patients with CF who are well and so is likely to be related to the primary defect in CF rather than a secondary effect of liver disease or pulmonary infection [21].

Adverse effects of treatment

As chronic infection and inflammation is a feature of cystic fibrosis, patients are uniquely vulnerable to the adverse effects related to cumulative exposure to antibiotics and anti-inflammatory drugs. Hearing loss in CF patients is associated with recurrent exposure to intravenous aminoglycosides [22]. Even an aminoglycoside given in nebulised form may cause adverse effects such as tinnitus [14]. Intravenous aminoglycosides have also been implicated as a cause of acute renal failure in CF patients [23]. Adverse effects also occur with the quinolones (cutaneous photosensitivity) and ceftazidime (urticarial rash). A number of trials have looked at the long term use of oral steroid treatment to break the cycle of infection and inflammation and prevent lung damage. However, although the decline in lung function was slower with steroid use, there were unacceptable adverse effects, with reduced linear growth, raised blood sugar and cataracts. The effect on growth was seen as early as 6 months into treatment on a dose of 2 mg/kg alternate daily [24].

9.2.4   CONCLUSION

Effective drug therapy in cystic fibrosis presents many challenges to the multidisciplinary team. Drug delivery may be impaired by malabsorption, variable lung deposition or loss of intravenous access after repeated therapy. Drugs may be less effective because of physiological limitations (as with pancreatic enzyme supplements) or organisms which are resistant to antibiotics or inaccessible within a biofilm. The need for multidrug regimens increases the risk of drug interactions and adverse effects, whilst long term therapy predisposes to cumulative toxicity. Finally, the therapeutic burden on patients and families may lead to non-adherence. Successful care depends on the judicious use of evidence based therapy, vigilance for drug side effects and long term support for patients and their families. With the prospect of gene therapy or novel treatments for the faulty CFTR protein, there is much optimism that the quality of life and prognosis for CF patients may improve further over the next decade.

REFERENCES

1. Vaisman, N., Pencharz, P. B., Corey, M., Canny, G. J., Hahn, E., *J. Pediatr., 111*, 496 (1987).
2. UK Cystic Fibrosis Trust Nutrition Working Group, *Nutritional management of cystic fibrosis*, Bromley, Cystic Fibrosis Trust, 2002.
3. Beverley, D. W., Kelleher, J., MacDonald, A., Littlewood, J. M., Robinson, T., Walters, M. P., *Arch. Dis. Child.*, 62, 564 (1987).

4. Taylor, C. J., Hillel, P. G., Ghosal, S., Frier, M., Senior, S., Tindale, W. B. et al., *Arch. Dis. Child.*, *80*,149 (1999).
5. Heijerman, H.G., *Neth. J. Med.*, *41*, 105 (1992).
6. Colombo, C., Battezzati, P. M., Podda, M., Bettinardi, N., Giunta, A., *Hepatology*, *23*, 1484 (1996).
7. Gross, K., Desanto, A., Grosfeld, J. L., West, K. W., Eigen, H., *J. Pediatr. Surg.*, *20*, 431 (1985).
8. O'Halloran, S. M., Gilbert, J., McKendrick, O. M., Carty, H. M., Heaf, D. P., *Arch. of Dis. Child.*, *61*, 1128 (1986).
9. Smyth, R. L., van Velzen, R., Smyth, A. R., Lloyd, D. A., Heaf, D. P., *Lancet*, *343*, 85 (1994).
10. Smyth, R. L., Ashby, D., O'Hea, U., et al., *Lancet*, *346*, 1247 (1995).
11. Armstrong, D. S., Grimwood, K., Carzino, R., et al., *B. M. J.*, *310*, 1571 (1995).
12. Smyth, A., Walters, S., "Prophylactic antibiotics for cystic fibrosis," (Cochrane Review). *The Cochrane Library*, *4*, Chichester, UK, John Wiley & Sons Ltd., (2003).
13. Frederiksen, B., Koch, C., Hoiby, N., *Pediatr. Pulmonol.*, *23*, 330 (1997).
14. Ramsey, B. W., Pepe, M. S., Quan, J. M., Otto, K. L., *N. Engl. J. Med .*, 340, 23 (1999).
15. Elborn, J. S., Prescott, R. J., Stack, B. H. R., et al., *Thorax*, *55*, 355 (2000).
16. Cheng, K., *Lancet*, *348*, 639 (1996).
17. Jones, A. P., Wallis, C. E., "Recombinant human deoxyribonuclease for cystic fibrosis," (Cochrane Review), *The Cochrane Library*, *4*, Chichester, UK, John Wiley & Sons Ltd., (2003).
18. Smyth, A., Tan, K., Hyman-Taylor, P., et al., *Pediatr. Pulmonol.*, *5 (suppl.)*, 292 (2003).
19. Rappaz, I., Decosterd, L. A., Bille, J., Pilet, M., Belaz, N., Roulet, M., *Eur. J. Pediatr.*, *159*, 919 (2000).
20. Touw, D. J., *Pharm. World Sci.*, *20*,149 (1998).
21. Parker, A. C., Pritchard, P., Preston, T., Smyth, R. L., Choonara, I., *Arch. Dis. Child.*, *77*, 239 (1997).
22. Mulheran, M., Degg, C., Burr, S., Morgan, D. W., Stableforth, D. E., *Antimicrob. Agents Chemother.*, *45*, 2502 (2001).
23. Drew, J. H., Watson, A. R., Evans, J. H. C., Smyth, A. R., *Paed. Perinat. Drug Ther.*, *5*, 65 (2002).
24. Cheng, K., Ashby, D., Smyth, R., "Oral steroids for cystic fibrosis" (Cochrane Review), *The Cochrane Library*, *4*, Chichester, UK: John Wiley & Sons Ltd, (2003).
25. UK Cystic Fibrosis Trust Antibiotic Group, Antibiotic Treatment for Cystic Fibrosis, "Report of the UK Cystic Fibrosis Trust Antibiotic Group," London, UK, Cystic Fibrosis Trust, (2002).
26. *Medicines for children*, London, Royal College of Paediatrics & Child Health, (2003).

9.3 Management of hypertensive emergencies in children[*]

Peter Houtman

Consultant Paediatrician, Children's Hospital, Leicester Royal Infirmary, UK

9.3.1 INTRODUCTION

Hypertensive emergencies are uncommon in paediatric practice. When faced with this problem, a working knowledge of the principles of management is more important than that of the specific features of the drugs used.

A background understanding of blood pressure (BP) variation in childhood will make management more appropriate. Genetic variability and environmental factors both have a role in determining an individual's BP, which is dependent on height more directly than age or other parameters of size. Measurement of BP is unfortunately relatively infrequently performed in children. Measuring equipment is often poorly maintained and the situation is now confounded by the withdrawal of mercury sphygmomanometers for safety and environmental concerns [1].

Technique is important. The child should be resting and the measuring cuff (bladder-size) should cover at least 2/3 of the upper arm length. Automated methods are often unreliable and machines must be well-calibrated. Always verify the reading yourself, if possible on more than one occasion. Invasive measurement is often very useful but not always appropriate. Beware of the "silent gap" if BP is very high. Mean BP can be used if available but, particularly in young children, the diastolic BP is often not easily measurable. Systolic BP is usually the most reliable parameter for monitoring purposes.

---

[*]Reprinted from *Paed. Perinatal Drug Ther.*, 5, 107 (2003)

9.3.2 HYPERTENSION

Charts of normal BP are available [2]. However "hypertension" in childhood is poorly defined: for example > 95th centile, 10 mmHg above 95th centile, and >99th centile are all used variably [3]. Table 1 shows values for "significant" and "severe" hypertension, which may be used in clinical practice as a rough guideline to BP levels which should be investigated and managed effectively [4].

A hypertensive emergency exists when there is organ damage, or impending organ damage, and is not defined in terms of the BP level, as this in itself cannot predict the severity of the problem alone [5,6]. The term hypertensive "urgency" is sometimes used [7,8] to distinguish those cases without organ damage, but with a possibility of such damage occurring in the next day or so. In practice this exact distinction is not often possible. The organs susceptible to damage include the brain, eyes, heart and kidney, with the major pathological process being fibrinoid necrosis of arterioles.

"Essential" hypertension is a poorly defined entity in children. Severe hypertension should always initially be considered to be "secondary" in cause. Causes of hypertensive emergencies include (in rough order of frequency) reflux nephropathy, obstructive uropathy, renovascular disease, glomerular disease, polycystic kidney disease, haemolytic-uraemic syndrome, coarctation, phaeochromocytoma, Wilm's tumour, renal dysplasia, intracranial disease and drugs [9]. Thus, renal causes predominate.

Table 1

Rough guide of hypertension values in childhood

Age	Significant hypertension		Severe hypertension	
	Systolic	Diastolic	Systolic	Diastolic
7 days	96		106	
<1 month	104		110	
<2 years	112	74	118	82
3-5 years	116	76	124	84
6-9 years	122	78	130	86
10-12 years	126	82	134	90
13-15 years	136	86	144	92
16-18 years	142	92	150	98

Often there are no symptoms even in severe hypertension [9]. If present, these are mainly neurological: visual symptoms, facial palsy, convulsions, hemiplegia and frank hypertensive encephalopathy. Cardiac failure is more commonly found in very young children in whom neurological complications are relatively rare.

9.3.3 MANAGEMENT

It is worth restating that the principles of the management of hypertensive emergencies are at least as important as the specifics of the drugs used. Appropriate intensive care facilities should be available. In the emergency evaluation one should assess cardiovascular status both clinically and with an ECG. Echocardiography is particularly helpful in assessing left ventricular function. Neurological status (conscious level, irritability, fits) will often determine the level of medical and nursing care required. Assessment of fluid balance and renal function will also influence treatment profoundly. An ophthalmological opinion is valuable. Thus, a multi-disciplinary approach is appropriate.

Beware of hypovolaemia. This can cause or exacerbate hypertension [10]. Also, it is sometimes difficult to decide whether fits in the presence of hypertension are the cause or an effect. In a post-ictal patient, the magnitude of BP increase has been found to be a useful clinical parameter to exclude a hypertensive crisis [11].

Treatment must be based on the cause if known or suspected. In renal disease intravascular fluid ("saline") overload is often under-estimated, and tissue oedema is not well correlated with intravascular volume. Thus the drug of choice may be a diuretic in the infant or child whose main problem is systemic fluid overload, with hypertension as part of the clinical picture. Renal function will profoundly influence management, particularly use of drugs. Urine output should be monitored and, if serum creatinine is outside the normal range, caution given in the use of drugs excreted though the renal route. Renal dialysis or filtration may be required. Renin-dependent causes will influence the use of ACE-inhibitors, particularly if renovascular. Catecholamine-driven hypertension needs specialist advice regarding both α-adrenergic and β-blockade, with particular attention to problems around surgery for phaeochromocytoma.

Consideration of the length of time over which the hypertension has developed is very important [12]. Cerebral autoregulation keeps cerebral blood flow relatively constant within systemic BP limits but over time this range is altered in hypertension. If BP is lowered too rapidly there is a real danger of relative hypotensive damage to watershed areas, particularly visual cortex, cerebellum and end-arteries (ciliary arteries to optic nerve and spinal arteries). Visual loss may be permanent. Thus, the principle of slow reduction of severe, chronic hypertension is paramount.

9.3.4   DRUG THERAPY

The specific drugs most widely recommended for initial emergency management are sodium nitroprusside and labetalol. Other drugs that have also been recommended include nifedipine, nicardipine and diazoxide. These are discussed individually below, and doses are given in Table 2. However, whichever drug is used, the principle of the slow reduction in blood pressure is the same. A helpful goal to strive for is to aim to reduce BP initially by only a third of the difference between the acute BP level and the appropriate normal value. Only after the first 24 hours or more should it be attempted to reduce BP further with the gradual addition of longer-acting drugs. Whichever agent is used, frequent and reliable BP monitoring is necessary. An intravenous saline infusion should be available to raise BP if it has dropped too low.

Sodium nitroprusside

Sodium nitroprusside is a potent vasodilator with an almost instant effect when given intravenously. It is also very short-acting (seconds) and therefore capable of exquisite BP control. Continuous BP monitoring is mandatory. The drug undergoes rapid photo-degradation so the solution must be covered with aluminium foil. *In vivo*, cyanide is produced from its local breakdown in smooth muscle, which is then metabolised to thiocyanate in the liver. This is excreted by the kidney. Therefore it must be used with caution in patients with renal or hepatic failure. In the absence of organ failure, a clinical problem should not be expected unless the drug is used for more than 48 hours or so. After this time a worsening metabolic acidosis can be anticipated. It may be possible to measure thiocyanate levels in these circumstances, but usually the patient will be suitable for longer-acting drugs by this time.

Table 2

Drug doses for hypertensive emergencies (start with the lower doses)

Drug	Route	Dose
sodium nitroprusside	IV	0.5 - 8 mcg/kg/min
labetalol	IV	0.25 - 1.5 mg/kg/h
nifedipine	Oral	0.1 - 0.25 mg/kg/dose (initial doses)
nicardipine	IV	1 - 3 mcg/kg/min
diazoxide	IV	1 - 3 mg/kg/dose

Labetalol

Labetalol is also commonly used in the acute situation. Most of its action is via β-adrenergic blockade but there is also α-adrenergic blockade (15%). It therefore reduces cardiac output and causes peripheral vasodilatation. It has a rapid onset but a longer duration of action (hours) and is therefore not capable of such sensitive BP regulation as nitroprusside. Its use is limited if there is clinical evidence of cardiac failure as it may further lower cardiac output. Like other β-blockers, there must be caution in asthmatics. It is metabolised in the liver and its use is not limited in renal failure. Initial incremental loading doses of 250 mcg/kg (up to 1 mg/kg) have been recommended followed by an infusion of 0.25 - 1.5 mg/kg/hour [13].

Nifedipine

The use of nifedipine in the acute phase is controversial [14]. In adults its use has been associated with extreme hypotension and neurological effects. However, in children this has rarely been reported and a recent series endorses its safety in children when used appropriately [15]. Its action is by vasodilatation. Cardiac output is maintained but there is often a reflex tachycardia, which in adults certainly may lead to myocardial or cerebral ischaemia. Headaches and flushing are common. Many paediatric centres use this drug in an otherwise well child and in small, frequent doses (initially 100-250 mcg/kg). Subsequently larger doses may be appropriate, when the child has demonstrated that initial doses have not caused a precipitate drop in BP. Small doses can be extracted from the capsules and the effect can be quite rapid, starting within 10 minutes. Unfortunately with the difficulty in accurately measuring the small doses of liquid from the capsules, dosing errors may occur and care needs to be taken [16]. The duration of effect and potency when used in this way is unpredictable, particularly when absorption is variably sublingual and from the stomach and small bowel, but the longer-acting (sustained-release) preparations are not suitable in the immediate emergency situation. Unfortunately nifedipine in this context illustrates the difficulty in administering medicines in small doses to children when the formulations available are really suitable only for larger individuals.

Nicardipine

Recently an intravenously administered calcium-channel antagonist, nicardipine, has become available. Although there is limited experience with its use in paediatric practice, reports of its suitability have been favourable [17], and it is increasingly being used for children. It has a very rapid onset of action, within a few minutes, and a relatively short duration of action (half-life 40 minutes). Advantages over nitroprusside include the ability to use it for more than a few days, as it does not produce toxic metabolites. In a recent study [18] when used in doses gradually increasing to 3 mcg/kg/min it only occasionally caused unwarranted hypotension and this was readily improved by stopping the drug. Intravenous calcium can also be used in situations of hypotension and reduced cardiac output (0.2 ml/kg i.v. in the form of 10% calcium chloride over 5 minutes) [19].

Diazoxide

For many years, diazoxide was the most frequently used drug for hypertensive emergencies. Its use has decreased as other drugs have become available. It has unpredictable potency and is associated with the complications relating to over-rapid reduction of blood pressure. However, it still can be useful if used in small, frequent doses (1-3 mg/kg/dose) rather than the large boluses employed in the past. There is a risk of hyperglycaemia and a maximum total dose of no more than 10 mg/kg per 24 hours is generally recommended.

Other drugs

Intravenous hydralazine is no longer indicated in the emergency situation now that drugs with more predictable potency and fewer side effects are available. Mention should be made of the ACE-inhibitors so as to emphasise that they are not usually suitable for use in an emergency – the magnitude of their effect is dependent on renin-status and they are particularly potent in renovascular disease. Their effects are therefore unpredictable, particularly if the cause of the hypertension is unknown. In unilateral renal vascular disease there is the very real risk of infarction of the affected kidney.

In the second phase of management there is gradual reduction of short-acting drugs and gradual introduction of oral longer-acting drugs. At this stage it is still important not to reduce BP too quickly. Usually more than one drug is used, for example a β-blocker, diuretic and vasodilator. Success depends on a good knowledge of the treatment of hypertension generally and will usually require an appropriate specialist.

9.3.5 CONCLUSION

The emergency management of severe hypertension in a child includes careful clinical assessment before embarking on specific drug treatment. In particular, consideration of the degree of systemic illness, and the relative likelihood of specific causes of the hypertension, will usually warrant multi-disciplinary care, often within the intensive therapy unit. Intravenous labetalol and sodium nitroprusside are commonly used in this situation, but more recently intravenous nicardipine has also been found to be valuable. Oral short-acting nifedipine can be given, but its use is controversial as the effect is less predictable. Whichever drug is prescribed, the aim is to reduce blood pressure slowly with careful blood pressure monitoring, as the complications of a precipitate drop in blood pressure are as serious as those of severe hypertension.

REFERENCES

1. Jones, D. W., Frohlich, E. D., Grim, C. M., Grim, C. E., Taubert, K. A., *Hypertension, 37*, 185 (2001).
2. National High Blood Pressure Education Program Working Group, *Pediatrics, 88*, 649 (1996).
3. Mentser, M., Bunchman, T., *Semin. Nephrol., 18*, 330 (1998).
4. National Heart, Lung, and Blood Institute, *Pediatrics, 79*, 1 (1987).
5. Groshong, T., *Pediatr. Ann., 25*, 368, 375 (1996).
6. Adelman, R. D., Coppo, R., Dillon, M. J., *Pediatr. Nephrol., 14*, 422 (2000).
7. Fivush, B., Neu, A., Furth, S., *Curr. Opin. Pediatr., 9*, 233 (1997).
8. Feld, L. G., Waz, W. R., "Treatment of hypertension" in *Pediatric Nephrology* (4th ed.), Barratt, T. M., Avner, E. D., Harmon, W. E. (eds.), Baltimore, Lippincott, Williams and Wilkins, pp. 1031-1049 (1999).
9. Deal, J. E., Barratt, T. M., Dillon, M. J., *Arch. Dis. Child., 67*, 1089 (1992).
10. Houtman, P. N., Shah, V., Barratt, T. M., Dillon, M. J., *Lancet, 336*, 1454 (1990).
11. Proulx, F., Lacroix, J., Farrell, C. A., Gauthier, M., *Crit. Care Med., 21*, 1541 (1993).
12. Houtman, P. N., Dillon, M. J., *Child. Nephrol. Urol., 12*, 154 (1992).
13. Bunchman, T. E., Lynch, R. E., Wood, E. G., *J. Pediatr., 120*, 140 (1992).
14. Sinaiko, A. R., Daniels, S. R., *J. Pediatr., 139*, 7 (2001).
15. Blaszak, R. T., Savage, J. A., Ellis, E. N., *J. Pediatr., 139*, 34 (2001).
16. Flynn, J. T., *J. Pediatr., 140* 787 (2002).
17. Michael, J., Groshong, T., Tobias, J. D., *Pediatr. Nephrol., 12*, 40 (1998).
18. Flynn, J. T., Mottes, T. A., Brophy, P. D., Kershaw, D. B., Smoyer, W. E., Bunchman, T. E., *J. Pediatr., 139*, 38 (2001).
19. Kenny, J., *BMJ, 308*, 992 (1994).

9.4 Heart failure

Beat Friedli

Paediatric Cardiac Unit, Hôpital Cantonal Universitaire, 1211 Genève, Switzerland

9.4.1 INTRODUCTION

Heart failure is a state in which the heart, as a pump, is unable to sufficiently perfuse the organs at rest or during exercise, and to meet their metabolic demands. The approach to this condition has changed considerably over the last two decades, in both adult medicine and paediatrics. Traditionally, the approach was mainly "haemodynamic", stimulating the failing heart muscle with inotropic drugs such as digitalis glycosides and vasoactive amines, or bipyridine derivatives such as amrinone and milrinone.

It became apparent that peripheral vasodilatation was effective in relieving heart failure; as peripheral resistance is increased in heart failure, lowering resistance reduces afterload and thus relieves the failing heart of some of its burden ("unloading").

More recently, it has been recognised that heart failure is accompanied by important neurohormonal activation: norepinephrine, angiotensin, atrial natriuretic peptide (ANP), B-type natriuretic peptide (BNP) and endothelin are all increased in chronic heart failure. Although this may be a beneficial compensatory mechanism initially, the excess presence of these hormones becomes detrimental to the myocardium over time and will worsen heart failure. Neutralising these hormones with angiotensin-converting enzyme (ACE) inhibitors, beta-blockers, and spironolactone not only improves symptoms of chronic heart failure, but can prolong survival.

Heart failure can be the result of two different mechanisms: 1) the heart muscle is weakened due to myocyte loss (ischaemia, inflammation, degeneration); 2) the heart muscle is able to pump vigorously, but the circulatory demands exceed its possibilities. Both types are seen in infants and children. Myocarditis, cardiomyopathy, anthracycline toxicity are typical examples of myocyte degeneration and loss. Massive left to right shunts, as seen in infants with septal defects, are typical of volume loads exceeding the possibilities of a normally functioning heart muscle.

The two mechanisms being different, the treatment modalities will also differ. We shall discuss the use of diuretics and digoxin, which remain the basic conventional treatment in paediatric heart failure, and then describe the use of ACE inhibitors, vasodilators, beta-blockers and spironolactone.

9.4.2 DIURETICS

Heart failure usually results in salt and water retention, as evidenced by oedema and liver congestion. Diuretics are effective in eliminating excess extracellular fluid, thus improving symptoms; therefore, they remain the mainstay treatment of congestive heart failure. Their doses are described in Table 1.

The two main families are the thiazides, acting on the distal tubule of the nephron, and the so-called loop diuretics, acting on the loop of Henle (furosemide, torasemide).

Spironolactone is only a weak diuretic when used alone and is best given as an adjunct to thiazides or loop diuretics; it helps avoid excessive potassium loss. Spironolactone has other positive effects in heart failure (see below).

Thiazides. Thiazide diuretics have been used in children for many years. Their action is on the distal tubule, where they inhibit sodium and chloride reabsorption. Compared with loop diuretics, they are more long-acting but less potent.

Hydrochlorothiazide is the most frequently used thiazide in children. Its half-life is 9.5 to 13 hours. The usual dose is 1 to 3 mg/kg /day, given in 2 doses.

Furosemide. Furosemide is a powerful, shorter-acting diuretic with its effect on the loop of Henle [1]. The half-life of furosemide is age dependent: 1.5 hours in adults and older children, 2 hours in infants, and 8 hours in the neonate, with large inter-individual variation. The volume of distribution in infants and children is twice as large as in adults.

The rapid action of furosemide is in part due to an extrarenal effect, namely the reduction of interstitial oedema. This makes furosemide the ideal drug for emergency measures, such as treating pulmonary oedema.

The dose is 1-2 mg/kg intravenously. Orally, the dose is 1 to 2 mg per kg per day, given in 2 to 4 doses (1 to 2 daily doses in neonates). The dose may be increased, in resistant cases, to a maximum of 6 mg/kg/day.

Torasemide. This is a new loop diuretic with a longer elimination half-life and possibly fewer side effects. Although torasemide has been studied in the immature animal [2], its use in paediatrics has not been properly evaluated, and the manufacturer therefore does not recommend its use in patients below 12 years of age.

Other diuretics

Metolazone acts on the proximal tubule of the nephron and the ascending part of the loop of Henle. It has been used in combination with furosemide for furosemide-resistant oedema in children.

Acetazolamide is a carbonic anhydrase inhibitor. It promotes bicarbonate excretion and therefore produces acidosis, contrary to other diuretics. Acetazolamide, therefore, is a useful adjunct in the presence of diuretic induced alkalosis.

Adverse effects

High doses of diuretics may cause hypovolaemia and a fall in cardiac output, particularly in newborns and small infants. Furosemide, as well as thiazide diuretics, affect potassium homeostasis, with resultant hypokalaemia. Therefore, potassium supplementation is often necessary. As an alternative, spironolactone, an aldosterone antagonist, may be used for its potassium-sparing effect. Spironolactone also has a weak diuretic effect and other positive effects in heart failure. Alkalosis occurs with chronic use of furosemide and thiazides. In high doses, furosemide is ototoxic and may cause deafness, especially in premature infants. Long-term diuretic treatments also cause hyponatraemia, which requires fluid intake restriction. Finally, renal calcification has been described with furosemide.

Table 1

Doses of diuretics

Diuretic	Dose mg/kg/day	Number of doses per day
chlorothiazide	20	2
hydrochlorothiazide	1-3	2
furosemide	1-2 max	2-4
metolazone	0.2-0.4	1-2
acetazolamide	5	1
spironolactone	1-3	2-3

9.4.3 DIGOXIN

Extracts of the digitalis plants (foxglove) have been used since 1785 for the treatment of heart failure. Digoxin is nowadays the only preparation routinely used in children. Although its usefulness has been repeatedly questioned, it is still widely prescribed. Indeed, recent studies have shown that it improves exercise tolerance in patients with heart failure, although it does not prolong life in adult patients [3].

Digoxin is a positive inotropic drug, increasing contractility of the heart muscle. Its main action at the cellular level is on the sodium pump; by blocking the sodium-potassium ATP-ase, it increases intracellular sodium and calcium, thus increasing contractility. In addition, digoxin also has a neurally mediated effect, decreasing sympathetic nervous system activity and enhancing vagal tone, thus slowing heart rate.

There is a debate about the usefulness of digoxin in the infant with a large left to right shunt [4,5]. Indeed contractility is good in this situation and does not necessarily need boosting. Nonetheless, clinical improvement is often demonstrated with digoxin in this situation [6], which may be due to the neurally mediated effect.

Digoxin has a long elimination half-life of 26 hours (with normal renal function). Without a loading dose, steady state is achieved within 7 days. Since digoxin is eliminated by the kidney, doses need to be reduced drastically in case of renal failure. The required dose can be calculated according to the creatinine clearance [7]. In order to obtain a rapid response, a loading dose is often given, divided into 3 doses over 24 hours (see Table 2). In older children and adults, digoxin is given once daily. In infants and small children it is given every 12

Table 2

Digoxin dosage [8]

	Loading, $\mu g/kg/day$, in 3 doses	Maintenance, $\mu g/kg/day$, in 2 doses, q12h
premature	20	5
neonate-infants (to 2 months)	30	8-10
infants < 2 years	40	10-12
children >2 years	30-40	8 -10

For children over 25 kg, the dose of 250 mcg should be given. For intravenous administration, doses must be reduced by 25 to 33%.

hours, since clearance is greater [8]. Clearance in the neonate, however, is reduced [9]. Drug interactions with quinidine and amiodarone result in an increase in the serum concentration of digoxin.

Adverse effects

Digoxin has a narrow therapeutic "window", and toxic effects of an overdose may be severe. Gastro-intestinal symptoms, mainly nausea, vomiting, anorexia, are common in adults but less so in infants, and difficult to evaluate. In infants and children, conduction disorders and bradycardia are more common than increased extrasystoles and tachycardias. In general, infants appear to tolerate digoxin better than adults.

Therapeutic drug monitoring

Determination of serum concentration is helpful and should be obtained mainly in two situations: when toxicity is suspected on clinical grounds, or when the effect is judged to be insufficient, before increasing the dose. Serum digoxin concentrations should be between 1.1 and 1.7 ng/ml (1.4 to 2.2 nmol/L).

9.4.4 ANGIOTENSIN CONVERTING ENZYME (ACE) INHIBITORS

ACE inhibitors are a mainstay therapy for heart failure, particularly when it is due to systemic ventricular dysfunction. The main effect of ACE inhibitors is to decrease afterload for the systemic ventricle, by reducing peripheral vascular resistance. Reduction of afterload in turn improves systolic function of the ventricle.

The use of these substances for the treatment of heart failure in children is undisputed in case of systemic ventricular dysfunction; they also have a positive effect in case of mitral or aortic regurgitation. ACE inhibitors are essential in the treatment of cardiomyopathies, but also in the failing single ventricle or failing systemic right ventricle (after Mustard or Senning procedure). They have been tested as a preventive measure after intra-atrial baffle procedures in the absence of overt heart failure, but no increase in exercise tolerance was found [10].

The use of ACE inhibitors in infants with large left to right shunts is more controversial. Because their effect is mainly on the systemic, not on the pulmonary circulation, they are expected to decrease left to right shunt and Qp/Qs. This has been demonstrated in the acute test in the catheterisation laboratory [11]. However, because systemic blood flow is low in large left to right shunts, and global cardiac output is at the upper limit, there is a risk of hypotension and renal failure in these infants [12,13]. ACE inhibitors may have a place in infants with large left to right shunts if conventional treatment with digoxin and diuretics are insufficient, but they have to be used with great care starting with very low doses.

The two ACE inhibitors mainly used in children are captopril and enalapril.

Captopril has a short duration of action, with an elimination half-life of 2 hours, and therefore must be given in 3 doses per day. Initial dose is 500 mcg/kg/day, mean dose 2 mg/kg/day, and maximum 4 mg/kg/day. An initial test dose of one tenth the mean daily dose should be given, especially in infants, and blood pressure must be measured at regular intervals thereafter.

Enalapril has a longer half-life of 8 to 11 hours; therefore the daily dose can be given in one or two doses. Recommended doses are 100 to 500 mcg/kg/day (mean dose 250 mcg/kg/day). Smaller doses must be given to neonates. Again, a test dose of one tenth the mean daily dose should be given to start with and blood pressure must be carefully monitored.

A large number of other ACE inhibitors have been developed for adult patients, but few have been tested in children in the context of heart failure. Table 3 lists doses.

Adverse effects

The main adverse reaction is hypotension, which occurs particularly after the first dose. Hypotension is a particular risk if ACE inhibitors are taken together

Table 3

ACE inhibitor dosage

Drug [Ref.]	Starting dose/kg/day	Mean dose/kg/day	Maximum dose/kg/day	Number of doses per day
captopril	500 μg	2 mg	4 mg	3
newborns	100 μg	400 μg		
enalapril	100 μg	250 μg	500 μg	1 or 2
lisinopril [14]*	20 μg	70 μg	600 μg	1
quinapril [15]		200 μg	400 μg	1 or 2
ramipril [16]*		1.5 mg/m^2		1
cilazapril [17]	10 μg	40 μg		2

* Tested in children with hypertension.

with diuretics. This is why a very small test dose should be administered at the onset of treatment. Hypotension may lead to oliguria and renal failure, especially in premature and term newborns. Cough is a well-known side effect in adults and is observed in children as well.

Allergic reactions may sometimes occur, such as leukopenia and rash. Angioedema is a rare but severe, life-threatening adverse effect.

9.4.5   OTHER VASODILATORS

If ACE inhibitors are not well tolerated, other vasodilators can be prescribed to decrease afterload. Vasodilators can be classified according to their site of action. Some have vasodilator effect predominantly on the venous side, such as nitroglycerine; others have an effect mainly on the arteriolar side, like hydralazine. Mixed arterial and venous vasodilators are prazosin, nitroprusside and the previously described ACE inhibitors.

All vasodilators have been shown to decrease systemic resistance and increase cardiac output acutely, in children presenting with heart failure. Nitroglycerin and nitroprusside are used by intravenous infusion essentially and are prescribed in acute situations, in particular after open-heart surgery. Organic nitrates, such as isosorbide dinitrate, are used in adult medicine to relieve angina, but are of little interest in children. The other vasodilators can be given orally to treat chronic heart failure. Hydralazine and prazosin are effective in the short term to treat chronic heart failure in children. Their doses are described in Table 4. However their effect diminishes after about one month, due to tachyphylaxis [18]. This is one reason why ACE inhibitors are generally preferred.

Adverse effects

Vasodilators can produce hypotension (in particular orthostatic hypotension) and tachycardia. Rarely, thromobocytopenia or agranulocytosis can complicate

Table 4
Dosage of additional vasodilators

	Initial dose/kg/day	Mean dose/kg/day	Maximum dose/kg/day	Doses per day
hydralazine	0.75 mg	2 mg	4 mg	3 to 4
prazosine	30 μg	100 μg	300 μg	3 to 4

hydralazine administration; a syndrome resembling systemic lupus erythematosus has been described in adults, again with hydralazine.

9.4.6 BETA-BLOCKERS

Beta-blockers have previously been considered contraindicated in heart failure. It is now recognised that the neurohormonal activation that occurs in chronic heart failure is deleterious for the heart muscle. By progressively blocking the sympathetic overactivity and noradrenaline, cardiac function can be improved, symptoms can be relieved, and it has been shown that life can be prolonged [19].

The two most used beta-blockers are metoprolol and carvedilol. Both have been evaluated in children suffering from heart failure, especially in the context of cardiomyopathy of various origins [20-22].

It is of utmost importance to start the treatment with very low doses to avoid the negative inotropic effect, and then to increase the dose very progressively over weeks, giving small increments every week, to finally reach the target dose, or the maximum tolerated dose. ("Start low, go slow."). The beneficial effect will be noted only after several weeks of treatment.

Beta-blockers are essentially beneficial in children with left ventricular dysfunction and decreased ejection fraction, as occurs in cardiomyopathies. They have been tried in infants with heart failure due to large left to right shunts in an uncontrolled study with some success [23]. The usefulness of beta-blockers in the context of large left to right shunts remains doubtful.

Metoprolol is a beta-1 specific blocker, so called cardiospecific beta-blocker. It has minimal effect on the bronchi. The starting dose is 100 to 200 mcg/kg given twice daily; it is then increased slowly to a mean dose of 1 mg/kg/day, maximum dose 2 mg/kg/day [20]. A slow release preparation is available in small, "paediatric" doses, so that it can be taken as a once-daily dose in the morning.

Carvedilol is a newer substance of particular interest in heart failure because it has both alpha-adrenergic and beta-adrenergic action (the latter is non-specific, in contrast to metoprolol which is cardiospecific). Carvedilol differs from other beta-blockers by being a vasodilator. It has been used successfully in children with heart failure secondary to left ventricular dysfunction [21,22]. Initial doses as low as 10 mcg/kg/day have been recommended, increasing slowly, by weekly increments, to 200 mcg/kg/day; the maximum dose is 700 mcg/kg/day.

Beta-blocker therapy should always be combined with administration of a diuretic to avoid fluid retention. In general, beta-blockers are added to a combined treatment of diuretics, ACE inhibitors and digoxin.

Table 5

Dosage of beta-blockers

	Initial dose/kg/day	Mean dose/kg/day	Maximum dose/kg/day	Doses per day
Metoprolol	200 mcg	1 mg	2 mg	2 (slow release:one)
Carvedilol	10 mcg	200-400 mcg	700 mcg	2

Adverse effects

The main side effects of beta-blockers are hypotension, dizziness, and bradycardia. Fatigue and headaches are also reported. Special attention must be given to fluid retention when beta-blockers are introduced. Patients should be weighed daily.

9.4.7 SPIRONOLACTONE

Spironolactone is an aldosterone antagonist. It has weak diuretic action and increases potassium reabsorption. Because of this potassium-sparing effect, it is often associated with other diuretics, such as thiazides or furosemide (see above).

More recently, other beneficial effects of spironolactone in heart failure patients have been discovered. Aldosterone participates in the remodelling of the left (systemic) ventricle, especially by promoting collagen deposition, which in turn deteriorates left ventricular function. A multicentre study with spironolactone in adults suffering from heart failure showed a favourable effect on mortality and the need for hospitalisation [24]. It is not known whether children with cardiomyopathy benefit in the same way by seeing their life prolonged.

If spironolactone is given in association with an ACE inhibitor, it is essential to closely monitor serum potassium, as both will raise the potassium level, with hyperkalaemia as a result. The dose in children is 1 to 3 mg/kg/day in two or three doses. The lower dosage is recommended if ACE inhibitors are prescribed at the same time.

REFERENCES

1. Shannan, K. E., Christensen M. L., *Pediatr. Nephrol., 12*, 603 (1998).
2. Dubourg, L., Mosig, D., Drukker, A., Guignard, J-P., *Pediatr. Nephrol., 14*, 476 (2000).
3. Hauptmann, P. J., Kelly, R. A. Digitalis, *Circulation, 99*, 1265 (1999).
4. Kimball, T. R., Daniels, S. R., Meyer, R. A., et al., *Am. J. Cardiol., 68*, 1377 (1991).
5. Seguchi, M., Nakazawa, M., Momma, K., *Am. J. Cardiol., 83*, 1408 (1999).
6. Berman, W., Yabek, S. M., Dillon, T., et al., *N. Engl. J. Med., 308*, 363 (1983).
7. Jelliffe, R. W., *Annals Intern. Med., 69*, 703 (1968).
8. Park, M. K., *J. Pediatr., 108*, 871 (1986).
9. Suematsu, E., Minemoto, M., Yukawa, E., et al., *J. Clin. Pharmac. Ther., 24*, 203 (1999).
10. Robinson, B., Heise, C. T., Moore, J. W., et al., *Pediatr. Cardiol., 23*, 618 (2002).
11. Sluysmans, T., Styns-Cailteux, M., Tremeroux-Wattiez, M., et al., *Am. J. Cardiol., 70*, 959 (1992).
12. Wessel, A., Buchhorn, R., Bursch, J., *Klin. Paediatr., 212*, 53 (2000).
13. Shaw, N. J., Wilson, N., Dickinson, D. F., *Arch. Dis. Child., 63*, 360 (1988).
14. Soffer, B., Zhang, Z., Miller, K. et al., *Am. J. Hypertens., 16*, 795 (2003).
15. Blumer, J. L., Daniels, S. R., Dreyer, W. J., et al., *J. Clin. Pharmacol., 43*, 128 (2003).
16. Soergel, M., Verho, M., Wuhl, E., et al., *Pediatr. Nephrol., 15*, 113 (2000).
17. Hazama, K., Nakazawa, M., Momma, K., *Am. J. Cardiol., 88*, 801 (2001).
18. Beekman, R. H., Rocchini, A. P., Dick, M., et al., *Pediatrics, 73*, 43 (1984).
19. Packer, M., Coats, A. J. S., Fowler, M. B., et al., *N. Engl. J. Med., 344*, 1651 (2001).
20. Shaddy, R. E., Tani, L. Y., Gidding, S. S., et al., *J. Heart Lung Transplant, 18*, 269 (1999).
21. Azeka, E., Franchini Ramirez, J. A., Valler, C., Alcides Bocchi, E., *J. Am. Coll. Cardiol., 40*. 2034 (2002).
22. Bruns, L. A., Kichuk Chrisant, M., Lamour, J. M., et al., *J. Pediatr., 138*, 505 (2001).
23. Buchhorn, R., Bartmus, D., Siekmeyer, W., et al., *Am. J. Cardiol., 81*. 1366 (1998).
24. Pitt, B., Zannad, F., Remme, W. J., et al., *N. Engl. J. Med., 341*, 709 (1999).

9.5 Diuretics

Jean-Pierre Guignard

Division of Nephrology, Department of Paediatrics, Lausanne University Medical Center, CH-1011 Lausanne, Switzerland

9.5.1 INTRODUCTION

Diuretics promote the excretion of water and electrolytes. They are primarily used in states of inappropriate salt and water retention. Diuretics are also used in a variety of clinical situations where an increase in sodium excretion is not the primary goal of treatment. Such conditions include acute renal failure, electrolyte disturbances (hypo- or hyperkalaemia, hypercalcaemia, hypercalciuria) and nephrogenic diabetes insipidus.

9.5.2 RENAL FUNCTION

Urine formation starts by the ultrafiltration of plasma through the glomerular capillary wall [1]. Reabsorption of filtered solutes is achieved by active or passive transport across the tubular cell membranes, using a transcellular or a paracellular route. Primary active transport requires a source of metabolic energy, provided by ATP hydrolysis. Secondary active transport of solutes along (*symport*) or against (*antiport*) the Na^+ gradient, created by its primary active transport, occurs via specific protein carriers molecules (*transporters*). Cell membranes contain channels allowing the rapid passage of specific ions (Na^+, K^+, Cl^-) across cellular membranes [1].

Reabsorption of sodium

The renal tubule can reabsorb up to 99% of the filtered load of sodium. Two thirds are reabsorbed in the proximal tubule, 25% in the ascending limb of Henle's loop, and 10% in the distal tubule and collecting duct (Fig. 1). The driving force for Na^+ reabsorption is the Na^+, K^+-ATPase in the basolateral membrane of the tubular cell [1].

In the early proximal tubule, the reabsorption of Na^+ is primarily coupled to that of HCO_3^- and a number of organic molecules. Sodium is reabsorbed mainly with Cl^- in the late proximal tubule. The thick ascending limb of Henle's loop reabsorbs approximately 25% of the filtered load of Na^+. It is impermeable to water. The movement of Na^+ across the luminal membrane is mediated by the Na^+, 2 Cl^-, K^+ symporter. The integrity of the Na^+, K^+-ATPase and of the Na^+, 2 Cl^-, K^+ cotransporter is necessary for the generation of a hypertonic medulla required to concentrate the urine. The late distal tubule and collecting duct are composed of two distinct cell types, the principal cells and the intercalated cells. The principal cells reabsorb Na^+ and water and secrete K^+, a process controlled by aldosterone. The intercalated cells reabsorb K^+ and secrete H^+ in the tubular lumen. These cells play an important role in regulating acid-base balance. In the distal tubule and collecting duct, part of sodium enters the luminal membrane via an epithelial Na^+ channel ($E_{Na}C$) [1,2].

Na^+ reabsorption along the nephron

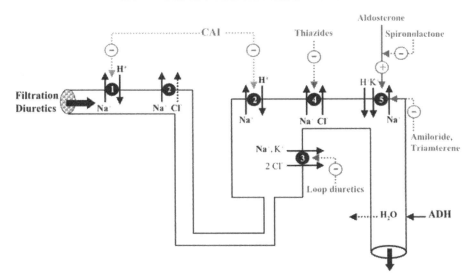

Figure 1. Sodium reabsorption along the nephron. The main transport systems involved in Na^+ transport along the nephron are as follows: ①Na^+-H^+ antiporter; ②Na^+-Cl^- symporter; ③Na^+-K^+-$2Cl^-$ symporter; ④Na^+-Cl^- symporter; ⑤E_{Na} channel or aldosterone receptor antgonists; ADH: Antidiuretic hormone; CAI Carbonic anhydrase inhibitors; (+): activation; (-): inhibition

The permeability of the collecting duct cells to water is under the influence of arginine vasopressin (AVP), also known as the antidiuretic hormone (ADH). This hormone increases the permeability of the cortical tubular cells by incorporating aquaporin-2 water channels in the apical membrane [3].

In addition to the glomerular tubular balance that, via Starling forces, adapts the reabsorption of Na^+ to its filtered load, Na^+ transport is regulated by various hormones and paracrine factors. Angiotensin II, aldosterone, the atrial natriuretic peptide, the prostaglandins, the catecholamines and dopamine play major roles.

9.5.3   BODY FLUID HOMEOSTASIS

The kidney is responsible for maintaining the ECF fluid volume and osmolality constant, in spite of large changes in salt and water intake.

Body fluid volume

NaCl, the major osmotically active solute in ECF, determines its volume. The balance between the intake and the renal excretion of Na^+ thus regulates ECF volume, and as a consequence cardiac output and blood pressure. Long-term changes in Na^+ excretion are regulated by aldosterone. More rapid changes in Na^+ excretion are achieved by changes in GFR, and/or by various intrarenal hormones that regulate Na^+ reabsorption [1].

Body fluid osmolality

The plasma osmolality is maintained within narrow limits. A 2-3% increase in P_{osm} stimulates the osmoreceptors with the consequent release of AVP, leading to the reabsorption of free-water in the collecting duct. Urine flow rate decreases and the osmolality of the urine increases. The opposite happens when P_{osm} decreases by 2-3% [1].

9.5.4   CLINICAL USE OF DIURETICS

Diuretics are used in various conditions associated with salt and water retention, with water and electrolytes imbalance, and with renal failure (Table 1) [4].

Oedematous states

Salt and water retention with oedema formation can occur as a primary event, or as a consequence of reduced effective circulating volume with secondary hyperaldosteronism [5]. The use of diuretics can be life-saving when Na retention is associated with an expansion of the ECF volume.

Table 1

Clinical indications for diuretics

filtration diuretics	oliguric renal failure
osmotic diuretics	oliguric acute renal failure cerebral oedema elevated intracranial pressure
carbonic anhydrase inhibitors	acute mountain sickness glaucoma refractory epilepsy production of alkaline diuresis
loop diuretics	oedematous states nephrotic syndrome oliguric states and prerenal failure severe hyponatraemia hypercalcaemia hyperkalaemia respiratory disorders in neonates
thiazides	oedematous states arterial hypertension hypercalciuria nephrogenic diabetes insipidus bronchopulmonary dysplasia
potassium sparing diuretics	adjonctive therapy with loop or thiazide diuretics prevention of hypokalaemia

Congestive heart failure

Congestive heart failure is associated with an increase in pressure in the venous circulation. The increased capillary pressure favors the movement of fluid into the interstitium, resulting in the formation of oedema. The treatment of this condition consists in restoring normal cardiac output. By mobilising the oedematous fluid, diuretics improve the symptoms of congestive heart failure. The pulmonary oedema, secondary to left heart failure, requires the urgent use of diuretics to reduce the life-threatening pulmonary congestion [6].

Nephrotic syndrome

The nephrotic syndrome is characterised by an increase in the permeability of the glomerular barrier to proteins, heavy proteinuria, hypoproteinaemia and the formation of oedema. Two pathogenic mechanisms have been proposed to explain the formation of oedema, the underfill and the overfill hypothesis [7]. Both hypovolaemia and hypervolaemia can be observed at some stage of the nephrotic syndrome. When present, the hypovolaemia is manifested by clinical signs such as tachycardia, abdominal pain and poorly perfused limbs. Elevated hematocrit and very low urinary sodium (<5 mmol/l) confirm the state of hypovolaemia. Diuretics should only be used with great caution in this condition, and only after correction of the hypovolaemia by albumin infusion.

In some patients, the oedema may also be caused by the primary retention of Na^+ and water, with increased plasma volume, increased hydrostatic capillary pressure, and leak of fluid into the interstitium. These patients benefit from diuretic administration [8].

Arterial hypertension

Arterial hypertension may be secondary to salt retention and expansion of the extracellular fluid volume. By leading to ECF volume contraction, diuretics decrease the blood pressure. They are often used as first-line drugs in the treatment of arterial hypertension [9].

Electrolyte imbalance

Diuretics can be used in various situations associated with electrolyte imbalance. They can increase K^+ excretion in hyperkalaemic states (loop diuretics, thiazides), increase calcium excretion in hypercalcaemia (loop diuretics) or decrease the rate of calcium excretion in hypercalciuric states (thiazides). Increased HCO_3^- excretion can be achieved by acetazolamide, while H^+ excretion can be stimulated by loop diuretics.

Nephrogenic diabetes insipidus

The patient with diabetes insipidus excretes large amounts of dilute urine. The thiazides can reduce the extracellular space volume by impairing sodium distal reabsorption, and increasing NaCl excretion. This decrease in ECF volume stimulates proximal salt and water reabsorption, thus eventually leading to reduced urine flow rate [10].

Acute renal failure

Because they have been shown to variably increase total renal blood flow, loop diuretics are often administered to patients with oliguric renal insufficiency, in the hope of promoting diuresis and improving GFR and renal perfusion. While diuretic administration may convert oliguric ARF to non-oliguric ARF, there is no evidence that this treatment can ameliorate renal function or improve the outcome of patients with ARF [11].

9.5.5   CLASSIFICATION OF DIURETICS

Diuretics can be classified according to their site and mode of action (Fig.1). They all increase sodium and water excretion and variably modify electrolyte excretion (Table 2). All diuretics have side effects which are actually an extension of their primary effects (Table 3). They also have non-electrolytic adverse effects (Table 4). The route and intervals of administration, as well as dosages of the main diuretic agents, are given in Table 5.

9.5.6   FILTRATION DIURETICS

Agents that increase diuresis by increasing GFR are sometimes called filtration diuretics. Such agents include the glucocorticoids and theophylline, as well as inotropic agents such as dopamine and dobutamine. The moderate natriuresis induced by these agents partly results from the increase in Na^+ filtered load, and partly from an inhibition of Na^+ tubular reabsorption. Dopamine has been claimed to improve diuresis and renal function in oliguric states. This claim remains, however, questionable [14].

Table 2

Acute effects of diuretics on electrolyte excretion *

	Na^+	K^+	Ca^{++}	H^+	HCO_3^-
osmotic diuretics	↑↑	↑	↑	?	↑
carbonic anhydrase inhibitors	↑	↑↑	=	↓	↑↑
loop diuretics	↑↑	↑↑	↑↑	↑	↑
thiazide diuretics	↑	↑↑	~↓	↑	↑
K^+-sparing diuretics	↑	↓	↓	↓	(↑)

Legend: ↑↑: marked increase; ↑: moderate increase; (↑): slight increase; ↓: decrease
=: no change; ~: variable effects; ?:insufficient data
* In the absence of significant volume depletion which would trigger complex adjustments
Adapted from Jackson, E.K. [12]

Table 3

Electrolyte disturbances induced by diuretics*

	Osmotic	CAI	Loop	Thiazides	K+-sparing
hypovolaemia	+++	-	+++	+	+
hyponatraemia	++	-	++	+++	-
hypokalaemia	++	+	+++	++	-
hyperkalaemia	-	-	-	-	++
hypercalciuria	+	-	++	-	-
hypercalcaemia	-	-	-	+	-
hypomagnesaemia	+	-	+	+	-
hypophosphataemia	+	-	+	+	-
hyperuricaemia	-	-	++	++	-
metabolic acidosis	+	++	-	-	+
metabolic alkalosis	+	-	++	++	-

CAI: carbonic anhydrase inhibitors
*Adapted from Sherbotie, J. R.; Kaplan, B. [13].

Table 4

Common adverse effects of diuretics

osmotic diuretics	congestive heart failure and pulmonary oedema nausea, vomiting
carbonic anhydrase inhibitors	drowsiness, fatigue, CNS depression, paresthesia, calculus formation
loop diuretics	ototoxicity (usually reversible), nephrocalcinosis in neonates, PDA in neonates, hyperuricaemia, hyperglycaemia, hyperlipidaemia, hypersensitivity
thiazides	hyperglycaemia, insulin resistance hyperlipidaemia, hypersensitivity (fever, rash, purpura, anaphylaxis, interstitial nephritis), hyperuricaemia
K+-sparing: miloride	diarrhoea, headache
triamterene	glucose intolerance, interstitial nephritis, blood dyscrasias
spironolactone	gynaecomastia, hirsutism, peptic ulcers, ataxia, headache

PDA: patent ductus arteriosus

Table 5

Dosages of common diuretics

Drug	Route / Interval (q/h)	Dosage (mg/kg/day)	Site of action	Comments
mannitol	iv/2-4	200-500	all nephrons	
acetazolamide	po/6-8	5	proximal + early dstal tubule	not effective at GFR < 20
furosemide	po/12-24 iv/12-24 civi*	1-2 0.5-1.5 100-200 microg/kg/h	Henle's Loop	effective at GFR < 10
torasemide	po	0.5-1	as above	effective at GFR < 10
ethacrynic acid	po/12-24	1-2	as above	effective at GFR < 10
bumetanide	po/12-24 iv/12-24 civi*	0.01-0.10 0.01-0.05 5-10 microg/kg/h	as above	effective at GFR < 10
hydrochlorothiazide	po/12-24	1-3	early distal tubule	not effective at GFR < 20
chlorthalidone	po/24-48	0.5-2.0	as above	not effective at GFR < 20

GFR: glomerular filtration rate (ml/min per 1.73 m^2)

*civi: constant iv infusion

Table 5 *(continued)*

Drug	Route / Interval (q/h)	Dosage (mg/kg/day)	Site of action	Comments
metolazone	po/12-24	0.2-0.4	as above	effective at GFR < 20
spironolactone	po/6-12	1-3	late distal tubule and collecting duct	delayed effect; do not use at GFR < 30
canreonate-K	iv/24	4-10	as above	do not use at GFR < 30
triamterene	po/12-24	2-4	as above	do not use at GFR < 30
amiloride	po/24	0.5	as above	do not use at GFR < 30

GFR: glomerular filtration rate (ml/min per 1.73 m^2)

9.5.7 OSMOTIC DIURETICS

Osmotic diuretics are agents that inhibit the reabsorption of solute and water by altering osmotic driving forces along the nephron. Osmotic diuretics include mannitol, glycerin, isosorbide and urea. Mannitol, a hexahydric alcohol related to mannose, with a molecular weight of 182 daltons, is most commonly used [15].

Mechanisms and site of action

Freely filtered and (mostly) not reabsorbed, osmotic diuretics increase the tubular fluid osmolality, thus impairing the diffusion of water out of the tubular lumen, as well as that of NaCl by a solvent drag effect. Osmotic diuretics increase, non-specifically, the excretion of all electrolytes.

Pharmacokinetics

Mannitol is excreted in the urine. The half-life of mannitol is 0.25-1.5 h; it is prolonged in patients with renal failure. Osmotic diuretics improve renal perfusion without significantly affecting GFR [11]. Mannitol is used to increase urine flow rate in patients with prerenal failure, to promote the excretion of toxic substances by forced diuresis, and to reduce elevated intracranial and intraocular pressures.

Adverse effects

Circulatory overload and congestive heart failure can occur as a consequence of inadequate expansion of ECFV.

Specific indications

Mannitol is infused in oliguric euvolaemic patients, at a dose of 2-5 ml/kg of 20% mannitol, iv over 5 min. The diuretic response appears within 1-3 h.

9.5.8 INHIBITORS OF CARBONIC ANHYDRASE

Acetazolamide, a sulfonamide derivative, is the main inhibitor of carbonic anhydrase used in humans.

Mechanisms and site of action

Carbonic anhydrase inhibitors are weak diuretics, at best producing the excretion of 5% of the Na^+ and water filtered load. The action of carbonic anhydrase inhibitors is self-limiting. The excretion of HCO_3^- decreases in parallel, with the development of metabolic acidosis in response to carbonic anhydrase inhibition.

Pharmacokinetics

Acetazolamide is readily absorbed and has a 100% oral bio-availability. Its half-life is 6-9 h. It is eliminated in the urine. Acetazolamide increases the urinary excretion of bicarbonate, sodium and potassium, promoting alkaline diuresis.

Adverse effects

The occurrence of metabolic acidosis is common if the urinary losses of HCO_3^- are not substituted. Paresthesias, drowsiness, rash and fever are not uncommon.

Specific indications

Acetazolamide can be used to alkalinise the urine, when necessary, as for instance when chemotherapy is given, or to treat glaucoma, to prevent mountain sickness, or as an adjunct in refractory epilepsy [16].

9.5.9 LOOP DIURETICS

Loop diuretics form a group of diuretics with diverse chemical structures. Furosemide and bumetanide are sulfonamide derivatives, torasemide is a sulfonylurea and ethacrynic acid a phenoxyacetic acid derivative.

Mechanisms and site of action

Loop diuretics block the Na^+, K^+, $2Cl^-$ symporter in the thick ascending limb of Henle's loop, where 25% of NaCl filtered load are reabsorbed [1]. They are consequently highly effective. The effect of loop diuretics is more closely related to their urinary excretion rate than to their plasma concentration. Loop diuretics interfere both with the diluting and the concentrating mechanism.

Pharmacokinetics

The diuretic response to loop diuretics appears within a few minutes after i.v. administration, and within 30-60 min after oral administration. The effect does not last over 2 h after IV injection, and 6 h after oral administration. The half-life is prolonged in patients with renal and liver insufficiency, and in premature and term neonates. Furosemide is the most commonly used diuretic in children. Torasemide has the same effect as furosemide, but its half-life is longer [19]. Loop diuretics cross the placental barrier and are secreted in breast milk. Loop diuretics are the most potent natriuretic agents, also markedly increasing Cl^-, K^+, Ca^{++} and Mg^{++} excretion [17]. They remain active in patients with advanced renal failure.

Continuous i.v. infusion of loop diuretics

Clinical trials in adults and children indicate that continuous infusion therapy can produce more efficient and better-controlled diuresis with less fluid shifts and greater hemodynamic stability [20].

Adverse effects

Because of their efficacy, common adverse effects including volume depletion, postural hypotension, dizziness and syncope, hyponatraemia and hypokalaemia are commonly observed. Hypochloraemic metabolic alkalosis occurs frequently as a consequence of direct stimulation by loop diuretics of H^+ secretion in the collecting tubule. Elevated Ca^{++} urinary losses may lead to nephrocalcinosis in premature infants, secondary hyperparathyroidism, bone resorption and rickets. The suggestion that, by stimulating prostaglandin synthesis, furosemide could promote PDA has not been confirmed. The beneficial renal effect of combining furosemide and indomethacin is still controversial [21]. The use of furosemide has been identified as an independent risk factor for sensorineural hearing loss in preterm infants. Hearing loss may be transient or permanent. By avoiding elevated peak concentrations of furosemide, continuous infusion decreases the risk of ototoxicity [21]. Pancreatitis, jaundice, deterioration of glucose intolerance, thrombocytopenia and serious skin disorders occur occasionally. The majority of adverse effects occur with the use of high doses. Drug interactions may occur with the co-administration of nephrotoxic antibiotics, NSAIDs, anticoagulants and cisplatin.

Specific Indications

Congestive heart failure (CHF) is the most common indication for the use of loop diuretics in neonates and infants.

In hypovolaemic patients with *nephrotic syndrome* and significant oedema, IV furosemide can be used to promote sodium and water excretion. Furosemide (1-2 mg/kg) should only be given after the expansion of the extracellular space with i.v. albumin (5 ml/kg of 20% albumin in 60 min). The effect is transient, but may be useful in patients with severe ascites and/or pulmonary oedema [22].

Furosemide is frequently used in *oliguric states* secondary to prerenal or renal failure, in the hope of promoting diuresis and improving renal function. There is as yet no clinical or experimental evidence that loop diuretics can prevent acute renal failure or improve the outcome of patients with acute renal failure [11].

A recent critical review of the literature [23] failed to support the routine administration of furosemide (or any diuretic) in *preterm infants with RDS*, and concluded that elective administration of furosemide should be weighed against the risk of precipitating hypovolaemia or of developing a symptomatic PDA by stimulating prostaglandin synthesis.

Loop diuretics have been given to *preterm infants with CLD,* in the hope of decreasing the need for oxygen or ventilatory support. Use cannot be recommended until randomised trials assessing effects on survival, duration of ventilatory support and oxygen administration, potential complications and long term outcome are available [24].

A critical review of three randomised trials in newborn infants with *posthaemorrhagic ventricular dilatation (PHVD)* concluded that acetazolamide and furosemide therapy is neither effective nor safe in treating preterm infants with PHVD [25]. A critical review concluded that there was, as yet, not enough evidence to support the administration of furosemide to preterm infants treated with indomethacin for PDA [26].

Loop diuretics can promote calcium excretion and decrease hypercalcaemia. Isotonic saline must be infused concomitantly to prevent volume depletion. Severe hyponatremia can be treated by loop diuretics and the concomitant isovolumetric infusion of hypertonic saline.

9.5.10 DISTAL DIURETICS

The benzothiadiazide derivatives are sulfonamides. They are weak diuretics that inhibit the reabsorption of NaCl at the diluting site in the early distal tubule. The main thiazides include chlorothiazide and hydrochlorothiazide. Thiazide-like agents, such as chlorthalidone and metolazone, belong to this group.

Mechanisms and sites of action

Thiazides decrease NaCl reabsorption in the distal convoluted tubule by inhibiting the Na^+-Cl^- apical symporter. The expression of this symporter, sometimes called ENCC1 or TSC, is upregulated by aldosterone [1]. Approximately 4-5% of the Na^+ filtered load is reabsorbed in the distal collecting duct and thus inhibition of Na^+ reabsorption at this site can only modestly increase NaCl excretion. The thiazides stimulate Ca^{++} reabsorption in the distal tubule, probably by opening the apical membrane Ca^{++} channels. The thiazides (but not metolazone) are ineffective at GFRs below 30 ml/min per 1.73 m^2.

Pharmacokinetics

Administration of thiazides initiates diuresis in 2 h, an effect that lasts for 12 h. The response to metolazone is somewhat more rapid (1 h) and lasts longer (12-24 h). The thiazides cross the placental barrier and are secreted in breast milk.

Because of the risk of inducing hypokalaemia, prophylactic co-administration of K^+-sparing diuretics may be indicated to prevent the occurrence of severe hypokalaemia.

Adverse effects

Thiazides may adversely affect water balance and induce electrolyte imbalances (Table 2). Other side effects include gastrointestinal disturbances, hypersensitivity reactions, cholestatic jaundice, pancreatitis, thrombocytopenia, and hyperglycemia in diabetic and susceptible patients.

Specific indications

The thiazides decrease calcium excretion and this effect may be useful in states of idiopathic *hypercalciuria*, as well as to prevent calcium losses in patients receiving glucocorticoids [27].

The thiazides have been successfully used in children with *nephrogenic diabetes insipidus*. By inducing volume contraction, they enhance the proximal tubular reabsorption of water and electrolytes, thus significantly decreasing urine output. While usefully decreasing urine output, volume contraction may inhibit growth in young children with nephrogenic diabetes insipidus.

While thiazides and thiazide-like diuretics are often used in the hope of improving the clinical outcome in *preterm infants with CLD*, there is little evidence to support any benefit on the need for ventilatory support, length of hospital stay or long-term outcome in infants receiving current therapy [28].

9.5.11 K+-SPARING DRUGS

Two types of diuretics form the group of K^+-sparing diuretics: a) the inhibitors of a renal epithelial Na^+ channels, and b) the antagonists of mineralocorticoid receptors. The overall effects of these two groups of diuretics differ only in their mode of action.

Mechanisms and site of action

a) The epithelial Na^+ channels blockers amiloride and triamterene block the entry of Na^+ into the cell through the Na^+-selective channels (ENaC) in the apical membrane.

b) The antagonists of the action of aldosterone on the principal cells of the collecting duct increase Na^+ excretion and decrease K^+ and H^+ secretion. Spironolactone, the main agent in this group, competitively inhibits the binding of aldosterone to the mineralocorticoid receptor, thus decreasing the synthesis of aldosterone-induced proteins (AIPs).

Pharmacokinetics

Spironolactone has a slow onset of action, requiring 2-3 days for maximum effect. Spironolactone crosses the placenta and is secreted into breast milk. Canreonate-potassium has similar actions to those of spironolactone. It is available for IV administration. Amiloride is incompletely absorbed from the gastrointestinal tract. It is excreted unchanged in the urine. Its half-life is prolonged in patients with hepatic or renal failure. Triamterene is unreliably absorbed. It is metabolised by hepatic conjugation.

The overall effects on electrolyte excretion are similar for spironolactone, amiloride and triamterene. They are weak natriuretic agents that reduce the excretion of potassium and hydrogen ions.

Adverse effects

The main adverse effects of K^+-sparing diuretics is to increase the K^+ plasma concentration to harmful levels. Significant adverse effects have been observed with spironolactone: gynaecomastia, hirsutism, impotence and menstrual irregularities can occur. Gynaecomastia in males is related to both the dose and duration of treatment.

Interactions

K^+-sparing diuretics should not be used in patients receiving angiotensin converting enzyme inhibitors, as the combination can worsen the risk of hyperkalaemia.

Specific indications

Refractory oedema, secondary to congestive heart failure, cirrhosis of the liver and the nephrotic syndrome represent the most common indications for the use of K^+-sparing diuretics. Because they induce K^+ retention, K^+-sparing diuretics should not be used in patients with impaired renal function, or in those receiving K^+ supplementation.

K^+-sparing diuretics are often used in association with thiazide diuretics in the management of preterm infants with CLD. While they certainly decrease the risk of hypokalaemia and facilitate the clinical management of the infants, there is, as yet, no definite proof that their association to thiazides improve the long-term outcome of preterm infants with CLD.

Amiloride has been successfully used in association with hydrochlorothiazide in patients with nephrogenic diabetes insipidus, obviating the need for using indomethacin [10].

9.5.12 NEW DEVELOPMENTS

Three categories of diuretics are under development and are discussed below.

Natriuretic peptides

The atrial natriuretic peptide (ANP) and the B-type natriuretic peptide (BNP) are two peptides with natriuretic and diuretic properties [29]. Enhancing the activity of the NPs could help patients presenting with inappropriate salt and water retention. Recent studies with BNP have produced promising results in situations of heart failure [30].

Adenosine A_1 receptor antagonists

In physiological conditions, adenosine participates in the regulation of GFR, renin secretion and sodium reabsorption. Renoprotection by theophylline, a non-specific adenosine receptor antagonist with natriuretic properties, has been demonstrated in both newborn animals and human neonates [31,32]. The use of A_1 adenosine receptor antagonists results in reduced proximal Na^+ reabsorption and blunting of the tubulo-glomerular feedback mechanism [33]. Experimental evidence indicates that A_1 receptors antagonists could also be cardioprotective by attenuating the A_1 mediated release of catecholamine, β-adrenergic mediated myocardial contraction and calcium overload [34].

Arginine-vasopressin antagonists

AVP acts on three type of receptors: 1) the V_{1A} receptors mediating vaso-constriction; 2) the V_{1B} receptors mediating the release of ACTH and 3) the V_2 receptors mediating free water reabsorption in the collecting duct. Orally active nonpeptide selective V_2-antagonists are being developed [35]. Experimental and adult human studies demonstrate that these agents increase the excretion of free-water and decrease the urine osmolality in a dose-dependent manner. They may thus prove useful in treating hyponatraemic states [36].

REFERENCES

1. Giebisch, G., Windhager, E., "The Urinary System, Part VI." in *Medical Physiology*, Boron, W. F., Boulpaep, E. L. (eds.), Saunders, p. 735 (2003).
2. Garty, H., Palmer, L. G., *Physiol. Rev.*, *77*, 359 (1997).
3. Yamamoto, T., Sasaki, S., *Kidney Int.*, *54*, 1041 (1998).
4. Wells, T. G., *Pediatr. Clin. North Am.*, *37*, 463 (1990).
5. Guignard, J. P., Gouyon, J. B., *Clin. Perinatol.*, *15*, 447 (1988).
6. Lowrie, L., *Prog. Pediatr. Cardiol.*, *12*, 45 (2000).
7. Schrier, R. W., Fassett, R. G., *Kidney Int.*, *53*, 1111 (1998).
8. Vande Walle, J. G., Donckerwolcke, R. A., *Pediatr. Nephrol.*, *16*, 283 (2001).
9. Reyes, A. J., *J. Hum. Hypertens.*, *16(suppl. 1)*, S78 (2002).
10. Kirchlechner, V., Koller, D. Y., Seidl, R., Waldhauser, F., *Arch. Dis. Child.*, *80*, 548 (1999).
11. Kellum, J. A., *Kidney Int. Suppl.*, *66*, S67 (1998).
12. Jackson, E. K., "Diuretics," in *Goodman and Gilman's, The Pharmacological basis of Therapeutics (*10th ed.), Hardman, J. G., Limbird, L. E. (eds.), McGraw-Hill Med. Publ. Div., p. 757 (2001).
13. Sherbotie, J. R., Kaplan, B., "Diuretics," in *Pediatric Pharmacology*, Aranda, J., Yaffé, S. (eds.), W.B. Saunders & Co., p. 524 (1992).
14. Gambaro, G., Bertaglia, G., *J. Nephrol.*, *15*, 213 (2002).
15. Better, O. S., Rubinstein, I., Winaver, J. M., Knochel, J. P., *Kidney Int.*, *52*, 886 (1997).
16. Reiss, W. G., Oles, K. S., *Ann. Pharmacother.*, *30*, 514 (1996).
17. Brater, D. C., *N. Engl. J. Med.*, *339*, 387 (1998).
18. Prandota, J., *Am. J. Ther.*, *8*, 275 (2001).
19. Knauf, H., Mutschler, E., *Clin. Pharmacokinet.*, *34*, 1 (1998).
20. Luciani, G. B., Nichani, S., Chang, A. C., et al., *Ann. Thorac. Surg.*, *64*, 1133 (1997).
21. Eades, S. K., Christensen, M. L., *Pediatr. Nephrol.*, *12*, 603 (1998).
22. Haws, R. M., Baum, M., *Pediatrics, 91*, 1142 (1993).
23. Brion, L. P., Soll, R. F., "Diuretics for respiratory distress syndrome in preterm infants," *Cochrane Database Syst. Rev.*, *(2)*, CD001454 (2001).
24. Brion, L. P., Primhak, R. A., *Cochrane Database Syst. Rev.(1)*, CD001453 (2002).
25. Whitelaw, A., Kennedy, C. R., Brion, L. P., *Cochrane Database Syst. Rev. (2)*, CD002270 (2001).
26. Brion, L. P., Campbell, D. E., *Pediatr. Nephrol.*, *13*, 212 (1999).
27. Lukert, B. P., Raisz, L. G., *Ann. Intern. Med.*, *112*, 352 (1990).
28. Brion, L. P., Primhak, R. A., Ambrosio-Perez, I., *Cochrane Database Syst. Rev. (1)*, CD001817 (2002).
29. Costello-Boerrigter, L. C., Boerrigter, G., Burnett, J. C., Jr., *Med. Clin. North Am.*, *87*, 475 (2003).
30. Colucci, W. S., Elkayam, U., Horton, D. P., et al., *N. Engl. J. Med.*, *343*, 246 (2000).
31. Gouyon, J. B., Guignard, J. P., *Kidney Int.*, *33*, 1078 (1988).
32. Huet, F., Semama, D., Grimaldi, M., et al., *Intensive Care Med.*, *21*, 511 (1995).
33. Pelleg, A., Porter, R. S., *Pharmacotherapy, 10*, 157 (1990).
34. Welch, W. J., *Expert Opin. Investig. Drugs, 11*, 1553 (2002).
35. Palm, C., Reimann, D., Gross, P., *Cardiovasc. Res.*, *51*, 403 (2001).
36. Udelson, J. E., Smith, W. B., Hendrix, G. H, et al., *Circulation, 104*, 2417 (2001).

9.6 Enuresis

Pierre Cochat[1], Behrouz Kassaï[2]

[1]*Department of Paediatrics and Inserm U499, Hôpital Edouard-Herriot & Université Claude-Bernard, 69437 Lyon cedex 03, France*

[2]*Department of Clinical Pharmacology, Université Claude-Bernard, 69376 Lyon cedex 08, France*

9.6.1 INTRODUCTION

Epidemiology

Bedwetting is a common, but socially disruptive and stressful condition, in school age children; with a prevalence rate of 15–20% at age 5, 8–10% at age 6, 3–4% at age 10 and 1–2% in young adults, without any significant difference between countries and cultures [1-5]. Two thirds experience monosymptomatic nocturnal enuresis (mainly boys) and one third suffer from voiding disorder (mainly girls), sometimes leading to persistent troubles toward adulthood. This chapter will focus on monosymptomatic nocturnal enuresis.

Definitions

Nocturnal enuresis is characterised by the frequent occurrence of complete uncontrolled micturition during sleep in children older than 5 years of age.

Voiding disorders include a wide number of functional problems, mainly bladder and urethral instability. Bladder instability is due to uninhibited contractions of the detrusor during bladder filling. Urethral instability is characterised by inadequate urethral sphincter relaxation during bladder filling in the absence of detrusor contraction.

Both nocturnal enuresis and bladder disorders may be of primary (absence of wetting-free period) or secondary (onset of wetting after a significant period of dryness) onset. Primary troubles are far more frequent than secondary ones, in an approximate ratio of 10:1.

Genetic factors

Heredity is one of the most important factors contributing to enuresis. The percentage of bedwetters increases from ~15% of children without any family history of enuresis to ~45% when one of the parents, and ~75% when both parents used to wet their bed. A strong family history of enuresis increases the probability of having enuresis. Recent evidence points to the involvement, in some cases, of a single autosomal dominant gene with reduced penetrance; linkage studies have shown that nocturnal enuresis is linked with regions on chromosomes 8, 12 and 13.

9.6.2 CLINICAL PRESENTATION

In order to better identify monosymptomatic enuresis and bladder disorders, both presentations will be summarised.

Voiding disorders – of which the most frequent is bladder instability – may mimic nocturnal enuresis. A simplified assistance for decision making is given in Figure 1.

Functional disorders may be clinically presented as:

- urge incontinence, which in young children is often due to them being too busy ("Nintendo enuresis") and inexperienced to get enough warning of a full bladder;

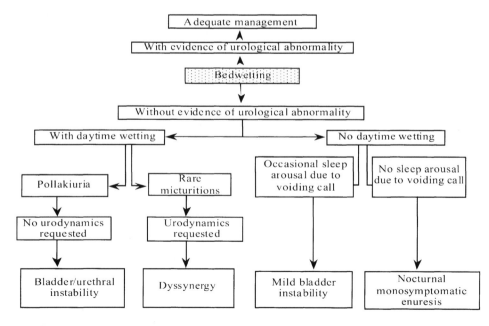

Figure 1. Clinical assistance for decision-making in the presence of bedwetting.

- stress incontinence on coughing or straining;
- in some children there is complete emptying of the bladder on giggling;
- others.

Primary nocturnal enuresis, long time regarded as a symptom of an underlying psychopathological problem, has been recently revisited. These surveys show that primary nocturnal enuresis is indeed a common biobehavioural problem without a psychiatric component, because the likelihood of significant psychopathology is less than 10%. Enuretic children tend to exhibit anxiety and problematic conducts, but the symptoms are more likely to be a result, than a cause, of enuresis; psychotherapy is indicated only when a psychopathological condition is present, or enuresis has major psychological consequences for the child or the family.

Practice point

In primary nocturnal enuresis, the likelihood of a significant psychopathology is less than 10%; however, enuresis causes significant distress and impairment of self-esteem [6].

In most cases, the diagnostic procedure of primary nocturnal enuresis is limited to the patient's history. Table 1 shows specific questions that should be asked systematically of the parents and the child.

No further investigation is required in the presence of a typical picture of primary monosymptomatic nocturnal enuresis if physical examination is normal. However, routine clinical parameters must be assessed, i.e. height and growth velocity, blood pressure and urinalysis using dipstick (glucose, leucocytes and nitrite, protein, pH, density) in order to exclude other specific diagnoses (e.g. diabetes mellitus, congenital tubular disorders, chronic renal insufficiency, central or nephrogenic diabetes insipidus, chronic tubulointerstitial nephritis, urinary tract infection). In selected cases, anatomical/physiological assessment, such as urinary tract ultrasonography, measurement of bladder capacity, evaluation of kidney and bladder function, has been suggested. In our experience, such indications are very limited.

Secondary nocturnal enuresis

Such a form of nocturnal enuresis may usually occur after personal or familial disturbance, e.g. moving, school problems, parental divorce, grandparent death, sudden illness in a family member or new arrival of a sibling. The clinical spectrum of secondary nocturnal enuresis is superimposable to the primary form, whereas bedwetting may be less frequent.

Table 1

Specific questions to be asked systematically to the parents and the child

family history of nocturnal enuresis?	present in 60% of the cases
daytime symptoms?	absent, unless it is associated with bladder disorder
night time wetting?	daily in most cases
environmental influence?	increased/decreased frequency of bedwetting
use of diapers?	frequent, because of daily night time wetting
sleep arousal because of voiding call?	absent; frequent if bladder instability
results of previous treatment(s)?	usually absent; frequent in bladder disorders
urinary tract infections?	absent; frequent in bladder disorder
constipation?	usually absent; frequent in bladder disorders
pollakiuria?	absent; frequent in bladder instability
polyuria?	should be excluded; symptom of tubular disorder
parasomnias ?	more frequent in enuretic children

9.6.3 PATHOPHYSIOLOGY

Nocturnal enuresis is a multifactorial condition and probably comprises several types [2].

Pathophysiological findings suggest 3 main interactive factors are involved:
- patient's inability to wake in response to signals from a full bladder;
- relative nocturnal polyuria, due to insufficient vasopressin release during sleep;
- reduced nocturnal functional bladder capacity especially in some cases that do not respond to desmopressin treatment.

Nocturnal enuresis, sometimes associated with diurnal enuresis, may be a part of psychiatric disorders, including attention-deficit hyperactivity disorder, autism and other severe disturbances. In such conditions, enuresis is a minor medical symptom but a major request with respect to the patient's quality of life, with an associated risk of punishment and abuse.

Practice points

Several factors may be associated in monosymptomatic nocturnal enuresis:
- genetic factors;
- a delay in the development of the arousal response to bladder distension;

• a maturation blunting of the normal vasopressin peak concentration during nocturnal sleep.

9.6.4 MANAGEMENT

Preliminary considerations

The medical approach to enuretic patients is challenging, since conventional treatment has to take place among others, such as psychotherapy, natural medicine, physiotherapy, hypnotherapy [1,2,7]. In addition, any kind of urinary dysfunction may have a psychosocial component (anxiety, guilty feeling, opposition) and their treatment has been shown to improve self-esteem. Some studies report a spontaneous cure rate of approximately 15% per year, whereas others report an apparent resistance to all available treatments. An overall 70 to 80% success rate of adequate treatments in compliant patients is usually expected.

Treatment of enuresis can have a pronounced impact on the family economy, since a bedwetting child represents a substantial expense to the family caused by addititonal washing and drying of bedclothes (100 to 450 € per year).

The access to the treatment of enuresis could be perceived as a medical privilege, because no treatment is available in developing countries. The lack of treatment leads to a wide use of traditional medicine based on religious beliefs and folk wisdom.

Management of nocturnal enuresis

An enuresis diary will allow determination of the severity of bedwetting and further interpretation of outcome under treatment. The main therapeutic options in children with nocturnal enuresis are summarised in Figure 2.

Behavioural and educational interventions

Nocturnal continence involves several skills which are generally not attained before the age of 4 to 5 years, i.e. awareness of urgency, ability to initiate urination, ability to inhibit urination while awake, and ability to inhibit urination while asleep. Teaching continence skills, such as bladder training and alarm systems, cure enuresis with minimal risk to the child's health [2-4,8-11].

Treatment should be adapted to the child's age. Children under the age of 6 years may be managed by reassurance. Explaining to the child and his family that he has different patterns of sleep and does not sense the need to urinate, can help to relieve the psychological consequences of enuresis. Preventing irregular sleep-

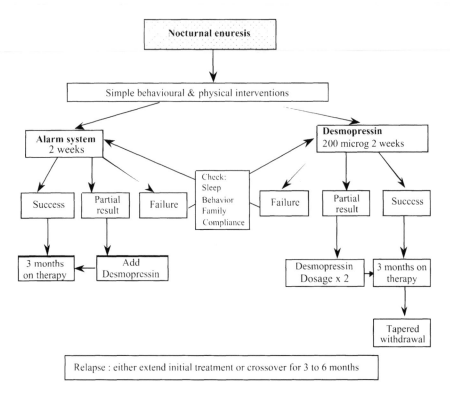

Figure 2. Therapeutic options in children with monosymptomatic nocturnal enurisis.

wake patterns and sleep deprivation may control bedwetting in young children. For older children, understanding and support are the most important attitudes towards helping the patient; but the child should not be left in diapers at night and should be assigned household responsibilities associated with the problem. Independent of the age of patients, reward systems (star charts given for dry nights, lifting or waking the child at night to urinate, retention control training to enlarge bladder capacity, fluid restriction) have been widely advised as a first-line method.

If no change in the nocturnal wetting pattern occurs after a period of about 3 months, and the child is motivated, the addition of an enuresis alarm system can be considered. Long-term pharmacological treatment should be limited to children older than 6 years who have not responded to behavioural approaches.

Several clinical trials have shown that alarm devices work in enuretics. In a randomised study enrolling 135 school age children, Faraj *et al.* [12] have shown that after 2 weeks, children with desmopressin achieved 80% dry nights compared to 50% in those with alarm treatment. The treatment was maintained for 3 months and the patients were evaluated 3 months after treatment withdrawal. Those who were under alarm treatment attained 94% dry nights *versus* 78% with desmopressin. Desmopressin offers better short-term results, but an enuresis alarm is significantly more efficient in the long-term. Unfortunately, in many countries, alarm devices are either not available or not reimbursed by national health services, and therefore such a strategy is difficult to implement. The cost of alarm devices ranges between 90 and 550 €.

Pharmacological treatment

• Tricyclics

The mechanism of action of tricyclics (e.g. imipramine) on nocturnal enuresis is unknown; it probably acts at several levels. The drug may work by its effect on sleep (arousal effects). It may also reduce premature contractions of the detrusor following partial filling of the bladder by its anticholinergic effect. Tricyclic agents have a positive effect on monosymptomatic nocturnal enuresis compared to placebo. However, their use should be limited to older children and adolescents, only for short periods of time and under selected conditions.

The daily dose of imipramine ranges between 1 and 2.5 mg/kg, or 25 to 75 mg depending on age and weight. Some patients do not respond, whereas others develop some kind of tolerance. In addition, the relapse rate after withdrawal is approximately 40 to 60%. Side effects include restlessness, sleep disturbance, attention disorders, headache, abdominal pain, constipation, weight loss, acrocyanosis. In addition, one should keep in mind that life threatening adverse events have been reported such as sudden death, coma, convulsions, heart or liver failure.

It is generally accepted that imipramine, a potentially dangerous drug, should not be recommended in a benign condition like monosymptomatic nocturnal enuresis.

• Desmopressin

Desmopressin is a synthetic agonist of the neuropeptide vasopressin with enhanced antidiuretic properties [13,14]. Thanks to its safety, it has replaced imipramine in the pharmacological approach to enuresis. Desmopressin mainly acts by reducing nocturnal urine production *via* direct vasopressin effects [15]. It may also act by improving the patient's ability to awaken and improve bladder instability; indeed studies have demonstrated that vasopressin facilitates arousal in humans and also appears to increase motor activity and stabilise unstable bladders in rats. Given orally at a dose of 200 to 400 microg (tablets 100 and 200 microg), or intranasally at a dose of 20 to 40 microg (2 to 4 sprays) for a period of ~3 months, desmopressin is effective in 50 to 80% of patients with monosymptomatic nocturnal enuresis. Unfortunately, such treatment is associated with a high relapse rate. Desmopressin has minor side effects including headache, mild abdominal cramp, nausea, nasal congestion, epistaxis, reversible body weight gain; water intoxication with hyponatremia and convulsions have been reported in a few cases. Children with a family history of nocturnal enuresis and a high nocturnal urine output seem to be good responders to desmopressin: Nørgaard *et al.* have shown that 91% of enuretic patients with a family history of enuresis compared with only 71% of those without are responders to desmopressin [16].

Conclusion from systematic reviews

Because a large number of individual therapeutic approaches in the management of nocturnal enuresis are based on opinions, evidence based medicine should be used in order to develop new guidelines [3,4,17-20].

- Simple behavioural methods (star charts given for dry nights, lifting or waking the child at night to urinate, retention control training to enlarge bladder capacity, fluid restriction) may be effective for some children and could therefore be tried as first line therapy before considering alarms or drugs.
- There is insufficient evidence to support the use of behavioural and educational interventions (dry bed training, full spectrum home training) without an alarm.
- Alarm interventions are highly effective for the treatment of nocturnal bedwetting.
- The combination of an alarm with behavioural and educational interventions offers better results than an alarm on its own.
- Only desmopressin, tricyclic drugs and alarm systems have been shown to be effective; however they are more demanding and may have side effects.
- Placebo is beneficial in around one third of the patients.
- Parents should be warned of the risk of serious and life threatening adverse effects of tricyclic overdose.
- Desmopressin and tricyclics have similar clinical effects, but alarm systems may produce more effective and sustained benefits. Indeed, most children relapse after stopping desmopressin or tricyclics, but only half relapse after alarm treatment.
- Desmopressin promptly reduces the number of wet nights, but there is some evidence that this is not sustained after drug discontinuation. It is recommended to limit fluid intake (< 250 mL) while desmopressin is effective.

- Giving extra fluids and avoiding penalties to the child after he has become dry using an alarm may further reduce the relapse rate.
- In some patients with partial response to either alarm or desmopressin, a combination of both treatments allows better results than single therapy.

Management of bladder disorders

The management of bladder disorders is based on a specific pathophysiological approach. Most children with bladder instability or bladder dyssynergy suffer from constipation, so that the first step is based on both dietary recommendations (hydration, fibres) and laxative therapy (lactulose, lactitol, macrogol). The second step will include anticholinergic agents (oxybutinine, tolderodine) and/or biofeedback pelvic floor training. In patients who resist to conventional treatment, constipation, vulvitis or recurrent urinary tract infection may have been underestimated and require a more agressive management. However, it must be noted that such troubles may sometimes be associated with congenital bladder abnormality, neurogenic bladder or child (sexual) abuse.

9.6.5 CONCLUSION

Monosymptomatic nocturnal enuresis is a very common finding in school-age children. The medical approach to enuresis needs further research: the pathophysiology is unknown, teaching depends on both the teacher and the country, and management depends on both cultural and financial abilities.

REFERENCES

1. Bourquia, A., Chihabeddine, K., *Saudi J. Kidney Dis. Transplant.*, *13*, 151 (2002).
2. Challamel, M. J., Cochat, P., *Sleep Med. Rev.*, *3*, 313 (1999).
3. Glazener, C.M.A., et al. http://www.cochrane.org/cochrane/revabstr/AB003637.htm (2004).
4. Glazener, C.M.A., et al. http://www.cochrane.org/cochrane/revabstr/AB004668.htm (2004).
5. Jarvelin, M. R., Vikeväinon-Tervonen, L., et al. *Acta Paediatr. Scand.*, *77*, 148 (1988).
6. Schulpen, T. W. J., *Acta Paediatr.*, *86*, 981 (1997).
7. Diseth, T. H., Vandvik, I. H., *Tidsskr. Nor. Laegeforen.*, *124*, 488 (2004).
8. Agence Nationale d'Accréditation et d'Evaluation en Santé, http://www.anaes.fr (2004).
9. Fritz, G., Rockney, R., et al., *J. Am. Acad. Child. Adolesc. Psychiatry*, *43*, 123 (2004).
10. Landgraf, J. M., Abidari, J., Cilento, B. G. Jr., et al., *Pediatrics*, *113*, 334 (2004).
11. Wille, S., *Acta Paediatr.*, *4*, 772 (1994).
12. Faraj, G., Cochat, P., Cavailles, M. L., Chevallier, C., *Arch. Pédiatr.*, *6*, 271 (1999).
13. Lackgren, G., Lilja, B., Nevus, T., Stenberg, A., *Br. J. Urol. 81* Suppl 3, 17 (1998).
14. Lehotska, V., Lichardus, B., Némethova, V., *Bratisl. Lek. Listy.*, *72*, 670 (1979).
15. Rittig, S., Knudsen, U. B., Nørgaard, J. P., et al. *Am. J. Physiol.*, *256*: 664 (1989).
16. Nørgaard, J. P., Rittig, S., Djurhuus, J. C., *J. Pediatr.*, *114*, 705 (1989).

17. Glazener, C. M. A., et al., http://www.cochrane.org/cochrane/revabstr/AB002112.htm (2004).
18. Glazener, C. M. A., et al., http://www.cochrane.org/cochrane/revabstr/AB002117.htm (2004).
19. Glazener, C. M. A., et al., http://www.cochrane.org/cochrane/revabstr/AB002238.htm (2004).
20. Glazener, C. M. A., et al., http://www.cochrane.org/cochrane/revabstr/AB002911.htm (2004).

9.7 Steroid sensitive nephrotic syndrome

Nicholas J. A. Webb

Consultant Paediatric Nephrologist, Royal Manchester Children's Hospital and Manchester Institute for Nephrology and Transplantation, Hospital Road, Pendlebury, Manchester, M27 4HA, UK

9.7.1 INTRODUCTION

Nephrotic syndrome develops when an abnormality of glomerular permeability results in the development of heavy proteinuria, hypoalbuminaemia and generalised oedema. Large studies in the 1970's reported that 80% of cases of childhood nephrotic syndrome are attributable to minimal change disease (MCD), so called because renal biopsy shows no significant abnormality at light microscopy level. This disorder is more common in males than females (ratio 3:2) and is predominantly a disease of younger children with a median age at presentation of 3 years. The incidence of MCD, in the UK, is around 2-4 per 100,000 children in the Caucasian population, but is approximately six times higher in children of Asian origin [1]. MCD is associated with response to oral steroid therapy in over 90% of cases and is usually associated with a favourable long-term outcome, although at least 70% of patients will experience a chronic, relapsing-remitting course [2, 3].

A heterogeneous group of conditions, including focal segmental glomerulosclerosis (FSGS) and mesangiocapillary glomerulonephritis (MCGN) account for the remaining 20% of cases of childhood nephrotic syndrome. In contrast to MCD, these diseases tend to present in an older age group of children and the majority of cases do not respond to oral steroid therapy alone. Their prognosis is correspondingly poorer.

As newly presenting cases of nephrotic syndrome have a high probability of having MCD, which is likely to respond to steroid therapy and up to 20% of those with alternative histological diagnoses may also respond to steroids, the large majority of children are treated with an empirical course of such therapy. Renal biopsy is performed only in those with atypical presenting features (age <12 m or >12 y, persistent hypertension or impaired renal function, gross hae-maturia, low plasma C3), or those who subsequently fail to respond to a standard course of steroids (see below). Hence, the majority of cases respond to empirical steroid therapy, without the underlying histological diagnosis being confirmed. The term steroid sensitive nephrotic syndrome (SSNS) has been adopted for this group of patients and recognises the important prognostic significance of res-ponse to steroid therapy. Reflecting changes in the use of routine biopsy, much of the early research in this field involved patients with a histological diagnosis of MCD, whereas later studies described patients with clinically defined SSNS. As a reflection of current clinical practice, the term SSNS will be used consistently throughout the rest of this chapter, including in discussion of older literature.

Those who do not respond to a standard initial course of steroids are defined as having steroid resistant nephrotic syndrome: outcome here is somewhat depen-dent upon histological diagnosis, though generally much poorer than SSNS with up to 50% developing end-stage renal failure. The management of this group falls outside of the remit of this chapter.

9.7.2　DRUG THERAPY OF THE PRESENTING EPISODE OF NEPHROTIC SYNDROME

The International Study for Kidney Diseases in Children (ISKDC) regimen, described in the late 1960s, remains in widespread use today. Prednisolone, or its active metabolite prednisone is given daily at a dose of 60 mg/m^2 (maximum dose 80 mg) for 28 days, followed by 40 mg/m^2 (maximum dose 60 mg) given on alternate days for a further 28 days. Undivided doses of predniso(lo)ne are used throughout, and there is no dose tapering following completion of the eight-week course.

A number of randomised controlled trials (RCTs) have investigated the effect of both longer and shorter steroid regimens on the subsequent clinical course. Meta-analysis has indicated that a longer duration of treatment at initial presentation may be associated with a higher rate of sustained remission and lower subsequent relapse rates, though further trials are required to help ascertain both the optimum duration of therapy and the impact of more intensive initial therapy on the inci-dence of side-effects [4]. One such RCT is shortly due to commence in the UK.

The child is generally hospitalised for the early part of this initial steroid regi-men to ensure that complications of the nephrotic state such, as hypovolaemia or infection, do not develop, and to educate the parents about the condition in general and the use of home urinalysis for the early detection of relapses (see below). Penicillin V (12.5 mg/kg bd) is commenced as prophylaxis against *Streptococcus pneumoniae* infection, and there should be a low threshold for the use of therapeutic dose antibiotics if infection is suspected. Oedema may be treated with fluid restriction and diuretics, unless there is evidence of hypovolaemia: where this is present, it should be corrected with the use of intravenous albumin solution. There is no evidence to support the use of bed-rest or alteration in dietary protein content. A no-added salt diet is a sensible measure given the presence of oedema and the administration of steroids.

A response to steroid therapy is indicated by the resolution of proteinuria, urinary remission being defined as three consecutive days of zero or trace proteinuria on Albustix®. Of those children who respond to steroids, approximately 80% will have entered remission within 14 days using this protocol [5]. By definition, children remitting during the first 28 days of therapy fall into the diagnostic category of SSNS described above. Children who have not entered remission by the end of 28 days of daily predniso(lo)ne treatment are regarded as steroid resistant. These children should be referred to a specialist paediatric nephrologist for further assessment: their management is complex and is outside the scope of this chapter.

9.7.3 MANAGEMENT OF INITIAL RELAPSES OF SSNS

Relapses (three consecutive days of 2+ or more proteinuria on Albustix®), which occur in 70-80% of children, are generally detected through routine urinalysis performed by the parents or older child at home. This allows the early commencement of treatment prior to the development of generalised oedema with its attendant complications. Treatment of relapses consists of predniso(lo)ne 60 mg/m^2 (maximum dose 80 mg) daily until urinary remission, followed by 40 mg/m^2 (maximum dose 60 mg) on alternate days for 14 doses over a 28 day period. Intensification of relapse therapy has not been shown to be of long-term clinical benefit [6].

9.7.4 MANAGEMENT OF FREQUENTLY RELAPSING AND STEROID DEPENDENT SSNS

At least 50% of patients will follow a frequently relapsing (2 relapses or more in the first six months after presentation or 4 relapses in any 12 month period) or steroid-dependent (relapses on or within 14 days of completion of steroid therapy) course. Such patients are generally managed by, or in consultation with, a paediatric nephrologist. The pattern of disease in the first six months is generally predictive of the complexity of the subsequent clinical course. Patients who relapse more than twice in the first six months after initial steroid response are at the highest risk of continuing to relapse frequently over the subsequent eighteen month period [7].

In the overwhelming majority of cases, relapses will remain responsive to steroids, though the repeated administration of courses of high dose steroid therapy is associated with the development of both short-term and long-term side-effects. Where these are present or anticipated, alternative therapy should be considered. An important general principle of management is that relapses should be prevented through the use of the smallest amount of immunosuppressive therapy possible, bearing in mind that over 80% of children will spontaneously 'grow out' of their disease and enter adult life free from relapses.

Long-term low dose maintenance steroid therapy

In most cases of frequently relapsing disease, a reasonable first measure is the use of low-dose alternate-day steroids, in an attempt to maintain the patient in stable remission. The side-effects of steroids are reduced by the use of the drug in a single alternate daily dose, and many patients can be kept free from relapse (and the requirement for high dose steroid therapy), using such an approach. After establishing remission using standard relapse therapy as described above, the predniso(lo)ne dose is gradually tapered, aiming for a dose in the range 100 – 500 microg/kg on alternate days to be given for 6 – 12 months. The smallest dose possible which maintains remission should be used. In steroid dependent patients, the physician will be aware of the dose of prednisolone on which the patient previously relapsed, and the maintenance dose of alternate day predniso(lo)ne should be targeted to be slightly above this figure.

The numerous short and long-term side-effects of steroids are well known and discussed in detail elsewhere in this textbook (see chapter on rheumatic disorders). Studies performed in children with nephrotic syndrome have shown the prevalence of these side-effects to be similar to those reported in children receiving steroids for other indications. Posterior subcapsular cataract occurs in 10-38% of children, though one large well conducted study failed to show any association between cataract development and the cumulative dose of prednisolone administered [8]. Data on growth on long-term alternate day steroid therapy in this condition are somewhat conflicting, and depend to some extent on the amount of steroid administered. Normal growth has been reported in pre-pubertal children [9, 10], though appears to fall off after 10 years of age, particularly in boys, in whom there is also some evidence of pubertal delay [11]. The small number of studies reporting final adult height data suggest only a minor reduction in this [12, 13]. Studies investigating bone mineral density have also reported conflicting results, partly due to variation in the methods and sites of measurement, though both reductions in both trabecular and cortical bone mineral density have been reported [14, 15].

The use of other agents to reduce steroid exposure

If the patient cannot be maintained in stable remission on an acceptably low dose of alternate day steroids, particularly where steroid side-effects have developed, the use of alternative agents should be considered. The three drugs, or classes of drug, which have an established place in the treatment of SSNS are levamisole, the alkylating agents (cyclophosphamide and chlorambucil) and ciclosporin [16]. All have the proven ability to both maintain remission and reduce or eliminate the requirement for steroids, though have different potential adverse effects. The balance of benefits and risks needs careful consideration for each individual patient, with detailed discussions being held with the family. Opinions differ as to the order in which these agents might be used, though a suggested algorithm is shown in Figure 1.

---

1. Initial Episode
   • Prednisolone 60 mg / m^2 daily for 28 days, followed by
   • Prednisolone 40 mg / m^2 on alternate days for 14 doses over 28 days

2. First Two Relapses
   • Prednisolone 60 mg / m^2 daily until remission, followed by
   • Prednisolone 40 mg / m^2 on alternate days for 14 doses over 28 days

3. Frequent Relapses
   • Maintenance Prednisolone: 0.1 – 0.5 mg / kg on alternate days for up to 12 months

4. Relapse on Prednisolone > 0.5 mg / kg / alternate days
   • Levamisole 2.5 mg / kg on alternate days for up to 12 months

5. Relapse on Prednisolone > 1.0 mg / kg / alternate days OR
   Relapse on Prednisolone > 0.5 mg / kg / alternate days plus adverse steroid effects / other risk factors
   • Cyclophosphamide 3mg / kg daily for 8 weeks

6. Post-cyclophosphamide relapses
   • As in 2 – 3 above

7. Relapse on Prednisolone > 0.5 mg / kg / alternate days
   • Ciclosporin 3 mg / kg per dose, given twice per day for up to 24 months

---

Figure 1. Typical management strategy for steroid sensitive nephrotic syndrome

9.7.5 LEVAMISOLE

Levamisole was developed as an antihelminthic agent, though isolated reports in the 1980s suggested that the drug was also effective for SSNS. Its immunomodulatory properties are not completely understood, but the drug appears to stimulate rather than suppress T-cell function: both T-cell and phagocyte function are restored to normal in immunosuppressed states. There are no anti-inflammatory properties or effects on B-cell function.

Three RCTs have shown levamisole alone, or with placebo, to be superior to steroids in preventing relapse in steroid dependent children; a meta-analysis has shown a lower relative risk for relapse at 6 to 12 months, RR 0.6 (95% CI 0.45-0.79%) [16]. Following the induction of remission with steroids, levamisole is commenced at a dose of 2.5 mg/kg on alternate days. This allows the tapering and possible discontinuation of concomitant alternate day steroids. The efficacy of levamisole is dependent upon its continuous administration, with a substantial risk of relapse occurring upon stopping therapy. When remission is successfully maintained, levamisole is typically continued for a period of up to 12-24 months, though there are no theoretical reasons why longer courses cannot be used.

The advantages of the drug lie in its potential to allow prolonged periods of remission off steroid therapy, its ease of administration and the low incidence of reported adverse effects. These include neutropenia, gastrointestinal upset and rash. Only one case of mild gastrointestinal upset was reported in the two largest RCTs, and there were no cases of neutropenia: the potential for this does, however, mandate that the patient's full blood count should be monitored on a regular basis.

9.7.6 ALKYLATING AGENTS

The alkylating agents possess alkyl chains which bind to purine bases and thus impair normal DNA transcription. They have both immunosuppressive and cytotoxic properties. The response of SSNS to mechlorethamine (nitrogen mustard), the parent compound, was first described over 50 years ago [17] though it has generally been superseded by two orally administered alkylating agents, cyclophosphamide and chlorambucil, the efficacies of which were first reported in the 1960s. These drugs retain an important role in the management of SSNS because of the demonstrated ability of a short course of therapy to induce long-term steroid-free remission or reduction of relapse frequency. Both have been shown by meta-analysis to significantly reduce the risk of relapse when compared with steroids alone (cyclophosphamide {RR 0.44; 95%CI 0.26-0.73}, chlorambucil {RR 0.13; 95%CI 0.03-0.57}) [16]. As with most therapies, children with frequently relapsing disease have a better outcome from alkylating agent therapy than those with steroid dependency, two year relapse free rates following cyclophosphamide therapy being approximately 70% and 25% respectively [18].

Cyclophosphamide is the most frequently prescribed oral alkylating agent, particularly in the UK, and there is a substantially larger volume of published literature relating to its use. For this reason, this section will focus on the use of this agent. Only one RCT has compared cyclophosphamide with chlorambucil, and whilst the study was underpowered, no difference was shown between the two in the risk of relapse at 12 and 24 months [19].

Following induction of remission with steroids, cyclophosphamide is commenced at a daily dose of 3 mg/kg for a total of eight weeks. No benefit has been shown by prolongation of this course to 12 weeks [18] and shorter courses have been shown to be less effective [20]. Once cyclophosphamide has been commenced, the dose of steroids can be reduced: it is the authors practice to give prednisolone 40 mg/m^2 for the first 4 weeks of therapy, then 20 mg/m^2 for the second 4 weeks with a subsequent taper, so that steroids are discontinued shortly after completion of cyclophosphamide.

The short-term adverse effects of cyclophosphamide include myelosuppression, nausea and gastrointestinal upset, temporary alopecia and haemorrhagic cystitis. Because of the risk of myelosuppression and the associated risk of infection, blood counts should be monitored on a weekly basis for 4 weeks, and then fortnightly throughout the course of cyclophosphamide: the dose should be reduced or stopped where neutropenia occurs. Once the neutrophil count has recovered, the drug can be recommenced at a lower dose. Medical advice should be sought promptly should the child develop a febrile illness, particularly in the presence of neutropenia. The alopecia is fully reversible and requires no specific therapy. Nausea and gastrointestinal upset generally responds well to anti-emetics such as ondansetron. Haemorrhagic cystitis is very rare at the doses of cyclophosphamide administered for SSNS.

In the longer-term, cyclophosphamide has the potential to induce both infertility and malignancy, though these side-effects appear to be related to higher dose therapy than that which is currently administered to children with nephrotic syndrome. A systematic review of eight studies, comprising 119 courses of cytotoxic therapy, reported a cumulative total dose-dependent effect of cyclophosphamide on sperm count in adult life, though those who had received a total dose of upto 168 mg/kg of cyclophosphamide (equivalent to 8 weeks at 3 mg/kg/day) had almost uniformly satisfactory sperm counts [21]. The risk of azoospermia increases with further courses of cyclophosphamide and may be greater in those given the drug in the peripubertal period. Because of slower cell division and turnover, the ovary is much more resistant to the effects of the alkylating agents and it is rare for females to suffer long-term effects of gonadal toxicity. Despite the mutagenic potential, no teratogenic effects have been observed in the offspring of adults treated with cyclophosphamide for SSNS in childhood.

This same systematic review pooled data from 38 studies, comprising 1573 courses of alkylating agent treatment for childhood nephrotic syndrome, and found a total of 14 cases of malignancy reported following such therapy [21]. The most recent case was reported in 1986, suggesting that treatment was given relatively early in the history of the use of the drug, and total doses administered were significantly in excess of those given today. At present, it has to be concluded that a single eight week course of therapy almost certainly does not raise the prevalence of malignancy above that of the general population (currently 0.17% by 15th birthday in the UK), though this risk may increase with further courses of alkylating agents.

9.7.7   CICLOSPORIN

The calcineurin inhibitor ciclosporin modifies T-cell function, inhibiting IL-2 production by activated T-cells. Developed as an immunosuppressant for use in renal transplant recipients, ciclosporin was first reported to be effective in nephrotic syndrome in 1986. Its efficacy has subsequently been confirmed in a number of further studies. Similar to levamisole, response to ciclosporin depends upon its continuous administration, with a high rate of relapse occurring upon discontinuation of the drug. This 'ciclosporin dependence' is well illustrated in a RCT comparing 24 weeks of ciclosporin with a single course of chlorambucil. At six months, there was no difference between the two study groups in maintenance of remission, though by 12 months, six months after completion of ciclosporin, significantly fewer in the ciclosporin group were in remission [22].

Ciclosporin may be prescribed in an attempt to control frequent relapses recurring after a previous course of cyclophosphamide. In selected circumstances, ciclosporin may be used earlier in preference to cyclophosphamide, for example in peri-pubertal boys possibly at increased risk of cyclophosphamide-related gonadal toxicity.

Much debate surrounds both the recommended dose and target trough blood levels of ciclosporin in SSNS. Most authorities would recommend a starting dose of 6 mg/kg/day in two divided doses, aiming for trough blood levels at the lower end of the range used for stable renal transplant recipients (75-125 μg/L). In practice, one should aim to use the smallest dose possible, which allows freedom from relapse, to avoid complications of therapy in children who have a condition which is likely to enter permanent remission in later childhood. When satisfactory therapeutic levels of ciclosporin have been achieved, alternate-day steroids may be tapered and possibly discontinued.

The short-term adverse effects of ciclosporin include hypertrichosis, gingival hyperplasia, hypertension, tremor and a rise in plasma creatinine mediated by glomerular vasoconstriction. The major factor limiting the prolonged use of ciclosporin is concern about potential chronic nephrotoxicity, which has been reported in SSNS patients taking ciclosporin continuously for more than two years [23]. Plasma creatinine concentration does not provide any indication of structural damage until a relatively advanced stage has been reached. For this reason, it is widely accepted practice to perform a renal biopsy after 18 – 24 months of therapy. Ciclosporin may be continued beyond two years in those patients with no histological evidence of chronic nephrotoxicity.

9.7.8   NEWER AGENTS

A number of newer immunosuppressive agents have been reported as showing efficacy in the management of SSNS. These include tacrolimus, which has a similar mechanism of action to ciclosporin, mycophenolate mofetil [24] and mizoribine, an imidazole nucleotide originally isolated from *Eupenicillium brefeldianum* [25].

9.7.9   PROGNOSIS OF SSNS

Despite the considerable therapeutic advances of the twentieth century, the four major series reporting long-term follow-up of patients with SSNS have shown mortality rates of 1-7.2 %, deaths being largely due to the acute complications of sepsis and vascular thrombosis [2,3,26,27]. Many of these deaths occurred in patients treated in the 1960s and 1970s, and current mortality rates are thought to be rather lower.

The long-term prognosis of SSNS in childhood is generally very good, with the large majority of children experiencing a decrease in the rate of relapses over the course of several years and entering long-term remission before adulthood. Up to 20% of SSNS patients continue to have relapses throughout the teenage years and beyond their twentieth birthday [3]. Most of these adults are thought to enter long-term remission in their third decade. A small number of publications have followed cohorts of SSNS patients into the third decade and have shown no excess incidence of hypertension and no cases of chronic renal impairment among patients who were fully steroid sensitive throughout the course of their illness [2,3,28]. There are no good studies documenting the outcome of these patients beyond this and it is not known whether this group of patients will have an excess of cardiovascular disease, diabetes or malignancy.

It is recognised that upto 4.5 % of initially corticosteroid sensitive patients will develop a steroid resistant pattern of illness in a subsequent episode [26,28-30]. Of these patients with "secondary" corticosteroid resistance, the majority will regain corticosteroid responsiveness following further immunosuppressive therapy, though a very small number, predominantly those subsequently shown to have FSGS histology, will remain corticosteroid resistant and develop chronic renal failure.

REFERENCES

1. Sharples, P. M., Poulton, J., *Arch. Dis. Child, 60,* 1014 (1985).
2. Trompeter, R. S., Lloyd, B. W., Hicks, J., White, R. H., Cameron, J. S., *Lancet,* 1368, (1985).
3. Lewis, M. A., Baildom, E. M., Davis, N., Houston, I. B., Postlethwaite, R. J., *Lancet,* 255-9 (1989).
4. Anonymous, *J. Pediatr., 98,* 561 (1981).
5. Hodson, E.M., Knight, J.F., Willis, N.S., Craig, J.C., *Arch. Dis. Child, 83,* 45 (2000).
6. Anonymous, *J. Pediatr., 95,* 239 (1979).
7. Anonymous, *J. Pediatr, 101,* 514 (1982).
8. Brocklebank, T. J., Harcourt, R. B., Meadow, S. R., *Arch. Dis. Child, 53,* 30 (1982).
9. Polito, C., Oporto, M. R., Totino, S. F., La Manna, A., Di Torro, R., *Acta Paediatr. Scand., 75,* 245 (1986).
10. Saha, M.-T., Laippala, P., Lenko, H. L., *Acta Paediatr., 87,* 545 (1998).
11. Rees, L., Greene, S. A., Adlard, P., et al., *Arch. Dis. Child, 63,* 484 (1988).
12. Foote, K. D., Brocklebank, T. J., Meadow, S. R., *Lancet,* 917 (1985).
13. Matsukura, H., Inaba, S., Shinozaki, K., et al., *Am. J. Nephrol., 21,* 362 (2001).
14. Lettgen, B., Jeken, C., Reiners, C., *Pediatr. Nephrol., 8,* 667 (1994).
15. Tenbrock, K., Kruppa, S., Mokov, E., Querfeld, U., Michalk, D., Schoenau, E., *Pediatr. Nephrol., 14,* 669 (2000).
16. Durkan, A. M., Hodson, E. M., Willis, N. S., Craig, J. C., *Kidney Int., 59,* 1919 (2001).
17. Kelley, V. C., Panos, T. C., *J. Pediatr. 41,* 507 (1952).
18. Ueda, N., Kuno, K., Ito, S., *Dis. Child, 65,* 1147 (1990).
19. Anonymous, *N. Engl. J. Med., 306,* 451 (1982).
20. Barratt, T. M., Cameron, J. S., Chantler, C. et al., *Arch. Dis. Child, 48,* 286 (1973).
21. Latta, K., von Schnakenburg, C., Ehrich, J. H., *Pediatr. Nephrol, 16,* 271 (2001).
22. Niaudet, P., *Pediatr. Nephrol., 6,* 1 (1992).
23. Niaudet, P., Habib, R., *J. Am. Soc. Nephrol., 5,* 1049 (1994).
24. Hogg, R., Fitzgibbons, L., Bruick, J., et al., *Nephrol. Dial. Transplant., 18,* 261 (2003).
25. Yoshioka, K., Ohashi, Y., Sakai, T., et al., *Kidney Int., 58,* 317-24 (2000).
26. Koskimies, O., Vilska, J., Rapola, J., Hallman, N., *Arch. Dis. Child, 57,* 544 (1982).
27. Anonymous, *Pediatrics, 73,* 497 (1984).
28. Neuhaus, T. J., Fay, J., Dillon, M. J., Trompeter, R. S., Barratt, T. M., *Arch. Dis. Child, 71,* 522 (1994).
29. Tarshish, P., Tobin, J. N., Bernstein, J., Edelmann, C. M., *J. Am. Soc. Nephrol, 8,* 769, (1997).
30. Webb, N. J. A., Iqbal, J., Smart, P. J., Lewis, M. A., Lendon, M., Houston, I. B., *Am. J. Kid. Dis., 27,* 484, (1996).

9.8 Cancer chemotherapy for paediatric malignancies

Gilles Vassal

Department of Paediatric Oncology and UPRES EA3535, Pharmacology and New Treatments of Cancers, Institut Gustave Roussy, Villejuif, France

The majority of malignant diseases in children are sensitive to cytotoxic chemotherapy. This chemosensitivity is generally attributed to the quickly proliferating character of these tumours and to their capacity to enter apoptosis. Chemotherapy has a major place in the treatment of paediatric cancers and contributes for a large part, to the current therapeutic successes. 75% of children can be cured with current chemotherapy. This raises the question of its long term toxicity and late effects in adults who recovered from a cancer during their childhood. This equally raises the issue of refractory malignant diseases (certain brain tumours, several metastatic diseases, high-risk leukaemias and some rare or complex mesenchymal tumours). Indeed, malignancy remains a major cause of death in children. There is an urgent need to better understand the biology of these refractory tumours and to develop new active treatments.

The classification of anticancer agents has been based on the knowledge of their mechanisms of action and/or of their cellular target. Thus, the therapeutic classes have not just been based on the chemical nature of the products, but on the mechanism of their effect. Anticancer agents currently used for the treatment of paediatric malignancies are described below. Other drugs are under evaluation for children, such as the topoisomerase 1 inhibitors (topotecan, irinotecan), fludarabine and tyrosine kinase inhibitors.

9.8.1   DACTINOMYCIN (ACTINOMYCIN D)

Mechanism of action

Dactinomycin is an antibiotic originating from the organism *streptomyces*. It is a cytotoxic agent which intercalates in between the DNA base pairs and inhibits the topoisomerases. It potentiates ionising radiations.

Pharmacology

Dactinomycin is administered intravenously. Very few studies have reported its pharmacokinetics. Metabolism is minimal with biliary excretion the major elimination pathway. Dactinomycin distributes rapidly into the tissues, the circulating nucleated cells and does not cross the blood brain barrier. Approximately 30% of the dose is found unchanged.

Use

Dactinomycin is used in the treatment of different solid tumours (nephroblastoma, sarcomas). The usual dose is 15 mcg/kg per day for 5 consecutive days. If irradiation is used, dactinomycin must be significantly reduced to avoid hepatotoxicity. Dactinomycin is toxic to the veins and any extravasation must lead to an immediate cessation of the administration, as well as to a local injection of sodium thiosulfate.

Toxicity

- Acute: myelosuppression, vomiting, radiosensitivity.
- Preventable: extravasation, hepatoxicity during concomitant irradiation.

9.8.2   ASPARAGINASE

Mechanism of action

Asparaginase is a bacterial enzyme extracted from *Escherichia coli* or from *Erwinia carotovora* which hydrolyses asparagine. By inducing asparagine deficiency, asparaginase becomes cytotoxic for the leukaemia cells which are incapable of synthesising asparagine.

Pharmacology

Asparaginase is administered intravenously or intramuscularly. The elimination half-life ranges from 8 to 30 hours. The bioavailability after an intramuscular injection is in the order of 50%. The appearance of antibodies brings about a rapid drop of the circulating concentrations of the enzyme [1].

Use

It is used for the specific treatment of acute lymphoblastic leukaemias and malignant non-Hodgkin's lymphomas.

Toxicity

- Acute: hypersensitivity including anaphylactic shock (reduced by IM administration), coagulation disorders, vomiting.
- Rare: pancreatitis.
- Long term: sterility.

9.8.3 ALKYLATING AGENTS

Busulfan

Mechanism of action

Busulfan is an alkylating agent and is toxic to cells in the G0 phase of the cell cycle, particularly haematopoietic stem cells.

Pharmacology

Busulfan is administered orally. An intravenous form exists for adults and will be available in the near future for children. Bioavailability is 80% with significant inter-individual variation. Busulfan distributes into the liver, the brain (up to 20% of the dose), the lungs and the bone marrow. The CSF to plasma concentration ratio is 1. The plasma half-life is 3 hours. The clearance is mainly metabolic, less than 1% is eliminated unchanged in the urine, and is higher in children than in adults. Busulfan, like other alkylating agents, is conjugated to glutathione by the glutathione-transferases [2].

Use

Busulfan is used before an autologous transplantation in the treatment of solid tumours (neuroblastoma, medulloblastoma, Ewing's tumour). It is also used before allogenic transplantation in the treatment of certain leukaemias and numerous genetic diseases like thalassaemia, immune deficiencies and lysosomal storage diseases. The usual total dose is 16 mg/kg or 600 mg/m^2, given every 6-12 hours over 4 consecutive days [3].

Toxicity

- Acute: myelosuppression, cutaneous pigmentation, veno-occlusive hepatic disease, mucositis, diarrhoea, pulmonary stenosis, seizures.
- Long term: sterility.

9.8.4 CYCLOPHOSPHAMIDE

Mechanism of action and metabolism

Cyclophosphamide is an alkylating agent, but is an inactive prodrug, which has to be metabolised into cytotoxic metabolites. Cytochromes P450 hydroxylate cyclophosphamide to 4 hydroxy-cyclophosphamide. Subsequent metabolites include mustard phosphoramide (cytotoxic) and acrolein. Part of the 4 hydroxy-cyclophosphamide and its tautomer, the adophosphamide, are transformed into inactive metabolites, mainly carboxyphosphamide, by aldehyde deshydrogenase. The mustard phosphoramide is the only metabolite which displays alkylating properties and is responsible for the cytotoxic antitumour activity. Acrolein is responsible for bladder toxicity.

Pharmacology

Cyclophosphamide is administered orally or intravenously. After intravenous injection, the elimination half-life of cyclophosphamide ranges from 4 to 10 hours. The peak of alkylating molecules is 1 to 2 hours after the cyclophosphamide peak. Their half-life is comparable to that of the prodrug. The 4 hydroxy-cyclophosphamide metabolite is weakly bound to plasma proteins. Brain distribution of cyclophosphamide and its metabolites is moderate with a CSF to plasma ratio of approximately 0.2. Renal excretion is the main route of elimination of cyclophosphamide and its metabolites.

Use

Cyclophosphamide is used for the treatment of numerous malignant diseases ,such as neuroblastoma and acute lymphoblastic leukaemias. It is used before allogenic haematopoietic stem cell transplantation in numerous malignant or non-malignant diseases. Cyclophosphamide is administered either at conventional doses of 100 to 300 mg/m^2, or at high doses (1.2 g/m^2 or 200 mg/kg) in combination with other cytotoxic agents. During the administration of doses over 1 g/m^2, prophylactic measures must be taken to avoid haemorrhagic cystitis. Acrolein is toxic to the bladder mucosa. The occurrence of bladder toxicity depends on the concentration of acrolein present in urine and the duration of the contact between acrolein and the mucosa. This toxicity is aggravated by urine stagnation and previous bladder irradiation. Hyperhydration, resulting in significant diuresis, along with the administration of mesna, which inactivates acrolein by conjugation, minimises toxicity.

Enzyme inducers, such as phenobarbital, increase the metabolism of cyclophosphamide. However, no clinical consequences of this drug interaction have been reported.

Toxicity

- Acute: myelosuppression, vomiting, alopecia. At high doses: syndrome of inappropriate antidiuretic hormone(SIADH) secretion, mucositis, diarrhoea.
- Preventable: haemorrhagic cystitis.
- Rare: cardiotoxicity (at high doses).
- Long term: sterility.

9.8.5 IFOSFAMIDE

Mechanism of action

Ifosfamide has a similar chemical structure to cyclophosphamide. It follows the same metabolic pathways of activation and inactivation. However, there are significant quantitative differences. Ifosfamide is hydroxylated in the liver by the cytochromes P450, and then transformed into mustard iso-phosphoramide (cytotoxic metabolite) and acrolein. Ifosfamide is partly transformed into inactive metabolites by the aldehyde dehydrogenase. Ifosfamide is also transformed in the liver by the cytochromes P450 into dechloroethylated metabolites. These metabolites are inactive but responsible for neurotoxicity.

Pharmacology

Ifosfamide can be administered orally and intravenously [4]. Following intravenous administration, the elimination half-life is in the order of 10 hours. Ifosfamide and its metabolites cross the normal blood brain barrier. Dose fractionation over several days is associated with autoinduction of ifosfamide metabolism and the formation of a greater quantity of active and inactive metabolites.

Use

Ifosfamide is used in the treatment of bone sarcomas, soft tissue sarcomas and nephroblastoma.

It is administered at doses of 6 to 12 g/m^2 fractionated over 2 to 5 days, usually in combination with other cytotoxic agents. Each administration of ifosfamide must be associated with hyperhydration and the administration of mesna to prevent bladder toxicity. Moderate to severe neurological toxicity may occur from headache to seizures and coma. It has been suggested that the administration of methylene blue may treat, or even prevent this neurotoxicity [5].

Toxicity

- Acute: myelosuppression, vomiting.
- Preventable: haemorrhagic cystitis.
- Long term: nephrotoxicity, sterility.

9.8.6   PLATINUM SALTS

Cisplatin

Mechanism of action

Cisplatin is an unstable salt of platinum in an aqueous environment, where it is spontaneously transformed into a reagent capable of inducing DNA damage by creating inter- and intra-strand crosslinks (adducts). Accumulation of these leads to cell death. It is stable in chlorinated environments such as saline.

Pharmacology

Cisplatin is administered intravenously [6]. The reactive forms are highly bound to plasma proteins. After intravenous administration, over 90% of platinum present in the blood is bound to proteins. The free platinum (presumed active forms) has a short half-life, 25 - 50 minutes. Cisplatin distributes largely with poor brain distribution (increased in brain tumours where the blood brain barrier is significantly altered). The main mechanism for the elimination of cisplatin is binding to plasma proteins. Free platinum is excreted by renal glomerular filtration [6].

Use

Cisplatin is used for the treatment of numerous paediatric malignant solid tumours, such as neuroblastoma, osteosarcoma, hepatoblastoma, germ cell tumours and brain tumours. Cisplatin is administered at doses of 80 to 120 mg/m^2, and in certain protocols 200 mg/m^2. The total cumulative dose must not exceed 600 mg/m^2. Because of its renal toxicity, cisplatin must be administered with hyperhydration (3 L/m²) associated with a forced diuresis by mannitol. At the highest doses of 200 mg/m², cisplatin can be administered in 3% hypertonic saline.

Toxicity

- Acute: myelosuppression (in particular anaemia), vomiting, hypomagnesaemia, hypocalcaemia, alopecia, peripheral neuropathy, anaphylaxis.
- Long term: ototoxicity, renal toxicity.

9.8.7 CARBOPLATIN

Mechanism of action

Carboplatin is an analogue of cisplatin with a similar cytotoxic mechanism of action and efficacy, but is less toxic.

Pharmacology

Carboplatin is administered intravenously. The pharmacokinetics of carboplatin are significantly different to those of cisplatin. Binding to plasma proteins is moderate. Renal glomerular filtration is the main route of excretion and accounts for 80% of total clearance. There is a good correlation between clearance of ultra filterable platinum and glomerular filtration rate, as measured by ⁵¹Cr EDTA clearance. It is, therefore, possible to estimate clearance for a patient and to calculate the dose to be administered. This allows one to control the inter-individual variability of clearance. Different formulae for calculating the dose have been proposed and validated, in particular for children [7,8]. The formulae for adults cannot be used for children.

Use

Carboplatin is used for the treatment of numerous malignant solid tumours such as neuroblastoma, nephroblastoma, germ cell tumours, brain tumours, and retinoblastoma. It is administered either at conventional doses of 300 to 800 mg/m² for several days in combination with other anticancer agents, or at high doses before haematopoietic stem cell transplantation. Unlike cisplatin, carboplatin does not require hyperhydration. When the dose is being calculated as a function of the glomerular filtration rate, the AUC target is, in general, 4 to 6 mg/min/ml in combination at conventional doses and in the order of 20 mg/min/ml during the administration of high doses. More recently, a weekly administration schedule has been developed for the treatment of brain tumours in young children.

Toxicity

- Acute: myelosuppression (particularly thrombocytopenia), vomiting, alopecia, nephrotoxicity.
- Rare toxicity: allergy (with a weekly schedule).
- Long term: deafness.

9.8.8 CYTARABINE

Mechanism of action

Cytarabine, or cytosine arabinoside (ara-C), is a cell-cycle-phase-specific antimetabolite and analogue of cytidine. Ara-C is transformed into a cytotoxic metabolite, ara-CTP, by successive intracellular phosphorylation. Ara-CTP is a competitive inhibitor of DNA polymerase. It also incorporates into DNA and alters its replication. Ara-C is rapidly inactivated (up to 80%) into ara-U by deamination in the liver, GI tract and leukocytes.

Pharmacology

Cytarabine can be administered intravenously, subcutaneously or intrathecally. After intravenous administration, the elimination half-life ranges from 1 to 3 hours. Binding to plasma proteins is weak (<15%). Cytarabine is widely distributed, in particular in the central nervous system. The CSF to plasma ratio ranges from 0.08 - 0.4 and depends on the duration of infusion. 90% of the administered dose is recovered in urine, mainly in the form of the inactive metabolite ara-U. Toxicity and efficacy correlate with the intracellular concentration of the active metabolite (ara-CTP) and not with the plasma levels of ara-C.

Use

Cytarabine is mainly used for the treatment of acute leukaemias and non-Hodgkin lymphoma. It is administered at doses of 2 to 3 g/m^2 every 12 hours for 6 consecutive days.

Toxicity

- Acute: myelosuppression, fever, vomiting, mucositis, diarrhoea, alopecia, skin rash; at high doses: conjunctivitis, cerebellar ataxia.
- Rare: encephalopathy, diffuse myalgias with fever.

9.8.9 ANTHRACYCLINES

Mechanism of action

Anthracyclines (doxorubicin, daunorubicin, epirubicin) are topoisomerase 2 inhibitors, which also intercalate in between the DNA base pairs. They induce cell death by apoptosis. Anthracyclines also generate free radicals toxic to the cells, in particular the striated muscular mononucleated fibres of the myocardium.

Pharmacology

Administration is intravenously. Doxorubicin distributes largely and rapidly into all tissues. After a very short distribution phase (5 min), the elimination is slow with a terminal half-life of 36 hours. Part of doxorubicin is transformed into doxorubicinol in the liver. Doxorubicin is mainly eliminated by biliary excretion as unchanged product and metabolites, representing 40 to 50% of the dose eliminated within 7 days. Urinary excretion is negligible. Doxorubicin does not cross the blood brain barrier. The cardiac toxicity of anthracyclines is not related to their cytotoxic mechanism of action.

Use

Doxorubicin and its analogue epirubicin are largely used in the treatment of paediatric solid malignancies. The doses differ from one tumour to another: from 15 to 60 mg/m^2 every 4 weeks for doxorubicin, from 40 to 100 mg/m^2 every 3 to 4 weeks for epirubicin. Daunorubicin is used in the treatment of acute leukaemias at doses of 30 to 60 mg/m^2 every 3 to 4 weeks.

Toxicity

- Acute: myelosuppression, vomiting, mucositis, diarrhoea, radiosensitivity, alopecia.
- Long term: cardiotoxicity, secondary acute leukaemias.

Cardiac toxicity

All anthracyclines are toxic to the heart and need regular echocardiograms (during and after treatment) measuring the systolic ejection fraction [9]. There are 3 types of cardiac toxicity:

- Acute during treatment.
- Non-acute during the year following treatment.
- Long term due to a rarefaction of myocardial fibres at the time of treatment. Cardiac failure occurs in adults during intensive muscular efforts, or during pregnancy.

Cardiac toxicity depends on the cumulated dose administered and on the duration of infusion (the risk is much higher for bolus injection, as compared to continuous infusion). The individual risk factors are yet to be identified. The guidelines to reduce the risk of cardiac toxicity are:

* Do not exceed the recommended maximum cumulative doses (doxorubicin < 550 mg/m^2).
* Do not administer anthracyclines as bolus injections, but rather by infusion of at least 6 hours, or even by continuous infusion of 24 to 48 hours. Infusion of more than 24 hours increases the risk of acute mucositis.

Agents to protect the myocardium, such as desrazoxane, are under evaluation for the child.

Anthracyclines must be injected intravenously. Any extravasation must lead to an immediate cessation of the injection.

9.8.10 ETOPOSIDE

Mechanism of action

Etoposide is a topoisomerase 2 inhibitor, which belongs to the chemical class of the epipodophyllotoxins. Inhibition of this enzyme induces DNA lesions, the accumulation of which triggers cell death by apoptosis. The cytotoxic effect depends on the concentration and the duration of exposure.

Pharmacology

Etoposide is administered orally or intravenously. Oral bioavailability ranges from 30 to 80%. Absorption is saturable at high doses. After intravenous administration, the elimination half-life is in the order of 7 hours. It is highly protein bound (94% binding) and the free active form is influenced by plasma albumin concentrations. Etoposide distributes largely in all tissues. Its distribution into the CSF is weak (CSF to plasma ratio < 0.01), but significant intratumour concentrations have been measured in brain tumours after intravenous administration. 30 to 60% of the administered dose is eliminated in urine with 25 to 50% in the form of unchanged product. Etoposide is also secreted in the saliva, which may explain the mucositis observed after high doses. Etoposide is partly metabolised by the cytochromes P450 and by glucuronidation.

Use

Etoposide is used in a large number of paediatric malignancies, such as acute leukaemias and lymphomas, neuroblastoma, nephroblastoma, sarcomas, germ cell tumors and brain tumours.

Three main modalities of administration are used in paediatric oncology:

- Sequential administration every 21 days, at doses of 150 to 500 mg/m², in combination with different anticancer agents.
- Continuous or semi-continuous administration, either for 21 consecutive days, or 3 days per week for 3 consecutive weeks [9].
- High doses (>1g/m²), either on its own, or in combination, before autologous or allogenic haematopoietic stem cell transplantation.

During the administration of high doses, the occurrence of severe mucositis is significantly reduced when etoposide is administered by continuous infusion, as compared to a repeated administration of short duration. The risk of secondary leukaemia is associated with the total dose and with the administration schedule [10]. The continuous and semi-continuous schedules have a higher risk of inducing secondary leukaemias.

Toxicity

- Acute: myelosuppression, diarrhoea, vomiting, alopecia.
- Rare: allergy.
- Long term: secondary leukaemias.

9.8.11 IMATINIB MESYLATE

Mechanism of action

Imatinib is an inhibitor of tyrosine kinase proteins. It inhibits bcr-abl, fusion protein expressed by the chronic myeloid leukaemia cells with Philadelphia chromosome. It also inhibits the PDGF (platelet-derived growth factor) receptors and c-kit, the SCF (stem cell factor) receptor [11].

Pharmacology

Administered orally, its bioavailability is 98% and its binding to plasma proteins 95%. The plasma half-life is 18 hours. There is a 1.5 to 2.5 fold accumulation at steady state. Imatinib is metabolised by CYP3A4 and its metabolism is inhibited by CYP3A4 inhibitors, such as erythromycin and fluconazole. Elimination is through the faeces, mostly as the unchanged form.

Use

Imatinib is used for the treatment of chronic myeloid leukaemias Philadelphia chromosome (bcr-abl) Positive (Ph+). Its activity is being studied in paediatric Ph+ acute leukaemias. Imatinib is administered orally to the child at doses of

240 mg/m^2 during the chronic phase (max dose 400 mg per day) and 340 mg/m^2 during the acute phase (max dose 600 mg per day) [4].

Toxicity

- Acute: myelosuppression, asthenia, facial oedema, hepatic cytolysis.
- Long term: still to be explored.

9.8.12 6-MERCAPTOPURINE (6-MP)

Mechanism of action and metabolism

6-mercaptopurine is an analogue of hypoxanthine. It is an inactive prodrug, which requires biotransformation to become cytotoxic, acting as an antimetabolite. Metabolism of 6-mercaptopurine is achieved through 3 different pathways. Activation is initiated by the hypoxanthine-guanine phophororibosyl transferase (HGPRT) with formation of cytotoxic metabolites, the 6-thioguanine nucleotides which are incorporated into the DNA and alter its replication. 6-MP is transformed into an inactive metabolite by thiopurine methyltransferase (TPMT) and by xanthine oxidase (XO) in the gastrointestinal tract (GIT) into 6-thiouric acid.

Pharmacology and pharmacogenetics

6-mercaptopurine is administered orally, but there is large inter- and intra-individual variability of 5 to 37% in bioavailability. The medicine is administered on an empty stomach. The peak concentration is reached after 2 hours. The plasma elimination half-life is in the order of 90 min, but the elimination half-life of the intra-cellular metabolites is very prolonged. Crossing of the blood brain barrier is poor. 6-mercaptopurine is excreted in the urine and in the bile.

TPMT exhibits gene polymorphism. Several mutations have been described. Three phenotypes can be described: one out of 300 individuals has an undetectable activity, 11% have an intermediate activity, whereas the remainder of the subjects have a high activity [5]. TPMT activity can be quantified within the red blood cells (check for the absence of a transfusion beforehand). This activity correlates with that observed in other tissues, such as lymphocytes, liver and kidney. Thus, metabolism of 6-mercaptopurine, and hence its efficacy and toxicity, is genetically determined. Patients having an undetectable activity of TPMT and receiving the usual dose are at high risk for developing severe toxicity [12].

Use

6-mercaptopurine is used for the treatment of acute lymphoblastic leukemias. It is administered orally daily at an initial dose of 75 mg/m². The daily dose is then adjusted as a function of haematological and hepatic tolerance. There is no correlation between the 6-mercaptopurine plasma levels and the observed effects. On the other hand, efficacy correlates with the red blood cell concentration of the active metabolites, namely the 6-thioguanine nucleotides. This is determined by the TPMT genotype of the patient, the administered dose and the duration of the treatment (a reversible increase of TPMT activity has been demonstrated during prolonged treatment). Children receiving maintenance therapy would theoretically benefit from red blood cell 6-thioguanine nucleotides monitoring, TPMT genotyping, and subsequent dose adjustment. However, the positive impact of such monitoring on survival has not yet been demonstrated [13]. Allopurinol significantly increases the peak concentration of 6-mercaptopurine.

Toxicity

- Acute: myelosuppression, hepatotoxicity, cholestasis, vomiting, mucositis.
- Rare: pancreatitis, fever with skin rash.
- Severe: for patients with TPMT deficiency.

9.8.13 METHOTREXATE

Mechanism of action

Methotrexate is a folic acid analogue that competitively inhibits dihydrofolate reductase (DHFR), depletes the intracellular pool of reduced folates, and finally inhibits DNA synthesis. It is an anti-metabolite, active in the S phase of the cell cycle.

Pharmacology

Methotrexate can be administered orally, intra-muscularly, intravenously, intrathecally and even intra-arterially. Following oral administration, the drug is rapidly absorbed and the peak plasma concentration is obtained at approximately 1 hour. Bioavailability is greater for small doses. Following intravenous administration, methotrexate distributes mainly into the liver, kidneys and, to a lesser degree, the GIT. Methotrexate is highly bound to plasma protein. It diffuses equally into the pleural effusions and ascites, from which it can be slowly released again into the circulation. Methotrexate does not distribute into the central nervous system, except following high doses. The CSF to plasma ratio ranges from 0.01 to 0.03.

Elimination of methotrexate is mainly renal through complex mechanisms of glomerular filtration, tubular secretion and tubular reabsorption. 40 to 80% of the administered dose is excreted unchanged in the urine. The elimination half-life ranges from 5 to 11 hours. Certain studies suggest that methotrexate clearance is higher in children than in adults.

Methotrexate is partly metabolised. Three metabolites have been identified in humans. DAMPA (4-deoxy-4-amino-N10-methyl-pteroic acid), an inactive metabolite, is formed by the bacterial flora of the gastrointestinal tract during the enterohepatic cycle of methotrexate. 7-hydroxymethotrexate is formed in the liver by P450 cytochromes. It exhibits a weak inhibiting activity on DHFR, which is approximately 200 times less than that of methotrexate. 7-hydroxy-methotrexate may interfere with the action of methotrexate by inhibiting its entrance into the cell and its biotransformation into polyglutamates. This metabolite is excreted in urine. 7-hydroxymethotrexate is less soluble in water than methotrexate and can precipitate in the renal tubules. Part of the intracellular methotrexate is converted into polyglutamates, which trap the drug in the cell, acting as intracellular storage for methotrexate. Methotrexate polyglutamates are powerful DHFR inhibitors. Methotrexate cytotoxicity correlates with the capacity of leukaemia blast cells to generate polyglutamates.

Following intrathecal administration, concentrations known to be cytotoxic _in vitro_ against leukaemia blast cells can be achieved in the CSF. However, the drug distribution within the upper cerebral spaces varies strongly from one patient to the other. That is why neuromeningeal prophylaxis calls for the administration of high doses of methotrexate (> 1.5 g/m^2). After intrathecal injection, methotrexate diffuses into systemic circulation and can induce systemic toxicity, in particular in case of repeated intrathecal injections. Polymorphism of the 5,10 methylene tetrahydrofolate gene may be associated with a higher toxicity risk.

Use

Methotrexate is used either at low doses for the maintenance therapy of acute lymphoblastic leukaemias (oral or IM), or at high doses (> 1.5 g/m^2) for the treatment of osteogenic sarcomas and malignant non-Hodgkin lymphomas. The duration of high dose methotrexate infusion ranges from 3 to 24 hours.
Administration of high dose methotrexate requires:

• An alkaline hyperdiuresis (pH > 7) to prevent the tubular precipitation of methotrexate and 7-hydroxymethotextrate that occurs in urine at pH < 7.

• The administration of folinic acid, that must always begin after the end of the perfusion, so as to ensure normal cell and tissue rescue by restoring the intracellular pool of reduced folates. Folinic acid administration must be extended until methotrexate plasma concentrations drop below 10^{-6} M.

• Strict clinical and biological monitoring focusing on the measurement of urine volumes and pH, monitoring of methotrexate plasma levels, and evaluation of the renal function (creatinine).

• Precise analysis of all the co-medications administered together with high dose methotrexate, since drug interaction may be responsible for severe toxicity.

In case of poisoning occurring during the administration of high doses of methotrexate, hyper-hydration is continued, the doses of folinic acid are increased, and one can use carboxypeptidase G2 [14,15], which is now commercially available (Voraxase®). This is an enzyme, which hydrolyses methotrexate into its non toxic DAMPA metabolite.

Several drugs inhibit the tubular secretion of methotrexate and/or of its metabolites and must not be used during the administration of methotextrate. These include salicylate derivates, penicillin, probenecid, sulphonamides and NSAIDs (indometacin, ketoprofen). The displacement of methotrexate bound to plasma proteins is another mechanism of drug interaction, which has been described in particular with sulfonamides and ketoconazole.

Toxicity

• Acute: myelosuppression, vomiting, mucositis, diarrhoea, alopecia.
• Preventable: acute renal insufficiency, hepatotoxicity.
• Rare: encephalopathy (potentialised by ionising radiations), allergy, anaphylaxis, interstitial pneumonitis, cerebrovascular stroke.

9.8.14 TEMOZOLOMIDE

Mechanism of action

Temozolomide is a methylating agent which is itself inactive. It undergoes a rapid chemical conversion at physiological pH to monomethyltriazenoimidazole (MTIC), the active compound. It alkylates DNA guanines.

Pharmacology

Temozolomide is administered orally. Its absorption is rapid, with a peak concentration reached within 30 minutes to 1.5 hours. The elimination half-life is short, 1.8 hours. Weakly bound to plasma proteins, temozolomide is eliminated mainly in the faeces in the form of inactive metabolites. It crosses the blood brain barrier and is largely distributed into the brain.

Use

Temozolomide is used in the treatment of relapsed malignant glial tumours. It is only administered orally, on an empty stomach, at a daily dose of 200 mg/m^2 for 5 consecutive days, every 28 days [16,17].

Toxicity

- Acute: myelosuppression (mainly thrombocytopenia), vomiting.
- Long term: not known yet.

9.8.15 VINCRISTINE AND OTHER SPINDLE POISONS

This group includes the vinca alkaloids (vincristine, vindesine, vinblastine), which are natural or synthetic alkaloids of the periwinkle plant.

Mechanism of action

Vinca alkaloids bind to tubulin and prevent the constitution of the mitotic spindle necessary for the migration of the chromosomes during the course of the metaphase and mitosis. These drugs are called spindle poisons and constitute the true antimitotic agent with anticancer therapeutic armantarium.

Pharmacology

Vinca alkaloids are only administered intravenously. Their elimination half-life is prolonged, from 9 to 155 hours. They distribute widely and are highly bound to plasma proteins and to the elements present in the blood (red blood cells, leukocytes, platelets). Elimination is essentially biliary, representing nearly 80% of the injected dose. 10 to 20% is eliminated in the urine. Diffusion into the CSF is poor.

Use

Vinca alkaloids are used for the treatment of acute lymphoblastic leukaemias and for many paediatric solid tumours. The dose of vincristine is 1.5 to 2 mg/m^2, without exceeding 2 mg, every 2 or 3 weeks. The dose of vindesine is 4 mg/m^2 every 7 to 10 days. The dose of vinblastine is 12 mg/m^2, usually by weekly administration.

In view of the major risk of cutaneous necrosis, injection of the vinca alkaloids must be intravenous. Any extravasation must lead to the immediate cessation of the injection, to a rinse of the vein and to the local subcutaneous injection of a mixture of sodium bicarbonate and hyaluronidase. One should consider urgent liposuction to remove the fatty tissues where the drug will accumulate.

Toxicity

- Acute: constipation, pain in the jaw, myelosuppression, alopecia.
- Rare: seizure, syndrome of inappropriate antidiuretic hormone (SIADH) secretion, cavernous sinus thrombosis, cutaneous toxicity.
- Chronic: peripheral neuropathy.

REFERENCES

1. Muller, H. J., Loning, L., Horn, A., et al., *Brit. J. Haematol., 110*, 379 (2000).
2. McCune, J. S., Gibbs, J. P., Slattery, J. T., *Clin. Pharmacokinet., 39*, 155 (2000).
3. Vassal, G., Deroussent, A., Challine, D., et al., *Blood, 79*, 2475 (1992).
4. Kerbusch, T., de Kraker, J., Keizer, H. J., et al., *Clin. Pharmacokinet., 40*, 41 (2001).
5. McLeod, H.L., Krynetski, E.Y., Relling, M.V., Evans, W.E., *Leukemia, 14*, 567-572 (2000).
6. Bin, P., Boddy A. V., English, M. W., Pearson, A. D., et al., *Anticancer Res., 14*, 2279 (1994).
7. Peng, B., Boddy, A. V., Cole, M., et al., *Eur. J. Cancer, 31A(11)*, 1804 (1995).
8. Chatelut, E., Boddy, A.V., Peng, B., et al., *Clin. Pharmacol. Ther., 59*, 436 (1996).
9. Pein, F., Vassal, G., Sakiroglu, C., et al., *Archives Pédiatriques, 2*, 988 (1995).
10. Le Deley, M. C., Leblanc, T., Shamsaldin, A., et al., *J. Clin. Oncol., 21*, 1074 (2003).
11. Champagne, M. A., Capdeville, R., Krailo, M., *Blood, 104*, 2655 (2004).
12. Relling, M. V., Hancock, M. L., Rivera, G. K., et al., *J. Natl. Cancer Inst., 91*, 2001 (1999).
13. Schmiegelow, K., Bjork, O., Glomstein, A., et al., *J. Clin. Oncol., 21*, 1332 (2003).
14. Widemann, B. C., Balis, F. M., Murphy, R. F., et al., *J. Clin. Oncol., 15*, 2125 (1997).
15. Widemann, B. C., Balis, F. M., Shalabi, A., et al., *J. Natl. Cancer Inst., 96*, 1557 (2004).
16. Estlin, E. J., Lashford, L. Ablett, S., et al., *Br. J. Cancer, 78*, 652 (1998).
17. Lashford, L. S., Thiesse, P., Jouvet, A., et al., *J. Clin. Oncol., 20*, 4684 (2002).

Index

A

absence, 65, 161, 238, 250, 291, 306, 314, 323-324, 327, 353, 355, 367, 450, 455, 457, 383, 437, 493, 508, 519, 523, 566, 570, 634, 712, 722, 731, 742, 755, 786

absorption, 10, 97, 117, 123-124, 136-137, 139, 148, 150-151, 155, 169-170, 180, 187, 192, 204, 207, 215, 228-229, 232, 236, 239, 284, 310, 351, 365-368, 374, 403, 406, 418, 421, 426, 440, 503, 529, 581, 596, 615, 623, 625, 634-635, 637, 644, 657, 662, 665, 667, 671-673, 676, 678-679, 687, 711-712, 723, 784, 789

abuse, 206, 227, 278, 604, 689, 758, 763

acceptability, 219-220, 222-224, 268

acetylation, 87, 98, 111, 125-126, 530

acetylcholine, 214, 321, 540, 554, 556, 592

acetylcholinesterase, 554

n-acetyltransferase type 2, 167

adaptive design, 60

addiction, 326, 348

adenocorticotropin, 315

adenosine receptors, 327

adhesion molecules, 639, 646

adolescent, 26, 36, 57, 80-81, 126, 270, 654, 687, 690, 697, 699-700

adrenocortical axis, 562

adverse drug reactions, 15, 118, 172, 191-192, 196, 246, 248, 562, 633-634, 637, 640

age groups, 21, 37, 51, 63, 74, 88-90, 222, 515, 523, 562, 592, 677, 708, 713

agonist, 227, 404, 561, 598, 600, 603-605, 615, 761

alarm system, 760

albumin, 125, 193, 207, 283-284, 350, 369-370, 429, 529, 548, 676, 741, 748, 767, 784

allergens, 703-704

allergy, 221, 273, 373, 446, 466, 483, 642, 662, 678, 710, 782, 785, 789

amnesic disorders, 692

amniotic fluid, 286, 311, 313, 338

amoebiasis, 668

anaesthesia, 119, 194, 199, 215, 226, 287, 303, 562, 581, 583-587, 589, 600, 602-603, 605-607, 610, 613, 617, 619, 624, 627, 644

analgesia, 104, 194, 228, 334, 371, 404-405, 414-415, 537, 561, 584-586, 589, 595-602, 604-605, 608, 613-617, 621-622, 625

angiotensin, 164, 336-337, 398, 727, 731, 739, 751

angiotensin converting enzyme, 336, 731, 751

animal model, 298, 540

animal study, 297

anorexia nervosa, 577, 578, 697, 731

antagonist, 313, 585, 590, 598, 604, 614, 648, 667, 723, 729, 735, 752

antibodies, 109-110, 215, 301, 313, 493, 504-506, 640, 658, 777

anticholinergic effects, 599, 691

antimicrobial agents, 67, 154, 484, 513

antipsychotic drugs, 77, 164, 694-695

antiretroviral agents, 183, 319

antituberculous medications, 183

antiviral agents, 155, 504

anuria, 317

anxiety, 277, 327, 582, 662, 678, 680, 688-689, 691-693, 697, 757, 759

anxiolysis, 561, 615

apnea, 181

apoptosis, 657, 775, 783-784

arachidonic acid, 338, 397, 646, 650

area under the concentration-time curve, 123, 149, 157, 187
area under the curve, 140, 183
arousability scale, 563
arthropathy, 279, 482
Asperger disorder, 696
astrocytes, 327
attention-deficit and hyperactivity disorder, 57, 77, 221, 269, 324, 325, 578, 583, 615, 687, 688, 758,
autistic disorder, 583, 696, 758

B

Baye's theorem, 61
bayesian analysis, 56, 60, 154
bedwetting, 755-757, 759-760, 762
benefit, 21, 31, 42-45, 47, 50-51, 58-59, 65, 140, 171, 179, 254, 318, 346, 353, 380, 382, 399-401, 427, 432, 509, 515, 523, 540, 575, 581, 647, 649, 679, 688, 696, 709, 712, 735, 741, 750, 767, 771, 787
benzo(a)pyrene, 301
best pharmaceuticals for children act, 23, 33, 65, 71
ß-blockers, 99, 164, 211, 213, 287, 723-724, 727-728, 734-735
ß-lactamases, 437, 469, 497, 498
bias, 56-57, 80, 293-294
bilirubin, 168, 193, 283, 334, 370, 453, 400, 429
binding sites, 334, 369-370, 475
bioavailability, 4, 90, 123-124, 135-136, 144, 148, 169, 171, 204, 222-223, 228-229, 239-242, 250, 271, 353, 356, 367-368, 400, 406, 421, 440, 507-508, 510, 542, 593, 598-599, 604-605, 614-616, 622, 626, 633, 648, 656, 706-707, 747, 776-777, 784-787
biotechnology, 20, 236, 292, 506
biotransformations, 283
birth defects, 330
bladder, 485-486, 488, 755-763, 778-780
blood brain barrier, 169, 367, 370, 776, 779-780, 783, 786, 789
blood pressure, 178, 245, 255, 335, 372, 394, 419, 421, 563, 615, 638, 650, 694, 719, 722, 724-725, 732, 739, 741, 757

bone marrow suppression, 442, 474, 479, 637
brain development, 359, 321-322, 324-325, 327-328, 331, 459
breast-feeding, 344-345, 349, 351, 353-356
breath test, 52, 101-105

C

caesarean section, 287, 303
calcineurin, 633-634, 636, 772
calmodulin, 633, 666
cancer, 95, 153, 156-158, 163, 167, 171, 176, 189, 267, 273, 598, 658, 775, 791
cardiotoxicity, 634, 691, 779, 783
catalepsia, 689
cerebral haemorrhage, 399-400
chemokines, 639
chemotherapy, 157, 167, 254, 267, 503, 747, 775
chlamydia trachomatis, 439
chlamydiae, 473, 474, 476, 482
chlamydiae and rickettsiae, 473, 476, 483
choroid plexus, 491
clearance, 88-94, 97, 104-108, 118, 126, 129-130, 133, 137-139, 141, 144, 148, 152-153, 155-157, 170, 177-178, 180-181, 183, 186, 205, 209-211, 360-361, 284, 301-302, 304, 311, 336, 340, 370-371, 458, 400-401, 404-406, 419, 426, 429, 440, 443, 519, 523, 547-548, 550, 560, 592-594, 596, 598-605, 614-616, 623, 625-626, 632-637, 661-662, 684, 706, 730-731, 777, 781, 788
cleft lip /palate, 295, 314, 329
clinical trial, 38-39, 49-52, 56-57, 64, 67-71, 184, 562, 570, 662
clostridium difficile, 471
clostridium spp, 478
Cochrane Reviews, 82, 346, 348, 382, 384-385, 389, 391, 395, 414-415, 432-434, 513, 580, 618, 663-664, 699, 718, 753, 763-764
collaboration, 23, 67, 69, 198, 415
colonisation, 435, 485, 491
colostrum, 350
comfort scale, 563

competence, 19, 39, 45, 344
compliance, 80, 177, 179, 183, 187, 223, 231, 236-237, 242, 265-270, 273, 360, 377, 389, 632, 657, 679, 688-689
computer programmes, 143
concentration, 37, 88, 117, 123, 125, 129-134, 136-145, 148-149, 151-153, 158, 178-182, 184-187, 204-207, 221, 226, 236-237, 281, 283-285, 287, 303-304, 310, 314, 329-330, 334, 350-353, 356, 366-370, 400, 418, 420-421, 426, 428, 441, 445, 494, 496, 499, 508, 518, 522-523, 547-549, 554, 556, 590, 595, 598-600, 602-603, 613-615, 622, 625-626, 632, 634-635, 637-638, 672, 676, 684, 687, 706, 716, 731, 747, 751, 759, 773, 777-778, 782, 784, 786-787, 789
concordance, 4, 223, 231, 265-266, 268, 270-273, 292, 297
conduct disorder, 696
congenital adrenal hyperplasia, 314-315
conjugation, 87, 92, 98, 165, 177, 193, 205, 214, 283, 371, 400, 426, 530, 602, 751, 779
consent, 3, 38-39, 43-47, 63, 71, 361
contra-indications, 6, 354, 382, 384, 431, 531, 658, 689
convulsions, 203, 228, 313, 344-347, 450, 453-456, 458, 460, 585, 709, 721, 761-762
corticosteroids, 75, 148, 227-228, 316, 335, 454-455, 377, 382, 385, 392, 509, 532-533, 632, 637, 639-640, 644, 646-647, 656-657, 704-710
cotyledon, 284
covariate, 151
cpmp, 20, 22, 24, 64, 235, 243
Crohn's disease, 653
cyclooxygenase, 316, 338, 372-373, 426, 428, 431, 639
cyclophilin, 633
cystitis, 485, 488, 771, 778-780
cytoarchitectonics, 324, 326
cytochrome P450, 87-88, 90, 98, 164, 177, 283, 299-300, 406, 443, 625, 716
cytokine receptors, 633

cytokines, 540, 633, 636, 639, 646, 650
CYP1A, 52, 87-92, 97, 99, 109, 164, 169, 172, 300, 301, 306, 311, 716
CYP2A, 166
CYP2C, 98, 300, 306
CYP2D, 306, 311
CYP2E, 300
CYP3A, 88-90, 98, 104-111, 116-117, 164-165, 170, 302, 306, 311, 371, 632-634, 637
cytotoxicity, 788
cytomegalovirus, 510
cytotoxic agents, 14, 136, 224, 637, 778, 780

D

data requirements, 150
database, 39, 69, 348, 513, 580, 618, 664, 699, 753
declaration of helsinki, 42, 47
dec-net, 69-70, 72
delivery, 19, 76, 205, 222, 224, 240, 242-243, 250, 270, 272, 287-289, 312, 329, 333-335, 340, 347, 351, 367, 380, 383-384, 389, 392-393, 431, 435-436, 518, 522, 540, 542-543, 548, 550, 664, 681, 709, 711, 717
demethylation, 90-91, 106, 125-126, 165, 400
dependence, 383, 604, 614, 772
depression, 77, 80, 82, 197, 278, 360, 313, 334-335, 404-405, 562, 579, 585, 589, 595-598, 600-601, 603-605, 614-615, 646, 690, 692, 743
design, 43, 46-47, 49-51, 53, 56-58, 60-62, 118, 150, 193, 246, 292, 506
detoxification, 98-99, 165, 194, 206
detrusor, 755, 761
diabetes mellitus, 650, 671, 757
diaphragmatic hernia, 317, 405, 541
diarrhea, 669
diffusion, 281-282, 286, 302, 350-351, 469, 507, 542, 548, 746, 790
dihydrofolate reductase, 480, 787
disposition, 32, 123, 161, 170, 175, 193, 360, 292, 302, 310-311, 315, 317, 351, 371, 465, 545-546, 554, 594, 632

distress, 284, 287, 309, 367, 377, 379-380, 384, 387, 395, 425, 559, 617, 692, 753, 757

Division of Pediatric Drug Development, 23

documentation, 252, 259, 306

dopamine, 170, 210, 306, 321, 324, 326, 373, 401, 574, 662, 687, 694, 739, 742

dopamine receptors, 694

dosage regimen, 148-149, 154, 156, 158, 161, 174, 178, 180

dosage requirements, 147, 151, 554

ductus arteriosus, 316, 339, 373, 377, 399, 425, 426, 429, 431, 539, 743

duplication, 39, 65, 67, 69, 169

duration, 51, 261, 268-269, 292, 304, 351-352, 455, 458, 405, 418, 427, 429, 439, 486, 488, 493, 500, 506-507, 510, 512, 519, 532, 557, 562, 591-592, 600-605, 616, 631, 658, 666-668, 672-673, 676-677, 693, 695, 716, 723-724, 732, 749, 751, 766, 778, 782, 784-785, 787-788

dysgenesis, 567

dyskinesia, 486, 506, 541, 694

dysmorphic features, 278, 296, 323, 324

dysplasia, 405, 425-426, 720

dysthymic disorder., 690

dysthyroidism, 315

dysuria, 485

E

e. coli., 436, 485, 667-668, 776

eclampsia, 317

education, 233, 246, 261, 269-270, 272-273, 575, 665, 703, 725

efficacy, 3, 14, 19-22, 34, 37, 41-42, 53, 56-59, 63-65, 69, 73, 77, 80, 136, 140, 142, 163-164, 167, 178, 183-184, 206, 235, 250, 312, 458, 403, 410, 413, 417-418, 420, 426-427, 431, 441, 444, 492, 494, 513, 518-519, 522, 532, 548, 586, 605, 613-614, 632, 634-635, 638, 640, 649, 651, 655-658, 662-663, 685, 688-689, 693-694, 696-697, 748, 770, 772-773, 781-782, 786-787

electroencephalogram, 346, 563, 569, 581, 589, 602

elimination, 37, 98, 126, 130-134, 137-138, 140-141, 145, 148, 153, 169-170, 180, 183, 194, 205-206, 208, 210, 225, 284, 286, 300, 310-311, 317, 335, 347, 351, 353, 356, 367, 372, 403, 406, 421, 426, 440, 444, 494, 512, 518-519, 550, 560, 594, 598-599, 601-605, 614, 616, 633, 637-638, 684, 716, 728, 730, 732, 776, 778-780, 782-786, 788-790

embryo, 165, 277, 291, 292-298, 304, 327, 329, 337, 338, 359

EMEA, 3, 19-22, 24, 29-30, 32, 35, 50, 64, 158, 235, 243

encephalopathy, 196, 453-454, 457, 459-460, 568, 667, 684, 721, 782, 789

endocytosis, 282, 504

endogenous digoxin-like substances, 181, 187

endothelin, 727

enterobacteriaceae, 437

enterococci, 437, 471, 482, 486

enuresis, 77, 228, 755-763

epilepsy, 13, 52, 164, 183, 197, 219, 267-268, 270, 309, 329-330, 449, 454-455, 459-461, 565-570, 575-576, 578-580, 689, 692, 696, 747

epileptic aura, 567

error, 58-59, 151, 223, 235, 246, 248-259, 262-263, 458, 595

erwinia, 776

ethanol, 202, 205-206, 211-212, 231, 237, 277, 323-325, 328

ethics, 23, 38-39, 42-43, 46-47, 50, 56, 69, 121, 291

escherichia coli, 436, 485, 667-671, 776

European Agency for the Evaluation of Medicinal Products, 20, 29

evaluation, 1, 3, 20, 23, 26, 29, 31, 33, 36, 38, 57, 64, 88, 123, 169, 178, 235, 284, 291-292, 299, 309, 312, 318, 452, 409, 549, 585, 636, 640, 659, 668, 685, 690-691, 697, 704, 721, 757, 763, 775, 784, 789

exacerbation, 270, 651, 706

excipient, 232

exclusivity, 26-30, 33, 64-65, 361

expertise, 20, 29, 49, 230
extemporaneous preparation, 223, 228-229, 239, 272
extemporaneously dispensed medicines, 4
extracorporeal membrane oxygenation, 157, 404, 539-541, 545

F

fasciculation, 201, 556
fast-dissolving drug formulations, 224, 272
FDA, 23, 25-28, 32-33, 35, 50, 64-65, 82, 150, 154, 158, 540
fetal liver, 88, 110, 164, 301-302, 311
fetal metabolism, 299
fetal safety, 277-279
fetal toxicity, 291-292, 294, 299
fever, 187, 485-487, 507, 510, 512, 527, 531-532, 534, 560, 621-622, 624-625, 650, 655, 657, 692, 743, 747, 782, 787
Finnegan score, 344-346
first-order kinetics, 132-133
flavin adenin dinucleotide, 540
flavin mononucleotide, 540
Food and Drug Administration, 22-23, 25-26, 35, 71, 82, 150, 154, 158, 540
fetus, 90, 104, 108, 109, 111-116, 126, 164, 227, 279, 281-289, 291-306, 309-316, 321-326, 328, 337, 338, 359, 360, 370, 373, 378, 382, 397, 398, 479, 515, 519, 520, 522, 593
folates, 787, 789
follow up, 314, 530, 654
formulation, 4, 21, 38, 124, 127, 135, 195-196, 204, 219, 224, 229, 232, 235, 238-240, 242, 255, 260, 269, 271-272, 365, 368, 421, 429, 431, 569, 596, 603, 624, 626, 663, 697-698
fungal infections, 435, 439

G

gastroesophageal reflux, 59, 419, 661-663
genotype, 105, 126, 155, 161, 163-164, 168, 170, 173-174, 178, 186, 522, 787
gingivitis, 634
gliogenesis, 324, 326

glomerular filtration, 126, 138, 153, 284, 302, 310-311, 317, 337-338, 351, 371-372, 398-400, 428, 440, 550, 594, 684, 744-745, 780-781, 788
glucocorticoids, 100, 169, 301, 315, 335, 336, 378-384, 639, 742, 750
glucuronidation, 87, 88, 92-94, 97-100, 108, 116, 119, 125, 168, 177, 301, 371, 404, 518-519, 596, 638, 784
glucuronosyltransferase, 193
glutathione, 87, 98, 203, 212, 214, 299, 625, 777
gluthathione S-tranferase, 165, 171
glycopeptides, 183, 441, 444, 470-471
good clinical practice, 21, 42, 64, 71, 436
goitre, 316
gram-negative, 370, 466, 469, 471, 473, 478-482
grapefruit juice, 135, 177
grey baby syndrome, 97, 125, 177, 193, 475
growth, 36-37, 73, 80, 226, 296-297, 300, 309, 316-317, 323-324, 326, 329, 331, 336-337, 372, 378, 380, 382, 400, 420, 465, 474, 504, 632, 638, 640, 654, 656-658, 667-668, 683, 686, 707-709, 712, 717, 750, 757, 768, 785
growth hormone, 226, 300, 683, 686
growth retardation, 296, 309, 316-317, 323-324, 326, 329, 640, 707-708
guidance, 22, 24, 26, 33, 35, 50-51, 64, 71, 158, 231, 652
guideline, 22, 35, 42, 64, 232, 720

H

haematuria, 485, 766
haemophilus, 483, 492, 529, 531
half-life, 93-94, 126-127, 130, 139, 141, 145, 170, 194, 360, 284, 286, 311, 315, 317, 327, 340, 347, 356, 372, 457-458, 378, 400, 406, 421, 426, 429, 440, 488, 494, 507-508, 518-519, 548-549, 560-562, 594, 596, 599, 601-605, 614-615, 623-625, 633, 637, 691, 724, 728, 730, 732, 746-747, 751, 776-780, 782-786, 788-790

hallucinations, 560, 689, 695
harmonisation, 19-20, 22, 35-36, 47, 50, 63-64, 74, 88, 291, 295, 298
healthy volunteer, 53, 168, 224, 560
hepatitis C, 345, 511
hepatotoxicity, 97, 193-194, 203, 214, 406, 442, 562, 576, 624-626, 637, 655, 776, 787, 789
high-frequency oscillatory ventilation, 541
hirsutism, 270, 633-634, 743, 751
holistic treatment, 578
human immunodeficiency virus, 187, 515-516
hydramnios, 316, 373, 399
hydroxylase deficiency, 314-315
hyperemesis gravidarum, 278
hypoglycaemia, 194, 202-203, 211, 360, 370, 453-454, 456, 672-673, 676, 678-681, 709
hyperlipidaemia, 637, 743
hypothalamo-pituitary-adrenal axis, 709
hypoxemia, 195
hypsarrhythmia, 568

I

immune system, 36, 118, 435
immunisation, 316, 505, 531-535, 651
immunoassays, 178, 181
immunosuppression, 532-533, 562, 632-633, 637, 640
importation, 5, 223, 271
in silico, 117-118, 173
incontinence, 756-757
individualisation, 157, 177-180, 183, 186, 641
infantile spasms, 452, 568, 570-571
infection, 183, 194-195, 312-313, 453, 380, 410, 422, 435-437, 439, 441, 444, 479-482, 485-486, 488, 491, 501, 503, 505-507, 510-513, 515, 523, 528-529, 534, 640, 650, 657-658, 668, 711, 713, 716-717, 757, 763, 766-767, 771
infectious agents, 273
inflammation, 398, 410, 486, 488, 646-647, 712-713, 717, 727

inflammatory bowel disease, 650
influenza virus, 505
inhalation, 202, 204, 222-223, 243, 294, 507, 705-706
inhaler, 76, 227, 231, 243, 261, 270-271, 705-706, 709
inhibitor, 13, 117, 177, 211-212, 314, 316, 327, 339, 399, 426, 494, 512, 518, 542, 621, 633, 638, 667, 693, 712, 729, 732, 735, 746, 772, 782, 784-785
inspection, 58
instability, 193, 208, 439-440, 755-756, 761, 763
insulin-like growth factor, 683
intestinal motility, 123, 284, 310, 366, 666
intracranial haemorrhage, 453, 568
intramuscular route, 124, 226
intranasal route, 222
intravenous route, 225, 252, 367, 400
intubation, 392-393, 405, 410, 553, 556-557, 601-602
investigators, 49, 51-52, 55-56, 58, 65, 103, 109, 300, 336, 339, 457, 548
ionisation, 123, 281, 310, 350, 366
IQ level, 325
isotope, 52, 102, 104
investigators, 49, 51-52, 55-56, 58, 65, 103, 109, 300, 336, 339, 457, 548

J - K

jacksonian attack, 450
klebsiella, 486
kwashiorkor, 529

L

laryngitis, 661
legislation, 4, 25-26, 29-30, 32-33, 36, 41, 64-65, 70-71, 311
leishmaniasis, 528
lennox-gastaut syndrome, 568
leukemias, 787
leukomalacia, 329
life-threatening disease, 50
limited sampling strategy, 632
lipid solubility, 281, 283, 310, 590-591, 597-598, 600, 602, 616

lipoatrophy, 678
lipohypertrophy, 678
liposolubility, 126, 281, 282, 310, 590-591, 597-598, 600, 602, 616
liposomes, 236
Lipshitz score, 345
lung maturation, 335, 378
lymphomas, 777, 784, 788

M

maintenance, 127, 194, 235, 267, 337, 456-459, 421, 427, 528, 548, 562, 570-571, 602, 625, 655-656, 688, 693, 716, 768-769, 772, 787-788
malaria, 527-529
malformations, 173, 295-296, 298, 305-306, 309, 311, 323-327, 329-330, 377, 486
malignancy, 640, 657, 663, 771-773, 775
malnutrition, 245, 527, 529, 624, 654
marasmus, 529
marketing authorisation, 4, 5, 19, 21-23, 41-42
maternal-fetal ratio, 303
maturation, 36, 90-91, 97, 103, 105, 118, 123-124, 126, 161, 163, 174, 181, 227, 296, 301, 304, 335, 337, 378, 403, 417, 444, 504, 592, 594, 623-624, 632, 707-708, 759
medication error, 246, 248-250, 257-259, 262-263, 595
medulloblastoma, 777
meningitis, 187, 255-256, 453, 455, 437, 439, 483, 491-500, 506, 527, 529
meningo-encephalitis, 453, 568
mental retardation, 313, 323-324
mesial temporal lobe epilepsy, 568
meta-analyses, 55, 345, 389, 488
metabolisers (extensive, poor), 155
metabolism, 52-53, 55, 87-91, 93-94, 97-98, 100, 103-104, 106, 108-109, 111, 117-119, 125, 135, 141, 145, 148, 155, 161-165, 168, 170-171, 173-175, 177, 180, 193-194, 198, 203, 205-207, 211-212, 214, 221, 287, 299-302, 306, 310, 351, 354, 365, 368, 371, 453, 458, 387-

388, 403, 405, 426, 440, 529, 550, 556, 596, 598-599, 602, 604-605, 613-614, 622, 634, 637, 650, 656-657, 683, 687, 706, 776, 778-779, 785-786
metanephros, 337
metaregister, 70, 72
methaemoglobinaemia, 192, 662
methodology, 38, 147-148, 150-151, 154-155, 158, 280, 488
methylation, 87, 98, 125, 476
methylxanthine, 418-419
Michaelis-Menten, 117, 132, 142
microarrays, 174
microassays, 37, 51
microcephaly, 326, 329
microsomes, 105, 108-110, 116, 165, 302
milk-plasma ratio, 356
minimal bactericidal concentration, 441, 494
minimal inhibitory concentration, 441, 496
minitablets, 240-241
mitrochondria, 563
model, 13, 109, 130, 133-134, 137, 139-140, 142-143, 151-153, 298, 373, 398, 494, 528, 530, 540, 549, 593, 622-623
modified release preparations, 225
molecular weight, 281-282, 285, 310, 350
monitoring, 15, 19, 53, 58, 67, 74, 140, 147, 150, 154-156, 177-178, 181, 183, 186-187, 198, 208, 210, 225, 245-246, 262, 302, 309, 450, 452, 456-457, 400, 404, 418, 420, 440, 444, 523, 528, 557, 563-564, 581-583, 632, 634-635, 637-638, 641, 649-650, 691, 694, 719, 722, 725, 731, 787, 789
monotherapy, 76, 184, 314, 330, 479, 569, 571, 645
Motherisk program, 277-280, 359
mucositis, 598, 778-779, 782-785, 787, 789
muscle relaxants, 288, 553
mutism, 693
mycobacteria, 482
mycobacterium tuberculosis, 529
mycoplasma, 473, 476

myelosuppression, 168, 771, 776, 778-783, 785-787, 789-791
myoclonic epilepsy, 568, 570, 576
myoglobinuria, 556

N

nanoparticle, 236
narcolepsy, 689
nasogastric route, 228, 421
nebuliser, 76, 219, 261, 705
needle-free injector, 226, 236
neisseria, 472, 492, 495, 497, 529
neisseria gonorrhoea, 529
neonatal abstinence syndrome, 343, 347
neonatal convulsions, 453-455, 458, 460
neonates, 7, 10, 51-52, 73-74, 88, 91, 93, 97, 103, 105, 107, 117, 123-126, 136, 138-139, 148, 150-153, 155-157, 168, 181, 183, 193, 195-197, 221, 226, 250-251, 279, 286, 302, 317, 334-335, 339, 343, 347, 363, 365-368, 370, 372, 449-454, 457-458, 461, 388, 397, 399-400, 403-406, 409-410, 413-415, 417-418, 421-422, 429-431, 435-436, 438, 440, 443-444, 475, 508, 515, 518-520, 522, 524, 545, 547-549, 590, 592, 594-595, 599-606, 614, 623, 625-626, 663, 728, 732, 743, 747-748, 752
nephroblastoma, 776, 779, 781, 784
nephrogenesis, 337-338, 397, 400, 440
nephrotoxicity, 440, 442, 633, 657, 716, 773, 780, 782
neural tube defects, 329-330
neuroblastoma, 777-778, 780-781, 784
neuroleptic malignant syndrome, 694
neuromuscular blockade, 553-554, 556-557
neuromuscular junction, 554
neuronal migration, 326
neuropathic pain, 615
neurotransmitter, 203, 329, 404, 589, 603
newborn, 26, 36, 43-44, 46-47, 88-90, 117, 148, 165, 168, 177, 191-193, 195, 224, 228, 286, 301, 304, 312, 324, 326, 331, 335, 340, 344-345, 347-349, 351, 353-356, 365, 368, 370, 372-373, 449-450,

459-460, 387-388, 409-410, 414, 435-436, 440, 446-447, 508, 510, 518, 522, 540-541, 554, 749, 752
Nissen's fundoplication, 663
nitric oxide, 337, 398, 425, 539-540, 542-544, 621
nitric oxide synthase, 394, 540
nitrosohaemoglobin, 542
non-invasive method, 52, 102, 104, 105, 181, 318, 563,
non-compartmental analysis, 130
non-invasive methods, 52
noradrenaline, 613, 691-692, 734
norepinephrine, 687, 693, 727
Nuremberg code, 42, 47

O

obsessive-compulsive disorder, 77, 80, 692
oesophagitis, 661, 663
off-label drug, 5, 15
oliguria, 339, 399, 427, 429, 733
ontogeny, 89, 92, 94, 97, 109, 111, 116-118, 125-126, 139, 167, 365, 371, 632
opioid receptor, 404, 590, 613
opportunistic infections, 435, 524, 650
opsonisation, 493
oral route, 223, 236, 421, 648
organ transplantation, 631, 636-637, 640
orodispersible tablet, 224
orphan drug, 33
osmolality, 224, 366, 661, 739, 746, 753
oxidation, 87, 92, 98, 105, 162, 168, 203, 205, 214, 301, 421, 551, 563, 576

P

paediatric board, 64
paediatric expert group, 22, 29
paediatric regulation, 23
paediatric research equity act, 28, 65, 71
paediatric rule, 27-29, 64
pain, 5, 13, 39, 44, 52, 124, 156, 225-226, 269, 403-406, 409-415, 485, 509, 583, 589, 591-592, 595, 598-599, 601, 603, 605-611, 615-618, 621-622, 626, 662-663, 688, 741, 761, 791

pain scores, 589, 622
palatability, 229, 268
pancreatitis, 637, 657, 748, 750, 777, 787
parallel-groups, 58
parameters, 51, 117-118, 129-130, 138-139, 147-148, 151-152, 158, 173, 250, 281-282, 284, 298, 355, 418, 426, 441, 523, 551, 594, 623-624, 632, 635-636, 684, 719, 757
parasitic infections, 439, 529
parasomnia, 758
parenteral administration, 38, 194, 195, 236, 294, 344, 365, 457, 487, 532, 550, 590, 596, 613, 615, 624, 648
participation, 71
passive diffusion, 281-282, 302, 350-351
patch, 227, 601, 615
patient-controlled analgesia, 334, 596, 598
pediatric rule, 27-29, 64
penicillin binding proteins, 465
perfused placenta, 284, 286
peridural anaesthesia, 287
perinatal transmission, 518, 519
persistent pulmonary hypertension, 539-540
pervasive developmental disorders, 696
p-glycoprotein, 103, 106, 125, 135, 169, 286, 361, 367, 369, 453, 633
pharmacodynamic, 37, 51-52, 63, 73, 142-143, 149, 151, 157-158, 365, 397, 403, 435, 440-441, 548, 589, 592, 597, 621-622, 632, 635, 638, 684
pharmacogenetics, 87, 97, 119, 122, 161-163, 169-173, 175-176, 309, 318, 406, 409, 578, 606-607, 631-632, 637, 641-642, 786
pharmacokinetic model, 549
pharmacokinetics, 51, 91-92, 94, 97, 108, 117, 127, 129-130, 145, 147-149, 152-153, 155-156, 158, 169-170, 174, 177, 187, 204, 250, 284, 289, 306, 310-312, 317, 340, 351, 400-401, 404, 418, 421, 426, 428-429, 440, 442-443, 465, 482, 515, 518-519, 522, 545, 547-549, 560, 592-594, 596, 599-600, 613, 615-616, 622, 632-634, 636-638, 641, 687, 706, 716, 746-747, 750-751, 776, 781

pharmacovigilance, 20, 22, 32, 53, 64, 172, 191, 196-198, 564
phase-1 reactions, 97, 125, 162
phase-2 reactions, 87, 98, 125, 162-163, 28
phenotype, 104-105, 126, 155, 161, 164-165, 167, 170, 173-174, 323
phosphatidylcholine, 387
pKa, 283, 310, 350, 366, 590, 602, 623
placebo, 51, 57, 77, 80, 360, 418, 420, 506-507, 509, 518, 600, 622, 640, 655, 657-658, 689-690, 693, 695-696, 761-762, 770
placenta, 126, 360, 282-288, 302-303, 310, 315-316, 321, 329, 333, 373, 378, 399, 751
placenta praevia, 329
placental transfer, 102, 281-289, 310, 312, 369, 599
plasmid, 437
plasmodium falciparum, 529
pneumococcus, 500
pneumocystis carinii, 480
polyglutamates, 788
polymorphism, 105, 115-116, 163-164, 167-169, 177, 597, 637, 657, 786, 788
polyuria, 339, 399, 419, 758
population pharmacokinetics, 51, 147
porins, 469, 473
potency, 142, 335, 404, 493, 548, 562, 592, 596-598, 600, 602, 604-605, 615-616, 646, 656, 723-724
power, 28, 55, 65, 153, 279, 563, 623, 716
Prader Willi syndrome, 685
pre-eclampsia, 173, 317, 373
pregnenolone-16α-carbonitrile, 301
premature, 26, 36, 107, 125, 133, 136, 138, 148, 155-157, 181, 195, 251, 255, 316, 329, 347, 371-373, 451-452, 383, 397, 399-401, 405-406, 409, 413, 417, 420-421, 425, 431, 440, 444, 475, 506, 519, 543, 576, 593-594, 602, 605, 623, 626, 729, 733, 747-748, 761
premature rupture of membranes, 329
premedication, 228, 581, 601, 616, 658
prescription, 13-14, 26, 33, 71, 74, 76-78, 80, 202, 253-255, 277, 400, 437, 507, 512, 578, 693-694, 709

presystemic clearance, 89-90
preterm labor, 340
prevention, 181, 198, 215, 262, 383, 387, 409-410, 428, 476, 500, 505, 507, 512, 518, 522, 527, 531, 543, 595, 637, 640
prior probability, 61
probe, 52, 90, 100, 102-110, 165
procedures, 4, 20, 22, 35, 44, 46, 61, 147, 154, 246, 248, 255, 271, 311-312, 318, 404, 410, 414, 436, 581, 584-586, 601, 603, 617, 664, 731
propofol infusion syndrome, 563
prostacyclin, 541
prostaglandins, 98, 280, 336, 338-339, 372, 397-398, 425, 646, 739
prostanoids, 316, 338, 339, 397-400
protease inhibitors, 169, 183, 286, 314, 516, 519, 522
protein binding, 53, 97, 106, 124-125, 137, 149, 171, 181, 192, 205, 360, 283, 286, 310, 334, 351, 369, 371, 374, 406, 418, 547, 594, 602-603, 687
pseudo-tumour cerebri, 686
pseudomonas, 439, 466, 479-480, 486, 711
pseudomonas aeruginosa, 466, 711
pseudocholinesterase, 554
psychotic episode, 689, 695
pulmonary hypoplasia, 393, 541
pulmonary route, 226
pulmonary vascular resistance, 394, 539
purines, 479
pyrimidines, 479

Q
questionnaire, 52

R
radiations, 776, 789
radiosensitivity, 776, 783
randomisation, 56
receptor, 142-143, 149, 164, 169, 173-174, 202, 311, 324, 335, 337-338, 373, 398, 403-404, 504, 506, 541, 548, 554, 581, 585, 589-590, 592, 596, 598, 600, 603-605, 613, 615-616, 634, 639-640, 649,

683-684, 738, 751-752, 785
recombination, 489
recommendation, 118, 534, 541
recruitment, 43, 52, 55, 61, 67, 69, 539
rectal route, 136, 222, 227-228, 623
red man syndrome, 444, 471
register, 33, 63, 67, 69-72, 650
regulation, 19-26, 29, 35, 64, 71, 111, 316, 331, 337, 622, 639, 723, 752
regulatory authorities, 23, 51, 150, 197-198, 692
rehydration, 665, 667-668
relapse, 167, 393-394, 427, 655, 657-658, 761-763, 766-772
renal blood flow, 138, 284, 310-311, 317, 335, 339, 372, 399, 646, 741
renal failure, 195, 203, 206, 212, 339, 372, 399, 486, 556, 602, 646, 717, 723, 730-733, 737-741, 746-751, 766
renal function, 105, 118, 138-139, 148, 151, 153, 156, 335, 372-373, 400, 404, 440, 508, 530, 593, 633, 651, 721, 730, 737, 742, 749, 752, 766, 789
renin-angiotensin system, 336, 550
residual, 151, 377, 393, 522, 541, 668, 672
resistance, 74, 136, 215, 327, 335, 338-339, 372, 394, 398, 400, 437, 439, 441, 444, 465, 469, 471-477, 479-481, 483, 486-488, 491, 497-498, 500, 508, 522, 528-529, 539, 542, 579, 656, 686, 716, 727, 731, 733, 743, 759, 774
respiratory depression, 197, 334-335, 404-405, 579, 589, 595-597, 600-601, 603-605, 614-615
respiratory distress syndrome, 367, 377, 379-380, 384, 387, 395, 753
response, 56, 58, 60, 80, 147-149, 152, 154, 156, 158, 161, 169, 171, 173-174, 178-179, 187, 204, 298, 312, 349, 380, 393, 404, 427, 435, 485-486, 512, 532, 534-535, 583, 589, 622, 638, 648-649, 656, 658, 668, 693, 695-696, 709, 730, 746-747, 750, 758, 763, 765-768, 770, 772
retinoblastoma, 781
retro-placental haematomaretro-placental haematoma, 329

reverse transcriptase, 109, 183, 314, 512, 516-519, 522
rhesus immunisation, 316
rickettsiae, 473-474, 476, 482
risk assessment, 257, 262-264, 291, 293-294, 297-298
risk factors, 262, 321, 703-704, 769, 784
route of administration, 6, 135, 219, 222, 225, 235, 252, 256, 261, 271, 284, 293, 351, 440, 486-487, 532
rubella vaccine, 279
rubella virus, 279
Rumack-Matthew nomogram, 625

S

s. pneumoniae, 492-500
safety, 3, 17, 19-21, 25, 29, 32, 35-36, 41-42, 50, 52-53, 58, 64-65, 69, 71, 73, 77, 80-82, 174, 232, 245, 247-251, 257-258, 260, 262-264, 277-279, 361, 291-294, 297-298, 314, 319, 325, 351, 354, 458, 403, 405, 410, 413, 421, 423, 510, 532, 562, 569, 571, 586, 597, 602, 641, 696-697, 713, 719, 723, 761
saliva sample, 37, 181
salmonellae typhi, 528
sample size, 57-59
sarcomas, 776, 779, 784, 788
saturable metabolism, 180
schizophrenia, 695
scientific advice, 19, 21
scintigraphy, 487-488
scores, 334, 344, 346, 589, 622, 690
sedation, 228, 404-405, 415, 557, 559-564, 575, 581-587, 590, 595-596, 601-603, 614-616, 644, 692, 694-696
seizure, 449-450, 452, 457-460, 482, 566-567, 569-571, 575-576, 578-580, 587, 791
sequelae, 313, 323, 329, 486, 493, 495, 497, 568, 689
sequencing techniques, 170, 174
sequential design, 57-58
septic shock, 437
septicaemia, 370, 436-437, 491, 527
seroconversion, 286

serotonin, 13, 82, 278, 321, 324, 329, 331, 613-614, 691-692, 694
serotoninergic effect, 574, 692
shigella dysenteriae, 528
Sigmoid E_{max} models, 621
skin test, 532
sleep pattern, 559
software, 152, 154, 432
solubility, 204, 236-238, 283, 310, 590-591, 597-598, 600, 602, 616
spina bifida, 317, 330
stability, 4, 38, 104, 123, 219-220, 223, 226, 229-230, 237-239, 250, 271, 333, 366, 405, 445, 505, 551, 600, 748
staphylococci, 436, 469, 472, 482-483, 486
status epilepticus, 329, 451, 454, 460, 579
stem cell transplantation, 778, 781, 785
sterility, 242, 777-780
steroid-nuclear receptor family, 639
Stevens-Johnson syndrome, 576
streptococcus agalactiae, 436-437
streptococcus pneumoniae, 482, 495, 528, 767
subcutaneous route, 136, 226, 236, 412, 413, 535, 581, 595, 597, 649, 650, 671-676, 680, 684, 782
substitution, 222, 343-344, 347, 355, 589, 676, 680
sudden infant death syndrome, 327-328
sulphation, 87-88, 92-94, 98, 125, 193, 203, 404, 530, 592
suppositories, 124, 222, 228, 240, 656
surfactant, 335, 377-378, 382-384, 387-396, 434, 539, 541
surgery, 265, 317, 403, 432, 491, 569, 579, 590, 594-595, 597-598, 603-604, 607, 614-616, 622, 654-655, 664, 713, 721, 733
sweetener, 221
syndrome of inappropriate antidiuretic hormone, 779, 791

T

tachyarrhythmia, 312
target, 141, 154, 162, 169-170, 179-184,

405, 440-441, 469, 475-477, 483, 517, 570, 590, 602, 622, 625-626, 632, 634-640, 683, 693, 695, 706, 734, 772, 775, 781

taste, 221-224, 237-240, 268

teenager, 270, 575, 680

teratogenicity, 279-280, 299, 306, 314, 318, 330

therapeutic drug monitoring, 147, 150, 154-155, 177-178, 181, 187, 302, 418, 420, 444, 632, 731

therapeutic range, 140-141, 161, 179, 184, 360, 421

thiopurine methyltransferase, 168, 637, 657, 786

thoracic rigidity, 602-603

thresholds, 44, 616

thrombocytopenia, 316, 637, 657, 748, 750, 782, 790

thromboxane A_2, 541

thyrotropin releasing factor, 383

tics, 688-689, 693, 696

timing, 31, 50, 179, 181, 316, 351, 457, 392, 534

tobacco, 245, 324, 328-329, 653, 703-704

tocolytic agent, 339, 399-400

tolerance, 312-313, 404, 506, 522, 556, 560, 562, 591-592, 595, 601, 603, 617, 640, 686, 696, 730-731, 761, 787

tonic-clonic convulsion, 567

tonsillectomy, 156, 595, 600, 615, 622-623

topical steroids, 271

Tourette's disorder, 689, 692, 696

toxic dose, 205, 298, 632

toxicity, 5, 14, 36, 38, 51, 53, 55, 80, 93, 104, 117, 140-141, 148, 155, 164, 171, 174, 178-180, 183-184, 186, 191-194, 198, 202-203, 205, 211, 215, 232, 239, 242, 291-299, 306, 313-314, 318-319, 321, 325, 329, 368, 370, 373, 397, 399, 418-419, 421, 432, 441, 444, 466, 471-472, 474-476, 478-480, 482-483, 506, 511, 519, 523, 530, 557, 562, 575, 577-578, 624-626, 632, 634-635, 646, 651, 655, 658, 662, 717, 728, 731, 771-772, 775-791

toxoids, 534

toxoplasma gondii, 313

toxoplasmosis, 286, 312-313, 476

transcutaneous route, 124, 226, 272, 557

transdermal route, 136, 227

transfer, 130, 139, 281-282, 284-287, 303, 310, 312, 350, 353, 357, 384, 473, 498, 504, 548, 599

transplantation, 109, 189, 598, 625, 631, 633-634, 636-638, 640-642, 765, 777-778, 781, 785

transporter, 118, 125, 173-174, 372, 444, 542, 637

trial, 38-39, 46, 49-52, 56-61, 64, 67-71, 77, 183-184, 360, 317, 456, 378-380, 382-384, 393, 427, 429, 431, 488, 512, 518, 522, 541, 562-563, 570, 585, 600, 647, 649-650, 662-663, 680, 682, 690, 716

triangular test, 58-59

trough concentrations, 150, 155, 182-183, 444-445, 632

trypanosomiasis, 528

tubular reabsorption, 126, 371, 742, 750, 788

tubular secretion, 126, 206, 371, 594, 788-789

Turner syndrome, 685

U

UDPGT, 167-172

UGT1A6, 93, 108

UGT1A9, 93, 108

UGT2B7, 92, 108, 371, 404, 592

ulcerative colitis, 653

unlicensed drug, 235

ureaplasma urealyticum, 439

V

vaccines, 3, 226, 500, 531-535, 651

variability, 117-118, 149, 151-152, 154, 158, 161, 165, 174, 177, 261, 289, 350, 366-367, 405, 418, 420, 429, 442, 493, 519, 523, 560, 562, 589, 592, 594-595, 599, 603, 622-623, 633, 641, 663, 719, 781, 786

varicella, 505, 512, 532, 534, 651
veno-occlusive hepatic disease, 778
ventilator dyssynchrony, 553
vesico-ureteric reflux, 486
video-recording, 450, 451, 566
viral replication, 504, 522
voiding disorders, 755-756
volume of distribution, 106, 125, 129, 133,
  136, 141, 145, 148, 152-153, 157, 180,
  205, 311, 351, 369-370, 400, 405-406,
  426, 440, 444, 547-549, 560, 594, 596,
  598, 605, 614, 616, 624-625, 633-634,
  637, 716, 728

W

website, 20, 69, 154
weight gain, 270, 294, 344, 679-680, 690,
  692, 694-695, 697, 713, 762
West syndrome, 568, 570
withdrawal syndrome, 46, 286, 325, 453,
  562, 575, 694
World Health Organization, 669

Z

zero-order kinetics, 132, 205